Texts in Computer Science

Series Editors

David Gries, Department of Computer Science, Cornell University, Ithaca, NY, USA

Orit Hazzan ⓘ, Faculty of Education in Technology and Science, Technion—Israel Institute of Technology, Haifa, Israel

More information about this series at http://www.springer.com/series/3191

Ze-Nian Li • Mark S. Drew •
Jiangchuan Liu

Fundamentals of Multimedia

Third Edition

Springer

Ze-Nian Li
School of Computing Science
Simon Fraser University
Burnaby, BC, Canada

Mark S. Drew
School of Computing Science
Simon Fraser University
Burnaby, BC, Canada

Jiangchuan Liu ⓘ
School of Computing Science
Simon Fraser University
Burnaby, BC, Canada

ISSN 1868-0941　　　　　　　　ISSN 1868-095X (electronic)
Texts in Computer Science
ISBN 978-3-030-62126-1　　　　ISBN 978-3-030-62124-7 (eBook)
https://doi.org/10.1007/978-3-030-62124-7

To my mom, and my wife Yansin.

Ze-Nian

To Noah, Ira, Eva and, especially, to Jenna.

Mark

To my wife Jill, and my children Kevin, Jerry, and Kathy.

Jiangchuan

Preface

In the 17 years since the first edition of *Fundamentals of Multimedia*, the field and applications of multimedia have flourished and are undergoing evermore rapid growth and evolution in various emerging interdisciplinary areas. However, a comprehensive textbook to aid the continuous learning and mastering of the fundamental concepts and knowledge in multimedia remains essential.

While the original edition was published by Prentice-Hall, starting from the second edition we have chosen Springer, a prestigious publisher that has a superb and rapidly expanding array of computer science textbooks, particularly the high-quality, dedicated, and established textbook series: Texts in Computer Science, of which this textbook forms a part. The second edition included considerable added depth to the networking aspect of the book. To this end, Dr. Jiangchuan Liu was added to the team of authors.

This third edition again constitutes a significant revision: the textbook has been thoroughly revised and updated to include recent developments in the field. For example, we updated the introduction to some of the current multimedia tools, we included current topics such as 360° video and the video coding standard H.266; new-generation social, mobile, and cloud computing for human-centric interactive multimedia, augmented reality and virtual reality; deep learning for multimedia processing; and their attendant technologies.

Multimedia is associated with a rich set of core subjects in Computer Science and Engineering, and we address those here. The book is not an introduction to simple design considerations and tools—it serves a more advanced audience than that. On the other hand, the book is not a reference work—it is more a traditional textbook. While we perforce may discuss multimedia tools, we would like to give a sense of the underlying issues at play in the tasks those tools carry out. Students who undertake and succeed in a course based on this text can be said to really understand fundamental matters in regard to this material, hence the title of the text.

In conjunction with this text, a full-fledged course should also allow students to make use of this knowledge to carry out interesting or even wonderful practical projects in multimedia; interactive projects that engage and sometimes amuse; and, perhaps, even teach these same concepts.

Who Should Read This Book?

This text aims at introducing the basic ideas used in multimedia, for an audience that is comfortable with technical applications, e.g., Computer Science students and Engineering students. The book aims to cover an upper level undergraduate multimedia course, but could also be used in more advanced courses. Indeed, a (quite long) list of courses making use of the first two editions of this text includes many undergraduate courses as well as use as a pertinent point of departure for graduate students who may not have encountered these ideas before in a practical way. As well, the book would be a good reference for anyone, including those in industry, who are interested in current multimedia technologies. The selection of material in the text addresses real issues that these learners will be facing as soon as they show up in the workplace. Some topics are simple, but new to the students; some are somewhat complex, but unavoidably so in this emerging area.

The text mainly presents concepts, not applications. A multimedia course, on the other hand, teaches these concepts, and tests them, but also allows students to utilize skills they already know, in coding and presentation, to address problems in multimedia. The accompanying website materials for the text include some code for multimedia applications along with some projects students have developed in such a course, plus other useful materials best presented in electronic form.

Have the Authors Used This Material in a Real Class?

Since 1996, we have taught a third-year undergraduate course in Multimedia Systems based on the introductory materials set out in this book. A one-semester course very likely could not include all the material covered in this text, but we have usually managed to consider a good many of the topics addressed, with mention made of a selected number of issues in Parts 3 and 4, within that time frame.

As well, over the same time period and again as a one-semester course, we have also taught a graduate-level course using notes covering topics similar to the ground covered by this text, as an introduction to more advanced materials. A fourth-year or graduate-level course would do well to discuss material from the first three parts of the book and then consider some material from the last part, perhaps in conjunction with some of the original research references included here along with results presented at topical conferences.

We have attempted to fill both needs, concentrating on an undergraduate audience but including more advanced material as well. Sections that can safely be omitted on a first reading are marked with an asterisk in the Table of Contents.

What is Covered in This Text?

In Part 1, Introduction and Multimedia Data Representations, we introduce some of the notions included in the term Multimedia, and look at its present as well as its history. Practically speaking, we carry out multimedia projects using software tools, so in addition to an overview of multimedia software tools we get down to some of the nuts and bolts of multimedia authoring. The representation of data is critical in the study of multimedia, and we look at the most important data representations for use in multimedia applications. Specifically, graphics and image data, video data, and audio data are examined in detail. Since color is vitally important in multimedia programs, we see how this important area impacts multimedia issues.

In Part 2, Multimedia Data Compression, we consider how we can make all this data fly onto the screen and speakers. Multimedia data compression turns out to be a very important enabling technology that makes modern multimedia systems possible. Therefore, we look at lossless and lossy compression methods, supplying the fundamental concepts necessary to fully understand these methods. For the latter category, lossy compression, arguably JPEG still-image compression standards, including JPEG 2000, are the most important, so we consider these in detail. But since a picture is worth 1,000 words, and so video is worth more than a million words per minute, we examine the ideas behind the MPEG standards MPEG-1, MPEG-2, MPEG-4, MPEG-7, and beyond into modern video coding standards H.264, H.265, and H.266. Audio compression is treated separately and we consider some basic audio and speech compression techniques and take a look at MPEG Audio, including MP3 and AAC.

In Part 3, Multimedia Communications and Networking, we consider the great demands multimedia communication and content sharing place on networks and systems. The Internet, however, was not initially designed for multimedia content distribution and there are significant challenges to be addressed. We discuss the wired Internet and wireless mobile network technologies and protocols, and the enhancements of them that make multimedia communications possible. We further examine state-of-the-art multimedia content distribution mechanisms, as well as modern cloud computing for highly scalable multimedia data processing. The discussion also includes the latest edge computing and serverless computing solutions toward fine-grained and flexible realtime multimedia.

In Part 4, Human-Centric Interactive Multimedia, we examine a number of technologies that form the heart of enabling the new Web 2.0 paradigm, with rich user interactions. Such popular Web 2.0-based social media sharing websites as YouTube, Facebook, Twitter, Twitch, and TikTok have drastically changed the content generation and distribution landscape, and indeed have become an integral part in people's daily life. The developments in the coding algorithms and hardware for sensing, communication, and interaction also empower virtual reality (VR) and augmented reality (AR), providing better immersive experiences beyond 3D. This part examines these new-generation interactive multimedia services and discusses their potential and challenges. The huge amount of multimedia content also militates for multimedia-aware search mechanisms, and we therefore consider the challenges and mechanisms for multimedia content search and retrieval.

Textbook Website

The book website is http://www.cs.sfu.ca/mmbook. There the reader will find general information about the book including previous editions, an errata sheet updated regularly, programs that help demonstrate concepts in the text, and a dynamic set of links for the "Further Exploration" section in some of the chapters. Since these links are regularly updated, and of course URLs change quite often, the links are online rather than within the printed text.

Instructors' Resources

The main text website has no ID and password, but access to sample student projects is at the instructor's discretion and is password-protected. For instructors, with a different password, the website also contains Course Instructor resources for adopters of the text. These include an extensive collection of online slides, solutions for the exercises in the text, sample assignments and solutions, sample exams, and extra exam questions.

Acknowledgments

We are most grateful to colleagues who generously gave of their time to review this text, and we wish to express our thanks to Edward Chang, Shu-Ching Chen, Qianping Gu, Mohamed Hefeeda, Rachelle S. Heller, Gongzhu Hu, S. N. Jayaram, Tiko Kameda, Joonwhoan Lee, Xiaobo Li, Jie Liang, Siwei Lu, Jiebo Luo, and Jacques Vaisey.

The writing of this text has been greatly aided by a number of suggestions and contributions from present and former colleagues and students. We would like to thank Mohamed Athiq, James Au, Yi Ching David Chou, Chad Ciavarro, Hossein Hajimirsadeghi, Hao Jiang, Mehran Khodabandeh, Steven Kilthau, Michael King, Tian Lan, Chenyu Li, Haitao Li, Cheng Lu, Minlong Lu, You Luo, Xiaoqiang Ma, Hamidreza Mirzaei, Peng Peng, Haoyu Ren, Ryan Shea, Chantal Snazel, Wenqi Song, Yi Sun, Dominic Szopa, Zinovi Tauber, Malte von Ruden, Fangxin Wang, Jian Wang, Jie Wei, Edward Yan, Osmar Zaïane, Cong Zhang, Lei Zhang, Miao Zhang, Wenbiao Zhang, Yuan Zhao, Ziyang Zhao, William Zhong, Qiang Zhu, and Yifei Zhu for their assistance. Yi Ching David Chou also helped with refreshing the companion website for the textbook. As well, Dr. Ye Lu made great contributions to Chaps. 8 and 9; Andy Sun contributed Chap. 20. Their valiant efforts are particularly appreciated. We are also most grateful for the students who generously made their course projects available for instructional use for this book.

Burnaby, Canada

Ze-Nian Li
Mark S. Drew
Jiangchuan Liu

Contents

Introduction and Multimedia Data Representations

As an introduction to multimedia, in Chap. 1, we consider the question of just what multimedia is. The components of multimedia are first introduced and then current multimedia research topics and projects are discussed to put the field into a perspective of what is actually at play at the edge of work in this field.

Since multimedia is indeed a practical field, Chap. 1 also supplies an overview of multimedia software tools, such as video editors and digital audio programs.

A Taste of Multimedia

As a "taste" of multimedia, in Chap. 2, we introduce a set of tasks and concerns that are considered in studying multimedia. Then issues in multimedia production and presentation on modern computing and communication platforms are discussed, followed by a further "taste" by considering how to produce sprite animation and "build-your-own" video transitions.

We then go on to review the current and future state of multimedia sharing and distribution, outlining later discussions of social media, video sharing, and new forms of online TV.

Finally, the details of some popular multimedia editing and authoring tools are set out for a quick start into the field.

Multimedia Data Representations

As in many fields, the issue of how to best represent the data is of crucial importance in the study of multimedia, and Chaps. 3–6 consider how this is addressed in this field. These chapters set out the most important data representations for use in multimedia applications. Since the main areas of concern are images, video, and audio, we begin investigating these in Chap. 3, Graphics and Image Data Representations. Before going on to look at Fundamental Concepts in Video in Chap. 5, we take a side trip in Chap. 4 to explore several issues in the use of color, since color is vitally important in multimedia programs.

Audio data has special properties and Chap. 6, Basics of Digital Audio, introduces methods to compress sound information, beginning with a discussion of digitization of audio, and linear and nonlinear quantization, including companding. MIDI is explicated, as an enabling technology to capture, store, and play back musical notes. Quantization and transmission of audio are discussed, including the notion of subtraction of signals from predicted values, yielding numbers that are easier to compress. Differential pulse code modulation (DPCM) and adaptive DPCM are introduced, and we take a look at encoder/decoder schema.

Introduction to Multimedia

1

1.1   What is Multimedia?

People who use the term "multimedia" may have quite different, even opposing, viewpoints. A consumer entertainment vendor, say a phone company, may think of multimedia as interactive TV with hundreds of digital channels or a cable-TV-like service delivered over a high-speed Internet connection. A hardware vendor might, on the other hand, like us to think of multimedia as a laptop that has good sound capability and perhaps the superiority of multimedia-enabled microprocessors that understand additional multimedia instructions.

A computer science or engineering student reading this book likely has a more application-oriented view of what multimedia consists of: applications that use multiple modalities to their advantage, including text, images, drawings, graphics, animation, video, sound (including speech), and, most likely, interactivity of some kind. This contrasts with media that use only rudimentary computer displays such as text-only or traditional forms of printed or hand-produced material.

The popular notion of "convergence" is one that inhabits the college campus as it does the culture at large. In this scenario, computers, smartphones, games, digital TV, multimedia-based search, and so on are converging in technology, presumably to arrive in the near future at a final and fully functional all-round, multimedia-enabled product. While hardware may indeed strive for such all-round devices, the present is already exciting—multimedia is part of some of the most interesting projects underway in computer science, with the keynote being *interactivity*. The convergence going on in this field is, in fact, a convergence of areas that have in the past been separated but are now finding much to share in this new application area. Graphics, visualization, HCI, artificial intelligence, computer vision, data compression, graph theory, networking, and database systems all have important contributions to make in multimedia at the present time.

© Springer Nature Switzerland AG 2021
Z.-N. Li et al., *Fundamentals of Multimedia*, Texts in Computer Science,
https://doi.org/10.1007/978-3-030-62124-7_1

1.1.1 Components of Multimedia

The multiple modalities of text, audio, images, drawings, animation, video, and interactivity in multimedia are put to use in ways as diverse as

- Videoconferencing.
- Tele-medicine.
- A web-based video editor that lets anyone create a new video by editing, annotating, and remixing professional videos on the cloud.
- Geographically based, real-time augmented reality, massively multiplayer online video games, making use of any portable device such as smartphones, laptops, or tablets, which function as GPS-aware mobile game consoles.
- Shapeshifting TV, where viewers vote on the plot path by phone text messages, which are parsed to direct plot changes in real time.
- A camera that suggests what would be the best type of next shot so as to adhere to good technique guidelines for developing storyboards.
- Cooperative education environments that allow schoolchildren to share a single educational game using two mice at once that pass control back and forth.
- Searching (very) large image and video databases for target visual objects, using semantics of objects.
- Compositing of artificial and natural video into hybrid scenes, placing real-appearing computer graphics and video objects into scenes so as to take the physics of objects and lights (e.g., shadows) into account.
- Visual cues of videoconferencing participants, taking into account gaze direction and attention of participants.
- Making multimedia components *editable*—allowing the user side to decide what components, video, graphics, and so on are actually viewed and allowing the client to move components around or delete them—making components distributed.
- Building "inverse-Hollywood" applications that can recreate the process by which a video was made, allowing storyboard pruning and concise video summarization.

From a computer science student's point of view, what makes multimedia interesting is that so much of the material covered in traditional computer science areas bears on the multimedia enterprise. In today's digital world, multimedia content is recorded and played, displayed, or accessed by digital information content-processing devices, ranging from smartphones, tablets, laptops, personal computers, smart TVs, and game consoles, to servers and data centers, over such distribution media as USB flash drives (keys), discs, and hard drives, or more popularly nowadays, wired and wireless networks. This leads to a wide variety of research topics:

- **Multimedia processing and coding**. This includes audio/image/video processing, compression algorithms, multimedia content analysis, content-based multimedia retrieval, multimedia security, and so on.

- **Multimedia system support and networking**. People look at such topics as network protocols, Internet and wireless networks, operating systems, servers and clients, and databases.
- **Multimedia tools, end systems, and applications**. These include hypermedia systems, user interfaces, authoring systems, multimodal interaction, and integration: "ubiquity"—web-everywhere devices, multimedia education, including computer-supported collaborative learning and design, and applications of virtual environments.

Multimedia research touches almost every branch of computer science. For example, data mining is an important current research area, and a large database of multimedia data objects is a good example of just what big data we may be interested in mining; telemedicine applications, such as "telemedical patient consultative encounters," are multimedia applications that place a heavy burden on network architectures. Multimedia research is also highly inter-disciplinary, involving such other research fields as electric engineering, physics, and psychology; signal processing for audio/video signals is an essential topic in electric engineering; color in image and video has a long history and solid foundation in physics; more importantly, all multimedia data are to be perceived by human beings, which is, certainly, related to medical and psychological research.

1.2 Multimedia: Past and Present

To place multimedia in its proper context, in this section, we briefly scan the history of multimedia, a relatively recent part of which is the connection between multimedia and hypermedia. We also show the rapid evolution and revolution of multimedia in the new millennium with the new generation of computing and communication platforms.

1.2.1 Early History of Multimedia

A brief history of the use of multimedia to communicate ideas might begin with newspapers, which were perhaps the *first* mass communication medium, using text, graphics, and images. Before still-image camera was invented, these graphics and images were generally hand-drawn.

Joseph Nicéphore Niépce captured the first natural image from his window in 1826 using a sliding wooden box camera [1,2]. It was made using an 8 h exposure on pewter coated with bitumen. Later, Alphonse Giroux built the first commercial camera with a double-box design. It had an outer box fitted with a landscape lens, and an inner box holding a ground glass focusing screen and image plate. Sliding the inner box makes objects at different distances be in focus. Similar cameras were used for exposing wet silver-surfaced copper plates, commercially introduced in 1839. In the 1870s, wet plates were replaced by the more convenient dry plates. Figure 1.1

Fig. 1.1 A vintage dry-plate camera. E&H T Anthony model Champion, circa 1890

(image from author's own collection) shows an example of a nineteenth-century dry-plate camera, with bellows for focusing. By the end of the nineteenth century, film-based cameras were introduced, which soon became dominant until replaced by digital cameras.

Thomas Alva Edison's phonograph, invented in 1877, was the first device that was able to record and reproduce sound. It originally recorded sound onto a tinfoil sheet phonograph cylinder [3]. Figure 1.2 shows an example of an Edison phonograph (Edison GEM, 1905; image from author's own collection).

Phonographs were later improved by Alexander Graham Bell. Most notable improvements include wax-coated cardboard cylinders, and a cutting stylus that moved from side to side in a "zig zag" pattern across the record. Emile Berliner further transformed the phonograph cylinders to gramophone records. Each side of such a flat disc has a spiral groove running from the periphery to near the center, which can be conveniently played by a turntable with a tonearm and a stylus. These components were improved over the time in the twentieth century, which eventually enabled quality sound reproducing that is very close to the original. The gramophone record was one of the dominant audio recording formats throughout much of the twentieth century. From the mid-1980s, gramophone use declined sharply because of the rise of audio tapes, and later the *compact disc* (CD) and other digital recording formats [4]. Figure 1.3 shows the evolution of audio storage media, starting from the Edison cylinder record, to the flat vinyl record, to magnetic tapes (reel-to-reel and cassette), and the modern digital CD.

Motion pictures were originally conceived of in the 1830s to observe motion too rapid for perception by the human eye. Edison again commissioned the invention of a motion picture camera in 1887 [5]. Silent feature films appeared from 1910 to 1927; the silent era effectively ended with the release of *The Jazz Singer* in 1927.

Fig. 1.2 An Edison phonograph, model GEM. Note the patent plate in the bottom picture, which suggests that the importance of patents had long been realized and also how serious Edison was in protecting his inventions. Despite the warnings in the plate, this particular phonograph was modified by the original owner, a good DIYer 100 years ago, to include a more powerful spring motor from an Edison Standard model and a large flower horn from the Tea Tray Company

In 1895, Guglielmo Marconi conducted the first wireless radio transmission at Pontecchio, Italy, and a few years later (1901), he detected radio waves beamed across the Atlantic [6]. Initially, invented for telegraph, radio is now a major medium for audio broadcasting. In 1909, Marconi shared the Nobel Prize for physics.[1]

---

[1] Reginald A. Fessenden, of Quebec, beat Marconi to human voice transmission by several years, but not all inventors receive due credit. Nevertheless, Fessenden was paid $2.5 million in 1928 for his purloined patents.

Fig. 1.3 Evolution of audio storage media. Left to right: an Edison cylinder record, a flat vinyl record, a reel-to-reel magnetic tape, a cassette tape, and a CD

Television, or TV for short, was the new medium for the twentieth century [7]. In 1884, Paul Gottlieb Nipkow, a 23-year-old university student in Germany, patented the first electromechanical television system, which employed a spinning disk with a series of holes spiralling toward the center. The holes were spaced at equal angular intervals such that, in a single rotation, the disk would allow light to pass through each hole and onto a light-sensitive selenium sensor which produced the electrical pulses. As an image was focused on the rotating disk, each hole captured a horizontal "slice" of the whole image. Nipkow's design would not be practical until advances in amplifier tube technology, in particular, the cathode ray tube (CRT), became available in 1907. Commercially available since the late 1920s, CRT-based TV established video as a commonly available medium and has since changed the world of mass communication.

All these media mentioned above are in the *analog* format, for which the time-varying feature (variable) of the signal is a continuous representation of the input, i.e., analogous to the input audio, image, or video signal. The connection between *computers* and *digital media*, i.e., media data represented using the discrete binary format, emerged actually only over a short period:

1945 As part of MIT's postwar deliberations on what to do with all those scientists employed on the war effort, Vannevar Bush wrote a landmark article [8] describing what amounts to a hypermedia system, called "Memex." Memex was meant to be a universally useful and personalized memory device that even included the concept of associative links—it really is the forerunner of the World Wide Web. After World War II, 6,000 scientists who had been hard at work on the war effort suddenly found themselves with time to consider other issues, and the Memex idea was one fruit of that new freedom.

1965 Ted Nelson started the Xanadu project and coined the term *hypertext*. Xanadu was the first attempt at a hypertext system—Nelson called it a "magic place of literary memory."

1967 Nicholas Negroponte formed the Architecture Machine Group at MIT.

1969 Nelson and van Dam at Brown University created an early hypertext editor called FRESS [9]. The present-day Intermedia project by the Institute for Research in Information and Scholarship (IRIS) at Brown is the descendant of that early system.

1976   The MIT Architecture Machine Group proposed a project entitled "Multiple Media." This resulted in the *Aspen Movie Map*, the first videodisc, in 1978.

1982   The *Compact Disc* (CD) was made commercially available by Philips and Sony, which was soon becoming the standard and popular medium for digital audio data, replacing the analog magnetic tape.

1985   Negroponte and Wiesner co-founded the MIT Media Lab, a leading research institution investigating digital video and multimedia.

1990   Kristina Hooper Woolsey headed the Apple Multimedia Lab, with a staff of 100. Education was a chief goal.

1991   MPEG-1 was approved as an international standard for digital video. Its further development led to newer standards, MPEG-2, MPEG-4, and further MPEGs, in the 1990s.

1991   The introduction of PDAs in 1991 began a new period in the use of computers in general and multimedia in particular. This development continued in 1996 with the marketing of the first PDA with no keyboard.

1992   JPEG was accepted as the international standard for digital image compression, which remains widely used today (say, by virtually every digital camera).

1992   The first audio multicast on the multicast backbone (MBone) was made.

1995   The JAVA language was created for platform-independent application development, which was widely used for developing multimedia applications.

1996   DVD video was introduced; high-quality, full-length movies were distributed on a single disc. The DVD format promised to transform the music, gaming, and computer industries.

1998   Handheld MP3 audio players were introduced to the consumer market, initially with 32 MB of flash memory.

1.2.2   Hypermedia, WWW, and Internet

The early studies laid a solid foundation for the capturing, representation, compression, and storage of each type of media. Multimedia, however, is not simply about putting different media together; rather, it focuses more on the integration of them so as to enable rich interaction among them, and as well between media and human beings.

We may think of a book as a *linear* medium, basically meant to be read from beginning to end. In contrast, a hypertext system is meant to be read nonlinearly, by following links that point to other parts of the document, or indeed to other documents. Figure 1.4 illustrates this familiar idea.

Douglas Engelbart, greatly influenced by Vannevar Bush's "As We May Think," demonstrated the *On-Line System* (NLS), another early hypertext program in 1968. Engelbart's group at Stanford Research Institute aimed at "augmentation, not automation," to enhance human abilities through computer technology. NLS consisted of such critical ideas as an outline editor for idea development, hypertext links, telecon-

Fig. 1.4 Hypertext is nonlinear

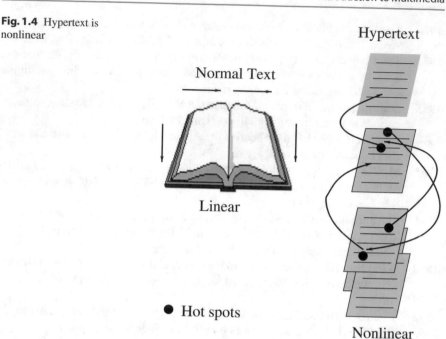

ferencing, word processing, and e-mail, and made use of the mouse pointing device, windowing software, and help systems [10].

Hypermedia, again first introduced by Ted Nelson around 1965, went beyond text-only. It includes a wide array of media, such as graphics, images, and especially the continuous media—sound and video, and links them together. The *World Wide Web* (WWW or simply Web) is the best example of a hypermedia application, which is also the largest.

Amazingly, this most predominant networked multimedia applications has its roots in nuclear physics! In 1990, Tim Berners-Lee proposed the World Wide Web to CERN (European Center for Nuclear Research) as a means for organizing and sharing their work and experimental results. With approval from CERN, he started developing a hypertext server, browser, and editor on a NeXTStep workstation. His team invented the *hypertext markup language* (HTML) and the *hypertext transfer protocol* (HTTP) for this purpose, too.

HyperText Markup Language (HTML)

It is recognized that documents need to have formats that are human-readable and that identify structure and elements. Charles Goldfarb, Edward Mosher, and Raymond Lorie developed the generalized markup language (GML) for IBM. In 1986, the ISO released a final version of the standard generalized markup language (SGML), mostly based on the earlier GML.

HTML is a language for publishing hypermedia on the web [11]. It is defined using SGML and derives elements that describe generic document structure and formatting. Since it uses ASCII, it is portable to all different (even binary incompatible) computer hardware, which allows for global exchange of information. As of 2020, the current version is HTML5.

HTML uses tags to describe document elements. The tags are in the format `<token params>` to define the start point of a document element and `</token>` to define the end of the element. Some elements have only inline parameters and don't require ending tags. HTML divides the document into a HEAD and a BODY part as follows:

```
<HTML>
<HEAD>
...
</HEAD>
<BODY>
...
</BODY>
</HTML>
```

The HEAD describes document definitions, which are parsed before any document rendering is done. These include page title, resource links, and meta-information the author decides to specify. The BODY part describes the document structure and content. Common structure elements are paragraphs, tables, forms, links, item lists, and buttons.

A very simple HTML page is as follows:

```
<HTML>
<HEAD>
  <TITLE>
  A sample web page.
  </TITLE>
  <META NAME = "Author" CONTENT = "Cranky Professor">
</HEAD> <BODY>
  <P>
  We can put any text we like here, since this is
  a paragraph element.
  </P>
</BODY>
</HTML>
```

Naturally, HTML has more complex structures and can be mixed with other standards. The standard has evolved to allow integration with script languages, dynamic manipulation of almost all elements and properties after display on the client side (*dynamic HTML*), and modular customization of all rendering parameters using a language called *cascading style sheets* (CSS). Nonetheless, HTML has rigid, non-descriptive structure elements, and modularity is hard to achieve.

Extensible Markup Language (XML)

There was also a need for a markup language for the web that has modularity of data, structure, and view. That is, we would like a user or an application to be able to *define* the tags (structure) allowed in a document and their relationship to each other, in one place, then define data using these tags in another place (the XML file) and, finally, define in yet another document how to render the tags.

Suppose we wanted to have stock information retrieved from a database according to a user query. Using XML, we would use a global *document-type definition* (DTD) we have already defined for stock data. Your server-side script will abide by the DTD rules to generate an XML document according to the query, using data from your database. Finally, we will send users your *XML style sheet* (XSL), depending on the type of device they use to display the information, so that our document looks best both on a computer with a 27-inch LED display and on a small-screen cellphone.

The original XML version was XML 1.0, approved by the W3C in February 1998, and was in its fifth edition as published in 2008. The original version is still recommended for general use. The second version XML 1.1 was introduced in 2004 and was in its second edition as published in 2006. XML syntax looks like HTML syntax, although it is much stricter. All tags are lowercase, and a tag that has only inline data has to terminate itself, for example, `<token params/>`. XML also uses namespaces, so that multiple DTDs declaring different elements but with similar tag names can have their elements distinguished. DTDs can be imported from URIs as well. As an example of an XML document structure, here is the definition for a small XHTML document:

```
<?xml version="1.0" encoding="iso-8859-1"?>
<!DOCTYPE html PUBLIC "-//W3C//DTD XHTML 1.0"
 "http://www.w3.org/TR/xhtml1/DTD/xhtml1-transition.dtd">
 <html xmlns="http://www.w3.org/1999/xhtml">
 ... [html that follows
     the above mentioned
     XML rules]
 </html>
```

All XML documents start with `<?xml version="ver"?>`. `<!DOCTYPE ...>` is a special tag used for importing DTDs. Since it is a DTD definition, it does not adhere to XML rules. `xmlns` defines a unique XML namespace for the document elements. In this case, the namespace is the XHTML specification website.

In addition to XML specifications, the following XML-related specifications are standardized:

- **XML Protocol**. Used to exchange XML information between processes. It is meant to supersede HTTP and extend it as well as to allow interprocess communications across networks.
- **XML Schema**. A more structured and powerful language for defining XML data types (tags). Unlike a DTD, XML schema uses XML tags for type definitions.
- **XSL**. This is basically CSS for XML. On the other hand, XSL is much more complex, having three parts: *XSL Transformations* (XSLT), *XML Path Language* (XPath), and *XSL Formatting Objects*.

The WWW quickly gained popularity, due to the amount of information available from web servers, the capacity to post such information, and the ease of navigating such information with a web browser, particularly after Marc Andreessen's introduction of the Mosaic browser in 1993 (this later became Netscape).

Today, web technology is maintained and developed by the World Wide Web Consortium (W3C), together with the Internet Engineering Task Force (IETF) to standardize the technologies. The W3C has listed the following three goals for the WWW: universal access of web resources (by everyone everywhere), effectiveness of navigating available information, and responsible use of posted material.

It is worth mentioning that the Internet serves as the underlying vehicle for the WWW and the multimedia content shared over it. Starting from the Advanced Research Projects Agency Network (ARPANET) with only two nodes in 1969, the Internet gradually became the dominating global network that interconnects numerous computer networks and their billions of users with the standard Internet protocol suite (TCP/IP). It evolved together with digital multimedia. On the one hand, the Internet carries much of the multimedia content. It has largely swept out optical discs as the storage and distribution media in the movie industry. It is currently reshaping the TV broadcast industry with an ever-accelerating speed. On the other hand, the Internet was not initially designed for multimedia data and was not quite friendly to multimedia traffic. Multimedia data, now occupying almost 90% of the Internet bandwidth, is the key driving force toward enhancing the existing Internet and toward developing the next generation of the Internet, as we will see in Chaps. 15 and 16.

1.2.3   Multimedia in the New Millennium

Entering the new millennium, we have witnessed the fast evolution toward a new generation of social, mobile, and cloud computing for multimedia processing and sharing. Today, the role of the Internet itself has evolved from the original use as a communication tool to providing easier and faster sharing of an infinite supply of information, and the multimedia content itself has been greatly enriched too. High-definition videos and even 3D/multiview videos can be readily captured and browsed by personal computing devices, and conveniently stored and processed with remote cloud resources. More importantly, users are now actively engaged to be part of a social ecosystem, rather than passively receiving media content. The revolution is being driven further by the deep penetration of 3G/4G/5G wireless networks and smart mobile devices. Coming with highly intuitive interfaces and exceptionally richer multimedia functionalities, they have been seamlessly integrated with online social networking for instant media content generation and sharing.

Below we list some important milestones in the development of multimedia in the new millennium. We believe that most of the readers of this textbook are familiar with them, as we are all in this Internet age, witnessing its dramatic changes; many readers, particularly the younger generation, would be even more familiar with the use of such multimedia services as YouTube, Facebook, Twitter, Twitch, and TikTok than the authors.

2000   WWW size was estimated at over 1 billion pages.
Sony unveiled the first Blu-ray Disc prototypes in October 2000, and the first prototype player was released in April 2003 in Japan.

2001   The first peer-to-peer file sharing (mostly MP3 music) system, Napster, was shut down by court order, but many new peer-to-peer file-sharing systems, e.g., Gnutella, eMule, and BitTorrent, were launched in the following years. Coolstreaming was the first large-scale peer-to-peer streaming system that was deployed in the Internet, attracting over 1 million in 2004. Later years saw the booming of many commercial peer-to-peer TV systems, e.g., PPLive, PPStream, and UUSee, particularly in East Asia.
NTT DoCoMo in Japan launched the first commercial 3G wireless network on October 1. 3G then started to be deployed worldwide, promising broadband wireless mobile data transfer for multimedia data.

2003   Skype was released for free peer-to-peer voice over the Internet.

2004   Web 2.0 was recognized as a new way to utilize software developers and end users use the web (and is not a technical specification for a new web). The idea is to promote user collaboration and interaction so as to generate content in a "virtual community," as opposed to simply passively viewing content. Examples include social networking, blogs, wikis, etc.
Facebook, the most popular online social network, was founded by Mark Zuckerberg.
Flickr, a popular photo hosting and sharing site, was created by Ludicorp, a Vancouver-based company founded by Stewart Butterfield and Caterina Fake.

2005   YouTube was created, providing an easy portal for video sharing, which was purchased by Google in late 2006.
Google launched the online map service, with satellite imaging, real-time traffic, and Streetview being added later.

2006   Twitter was created, and rapidly gained worldwide popularity, with 500 million registered users in 2012, who posted 340 million tweets per day.
Amazon launched its cloud computing platform, Amazon's Web Services (AWS). The most central and well known of these services are Amazon EC2 and Amazon S3.

Nintendo introduced the Wii home video game console, whose remote controller can detect movement in three dimensions.

2007 Apple launched the first generation of iPhone, running the iOS mobile operating system. Its touch screen enabled very intuitive operations, and the associated App Store offered numerous mobile applications.

Google unveiled Android mobile operating system, along with the founding of the Open Handset Alliance: a consortium of hardware, software, and telecommunication companies devoted to advancing open standards for mobile devices. The first Android-powered phone was sold in October 2008, and Google Play, Android's primary app store, was soon launched.

In the following years, tablet computers using iOS, Android, and Windows with larger touch screens joined the ecosystem, too.

2009 The first LTE (long-term evolution) network was set up in Oslo, Norway, and Stockholm, Sweden, making an important step toward 4G wireless networking.

James Cameron's film, Avatar, created a surge of the interest in 3D video.

2010 Netflix, which used to be a DVD rental service provider, migrated its infrastructure to the Amazon AWS cloud computing platform, and became a major online streaming video provider. Master copies of digital films from movie studios are stored on Amazon S3, and each film is encoded into over 50 different versions based on video resolution, audio quality using machines on the cloud. In total, Netflix has over 1 petabyte of data stored on Amazon's cloud.

Microsoft introduced Kinect, a horizontal bar with full-body 3D motion capture, facial recognition, and voice recognition capabilities, for its game console Xbox 360.

2012 HTML5 subsumes the previous version, HTML4, which was standardized in 1997. HTML5 is a W3C "Candidate Recommendation." It is meant to provide support for the latest multimedia formats while maintaining consistency for current web browsers and devices, along with the ability to run on low-powered devices such as smartphones and tablets.

Twitch started offering video live streaming service, in particular, eSports competitions.

2013 Twitter offered Vine, a mobile app that enables its users to create and post short video clips.

Sony released its PlayStation 4, a video game console, which is to be integrated with Gaikai, a cloud-based gaming service that offers streaming video game content.

4K resolution TV started to be available in the consumer market.

Google started selling Google Glass, an optical head-mounted display designed in the shape of a pair of eyeglasses.

2015 YouTube launched support for publishing and viewing 360° videos, with playback on its website and its Android mobile apps.

AlphaGo, a computer program that plays the board game Go, became the first program to beat a human professional player. The event attracted significant attention from the general public, particular on its core technology, deep learning, which, in the coming years, has seen success in multimedia content understanding and generation.

2016 HoloLens, a pair of mixed reality smartglasses developed and manufactured by Microsoft, started to be available in the market.

Pokémon Go, an augmented reality (AR) mobile game, was released and credited with popularizing location-based and AR technologies.

Netflix completely migrated to the Amazon AWS cloud platform and Skype moved to the Microsoft Azure platform.

2017 TikTok, a video-sharing social networking service for creating and sharing short lip-sync, comedy, and talent videos, was launched for the global market (its Chinese version, Douyin, was launched in 2016). Both soon became the most popular app in this category.

2018 The world's first 16K ultra-high definition (UHD) short video film, Prairie Wind, was created.

2019 5G cellular systems started deployment, providing enhanced mobile broadband and ultra-low latency access.

The Wi-Fi 6 (802.11ax) standard was released, offering theoretical maximum throughput of 1 Gbps.

2020 Due to the outbreak of corona virus (COVID-19) around the world, work/study from home became a norm in early 2020. Multimedia-empowered online meeting and teaching tools, e.g., Zoom, Google Class, and Microsoft Teams, saw booming use during this period.

1.3 Multimedia Software Tools: A Quick Scan

For a concrete appreciation of the current state of multimedia software tools available for carrying out tasks in multimedia, we now include a quick overview of software categories and products.

These tools are really only the beginning—a fully functional multimedia project can also call for stand-alone programming as well as just the use of predefined tools to fully exercise the capabilities of machines and the Internet.[2]

In courses we teach using this text, students are encouraged to try these tools, producing full-blown and creative multimedia productions. Yet this textbook is not a "how-to" book about using these tools—it is about understanding the fundamental design principles behind these tools! With a clear understanding of the key multimedia data structures, algorithms, and protocols, a student can make smarter and advanced use of such tools, so as to fully unleash their potentials, and even improve the tools themselves or develop new tools.

---

[2] In a typical computer science course in multimedia, the tools described here might be used to create a small multimedia production as a first assignment. Some of the tools are powerful enough that they might also form part of a course project.

The categories of software tools we examine here are

- Music sequencing and notation,
- Digital audio,
- Graphics and image editing,
- Video editing,
- Animation,
- Multimedia authoring, and
- Multimedia broadcasting.

1.3.1 Music Sequencing and Notation

Cakewalk by Bandlab

Cakewalk by Bandlab is a very straightforward music-notation program for "sequencing." The term *sequencer* comes from older devices that stored sequences of notes in the MIDI music language (*events*, in MIDI, see Sect. 6.2).

Finale, Avid Sibelius

Finale and Sibelius are two composer-level notation systems; these programs likely set the bar for excellence, but their learning curve is fairly steep.

1.3.2 Digital Audio

Digital Audio tools deal with accessing and editing the actual sampled sounds that make up audio.

Adobe Audition

Adobe Audition (formerly Cool Edit) is a powerful, popular digital audio toolkit that emulates a professional audio studio, including multitrack productions and sound file editing, along with digital signal processing effects.

Sound Forge

Like Audition, Sound Forge is a sophisticated PC-based program for editing WAV sound files. Sound can be captured through the sound card, and then mixed and edited. It also permits adding complex special effects.

Avid Pro Tools

Pro Tools is a high-end integrated audio production and editing environment that runs on Macintosh computers as well as Windows. Pro Tools offers easy MIDI creation and manipulation as well as powerful audio mixing, recording, and editing software. Full effects depend on purchasing a dongle.

1.3.3 Graphics and Image Editing

Adobe Illustrator

Illustrator is a powerful publishing tool for creating and editing vector graphics, which can easily be exported to use on the web. It is especially useful for logo graphics design.

Adobe Photoshop

Photoshop is the standard in a tool for graphics, image processing, and image manipulation. Layers of images, graphics, and text can be separately manipulated for maximum flexibility, and its set of filters permits creation of sophisticated lighting effects.

GIMP

GIMP is a free and open-source graphics editor alternative to Photoshop. It supports many bitmap formats, such as GIF, PNG, and JPEG. These are *pixel-based* formats, in that each pixel is specified. It also supports *vector-based* formats, in which endpoints of lines are specified instead of the pixels themselves, such as SWF (Adobe Flash). It can also read Photoshop format.

1.3.4 Video Editing

Adobe Premiere

Premiere is a simple, intuitive video editing tool for *nonlinear* editing—putting video clips into any order. Video and audio are arranged in *tracks*, like a musical score. It provides a large number of video and audio tracks, superimpositions, and virtual clips. A large library of built-in transitions, filters, and motions for clips allows easy creation of effective multimedia productions.

CyberLink PowerDirector

PowerDirector, produced by CyberLink Corp., is another popular nonlinear video editing software. It provides a rich selection of audio and video features and special effects and is easy to use. It supports modern video formats including AVCHD 2.0, 4K Ultra HD, and 3D video. It also supports end-to-end 360° video editing, as well as 64-bit video processing, graphics card acceleration, and multiple CPUs. Its processing and preview are known to be faster than Premiere. However, it is not as "programmable" as Premiere.

Adobe After Effects

After Effects is a powerful video editing tool that enables users to add and change existing movies with effects such as lighting, shadows, and motion blurring. It also allows layers, as in Photoshop, to permit manipulating objects independently.

iMovie

iMovie is a video editing tool for MacOS and iOS devices. It is versatile, convenient for video editing and creation of movie trailers. iMovie on iPhones is especially handy and popular. Later versions of iMovie also support 4K UHD video editing.

Final Cut Pro

Final Cut Pro is a professional video editing tool offered by Apple for MacOS. It allows the input of video and audio from numerous sources, and provides a complete environment, from editing and color correction to the final output of a video file.

1.3.5 Animation

Multimedia APIs

Java3D is an API used by Java to construct and render 3D graphics, similar to the way Java Media Framework handles media files. It provides a basic set of object primitives (cube, splines, etc.) upon which the developer can build scenes. It is an abstraction layer built on top of OpenGL or DirectX (the user can select which), so the graphics are accelerated.

DirectX, a Windows API that supports video, images, audio, and 3D animation, is a common API used to develop multimedia Windows applications such as computer games.

OpenGL was created in 1992 and is still a popular 3D API today. OpenGL is highly portable and will run on all popular modern operating systems, such as UNIX, Linux, Windows, and MacOS.

Animation Software

Autodesk 3ds Max (formerly 3D Studio Max) includes a number of high-end professional tools for character animation, game development, and visual effects production. Models produced using this tool can be seen in several consumer games, such as for the Sony Playstation.

Autodesk Maya is a complete modeling, animation, and rendering package. It features a wide variety of modeling and animation tools, such as to create realistic clothes and fur. Autodesk Maya runs on Windows, MacOS, and Linux.

Blender is a free and open-source alternative to the paid Autodesk suite of tools. It also offers a complete modeling, animation, and rendering feature set, as well as Python scripting capabilities.

GIF Animation Packages

For a much simpler approach to animation that also allows quick development of effective small animations for the web, many shareware and other programs permit creating animated GIF images. GIFs can contain several images, and looping through them creates a simple animation. Linux also provides some simple animation tools, such as `animate`.

1.3.6   Multimedia Authoring

Tools that provide the capability for creating a complete multimedia presentation, including interactive user control, are called *authoring* programs.

Adobe Animate

Adobe Animate (formerly Adobe Flash[3]) allows users to create interactive presentations for many different platforms in many different formats, such as HTML5 and WebGL. The content creation process in Animate follows the score metaphor—a timeline arranged in parallel event sequences, much like a musical score consisting

---

[3] Flash and Shockwave were the dominant online multimedia platforms from the late 1990s up until the early 2010s. It fell out of flavor due to its many inherit security flaws and resources required to support the ecosystem. HTML5 has largely overtaken this new role.

of musical notes. Elements in Animate are called *symbols*. The symbols are added to a central repository, called a library, and can be added to the multimedia timeline. Once the symbols are present at a specific time, they appear on the stage, which represents what the movie looks like at a certain time, and can be manipulated and moved by the tools built into Animate.

Adobe Director

Adobe Director (formerly Macromedia Director) is a multimedia application authoring platform. It uses a movie metaphor to create interactive presentations. It includes a built-in scripting language, Lingo, that allows creation of complex interactive movies. Adobe has discontinued its support to Director since 2017. However, as a powerful tool, Director is still being used in the community to date.

Adobe Dreamweaver

Dreamweaver is a webpage authoring tool that allows users to produce web-based multimedia presentations without learning any HTML.

Software Development Kits

Unity Engine is a multimedia engine development kit targeting novice or small development teams for producing video games and other interactive off-the-shelf experiences. Unity can be deployed to many platforms, such as a PC, smartphone, or a console window. It is free for personal use and can also be used professionally with varying subscription costs and no royalties.

Unreal Engine offers high-fidelity visuals out of the box, unlike Unity, but has a steep learning curve. It is commonly used by and for large game development studios targeting high-quality visual effects with the latest hardware.

1.3.7 Multimedia Broadcasting

OBS, XSplit

OBS and XSplit are two widely used broadcasting tools. OBS is free and open source, while XSplit is proprietary and paid. These tools can be thought of as an entire broadcasting production studio in digital form. They offer built-in support for switching between different cameras and other multimedia sources for real-time broadcasting. Users can broadcast live video feeds to websites like YouTube Live, Mixer, Twitch, and various other live streaming websites.

1.4   The Future of Multimedia

This textbook emphasizes the *fundamentals* of multimedia, focusing on the basic and mature techniques that collectively form the foundation of today's multimedia systems. It is worth noting, however, that multimedia research remains young and is vigorously growing. It brings many exciting topics together, and we will certainly see great innovations that will dramatically change our life in the near future [12, 13].

For example, researchers are interested in camera-based object tracking technology. But while face detection is ubiquitous, with camera software doing a reasonable job of identifying faces in images and video, face recognition, object tracking, and action/event analysis are by no means solved problems today. While shot detection—finding where scene changes exist in video—and video classification have for some time been of interest, new challenges have now arisen in these old subjects due to the abundance of online video that is not professionally edited.

Extending conventional 2D video, today's 3D capture technology is fast enough to allow acquiring dynamic characteristics of the human facial expression during speech, to synthesize highly realistic facial animation from speech for low-bandwidth applications. Beyond this, multiple views from several cameras or from a single camera under differing lighting can accurately acquire data that gives both the shape and surface properties of materials, thus automatically generating synthetic graphics models. This allows photo-realistic (and video quality) synthesis of virtual actors. Multimedia applications aimed at handicapped persons, particularly those with poor vision and the elderly, are a rich field of endeavor in current research, too. Another related example is *Google Glass*, which, equipped with an optical head-mounted display, enables interactive, smartphone-like information display for its users. Wirelessly connected to the Internet, it can also communicate using natural language voice commands. All these scenarios for exciting new technologies make a good step forward toward *wearable computing*, with great potential—arguably in the near-future.

The past decade has also witnessed an explosion of new-generation communication technologies for multimedia processing and delivery. Without a doubt, networked services in the future will focus on the user experience and participation with rich media. Online social media, such as YouTube, Facebook, Twitter, Twitch, and TikTok, are rapidly changing mechanisms for information generation and sharing, even in our daily life. Research on online social media is likely one of the most important areas under scrutiny, with some 100,000 academic articles produced per year in this area.

The vast coverage of social media sharing, however, presents unprecedented scale challenges. Highly diversified content types, origins, and distribution channels further impose complex interactions among the individual participants, particularly for advanced immersive applications such as interactive livecast/gamecast, multi-party media production, online gaming, and augmented reality (AR) and virtual reality (VR). They make very high demands on quality and responsiveness, much higher than conventional voice communications or Internet TV (e.g., latency on the order of 100 ms or even below 10 ms). For emerging applications such as video analytics and assisted/autonomous driving, the consumers have largely become intelligent machines, which, without a human-in-the-loop, impose even more stringent demands. In light of this, new networking and distributed computing architectures have been advocated, e.g., 5G cellular networks, as well as cloud and edge computing. The 6G cellular network standard is still under development, and ultra-low latency and high-bandwidth support for massive scale multimedia end devices have been the primary focus.

Crowdsourcing for multimedia: This concept, making use of the input from a large number of human contributors in multimedia projects, has experienced a large growth in attention. For example, having people provide tags to aid in understanding the visual content of images and video, such as Amazon's "Mechanical Turk," to outsource such time-consuming tasks as semantic video annotation to a large number of workers who are willing to work for small reward or just for fun. A straightforward use of such large populations is to analyze "sentiment," such as the popularity of a particular brand name as evidenced by reading several thousand tweets on the subject. Another example is "Digital fashion," which aims to develop smart clothing that can communicate with other such enhanced clothing using wireless communication, so as to artificially enhance human interaction in a social setting. The vision here is to use technology to allow individuals to allow certain thoughts and feelings to be broadcast automatically, for exchange with others equipped with similar technology.

Executable academic papers: In science and engineering, one traditional way to communicate findings is by publication of papers in academic journals. A new idea that exploits the completely digital pathway for broadcast of information is called "Executable papers." The idea here is that results discussed in a published paper are often difficult to reproduce. The reason is that datasets being used and programming code working on that data are typically not supplied as part of the publication. The executable papers approach allows the "reader" to interact with and interactively manipulate the data and code, to further understand the findings being presented. Moreover, the concept includes allowing the reader to rerun the code, change parameters, or upload different data.

Animated lifelike virtual agents: For example, virtual educators, in particular, as social partners for special needs children, and various other roles that are designed to demonstrate emotion and personality, with a variety of embodiments. The objective is flexibility, as opposed to a fixed script.

Behavioral science models can be brought into play to model interaction between people, which can then be extended to enable natural interaction by virtual characters. Such "augmented interaction" applications can be used to develop interfaces between real and virtual humans for tasks such as augmented storytelling.

Each of these application areas pushes the development of computer science generally, stimulates new applications, and fascinates practitioners. The chief leaders of multimedia research have generated several overarching "grand challenge" problems, which act as type of state of the art for multimedia interests. At present, some of these consist of the following:

- Social event detection for social multimedia: discovering social events planned and attended by people, as indicated by collections of multimedia content that was captured by people and uploaded to social-media sites.

- Sports video annotation: using video classification to label video segments with certain actions such as strokes in table tennis, penalty kicks in soccer games, etc.
- GameStory: a video game analytics challenge in which e-sport games often involving millions of players and viewers are analyzed. Audio and video streams, commentaries, game data and statistics, interaction traces, and other multimedia data are used to summarize and predict the development and evolution of the games.
- Live video streaming: requiring ultra-low end-to-end latency for real-time interaction between the broadcaster and the viewers. The main challenge is the QoE (quality of experience), due to the latency constraint.
- Violent scenes detection in film: automatically detecting portions of movies depicting violence. Again, all aspects available such as text and audio could be brought into play.
- Preserving privacy in surveillance videos: methods obscuring private information (such as faces on Google Earth), so as to render privacy-sensitive elements of video unrecognizable, while at the same time allowing the video to still be viewable by people and also allow computer vision tasks such as object tracking.
- Deep video understanding: understanding the relationships between different entities from a long duration movie. The relations can be family, work, social, and other types. All modalities of multimedia will be exploited and a knowledge base will be constructed and consulted.
- Large-scale human-centric video analysis: analyzing various crowd and complex events such as getting off a train, dining in a busy restaurant, earthquake escape, etc. Issues include multi-person tracking, crowd pose estimation and tracking, and person-level action recognition.
- Searching and question answering for the spoken web: searching for audio content within audio content by using an audio query, matching spoken questions with a collection of spoken answers.
- Multimedia recommender systems: improving the quality of recommender systems to produce items more relevant to users' interests. Applications include movie recommendation, news recommendation, etc.

Solutions to these challenges can be difficult, but the impact can be enormous, not only to the IT industry, but also to everyone, as we all live in a digital multimedia world. We want this textbook to bring valuable knowledge about multimedia to you, and hope you enjoy it and perhaps even contribute to this promising field (maybe for some of the topics listed above or beyond) in your future career!

1.5 Exercises

1. Using your own words, describe what "multimedia" is. Is multimedia simply a collection of different types of media?
2. Identify three novel multimedia applications. Discuss why you think these are novel and their potential impact.
3. Discuss the relation between multimedia and hypermedia.

4. Briefly explain, in your own words, "Memex" and its role regarding hypertext. Could we carry out the Memex task today? How do you use Memex ideas in your own work?
5. Discover a current media input, storage, or playback device that is analog. Is it necessary to convert to digital? What are the pros and cons of being analog or digital?
6. Your task is to think about the transmission of smell over the Internet. Suppose we have a smell sensor at one location and wish to transmit the *Aroma Vector* (let us say) to a receiver, to reproduce the same sensation. You are asked to design such a system. List three key issues to consider and two applications of such a delivery system. *Hint*: Think about medical applications.
7. Tracking objects or people can be done by both sight and sound. While vision systems are precise, they are relatively expensive; on the other hand, a pair of microphones can detect a person's *bearing* inaccurately but cheaply. Sensor *fusion* of sound and vision is thus useful. Surf the web to find out who is developing tools for videoconferencing using this kind of multimedia idea.
8. *Non-photorealistic* graphics means computer graphics that do well enough without attempting to make images that look like camera images. An example is conferencing. For example, if we track lip movements, we can generate the right animation to fit our face. If we don't much like our own face, we can substitute another one—facial-feature modeling can map correct lip movements onto another model. See if you can find out who is carrying out research on generating avatars to represent conference participants' bodies.
9. Watermarking is a means of embedding a hidden message in data. This could have important legal implications: Is this image copied? Is this image doctored? Who took it? Where? Think of "messages" that could be sensed while capturing an image and secretly embedded in the image, so as to answer these questions. (A similar question is derived from the use of cell phones. What could we use to determine who is putting this phone to use, and where, and when? This could eliminate the need for passwords or others using the phone you lost.).

References

1. B. Newhall, *The History of Photography: from 1839 to the Present*. (The Museum of Modern Art, 1982)
2. T. Gustavson, G.E. House, *Camera: a History of Photography from Daguerreotype to Digital*. (Sterling Signature, 2012)
3. A. Koenigsberg, *The Patent History of the Phonograph, 1877–1912*. (APM Press, 1991)
4. D.L. Morton, Jr., *Sound Recording: the Life Story of a Technology*. (Johns Hopkins University Press, 2006)
5. Q.D. Bowers, K. Fuller-Seeley, *One Thousand Nights at the Movies: an Illustrated History of Motion Pictures, 1895–1915*. (Whitman Publishing, 2012)
6. T.K. Sarkar, R. Mailloux, A. Arthur, M. Salazar-Palma, D.L. Sengupta, *History of Wireless*. (Wiley-IEEE Press, Oliner, 2006)

7. M. Hilmes, J. Jacobs, *The Television History Book (Television, Media and Cultural Studies)*. (British Film Institute, 2008)
8. V. Bush, As we may think, in *The Atlantic Monthly* (1945)
9. N. Yankelovitch, N. Meyrowitz, A. van Dam, Reading and writing the electronic book, in *Hypermedia and Literary Studies*, ed. by P. Delany, G.P. Landow (MIT Press, 1991)
10. D. Engelbart, H. Lehtman, Working together, in *BYTE Magazine* (1988), pp. 245–252
11. J. Duckett, *HTML and CSS: design and Build Websites* (Wiley, 2011)
12. K. Nahrstedt, R. Lienhart, Mal. Slaney. *Special Issue on the 20th Anniversary of ACM SIGMM*. ACM Trans. Multimed. Comput. Commun. Appl. (TOMCCAP) (2013)
13. A. Nauman et al., Multimedia Internet of things: a comprehensive survey. IEEE Access **8**, 8202–8250 (2020)

A Taste of Multimedia

2

2.1 Multimedia Tasks and Concerns

Multimedia content is ubiquitous in software all around us, including in our phones, of course. We are interested in this subject from a computer science and engineering point of view, as well as being interested in making interactive applications (or "presentations"), using video editors such as Adobe Premiere and still-image editors such as Adobe Photoshop in the first instance, but then combining the resulting resources into interactive programs by making use of "authoring" tools such as Flash and Director that can include sophisticated programming. Multimedia often generates problems and considerations that have a more general computer science flavor. For example, most cameras now are smart enough to find faces (with reasonable success)—but just recently such a task was firmly in the domain of Computer Vision, i.e., a branch of Artificial Intelligence dealing with trying to *understand* image content. So such more basic concerns do impact multimedia as it now appears in products, and will tend to increasingly influence the field. Continuing in the Computer Vision direction, a camera owner might be encouraged to think like a computer scientist and ask "What is going on in an image?" A less high-level question is "Where has this image been taken?" (scene recognition) or "Does the image contain a particular object?" (object classification). A still quite difficult question is "Where is an object of interest?" (object detection). And a lower level question might be "Which object does each pixel belong to?" (image segmentation). Thus, it does not take long before we find ourselves fully engaged in a classic Computer Vision hierarchy of high level to detailed description of an image, with scene recognition at the top and image segmentation at the bottom.

In this text, we take a very moderate approach to difficulty level, and don't presume to answer such sophisticated questions as those posed above. Nonetheless, studying the fundamentals of the multimedia problem is indeed a fruitful concern and we

© Springer Nature Switzerland AG 2021
Z.-N. Li et al., *Fundamentals of Multimedia*, Texts in Computer Science,
https://doi.org/10.1007/978-3-030-62124-7_2

aim the book at giving readers the tools they would need to eventually tackle such difficult questions, for example, in a work situation.

2.2 Multimedia Presentation

In this section, we briefly outline some effects to keep in mind for presenting multimedia content as well as some useful guidelines for content design [1,2].

Graphics Styles

Careful thought has gone into combinations of color schemes and how lettering is perceived in a presentation. Many presentations are meant for business projected displays, rather than appearing on a screen close to the eye. Human visual dynamics are considered in regard to how such presentations must be constructed. Most of the observations here are drawn from Vetter et al. [3], as shown in Fig. 2.1.

Color Principles and Guidelines

Some *color schemes* and *art styles* are best combined with a certain theme or style. Color schemes could be, for example, natural and floral for outdoor scenes and solid colors for indoor scenes. Examples of art styles are oil paints, watercolors, colored pencils, and pastels.

A general hint is to not use too many colors, as this can be distracting. It helps to be consistent with the use of color—then color can be used to signal changes in theme.

Fonts

For effective visual communication, large fonts (18–36 points) are best, with no more than six to eight lines per screen. As shown in Fig. 2.1, sans serif fonts work better than serif fonts (serif fonts are those with short lines stemming from and at an angle to the upper and lower ends of a letter's strokes). Figure 2.1 shows a comparison of two screen projections, (Figs. 2 and 3 from Vetter et al. [3]).

The top figure shows *good* use of color and fonts. It has a consistent color scheme, uses large and all sans serif (Arial) fonts. The bottom figure is *poor*, in that too many colors are used, and they are inconsistent. The red adjacent to the blue is hard to focus on, because the human retina cannot focus on these colors simultaneously. The serif (Times New Roman) font is said to be hard to read in a darkened, projection setting. Finally, the lower right panel does not have enough contrast—pretty pastel colors are often usable only if their background is sufficiently different.

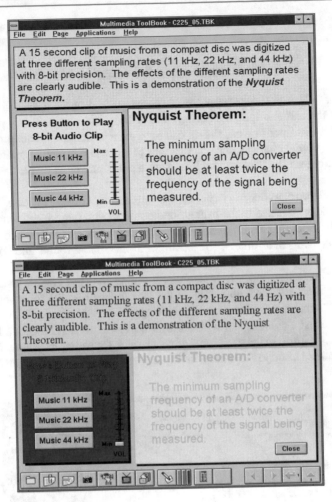

Fig. 2.1 Colors and fonts. Courtesy of Ron Vetter

A Color Contrast Program

Seeing the results of research of Vetter et al., we constructed a small Java program[1] to investigate how readability of text depends on the text color and the color of the background.

The simplest approach to making readable colors on a screen is to use the principal complementary color as the background for text. For color values in the range 0–1 (or, effectively, 0–255), if the text color is some triple (red, green, blue), or (R, G, B)

---

[1] See http://www.cs.sfu.ca/mmbook. There both the executable and the program source are given.

for short, a legible color for the background is likely given by that color subtracted from the maximum:

$$(R, G, B) \Rightarrow (1 - R, 1 - G, 1 - B) \tag{2.1}$$

That is, not only is the color "opposite" in some sense (not the same sense as artists use), but if the text is bright, the background is dark, and vice versa.

In the Java program given, sliders can be used to change the background color. As the background changes, the text changes to equal the principal complementary color. Clicking on the background brings up a color picker as an alternative to the sliders.

If you feel you can choose a better color combination, click on the text. This brings up a color picker not tied to the background color, so you can experiment. (The text itself can also be edited.) A little experimentation shows that some color combinations are more pleasing than others, for example, a pink background and forest green foreground or a green background and mauve foreground. Figure 2.2 shows this small program in operation. It is also worth noting that the graphic user interface (GUI) of this program is with a modern flat design (see Fig. 2.1 for comparison).

Figure 2.3 shows a "color wheel," with opposite colors equal to $(1 - R, 1 - G, 1 - B)$. An artist's color wheel will not look the same, as it is based on feel rather than on an algorithm. In the traditional artist's wheel, for example, yellow is opposite magenta, instead of opposite blue as in Fig. 2.3, and blue is instead opposite orange.

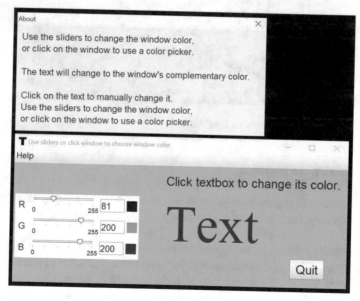

Fig. 2.2 Program to investigate colors and readability

Fig. 2.3 Color wheel

Sprite Animation

Sprites are often used in animation. For example, in Adobe Director, the notion of a sprite is expanded to an instantiation of any resource. However, the basic idea of sprite animation is simple. Suppose we have produced an animation figure, as in Fig. 2.4a. Then it is a simple matter to create a 1-bit mask M, as shown in Fig. 2.4b, black on white, and the accompanying sprite S, as shown in Fig. 2.4c.

Now we can overlay the sprite on a colored background B, as shown in Fig. 2.5a, by first ANDing B and M, then ORing the result with S, with the final result as in Fig. 2.5e. Operations are available to carry out these simple compositing manipulations at frame rate and so produce a simple 2D animation that moves the sprite around the frame but does not change the way it looks.

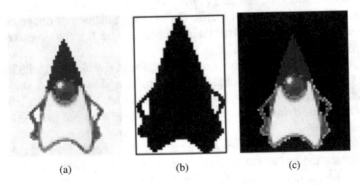

      (a)   (b)   (c)

Fig. 2.4 Sprite creation: **a** original; **b** mask image M; and **c** sprite S. *"Duke" figure courtesy of Sun Microsystems*

Fig. 2.5 Sprite animation: **a** Background B; **b** mask M; **c** B AND M; **d** sprite S; **e** B AND M OR S

Video Transitions

Video *transitions* can be an effective way to indicate a change to the next section. Video transitions are syntactic means to signal "scene changes" and often carry semantic meaning. Many different types of transitions exist; the main types are *cuts, wipes, dissolves, fade-ins,* and *fade-outs.*

A cut, as the name suggests, carries out an abrupt change of image contents in two consecutive video frames from their respective clips. It is the simplest and most frequently used video transition.

A wipe is a replacement of the pixels in a region of the viewport with those from another video. If the boundary line between the two videos moves slowly across the screen, the second video gradually replaces the first. Wipes can be left-to-right, right-to-left, vertical, horizontal, like an iris opening, swept out like the hands of a clock, and so on.

A dissolve replaces every pixel with a mixture over time of the two videos, gradually changing the first to the second. A fade-out is the replacement of a video by black (or white), and fade-in is its reverse. Most dissolves can be classified into two types, corresponding, for example, to *cross dissolve* and *dither dissolve* in Adobe Premiere video editing software.

In Type I (cross dissolve), every pixel is affected gradually. It can be defined by

$$\mathbf{D} = (1 - \alpha(t)) \cdot \mathbf{A} + \alpha(t) \cdot \mathbf{B} \tag{2.2}$$

(a)   (b)   (c)

Fig. 2.6 **a** Video$_L$; **b** Video$_R$; **c** Video$_L$ sliding into place and pushing out Video$_R$

where **A** and **B** are the color 3-vectors for video A and video B. Here, $\alpha(t)$ is a transition function, which is often linear with time t:

$$\alpha(t) = kt, \quad \text{with} \quad kt_{\max} \equiv 1 \tag{2.3}$$

Type II (dither dissolve) is entirely different. Determined by $\alpha(t)$, increasingly more and more pixels in video A will abruptly (instead of gradually, as in Type I) change to video B. The positions of the pixels subjected to the change can be random or sometimes follow a particular pattern.

Obviously, fade-in and fade-out are special types of a Type I dissolve, in which video A or B is black (or white). Wipes are special forms of a Type II dissolve, in which changing pixels follow a particular geometric pattern.

Despite the fact that many digital video editors include a preset number of video transitions, we may also be interested in building our own. For example, suppose we wish to build a special type of wipe that slides one video out while another video slides in to replace it. The usual type of wipe does not do this. Instead, each video stays in place, and the transition line moves across each "stationary" video, so that the left part of the viewport shows pixels from the left video, and the right part shows pixels from the right video (for a wipe moving horizontally from left to right).

Suppose we would like to have each video frame not held in place, but instead move progressively farther into (out of) the viewport: we wish to slide Video$_L$ in from the left and push out Video$_R$. Figure 2.6 shows this process. Each of Video$_L$ and Video$_R$ has its own values of R, G, and B. Note that R is a function of position in the frame, (x, y), as well as of time t. Since this is video and not a collection of images of various sizes, each of the two videos has the same maximum extent, $x_{\max}$. (Premiere actually makes all videos the same size—the one chosen in the preset selection—so there is no cause to worry about different sizes.)

As time goes by, the horizontal location x_T for the transition boundary moves across the viewport from $x_T = 0$ at $t = 0$ to $x_T = x_{\max}$ at $t = t_{\max}$. Therefore, for a transition that is linear in time, $x_T = (t/t_{\max})x_{\max}$.

So for any time t, the situation is as shown in Fig. 2.7a. The viewport, in which we shall be writing pixels, has its own coordinate system, with the x-axis from 0 to $x_{\max}$. For each x (and y), we must determine (a) from which video we take RGB, i.e., (red, green, blue) values, and (b) from what x-position in the *unmoving* video we take pixel values, that is, from what position x from the left video, say, in its own

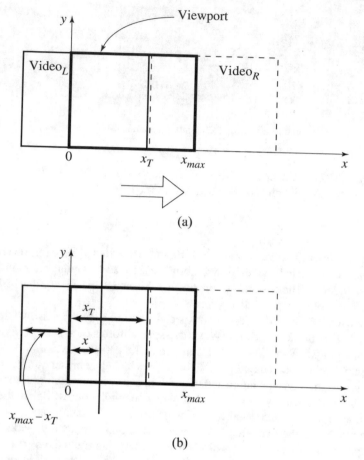

Fig. 2.7 **a** Geometry of Video$_L$ pushing out Video$_R$; **b** Calculating position in Video$_L$ from where pixels are copied to the viewport

coordinate system. It is a video, so of course the image in the left video frame is changing in time.

Let's assume that dependence on y is implicit. In any event, we use the same y as in the source video. Then for the red channel (and similarly for the green and blue), $R = R(x, t)$. Suppose we have determined that pixels should come from Video$_L$. Then the x-position x_L in the *unmoving* video should be $x_L = x + (x_{max} - x_T)$, where x is the position we are trying to fill in the viewport, x_T is the position in the viewport that the transition boundary has reached, and x_{max} is the maximum pixel position for any frame.

To see this, we note from Fig. 2.7b that we can calculate the position x_L in Video$_L$'s coordinate system as the sum of the distance x, in the viewport, and the difference $x_{max} - x_T$.

```
for  t  in  0..t_max
    for  x  in  0..x_max
        if  ( x/x_max  <  t/t_max )
            R = R_L ( x + x_max * [1 - t/t_max], t)
        else
            R = R_R ( x - x_max * t/t_max, t)
```

Fig. 2.8 Pseudocode for slide video transition

Substituting the fact that the transition moves linearly with time, $x_T = x_{max}(t/t_{max})$, we can set up a pseudocode solution as shown in Fig. 2.8. In Fig. 2.8, the slight change in formula if pixels are actually coming from Video$_R$ instead of from Video$_L$ is easy to derive.

The exercise section in the end of the chapter contains suggestions for further such video transitions. As a computer scientist or engineer, you should be easily capable of constructing your own video transitions, rather than relying on simply choosing items from a menu. A career in multimedia involves addressing interesting and sometimes challenging tasks that no-one actually solved before!

2.3   Data Compression

One of the most evident and important challenges of using multimedia is the necessity to compress data. Table 2.1 shows some values for standard definition and for high-definition broadcast video. Clearly, we need excellent and fast data compression in order to avoid such high data rates that cause problems for storage and networks, if we tried to share such data and also for disk I/O.

Table 2.1 Uncompressed video sizes

Standard definition video	
640×480 full color	= 922 kB/frame
@ 30 frames/s	= 28 MB/s
	= 221 Mb/s
× 3,600 s/h	= 100 GB/h
High definition video	
1,920×1,080 full color	= 6.2 MB/frame
@ 30 frames/s	= 187 MB/s
	= 1.5 Gb/s
× 3,600 s/h	= 672 GB/h

How much compression is required? In effect, this depends on the application, on the capability of the viewing computer and display, and on the bandwidth (in bits per second) available to perhaps stream and certainly to view the decompressed result.

In the JPEG image compression standard, the amount of compression is controlled by a value qf (*Quality Factor*) in the range 1–100 (see Sect. 9.1 for details). The "quality" of the resulting image is best for $qf = 100$ and worst for $qf = 1$.

Figure 2.9a shows an original, uncompressed image (Český Krumlov, Czech Republic) taken by a digital camera that allows full-accuracy images to be captured, with no data compression at all. The original image resolution is 3,232 by 2,424. It is reduced to 485 columns and 364 rows of pixel data to better illustrate the effect of qf. With 8-bit accuracy in each of red, green, and blue pixel values, the total file size is $485 \times 364 \times 3 = 529,620$ bytes (not including file-header information, which stores such values as the column and row size).

In Table 2.2, we show results using different quality factors in JPEG compression. Indeed, we can greatly shrink the file size down, but for small values of qf the resulting image is poor.

We can see in Fig. 2.9 that while $qf = 25$ is not terrible, if we insist on going down to a quality factor of $qf = 5$ we do end up with an unusable image. However,

(a)   (b)

(c)   (d)

Fig. 2.9 JPEG compression: **a** Original uncompressed image; **b** JPEG compression with quality factor $qf = 75$ (the typical default); **c, d** Quality factors $qf = 25$ and $qf = 5$

Table 2.2 JPEG file sizes (bytes) and percentage size of data for JPEG compression with quality factor $qf = 75, 25$, and 5

Quality factor (qf)	Compressed file size	Percentage of original (%)
–	529,620	100
75	37,667	7.11
25	16,560	3.13
5	5,960	1.13

this exercise does show us something interesting: the color part, as opposed to the black and white (i.e., the grayscale) may well be the less noticeable problem for high compression ratios (i.e., low amounts of data surviving compression). We'll see how color and grayscale are, in fact, treated differently in Chap. 9.

Compression indeed saves the day, but at a price too. JPEG compression can effect a compression ratio of 15:1 with little loss of quality. For video compression, the MPEG video compression standard, set out in Chap. 11, can produce a compression ratio of 100:1 while retaining reasonable quality.

However, let's look at how expensive image and video processing is in terms of processing in the CPU. Suppose we have an image whose pixels we wish to darken, by a factor of 2. The following code fragment is pseudocode for such an operation:

```
for x = 1 to columns
{
    for y = 1 to rows
    {
        image[x,y].red    /= 2;
        image[x,y].green /= 2;
        image[x,y].blue  /= 2;
    }
}
```

On a RISC machine, the loop amounts to one increment, one check, and one branch instruction. There are also three loads, three shifts, and three stores. This makes a total of 12 instructions per pixel, i.e., per 3 bytes. So, we have four instructions per image byte. For standard-definition video, we have 28 MB/s, meaning 28 × 4=112 mega instructions per second. For high definition, at 187 MB/s, we need 748 mega instructions per second.

This is certainly possible. However, JPEG compression takes some 300 instructions per pixel or, in other words, 100 instructions per image byte. This yields numbers of 2.8 billion instructions per second for standard definition and 19 billion instructions per second for high definition, which begins to be a real constraint! Clearly, clever techniques are required, and we will view these in later chapters.

Other issues arise from trying to interact with multiple streams of data, e.g., what happens if we tried to show video of a news interview, plus some video of background information, plus data streams of additional information, etc. Is compositing (putting together) such information first, and then compressing the best way forward? Or is compositing at the receiver end? Multimedia tends to open up new questions on computer science itself. Multiple data streams place new burdens on operating systems, in terms of scheduling and resource management.

As well, new capabilities can imply new demands: what happens if the rock band needs to rehearse music together, but they are not in the same place (a distributed music problem). The question becomes one of how much latency (time lag) is acceptable when we're doing compression, for various applications. For music rehearsal, all the band members have to hit the lead note at very close to the same time!

2.4   Multimedia Production

A multimedia project can involve a host of people with specialized skills. In this book, we emphasize technical aspects, but as well multimedia production can easily involve an art director, graphic designer, production artist, producer, project manager, writer, user interface designer, sound designer, videographer, and 3D and 2D animators, as well as programmers.

The production timeline would likely only involve programming when the project is about 40% complete, with a reasonable target for an alpha version (an early version that does not contain all planned features) being perhaps 65–70% complete. Typically, the design phase consists of storyboarding, flowcharting, prototyping, and user testing, as well as a parallel production of media. Programming and debugging phases would be carried out in consultation with marketing, and the distribution phase would follow.

A storyboard depicts the initial idea content of a multimedia concept in a series of sketches. These are like "keyframes" in a video—the story hangs from these "stopping places." A flowchart organizes the storyboards by inserting navigation information—the multimedia concept's structure and user interaction. The most reliable approach for planning navigation is to pick a traditional data structure. A hierarchical system is perhaps one of the simplest organizational strategies.

Multimedia is not really like other presentations, in that careful thought must be given to organization of movement between the "rooms" in the production. For example, suppose we are navigating an African safari, but we also need to bring specimens back to our museum for close examination—just how do we effect the transition from one locale to the other? A flowchart helps to imagine the solution.

The flowchart phase is followed by development of a detailed functional specification. This consists of a walk-through of each scenario of the presentation, frame by frame, including all screen action and user interaction. For example, during a mouseover for a character, the character reacts, or a user clicking on a character results in an action.

The final part of the design phase is prototyping and testing. Some multimedia designers use a specialized multimedia authoring tool at this stage already, even if the intermediate prototype will not be used in the final product or continued in another tool. User testing is, of course, extremely important before the final development phase.

2.5   Multimedia Sharing and Distribution

Multimedia content, once produced, needs to be published and then shared among users. In recent years, traditional storage and distribution media, such as optical discs, have been largely replaced by USB flash drives or solid-state drives (SSD), or more conveniently, the Internet.

Consider YouTube, the most popular video sharing site over the Internet, as an example. A user can easily create a Google account and channel (as YouTube is now owned by Google), and then upload a video, which will be shared to everyone or to selected users. YouTube further enables titles and tags that are used to classify the videos and link similar videos together (shown as a list of related videos). Figure 2.10 shows the webpage for uploading a video from a local computer to YouTube. The video, captured by us for a 1905 Edison Fireside phonograph with a cygnet horn playing a cylinder record, can be searched from YouTube's homepage with "Edison Phonograph Multimedia Textbook 3rd Edition," which was the title and tag supplied by us.

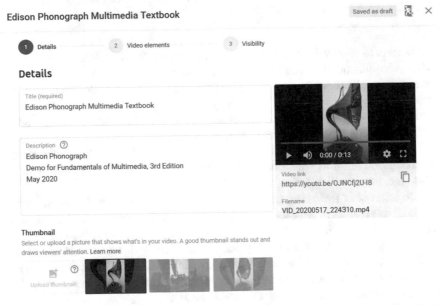

Fig. 2.10 The webpage for uploading a YouTube video. The video, titled "Edison Phonograph Multimedia Textbook 3rd Edition," is open to all users (Privacy settings: Public) and can be searched in the YouTube homepage using the title or tags. Note that the video thumbnails are automatically generated by YouTube

Fig. 2.11 The YouTube page for the video uploaded. The list of related videos are shown on the right side and users can post their comments, too

Figure 2.11 shows the YouTube page of this video. It also offers a list of related videos, recommended by YouTube. Ideally, we expect that it is linked to other videos about Edison phonographs or multimedia textbooks. The results shown in Fig. 2.11 are not exactly what we would expect, but they do relate to a certain degree. Indeed, multimedia content retrieval and recommendation remains quite difficult, and we will review some of the basic techniques in Chap. 21.

The link to this video can be fed into such other social networking sites such as Facebook or Twitter as well, potentially propagating to many users of interest in a short time, as we will examine in Chap. 19.

The Internet is reshaping traditional TV broadcasting as well. Netflix, which builds all its services in Amazon's cloud platform, has become one of the largest TV content providers. With a customer base over 137 million worldwide (Year 2018), it has surpassed most conventional TV broadcasters. Twitch.tv, a gamecast and livecast platform, has 2.2 million broadcasters monthly and 15 million daily active users, with around a million average concurrent users. The users are mostly young, who don't have strong interest in traditional TV broadcast at all.

Users' viewing habits are also changing. Compelling content is the core foundation of any *Internet Protocol TV* (IPTV) proposition, which remains true today; yet IPTV services are becoming highly personalized, integrated, portable, and on-demand. Most service providers are moving beyond basic video offerings toward richer user experiences, particularly with the support for multi-screen viewing across TVs, PCs, tablets, and smartphones. Meanwhile, 3D, multi-view, and multi-streaming are being developed, in which multiple video streams from the same event are delivered to a user, who will be able to switch between camera views. This is a real recognition by service providers of what is happening in homes across the planet—that families are voraciously and simultaneously consuming streamed high-definition video on devices other than the traditional set-top box/TV pairs.

The scaling challenge for such multimedia content distribution is enormous [4]. To reach 100 million viewers, delivery of TV quality video encoded in MPEG-4 (1.5 Mbps) may require an aggregate capacity of 150 Tbps. To put things into perspective, consider two large-scale Internet video broadcasts: the NBC's live telecast of Super Bowl XLIX in 2015, which has a record-high viewership within the United States with 114.4 million viewers, and the 2018 FIFA World Cup Final (France versus Croatia), which has around 1 billion viewers worldwide. Even with low-bandwidth Internet video of 500 Kbps, the NBC broadcast needs more than 57 Tbps server and network bandwidth, and the FIFA broadcast worldwide would be an order of magnitude higher. These can hardly be handled by any single server. Later, in Chaps. 16 and 18, we will see effective solutions using peer-to-peer, content distribution networks, or the cloud to deal with such challenges.

2.6   Some Useful Editing and Authoring Tools

This textbook is primarily concerned with principles of multimedia—the fundamentals to be grasped for a real understanding of this subject. Nonetheless, we need real vehicles for showing this understanding, and straight programming in C++ or Java is not always the best way of showing your knowledge. Most introductory multimedia courses ask you to at least start off delivering some multimedia product (e.g., see Exercise 9).

Therefore, we'll consider some popular authoring tools. Since the first step in creating a multimedia application is probably creation of interesting video clips, we start off looking at a video editing tool. This is not really an authoring tool, but video creation is so important that we include a small introduction to one such program.

The tools we look at are the following:

- Premiere,
- HTML5 canvas,
- Director, and
- XD.

While this is, of course, by no means an exhaustive list, these tools are often used in creating multimedia content. Also note that many of them are now in Adobe's Creative Cloud, which gives subscribers access to a collection of software used for graphic design, video editing, web development, and photography, along with a set of mobile applications and also some optional cloud services.

2.6.1 Adobe Premiere

Premiere Basics

Adobe Premiere is a very simple yet powerful video editing program that allows you to quickly create a simple digital video by assembling and merging multimedia components. It effectively uses a "score" authoring metaphor, in that components are placed in "tracks" horizontally, in a Timeline window that in a sense resembles a musical score.

The `File > New Project` command opens a window that displays a series of "presets"—assemblies of values for frame resolution, compression method, and frame rate. Start by importing resources, such as AVI (Audio Video Interleave) video files and WAV sound files and dragging them from the Project window onto tracks 1 or 2.

Usually, you see only three tracks: Video 1, Video 2, and Video 3. Video transitions, meaning how we change from one video segment to another, are in the Effects window. Transitions are dragged into the Transitions track from the Transition window, such as a gradual replacement of Video 1 by Video 2 (a dissolve), sudden replacement of random pixels in a checkerboard (a dither dissolve), or a wipe, with one video sliding over another. There are many other transitions to choose from.

You can import WAV sound files by dragging them to Audio 1 or Audio 2 of the Timeline window or to any additional sound tracks. You can edit the properties of any sound track by right-clicking on it.

Figure 2.12 shows what a typical Premiere screen might look like. The yellow ruler at the top of the Timeline window delineates the working timeline—drag it to the right amount of time. The 1-second dropdown box at the bottom represents showing one video keyframe per 1 second.

To "compile" the video, go to `Sequence > Render Work Area` to have a look at the product you're making, and save the project as a `.prproj` file. To save the movie, select `File > Export > Movie`. Now it gets interesting, because you must make some choices here, involving how and in what format the movie is to be saved. Figure 2.13 shows the project options. The dialogs that tweak each codec are provided by the codec manufacturer; bring these up by clicking on the parts of the project being controlled in a panel or via a configure button. Compression codecs (compression–decompression protocols) are often in hardware on the video capture card. If you choose a codec that requires hardware assistance, someone else's system may not be able to play your brilliant digital video, and all is in vain!

Images can also be inserted into tracks. We can use transitions to make the images gradually appear or disappear in the final video window. To do so, set up a "mask" image, as in Fig. 2.14. Here, we have imported an Adobe Photoshop layered image, with accompanying alpha channel made in Photoshop.

Then in Premiere, we click on the image, which has been placed in its own video track, and in the Effect Controls window, click the triangle next to the Opacity property to enter a new opacity value, making the image partially transparent. Premiere controls making the face shown in the image have a transparent background using

Fig. 2.12 Adobe Premiere screen

Fig. 2.13 Adobe Premiere export settings screen

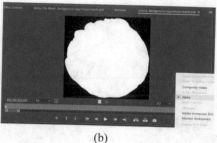

(a) (b)

Fig. 2.14 Adobe Premiere preview clip viewer, for an Adobe Photoshop image with an alpha-channel layer. **a**: RGB channels: color on black background. **b**: Alpha channel: white on black background

the alpha channel. It is also simple to use Motion effect to have the image fly in and out of the frame.

In Photoshop, we set up an alpha channel as follows:

1. Use an image you like—a `.JPG`, say.
2. Make the background some solid color—white, say.
3. Make sure you have chosen `Image > Mode > RGB Color`.
4. Select that background area (you want it to remain opaque in Premiere)—use the magic wand tool.
5. Go to `Select > Save Selection....`
6. Ensure that `Channel = New`. Press `OK`.
7. Go to `Window > Show Channels`, double-click the new channel, and rename it Alpha; make its color (0, 0, 0).
8. Save the file as a `PSD`.

If the alpha channel you created in Photoshop (the "key" we are using) has a white background, you'll need to reverse the key in Premiere.

Premiere has its own simple method of creating titles (to give credit where credit is due) for your digital video.

Another nice feature of Premiere is that it is simple to use in capturing video from old analog sources. To form a digital video from a camcorder input, go to `File > Capture`. (The menu for video/audio capture options appears by right-clicking the capture window.)

You can also set the compression and format you would like in the export settings by going to `File > Export > Media` or pressing `Ctrl+M`. Here, you have access to a drop down of preset compression options such as H.265, MPEG4, Quick-Time, and more. Alternatively, if you would like custom compression settings, you may go to the video tab to control the frame size, frame rate, pixel ratio, and more.

2.6.2 HTML5 Canvas

HTML5 is the latest evolution of the standard that defines HTML (Hypertext Markup Language), as recommended by the World Wide Web Consortium. It comes with new elements, attributes, and behaviors, and a larger set of technologies that allows the building of more diverse and powerful websites and applications, particularly with improved support for multimedia. From the late 1990s up until the early 2010s, Adobe Flash and Shockwave were the dominant tools for the creation of interactive animation. They fell out of favor due to the many inherent security flaws and resources required to support the ecosystem. HTML5 has largely overtaken this role.

HTML5 canvas is an element used to draw animated graphics on a webpage. Here, we give a brief introduction to HTML5 canvas and provide some simple examples of its use. More can be found in the Mozilla Developer Network (MDN) Web Docs and other sources.

In HTML5, a canvas is a rectangular area that is generated by HTML code with width and height attributes. Its background color is transparent by default. In the sample code below, we set the page body background color to be gray, and the canvas background color to be red, so as to make it visible.

```
<canvas id="canvas" width="300" height="300" style
="background-color: red">
 </canvas>
```

The <canvas> element is only a container for graphics. You need to use JavaScript to actually draw the graphical objects. When you create a new canvas, it generates one or more rendering context to display the content. Below is the JavaScipt code to get a 2D context. For your interest, WebGL (Web Graphics Library) can be used to render 3D context.

```
var canvas = document.getElementById("canvas");
var ctx = canvas.getContext("2d");
```

Draw on Canvas

You can draw simple shapes, text, images, path, or other objects on the canvas. The canvas is filled by grids, acting as an x-y coordinate system. The code below draws a rectangle, where x and y are the coordinates of top-left corner of the rectangle.

```
// a rectangle with filled color
ctx.fillRect(x, y, width, height);

// a rectangle with border
ctx.strokeRect(x, y, width, height);
```

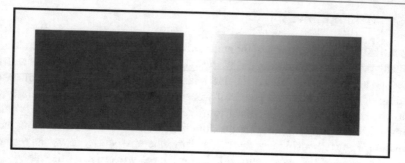

Fig. 2.15 A rectangle with pure color (left) and a rectangle with gradient color (right)

```
// clear a rectangle area
ctx.clearRect(x, y, width, height);
```

The shape can be filled up by the `fillStyle` method. Figure 2.15 shows two examples with a pure color and gradient color, respectively.

Transform and Animation

The shapes of the graphics can be transformed by `translate`, `rotate`, and `scale`.

```
ctx.translate(x,y);
ctx.rotate(degree);
ctx.scale(x_scaled_factor,y_scaled_factor);
```

To create an animation, you can take the following steps:

1. Clear the canvas: `ctx.clearRect(x, y, width, height)`.
2. Save the current state of canvas: `ctx.save()`.
3. Draw the graphics.
4. Restore the canvas state: `ctx.restore()`.

2.6.3 Adobe Director

Director is a complete environment (see Fig. 2.16) for creating interactive "movies." The movie metaphor is used throughout Director, and the windows used in the program reflect this. The main window, on which the action takes place, is the Stage.

The other two main windows are Cast and Score. A Cast consists of resources a movie may use, such as bitmaps, sounds, vector-graphics shapes, Flash movies,

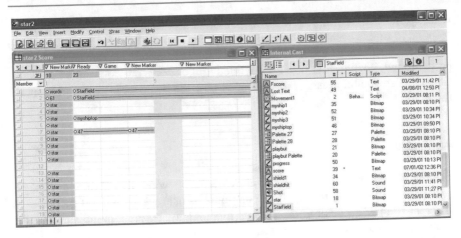

Fig. 2.16 Director: main windows

digital videos, and scripts. Cast members can be created directly or simply imported. Typically, you create several casts, to better organize the parts of a movie. Cast members are placed on the Stage by dragging them there from the Cast window. Because several instances may be used for a single cast member, each instance is called a sprite. Typically, cast members are raw media, whereas sprites are objects that control where, when, and how cast members appear on the stage and in the movie.

Sprites can become interactive by attaching "behaviors" to them (for example, make the sprite follow the mouse) either prewritten or specially created. Behaviors are in the internal script language of Director, called Lingo. Director is a standard event-driven program that allows easy positioning of objects and attachment of event procedures to objects. A very useful part of Lingo is called Imaging Lingo, which can directly manipulate images from within Director. This means that image manipulation can be carried out in code, making for code-based visual effects.

The set of predefined events is rich and includes mouse events as well as network events (an example of the latter would be testing whether cast members are downloaded yet). The type of control achievable might be to loop part of a presentation until a video is downloaded, then continue or jump to another frame. Bitmaps are used for buttons, and the most typical use would be to jump to a frame in the movie after a button-click event.

The Score window is organized in horizontal lines, each for one of the sprites, and vertical frames. Thus, the Score looks somewhat like a musical score, in that time is from left to right, but it more resembles the list of events in a MIDI file (see Chap. 6).

Both types of behaviors, prewritten and user-defined, are in Lingo. The Library palette provides access to all prewritten behavior scripts. You can drop a behavior onto a sprite or attach behaviors to a whole frame.

If a behavior includes parameters, a dialog box appears. For example, navigation behaviors must have a specified frame to jump to. You can attach the same behavior to many sprites or frames and use different parameters for each instance. Most behaviors respond to simple events, such as a click on a sprite or the event triggered when the "playback head" enters a frame. Most basic functions, such as playing a sound, come prepackaged. Writing your own user-defined Lingo scripts provides more flexibility. Behaviors are modified using *Inspector* windows: the Behavior Inspector or Property Inspector.

Animation

Traditional animation (cel animation) is created by showing slightly different images over time. In Director, this approach amounts to using different cast members in different frames. To control this process more easily, Director permits combining many cast members into a single sprite.

A less sophisticated-looking but simple animation is available with the *tweening* feature of Director. Here, you specify a particular image and move it around the stage without altering the original image. "Tweening" refers to the job of minor animators, who used to have to fill in between the keyframes produced by more experienced animators—a role Director fulfills automatically.

To prepare such an animation, specify the path on the stage for the tweened frames to take. You can also specify several keyframes and the kind of curve for the animation to follow between keyframes. You also specify how the image should accelerate and decelerate at the beginning and end of the movement ("ease-in" and "ease-out"). Figure 2.17 shows a tweened sprite.

Lingo Scripts and 3D Sprites

Director uses four types of scripts: behaviors, scripts attached to cast members, movie scripts, and parent scripts. Behaviors, movie scripts, and parent scripts all appear as cast members in the Cast window.

Another sophisticated feature in Director is the ability to create, import, and manipulate 3D objects on the stage. For example, a simple 3D object that can be added in Director is 3D text.

2.6.4   Adobe XD

Adobe XD is a program to create mockups and prototypes of mobile apps, websites, and software using a series of artboards and links between them. It is primarily for user interface or user experience designers to test the visual appearance of their interface designs before it goes to a developer to build in code. This way, designers can finalize the layout and user flow of their creation without coding knowledge and make design adjustments and changes easily.

Fig. 2.17 A tweened sprite

There are three main modes in Adobe XD, namely, *design*, *prototype*, and *share*. In the design mode (Fig. 2.18), a user can make artboards to place images, text, vector icons, and more as if they were on a screen. There are also important design controls such as adding layout columns and adding component states (Fig. 2.19).

Fig. 2.18 The design mode in Adobe XD

Fig. 2.19 Component states are used to change the look of an element when it is triggered

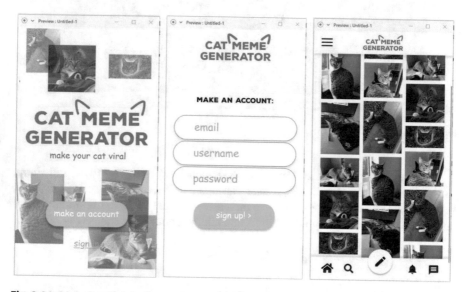

Fig. 2.20 Linked screens in the preview mode. Once linked in the prototype mode, the elements are clickable, allowing users to move between screens

Once a designer has completed the interface, the screens can be linked together through interactions in the prototype mode. Therefore, any element can be used to trigger a transition to a new screen, giving the sense of a real app when the design is previewed in (Fig. 2.20). When the design is ready, the user can send a link to the working prototype in the share mode.[2]

Working prototypes made in Adobe XD and other prototyping programs have many uses in the interface design process. They can be used to obtain design critique, run user testing on a design, show developers how and what to build, and show the app in presentations, either live or in recordings. Working prototypes can be played

---

[2]See http://www.cs.sfu.ca/mmbook for a demo.

on desktop as well as on a mobile device using the Adobe XD mobile app, through which designers can get a good sense of scale and how their design sits on a real smartphone screen.

2.7 Exercises

1. What extra information is multimedia good at conveying?

    (a) What can spoken text convey that written text cannot?
    (b) When might written text be better than spoken text?

2. Find and learn Autodesk 3ds Max in your local lab software. Read the online tutorials to see this software's approach to a 3D modeling technique. Learn texture mapping and animation using this product. Make a 3D model after carrying out these steps.

3. Suppose we wish to create a simple animation, as in Fig. 2.21. Note that this image is exactly what the animation looks like at some time, not a figurative representation of the *process* of moving the fish; the fish is repeated as it moves. State what we need to carry out this objective, and give a simple pseudocode solution for the problem. Assume we already have a list of (x, y) coordinates for the fish path that we have available a procedure for centering images on path positions, and that the movement takes place on top of a video.

4. For the slide transition in Fig. 2.8, explain how we arrive at the formula for x in the unmoving right video R_R.

5. Suppose we wish to create a video transition such that the second video appears under the first video through an opening circle (like a camera iris opening), as shown in Fig. 2.22. Write a formula to use the correct pixels from the two videos to achieve this special effect. Just write your answer for the red channel.

6. Now suppose we wish to create a video transition such that the second video appears under the first video through a moving radius (like a clock hand), as shown in Fig. 2.23. Write a formula to use the correct pixels from the two videos to achieve this special effect for the red channel.

7. Suppose you wish to create a wavy effect, as shown in Fig. 2.24. This effect comes from replacing the image x value by an x value offset by a small amount. Suppose the image size is 160 rows × 120 columns of pixels.

    (a) Using float arithmetic, add a sine component to the x value of the pixel such that the pixel takes on an RGB value equal to that of a different pixel in the original image. Make the maximum shift in x equal to 16 pixels.

Fig. 2.21 Sprite, progressively taking up more space

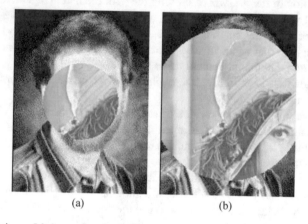

(a)   (b)

Fig. 2.22 Iris wipe: **a** Iris is opening; **b** at a later moment

(b) In Premiere and other packages, only integer arithmetic is provided. Functions such as `sin` are redefined so as to take an `int` argument and return an `int`. The argument to the `sin` function must be in 0 .. 1, 024, and the value of `sin` is in −512 .. 512: `sin(0)` returns 0, `sin(256)` returns 512, `sin(512)` returns 0, `sin(768)` returns −512, and `sin(1,024)` returns 0. Rewrite your expression in part (a) using integer arithmetic.

(c) How could you change your answer to make the waving time-dependent?

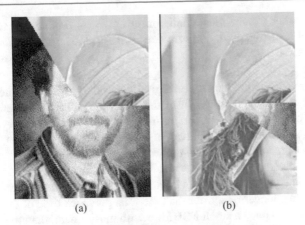

(a)   (b)

Fig. 2.23 Clock wipe: **a** Clock hand is sweeping out; **b** at a later moment

Fig. 2.24 Filter applied to video

8. How would you create the color wheel image in Fig. 2.3? Write a small program to make such an image. *Hint:* Place R, G, and B at the corners of an equilateral triangle inside the circle. It's best to go over *all columns and rows* in the output image rather than simply going around the disk and trying to map results back to (x, y) pixel positions.

9. As a longer exercise for learning existing software for manipulating images, video, and music, make a 1-min digital video. By the end of this exercise, you should know how to use a video editor (e.g., Adobe Premiere), an image editor (espe-

cially Photoshop), some music notation program for producing MIDI, and perhaps digital-audio manipulation software such as Adobe Audition, as well as other multimedia software.

(a) Acquire (or find) at least three digital video files. You can either use a camcorder or download some from the net, or use the video setting on still-image camera, phone, etc. (or, for interesting legacy video, use video-capture through Premiere or an equivalent product to make your own, from an old analog Camcorder or VCR—this is challenging, and fun).

(b) Try to upload one of the videos to YouTube. Check the time that is taken to upload the video, and discuss its relation with your video's quality and size. Is this time longer or shorter than the total playback time of the video?

(c) Compose (or edit) a small MIDI file with music-manipulation software.

(d) Create (or find) at least one WAV file (ok—could be MP3 or OPUS). You may either digitize your own or find some on the net, etc. You might like to edit this digital-audio file using software such as Audition, Audacity, etc.

(e) Use Photoshop to create a title and an ending. This is not trivial; however, you cannot say you know about multimedia without having worked with Photoshop.

As stated above, a useful feature to know in Photoshop is how to create an alpha channel:

- Use an image you like: a .JPG, say.
- Make the background some solid color, white, say.
- Make sure that you have chosen `Image > Mode > RGB Color`.
- Select that background area (you want it to remain opaque in Premiere): MagicWandTool.
- `Select > Save Selection > Channel=New; OK`.
- `Window > ShowChannels;` Double click the new channel and rename it Alpha; make its color (0, 0, 0).
- Save the file as a .PSD.

If the alpha channel you created in Photoshop has a white background, you'll need to choose ReverseKey in Premiere when you choose `Transparency > Alpha`.

(f) Combine all of the above to produce a movie about 60 seconds long, including a title, some credits, some soundtracks, and at least three transitions. The plotline of your video should be interesting, to you!

(g) Experiment with different compression methods; you are encouraged to use MPEG for your final product. We are very interested in seeing how the concepts in the textbook go over into the production of actual video. Adobe Premiere can use the DivX codec to generate movies, with the output movie actually playable on (that) machine, but wouldn't it be interesting to try various codecs?

(h) The above constitutes a minimum statement of the exercise. You may be tempted to get very creative, and that's fine, but don't go overboard and take too much time away from the rest of your life!

References

1. A.C. Luther, *Authoring Interactive Multimedia (The IBM Tools Series).* (AP Professional, 1994)
2. D.E. Wolfgram, *Creating Multimedia Presentations.* (QUE, 1994)
3. R. Vetter, C. Ward, S. Shapiro, Using color and text in multimedia projections. IEEE Multimed. **2**(4), 46–54 (1995)
4. Jiangchuan Liu, Sanjay G. Rao, Bo Li, Hui Zhang, Opportunities and challenges of peer-to-peer Internet video broadcast. Proc. of the IEEE **96**(1), 11–24 (2008)

Graphics and Image Data Representations

3

In this chapter, we look at images, starting with 1-bit images, then 8-bit grayscale images and how to print them, then 24-bit color images and 8-bit versions of color images. The specifics of file formats for storing such images will also be discussed.

We consider the following topics:

- Graphics/image data types,
- Popular file formats.

3.1 Graphics and Image Data Types

The number of file formats used in multimedia [1] continues to proliferate. For example, Table 3.1 shows a list of some of the file formats used in the popular product Adobe Premiere. In this text, we shall study just a few popular file formats to develop a sense of how they operate. We shall concentrate on GIF and JPG image file formats, since the GIF file format is one of the simplest and it contains several fundamental features, and the JPG file format is arguably the most important overall.

To begin, we shall discuss the features of various types of images in general.

3.1.1 1-Bit Images

Images consist of *pixels*—picture elements in digital images. A 1-bit image consists of on and off bits only and thus is the simplest type of image. Each pixel is stored as a single bit (0 or 1). Hence, such an image is also referred to as a *binary image*.

It is also sometimes called a 1-bit *monochrome* image, since it contains no color. Figure 3.1 shows a 1-bit monochrome image (called "Lena" by multimedia

© Springer Nature Switzerland AG 2021
Z.-N. Li et al., *Fundamentals of Multimedia*, Texts in Computer Science,
https://doi.org/10.1007/978-3-030-62124-7_3

Table 3.1 Some popular Adobe Premiere file formats

Image	Audio	Video
BMP, DIB,	AIFF, AAC,	AVI, DV,
GIF, HEIF,	AC3, BWF,	FLV, HEVC,
JPG, PICT,	MP3, M4A,	M4V, MOV, MP4,
PNG, PSD,	WAV, WMA	MPG, MTS, MXF,
TGA, TIF		SWF, WMV

Fig. 3.1 Monochrome 1-bit
Lena image

scientists—this is a standard image used to illustrate many algorithms). A 640×480 monochrome image requires 38.4 kB of storage ($= 640 \times 480/8$). Monochrome 1-bit images can be satisfactory for pictures containing only simple graphics and text. Moreover, fax machines use 1-bit data, so in fact 1-bit images are still important even though storage capacities have increased enough to permit the use of imaging that carries more information.

3.1.2 8-Bit Gray-Level Images

Now consider an 8-bit image—that is, one for which each pixel has a *gray value* between 0 and 255. Each pixel is represented by a single byte—for example, a dark pixel might have a value of 10, and a bright one might be 230.

The entire image can be thought of as a two-dimensional array of pixel values. We refer to such an array as a *bitmap*—a representation of the graphics/image data that parallels the manner in which it is stored in video memory.

Image resolution refers to the number of pixels in a digital image (higher resolution always yields better quality). Fairly high resolution for such an image might be $1,600 \times 1,200$, whereas lower resolution might be 640×480. Notice that here we are using an *aspect ratio* of 4:3. We don't have to adopt this ratio, but it has been found to look natural. For this reason, 4:3 was adopted for early TV (see Chap. 5) and most early laptop screens. Later displays typically use an aspect ration of 16:9 to match high-definition video and TV.

Fig. 3.2 Bitplanes for 8-bit
grayscale image

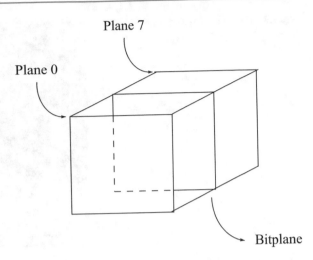

Plane 7

Plane 0

Bitplane

Such an array must be stored in hardware; we call this hardware a *frame buffer*. Special (relatively expensive) hardware called a "video" card (actually a graphics card) is used for this purpose. The resolution of the video card does not have to match the desired resolution of the image, but if not enough video card memory is available, the data has to be shifted around in RAM for display.

We can think of the 8-bit image as a set of 1-bit *bitplanes*, where each plane consists of a 1-bit representation of the image: a bit is turned on if the image pixel has a nonzero value at that bit level.

Figure 3.2 displays the concept of bitplanes graphically. Each bitplane can have a value of 0 or 1 at each pixel but, together, all the bitplanes make up a single byte that stores values between 0 and 255 (in this 8-bit situation). For the least significant bit, the bit value translates to 0 or 1 in the final numeric sum of the binary number. Positional arithmetic implies that for the next, second, bit each 0 or 1 makes a contribution of 0 or 2 to the final sum. The next bits stand for 0 or 4, 0 or 8, and so on, up to 0 or 128 for the most significant bit. Video cards can refresh bitplane data at video rate but, unlike RAM, do not hold the data well. Raster fields are refreshed at 60 cps (cycles per second) in North America and 50 cps in Europe.

Each pixel is usually stored as a byte (a value between 0 and 255), so a 640×480 grayscale image requires 300 kB of storage ($640 \times 480 = 307{,}200$). Figure 3.3 shows the Lena image again, this time in grayscale.

If we wish to *print* such an image, things become more complex. Suppose we have available a 600 dot-per-inch (dpi) laser printer. Such a device can usually only print a dot or not print it. However, a 600×600 image will be printed in a 1-inch space and will thus not be very pleasing. Instead, *dithering* is used. The basic strategy of dithering is to trade *intensity resolution* for *spatial resolution*. (See [2] for a good discussion of dithering.).

Fig. 3.3 Grayscale image of
Lena

Dithering

For printing on a 1-bit printer, dithering is used to calculate larger patterns of dots, such that values from 0 to 255 correspond to pleasing patterns that correctly represent darker and brighter pixel values. The main strategy is to replace a pixel value by a larger pattern, say 2×2 or 4×4, such that the number of printed dots approximates the varying-sized disks of ink used in *halftone printing*. Halftone printing is an analog process that uses smaller or larger filled circles of black ink to represent shading, for newspaper printing, say.

If instead we use an $n \times n$ matrix of on-off 1-bit dots, we can represent $n^2 + 1$ levels of intensity resolution—since, for example, three dots filled in any way counts as one intensity level. The dot patterns are created heuristically. For example, if we use a 2×2 "dither matrix":

$$\begin{pmatrix} 0 & 2 \\ 3 & 1 \end{pmatrix}$$

we can first remap image values in $0..255$ into the new range $0..4$ by (integer) dividing by 256/5. Then, for example, if the pixel value is 0, we print nothing in a 2×2 area of printer output. But if the pixel value is 4, we print all four dots. So the rule is

> If the intensity is greater than the dither matrix entry, print an on dot at that entry location: replace each pixel by an $n \times n$ matrix of such on or off dots.

However, we notice that the number of levels, so far, is small for this type of printing. If we increase the number of effective intensity levels by increasing the dither matrix size, we also increase the size of the output image. This reduces the amount of detail in any small part of the image, effectively reducing the spatial resolution.

Note that the image size may be much larger for a dithered image, since replacing each pixel by a 4×4 array of dots, say, makes an image $4^2 = 16$ times as large.

(a)   (b)   (c)

Fig. 3.4 Dithering of grayscale images. **a** 8-bit grayscale image `lenagray.bmp`; **b** dithered version of the image; **c** detail of dithered version

However, a clever trick can get around this problem. Suppose we wish to use a larger, 4×4 dither matrix, such as

$$\begin{pmatrix} 0 & 8 & 2 & 10 \\ 12 & 4 & 14 & 6 \\ 3 & 11 & 1 & 9 \\ 15 & 7 & 13 & 5 \end{pmatrix}$$

Then suppose we slide the dither matrix over the image four pixels in the horizontal and vertical directions at a time (where image values have been reduced to the range $0..16$). An "ordered dither" consists of turning on the printer output bit for a pixel if the intensity level is greater than the particular matrix element just at that pixel position. Figure 3.4a shows a grayscale image of Lena. The ordered-dither version is shown as Fig. 3.4b, with a detail of Lena's right eye in Fig. 3.4c. An algorithm for ordered dither, with $n \times n$ dither matrix, is as follows:

Algorithm 3.1   Ordered Dither

begin
    for $x = 0$ to x_{max} // columns
        for $y = 0$ to y_{max} // rows
            $i = x \bmod n$
            $j = y \bmod n$
            // $I(x, y)$ is the input, $O(x, y)$ is the output, D is the dither matrix.
            if $I(x, y) > D(i, j)$
                $O(x, y) = 1$;
            else
                $O(x, y) = 0$;
end

Foley et al. [2] provides more details on ordered dithering.

3.1.3 Image Data Types

The next sections introduce some of the most common data types for graphics and image file formats: 24-bit color and 8-bit color. We then discuss file formats. There are some formats that are restricted to particular hardware/operating system platforms (e.g., MS Windows or macOS), while many others are *platform-independent*, or *cross-platform*, formats. Even if some formats are not cross-platform, conversion applications can recognize and translate formats from one system to another.

Most image formats incorporate some variation of a *compression* technique due to the large storage size of image files. Compression techniques can be classified as either *lossless* or *lossy*. We will study various image, video, and audio compression techniques in Chaps. 7 through 14.

3.1.4 24-Bit Color Images

In a color 24-bit image, each pixel is represented by three bytes, usually representing RGB. Since each value is in the range 0–255, this format supports $256 \times 256 \times 256$, or a total of 16,777,216, possible combined colors. However, such flexibility does result in a storage penalty: a 640×480 24-bit color image would require 921.6 kB of storage without any compression.

An important point to note is that many 24-bit color images are actually stored as 32-bit images, with the extra byte of data for each pixel storing an α (*alpha*) value representing special-effect information. (See [2] for an introduction to use of the α-channel for compositing several overlapping objects in a graphics image. The simplest use is as a transparency flag.)

Figure 3.5 shows the image `forestfire.bmp`, a 24-bit image in Microsoft Windows BMP format (discussed later in the chapter). Also shown are the grayscale images for just the red, green, and blue channels for this image. Taking the byte values 0 .. 255 in each color channel to represent intensity, we can display a grayscale image for each color separately.

3.1.5 Higher Bit-Depth Images

Among image formats that are usually *not* compressed if possible are ones that require the maximum faithfulness to the viewed scene for various reasons such as medical liability. For example, an image of a patient's liver had better represent the colors red and purple, say, very accurately!

Other image formats recognize that more information about the scene being imaged can be gained by using special cameras that view more than just three colors, i.e., RGB. Here the idea might be to use invisible light (e.g., infrared, ultraviolet) for security cameras, say, or to produce medical images of skin that can utilize the additional colors to better diagnose skin ailments such as carcinoma. Another reason

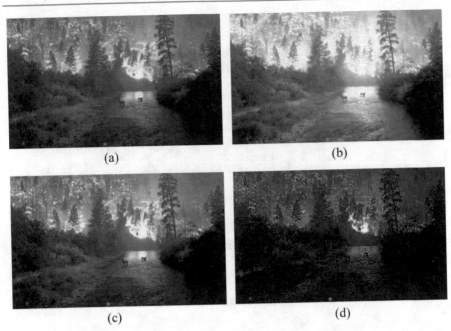

Fig. 3.5 High-resolution color and separate R, G, B color channel images. **a** example of 24-bit color image `forestfire.bmp`; **b, c, d** R, G, and B color channels for this image

for using high bit-depth is in satellite imaging, where extra information can give an indication of types of crop growth, etc.: here, the cost of simply lifting the camera into high altitude or into space motivates the idea of obtaining as much information as possible, perhaps even if we cannot as yet make use of all the data.

Such images are called *multispectral* (more than 3 colors) or hyperspectral (a great many image planes, say 224 colors for satellite imaging).

In this text, we shall stick to grayscale or RGB color images.

3.1.6 8-Bit Color Images

If space is a concern (and it almost always is—e.g., we don't want to fill up our smartphone memory needlessly), reasonably accurate color images can be obtained by *quantizing* the color information to collapse it. Many systems can utilize color information stored with only 8 bits of information (the so-called "256 colors") in producing a screen image. Even if a system has the electronics to actually use 24-bit information, backward compatibility demands that we understand 8-bit color image files, as well as their being smaller in size and quite useful. We shall also see that there are tricks that can be used only for such imagery.

8-bit color image files use the concept of a *lookup table* to store color information. Basically, the image stores not color but instead just a set of bytes, each of which is an index into a table with 3-byte values that specify the 24-bit color for a pixel with

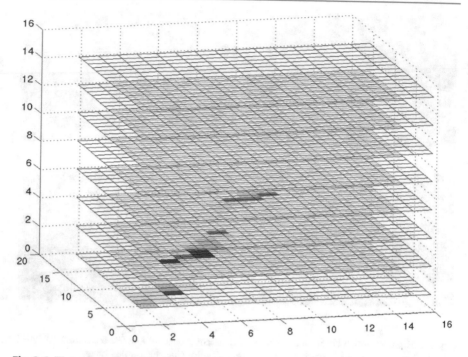

Fig. 3.6 Three-dimensional histogram of RGB colors in `forestfire.bmp`

that lookup table index. In a way, it's a bit like a paint-by-number children's art set, with number 1 perhaps standing for orange, number 2 for green, and so on—there is no inherent pattern to the set of actual colors.

It makes sense to carefully choose just which colors to represent best in the image: if an image is mostly red sunset, it's reasonable to represent red with precision and store only a few greens.

Suppose all the colors in a 24-bit image were collected in a $256 \times 256 \times 256$ set of cells, along with the count of how many pixels belong to each of these colors stored in that cell. For example, if exactly 23 pixels have RGB values $(45, 200, 91)$ then store the value 23 in a three-dimensional array, at the element indexed by the index values [45, 200, 91]. This data structure is called a *color histogram* (see, e.g., [3,4]). It is a very useful tool for image transformation and manipulation in Image Processing.

Figure 3.6 shows a 3D histogram of the RGB values of the pixels in `forestfire.bmp`. The histogram has $16 \times 16 \times 16$ bins and shows the count in each bin in terms of intensity and pseudocolor. We can see a few important clusters of color information, corresponding to the reds, yellows, greens, and so on, of the `forestfire` image. Clustering in this way allows us to pick the most important 256 groups of color.

Basically, large populations in 3D histogram bins can be subjected to a split-and-merge algorithm to determine the "best" 256 colors. Figure 3.7 shows the resulting 8-bit image in GIF format (discussed later in this chapter). Notice that the difference

Fig. 3.7 Example of an 8-bit color image

between Fig. 3.5a, the 24-bit image, and Fig. 3.7, the 8-bit image, is reasonably small. This is not always the case. Consider the field of medical imaging: would you be satisfied with only a "reasonably accurate" image of your brain for potential laser surgery? Likely not—and that is why consideration of 64-bit imaging for medical applications is not out of the question.

Note the great savings in space for 8-bit images over 24-bit ones: a 640 × 480 8-bit color image requires only 300 kB of storage, compared to 921.6 kB for a color image (again, without any compression applied).

3.1.7 Color Lookup Tables (LUTs)

Again, the idea used in 8-bit color images is to store only the index, or code value, for each pixel. Then, if a pixel stores, say, the value 25, the meaning is to go to row 25 in a color lookup table (LUT). While images are displayed as two-dimensional arrays of values, they are usually *stored* in row-column order as simply a long series of values. For an 8-bit image, the image file can store in the file header information just what 8-bit values for R, G, and B correspond to each index. Figure 3.8 displays this idea. The LUT is often called a *palette*.

A *color picker* consists of an array of fairly large blocks of color (or a semicontinuous range of colors) such that a mouse click will select the color indicated. In reality, a color picker displays the palette colors associated with index values from 0 to 255. Figure 3.9 displays the concept of a color picker: if the user selects the color block with index value 2, then the color meant is cyan, with RGB values (0, 255, 255).

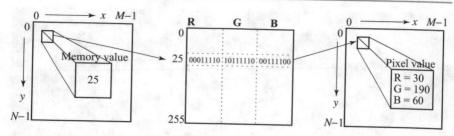

Fig. 3.8 Color LUT for 8-bit color images

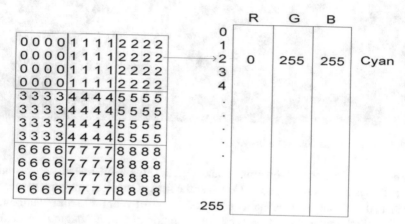

Fig. 3.9 Color picker for 8-bit color: each block of the color picker corresponds to one row of the color LUT

A simple animation process is possible via simply changing the color table: this is called *color cycling* or *palette animation*. Since updates from the color table are fast, this can result in a simple, pleasing effect, such as a marquee around a textbox with colored balls that appear to move around the border of the box. Because of the simplicity of changing the small palette data, tricks like this are possible for 8-bit color images.

Dithering can also be carried out for color printers, using 1 bit per color channel and spacing out the color with R, G, and B dots. Alternatively, if the printer or screen can print only a limited number of colors, say using 8 bits instead of 24, color can be made to seem printable, even if it is not available in the color LUT. The apparent color resolution of a display can be increased without reducing spatial resolution by averaging the intensities of neighboring pixels. Then it is possible to trick the eye into perceiving colors that are not available, because it carries out a spatial blending that can be put to good use. Figure 3.10a shows a 24-bit color image of Lena, and Fig. 3.10b shows the same image reduced to only 5 bits via dithering. Figure 3.10c shows a detail of the left eye.

Fig. 3.10 a 24-bit color image `lena.bmp`; **b** version with color dithering; **c** detail of dithered version

How to Devise a Color Lookup Table

In Sect. 3.1.6, we briefly discussed the idea of *clustering* to generate the most important 256 colors from a 24-bit color image. However, in general, clustering is an
expensive and slow process. But we need to devise color LUTs somehow—how
shall we accomplish this?

The most straightforward way to make 8-bit lookup color out of 24-bit color would
be to divide the RGB cube into equal slices in each dimension. Then the centers of
each of the resulting cubes would serve as the entries in the color LUT, and simply
scaling the RGB ranges 0 .. 255 into the appropriate ranges would generate the 8-bit
codes.

Since humans are more sensitive to R and G than to B, we could shrink the R
range and G range 0 .. 255 into the 3-bit range 0 .. 7 and shrink the B range down to
the 2-bit range 0 .. 3, making a total of 8 bits. To shrink R and G, we could simply
divide the R or G byte value by $(256/8 =)$ 32 and then truncate. Then each pixel
in the image gets replaced by its 8-bit index, and the color LUT serves to generate
24-bit color.

However, what tends to happen with this simple scheme is that edge artifacts
appear in the image. The reason is that if a slight change in RGB results in shifting
to a new code, an edge appears, and this can be quite annoying perceptually.

A simple alternate solution for this color reduction problem called the *median-
cut algorithm* does a better job (and several other competing methods do as well or
better). This approach derives from computer graphics [5]; here, we show a much
simplified version. The method is a type of adaptive partitioning scheme that tries to
put the most bits, the most discrimination power, where colors are most clustered.

The idea is to sort the R byte values and find their median. Then values smaller
than the median are labeled with a 0 bit and values larger than the median are labeled
with a 1 bit. The median is the point where half the pixels are smaller and half are
larger.

Suppose we are imaging some apples, and most pixels are reddish. Then the
median R byte value might fall fairly high on the red 0 .. 255 scale. Next, we consider
only pixels with a 0 label *from the first step* and sort their G values. Again, we label
image pixels with another bit (and this is the second bit assigned), 0 for those less

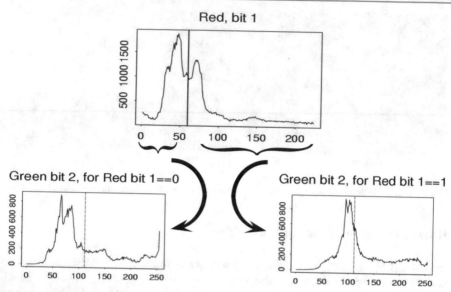

Fig. 3.11 Histogram of R bytes for the 24-bit color image forestfire.bmp results in a 0 or 1-bit label for every pixel. For the second bit of the color table index being built, we take R values less than the R median and label just those pixels as 0 or 1 according to their G value is less or greater than the median of the G value. Continuing over R, G, B for 8 bits gives a color LUT 8-bit index

than the median in the greens and 1 for those greater. Now applying the same scheme to pixels that received a 1 bit for the red step, we have arrived at 2-bit labeling for all pixels.

Carrying on to the blue channel, we have a 3-bit scheme. Repeating all steps, R, G, and B, results in a 6-bit scheme, and cycling through R and G once more results in 8 bits. These bits form our 8-bit color index value for pixels, and corresponding 24-bit colors can be the centers of the resulting small color cubes.

You can see that in fact this type of scheme will indeed concentrate bits where they most need to differentiate between high populations of close colors. We can most easily visualize finding the median by using a histogram showing counts at position 0..255. Figure 3.11 shows a histogram of the R byte values for the forestfire.bmp image along with the median of these values, depicted as a vertical line.

The 24-bit color image resulting from replacing every pixel by its corresponding color LUT 24-bit color is only an approximation to the original 24-bit image, of course, but the above algorithm does a reasonable job of putting most discriminatory power where it is most needed—where small color shading differences will be most noticeable. It should also be mentioned that several methods exist for distributing the approximation errors from one pixel to the next. This has the effect of smoothing out problems in the 8-bit approximation.

The more accurate version of the median-cut algorithm proceeds via the following steps:

1. Find the smallest box that contains all the colors in the image.
2. Sort the enclosed colors along the longest dimension of the box.
3. Split the box into two regions at the median of the sorted list.
4. Repeat the above process in steps (2) and (3) until the original color space has been divided into, say, 256 regions.
5. For every box, call the mean of R, G, and B in that box the representative (the center) color for the box.
6. Based on the Euclidean distance between a pixel RGB value and the box centers, assign every pixel to one of the representative colors. Replace the pixel by the code in a lookup table that indexes representative colors (in the table, each representative color is 24-bits—8 bits each for R, G, and B.).

This way, we might have a table of 256 rows, each containing three 8-bit values. The row indices are the codes for the lookup table, and these indices are what are stored in pixel values of the new, *color quantized* or *palettized* image.

3.2 Popular File Formats

Some popular image file formats are described below. One of the simplest is the 8-bit GIF format, and we study it because it is easily understood, and as well because of its historical connection to the WWW and HTML markup language as the first image type recognized by net browsers. However, currently the most important common file format is JPEG, which will be explored in great depth in Chap. 9.

3.2.1 GIF

Graphics Interchange Format (GIF) was devised by UNISYS Corporation and CompuServe, initially for transmitting graphical images over phone lines via modems. The GIF standard uses the Lempel-Ziv-Welch algorithm (a form of compression—see Chap. 7), modified slightly for image scanline packets to use the line grouping of pixels effectively.

The GIF standard is limited to 8-bit (256) color images only. While this produces acceptable color, it is best suited for images with few distinctive colors (e.g., graphics or drawing).

The GIF image format has a few interesting features, notwithstanding the fact that it has been largely supplanted. The standard supports *interlacing*—the successive display of pixels in widely spaced rows by a four-pass display process.

In fact, GIF comes in two flavors. The original specification is GIF87a. The later version, GIF89a, supports simple *animation* via a Graphics Control Extension block in the data. This provides simple control over *delay time*, a *transparency index*, and so on. Most image editing software supports GIF, and it is widely used in today's webpages.

Fig. 3.12 GIF file format

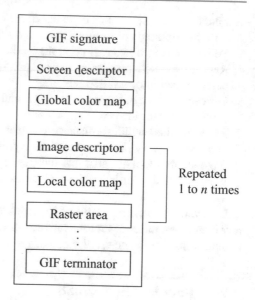

It is worthwhile examining the file format for GIF87 in more detail, since many such formats bear a resemblance to it but have grown a good deal more complex than this "simple" standard. For the standard specification, the general file format is as in Fig. 3.12. The *Signature* is 6 bytes: GIF87a; the *Screen Descriptor* is a 7-byte set of flags. A GIF87 file can contain more than one image definition, usually to fit on several different parts of the screen. Therefore, each image can contain its own color lookup table, a *Local Color Map*, for mapping 8 bits into 24-bit RGB values. However, it need not, and a global color map can instead be defined to take the place of a local table if the latter is not included.

The Screen Descriptor comprises a set of attributes that belong to every image in the file. According to the GIF87 standard, it is defined as in Fig. 3.13. *Screen Width* is given in the first 2 bytes. Since some machines invert the order MSB/LSB (most significant byte/least significant byte—i.e., byte order), this order is specified. *Screen Height* is the next 2 bytes. The "m" in byte 5 is 0 if no global color map is given. Color resolution, "cr", is 3 bits in 0 .. 7. Since this is an old standard meant to operate on a variety of low-end hardware, "cr" is *requesting* this much color resolution.

The next bit, shown as "0", is extra and is not used in this standard. "Pixel" is another 3 bits, indicating the number of bits per pixel in the image, as stored in the file. Although "cr" usually equals "pixel", it need not. Byte 6 gives the color table index byte for the background color, and byte 7 is filled with zeros. For present usage, the ability to use a small color resolution is a good feature, since we may be interested in very low-end devices such as web-enabled wristwatches, say.

A *color map* is set up in a simple fashion, as in Fig. 3.14. However, the actual length of the table equals $2^{\text{pixel}+1}$ as given in the screen descriptor.

Each image in the file has its own Image Descriptor, defined as in Fig. 3.15. Interestingly, the developers of this standard allowed for future extensions by ignoring any bytes between the end of one image and the beginning of the next, identified by a

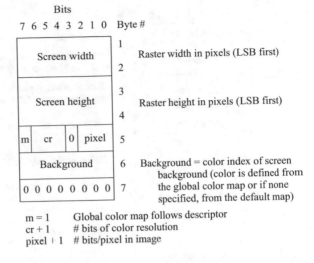

Fig. 3.13 GIF screen descriptor

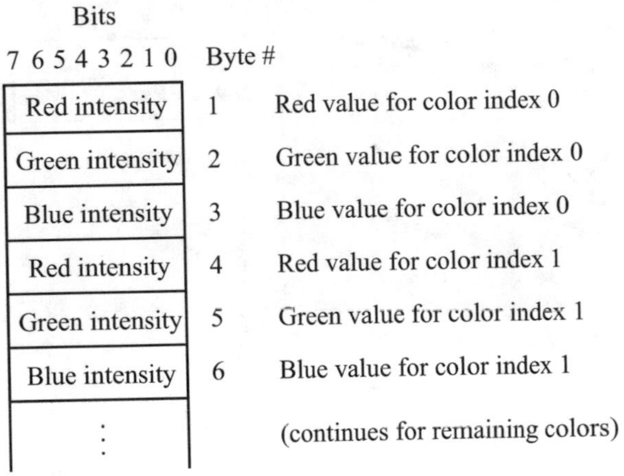

m = 1   Global color map follows descriptor
cr + 1   # bits of color resolution
pixel + 1   # bits/pixel in image

Fig. 3.14 GIF color map

comma character. In this way, future enhancements could have been simply inserted in a backward-compatible fashion.

If the *interlace* bit is set in the local Image Descriptor, the rows of the image are displayed in a four-pass sequence, as in Fig. 3.16. Here, the first pass displays rows 0 and 8, the second pass displays rows 4 and 12, and so on. This allows for a quick sketch to appear when a web browser displays the image, followed by more detailed fill-ins. The JPEG standard (below) has a similar display mode, denoted *progressive mode*.

The *actual raster data* itself is first compressed using the LZW compression scheme (see Chap. 7) before being stored.

The GIF87 standard also set out, for future use, how Extension Blocks could be defined. Even in GIF87, simple animations can be achieved, but no delay was defined between images, and multiple images simply overwrite each other with no screen clears.

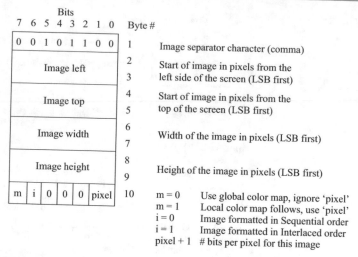

Fig. 3.15 GIF image descriptor

Fig. 3.16 GIF four-pass interlace display row order

Image row	Pass 1	Pass 2	Pass 3	Pass 4	Result
0	*1a*				*1a*
1				*4a*	*4a*
2			*3a*		*3a*
3				*4b*	*4b*
4		*2a*			*2a*
5				*4c*	*4c*
6			*3b*		*3b*
7				*4d*	*4d*
8	*1b*				*1b*
9				*4e*	*4e*
10			*3c*		*3c*
11				*4f*	*4f*
12		*2b*			*2b*
:					

GIF89 introduced a number of Extension Block definitions, especially those to assist animation: transparency and delay between images. A quite useful feature introduced in GIF89 is the idea of a sorted color table. The most important colors appear first, so that if a decoder has fewer colors available, the most important ones are chosen. That is, only a segment of the color lookup table is used, and nearby colors are mapped as well as possible into the colors available.

We can investigate how the file header works in practice by having a look at a particular GIF image. Figure 3.7 is an 8-bit color GIF image. To see how the file header looks, we can simply use everyone's favorite command in the Unix/Linux/macOS operating system: od (octal dump). In Unix,[1] then, we issue the command

```
od -c forestfire.gif | head -2
```

and we see the first 32 bytes interpreted as characters:

```
G   I    F   8   7   a    \208 \2    \188 \1    \247 \0   \0    \6    \3 \5
J \132 \24 |     )    \7   \198 \195 \      \128 U      \27 \196 \166 &   T
```

To decipher the remainder of the file header (after GIF87a), we use hexadecimal:

```
od -x forestfire.gif | head -2
```

with the result

```
4749 4638 3761 d002 bc01 f700 0006 0305
ae84 187c 2907 c6c3 5c80 551b c4a6 2654
```

The d002 bc01 following the Signature are Screen Width and Height; these are given in least-significant-byte-first order, so for this file in decimal the Screen Width is $0 + 13 \times 16 + 2 \times 16^2 = 720$, and Screen Height is $11 \times 16 + 12 + 1 \times 16^2 = 444$. Then the f7 (which is 247 in decimal) is the fifth byte in the Screen Descriptor, followed by the background color index, 00, and the 00 delimiter. The set of flags, f7, in bits, reads 1, 111, 0, 111, or in other words: global color map is used, 8-bit color resolution, 0 separator, 8-bit pixel data.

3.2.2 JPEG

The most important current standard for image compression is JPEG [6]. This standard was created by a working group of the International Organization for Standardization (ISO) that was informally called the Joint Photographic Experts Group and is therefore so named. We shall study JPEG in a good deal more detail in Chap. 9, but a few salient features of this compression standard can be mentioned here.

The human vision system has some specific limitations, which JPEG takes advantage of to achieve high rates of compression. The eye–brain system cannot see extremely fine detail. If many changes occur within a few pixels, we refer to that

---

[1]CentOS version; older versions use slightly different syntax.

Fig. 3.17 JPEG image with the low quality specified by user

image segment as having *high spatial frequency*—that is, a great deal of change in (x, y) space. This limitation is even more conspicuous for color vision than for grayscale (black and white). Therefore, color information in JPEG is *decimated* (partially dropped, or averaged) and then small blocks of an image are represented in the spatial frequency domain (u, v), rather than in (x, y). That is, the speed of changes in x and y is evaluated, from low to high, and a new "image" is formed by grouping the coefficients or weights of these speeds.

Weights that correspond to slow changes are then favored, using a simple trick: values are divided by some large integer and truncated. In this way, small values are zeroed out. Then a scheme for representing long runs of zeros efficiently is applied, and *voila!*—the image is greatly compressed.

Since we effectively throw away a lot of information by the division and truncation step, this compression scheme is "lossy" (although a lossless mode exists). What's more, since it is straightforward to allow the user to choose how large a denominator to use and hence how much information to discard, JPEG allows the user to set a desired level of quality, or compression ratio (input divided by output).

As an example, Fig. 3.17 shows our `forestfire` image with a quality factor $qf = 10$. (The usual default quality factor is $qf = 75$.).

This image is a mere 1.5% of the original size. In comparison, a JPEG image with $qf = 75$ yields an image size 5.6% of the original, whereas a GIF version of this image compresses down to 23.0% of the uncompressed image size.

3.2.3 PNG

One interesting development stemming from the popularity of the Internet is efforts toward more system-independent image formats. One such format is *Portable Network Graphics* (PNG). This standard is meant to supersede the GIF standard and extends it in important ways. The motivation for a new standard was in part the patent held by UNISYS and CompuServe on the LZW compression method. (Interestingly, the patent covers only compression, not decompression: this is why the Unix `gunzip` utility and others can decompress LZW-compressed files.)

Special features of PNG files include support for up to 16 bits per pixel in each color channel, i.e., 48-bit color—a large increase. Files may also contain gamma-correction information (see Sect. 4.1.6) for the correct display of color images and α-channel information (up to 16 bits) for such uses as control of transparency. Instead of a progressive display based on row-interlacing as in GIF images, the display progressively displays pixels in a two-dimensional interlacing over seven passes through each 8×8 block of an image. It supports both lossless and lossy compression with performance better than GIF. PNG is widely supported by various web browsers and imaging software.

3.2.4 TIFF

Tagged Image File Format (TIFF) is another popular image file format. Developed by the Aldus Corporation in the 1980s, it was later supported by Microsoft and other companies, being popular among graphic artists, photographers, and the publishing industry. Its support for attachment of additional information (referred to as "tags") provides a great deal of flexibility. The most important tag is a format signifier: what type of compression, etc., is in use in the stored image. For example, TIFF can store many different types of images: 1-bit, grayscale, 8-bit, 24-bit RGB, and so on. TIFF was originally a lossless format, but an added tag allows you to opt for JPEG, JBIG, and even JPEG 2000 compressions. Since TIFF is not as user-controllable as JPEG, it does not provide any major advantages over the latter for lossy compression. It is quite common to use TIFF files to store *uncompressed* data. TIFF files are divided into sections, each of which can store a bitmap image, a vector- or stroke-based image (see PostScript below), or other types of data. Each section's data type is specified in its tag.

3.2.5 Windows BMP

BitMap (BMP), also known as *Device Independent Bitmap* (DIB), is a major system standard image file format for Microsoft Windows. It uses raster graphics. BMP supports many pixel formats, including indexed color (up to 8 bits per pixel), and 16, 24, and 32-bit color images. It makes use of Run-Length Encoding (RLE) compression (see Chap. 7) and can fairly efficiently compress 24-bit color images due to its 24-bit RLE algorithm. BMP images can also be stored uncompressed. In particular, the 16-bit and 32-bit color images (with α-channel information) are always uncompressed.

3.2.6 Windows WMF

Windows MetaFile (WMF) is the native vector file format for the Microsoft Windows operating environment. WMF files actually consist of a collection of *Graphics Device Interface* (GDI) function calls, also native to the Windows environment. When a WMF file is "played" (typically using the Windows `PlayMetaFile()` function) the described graphic is rendered. WMF files are ostensibly device-independent and unlimited in size. The later Enhanced Metafile Format Plus Extensions (EMF+) format is device independent.

3.2.7 Netpbm Format

PPM (Portable PixMap), *PGM* (Portable GrayMap), and *PBM* (Portable BitMap) belong to a family of open-source Netpbm formats. These formats are mostly common in the Linux/Unix environments. They are sometimes also collectively known as PNM or PAM (Portable AnyMap). These are either ASCII files or raw binary files with an ASCII header for images. Because they are so simple, they can always be used for cross-platform applications. They are widely supported by various software, e.g., GIMP in Linux and mac OS, and work in Windows as well.

3.2.8 EXIF

Exchangeable Image File (EXIF) is an image format for digital cameras. It enables the recording of image metadata (exposure, light source/flash, white balance, type of scene, etc.) for the standardization of image exchange. A variety of tags (many more than in TIFF) is available to facilitate higher-quality printing, since information about the camera and picture-taking conditions can be stored and used, e.g., by printers for possible color-correction algorithms. The EXIF format is incorporated in the JPEG software in most digital cameras. It also includes specification of file format for audio that accompanies digital images.

3.2.9 HEIF

High Efficiency Image File (HEIF) is the still-image version of the HEVC (or H. 265) video format. It has been adopted by Apple to replace JPEG in all of its iDevices. HEIF is said to be a container to wrap still images compressed with the HEVC codec. The container is very flexible: it can be used to store image sequences, depth information, thumbnail images, audio, etc. HEIF is more efficient than JPEG, i.e., the image size can be reduced by half with almost imperceptible changes to image quality. HEIF supports up to 14-bit color depth, which is higher than the color depth in JPEG (usually 8-bit, can be up to 12-bit). As in GIF, HEIF also supports transparency.

3.2.10 PS and PDF

PostScript is an important language for typesetting, and many high-end printers have a PostScript interpreter built into them. PostScript is a vector-based, rather than pixel-based, picture language: page elements are essentially defined in terms of vectors. With fonts defined this way, PostScript includes vector/structured graphics as well as text; bit-mapped images can also be included in output files. Encapsulated PostScript files add some information for including PostScript files in another document.

Several popular graphics programs, such as Adobe Illustrator, use PostScript. Note, however, that the PostScript page description language does not provide compression; in fact, PostScript files are just stored as ASCII. Therefore files are often large, and in academic settings, it is common for such files to be made available only after compression by some Unix utility, such as `compress` or `gzip`.

Therefore, another text + figures language has largely superseded PostScript in nonacademic settings: Adobe Systems Inc. includes LZW (see Chap. 7) compression in its *Portable Document Format* (PDF) file format. As a consequence, PDF files that do not include images have about the same compression ratio, 2:1 or 3:1, as do files compressed with other LZW-based compression tools, such as the Unix `compress` or `gzip`, or the PC-based `winzip` (a variety of `pkzip`) or WinRAR. For files containing images, PDF may achieve higher compression ratios by using separate JPEG compression for the image content (depending on the tools used to create original and compressed versions). A useful feature of the Adobe Acrobat PDF reader is that it can be configured to read documents structured as linked elements, with clickable content and handy summary tree-structured link diagrams provided.

It is interesting for computer science and engineering students to know that the name PostScript arose because its language is based on the stack data structure, with *postfix* notation, where an operator follows its operands. The *stroke-based* graphics feature in PostScript means that diagrams should appear with crisp lines on any output device and, more importantly, at any zoom level (this said, low-resolution printers will still produce low-resolution output). The idea is that the PostScript engine in the output device (say, a screen) renders the stroke command as neatly as possible. For example, if we execute a command `100 200 moveto`, the PostScript interpreter pushes an x and y position onto the stack; if we follow with `250 75 lineto` and `stroke` we get a line to the next point.

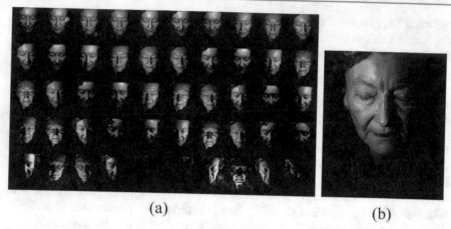

Fig. 3.18 a 50 input images for PTM: lights individually from 50 different directions $e^i, i = 1 \ldots 50$; **b** interpolated image under new light e

3.2.11 PTM

PTM (Polynomial Texture Mapping) is a technique for storing a representation of a camera scene that contains information about a set of images taken under a set of lights that each have the same spectrum, but with each placed at a different direction from the scene [7].

Suppose we have acquired n images of a scene, taken with a fixed-position camera but with lighting from $i = 1 \ldots n$ different lighting directions $e^i = (u^i, v^i, w^i)^T$. For example, a hemispherical lighting frame could be used with, say, 40 or 50 lights, one at each vertex of a geodesic dome. The objective of PTM is in part to find out the surface properties of the object being imaged—this has been used for imaging museum artifacts and paintings, for instance. But the main task for PTM is being able to interpolate the lighting directions, so as to generate new images not seen before. The file size for PTM image collections is kept small by packing multiple interpolation coefficients into integer data.

Figure 3.18a shows what a typical set of input images look like, here for a set of 50 input images.[2] Figure 3.18b shows an interpolated image, for a light direction not in the input set [8]: here the light is coming from $\theta = 42.6°$, $\phi = -175.4°$, where θ is the polar angle from the camera to the object and ϕ is the angle on the x, y plane perpendicular to that axis.

How PTM proceeds is to assume a polynomial model for generating Luminance $L = R + G + B$ (or R, G, B separately) and forming regression coefficients from the set of light-directions e^i to the $i = 1 \ldots n$ values L^i [7]. An additional level using Radial Basis Function (RBF) interpolation is used to interpolate non-smooth phenomena such as shadows and specularities [8].

---

[2]Dataset courtesy of Tom Malzbender, Hewlett-Packard.

3.3 Exercises

1. Briefly explain why we need to be able to have less than 24-bit color and why this makes for a problem. Generally, what do we need to do to adaptively transform 24-bit color values to 8-bit ones?
2. Suppose we decide to quantize an 8-bit grayscale image down to just 2 bits of accuracy. What is the simplest way to do so? What ranges of byte values in the original image are mapped to what quantized values?
3. Suppose we have a 5-bit grayscale image. What size of Ordered-dither matrix do we need to display the image on a 1-bit printer?
4. Suppose we have available 24 bits per pixel for a color image. However, we notice that humans are more sensitive to R and G than to B—in fact, 1.5 times more sensitive to R or G than to B. How could we best make use of the bits available?
5. At your job, you have decided to impress the boss by using up more disk space for the company's grayscale images. Instead of using 8 bits per pixel, you'd like to use 48 bits per pixel in RGB. How could you store the original grayscale images so that in the new format they would appear the same as they used to, visually?
6. Suppose an 8-bit grayscale image appears as in Fig. 3.19a; i.e., linear shading goes from 0 to 255 from left to right, illustrated in Fig. 3.19b.
   The image is 100 rows by 100 columns.
   For the most significant bitplane, please draw an image showing the 1's and 0's. How many 1's are there?.
   For the next-most significant bitplane, please draw an image showing the 1's and 0's.
   How many 1's are there?

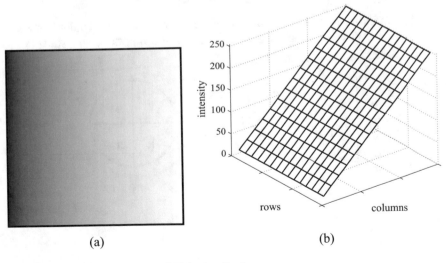

(a)   (b)

Fig. 3.19 **a** Grayscale image in $0..255$; **b** visualized as a ramp

7. For the color LUT problem, try out the median-cut algorithm on a sample image. Explain briefly why it is that this algorithm, carried out on an image of red apples, puts more color gradation in the resulting 24-bit color image where it is needed, among the reds.

8. In regard to non-ordered dithering, a standard graphics text [2] states, "Even larger patterns can be used, but the spatial versus intensity resolution trade-off is limited by our visual acuity (about one minute of arc in normal lighting)."

   (a) What does this sentence mean?

   (b) If we hold a piece of paper out at a distance of 1 foot, what is the approximate linear distance between dots? (*Information*: One minute of arc is 1/60 of one degree of angle. Arc length on a circle equals angle (in radians) times radius.) Could we see the gap between dots on a 300 dpi printer?

9. Write down an algorithm (pseudocode) for calculating a color histogram for RGB data.

10. (a) Describe in detail how to use a **single** image and several color lookup tables to realize a simple animation—a rotating color wheel, in which a sequence of 4 snapshots will appear repetitively. The wheel rotates $\theta = 90°$ clockwise each time; the first three snapshots are shown below. C—Cyan, M—Magenta, Y—Yellow, R—Red.

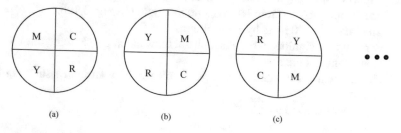

(a)   (b)   (c)

   (b) Similarly, how can you realize a sequence of 8 snapshots so that the wheel rotates $\theta = 45°$ clockwise each time, and the first three snapshots are as below?

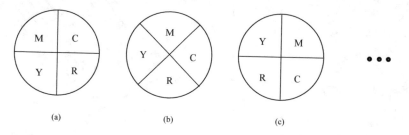

(a)   (b)   (c)

(c) Apparently, it is trivial to make a smoother animation of the rotating color wheel by simply decreasing the value of the rotation angle θ each time. If each pixel in the single image is limited to 8 bits, what is the maximum number of snapshots you can produce if your animation is similar to case (b) above? Why?

References

1. J. Miano, *Compressed Image File Formats: JPEG, PNG, GIF, XBM, BMP* (Addison Wesley Professional, 1999)
2. J.F. Hughes, A. van Dam, M. McGuire, D.F. Sklar, J.D. Foley, S.K. Feiner, K. Akeley, *Computer Graphics: Principles and Practice*, 3rd edn. in C (Addison-Wesley, 2013)
3. M. Sonka, V. Hlavac, R. Boyle, *Image processing, Analysis, and Machine Vision*, 4th edn. (PWS Publishing, 2014)
4. L.G. Shapiro, G.C. Stockman, *Computer Vision* (Prentice-Hall, 2001)
5. P. Heckbert, Color image quantization for frame buffer display. SIGGRAPH Proc. **16**, 297–307 (1982)
6. W.B. Pennebaker, J.L. Mitchell, *The JPEG Still Image Data Compression Standard* (Van Nostrand Reinhold, New York, 1992)
7. T. Malzbender, D. Gelb, H. Wolters. Polynomial texture maps, in *Computer Graphics, SIGGRAPH 2001 Proceedings* (2001), pp. 519–528
8. M.S. Drew, Y. Hel-Or, T. Malzbender, N. Hajari, Robust estimation of surface properties and interpolation of shadow/specularity components. Image Vis. Comput. **30**(4–5), 317–331 (2012)

Color in Image and Video

4

Color images and videos are ubiquitous on the web and in multimedia productions. Increasingly, we are becoming more aware of the discrepancies between color as seen by people and the sometimes very different color displayed on our screens. The latest version of the HTML standard attempts to address this issue by specifying color in terms of a standard, "sRGB," arrived at by color scientists.

To become aware of the simple, yet strangely involved world of color, in this chapter, we shall consider the following topics:

- Color science,
- Color models in images, and
- Color models in video.

4.1 Color Science

4.1.1 Light and Spectra

Recall from high school that light is an electromagnetic wave, and that its color is characterized by the wavelength of the wave. Laser light consists of a single wavelength, e.g., a ruby laser produces a bright, scarlet red beam. So if we were to make a plot of the light intensity versus wavelength, we would see a spike at the appropriate red wavelength, and no other contribution to the light.

In contrast, most light sources produce contributions over many wavelengths. However, humans cannot detect all light, but just contributions that fall in the "visible wavelengths." Short wavelengths produce a blue sensation and long wavelengths produce a red one.

We measure visible light using a device called a spectrophotometer, by reflecting light from a diffraction grating (a ruled surface) that spreads out the different wavelengths much as a prism does. Figure 4.1 shows the phenomenon that white light

© Springer Nature Switzerland AG 2021
Z.-N. Li et al., *Fundamentals of Multimedia*, Texts in Computer Science,
https://doi.org/10.1007/978-3-030-62124-7_4

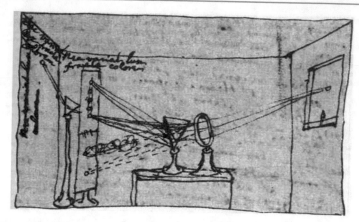

Fig. 4.1 Sir Isaac Newton's experiments. By permission of the Warden and Fellows, New College, Oxford

Fig. 4.2 Spectral power distribution of daylight

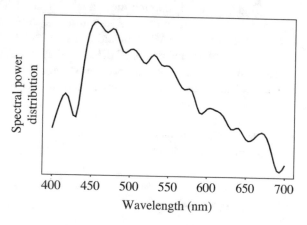

contains all the colors of a rainbow. If you have ever looked through a toy prism, you will have noticed that a rainbow effect is generated—the effect due to a natural phenomenon called dispersion. You see a similar effect on the surface of a soap bubble.

Visible light is an electromagnetic wave in the range 400–700 nm (where nm stands for nanometer, or 10^{-9} m). Figure 4.2 shows the relative power in each wavelength interval for typical outdoor light on a sunny day. This type of curve is called a spectral power distribution (SPD) or a *spectrum*. It shows the relative amount of light energy (electromagnetic signal) at each wavelength. The symbol for wavelength is λ (*lambda*) so this type of curve might be called $E(\lambda)$. In practice, measurements are used that effectively sum up voltage in a small wavelength range, say 5 or 10 nm, so such plots usually consist of segments joining function values every 10 nm. That means, also, that such profiles are actually stored as vectors; however, below we show equations that treat $E(\lambda)$ as a continuous function—in reality, integrals are calculated using sums over elements of vectors.

4.1.2 Human Vision

The eye works like a camera, with the lens focusing an image onto the retina (upside-down and left-right reversed). The retina consists of an array of *rods* and three kinds of *cones* . These receptors are called such because they are shaped like cones and rods, respectively. The rods come into play when light levels are low and produce a image in shades of gray ("all cats are gray at night!"). For higher light levels, the cones each produce a signal. Because of their differing pigments, the three kinds of cones are most sensitive to red (R), green (G), and blue (B) light.

Higher light levels result in more neurons firing; the issue of just what happens in the brain further down the pipeline is the subject of much debate. However, it seems likely that the brain makes use of *differences* $R - G$, $G - B$, and $B - R$, as well as combining all of R, G, and B into a high-light-level achromatic channel (and thus we can say that the brain is good at algebra).

4.1.3 Spectral Sensitivity of the Eye

The eye is most sensitive to light in the middle of the visible spectrum. Like the SPD profile of a light source, as in Fig. 4.2, for receptors we show the relative sensitivity as a function of wavelength. The blue receptor sensitivity is not shown to scale because, in fact, it is much smaller than the curves for red or green—blue is a late addition, in evolution (and, statistically, is the favorite color of humans, regardless of nationality—perhaps for this reason, blue is a bit surprising!). Figure 4.3 shows the overall sensitivity as a dashed line. This dashed curve in Fig. 4.3 is important and is called the luminous-efficiency function. It is usually denoted $V(\lambda)$ and is formed as the sum of the response curves to red, green, and blue [1,2].

The rods are sensitive to a broad range of wavelengths, but produce a signal that generates the perception of the black–white scale only. The rod sensitivity curve looks like the luminous-efficiency function $V(\lambda)$ but is shifted somewhat to the red end of the spectrum [1].

The eye has about 6 million cones, but the proportions of R, G, and B cones are different. They likely are present in the ratios 40:20:1 (see [3] for a complete explanation). So the achromatic channel produced by the cones is thus something like $2R + G + B/20$.

These spectral sensitivity functions are usually denoted by some other letters than "R, G, B," so here let us denote them by the vector function $q(\lambda)$, with components

$$q(\lambda) = (q_R(\lambda), q_G(\lambda), q_B(\lambda))^T \qquad (4.1)$$

That is, there are three sensors (a vector index $k = 1 .. 3$ therefore applies), and each is a function of wavelength.

The response in each color channel in the eye is proportional to the number of neurons firing. For the red channel, any light falling anywhere in the nonzero part of the red cone function in Fig. 4.3 will generate some response. So the total response

Fig. 4.3 R, G, and B cones, and luminous-efficiency curve V(λ)

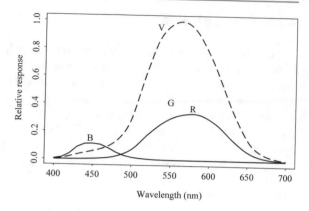

of the red channel is the sum over all the light falling on the retina that the red cone is sensitive to, weighted by the sensitivity at that wavelength. Again thinking of these sensitivities as continuous functions, we can succinctly write down this idea in the form of an integral:

$$R = \int E(\lambda)\, q_R(\lambda)\, d\lambda$$

$$G = \int E(\lambda)\, q_G(\lambda)\, d\lambda \qquad (4.2)$$

$$B = \int E(\lambda)\, q_B(\lambda)\, d\lambda$$

Since the signal transmitted consists of three numbers, colors form a three-dimensional vector space.

4.1.4   Image Formation

Equation 4.2 actually only applies when we view a self-luminous object (i.e., a light). In most situations, we actually image light that is reflected from a surface. Surfaces reflect different amounts of light at different wavelengths, and dark surfaces reflect less energy than light surfaces. Figure 4.4 shows the surface spectral reflectance from (1) orange sneakers and (2) faded bluejeans [4]. The reflectance function is denoted $S(\lambda)$.

The image formation situation is thus as follows: light from the illuminant with SPD $E(\lambda)$ impinges on a surface, with surface spectral reflectance function $S(\lambda)$, is reflected, and then is filtered by the eye's cone functions $q(\lambda)$. The basic arrangement is as shown in Fig. 4.5. The function $C(\lambda)$ is called the *color signal* and consists of the product of the illuminant $E(\lambda)$ times the reflectance $S(\lambda)$: $C(\lambda) = E(\lambda)\, S(\lambda)$.

Fig. 4.4 Surface spectral reflectance functions $S(\lambda)$ for objects

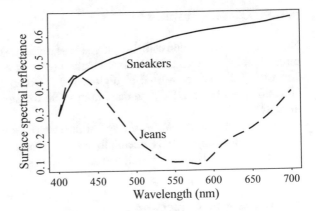

Fig. 4.5 Image formation model

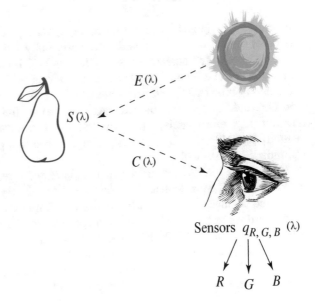

The equations similar to Eq. (4.2), then, that take into account the image formation model are

$$R = \int E(\lambda)\ S(\lambda)\ q_R(\lambda)\ d\lambda$$

$$G = \int E(\lambda)\ S(\lambda)\ q_G(\lambda)\ d\lambda \qquad (4.3)$$

$$B = \int E(\lambda)\ S(\lambda)\ q_B(\lambda)\ d\lambda$$

4.1.5   Camera Systems

Now, we humans develop camera systems in a similar fashion, and a good camera has three signals produced at each pixel location (corresponding to a retinal position). Analog signals are converted to digital, truncated to integers, and stored. If the precision used is 8 bit, then the maximum value for any of R, G, B is 255, and the minimum is 0.

However, the light entering the eye of the computer user is that which is emitted by the screen—the screen is essentially a self-luminous source. Therefore, we need to know the light $E(\lambda)$ entering the eye.

4.1.6   Gamma Correction

Modern displays (e.g., LCD, LED, OLED, etc.) attempt to mimic older CRT (cathode ray tube) displays, since standards were originally built on these displays. Thus, knowing about the characteristics of CRTs is still important. The RGB numbers in an image file are converted back to analog, and drive the electron guns in the cathode ray tube (CRT). Electrons are emitted proportional to the driving voltage, and we would like to have the CRT system produce light that is linearly related to the voltage. Unfortunately, it turns out that this is not the case. The light emitted is actually roughly proportional to the voltage raised to a power; this power is called "gamma," with symbol γ.

Thus, if the file value in the red channel is R, the screen emits light proportional to R^γ, with SPD equal to that of the red phosphor paint on the screen that is the target of the red channel electron gun. The value of gamma is around 2.2.

Since the mechanics of a television receiver are the same as those for the old standard computer CRT displays, TV systems, regardless of the actual display used, pre-correct for this situation by actually applying the inverse transformation before generating TV voltage signals. It is customary to append a prime to signals that are "gamma-corrected" by raising to the power $(1/\gamma)$ before transmission. Thus, we have

$$R \;\rightarrow\; R' = R^{1/\gamma} \;\Rightarrow\; (R')^\gamma \;\rightarrow\; R, \tag{4.4}$$

and we arrive at "linear signals." Again, cameras store gamma-corrected values R', so that when the exponentiation by γ occurs in the display, images will look right.

Voltage is often normalized to maximum 1, and it is interesting to see what effect these gamma transformations have on signals. Figure 4.6a shows the light output with no gamma-correction applied. We see that darker values are displayed too dark. This is also shown in Fig. 4.7a, which displays a linear ramp from left to right.

Figure 4.6b shows the effect of pre-correcting signals by applying the power law $R^{1/\gamma}$, where it is customary to normalize voltage to the range 0–1. We see that applying first the correction in Fig. 4.6b, followed by the effect of the display (an ostensible CRT) system, in Fig. 4.6a, would result in linear signals. The combined

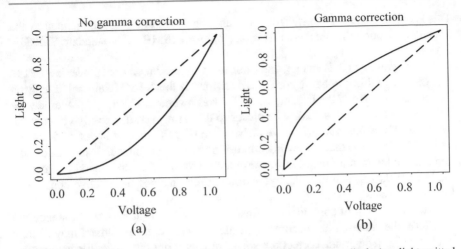

Fig. 4.6 a Effect of putative standard CRT (mimicked by an actual modern display) on light emitted from screen (voltage is normalized to range 0–1). **b** Gamma correction of signal

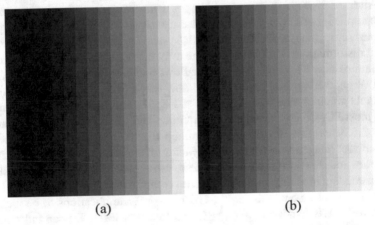

Fig. 4.7 a Display of ramp from 0 to 255, with no gamma correction. **b** Image with gamma correction applied

effect is shown in Fig. 4.7b. Here, a ramp is shown in 16 steps from gray level 0 to gray level 255.

A more careful definition of gamma recognizes that a simple power law would result in an infinite derivative at zero voltage—and this made constructing a circuit to accomplish gamma correction difficult to devise in analog. In practice, a more general transform, such as $R \rightarrow R' = a \times R^{1/\gamma} + b$, is used, along with special care at the origin:

$$V_{out} = \begin{cases} 4.5 \times V_{in} & V_{in} < 0.018 \\ 1.099 \times (V_{in}^{0.45} - 0.099), & V_{in} \geq 0.018 \end{cases} \tag{4.5}$$

This is called a camera transfer function, and the above law is recommended by the Society of Motion Picture and Television Engineers (SMPTE) as standard SMPTE–170M.

Why a gamma of 2.2? In fact, this value does *not* produce a final power law of 1.0. The history of this number is buried in decisions of the NTSC (National Television System Committee of the U.S.A.) when TV was invented. The actual power law for color receivers may in actuality be closer to 2.8. However, if we only compensate for about 2.2 of this power law, we arrive at an overall value of about 1.25 instead of 1.0. The idea was that in viewing conditions with a dim surround, such an overall gamma produces more pleasing images albeit with color errors—darker colors are made even darker, and also the eye–brain system changes the relative contrast of light and dark colors [5].

With the advent of computer systems, ostensibly built (at least in the standards) on CRTs, the situation has become even more interesting: the camera may or may not have inserted gamma correction; software may write the image file using some gamma; software may decode expecting some (other) gamma; the image is stored in a frame buffer; and it is common to provide a lookup table for gamma in the frame buffer. After all, if we have generated images using computer graphics, no gamma has been applied, but a gamma is still necessary to pre-compensate for the display.

It makes sense, then, to define an overall "system" gamma that takes into account all such transformations. Unfortunately, one must often simply guess what the actual overall gamma may be. Adobe Photoshop allows one to try different gamma values. For WWW publishing, it is important to know that a Macintosh does gamma correction in its graphics card with a gamma of 1.8, SGI machines expect a gamma of 1.7, and most PCs do no extra gamma correction and likely have a display gamma of about 2.5. Therefore, for the most common machines, it might make sense to gamma-correct images at the average of Macintosh and PC values, or about 2.1.

However, most practitioners might use a value of 2.4, adopted by the sRGB group—a standard modeling of typical light levels and monitor conditions is included in the definition of a new "standard" RGB for WWW applications to be included in all future HTML standards called sRGB: a (more or less) "Device Independent Color Space for the Internet."

An issue related to gamma correction is the decision of just what intensity levels will be represented by what bit patterns in the pixel values in a file. The eye is most sensitive to *ratios* of intensity levels, rather than absolute intensities: this means that the brighter the light, the greater must be changed in light level in order for the change to be perceived. If we had precise control over what bits represented what intensities, then it would make sense to code intensities logarithmically for maximum usage of the bits available, and then include that coding in an inverse of the $(1/\gamma)$ power law transform, as in Eq. (4.4), or perhaps a lookup table implementation of such an inverse function (see [6], p. 564). However, it is most likely that images or videos we encounter have no nonlinear encoding of bit levels, but have indeed been produced by a camcorder or are for broadcast TV. These images will have been gamma-corrected according to Eq. (4.4). The CIE-sponsored CIELAB perceptually based color difference metric (see page 100) provides a careful algorithm for including the nonlinear aspect of human brightness perception.

4.1.7 Color-Matching Functions

Practically speaking, many color applications involve specifying and recreating a particular desired color. For example, suppose one wishes to duplicate a particular shade on the screen, or a particular shade of dyed cloth. Over many years, even before the eye-sensitivity curves of Fig. 4.3 were known, a technique evolved in psychology for matching a combination of basic R, G, and B lights to a given shade. A particular set of three basic lights was available: these are called the set of *color primaries*. Then to match a given shade, a set of observers were asked to separately adjust the brightness of the three primaries using a set of controls until the resulting spot of light most closely matched the desired color. The basic situation is shown in Fig. 4.8. A device for carrying out such an experiment is called a *colorimeter*.

The international standards body for color, the CIE (Commission Internationale de L'Eclairage) pooled all such data in 1931, in a set of curves called the *color-matching functions*. They used color primaries with peaks at 440, 545, and 580 nm. Suppose instead of a swatch of cloth one was interested in matching a given wavelength of laser light (i.e., monochromatic light). Then the color-matching experiments are summarized by a statement of what proportion of the color primaries are needed for each individual narrowband wavelength light. General lights are then matched by a linear combination of single wavelength results.

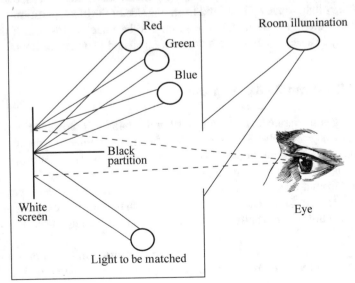

Fig. 4.8 Colorimeter experiment

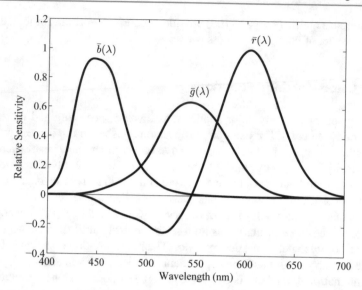

Fig. 4.9 CIE color-matching functions $\bar{r}(\lambda)$, $\bar{g}(\lambda)$, $\bar{b}(\lambda)$

Figure 4.9 shows the CIE color-matching curves, which are denoted $\bar{r}(\lambda)$, $\bar{g}(\lambda)$, and $\bar{b}(\lambda)$. In fact, such curves are a linear matrix multiplication away from the eye sensitivities in Fig. 4.3.

Why are some parts of the curves negative? This indicates that some colors cannot be reproduced by a linear combination of the primaries. For such colors, one or more of the primary lights have to be shifted from one side of the black partition in Fig. 4.8 to the other—then they illuminate the sample to be matched instead of the white screen. Thus, in a sense, such colors are being matched by negative lights.

4.1.8   CIE Chromaticity Diagram

In times long past, engineers found it upsetting that one CIE color-matching curve in Fig. 4.9 has a negative lobe. Therefore, a set of fictitious primaries were devised that lead to color-matching functions with only positives values. The resulting curves are shown in Fig. 4.10; these are usually referred to as *the* color-matching functions. They result from a linear (3 × 3 matrix) transform from the $\bar{r}$, $\bar{g}$, $\bar{b}$ curves, and are denoted $\bar{x}(\lambda)$, $\bar{y}(\lambda)$, $\bar{z}(\lambda)$. The matrix is chosen such that the middle standard color-matching function $\bar{y}(\lambda)$ exactly equals the luminous-efficiency curve $V(\lambda)$ shown in Fig. 4.3.

For a general SPD, $E(\lambda)$, the essential "colorimetric" information required to characterize a color is the set of *tristimulus values* X, Y, Z defined in analogy to Eq. 4.2 as

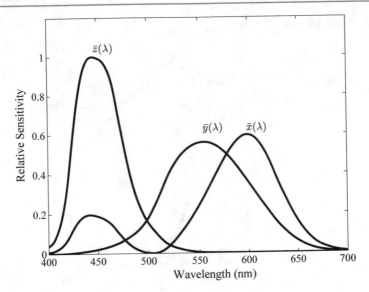

Fig. 4.10 CIE standard color-matching functions $\bar{x}(\lambda)$, $\bar{y}(\lambda)$, $\bar{z}(\lambda)$

$$X = \int E(\lambda)\,\bar{x}(\lambda)\,d\lambda$$

$$Y = \int E(\lambda)\,\bar{y}(\lambda)\,d\lambda \qquad (4.6)$$

$$Z = \int E(\lambda)\,\bar{z}(\lambda)\,d\lambda$$

The middle value, Y, is called the *luminance*. All color information and transforms are tied to these special values. They incorporate substantial information about the human visual system. However, 3D data is difficult to visualize, and consequently the CIE devised a 2D diagram based on the values of (X, Y, Z) triples implied by the curves in Fig. 4.10; for each wavelength in the visible, the values of X, Y, Z given by the three curve values form the limits of what humans can see. However, from Eq. (4.6), we observe that increasing the brightness of illumination (turning up the light bulb) increases the tristimulus values by a scalar multiple. Therefore, it makes sense to devise a 2D diagram by somehow factoring out the magnitude of vectors (X, Y, Z). In the CIE system, this is accomplished by dividing by the sum $X + Y + Z$:

$$
\begin{aligned}
x &= X/(X+Y+Z) \\
y &= Y/(X+Y+Z) \\
z &= Z/(X+Y+Z)
\end{aligned}
\qquad (4.7)
$$

This effectively means that one value out of the set (x, y, z) is redundant, since we have

$$x + y + z = \frac{X+Y+Z}{X+Y+Z} \equiv 1 \qquad (4.8)$$

Fig. 4.11 CIE chromaticity
diagram

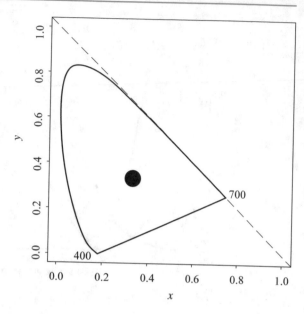

so that

$$z = 1 - x - y \ . \tag{4.9}$$

Values x, y are called *chromaticities*.

Effectively, we are projecting each tristimulus vector (X, Y, Z) onto the plane connecting points $(1, 0, 0)$, $(0, 1, 0)$, and $(0, 0, 1)$. Usually, this plane is viewed projected onto the $z = 0$ plane, as a set of points inside the triangle with vertices having (x, y) values $(0, 0)$, $(1, 0)$, and $(0, 1)$.

Figure 4.11 shows the locus of points for monochromatic light, drawn on this CIE "chromaticity diagram." The straight line along the bottom of the "horseshoe" joins points at the extremities of the visible spectrum, 400 nm and 700 nm (from blue through green to red). That straight line is called the "line of purples." The horseshoe itself is called the "spectrum locus" and shows the (x, y) chromaticity values of monochromatic light at each of the visible wavelengths.

The color-matching curves are devised so as to add up to the same value (the area under each curve is the same for each of $\bar{x}(\lambda)$, $\bar{y}(\lambda)$, $\bar{z}(\lambda)$). Therefore, for a white illuminant with all SPD values equal to 1—an "equi-energy white light"— the chromaticity values are $(1/3, 1/3)$. Figure 4.11 displays this white point in the middle of the diagram. Finally, since we must have $x, y \leq 1$ and $x + y \leq 1$, all possible chromaticity values must necessarily lie below the dashed diagonal line in Fig. 4.11.

Note that one may choose different "white" spectra as the standard illuminant. The CIE defines several of these, such as illuminant A, illuminant C, and standard daylights D65 and D100. Each of these will display as a somewhat different white spot on the CIE diagram: D65 has a chromaticity equal to $(0.312713, 0.329016)$ and illuminant C has chromaticity $(0.310063, 0.316158)$. Figure 4.12 displays the SPD

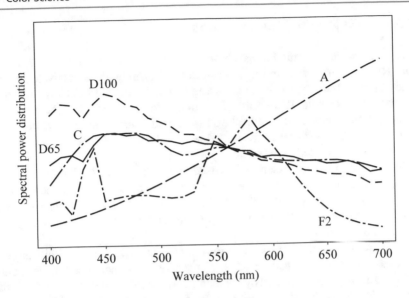

Fig. 4.12 Standard illuminant SPDs

curves for each of these standard lights. Illuminant A is characteristic of incandescent lighting, with an SPD typical of a tungsten bulb, and is quite red. Illuminant C is an early attempt to characterize daylight, while D65 and D100 are, respectively, a mid-range and a bluish commonly used daylight. Figure 4.12 also shows the much more spiky SPD for a standard fluorescent illumination, called F2 [2].

Colors with chromaticities on the spectrum locus represent "pure" colors. These are the most "saturated": think of paper becoming more and more saturated with ink. In contrast, colors close to the white point are more unsaturated.

The chromaticity diagram has the nice property that, for a mixture of two lights, the resulting chromaticity lies on the straight line joining the chromaticities of the two lights. Here, we are being slightly cagey in not saying that this is the case for *colors*, in general, but just for "lights." The reason is that so far we have been adhering to an *additive* model of color mixing. This model holds good for lights, or, as a special case, for monitor colors. However, as we shall see below, it does not hold for printer colors (see p. 106).

For any chromaticity on the CIE diagram, the "dominant wavelength" is the position on the spectrum locus intersected by a line joining the white point to the given color, and extended through it. (For colors that give an intersection on the line of purples, a complement dominant wavelength is defined by extending the line backward through the white point.) Another useful definition is the set of complementary colors for some given color, which is given by all the colors on the line through the white spot. Finally, the "excitation purity" is the ratio of distances from the white spot to the given color and to the dominant wavelength, expressed as a percentage.

4.1.9   Color Monitor Specifications

Color monitors are specified in part by the white point chromaticity that is desired
if the RGB electron guns in the ostensible CRT display are all activated at their
highest power. Actually, we are likely using gamma-corrected values R', G', B', as
supplied by the camera. If we normalize voltage to the range 0 to 1, then we wish
to specify a monitor such that it displays the desired white point when $R'=G'=B'=1$
(abbreviating the transform from file value to voltage by simply stating the pixel
color values, normalized to maximum 1).

However, the phosphorescent paints used on the inside of a CRT monitor screen,
in fact, have their own chromaticities, so that at first glance it would appear that one
could not independently control the monitor white point. However, this is remedied
by setting the gain control for each electron gun such that at maximum voltages the
desired white appears.

There are several monitor specifications in current use. Monitor specifications in
the standards we are still using consist of the fixed chromaticities for the monitor
phosphors as they were specified by the manufacturers, along with the standard white
point needed. Table 4.1 shows these values for three common specification state-
ments. NTSC is the standard North American and Japanese specification; SMPTE
is a more modern version of this, wherein the standard illuminant is changed from
illuminant C to illuminant D65 and the phosphor chromaticities are more in line
with later machines. Digital video specifications use a similar specification in North
America. The EBU system is derived from the European Broadcasting Union and is
used in PAL and SECAM video systems.

4.1.10   Out-of-Gamut Colors

For the moment, let's not worry about gamma correction. Then the really basic problem for displaying color is how to generate *device-independent* color, by agreement
taken to be specified by (x, y) chromaticity values, using *device-dependent* color
values RGB.

For any (x, y) pair, we wish to find that RGB triple giving the specified (x, y, z);
therefore, we form the z values for the phosphors, via $z = 1 - x - y$ and solve for
RGB from the manufacturer-specified chromaticities. Since, if we had no green or
blue value (i.e., file values of zero) we would simply see the red-phosphor chromaticities, and similarly for G and B, we combine nonzero values of R, G, and B via

Table 4.1 Chromaticities and white points for monitor specifications

System	Red		Green		Blue		White point	
	x_r	y_r	x_g	y_g	x_b	y_b	x_W	y_W
NTSC	0.67	0.33	0.21	0.71	0.14	0.08	0.3101	0.3162
SMPTE	0.630	0.340	0.310	0.595	0.155	0.070	0.3127	0.3290
EBU	0.64	0.33	0.29	0.60	0.15	0.06	0.3127	0.3290

$$\begin{bmatrix} x_r & x_g & x_b \\ y_r & y_g & y_b \\ z_r & z_g & z_b \end{bmatrix} \begin{bmatrix} R \\ G \\ B \end{bmatrix} = \begin{bmatrix} x \\ y \\ z \end{bmatrix} \tag{4.10}$$

If (x, y) is *specified* instead of derived from the above, then we have to invert the matrix of phosphor (x, y, z) values to obtain the correct RGB values to use in order to obtain the desired chromaticity.

But what do we do if any of the RGB numbers is *negative*? The problem in this case is that, while humans are able to perceive the color, it is not representable on the device being used. We say in that case the color is "out of gamut," since the set of all possible displayable colors constitutes the gamut of the device.

One method that is used to deal with this situation is to simply use the closest in-gamut color available. Another common approach is to select the closest complementary color.

For a CRT monitor, every displayable color is within a *triangle*; this follows from so-called "Grassman's Law," describing human vision—Grassman's law states that "color matching is linear," and this means that linear combinations of lights made up of three primaries are just the linear set of weights used to make the combination times those primaries. That is, if we compose colors from a linear combination of the three "lights" available from the three phosphors, then we can only create colors from the convex set derived from the lights—in this case, a triangle. (We'll see below that for printers this convexity no longer holds.)

Figure 4.13 shows the triangular gamut for the NTSC system, drawn on the CIE diagram. Now suppose the small triangle represents a given, desired color. Then the in-gamut point on the boundary of the NTSC monitor gamut is taken to be the intersection of the line (labeled (a) in the figure) connecting the desired color to the white point with the nearest line (labeled (b) in the figure) forming the boundary of the gamut triangle.

4.1.11 White Point Correction

One deficiency in what we have done so far is that in reality we need to be able to map tristimulus values XYZ to device RGB's, and not just deal with chromaticity xyz. The difference is that XYZ values include the magnitude of the color. As well, we need to be able to alter matters such that when each of R, G, B are at maximum value we obtain the white point.

But so far, Table 4.1 would produce incorrect values. For example, consider the SMPTE specifications. Setting $R = G = B = 1$ results in a value of X that equals the sum of the x values, or $0.630 + 0.310 + 0.155$, which is 1.095. Similarly, the Y and Z values come out to 1.005 and 0.9. Now, dividing by $(X + Y + Z)$ this results in a chromaticity of $(0.365, 0.335)$, rather than the desired values of $(0.3127, 0.3290)$.

Fig. 4.13 Approximating an
out-of-gamut color by an
in-gamut one. The
out-of-gamut color shown by
a triangle is approximated by
the intersection of the line **a**
from that color to the white
point with the boundary **b** of
the device color gamut

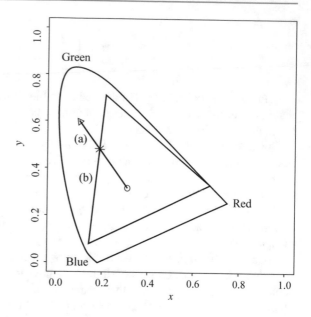

The method used to correct both deficiencies is to first take the white point magnitude of Y as unity:

$$Y \text{ (white point)} \; = \; 1. \tag{4.11}$$

Now we need to find a set of three correction factors such that if the gains of the three electron guns are multiplied by these values we get exactly the white point XYZ value at $R = G = B = 1$. Suppose the matrix of phosphor chromaticities $x_r, x_g, \ldots$, etc. in Eq. (4.10) is called M. We can express the correction as a diagonal matrix $D = diag(d_1, d_2, d_3)$ such that

$$XYZ_{\text{white}} \; \equiv \; M \, D \, (1, 1, 1)^T \tag{4.12}$$

where $(\;)^T$ means transpose.

For the SMPTE specification, we have $(x, y, z) = (0.3127, 0.3290, 0.3582)$ or dividing by the middle value, $XYZ_{\text{white}} = (0.95045, 1, 1.08892)$. We note that multiplying D by $(1, 1, 1)^T$ just gives $(d_1, d_2, d_3)^T$ and we end up with an equation *specifying* $(d_1, d_2, d_3)^T$:

$$\begin{bmatrix} X \\ Y \\ Z \end{bmatrix}_{\text{white}} = \begin{bmatrix} 0.630 & 0.310 & 0.155 \\ 0.340 & 0.595 & 0.070 \\ 0.03 & 0.095 & 0.775 \end{bmatrix} \begin{bmatrix} d_1 \\ d_2 \\ d_3 \end{bmatrix} \tag{4.13}$$

Inverting, with the new values XYZ_{white} specified as above, we arrive at

$$(d_1, d_2, d_3) \; = \; (0.6247, 1.1783, 1.2364). \tag{4.14}$$

4.1.12 XYZ to RGB Transform

Now the 3 × 3 transform matrix from XYZ to RGB is taken to be

$$T = MD \tag{4.15}$$

even for points other than the white point:

$$\begin{bmatrix} X \\ Y \\ Z \end{bmatrix} = T \begin{bmatrix} R \\ G \\ B \end{bmatrix}. \tag{4.16}$$

For the SMPTE specification, we arrive at

$$T = \begin{bmatrix} 0.3935 & 0.3653 & 0.1916 \\ 0.2124 & 0.7011 & 0.0866 \\ 0.0187 & 0.1119 & 0.9582 \end{bmatrix}. \tag{4.17}$$

Written out, this reads

$$\begin{aligned} X &= 0.3935 \cdot R + 0.3653 \cdot G + 0.1916 \cdot B \\ Y &= 0.2124 \cdot R + 0.7011 \cdot G + 0.0866 \cdot B \\ Z &= 0.0187 \cdot R + 0.1119 \cdot G + 0.9582 \cdot B \end{aligned} \tag{4.18}$$

4.1.13 Transform with Gamma Correction

The above calculations assume that we are dealing with linear signals. However, it is most likely that instead of linear R, G, B we actually have nonlinear, gamma-corrected R', G', B' camera values.

The best way of carrying out an XYZ to RGB transform is to calculate the linear RGB required, by inverting Eq. (4.16), and then create nonlinear signals via gamma correction.

Nevertheless, this is not usually done as stated. Instead, the equation for the Y value is used as is, but applied to nonlinear signals. This does not imply much error, in fact, for colors near the white point. The only concession to accuracy is to give the new name Y' to this new Y value created from R', G', B'. The significance of Y' is that it codes a descriptor of brightness for the pixel in question.[1]

The most used set of transform equations are those for the original NTSC system, based upon an Illuminant C white point, even though these are outdated. Following

---

[1] In the color FAQ file [7], this new value Y' is called "luma."

the procedure outlined above, but with the values in the top row of Table 4.1, we arrive at the following transform:

$$
\begin{aligned}
X &= 0.607 \cdot R + 0.174 \cdot G + 0.200 \cdot B \\
Y &= 0.299 \cdot R + 0.587 \cdot G + 0.114 \cdot B \\
Z &= 0.000 \cdot R + 0.066 \cdot G + 1.116 \cdot B
\end{aligned}
\tag{4.19}
$$

Thus, coding for nonlinear signals begins with encoding the nonlinear signal correlate of luminance:

$$
Y' = 0.299 \cdot R' + 0.587 \cdot G' + 0.114 \cdot B' \tag{4.20}
$$

(See Sect. 4.3 for more discussion on encoding of nonlinear signals.)

4.1.14 L*a*b* (CIELAB) Color Model

The discussion above of how best to make use of the bits available to us (see p. 65) touched on the issue of how well human vision sees changes in light levels. This subject is actually an example of *Weber's Law*, from psychology: the more there is of a quantity, the more change there must be to perceive a difference. For example, it's relatively easy to tell the difference in weight between picking up your 4-year-old sister and your 5-year-old brother (aside from their other attributes). However, it is more difficult to tell the difference in weight between two heavy objects. Another example is that to see a change in a bright light the difference must be much larger than to see a change in a dim light. A rule of thumb for this phenomenon states that equally perceived changes must be relative—changes are about equally perceived if the ratio of the change is the same, whether for dark or bright lights, etc. Some thought using this idea leads one to a logarithmic approximation to perceptually equally spaced units.

For human vision, however, the CIE arrived at a somewhat more involved version of this kind of rule, called the CIELAB space. What is being quantified in this space is, again, *differences* perceived in color and brightness. This makes sense, in fact, since, practically speaking, color differences are most useful for the comparison of source and target colors, as it were: one would be interested, e.g., in whether a particular batch of dyed cloth has the same color as an original swatch. Figure 4.14 shows a cutaway into a 3D solid of the coordinate space associated with this color difference metric.

CIELAB (also known as L*a*b*) uses a power law of 1/3 instead of a logarithm. It uses three values that correspond roughly to (a) luminance; plus (b) a pair that combine to make colorfulness and hue. The color difference, for two colors in this color space, is simply defined as a Euclidean distance:

$$
\Delta E = \sqrt{(L_1^* - L_2^*)^2 + (a_1^* - a_2^*)^2 + (b_1^* - b_2^*)^2} \tag{4.21}
$$

Fig. 4.14 CIELAB model

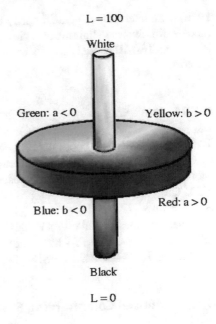

where

$$L^* = 116 \left(\frac{Y}{Y_n}\right)^{(1/3)} - 16$$

$$a^* = 500 \left[\left(\frac{X}{X_n}\right)^{(1/3)} - \left(\frac{Y}{Y_n}\right)^{(1/3)}\right]$$

$$b^* = 200 \left[\left(\frac{Y}{Y_n}\right)^{(1/3)} - \left(\frac{Z}{Z_n}\right)^{(1/3)}\right] \tag{4.22}$$

with X_n, Y_n, Z_n being the XYZ values of the white point. Auxiliary definitions are

$$\text{chroma} = c^* = \sqrt{(a^*)^2 + (b^*)^2}$$

$$\text{hue angle} = h^* = \arctan \frac{b^*}{a^*} \tag{4.23}$$

Roughly, the maximum and minimum of value a^* correspond to red and green, while b^* ranges from yellow to blue. The *chroma* is a scale of colorfulness, with more colorful (more saturated) colors occupying the outside of the CIELAB solid at each L^* brightness level, and more washed-out (desaturated) colors nearer the central achromatic axis. The hue angle expresses more or less what most people mean by "the color," i.e., would one describe the color as red, orange, etc.

The development of such color differences models is a very active field of research, and there are a plethora of other human-perception-based formulae (the other competitor of the same vintage as CIELAB is called CIELUV—both were devised in

1976). The interest is generated partly because such color metrics impact how one models differences in lighting and viewing across device and/or network boundaries [8]. The CIELAB model is used by several high-end products, including Adobe Photoshop.

4.1.15 More Color Coordinate Schemes

There are several other coordinate schemes in use to describe color as humans perceive it, with some confusion in the field as to whether gamma correction should or should not be applied. Here we are describing device-independent color—based on XYZ and correlated to what humans see. However, generally users make free use of RGB or R', G', B'.

Other schemes include: CMY (described below on p. 105); HSL—Hue, saturation, and lightness; HSV—hue, saturation, and value; HSI—and intensity; HCI–C=chroma; HVC—V=value; HSD—D=darkness: the beat goes on!

4.1.16 Munsell Color Naming System

Accurate *naming* of colors is also an important consideration. One time-tested standard system was devised by Munsell in the early 1900s and revised many times (the last one is called the Munsell renotation) [9]. The idea is to set up (yet another) approximately perceptually uniform system of three axes to discuss and specify color. The axes are value (black-white), hue, and chroma. Value is divided into 9 steps, Hue is in 40 steps around a circle, and chroma (saturation) has a maximum of 16 levels, but the circle radius varies with value.

The main idea is a fairly invariant specification of color for any user, including artists, and the Munsell corporation therefore sells books of all these patches of paint made up with proprietary paint formulae (the book is quite expensive!). It has been asserted that this is the most often used uniform scale.

4.2   Color Models in Images

We now have seen an introduction to color science and an introduction to some of the problems that crop up with respect to color for image displays. But how are colors models and coordinate systems really used for stored, displayed, and printed images?

4.2.1   RGB Color Model for Displays

According to Sect. 4.3, we usually store color information directly in RGB form. However, we note from Sect. 4.1 that such a coordinate system is in fact device dependent.

We expect to be able to use 8 bits per color channel for color that is accurate enough. However, in fact, we have to use about 12 bits per channel to avoid an aliasing effect in dark image areas—contour bands that result from gamma correction since gamma correction results in many fewer available integer levels (see Exercise 9.)

For images produced from computer graphics, we store integers proportional to intensity in the frame buffer; then, we should have a gamma correction LUT between the frame buffer and the display. If gamma correction is applied to floats before quantizing to integers, before storage in the frame buffer, then, in fact, we can use only 8 bits per channel and still avoid contouring artifacts.

4.2.2 Multi-sensor Cameras

More accurate color can be achieved by using cameras with more than three sensors, i.e., more than three color filters. One way of doing this is by using a rotating filter, which places a different color filter in the light path over a quick series of shots. In work on capture of artwork at the Museum of Modern Art in New York City, a six-channel camera [10] has been used to accurately capture images of important artworks, such that images are closer to full spectrum; this work uses an altered color filter checkerboard array, or set of these, built into the camera ("Art Spectral Imaging"). Part of work in this direction also has included removing the near-infrared filter typically placed in a camera, so as to extend the camera's sensitivity into the infrared [11].

4.2.3 Camera-Dependent Color

The values of R, G, B at a pixel depend on what camera sensors are used to image a scene. Here we look at other two other camera-dependent color spaces that are commonly used: HSV and sRGB.

First, recall that in Sect. 4.1.14 on CIELAB, we defined color coordinates that are meant to be *camera independent*, i.e., human-perception oriented. There, the proposed set of axes L^*, a^* (redness-greenness), b^* (yellowness-blueness) are associated with human visual system percepts of *Lightness* L^*; hue h^*, meaning a magnitude-independent notion of color; and chroma c^*, meaning the purity (vividness) of a color.

HSV

Along this same direction, in order to tie such perceptual concepts into camera-dependent color, the HSV color system tries to generate similar quantities. While there are many commonly used variants, HSV is by far the most common. H stands for hue; S stands for "saturation" of a color, defined by chroma divided by its luminance—the more desaturated the color is the closer it is to gray; and V stands for "value," meaning a correlate of brightness as perceived by humans. The HSV color model is commonly used in image processing and editing software.

RGB data is converted into the HSV color space as follows: assuming R, G, B are in $0..255$,

$$M = \max\{R, G, B\}$$

$$m = \min\{R, G, B\}$$

$$V = M$$

$$S = \begin{cases} 0 & \text{if } V = 0 \\ (V - m)/V & \text{if } V > 0 \end{cases}$$

$$H = \begin{cases} 0 & \text{if } S = 0 \\ 60(G - B)/(M - m) & \text{if } (M = R \text{ and } G \geq B) \\ 60(G - B)/(M - m) + 360 & \text{if } (M = R \text{ and } G < B) \\ 60(B - R)/(M - m) + 120 & \text{if } M = G \\ 60(R - G)/(M - m) + 240 & \text{if } M = B \end{cases}$$

(4.24)

where M and m denote the maximum and minimum of the (R, G, B) triplet.

sRGB

As a balance between human color perception and device-dependent color, the sRGB color space was devised as tied to the color space of a particular reference display device. sRGB has very generally been adopted as a reference color space on the web, in the sense that unless otherwise stated the color space for encoded/transmitted values is assumed to be sRGB.

Originally, sRGB was proposed by Hewlett-Packard and Microsoft, and was later standardized by the International Electrotechnical Commission (IEC) [12]. sRGB presupposes certain standard viewing conditions, typical of use for computer monitors (details may be found in [13,14]). As well, it specifies a transform for going from a standard gamma-corrected image to one which is linear with light intensity, as follows (with each color channel I in (R, G, B) now normalized into the range $[0, 1]$): For $I = R, G, B$, we apply a function

$$\begin{cases} I = I'/12.92, & \text{if } I' < 0.04045; \\ I = ((I' + 0.055)/1.055)^{2.4} & \text{otherwise} \end{cases}$$

(4.25)

Taking into account the whole curve shape, this is approximately a γ value of 2.2.

As well, the sRGB standard specifies a colorimetric transform to go from such linear sRGB values to human-centered CIEXYZ tristimulus color space values:

$$\begin{pmatrix} X \\ Y \\ Z \end{pmatrix} = \begin{pmatrix} 0.4124 & 0.3576 & 0.1805 \\ 0.2126 & 0.7152 & 0.0722 \\ 0.0193 & 0.1192 & 0.9505 \end{pmatrix} \begin{pmatrix} R \\ G \\ B \end{pmatrix}$$

(4.26)

With this definition, when white is $(R, G, B) = (1, 1, 1)$ the XYZ triple becomes that of standard light D65 (divided by 100): $(X, Y, Z) = (0.9505, 1.0000, 1.0890)$.

4.2.4   Subtractive Color: CMY Color Model

So far, we have effectively been dealing only with *additive color*. Namely, when two light beams impinge on a target, say from two color projectors sending light to a white screen, their colors add. For theoretical CRTs, still the displays upon which standards are built, when two phosphors on a CRT screen are turned on, their colors add. So, for example, red phosphor + green phosphor makes yellow light.

But for ink deposited on paper, in essence the opposite situation holds: yellow ink *subtracts* blue from white illumination, but reflects red and green, and that is why it appears yellow!

So, instead of red, green, and blue primaries, we need primaries that amount to −red, −green, and −blue; we need to *subtract* R or G or B. These subtractive color primaries are cyan (C), magenta (M), and yellow (Y) inks. Figure 4.15 shows how the two systems, RGB and CMY, are connected. In the additive (RGB) system, black is "no light," $RGB = (0, 0, 0)$. But in the subtractive CMY system, black arises from subtracting all the light by laying down inks with $C = M = Y = 1$.

4.2.5   Transformation from RGB to CMY

Given our identification of the role of inks in subtractive systems, the simplest model we can invent to specify what ink density to lay down on paper, to make a certain

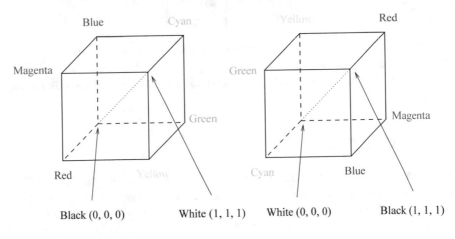

The RGB Cube   The CMY Cube

Fig. 4.15 RGB and CMY color cubes

desired RGB color, is as follows:

$$\begin{bmatrix} C \\ M \\ Y \end{bmatrix} = \begin{bmatrix} 1 \\ 1 \\ 1 \end{bmatrix} - \begin{bmatrix} R \\ G \\ B \end{bmatrix} \tag{4.27}$$

Then the inverse transform is

$$\begin{bmatrix} R \\ G \\ B \end{bmatrix} = \begin{bmatrix} 1 \\ 1 \\ 1 \end{bmatrix} - \begin{bmatrix} C \\ M \\ Y \end{bmatrix} \tag{4.28}$$

4.2.6   Undercolor Removal: CMYK System

C, M, and Y are supposed to mix to black. However, more often they mix to a muddy brown (we all know this, from kindergarten!). Truly "black" black ink is, in fact, cheaper than mixing colored inks to make black, so a simple approach to producing sharper printer colors is to calculate that part of the three-color mix that would be black, remove it from the color proportions, and add it back as real black. This is called "undercolor removal."

With K representing the amount of black, the new specification of inks is thus

$$K \equiv \min\{C, M, Y\} \tag{4.29}$$

$$\begin{bmatrix} C \\ M \\ Y \end{bmatrix} \Rightarrow \begin{bmatrix} C - K \\ M - K \\ Y - K \end{bmatrix}$$

Figure 4.16 depicts the color combinations that result from combining primary colors available in the two situations, additive color, in which one usually specifies color using RGB, and subtractive color, in which one usually specifies color using CMY or CMYK.

4.2.7   Printer Gamuts

In a common model of the printing process, printers lay down transparent layers of ink onto a (generally white) substrate. If we wish to have a cyan printing ink truly equal to minus-red, then our objective is to produce a cyan ink that completely blocks red light but also completely passes all green and blue light. Unfortunately, such "block dyes" are only approximated in industry and in reality transmission curves overlap for the C, M, Y inks. This leads to "crosstalk" between the color channels and difficulties in predicting colors achievable in printing.

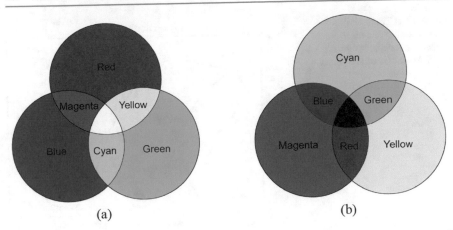

(a)   (b)

Fig. 4.16 Additive and subtractive color. **a** RGB is used to specify additive color. **b** CMY is used to specify subtractive color

Figure 4.17a shows typical transmission curves for real block dyes, and Fig. 4.17b shows the resulting color gamut for a color printer that uses such inks. We see that the gamut is smaller than that of an NTSC monitor, and can overlap it.

Such a gamut arises from the model used for printer inks. Transmittances are related to *optical density* D via a logarithm: $D = -\log T$, where T is one of the curves in Fig. 4.17a. An actual color is formed by a linear combination D of inks, with D a combination of the three densities weighted by weights w_i, $i = 1..3$, and w_i can be in the range from zero to the maximum allowable without smearing. So the overall transmittance T is formed as a product of exponentials of the three weighted densities—light is extinguished exponentially as it travels through a "sandwich" of transparent dyes. The light reflected from paper (or through a piece of slide film) is $T E = e^{-D} E$, where E is the illuminating light. Forming colors XYZ with Eq. (4.6) leads to the printer gamut in Fig. 4.17b.

The center of the printer gamut is the white–black axis; the six boundary vertices correspond to C, M, Y, and the three combinations CM, CY, and MY laid down at full density. Lesser ink densities lie more in the middle of the diagram. Full density for all inks correspond to the black/white point, which lies in the center of the diagram at the point marked "o." For these particular inks, that point has chromaticity $(x, y) = (0.276, 0.308)$.

4.2.8 Multi-ink Printers

Increasingly, printers are being manufactured with more than four (CMYK) inks, i.e., printing systems with many more colorants. An example is a CMYKRGB printer. The objective is to greatly increase the size of the printer gamut [15].

Fig. 4.17 a Transmission curves for block dyes. **b** Spectrum locus, triangular NTSC gamut, and six-vertex printer gamut

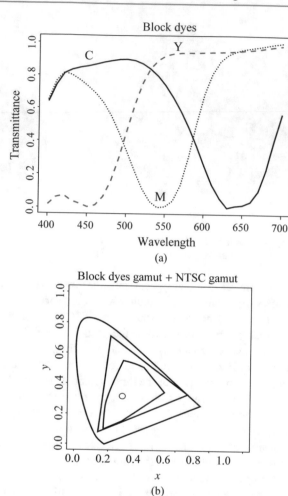

4.3   Color Models in Video

4.3.1   Video Color Transforms

Methods of dealing with color in digital video are largely derived from older analog methods of coding color for TV. Typically, some version of the luminance is combined with color information in a single signal. For example, a matrix transform method similar to Eq. (4.19) called YIQ is used to transmit TV signals in North America and Japan.

In Europe, video tape used the PAL or SECAM codings, which are based on TV that uses a matrix transform called YUV.

Finally, digital video mostly uses a matrix transform called YCbCr that is closely related to YUV.[2]

4.3.2 YUV Color Model

Initially, YUV coding was used for PAL analog video. A version of YUV is now also used in the CCIR 601 standard for digital video.

Firstly, it codes a luminance signal (for gamma-corrected signals) equal to Y' in Eq. (4.20). (Recall that Y' is often called the "luma.") The luma Y' is similar, but not exactly the same as, the CIE luminance value Y, gamma-corrected. In multimedia, practitioners often blur the difference and simply refer to both as the luminance.

As well as magnitude or brightness we need a colorfulness scale, and to this end *chrominance* refers to the difference between a color and a reference white at the same luminance. It can be represented by the color *differences* U, V:

$$U = B' - Y'$$
$$V = R' - Y' \tag{4.30}$$

From Eqs. (4.20) to (4.30) reads

$$\begin{bmatrix} Y' \\ U \\ V \end{bmatrix} = \begin{bmatrix} 0.299 & 0.587 & 0.114 \\ -0.299 & -0.587 & 0.886 \\ 0.701 & -0.587 & -0.114 \end{bmatrix} \begin{bmatrix} R' \\ G' \\ B' \end{bmatrix} \tag{4.31}$$

One goes backward, from (Y', U, V) to (R', G', B'), by inverting the matrix in Eq. (4.31).

Note that for a gray pixel, with $R' = G' = B'$, the luminance Y' is equal to that same gray value, R', say, since the sum of the coefficients in Eq. (4.20) is $0.299 + 0.587 + 0.114 = 1.0$. Also, for such a gray ("black and white") image, the chrominance (U, V) is zero since the sum of coefficients in each of the lower two equations in (4.31) is zero. Hence, color TV could be displayed on a precursor black and white television by just using the Y' signal.[3] And for backward compatibility, color TV uses old black and white signals with no color information by identifying the signal with Y'.

Finally, in the actual implementation, U and V are rescaled for purposes of having a more convenient maximum and minimum. For analog video, the scales were chosen such that each of U or V is limited to the range between ± 0.5 times the maximum of Y' [16]. (Note that actual voltages are in another, non-normalized range—for analog

---

[2]The luminance–chrominance color models (YIQ, YUV, YCbCr) are proven effective. Hence, they are also adopted in image compression standards such as JPEG and JPEG2000.

[3]It should be noted that many authors and users simply use these letters with no primes, and (perhaps) mean them as if they were with primes!

Y' is often in the range 0 to 700 mV and so rescaled U and V, called P_B and P_R in that context, range over ± 350 mV.)

Such a scaling reflects how to deal with component video—three separate signals. However, for dealing with *composite* video, in which we want to compose a single *chrominance* signal out of both U and V at once, it turns out to be convenient to contain U, V within the range $-1/3$ to $+4/3$. This is so that the composite signal magnitude $Y' \pm \sqrt{U^2 + V^2}$ will remain within the amplitude limits of the recording equipment. For this purpose, U and V are rescaled as follows:

$$
\begin{aligned}
U &= 0.492111(B' - Y') \\
V &= 0.877283(R' - Y')
\end{aligned}
\tag{4.32}
$$

(with multipliers sometimes rounded to three significant digits). Altogether, this makes the transform from R', G', B' to Y', U, V as follows:

$$
\begin{bmatrix} Y' \\ U \\ V \end{bmatrix} =
\begin{bmatrix}
0.299 & 0.587 & 0.114 \\
-0.14713 & -0.28886 & 0.436 \\
0.615 & -0.51499 & -0.10001
\end{bmatrix}
\begin{bmatrix} R' \\ G' \\ B' \end{bmatrix}
\tag{4.33}
$$

One goes backward, from (Y', U, V) to (R', G', B'), by inverting the matrix in Eq. (4.33), as follows:

$$
\begin{bmatrix} R' \\ G' \\ B' \end{bmatrix} =
\begin{bmatrix}
1.0000 & 0.0000 & 1.13983 \\
1.0000 & -0.39465 & -0.58059 \\
1.0000 & 2.03211 & 0.0000
\end{bmatrix}
\begin{bmatrix} Y' \\ U \\ V \end{bmatrix}
\tag{4.34}
$$

Then the chrominance signal is composed of U and V as the composite signal

$$
C = U \cdot \cos(\omega t) + V \cdot \sin(\omega t)
\tag{4.35}
$$

where ω represents the NTSC color frequency.

From Eq. (4.33), we note that zero is not the minimum value for U, V. In terms of real, positive colors, U is approximately from blue ($U > 0$) to yellow ($U < 0$) in the RGB cube; V is approximately from red ($V > 0$) to cyan ($V < 0$).

Figure 4.18 shows the decomposition of a typical color image into its Y', U, V components. Since both U and V go negative, for display, in fact, the images are shifted, rescaled versions of the actual signals.

Since the eye is most sensitive to black and white variations, in terms of spatial frequency (e.g., the eye can see a grid of fine gray lines more clearly than fine colored lines), in the analog PAL signal a bandwidth of only 1.3 MHz was allocated to each of U and V, while 5.5 MHz was reserved for the Y' signal. In fact, color information that is transmitted for color TV is actually very blocky but we don't perceive this low level of color information.

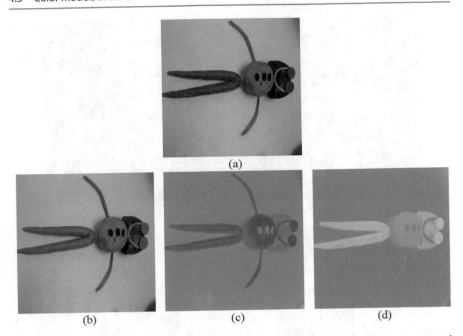

Fig. 4.18 $Y'UV$ decomposition of color image. Top image **a** is original color image; **b** is Y'; **c, d** are (U, V)

4.3.3 YIQ Color Model

YIQ (actually, $Y' I Q$) is used in NTSC color TV broadcasting. Again, gray pixels generate zero (I, Q) chrominance signal. The original meanings of these names came from combinations of analog signals, I for "in-phase chrominance" and Q for "quadrature chrominance" signal—these names can now be safely ignored.

It is thought that, although U and V are more simply defined, they do not capture the most-to-least hierarchy of human vision sensitivity. Although U and V nicely define the color differences, they do not best correspond to actual human perceptual color sensitivities. In NTSC, I and Q are used instead.

YIQ is just a version of YUV, with the same Y', but with U and V rotated by $33°$:

$$I = 0.492111(R' - Y') \cos 33° - 0.877283(B' - Y') \sin 33° \quad (4.36)$$
$$Q = 0.492111(R' - Y') \sin 33° + 0.877283(B' - Y') \cos 33°$$

This leads to the following matrix transform:

$$\begin{bmatrix} Y' \\ I \\ Q \end{bmatrix} = \begin{bmatrix} 0.299 & 0.587 & 0.114 \\ 0.595879 & -0.274133 & -0.321746 \\ 0.211205 & -0.523083 & 0.311878 \end{bmatrix} \begin{bmatrix} R' \\ G' \\ B' \end{bmatrix} \quad (4.37)$$

I is roughly the orange–blue direction and Q roughly corresponds to the purple–green direction.

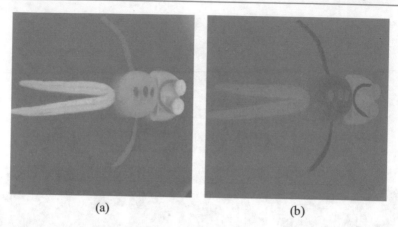

Fig. 4.19 I and Q components of color image

To go back from (Y', I, Q) to (R', G', B'), one inverts Eq. (4.37):

$$\begin{bmatrix} R' \\ G' \\ B' \end{bmatrix} = \begin{bmatrix} 1.0000 & 0.95630 & 0.62103 \\ 1.0000 & -0.27256 & -0.64671 \\ 1.0000 & -1.10474 & 1.70116 \end{bmatrix} \begin{bmatrix} Y' \\ I \\ Q \end{bmatrix} \quad (4.38)$$

Figure 4.19 shows the decomposition of the same color image as above into YIQ components. Only the I and Q components are shown since the original image and the Y' component are the same as in Fig. 4.18.

For this particular image, most of the energy is captured in the Y' component; this is typical. However, in this case, the YIQ decomposition does a better of job of forming a hierarchical sequence of images; for the 8-bit Y' component, the root-mean-square value is 146 (with 255 the maximum possible). The U, V components have RMS values 19 and 21. For the YIQ decomposition, the I and Q components, on the other hand, have RMS values 20 and 5, and so better prioritize color values.

4.3.4 YCbCr Color Model

The international standard for component (three-signal, studio quality) *digital* video is officially "Recommendation ITU-R BT.601-4" (known as Rec. 601). This standard uses another color space, YC_bC_r, often simply written YCbCr. The YCbCr transform is closely related to the YUV transform. YUV is changed by scaling such that C_b is U, but with a coefficient of 0.5 multiplying B'. In some software systems, C_b and C_r are also shifted such that values are between 0 and 1. This makes the equations as follows:

$$C_b = ((B' - Y')/1.772) + 0.5 \quad (4.39)$$
$$C_r = ((R' - Y')/1.402) + 0.5$$

Written out, we then have

$$\begin{bmatrix} Y' \\ C_b \\ C_r \end{bmatrix} = \begin{bmatrix} 0.299 & 0.587 & 0.114 \\ -0.168736 & -0.331264 & 0.5 \\ 0.5 & -0.418688 & -0.081312 \end{bmatrix} \begin{bmatrix} R' \\ G' \\ B' \end{bmatrix} + \begin{bmatrix} 0 \\ 0.5 \\ 0.5 \end{bmatrix}$$
(4.40)

In practice, however, Recommendation 601 specifies 8-bit coding, with a maximum Y' value of only 219, and a minimum of $+16$. Values below 16 and above 235 are reserved for other processing (these are denoted "headroom" and "footroom"). Cb and Cr have a range of ± 112 and offset of $+128$ (in other words, a maximum of 240 and a minimum of 16). If R', G', B' are floats in $[0..1]$, then we obtain Y', C_b, C_r in $[0..255]$ via the transform [16]

$$\begin{bmatrix} Y' \\ C_b \\ C_r \end{bmatrix} = \begin{bmatrix} 65.481 & 128.553 & 24.966 \\ -37.797 & -74.203 & 112 \\ 112 & -93.786 & -18.214 \end{bmatrix} \begin{bmatrix} R' \\ G' \\ B' \end{bmatrix} + \begin{bmatrix} 16 \\ 128 \\ 128 \end{bmatrix}$$
(4.41)

In fact, the output range is also clamped to $[1..254]$ since the Rec. 601 synchronization signals are given by codes 0 and 255.

The inverse transform to Eq. (4.41) is as follows [16]:

$$\begin{bmatrix} R' \\ G' \\ B' \end{bmatrix} = \begin{bmatrix} 0.00456621 & 0.0000 & 0.00625893 \\ 0.00456621 & -0.00153632 & -0.00318811 \\ 0.00456621 & 0.00791071 & 0.0000 \end{bmatrix} \begin{bmatrix} Y' - 16 \\ C_b - 128 \\ C_r - 128 \end{bmatrix}$$
(4.42)

The YCbCr transform is used in JPEG image compression and MPEG video compression.

4.4 Exercises

1. Consider the following set of color-related terms:

   (a) wavelength,
   (b) color level,
   (c) brightness, and
   (d) whiteness.

   How would you match each of the following (more vaguely stated) characteristics to each of the above terms?

   (a) luminance,
   (b) hue,
   (c) saturation, and
   (d) chrominance.

2. What color is outdoor light? That is, around what wavelength would you guess the peak power is for a red sunset? For blue sky light?

3. "The LAB gamut covers all colors in the visible spectrum."
   What does that statement mean? Briefly, how does LAB relate to color?—just be descriptive.
   What are (roughly) the relative sizes of the LAB gamut, the CMYK gamut, and a monitor gamut?

4. Prove that straight lines in (X, Y, Z) space project to straight lines in (x, y) chromaticity space. That is, let $C_1 = (X_1, Y_1, Z_1)$ and $C_2 = (X_2, Y_2, Z_2)$ be two different colors, and let $C_3 = (X_3, Y_3, Z_3)$ fall on a line connecting C_1 and C_2: $C_3 = \alpha C_1 + (1 - \alpha)C_2$. Then show that $(x_3, y_3) = \beta(x_1, y_1) + (1 - \beta)(x_2, y_2)$ for some β.

5. Where does the chromaticity "horseshoe" shape in Fig. 4.11 come from? Can we calculate it?
   Write a small pseudocode solution for the problem of finding this so-called "spectrum locus."
   Hint: Figure 4.20a shows the color-matching functions in Fig. 4.10 drawn as a set of points in 3-space. And Fig. 4.20b shows these points mapped into another 3D set of points.
   Hint: Try a programming solution for this problem, to help you answer it more explicitly.

6. Suppose we use a new set of color-matching functions $\bar{x}^{new}(\lambda), \bar{y}^{new}(\lambda), \bar{z}^{new}(\lambda)$ with values

λ (nm)	$\bar{x}^{new}(\lambda)$	$\bar{y}^{new}(\lambda)$	$\bar{z}^{new}(\lambda)$
450	0.2	0.1	0.5
500	0.1	0.4	0.3
600	0.1	0.4	0.2
700	0.6	0.1	0.0

In this system, what are the chromaticity values (x, y) of equi-energy white light $E(\lambda)$ where $E(\lambda) \equiv 1$ for all wavelengths λ? Explain.

7. Repeat the steps leading up to Eq. (4.18), but this time using the NTSC standard— if you use the number of significant digits as in Table 4.18 you will end up with the transform in Eq. (4.19).

8. (a) Suppose images are *not* gamma-corrected by a camcorder. Generally, how would they appear on a screen?
   (b) What happens if we artificially increase the output gamma for stored image pixels? (One can do this in Photoshop.) What is the effect on the image?

9. Suppose image file values are in $0..255$ in each color channel. If we define $\bar{R} = R/255$ for the red channel, we wish to carry out gamma correction by passing a new value $\bar{R}'$ to the display device, with $\bar{R}' \simeq \bar{R}^{1/2.0}$.

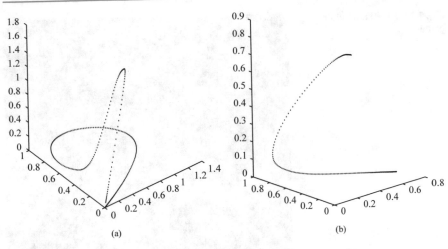

Fig. 4.20 **a** Color-matching functions; **b** transformed color-matching functions

It is common to carry out this operation using integer math. Suppose we approximate the calculation as creating new integer values in $0..255$ via

$$(int)\,(255 \cdot (\overline{R}^{1/2.0}))$$

(a) Comment (very roughly) on the effect of this operation on the number of actually available levels for display. Hint: Coding this up in any language will help you understand the mechanism at work better—and as well, then you can simply count the output levels.

(b) Which end of the levels $0..255$ is affected most by gamma correction, the low end near 0 or the high end near 255? Why? How much at each end?

10. In many computer graphics applications, γ correction is performed only in the color LUT (lookup table). Show the first five entries of the color LUT if it is meant for use in γ correction.
    Hint: Coding this up saves you the trouble of using a calculator for this question.

11. Devise a program to produce Fig. 4.21, showing the color gamut of a monitor that adheres to SMPTE specifications.

12. The "hue" is the color, independent of brightness and how much pure white has been added to it. We can make a *simple* definition of hue as the set of ratios R:G:B.

    (a) Suppose a color (i.e., an RGB) is divided by 2.0, so that the RGB triple now has values 0.5 times its former values.
        Explain using numerical values:

Fig. 4.21 SMPTE monitor gamut

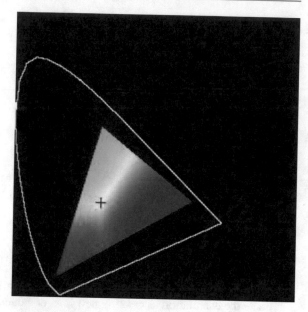

(i) If gamma correction is applied after the division by 2.0 and before the color is stored, does the darker RGB have the same hue as the original in the sense of having the same ratios R:G:B of light emanating from the display device? (we're not discussing any psychophysical effects that change our perception—here we're just worried about the machine itself).

(ii) If gamma correction is *not* applied, does the second RGB above, = RGB/2, have the same hue as the first RGB, when displayed? And are these the same hues as for the original color as *stored*, not the light as displayed?

(b) Assuming no gamma correction is applied, for what color triples is the hue as displayed the same as for the original color as stored?

13. We wish to produce a graphic that is pleasing and easily readable. Suppose we make the background color pink. What color text font should we use to make the text most readable? Justify your answer.

14. To make matters simpler for eventual printing, we buy a camera equipped with CMY sensors, as opposed to RGB sensors (CMY cameras are, in fact, available).

(a) Draw spectral curves roughly depicting what such a camera's sensitivity to wavelength might look like.

(b) Could the output of a CMY camera be used to produce ordinary RGB pictures? How?

15. Color ink-jet printers use the CMYK model. When the color ink *cyan* is sprayed onto a sheet of white paper,

   (i) why does it look cyan under daylight?
   (ii) what color would it appear to be under a *blue* light. Why?

References

1. D.H. Pritchard, U.S. color television fundamentals—A review. IEEE Trans. Consum. Electron. **23**(4), 467–478 (1977)
2. G. Wyszecki, W.S. Stiles, *Color Science: concepts and Methods, Quantitative Data and Formulas*, 2nd edn. (Wiley, New York, 2000)
3. R.W.G. Hunt, Color reproduction and color vision modeling, in *1st Color Imaging Conference on Transforms & Transportability of Color*, pp. 1–5. Society for Imaging Science & Technology (IS&T)/Society for Information Display (SID) joint conference (1993)
4. M.J. Vrhel, R. Gershon, L.S. Iwan, Measurement and analysis of object reflectance spectra. Color Res. Appl. **19**, 4–9 (1994)
5. R.W.G. Hunt, *The Reproduction of Color*, 6th edn. (Fountain Press, England, 2004)
6. J.F. Hughes, A. van Dam, M. McGuire, D.F. Sklar, J.D. Foley, S.K. Feiner, K. Akeley, *Computer Graphics: Principles and Practice*, 3rd edn. (Addison-Wesley, 2013)
7. C.A. Poynton, Color FAQ—frequently asked questions color (2006). http://www.poynton.com/notes/colour_and_gamma/ColorFAQ.html
8. M.D. Fairchild, *Color Appearance Models*, 3rd edn. (Addison-Wesley, 2013)
9. D. Travis, *Effective Color Displays*. (Academic Press, 1991)
10. R.S. Berns, L.A. Taplin, Practical spectral imaging using a color-filter array digital camera. art-si.org/PDFs/Acquisition/TR_Practical_Spectral_Imaging.pdf
11. C. Fredembach, S. Susstrunk, Colouring the near infrared, in *16th Color Imaging Conference*, pp. 176–182 (2008)
12. International Electrotechnical Commission, *IEC 61966-2-1: multimedia systems and equipment—Colour measurement and management—Part 2-1: colour management—Default RGB colour space—sRGB*. (IEC, Geneva, 2007)
13. M. Stokes, M. Anderson, S. Chandrasekar, R. Motta, A standard default color space for the internet: sRGB (1996). http://www.color.org/sRGB.html
14. Microsoft Corporation, Colorspace Interchange Using sRGB (2001). http://www.microsoft.com/whdc/device/display/color/sRGB.mspx
15. P. Urban, Ink limitation for spectral or color constant printing, in *11th Congress of the International Colour Association* (2009)
16. C.A. Poynton, *A Technical Introduction to Digital Video* (Wiley, New York, 1996)

Fundamental Concepts in Video

5

In this chapter, we introduce the principal notions needed to understand video. Digital video compression is explored separately, in Chaps. 10 through 12.

Here we consider the following aspects of video and how they impact multimedia applications:

- Analog video,
- Digital video,
- Video display interfaces,
- 360° video,
- 3D video and TV, and
- Video quality assessment.

Since video is created from a variety of sources, we begin with the signals themselves. Analog video is represented as a continuous (time-varying) signal, and the first part of this chapter discusses how it is created and measured. Digital video is represented as a sequence of digital images. Nowadays, it is omnipresent in many types of multimedia applications. Therefore, the second part of the chapter focuses on issues in digital video including HDTV, UHDTV, and 3D TV.

5.1 Analog Video

Up until last decade, most TV programs were sent and received as an analog signal. Once the electrical signal is received, we may assume that brightness is at least a monotonic function of voltage, if not necessarily linear, because of gamma correction (see Sect. 4.1.6).

An analog signal $f(t)$ samples a time-varying image. So-called *progressive* scanning traces through a complete picture (a frame) row-wise for each time interval. A high-resolution computer monitor typically uses a time interval of 1/72 s.

© Springer Nature Switzerland AG 2021
Z.-N. Li et al., *Fundamentals of Multimedia*, Texts in Computer Science,
https://doi.org/10.1007/978-3-030-62124-7_5

Fig. 5.1 Interlaced raster scan

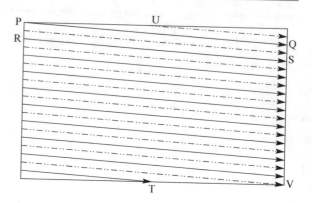

In TV and in some monitors and multimedia standards, another system, *interlaced* scanning, is used. Here, the odd-numbered lines are traced first, then the even-numbered lines. This results in "odd" and "even" *fields*—two fields make up one frame.

In fact, the odd lines (starting from 1) end up at the middle of a line at the end of the odd field, and the even scan starts at a half-way point. Figure 5.1 shows the scheme used. First, the solid (odd) lines are traced—P to Q, then R to S, and so on, ending at T—then the even field starts at U and ends at V. The scan lines are not horizontal because a small voltage is applied, moving the electron beam down over time.

Interlacing was invented because, when standards were being defined, it was difficult to transmit the amount of information in a full frame quickly enough to avoid flicker. The double number of fields presented to the eye reduces perceived flicker.

Because of interlacing, the odd and even lines are displaced in time from each other. This is generally not noticeable except when fast action is taking place onscreen, when blurring may occur. For example, in the video in Fig. 5.2, the moving helicopter is blurred more than the still background.

Since it is sometimes necessary to change the frame rate, resize, or even produce stills from an interlaced source video, various schemes are used to *de-interlace* it. The simplest de-interlacing method consists of discarding one field and duplicating the scan lines of the other field, which results in the information in one field being lost completely. Other, more complicated methods retain information from both fields.

Traditional CRT (cathode ray tube) displays are built like fluorescent lights and must flash 50–70 times per second to appear smooth. In Europe, this fact is conveniently tied to their 50 Hz electrical system, and they use video digitized at 25 frames per second (fps); in North America, the 60 Hz electric system dictates 30 fps.

The jump from Q to R and so on in Fig. 5.1 is called the *horizontal retrace*, during which the electronic beam in the CRT is blanked. The jump from T to U or V to P is called the *vertical retrace*.

(a)

(b)   (c)   (d)

Fig. 5.2 Interlaced scan produces two fields for each frame: **a** The video frame; **b** Field 1; **c** Field 2; **d** difference of fields

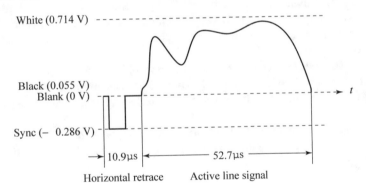

White (0.714 V)

Black (0.055 V)
Blank (0 V)

Sync (− 0.286 V)

10.9μs   52.7μs

Horizontal retrace   Active line signal

Fig. 5.3 Electronic signal for one NTSC scan line

Since voltage is one dimensional—it is simply a signal that varies with time—how do we know when a new video line begins? That is, what part of an electrical signal tells us that we have to restart at the left side of the screen?

The solution used in analog video is a small voltage offset from zero to indicate black and another value, such as zero, to indicate the start of a line. Namely, we could use a "blacker-than-black" zero signal to indicate the beginning of a line.

Figure 5.3 shows a typical electronic signal for one scan line of NTSC composite video. "White" has a peak value of 0.714 V, "Black" is slightly above zero at 0.055 V, whereas Blank is at zero volts. As shown, the time duration for blanking pulses in the signal is used for synchronization as well, with the tip of the sync signal at approximately −0.286 V. In fact, the problem of reliable synchronization is so important that special signals to control sync take up about 30% of the signal!

The vertical retrace and sync ideas are similar to the horizontal one, except that they happen only once per field. Tekalp [1] presents a good discussion of the details of analog (and digital) video. The handbook [2] considers many fundamental problems in video processing in great depth.

5.1.1 NTSC Video

The NTSC TV standard was mostly used in North America and Japan. It uses a familiar 4:3 *aspect ratio* (i.e., the ratio of picture width to height) and 525 scan lines per frame at 30 frames per second.

More exactly, for historical reasons NTSC uses 29.97 fps—or, in other words, 33.37 ms per frame. NTSC follows the interlaced scanning system, and each frame is divided into two fields, with 262.5 lines/field. Thus, the horizontal sweep frequency is $525 \times 29.97 \approx 15,734$ lines/s, so that each line is swept out in $1/15,734$ s $\approx 63.6\,\mu$s. Since the horizontal retrace takes $10.9\,\mu$s, this leaves $52.7\,\mu$s for the active line signal, during which image data is displayed (see Fig. 5.3).

Figure 5.4 shows the effect of "vertical retrace and sync" and "horizontal retrace and sync" on the NTSC video raster. Blanking information is placed into 20 lines reserved for control information at the beginning of each field. Hence, the number of *active video lines* per frame is only 485. Similarly, almost 1/6 of the raster at the left side is blanked for horizontal retrace and sync. The nonblanking pixels are called *active pixels*.

Pixels often fall between scan lines. Therefore, even with noninterlaced scan, NTSC TV is capable of showing only about 340 (visually distinct) lines—about 70% of the 485 specified active lines. With interlaced scan, it could be as low as 50%.

Image data is not encoded in the blanking regions, but other information can be placed there, such as V-chip information, stereo audio channel data, and subtitles in many languages.

Fig. 5.4 Video raster, including retrace and sync data

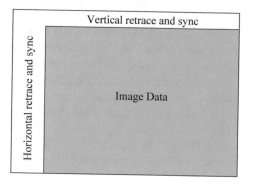

Table 5.1 Samples per line for various analog video formats

Format	Samples per line
VHS	240
S-VHS	400–425
Beta-SP	500
Standard 8 mm	300
Hi-8 mm	425

NTSC video is an analog signal with no fixed horizontal resolution. Therefore, we must decide how many times to sample the signal for display. Each sample corresponds to one pixel output. A *pixel clock* divides each horizontal line of video into samples. The higher the frequency of the pixel clock, the more samples per line.

Different video formats provide different numbers of samples per line, as listed in Table 5.1. Laser disks had about the same resolution as Hi-8. (In comparison, miniDV 1/4-inch tapes for digital video were 480 lines by 720 samples per line.)

NTSC uses the YIQ color model. We employ the technique of *quadrature modulation* to combine (the spectrally overlapped part of) I (in-phase) and Q (quadrature) signals into a single chroma signal C [1,3]:

$$C = I \cos(F_{sc}t) + Q \sin(F_{sc}t) \qquad (5.1)$$

This modulated chroma signal is also known as the *color subcarrier*, whose magnitude is $\sqrt{I^2 + Q^2}$ and phase is $\tan^{-1}(Q/I)$. The frequency of C is $F_{sc} \approx 3.58$ MHz.

The I and Q signals are multiplied in the time domain by cosine and sine functions with the frequency F_{sc} (Eq. (5.1)). This is equivalent to convolving their Fourier transforms in the frequency domain with two impulse functions at F_{sc} and $-F_{sc}$. As a result, a copy of I and Q frequency spectra is made which are centered at F_{sc} and $-F_{sc}$, respectively.[1]

The NTSC composite signal is a further composition of the luminance signal Y and the chroma signal, as defined below:

$$\text{composite} = Y + C = Y + I \cos(F_{sc}t) + Q \sin(F_{sc}t) \qquad (5.2)$$

NTSC assigned a bandwidth of 4.2 MHz to Y but only 1.6 MHz to I and 0.6 MHz to Q, due to humans' insensitivity to color details (high-frequency color changes). As Fig. 5.5 shows, the picture carrier is at 1.25 MHz in the NTSC video channel, which has a total bandwidth of 6 MHz. The chroma signal is being "carried" by $F_{sc} \approx 3.58$ MHz toward the higher end of the channel and is thus centered at 1.25 +

---

[1] Negative frequency $(-F_{sc})$ is a mathematical notion needed in the Fourier transform. In the physical spectrum, only positive frequency is used.

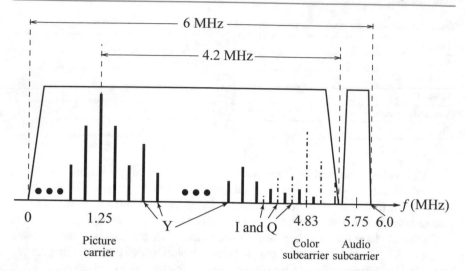

Fig. 5.5 Interleaving Y and C signals in the NTSC spectrum

$3.58 = 4.83$ MHz. This greatly reduces the potential interference between the Y (luminance) and C (chrominance) signals, since the magnitudes of higher frequency components of Y are significantly smaller than their lower frequency counterparts.

Moreover, as Blinn[3] explains, great care is taken to interleave the discrete Y and C spectra so as to further reduce the interference between them. The "interleaving" is illustrated in Fig. 5.5, where the frequency components for Y (from the discrete Fourier transform) are shown as solid lines, and those for I and Q are shown as dashed lines. As a result, the 4.2 MHz band of Y is overlapped and interleaved with the 1.6 MHz to I and 0.6 MHz to Q.

The first step in decoding the composite signal at the receiver side is to separate Y and C. Generally, low-pass filters can be used to extract Y, which is located at the lower end of the channel. TV sets with higher quality also use comb filters [3] to exploit the fact that Y and C are interleaved.

After separation from Y, the chroma signal C can be demodulated to extract I and Q separately.

To extract I:

1. Multiply the signal C by $2\cos(F_{sc}t)$

$$
\begin{aligned}
C \cdot 2\cos(F_{sc}t) &= I \cdot 2\cos^2(F_{sc}t) + Q \cdot 2\sin(F_{sc}t)\cos(F_{sc}t) \\
&= I \cdot (1 + \cos(2F_{sc}t)) + Q \cdot 2\sin(F_{sc}t)\cos(F_{sc}t) \\
&= I + I \cdot \cos(2F_{sc}t) + Q \cdot \sin(2F_{sc}t)
\end{aligned}
$$

2. Apply a low-pass filter to obtain I and discard the two higher frequency ($2F_{sc}$) terms.

Similarly, extract Q by first multiplying C by $2\sin(F_{sc}t)$ and then applying low-pass filtering.

The NTSC bandwidth of 6 MHz is tight. Its audio subcarrier frequency is 4.5 MHz, which places the center of the audio band at $1.25 + 4.5 = 5.75$ MHz in the channel (Fig. 5.5). This would actually be a bit too close to the color subcarrier—a cause for potential interference between the audio and color signals. It was due largely to this reason that NTSC color TV slowed its frame rate to $30 \times 1,000/1,001 \approx 29.97$ fps [4]. As a result, the adopted NTSC color subcarrier frequency is slightly lowered, to

$$f_{sc} = 30 \times 1,000/1,001 \times 525 \times 227.5 \approx 3.579545 \text{ MHz}$$

where 227.5 is the number of color samples per scan line in NTSC broadcast TV.

5.1.2 PAL Video

PAL (phase alternating line) is a TV standard originally invented by German scientists. It uses 625 scan lines per frame, at 25 frames per second (or 40 ms/frame), with a 4:3 aspect ratio and interlaced fields. Its broadcast TV signals are also used in composite video. This important standard is widely used in Western Europe, China, India, and many other parts of the world. Because it has higher resolution than NTSC (625 versus 525 scan lines), the visual quality of its pictures is generally better.

PAL uses the YUV color model with an 8 MHz channel, allocating a bandwidth of 5.5 MHz to Y and 1.8 MHz each to U and V. The color subcarrier frequency is $f_{sc} \approx 4.43$ MHz. To improve picture quality, chroma signals have alternate signs (e.g., $+$U and $-$U) in successive scan lines; hence, the name "Phase Alternating Line."[2] This facilitates the use of a (line-rate) comb filter at the receiver—the signals in consecutive lines are averaged so as to cancel the chroma signals (which always carry opposite signs) for separating Y and C and obtain high-quality Y signals.

5.1.3 SECAM Video

SECAM, which was invented by the French, is the third major broadcast TV standard. SECAM stands for *Systeme Electronique Couleur Avec Memoire*. SECAM also uses 625 scan lines per frame, at 25 frames per second, with a 4:3 aspect ratio and interlaced fields. The original design called for a higher number of scan lines (over 800), but the final version settled for 625.

SECAM and PAL are similar, differing slightly in their color coding scheme. In SECAM, U and V signals are modulated using separate color subcarriers at 4.25 MHz and 4.41 MHz, respectively. They are sent in alternate lines, that is, only one of the U or V signals will be sent on each scan line.

---

[2]According to Blinn [3], NTSC selects a half integer (227.5) number of color samples for each scan line. Hence, its chroma signal also switches sign in successive scan lines.

Table 5.2 Comparison of analog broadcast TV systems

Series TV system	Frame rate (fps)	Series number of scan lines	Total channel series width (MHz)	Bandwidth series allocation (MHz)		
				Y	Series I or U	Series Q or V
NTSC	29.97	525	6.0	4.2	1.6	0.6
PAL	25	625	8.0	5.5	1.8	1.8
SECAM	25	625	8.0	6.0	2.0	2.0

Table 5.2 gives a comparison of the three major analog broadcast TV systems. It is worth noting that most countries have ceased their operation (e.g., NTSC broadcast was stopped in Canada and the United States in 2011 and 2015, respectively) and have switched to digital TV broadcast.

5.2 Digital Video

The advantages of digital representation for video are many. It permits

- Storing video on digital devices or in memory, ready to be processed (noise removal, cut and paste, and so on) and integrated into various multimedia applications.
- Direct access, which makes nonlinear video editing simple.
- Repeated recording without degradation of image quality.
- Ease of encryption and better tolerance to channel noise.

In earlier Sony or Panasonic recorders, digital video was in the form of composite video. Modern digital video generally uses component video, although RGB signals are first converted into a certain type of color opponent space. The usual color space is YCbCr [5].

5.2.1 Chroma Subsampling

Since humans see color with much less spatial resolution than black and white, it makes sense to decimate the chrominance signal. Interesting but not necessarily informative names have arisen to label the different schemes used. To begin with, numbers are given stating how many pixel values, per four original pixels, are actually sent. Thus, the chroma subsampling scheme "4:4:4" indicates that no chroma subsampling is used. Each pixel's Y, Cb, and Cr values are transmitted, four for each of Y, Cb, and Cr.

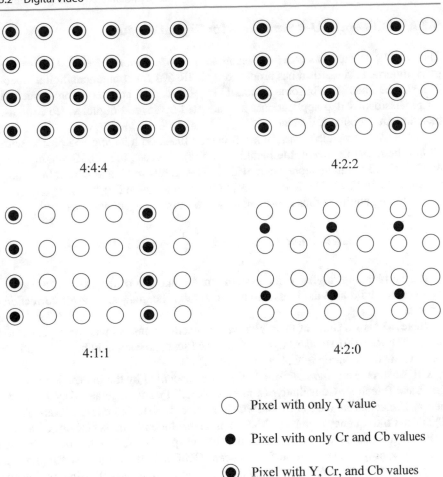

4:4:4   4:2:2

4:1:1   4:2:0

○ Pixel with only Y value

● Pixel with only Cr and Cb values

◉ Pixel with Y, Cr, and Cb values

Fig. 5.6 Chroma subsampling

The scheme "4:2:2" indicates horizontal subsampling of the Cb and Cr signals by a factor of 2. That is, of four pixels horizontally labeled 0–3, all four Ys are sent, and every two Cbs and two Crs are sent, as $(Cb0, Y0)(Cr0, Y1)(Cb2, Y2)(Cr2, Y3)(Cb4, Y4)$, and so on.

The scheme "4:1:1" subsamples horizontally by a factor of 4. The scheme "4:2:0" subsamples in both the horizontal and vertical dimensions by a factor of 2. Theoretically, an average chroma pixel is positioned between the rows and columns, as shown in Fig. 5.6. We can see that the scheme 4:2:0 is, in fact, another kind of 4:1:1 sampling, in the sense that we send 4, 1, and 1 values per 4 pixels. Therefore, the labeling scheme is not a very reliable mnemonic!

Scheme 4:2:0, along with others, is commonly used in JPEG and MPEG (see later chapters in Part II).

5.2.2 CCIR and ITU-R Standards for Digital Video

The CCIR is the *Consultative Committee for International Radio*. One of the most important standards it has produced is CCIR-601 for component digital video. This standard has since become standard ITU-R Rec. 601, an international standard for professional video applications. It is adopted by several digital video formats, including the popular DV video.

The NTSC-compatible version has 525 scan lines, each having 858 pixels (with 720 of them visible, not in the blanking period). Because the NTSC version uses 4:2:2, each pixel can be represented with 2 bytes (8 bits for Y and 8 bits alternating between Cb and Cr). The Rec. 601 (NTSC) data rate (including blanking and sync but excluding audio) is thus approximately 216 Mbps (megabits per second):

$$525 \times 858 \times 30 \times 2 \text{ bytes} \times 8 \frac{\text{bits}}{\text{byte}} \approx 216 \text{ Mbps}$$

During blanking, digital video systems may make use of the extra data capacity to carry audio signals, translations into foreign languages, or error-correction information.

Table 5.3 shows some of the digital video specifications, all with an aspect ratio of 4:3. The Rec. 601 standard uses an interlaced scan, so each field has only half as much vertical resolution (e.g., 240 lines in NTSC).

CIF stands for *common intermediate format*, specified by the International Telegraph and Telephone Consultative Committee (CCITT), now superseded by the International Telecommunication Union, which oversees both telecommunications (ITU-T) and radio frequency matters (ITU-R) under one United Nations body. The idea of CIF, which is about the same as VHS quality, is to specify a format for lower bitrate. CIF uses a progressive (noninterlaced) scan. QCIF stands for Quarter-CIF, and is for even lower bitrate. All the CIF/QCIF resolutions are evenly divisible by 8, and all except 88 are divisible by 16; this is convenient for block-based video coding in H.261 and H.263, discussed in Chap. 10.

Table 5.3 ITU-R digital video specifications

	Rec. 601 525/60 NTSC	Rec. 601 625/50 PAL/SECAM	CIF	QCIF
Luminance resolution	720 × 480	720 × 576	352 × 288	176 × 144
Chrominance resolution	360 × 480	360 × 576	176 × 144	88 × 72
Color subsampling	4:2:2	4:2:2	4:2:0	4:2:0
Aspect ratio	4:3	4:3	4:3	4:3
Fields/s	60	50	30	30
Interlaced	Yes	Yes	No	No

CIF is a compromise between NTSC and PAL, in that it adopts the NTSC frame rate and half the number of active lines in PAL. When played on existing TV sets, NTSC TV will first need to convert the number of lines, whereas PAL TV will require frame-rate conversion.

5.2.3   High Definition TV (HDTV)

The introduction of wide-screen movies brought the discovery that viewers seated near the screen enjoyed a level of participation (sensation of immersion) not experienced with conventional movies. Apparently, the exposure to a greater field of view, especially the involvement of peripheral vision, contributes to the sense of "being there." The main thrust of high-definition TV (HDTV) is not to increase the "definition" in each unit area, but rather to increase the visual field, especially its width.

The first-generation HDTV was based on an analog technology developed by Sony and NHK in Japan in the late 1970s. HDTV successfully broadcasts the 1984 Los Angeles Olympic Games in Japan. MUltiple sub-Nyquist sampling encoding (MUSE) was an improved NHK HDTV with hybrid analog/digital technologies that was put in use in the 1990s. It has 1,125 scan lines, interlaced (60 fields per second), and a 16:9 aspect ratio. It uses satellite to broadcast—quite appropriate for Japan, which can be covered with one or two satellites. The direct broadcast satellite (DBS) channels used have a bandwidth of 24 MHz.

In general, terrestrial broadcast, satellite broadcast, cable, and broadband networks are all feasible means for transmitting HDTV as well as conventional TV. Since uncompressed HDTV will easily demand more than 20 MHz bandwidth, which will not fit in the current 6 or 8 MHz channels, various compression techniques are being investigated. It is also anticipated that high-quality HDTV signals will be transmitted using more than one channel, even after compression.

In 1987, the FCC decided that HDTV standards must be compatible with the existing NTSC standard and must be confined to the existing very high-frequency (VHF) and ultra-high-frequency (UHF) bands. This prompted a number of proposals in North America by the end of 1988, all of them analog or mixed analog/digital.

In 1990, the FCC announced a different initiative—its preference for full-resolution HDTV. They decided that HDTV would be simultaneously broadcasted with existing NTSC TV and eventually replace it. The development of digital HDTV immediately took off in North America.

Witnessing a boom of proposals for digital HDTV, the FCC made a key decision to go all digital in 1993. A "grand alliance" was formed that included four main proposals, by General Instruments, MIT, Zenith, and AT&T, and by Thomson, Philips, Sarnoff, and others. This eventually led to the formation of the Advanced Television Systems Committee (ATSC), which was responsible for the standard for TV broadcasting of HDTV. In 1995, the U.S. FCC Advisory Committee on Advanced Television Service recommended that the ATSC digital television standard be adopted.

Table 5.4 Advanced digital TV formats supported by ATSC

Number of active pixels per line	Number of active lines	Aspect ratio	Frame rate
1,920	1,080	16:9	60P 60I 30P 24P
1,280	720	16:9	60P 30P 24P
720	480	16:9 or 4:3	60P 60I 30P 24P
640	480	4:3	60P 60I 30P 24P

Table 5.4 lists some of the standard supported video scanning formats. (For the 50 Hz systems, 60P becomes 50P, 30P becomes 25P, etc.) In the table, "I" means interlaced scan and "P" means progressive (noninterlaced) scan. The frame rates supported are both integer rates and the NTSC rates, that is, 60.00 or 59.94, 30.00 or 29.97, 24.00 or 23.98 fps.

For video, MPEG-2 was initially chosen as the compression standard. As will be seen in Chap. 11, it uses main level to high level of the main profile of MPEG-2. For audio, AC-3 is the standard. It supports the so-called 5.1 channel Dolby surround sound—five surround channels plus a subwoofer channel. In 2008, ATSC was updated to adopt the H.264 video compression standard.

The salient difference between conventional TV and HDTV [4,6] is that the latter has a much wider aspect ratio of 16:9 instead of 4:3. (Actually, it works out to be exactly one-third wider than current TV). Another feature of HDTV is its move toward progressive (noninterlaced) scan. The rationale is that interlacing introduces serrated edges to moving objects and flickers along horizontal edges.

Consumers with analog TV sets are still able to receive signals via an 8-VSB (8-level vestigial sideband) demodulation box. The services provided include the following:

- **Standard-definition TV (SDTV)**—the NTSC TV or higher.
- **Enhanced-definition TV (EDTV)**—480 active lines or higher—the third and fourth rows in Table 5.4.
- **High-definition TV (HDTV)**—720 active lines or higher. Popular choices include 720P (1,280 × 720, progressive scan, 30 fps), 1080I (1,920 × 1,080, interlaced, 30 fps), and 1080P (1,920 × 1,080, progressive scan, 30 or 60 fps), where the last one has become the mainstream nowadays.

5.2.4 Ultra-High-Definition TV (UHDTV)

Ultra-High-Definition TV (UHDTV), also known as Ultra HD or UHD, is a new generation of HDTV. Effort toward UHDTV standards was initialized in 2012 and the ATSC called for proposals to support the 4K UHDTV (2160P) at 60 fps in 2013. As of today, UHDTV mainly includes 4K UHDTV: 2160P (3, 840 × 2, 160, progressive scan) and 8K UHDTV: 4320P (7, 680 × 4, 320, progressive scan). The aspect ratio

Table 5.5 A summary of ultra-high-definition TV (UHDTV)

Type of UHDTV	Resolution	Bit depth (bits)	Aspect ratio	Frame rate
4K UHD (2160P)	3,840 × 2,160	10 or 12	16:9	up to 120P
8K UHD (4320P)	7,680 × 4,320	10 or 12	16:9	up to 120P
16K UHD (8640P)	15,360 × 8,640	10 or 12	16:9	up to 240P

is 16:9. The bit depth is 10 or 12 bits per sample, and the chroma subsampling can be 4:2:0, 4:2:2, or 4:4:4. The supported maximal frame rate has been increased to 120 fps. The UHDTV provides a superior picture quality, close to IMAX movies. Compared to HDTV, it requires a much higher bandwidth/bitrate.

16K UHDTV has also been demonstrated in 2018, targeting applications such as virtual reality with true immersion. Its resolution is 15,360 × 8,640 for a total of 132.7 megapixels.

Table 5.5 summarizes the main specifications of 4K, 8K, and 16K UHDTV.

5.3   Video Display Interfaces

We now discuss the interfaces for video signal transmission from some output devices (e.g., set-top box, video player, video card, etc.) to a video display (e.g., TV, monitor, projector, etc.). There have been a wide range of video display interfaces, supporting video signals of different formats (analog or digital, interlaced or progressive), different frame rates, and different resolutions [7]. We start our discussion with analog interfaces, including Component Video, Composite Video, and S-Video, and then digital interfaces, including DVI, HDMI, and DisplayPort.

5.3.1   Analog Display Interfaces

Analog video signals are often transmitted in one of three different interfaces: *component video*, *composite video*, and *S-video*. Figure 5.7 shows the typical connectors for them.

Fig. 5.7 Connectors for typical analog display interfaces. From left to right: component video, composite video, S-video, and VGA

Component Video

Higher end video systems, such as for studios, make use of three separate video signals for the red, green, and blue image planes. This is referred to as *component video*. This kind of system has three wires (and connectors) connecting the camera or other devices to a TV or monitor.

Color signals are not restricted to always being RGB separations. Instead, as we saw in Chap. 4 on color models for images and video, we can form three signals via a luminance–chrominance transformation of the RGB signals—for example, YIQ or YUV.

For any color separation scheme, component video gives the best color reproduction, since there is no "crosstalk" between the three different channels, unlike composite video or S-video. Component video, however, requires more bandwidth and good synchronization of the three components.

Composite Video

In *composite video*, color ("chrominance") and intensity ("luminance") signals are mixed into a *single* carrier wave. Chrominance is a composite of two color components (I and Q or U and V). This is the type of signal used by broadcast color TV; it is downward compatible with black-and-white TV.

In NTSC TV, for example [3], I and Q are combined into a chroma signal, and a color subcarrier then puts the chroma signal at the higher frequency end of the channel shared with the luminance signal. The chrominance and luminance components can be separated at the receiver end, and the two color components can be further recovered.

When connecting to TVs or VCRs, composite video uses only one wire (and hence one connector, such as a BNC connector at each end of a coaxial cable or an RCA plug at each end of an ordinary wire), and video color signals are mixed, not sent separately. The audio signal is another addition to this one signal. Since color information is mixed and both color and intensity are wrapped into the same signal, some interference between the luminance and chrominance signals is inevitable.

S-Video

As a compromise, *S-video* (separated video or super video, e.g., in S-VHS) uses two wires: one for luminance and another for a composite chrominance signal. As a result, there is less crosstalk between the color information and the crucial grayscale information.

The reason for placing luminance into its own part of the signal is that black-and-white information is most important for visual perception. As noted in the previous chapter, humans are able to differentiate spatial resolution in the grayscale ("black-and-white") part much better than for the color part of RGB images. Therefore, color information transmitted can be much less accurate than intensity information. We can see only fairly large blobs of color, so it makes sense to send less color detail.

Video Graphics Array (VGA)

The *video graphics array* (VGA) is a video display interface that was first introduced by IBM in 1987, along with its PS/2 personal computers. It has since been widely used in the computer industry with many variations, which are collectively referred to as VGA.

The initial VGA resolution was 640×480 using the 15-pin D-subminiature VGA connector. Later extensions can carry resolutions ranging from 640×400 pixels at 70 Hz (24 MHz of signal bandwidth) to $1,280 \times 1,024$ pixels (SXGA) at 85 Hz (160 MHz) and up to $2,048 \times 1,536$ (QXGA) at 85 Hz (388 MHz).

The VGA video signals are based on analog component RGBHV (red, green, blue, horizontal sync, vertical sync). It also carries the *display data channel* (DDC) data defined by *Video Electronics Standards Association* (VESA). Since the video signals are analog, it will suffer from interference, particularly when the cable is long.

5.3.2 Digital Display Interfaces

Given the rise of digital video processing and the monitors that directly accept digital video signals, there is a great demand toward video display interfaces that transmit digital video signals. Such interfaces emerged in 1980s (e.g., color graphics adapter (CGA) with the D-subminiature connector), and evolved rapidly. Today, the most widely used digital video interfaces include digital visual interface (DVI), high-definition multimedia interface (HDMI), and DisplayPort, as shown in Fig. 5.8.

Digital Visual Interface (DVI)

Digital visual interface (DVI) was developed by the *Digital Display Working Group* (DDWG) for transferring digital video signals, particularly from a computer's video

Fig. 5.8 Connectors of different digital display interfaces. from left to right: DVI, HDMI, Display-Port

card to a monitor. It carries uncompressed digital video and can be configured to support multiple modes, including DVI-D (digital only), DVI-A (analog only), or DVI-I (digital and analog). The support for analog connections makes DVI backward compatible with VGA (though an adapter is needed between the two interfaces).

DVI's digital video transmission format is based on *PanelLink*, a high-speed serial link technology using *transition minimized differential signaling* (TMDS). Through DVI, a source, e.g., a video card, can read the display's *extended display identification data* (EDID), which contains the display's identification, color characteristics (such as gamma level), and table of supported video modes. When a source and a display are connected, the source first queries the display's capabilities by reading the monitor's EDID block. A preferred mode or native resolution can then be chosen.

In a single-link mode, the maximum pixel clock frequency of DVI is 165 MHz, which supports a maximum resolution of 2.75 megapixels at the 60 Hz refresh rate. This allows a maximum 16:9 screen resolution of 1,920 × 1,080 at 60 Hz. The DVI specification also supports dual link, which achieves even higher resolutions up to 2,560 × 1,600 at 60 Hz.

High-Definition Multimedia Interface (HDMI)

HDMI is a newer digital audio/video interface developed to be backward compatible with DVI. It was promoted by the consumer electronics industry, and has been widely used in the consumer market since 2002. The HDMI specification defines the protocols, signals, electrical interfaces, and mechanical requirements. Its electrical specifications, in terms of TMDS and VESA/DDC links, are identical to those of DVI. As such, for the basic video, an adapter can convert their video signals losslessly. HDMI, however, differs from DVI in the following aspects:

1. HDMI doesn't carry analog signal and hence is not compatible with VGA.
2. DVI is limited to the RGB color range (0–255). HDMI supports both RGB and YCbCr 4:4:4 or 4:2:2. The latter are more common in application fields other than computer graphics.
3. HDMI supports digital audio, in addition to digital video.

The maximum pixel clock rate for HDMI 1.0 is 165 MHz, which is sufficient to support 1080P and WUXGA (1,920 × 1,200) at 60 Hz. HDMI 1.3 increases that to 340 MHz, which allows for higher resolution (such as WQXGA, 2,560 × 1,600) over a single digital link. HDMI 2.1 was released in 2017, which supports for higher resolutions and higher refresh rates. HDMI 2.1 also introduces a new HDMI cable category called ultra-high speed (up to 48 Gbit/s bandwidth), which is enough for 8K resolution at approximately 50 Hz. Using display stream compression with a compression ratio of up to 3:1, formats up to 8K (7,680 × 4,320) at 120 Hz or 10K (10,240 × 4,320) at 100 Hz are possible.

DisplayPort

DisplayPort is a digital display interface developed by VESA, starting from 2006. It is the first display interface that uses packetized data transmission, like the Internet or Ethernet (see Chap. 15). Specifically, it is based on small data packets known as *micro-packets*, which can embed the clock signal within the data stream. As such, DisplayPort can achieve a higher resolution yet with fewer pins than the previous technologies. The use of data packets also allows DisplayPort to be extensible, i.e., new features can be added over time without significant changes to the physical interface itself.

DisplayPort can be used to transmit audio and video simultaneously, or either of them. The video signal path can have 6–16 bits per color channel, and the audio path can have up to 8 channels of 24-bit 192 kHz uncompressed PCM audio or carry compressed audio. A dedicated bidirectional channel carries device management and control data.

VESA designed DisplayPort to replace VGA and DVI. To this end, it has a much higher video bandwidth, enough for four simultaneous 1080P 60 Hz displays or 4K video at 60 Hz. Backward compatibility to VGA and DVI is achieved by using active adapters. Compared with HDMI, DisplayPort has slightly more bandwidth, which also accommodates multiple streams of audio and video to separate devices. Furthermore, the VESA specification is royalty-free, while HDMI charges an annual fee to manufacturers. These points make DisplayPort a strong competitor to HDMI in the consumer electronics market, as well.

DisplayPort cables and ports may have either a full-size connector or a mini connector. The latter was first developed and used by Apple in its computer products. Despite the different shapes, they are identical in performance.

In 2019, VESA formally released the DisplayPort 2.0 standard, which expands the data rate to 77.37 Gbit/s. It also incorporates a number of advanced features, such as beyond-8K resolutions, higher refresh rates and high dynamic range (HDR) support at higher resolutions, improved support for multiple display configurations, as well as improved user experience with augmented/virtual reality (AR/VR) displays.

5.4 360° Video

360° *video* is also known as *Omnidirectional Video*, *Spherical Video*, or *Immersive Video*. Due to the rapid development of the VR (virtual reality) technology, it is becoming very popular [8,9]. An intuitive way to envision a 360° video is to take a roller coaster ride when one is looking left and right, front and behind, up and down (and often upside-down), while enduring various irregular movements. As the name implies, 360° video can span 360° horizontally and 180° vertically. It is captured by cameras at (almost) all possible viewing angles. In actual implementations, the video is usually captured by an omnidirectional camera or a collection of cameras with a wide field of view. Although theoretically possible, most 360° videos have a limited range of views vertically, i.e., they may not include the top and bottom of the sphere.

When multiple cameras are employed, they usually have overlapping views so that they can be merged to obtain the spherical video. *Video stitching* is performed for generating the merged footage, which is similar to stitching for panoramic images. Since the color and contrast of the shots often vary significantly, their calibration/adjustment is a challenging task.

360° videos can be monoscopic or stereoscopic. The latter is especially suitable for VR applications, often viewed on smart devices or with a VR headset.

5.4.1 Equirectangular Projection (ERP)

In order to view the 360° (spherical) video on flat 2D screens, some special projections are required. One of the most common projections is the *equirectangular projection (ERP)*.

If we adopt the spherical coordinate system, then a point on the sphere will be represented by (R, θ, ϕ), where R is the radius of the sphere, θ indicates the longitude, and ϕ the latitude. The angle θ has a range of $[-180°, 180°]$, commonly known as longitude 180°W to 180°E, and ϕ has a range of $[-90°, 90°]$, i.e., latitude 90°S to 90°N. At the prime meridian (aka the Greenwich meridian), $\theta = 0°$, and at the equator, $\phi = 0°$.

In this discussion of the ERP, we will always use the equator as the central parallel, where $y = 0$. Let θ_0 be the longitude of the central meridian, where $x = 0$, then the equations for the ERP are

$$x = R \cdot (\theta - \theta_0) \tag{5.3}$$
$$y = R \cdot \phi \tag{5.4}$$

where (x, y) are the projection coordinates after the ERP, and θ and ϕ are in radians.

If $\theta_0 = 0°$, i.e., we use the prime meridian as the central meridian, where $x = 0$, then the equations for the ERP are simply

$$x = R \cdot \theta \tag{5.5}$$
$$y = R \cdot \phi \tag{5.6}$$

Figure 5.9 illustrates the world map after the ERP, where $\theta_0 = 0°$. As shown in the figure by the distorted "circles," the distortion caused by the ERP is severe, especially when it is near the poles (not to mention at the poles). In many applications, a rectilinear correction is required.

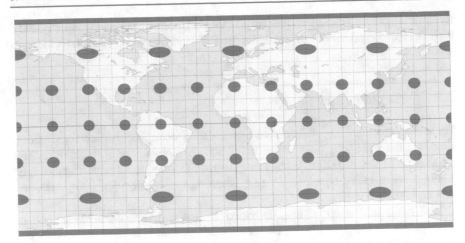

Fig. 5.9 The equirectangular projection (ERP) of the world map. (Image created by Justin Kunimune, github.com/jkunimune15/Map-Projections. © CC BY-SA 4.0)

5.4.2   Other Projections

The following are three of the other popular projection models:

- Cubemap projection (CMP)—Instead of projecting onto a single plane as in ERP, projections are made onto the six faces of a cube, i.e., left, front, right, behind, top, and bottom. As a result, the distortion caused by the projection is much reduced. CMP can be realized by a *radial projection*, in which a sphere is embedded in a cube. A ray connecting the center of the sphere and the point on the sphere is extended to get the projection image on one of the faces of the cube.
- Equi-angular cubemap (EAC)—This is a variation of CMP adopted by Google for streaming VR videos. In the conventional CMP, although the projection rays are equally spaced over the circle, i.e., with an identical $\Delta\theta$, the projection intervals on the cube surface are not equal. If the two rays emanating around the center of the view form an interval of length d_0, then $d_0 < d_i$, when $i \neq 0$. The difference $|d_i - d_0|$ increases when $|i|$ increases, this causes undesirable nonuniform sampling. To avoid this distortion, in EAC, these intervals are remapped to new intervals of equal lengths.
- Pyramid format—This format was advocated by Facebook. In the so-called *pyramid projection*, the base of the pyramid provides the full resolution image of the front view. The sides of the pyramid provide the side view images with gradually reduced resolution. The main advantage of this multiresolution approach is to reduce the bitrate required for video transmission, i.e., to use the higher resolutions only when needed.

Ye et al. [9] presented a good summary of various projection formats for 360° videos included in the *360Lib platform*, which was established by the Joint Video

Exploration Team (JVET) of the ITU-T Video Coding Experts Group (VCEG). 360Lib [10] also provides tools for converting 360° videos among various projection models.

5.5   3D Video and TV

Three-dimensional (3D) pictures and movies have been in existence for decades. However, the rapid progress in the research and development of 3D technology and the success of the 2009 film Avatar have pushed 3D video to its peak. Increasingly, it is in movie theaters, broadcast TV (e.g., sporting events), personal computers, and various handheld devices.

The main advantage of the 3D video is that it enables the experience of immersion—be there, and really Be there! Besides Avatar, the movies such as Up, Hugo, Toy Story 3, and recently, Toy Story 4 are some of the successful examples.

We will start with an introduction to the fundamentals of 3D vision or 3D percept, emphasizing stereo vision (or stereopsis) since most modern 3D video and 3D TV are based on the stereoscopic vision.

5.5.1   Cues for 3D Percept

The human vision system is capable of achieving a 3D percept by utilizing multiple cues. They are combined to produce optimal (or nearly optimal) depth estimates. When the multiple cues agree, this enhances the 3D percept. When they conflict with each other, the 3D percept can be hindered. Sometimes, illusions can arise.

Monocular Cues

Monocular cues that do not necessarily involve both eyes include the following:

- Shading—depth perception by shading and highlights.
- Perspective scaling—converging parallel lines with distance and at infinity.
- Relative size—distant objects appear smaller compared to known same-sized objects not in distance.
- Texture gradient—the appearance of textures changing when they recede in distance.
- Blur gradient—objects appear sharper at the distance where the eyes are focused, whereas nearer and farther objects are gradually blurred.
- Haze—due to light scattering by the atmosphere, objects at distance have lower contrast and lower color saturation.
- Occlusion—a far object occluded by nearer object(s).
- Motion parallax—induced by object movement and head movement, such that nearer objects appear to move faster.

Among the above monocular cues, it has been said that occlusion and motion parallax are most effective.

Binocular Cues

The human vision system utilizes effective binocular vision, i.e., *stereo vision*, aka *stereopsis*. Our left and right eyes are separated by a small distance, on average approximately two and half inches, or 65 mm. This is known as the *interocular distance*. As a result, the left and right eyes have slightly different views, i.e., images of objects are shifted horizontally. The amount of the shift, or *disparity*, is dependent on the object's distance from the eyes, i.e., its *depth*, thus providing the binocular cue for the 3D percept. The horizontal shift is also known as *horizontal parallax*. The fusion of the left and right images into single vision occurs in the brain, producing the 3D percept.

Current 3D video and TV systems are almost all based on stereopsis, which is believed to be the most effective cue.

5.5.2 3D Camera Models

Simple Stereo Camera Model

We can design a simple (artificial) stereo camera system in which the left and right cameras are identical (same lens, same focal length, etc.); the cameras' optical axes are parallel, pointing at the Z-direction, the scene depth. The cameras are placed at $(-b/2, 0, 0)$ and $(b/2, 0, 0)$ in the world coordinate system (as opposed to a local coordinate system based on the camera axes), where b is camera separation, or the length of the *baseline*. Given a point $P(X, Y, Z)$ in the 3D space, and x_l and x_r being the x-coordinates of its projections on the left and right camera image planes, the following can be derived:

$$d = fb/Z \qquad (5.7)$$

where f is the focal length, $d = x_l - x_r$ is the *disparity* or *horizontal parallax*.

This suggests that disparity d is inversely proportional to the depth Z of the point P. Namely, objects near the cameras yield large disparity values, and far objects yield small disparity values. When the point is very far, approaching infinity, $d \to 0$.

Almost all amateur and professional stereo video cameras use the above simple stereo camera model where the camera axes are in parallel. The obvious reason is that it is simple and easy to manufacture. Moreover, objects at the same depth in the scene will have the same disparity d according to Eq. (5.7). This enables us to depict the 3D space with a stack of *depth planes*, or equivalently, *disparity planes*, which is handy in camera calibration, video processing, and analysis.

Toed-in Stereo Camera Model

Human eyes are known to behave differently from the simple camera model above. When humans focus on an object at a certain distance, our eyes rotate around a vertical axis in opposite directions in order to obtain (or maintain) single binocular vision. As a result, disparity $d = 0$ at the object of focus, and at the locations that have the same distance from the observer as the object of focus; $d > 0$ for objects farther than the object of focus (the so-called *positive parallax*); and $d < 0$ for nearer objects (*negative parallax*).

Human eyes can be emulated by so-called toed-in stereo cameras, in which the camera axes are usually converging and not in parallel.

One of the complications of this model is that objects at the same depth (i.e., the same Z) in the scene no longer yield the same disparity. In other words, the "disparity planes" are now curved. Objects on both sides of the view appear farther away than the objects in the middle, even when they have the same depth Z.

5.5.3   3D Movie and TV Based on Stereo Vision

3D Movie Using Colored Glasses

In the early days, most movie theaters offering a 3D experience provided glasses tinted with complementary colors, usually red on the left and cyan on the right. This technique is called *Anaglyph 3D*. Basically, in preparing the stereo pictures, the left image is filtered to remove blue and green, and the right image is filtered to remove red. They are projected onto the same screen with good alignment and proper disparities. After the stereo pictures pass through the colored glasses, they are mentally combined (fused) and the color 3D picture is reproduced in the viewer's brain.

Anaglyph 3D movies are easy to produce. However, due to the color filtering, the color quality is not necessarily the best. Anaglyph 3D is still widely used in scientific visualization and various computer applications.

3D Movies Using Circularly Polarized Glasses

Nowadays, the dominant technology in 3D movie theaters is the RealD Cinema System. Movie-goers are required to wear polarized glasses in order to see the movie in 3D. Basically, the lights from the left and right pictures are polarized in different directions. They are projected and superimposed on the same screen. The left and right polarized glasses that the audience wears are polarized accordingly, which allows one of the two polarized pictures to pass through while blocking the other. To cut costs, a single projector is used in most movie theaters. It has a Z screen polarization switch to alternatively polarize the lights from the left and right pictures before projecting onto the screen. The frame rate is said to be 144 fps.

Circularly (as opposed to linearly) polarized glasses are used so that users can tilt their heads and look around a bit more freely without losing the 3D percept.

3D TV with Shutter Glasses

Most TVs for home entertainment, however, use *shutter glasses*. Basically, the liquid crystal layer on the glasses that the user wears becomes opaque (behaving like a shutter) when some voltage is applied. It is otherwise transparent. The glasses are actively (e.g., via infrared) synchronized with the TV set that alternately shows left and right images (e.g., 120 Hz for the left and 120 Hz for the right) in a time sequential manner.

3D vision with shutter glasses can readily be realized on desktop computers or laptops with a modest addition of specially designed hardware and software. The NVIDIA GeForce 3D Vision Kit is such an example.

5.5.4   The Vergence–Accommodation Conflict

Current stereoscopic technology for 3D video has many drawbacks. It is reported that a large number of viewers have difficulties watching 3D movies and/or TVs. 3D objects can appear darker, smaller, and flattened compared to their appearance in the real world. Moreover, they cause eye fatigue and strain. They can make viewers dizzy, causing headache and even nausea.

Beside many obvious technical challenges in making the left and right images undistorted, synchronized, and separated, there is a more fundamental issue, i.e., the *Vergence–Accommodation Conflict* [11, 12].

The word "accommodation" here refers to the physical act of the eye required to maintain a clear (focused) image on an object when its distance changes. As depicted in Fig. 5.10a, human eyes harmonize accommodation and vergence. When we keep our focus on an object of interest, our eyes also converge at the same spot. As a result, Focal distance = Vergence distance. The system is, of course, dynamic: we change our focus of attention when needed, and adjust our vergence and accommodation accordingly.

In a 3D movie theater, or when we gaze at a 3D display device, the situation is different. We are naturally focusing on the screen at a fixed distance. When our brain processes and fuses the left and right images, we are supposed to decouple our vergence from accommodation. This is the *vergence–accommodation conflict*. When the object is supposed to be behind the screen (with positive parallax) as indicated in Fig. 5.10b, Focal distance < Vergence distance, and vice versa.

Most of us seem capable of doing so, except it demands a heavy cognitive load. This explains why we quickly feel visual fatigue and so on. To cite Walter Murch, a distinguished film editor and sound designer, in one of his communications with Roger Ebert, the legendary film critic: "The biggest problem with 3-D is the 'convergence/focus' issue. …3-D films require us to focus at one distance and converge at

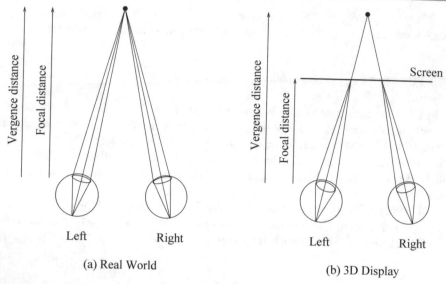

Fig. 5.10 The vergence–accommodation conflict

another, and 600 million years of evolution has never presented this problem before. All living things with eyes have always focused and converged at the same point."

The movie industry has invented many techniques to alleviate this conflict [13]. For example, a common practice is to avoid depth discontinuity between cuts. Within clips, efforts are made to keep the main object of interest at roughly the screen depth, and to keep its average depth at that level when there must be movements causing depth changes.

5.5.5 Autostereoscopic (Glasses-Free) Display Devices

Wearing glasses while watching 3D video/TV/movies itself is another major drawback. It is uncomfortable, especially for those who already wear prescription eye glasses. The filters in the glasses inevitably dim the picture by reducing its brightness and contrast, not to mention the color distortion. Figure 5.11 shows two popular glasses-free, so-called *autostereoscopic display devices*.

Figure 5.11a depicts the technique of *parallax barrier*, in which a layer of opaque material with slits is placed in front of the normal display device, e.g., an LCD. As a result, each eye only sees half of the columns on the display. By properly arranging the stereo left–right images, separate viewing of the two images in the left and right eyes is realized.

A number of commercial products use the parallax barrier technology, such as the portable Nintendo 3DS game console, the screen on the Fujifilm 3D camera FinePix Real 3D W3, and several smartphones.

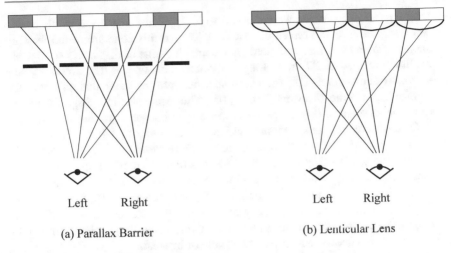

Fig. 5.11 Autostereoscopic display devices

In order to allow a larger viewing angle so that the device can be used from multiple positions, and potentially by multiple users, more than one pair of stereo images can be used, e.g., in one of Toshiba's glasses-free 3D TVs.

Figure 5.11b depicts the technique of using a *lenticular lens* . Instead of barriers, columns of magnifying lenses can be placed in front of the display to direct lights properly to the left and right eyes. The same technology has been also applied to *lenticular printing* to generate various 3D pictures and/or animations.

The lenticular technology is a type of *integral imaging*, originally proposed by Gabriel Lippmann in 1908 [14]. Instead of cylindrical lenses as shown above, an array of spherical convex microlenses can be used to generate a large number of distinct micro-images. These are computer-generated 2D views of the 3D scene, with one view per micro-lens. Therefore, this technique enables the rendering of multiple views from any directions. The Lytro camera, based on the technology of the *4D light field* [15], was an attempt toward this goal.

5.5.6 Disparity Manipulation in 3D Content Creation

The creation of 3D video content is a major challenge technically, perceptually, and artistically. In post-production, disparity values are manipulated to create the best 3D percept. Below we will focus on various methods of disparity manipulation where the geometry will be altered simply. The disparity here is the *image disparity* measured in pixels.

As summarized by Lang et al. in their SIGGRAPH 2010 paper on nonlinear disparity mapping [16], the following are essential concepts:

- **Disparity range**—When we are asked to look (focus) at the screen, there is a *comfort zone* near the screen distance. Objects in this zone are in the viewing angles of both eyes, and will yield an acceptable range of disparities so they are readily perceived in 3D. In creating 3D video contents, it is a common practice to map (often suppress) the original disparities into the range that will fit in the comfort zone of most viewers. To cite [16], "Practical values for disparity on a 30 foot cinema screen, are between +30 (appears behind screen) and −100 (appears in front of screen) pixels, assuming video with a width of 2,048 pixels."
- **Disparity sensitivity**—Our vision system is more capable of discriminating different depths when they are nearby. The sensitivity to depth drops rapidly with increased viewing distance: it is said to be inversely proportional to the square of the distance. This justifies a nonlinear disparity mapping [16] in which more disparity compression takes place at larger viewing distances. Since the disparity range of nearby objects is better preserved, this alleviates the problem of flattening of foreground objects, which are often more of interest.
- **Disparity gradient**—This measures the rate of disparity changes within a distance in the stereoscopic images. For example, two points on a frontal surface in the 3D world will yield (approximately) the same disparity in the left and right images due to their identical depth; this will yield a disparity gradient of (near) zero. On the other hand, two points on an oblique surface may yield different disparity values due to their difference in depth, and hence yield a nonzero disparity gradient. Burt and Julesz [17] pointed out that human vision has a limit of disparity gradient in binocular fusion. Beyond this limit, fusion into a single vision is difficult and mostly impossible, which is thus avoided in disparity gradient editing.
- **Disparity velocity**—When consecutive scenes present little disparity change, we can process the stereoscopic information very quickly. When there are large accommodation and vergence changes (i.e., disparity changes), we will slow down considerably. This is due to the limit of temporal modulation frequency of disparity. As discussed earlier, while focusing on the screen watching 3D video, the rapid change in vergence is a main cause for visual fatigue and must be restricted. We can tolerate some changes of convergence (i.e., disparity) as long as the speed of the changes is moderate.

Some additional technical issues are as follows:

- Most stereoscopic cameras adopt the simple camera model where the camera optical axes are in parallel. This yields near-zero disparity for far objects and very large disparity for nearby objects, and is very different from the toed-in camera model which better emulates the human vision system. In this case, a conversion of the image disparity values is necessary in the 3D video post-production stage. A variety of techniques are described in [13,18], among them *floating window* where the screen distance can be artificially shifted.
- As stated above, the average *interocular distance* of viewers is approximately 2.5 inches. As a result, in the toed-in camera model, the projected images of a

very far object (near infinity), for example, should be about 2.5 inches apart in the left and right images on the screen in order to generate the required positive parallax. Depending on the screen size and screen resolution, a very different *image disparity* will be required. It is therefore a common practice to produce different 3D contents with very different image disparity values targeted for different purposes (large cinema screens versus small PC or smartphone screens, high resolution versus low resolution).

The multimedia and movie industries are keenly interested in converting the vast amount of 2D contents into 3D. Zhang et al. [19] provide a good survey on the issues involved in manually and (semi-)automatically converting such videos and films.

5.6 Video Quality Assessment (VQA)

In order to reduce the storage and transmission overhead, it is common for digital video to be compressed. When comparing the coding efficiency of different video compression methods, a common practice is to compare the bitrates of the coded video bitstreams at the same *quality*. The video quality assessment approaches can be *objective* or *subjective*: the former is done automatically by computers and the latter requires human judgement.

5.6.1 Objective Assessment

The most common criterion used for the objective assessment is PSNR (peak signal-to-noise ratio). For images it is

$$PSNR = 10\log_{10}\frac{I_{max}^2}{MSE},\qquad(5.8)$$

where I_{max} is the maximum intensity value, e.g., 255 for 8-bit images, and MSE is the *mean squared error* between the original image I and the compressed image $\tilde{I}$. For videos, the PSNR is usually the average of the PSNRs of the images in the video sequence.

In recent literature, *BD-rate* is often adopted for the comparison of video coding efficiency. BD-rate (short for *Bjontegaard's delta-rate*) [20] is based on the difference of bitrates between two compression methods, usually the new versus the baseline, when the same PSNR is maintained. In actual implementations, it is the average bitrate difference over a range of PSNRs (or quality levels). A negative BD-rate indicates an improvement in coding efficiency (saving of bitrate) achieved by the new method. The Luma BD-rate is generally viewed as more reliable than the Chroma BD-rate, although the latter is sometimes also measured.

5.6.2   Subjective Assessment

The main advantage of PSNR is that it is easy to calculate. However, it does not necessarily reflect the quality as perceived by humans, i.e., visual quality. An obvious example would be to add (or subtract) a small and fixed amount to the intensity values of all the pixels in the picture. Visually (subjectively), we may not notice any quality change. On the other hand, the PSNR will certainly be affected.

The methodology of subjective assessment of television pictures is specified in ITU-R Recommendation BT.500. Its latest version is BT.500-13 [21] revised in 2012.

For example, in the experiment of Ohm et al. [22], the original and compressed video clips are shown in succession to human subjects (the so-called *double stimulus* method). The subjects are asked to grade the compressed videos by their quality, 0—lowest, 10—highest. The MOS (*mean opinion score*), which is the arithmetic mean of their scores, is used as the measure for the subjective quality of different video compression methods.

The above subjective assessment method is extensively employed in the development of the video compression standards such as H.265 and H.266, especially for HDR videos [23], where the MOS scores are used to compare the effectiveness of different techniques and algorithms so as to guide the development of the coding tools.

5.6.3   Other VQA Metrics

Video quality assessment (VQA) is an active research area [24]. The main efforts are to find better metrics than simple measures such as PSNR, so that assessments can be conducted objectively (by computer) and their results will be comparable to those of human subjects. Many VQA algorithms are derived from their counterparts in image quality assessment (IQA). Wang et al. [25] presented the *structural similarity (SSIM) index* that captures some simple image structural information (e.g., luminance and contrast). It has become very popular in image and video quality assessments. Peng et al. [26] presented a brief survey of VQA and a good metric based on a novel spacetime texture representation. Wang et al. [27] presented a VQA model for asymmetrically compressed stereoscopic 3D videos, where the left and right stereo images can have different resolutions and asymmetric transform-domain quantization coding may have been applied.

5.7   Exercises

1. NTSC video has 525 lines per frame and 63.6 µs per line, with 20 lines per field of vertical retrace and 10.9 µs horizontal retrace.

   (a) Where does the 63.6 µs come from?
   (b) Which takes more time, horizontal retrace or vertical retrace? How much more time?

2. Which do you think has less detectable flicker, PAL in Europe or NTSC in North America? Justify your conclusion.
3. Sometimes the signals for television are combined into fewer than all the parts required for TV transmission.

    (a) Altogether, how many and what are the signals used for studio broadcast TV?
    (b) What does S-video stand for? How many and what signals are used in S-video?
    (c) How many signals are actually broadcasted for standard analog TV reception? What kind of video is that called?

4. Show how the Q signal can be extracted from the NTSC chroma signal C (Eq. (5.1)) during demodulation.
5. One sometimes hears that the old Betamax format for videotape, which competed with VHS and lost, was actually a better format. How would such a statement be justified?
6. We don't see flicker on a workstation screen when displaying video at NTSC frame rate. Why do you think this might be?
7. Digital video uses *chroma subsampling*. What is the purpose of this? Why is it feasible?
8. What are the most salient differences between ordinary TV and HDTV/UHDTV? What was the main impetus for the development of HDTV/UHDTV?
9. What is the advantage of interlaced video? What are some of its problems?
10. One solution that removes the problems of interlaced video is to de-interlace it. Why can we not just overlay the two fields to obtain a de-interlaced image? Suggest some simple de-interlacing algorithms that retain information from both fields.
11. Assuming the bit depth of 12 bits per sample, 120 fps, and 4:2:2 chroma subsamplng, what are the bitrates of 4K, 8K, and 16K UHDTV videos if they are uncompressed?
12. In equirectangular projection (ERP), if we use the prime meridian as the central meridian where $x = 0$, and the equator as the central parallel where $y = 0$, what are the projected coordinates for a city at 31.23°N, 121.47°E ?
13. Assuming we use the toed-in stereo camera model, the interocular distance is I, and the screen is D meters away, (a) At what distance will a point P generate a positive parallax equal to I on the screen? (b) At what distance will a point P generate a negative parallax equal to $-I$?
14. The most common criterion used for the objective image quality assessment is PSNR (peak signal-to-noise ratio). For 8-bit images, it is

$$PSNR = 10\log_{10}\frac{255^2}{MSE} \tag{5.9}$$

where MSE stands for the *mean squared error* between the original image and the image after compression.

If the original image and the lossy image after compression are as in (a) and (b) given below:

50	50	50	50	200	200	200	200
50	50	50	50	200	200	200	200
50	50	50	50	200	200	200	200
50	50	50	50	200	200	200	200
50	50	50	50	200	200	200	200
50	50	50	50	200	200	200	200
50	50	50	50	200	200	200	200
50	50	50	50	200	200	200	200

50	50	60	80	170	190	200	200
50	50	60	80	170	190	200	200
50	50	60	80	170	190	200	200
50	50	60	80	170	190	200	200
50	50	60	80	170	190	200	200
50	50	60	80	170	190	200	200
50	50	60	80	170	190	200	200
50	50	60	80	170	190	200	200

(a) Original image   (b) Lossy image after compression

(a) Explain what is the main reason of the loss of the sharp edge in the lossy image after compression.

(b) What is the PSNR for this lossy image?

(c) If we arbitrarily add an intensity value of 10 to all pixels in the lossy image in (b), what would be the PSNR?

(d) The changes (additional errors) in (c) are hardly noticeable visually, i.e., they barely affect the perceived image quality. Hence, it is often argued that PSNR is not the best measure for image quality assessment.

Suggest a method that will alleviate the impact of such changes to the measure of PSNR.

15. Give at least three reasons to argue that PSNR is not necessarily a good metric for video quality assessment.

References

1. A.M. Tekalp, *Digital Video Processing*, 2nd edn. (Prentice Hall, 2015)
2. A. Bovik (ed.), *Handbook of Image and Video Processing*, 2nd edn. (Academic Press, 2010)
3. J.F. Blinn, NTSC: nice technology, super color. IEEE Comput. Graph. Appl. **13**(2), 17–23 (1993)
4. C.A. Poynton, *A Technical Introduction to Digital Video* (Wiley, New York, 1996)
5. J.F. Blinn, The world of digital video. IEEE Comput. Graph. Appl. **12**(5), 106–112 (1992)
6. C.A. Poynton, *Digital Video and HD: Algorithms and Interfaces*, 2nd edn. (Elsevier, 2012)
7. R.L. Myers, *Display Interfaces: Fundamentals and Standards* (Wiley, 2002)
8. C.L. Fan, et al., A survey on 360° video streaming: acquisition, transmission, and display. ACM Comput. Surv. **52**(4), 71:1–71:36 (2019)
9. Y. Ye, J.M. Boyce, P. Hanhart, Omnidirectional 360° video coding technology in responses to the joint call for proposals on video compression with capability beyond HEVC. IEEE Trans. Circuits Syst. Video Technol. **30**(5), 1241–1252 (2020)
10. ISO/IEC JTC1/SC29/WG11/N17197, *Algorithm Descriptions of Projection Format Conversion and Video Quality Metrics in 360Lib*. (ITU and ISO/IEC, 2017)
11. D.M. Hoffman, A.R. Girshick, K. Akeley, M.S. Banks, Vergence-accommodation conflicts hinder visual performance and cause visual fatigue. J. Vis. **8**(3) (2008)

12. T. Shibata, J. Kim, D.M. Hoffman, M.S. Banks, The zone of comfort: predicting visual discomfort with stereo displays. J. Vis. **11**(8) (2011)
13. B. Mendiburu, *3D Movie Making: Stereoscopic Digital Cinema from Script to Screen*. (Elsevier Science, 2009)
14. G. Lippmann, La Photographie Integrale. Comptes Rendus Academie des Sciences **146**, 446–451 (1908)
15. M. Levoy, P. Hanrahan, Light field rendering, in *Proceedings of International Conference on Computer Graphics and Interactive Techniques (SIGGRAPH)* (1996)
16. M. Lang, A. Hornung, O. Wang, S. Poulakos, A. Smolic, M. Gross, Nonlinear Disparity mapping for stereoscopic 3D. ACM Trans. Graph. **29**(4) (2010)
17. P. Burt, B. Julesz, A disparity gradient limit for binocular fusion. Science **208**(4444), 615–617 (1980)
18. R. Ronfard, G. Taubin (eds.), *Image and Geometry Processing for 3-D Cinematography: an Introduction*. (Springer, 2010)
19. L. Zhang, C. Vazquez, S. Knorr, 3D-TV content creation: automatic 2D-to-3D video conversion. IEEE Trans. Broadcast. **57**(2), 372–383 (2011)
20. G. Bjontegaard, Calculation of average PSNR differences between RD-curves. Technical Report VCEG-M33 (2001)
21. ITU-R Rec. BT.500-13. *Methodology for the Subjective Assessment of the Quality of Television Pictures*. ITU-R (2012)
22. J.R. Ohm, et al., Comparison of the coding efficiency of video coding standards—including high efficiency video coding (HEVC). IEEE Trans. Circ. Syst. Video Technol. **22**(12), 1669–1684 (2012)
23. E. François, et al., High dynamic range video coding technology in responses to the joint call for proposals on video compression with capability beyond HEVC. IEEE Trans. Circ. Syst. Video Technol. **30**(5), 1253–1266 (2020)
24. Y. Chen, K. Wu, Q. Zhang, From QoS to QoE: a tutorial on video quality assessment. IEEE Commun. Surv. Tutor. **17**(2), 1126–1165 (2015)
25. Z. Wang et al., Image quality assessment: from error visibility to structural similarity. IEEE Trans. Image Process. **13**(4), 600–612 (2004)
26. P. Peng, D. Liao, Z.N. Li, An efficient temporal distortion measure of videos based on spacetime texture. Pattern Recogn. **70**, 1–11 (2017)
27. J. Wang et al., Asymmetrically compressed stereoscopic 3D videos: quality assessment and rate-distortion performance evaluation. IEEE Trans. Image Process. **26**(3), 1330–1343 (2017)

Basics of Digital Audio

6

Audio information is crucial for multimedia presentations and, in a sense, is the simplest type of multimedia data. However, some important differences between audio and image information cannot be ignored. For example, while it is customary and useful to occasionally drop a video frame from a video stream, to facilitate viewing speed, we simply cannot do the same with sound information or all sense will be lost from that dimension. We introduce basic concepts for sound in multimedia in this chapter and examine the arcane details of compression of sound information in Chaps. 13 and 14. The digitization of sound necessarily implies sampling and quantization of signals, so we introduce these topics here.

We begin with a discussion of just what makes up sound information, then we go on to examine the use of MIDI as an enabling technology to capture, store, and play back musical notes. We go on to look at some details of audio quantization, and give some introductory information on how digital audio is dealt with for storage or transmission. This entails a first discussion of how subtraction of signals from predicted values yields numbers that are close to zero, and hence easier to deal with.

6.1 Digitization of Sound

6.1.1 What Is Sound?

Sound is a wave phenomenon like light, but it is macroscopic and involves molecules of air being compressed and expanded under the action of some physical device. For example, a speaker in an audio system vibrates back and forth and produces a longitudinal pressure wave that we perceive as sound. (As an example, let us use a Slinky and place it on the table. We get a longitudinal wave by vibrating the Slinky along its length; in contrast, we get a transverse wave by waving the Slinky back and forth perpendicular to its length.)

© Springer Nature Switzerland AG 2021
Z.-N. Li et al., *Fundamentals of Multimedia*, Texts in Computer Science,
https://doi.org/10.1007/978-3-030-62124-7_6

Without air there is no sound—for example, in space. Since sound is a pressure wave, it takes on continuous values, as opposed to digitized ones with a finite range. Nevertheless, if we wish to use a digital version of sound waves, we must form digitized representations of audio information.

Even though such pressure waves are longitudinal, they still have ordinary wave properties and behaviors, such as reflection (bouncing), refraction (change of angle when entering a medium with a different density), and diffraction (bending around an obstacle). This makes the design of "surround sound" possible.

Since sound consists of measurable pressures at any 3D point, we can detect it by measuring the pressure level at a location, using a transducer to convert pressure to voltage levels.

In general, any signal can be decomposed into a sum of sinusoids, if we are willing to use enough sinusoids. Figure 6.1 shows how weighted sinusoids can build up quite a complex signal. Whereas frequency is an absolute measure, pitch is a perceptual, subjective quality of sound. Generally, pitch is relative. Pitch and frequency are linked by setting the note A above middle C to exactly 440 Hz. An *octave* above that note corresponds to doubling the frequency and takes us to another A note. Thus, with the middle A on a piano ("A4" or "A440") set to 440 Hz, the next A up is 880 Hz, one octave above, and so on.

Given a sound with fundamental frequency f, we define *harmonics* as any musical tones whose frequencies are integral multiples of the fundamental frequency, i.e., $2f, 3f, 4f, ..,$ etc. For example, if the fundamental frequency (also known as *first harmonic*) is $f = 100$ Hz, the frequency of the second harmonic is 200 Hz. For the third harmonic, it is 300 Hz and so on. The harmonics can be linearly combined to form a new signal. Because all the harmonics are periodic at the fundamental frequency, their linear combinations are also periodic at the fundamental frequency. Figure 6.1 shows the appearance of the combination of these harmonics.

Now, if we allow any (integer and noninteger) multiples higher than the fundamental frequency, we allow *overtones* and have a more complex and interesting resulting sound. Together, the fundamental frequency and overtones are referred to as *partials*. The harmonics we discussed above can also be referred to as *harmonic partials*.

6.1.2 Digitization

Figure 6.2 shows the "one-dimensional" nature of sound. Values change over time in *amplitude*: the pressure increases or decreases with time [1]. Since there is only one independent variable, time, we call this a 1D signal—as opposed to images, with data that depends on two variables, x and y, or video, which depends on three variables, x, y, t. The amplitude value is a continuous quantity. Since we are interested in working with such data in computer storage, we must *digitize* the *analog signals* (i.e., continuous-valued voltages) produced by microphones. Digitization means conversion to a stream of numbers—preferably *integers* for efficiency.

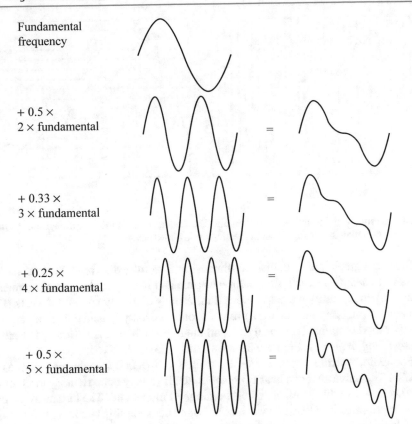

Fundamental
frequency

+ 0.5 ×
2 × fundamental

+ 0.33 ×
3 × fundamental

+ 0.25 ×
4 × fundamental

+ 0.5 ×
5 × fundamental

Fig. 6.1 Building up a complex signal by superposing sinusoids

Fig. 6.2 An analog signal: continuous measurement of pressure wave

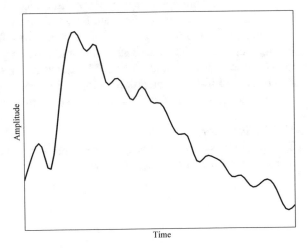

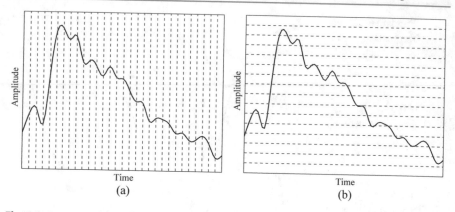

Fig. 6.3 Sampling and quantization: **a** Sampling the analog signal in the time dimension; **b** quantization is sampling the analog signal in the amplitude dimension

Since the graph in Fig. 6.2 is two dimensional, to fully digitize the signal shown we have to *sample* in each dimension—in time and in amplitude. Sampling means measuring the quantity we are interested in, usually at evenly spaced intervals. The first kind of sampling—using measurements only at evenly spaced *time* intervals—is simply called *sampling* (surprisingly), and the rate at which it is performed is called the *sampling frequency*. Figure 6.3a shows this type of digitization.

For audio, typical sampling rates are from 8 kHz (8,000 samples per second) to 48 kHz. The human ear can hear from about 20 Hz (a very deep rumble) to as much as 20 kHz; above this level, we enter the range of ultrasound. The human voice can reach approximately 4 kHz and we need to bound our sampling rate from below by at least double this frequency (see the discussion of the Nyquist sampling rate, below). Thus, we arrive at the useful range of about 8–40 or so kHz.

Sampling in the amplitude or voltage dimension is called *quantization*, shown in Fig. 6.3b. While we have discussed only uniform sampling, with equally spaced sampling intervals, nonuniform sampling is possible. This is not used for sampling in time but is used for quantization (see the μ-law rule, below). Typical uniform quantization rates are 8 and 16 bits; 8-bit quantization divides the vertical axis into 256 levels and 16-bit quantization divides it into 65,536 levels.

To decide how to digitize audio data, we need to answer the following questions:

1. What is the sampling rate?
2. How finely is the data to be quantized, and is the quantization uniform?
3. How is audio data formatted (i.e., what is the file format)?

6.1.3 Nyquist Theorem

As we know now, each sound is just made from sinusoids. As a simple illustration, Fig. 6.4a shows a single sinusoid: it is a single, pure, frequency (only electronic instruments can create such boring sounds).

Now if the sampling rate just equals the actual frequency, we can see from Fig. 6.4b that a false signal is detected: it is simply a constant, with zero frequency. If, on the other hand, we sample at 1.5 times the frequency, Fig. 6.4c shows that we obtain an incorrect (*alias*) frequency that is lower than the correct one—it is half the correct one (the wavelength, from peak to peak, is double that of the actual signal). In computer graphics, much effort is aimed at masking such alias effects by various methods of

Fig. 6.4 Aliasing: **a** A single frequency; **b** sampling at exactly the frequency produces a constant; **c** sampling at 1.5 times per cycle produces an *alias* frequency that is perceived

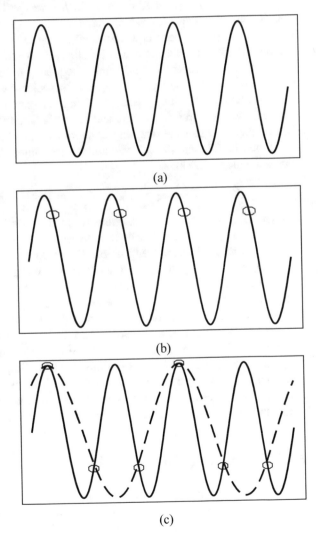

(a)

(b)

(c)

anti-aliasing. An alias is any artifact that does not belong to the original signal. Thus, for correct sampling, we must use a sampling rate equal to at least twice the maximum frequency content in the signal. This is called the *Nyquist rate*.

The Nyquist theorem is named after Harry Nyquist, a famous mathematician who worked at Bell Labs. More generally, if a signal is *band limited*—that is, if it has a lower limit f_1 and an upper limit f_2 of frequency components in the signal—then we need a sampling rate of at least $2(f_2 - f_1)$.

Suppose we have a *fixed* sampling rate. Since it would be impossible to recover frequencies higher than half the sampling rate in any event, most systems have an *anti-aliasing filter* that restricts the frequency content of the sampler's input to a range at or below half the sampling frequency. Confusingly, the frequency equal to half the Nyquist rate is called the *Nyquist frequency*. Then for our fixed sampling rate, the Nyquist frequency is half the sampling rate. The highest possible signal frequency component has frequency equal to that of the sampling itself.

Note that the true frequency and its alias are located symmetrically on the frequency axis with respect to the Nyquist frequency pertaining to the sampling rate used. For this reason, the Nyquist frequency associated with the sampling frequency is often called the "folding" frequency. That is to say, if the sampling frequency is less than twice the true frequency, and is greater than the true frequency, then the alias frequency equals the sampling frequency minus the true frequency. For example, if the true frequency is 5.5 kHz and the sampling frequency is 8 kHz, then the alias frequency is 2.5 kHz:

$$f_{\text{alias}} = f_{\text{sampling}} - f_{\text{true}}, \quad \text{for} \quad f_{\text{true}} < f_{\text{sampling}} < 2 \times f_{\text{true}} \tag{6.1}$$

As well, a frequency at double any frequency could also fit sample points. In fact, adding any positive or negative multiple of the sampling frequency to the true frequency always gives another possible alias frequency, in that such an alias gives the same set of samples when sampled at the sampling frequency.

So, if again the sampling frequency is less than twice the true frequency and is less than the true frequency, then the alias frequency equals n times the sampling frequency minus the true frequency, where the n is the lowest integer that makes n times the sampling frequency larger than the true frequency. For example, when the true frequency is between 1.0 and 1.5 times the sampling frequency, the alias frequency equals the true frequency minus the sampling frequency.

In general, the apparent frequency of a sinusoid is the lowest frequency of a sinusoid that has exactly the same samples as the input sinusoid. Figure 6.5 shows the relationship of the apparent frequency to the input (true) frequency.

6.1.4 Signal-to-Noise Ratio (SNR)

In any analog system, random fluctuations produce *noise* added to the signal, and the measured voltage is thus incorrect. The ratio of the power of the correct signal to the noise is called the *signal-to-noise ratio* (*SNR*). Therefore, the SNR is a measure of the quality of the signal.

Fig. 6.5 Folding of sinusoid frequency sampled at 8,000 Hz. The folding frequency, shown dashed, is 4,000 Hz

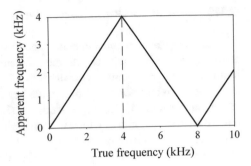

The SNR is usually measured in *decibels* (*dB*), where 1 dB is a tenth of a *bel*. The SNR value, in units of dB, is defined in terms of base-10 logarithms of squared voltages:

$$SNR = 10 \log_{10} \frac{V_{signal}^2}{V_{noise}^2} = 20 \log_{10} \frac{V_{signal}}{V_{noise}} \tag{6.2}$$

The power in a signal is proportional to the square of the voltage. For example, if the signal voltage V_{signal} is ten times the noise, the SNR is $20 \times \log_{10}(10) = 20$ dB.

In terms of power, if the squeaking we hear from ten violins playing is ten times the squeaking we hear from one violin playing, then the ratio of power is given in terms of decibels as 10 dB, or, in other words, 1 Bel. Notice that decibels are always defined in terms of a ratio. The term "decibels" as applied to sounds in our environment usually is in comparison to a just-audible sound with frequency 1 kHz. The levels of sound we hear around us are described in terms of decibels, as a ratio to the quietest sound we are capable of hearing. Table 6.1 shows approximate levels for these sounds.

Table 6.1 Magnitudes of common sounds, in decibels

Threshold of hearing	0
Rustle of leaves	10
Very quiet room	20
Average room	40
Conversation	60
Busy street	70
Loud radio	80
Train through station	90
Riveter	100
Threshold of discomfort	120
Threshold of pain	140
Damage to eardrum	160

6.1.5 Signal-to-Quantization-Noise Ratio (SQNR)

For digital signals, we must take into account the fact that only quantized values are stored. For a digital audio signal, the precision of each sample is determined by the number of bits per sample, typically 8 or 16.

Aside from any noise that may have been present in the original analog signal, additional error results from quantization. That is, if voltages are in the range of 0–1 but we have only 8 bits in which to store values, we effectively force all continuous values of voltage into only 256 different values. Inevitably, this introduces a roundoff error. Although it is not really "noise," it is called *quantization noise* (or *quantization error*). The association with the concept of noise is that such errors will essentially occur randomly from sample to sample.

The quality of the quantization is characterized by the *signal-to-quantization-noise ratio (SQNR)*. Quantization noise is defined as the difference between the value of the analog signal, for the particular sampling time, and the nearest quantization interval value. At most, this error can be as much as half of the interval.

For a quantization accuracy of N bits per sample, the range of the digital signal is -2^{N-1} to $2^{N-1} - 1$. Thus, if the actual analog signal is in the range from $-V_{max}$ to $+V_{max}$, each quantization level represents a voltage of $2V_{max}/2^N$, or $V_{max}/2^{N-1}$. SQNR can be simply expressed in terms of the peak signal, which is mapped to the level V_{signal} of about 2^{N-1}, and the SQNR has as denominator the maximum V_{quan_noise} of $1/2$. The ratio of the two is a simple definition of the SQNR[1]:

$$SQNR = 20 \log_{10} \frac{V_{signal}}{V_{quan\_noise}} = 20 \log_{10} \frac{2^{N-1}}{\frac{1}{2}}$$
$$= 20 \times N \times \log 2 = 6.02N \, (\text{dB}) \qquad (6.3)$$

In other words, each bit adds about 6 dB of resolution, so 16 bits provide a maximum SQNR of 96 dB.

We have examined the worst case. If, on the other hand, we assume that the input signal is sinusoidal, that is, quantization error is statistically independent, and that its magnitude is uniformly distributed between 0 and half the interval, we can show [2] that the expression for the SQNR becomes

$$SQNR = 6.02N + 1.76 \, (\text{dB}) \qquad (6.4)$$

Since larger is better, this shows that a more realistic approximation gives a better characterization number for the quality of a system.

We can simulate quantizing samples, e.g., drawing values from a sinusoidal probability function, and verify Eq. (6.4). Defining the SQNR in terms of the RMS (root-mean-square) value of the signal versus the RMS value of the quantization noise, the following MATLAB fragment does indeed comply with Eq. (6.4):

---

[1]This ratio is actually the *peak* signal-to-quantization-noise ratio, or PSQNR.

```
% sqnr_sinusoid.m
%
% Simulation to verify SQNR for sinusoidal
%   probability density function.
b = 8; % 8-bit quantized signals
q = 1/10000; % make sampled signal with interval size 1/10001
seq = [0 : q : 1];
x = sin(2*pi*seq); % analog signal --> 10001 samples
% Now quantize:
x8bit = round(2^(b-1)*x) / 2^(b-1); % in [-128,128]/128=[-1,+1]
quanterror = x - x8bit;
%
SQNR = 20*log10( sqrt(mean(x.^2))/sqrt(mean(quanterror.^2)) ) %
% 50.0189dB
SQNRtheory = 6.02*b + 1.76 % 1.76=20*log10(sqrt(3/2))
% 49.9200dB
```

The more careful Eq. (6.4) can actually be proved analytically if wish to do so: if the error obeys a uniform-random probability distribution in the range $[-0.5, 0.5]$, then its RMS (root-mean-square) value is $\sqrt{\int_{-0.5}^{0.5} x^2 dx} = 1/\sqrt{12}$. Now assume the signal itself is a sinusoid, $\sin(2\pi x) = \sin\theta$. This has to be multiplied by a scalar value D, giving the range over which the sinusoid travels, i.e., the max minus the min: $D = [(2^{N-1} - 1) - (-2^{N-1})] \simeq 2^N$. The sine curve is multiplied by the factor $D/2$.

Then the RMS value of the signal is $\sqrt{1/(2\pi) \int_0^{2\pi} (\frac{D}{2} \sin\theta)^2 d\theta} = D/(2\sqrt{2})$. Forming the ratio of the RMS signal over the RMS quantization noise, we get $20 \log_{10}(\sqrt{12} D/(2\sqrt{2})) = 20 \log_{10}(D\sqrt{3/2}) = 20 \log_{10}(D) + 20 \log_{10}(\sqrt{3/2}) = 20 \log_{10}(2^N) + 20 \log_{10}(\sqrt{3/2})$, which just gives Eq. (6.4).

Typical digital audio sample precision is either 8 bits per sample, equivalent to about telephone quality, or 16 bits, for CD quality. In fact, 12 bits or so would likely do fine for adequate sound reproduction.

6.1.6 Linear and Nonlinear Quantization

We mentioned above that samples are typically stored as uniformly quantized values. This is called *linear format*. However, with a limited number of bits available, it may be more sensible to try to take into account the properties of human perception and set up nonuniform quantization levels that pay more attention to the frequency range over which humans hear best.

Remember that here we are quantizing magnitude, or amplitude—how loud the signal is. In Chap. 4, we discussed an interesting feature of many human perception subsystems (as it were)—Weber's law—which states that the more there is, proportionately more must be added to discern a difference. Stated formally, Weber's law says that equally perceived differences have values proportional to absolute levels:

$$\Delta\text{Response} \propto \Delta\text{Stimulus}/\text{Stimulus} \qquad (6.5)$$

This means that, for example, if we can feel an increase in weight from 10 to 11 pounds, then if instead we start at 20 pounds, it would take 22 pounds for us to feel an increase in weight.

Inserting a constant of proportionality k, we have a differential equation that states

$$dr = k(1/s)\,ds \qquad (6.6)$$

with response r and stimulus s. Integrating, we arrive at a solution

$$r = k \ln s + C \qquad (6.7)$$

with constant of integration C. Stated differently, the solution is

$$r = k \ln(s/s_0) \qquad (6.8)$$

where s_0 is the lowest level of stimulus that causes a response ($r = 0$ when $s = s_0$).

Thus, nonuniform quantization schemes that take advantage of this perceptual characteristic make use of logarithms. The idea is that in a log plot derived from Eq. (6.8), if we simply take uniform steps along the s-axis, we are not mirroring the nonlinear response along the r-axis.

Instead, we would like to take uniform steps along the r-axis. Thus, nonlinear quantization works by first transforming an analog signal from the raw s space into the theoretical r space, then uniformly quantizing the resulting values. The result is that for steps near the low end of the signal, quantization steps are effectively more concentrated on the s-axis, whereas for large values of s, one quantization step in r encompasses a wide range of s values.

Such a law for audio is called μ-*law* encoding, or *u-law*, since it's easier to write. A very similar rule, called *A-law*, is used in telephony in Europe.

The equations for these similar encoding methods are as follows:
μ-law:

$$r = \frac{\text{sign}(s)}{\ln(1+\mu)} \ln \left\{ 1 + \mu \left| \frac{s}{s_p} \right| \right\}, \qquad \left| \frac{s}{s_p} \right| \le 1 \qquad (6.9)$$

A-law:

$$r = \begin{cases} \frac{A}{1+\ln A} \left(\frac{s}{s_p} \right), & \left| \frac{s}{s_p} \right| \le \frac{1}{A} \\[2mm] \frac{\text{sign}(s)}{1+\ln A} \left[1 + \ln A \left| \frac{s}{s_p} \right| \right], & \frac{1}{A} \le \left| \frac{s}{s_p} \right| \le 1 \end{cases} \qquad (6.10)$$

$$\text{where } \text{sign}(s) = \begin{cases} 1 & \text{if } s > 0, \\ -1 & \text{otherwise} \end{cases}$$

Figure 6.6 depicts these curves. The parameter of the μ-law encoder is usually set to $\mu = 100$ or $\mu = 255$, while the parameter for the A-law encoder is usually set to $A = 87.6$.

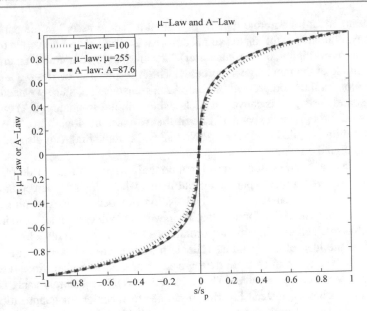

Fig. 6.6 Nonlinear transform for audio signals

Here, s_p is the *peak signal value* and s is the current signal value. So far, this simply means that we wish to deal with s/s_p, in the range -1 to 1.

The idea of using this type of law is that if s/s_p is first transformed to values r as above and then r is quantized uniformly before transmitting or storing the signal, most of the available bits will be used to store information where changes in the signal are most apparent to a human listener, because of our perceptual nonuniformity.

To see this, consider a small change in $|s/s_p|$ near the value 1.0, where the curve in Fig. 6.6 is flattest. Clearly, the change in s has to be much larger in the flat area than near the origin to be registered by a change in the quantized r value. And it is at the quiet, low end of our hearing that we can best discern small changes in s. The μ-law transform concentrates the available information at that end.

First, we carry out the μ-law transformation, then we quantize the resulting value, which is a nonlinear transform away from the input. The logarithmic steps represent low-amplitude, quiet signals with more accuracy than loud, high-amplitude ones. What this means for signals that are then encoded as a fixed number of bits is that for low-amplitude, quiet signals, the amount of noise—the error in representing the signal—is a smaller number than for high-amplitude signals. Therefore, the μ-law transform effectively makes the signal-to-noise ratio more uniform across the range of input signals.

This technique is based on human perception—a simple form of "perceptual coder." Interestingly, we have in effect also made use of the statistics of sounds we are likely to hear, which are generally in the low-volume range. In effect, we are asking for most bits to be assigned where most sounds occur—where the probability density is highest. So this type of coder is also one that is driven by statistics.

In summary, a logarithmic transform, called a "compressor" in the parlance of telephony, is applied to the analog signal before it is sampled and converted to digital (by an analog-to-digital, or AD, converter). The amount of compression increases as the amplitude of the input signal increases. The AD converter carries out a uniform quantization on the "compressed" signal. After transmission, since we need analog to hear sound, the signal is converted back, using a digital-to-analog (DA) converter, then passed through an "expander" circuit that reverses the logarithm. The overall transformation is called *companding*. Nowadays, companding can also be carried out in the digital domain.

The μ-law in audio is used to develop a nonuniform quantization rule for sound. In general, we would like to put the available bits where the most perceptual acuity (sensitivity to small changes) is. Ideally, bit allocation occurs by examining a curve of stimulus versus response for humans. Then we try to allocate bit levels to intervals for which a small change in stimulus produces a large change in response.

That is, the idea of companding reflects a less specific idea used in assigning bits to signals: put the bits where they are most needed to deliver finer resolution where the result can be perceived. This idea militates against simply using uniform quantization schemes, instead favoring nonuniform schemes for quantization. The μ-law (or A-law) for audio is an application of this idea.

Savings in bits can be gained by transmitting a smaller bit depth for the signal, if this is indeed possible without introducing too much error. Once telephony signals became digital, it was found that the original continuous-domain μ-law transform could be used with a substantial reduction of bits during transmission and still produce reasonable-sounding speech upon expansion at the receiver end. The μ-law often starts with a bit depth of 16 bits, but transmits using 8 bits, and then expands back to 16 bits at the receiver.

Suppose we use the μ-law Eq. (6.9) with $\mu=255$. Here the signal s is normalized into the range $[-1, 1]$. If the input is in -2^{15} to $(+2^{15} - 1)$, we divide by 2^{15} to normalize. Then the μ-law is applied to turn s into r; this is followed by reducing the bit depth down to 8-bit samples, using $\hat{r} = \text{sign}(s) * \text{floor}(128 * r)$.

Now the 8-bit signal $\hat{r}$ is transmitted.

Then, at the receiver side, we normalize $\hat{r}$ by dividing by 2^7, and then apply the inverse function to Eq. (6.9), which is as follows:

$$\hat{s} = \text{sign}(s) \left(\frac{(\mu + 1)^{|\hat{r}|} - 1}{\mu} \right) \tag{6.11}$$

Finally, we expand back up to 16 bits: $\tilde{s} = ceil(2^{15} * \hat{s})$. Below we show a MATLAB function for these operations.

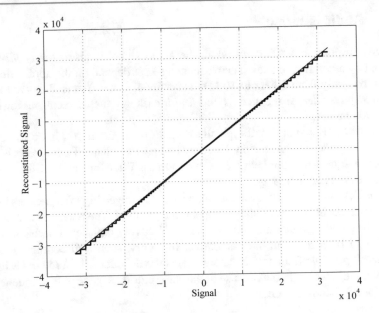

Fig. 6.7 Nonlinear quantization by companding

```
function x_out = mu_law_8bitsf(x)
% signal x  is 16-bit
mu=255;
xnormd = x/2^15;
y=sign(x)*( (log(1+mu*abs(xnormd)))/log(1+mu) );

y8bit = floor(128*y);
  % TRANSMIT
y8bitnormd = y8bit/2^7;
x_hat = sign(x)*( ( (mu+1)^abs(y8bitnormd)-1 )/mu );

% scale to 16 bits:
x_out = ceil(2^15*x_hat);
```

For the 2^{16} input values, shown as a solid line in Fig. 6.7, the companded output values are shown as the staircase steps, in a thicker line. Indeed, we see that the companding puts the most accuracy at the quiet end nearest zero.

6.1.7 Audio Filtering

Prior to sampling and AD conversion, the audio signal is also usually *filtered* to remove unwanted frequencies. The frequencies kept depend on the application. For speech, typically from 50 Hz to 10 kHz is retained. Other frequencies are blocked by a *band-pass filter*, also called a *band-limiting* filter, which screens out lower and higher frequencies.

An audio music signal will typically contain from about 20 Hz up to 20 kHz. (Twenty Hz is the low rumble produced by an upset elephant. Twenty kHz is about the highest squeak we can hear.). So the band-pass filter for music will screen out frequencies outside this range.

At the DA converter end, even though we have removed high frequencies that are likely just noise in any event, they reappear in the output. The reason is that because of sampling and then quantization, we have effectively replaced a perhaps smooth input signal by a series of step functions. In theory, such a discontinuous signal contains all possible frequencies. Therefore, at the decoder side, a *low-pass* filter is used after the DA circuit, making use of the same cutoff as at the high-frequency end of the coder's band-pass filter.

6.1.8 Audio Quality versus Data Rate

The uncompressed data rate increases as more bits are used for quantization. Stereo information, as opposed to mono, doubles the *bitrate* (in bits per second) needed to transmit a digital audio signal. Table 6.2 shows how audio quality is related to bitrate and bandwidth.

The term *bandwidth*, derived from analog devices in signal processing, refers to the part of the response or transfer function of a device that is approximately constant, or flat, with the x-axis being the frequency and the y-axis equal to the transfer function. *Half-power bandwidth* (*HPBW*) refers to the bandwidth between points when the power falls to half the maximum power. Since $10 \log_{10}(\frac{1}{2}) \approx -3.0$, the term -3 dB bandwidth is also used to refer to the HPBW.

So for analog devices, the bandwidth was expressed in the frequency unit, called *Hertz* (Hz), which is cycles per second (for example, heartbeats per second). For digital devices, on the other hand, the amount of data that can be transmitted in a fixed bandwidth is usually expressed in bitrate, i.e., bits per second (bps) or bytes per amount of time.

In contrast, in computer networking, the term *bandwidth* refers to the data rate (bps) that the network or transmission link can deliver. We will examine this issue in detail in later chapters on multimedia networks.

Telephony uses μ-law (which may be written "u-law") encoding, or A-law in Europe. The other formats use linear quantization. Using the μ-law rule shown in Eq. (6.9), the dynamic range—the ratio of highest to lowest nonzero value, expressed in dB for the value 2^n for an n-bit system, or simply stated as the number of bits—of digital telephone signals is effectively improved from 8 bits to 12 or 13.

Table 6.2 Bitrate and bandwidth in sample audio applications

Quality	Sampling rate (kHz)	Bits per sample	Mono/Stereo	Bitrate (if uncompressed) (kB/s)	Signal bandwidth (Hz)
Telephone	8	8	Mono	8	200–3,400
AM radio	11.025	8	Mono	11.0	100–5,500
FM radio	22.05	16	Stereo	88.2	20–11,000
CD	44.1	16	Stereo	176.4	5–20,000
DVD audio	192 (max)	24 (max)	Up to 6 channels	1,200.0 (max)	0–96,000 (max)

The standard sampling frequencies used in audio are 5.0125, 11.025, 22.05, and 44.1 kHz, with some exceptions, and these frequencies are supported by most sound cards.

Sometimes it is useful to remember the kinds of data rates in Table 6.2 in terms of bytes per minute. For example, the uncompressed digital audio signal for CD-quality stereo sound is 10.6 megabytes per minute—roughly 10 megabytes—per minute.

6.1.9 Synthetic Sounds

Digitized sound must still be converted to analog, for us to hear it. There are two fundamentally different approaches to handle stored sampled audio. The first is termed *FM*, for *frequency modulation*. The second is called *Wave Table*, or just *Wave*, sound.

In the first approach, a carrier sinusoid is changed by adding another term involving a second, modulating frequency. A more interesting sound is created by changing the argument of the main cosine term, putting the second cosine inside the argument itself—then we have a cosine of a cosine. A time-varying amplitude "envelope" function multiplies the whole signal, and another time-varying function multiplies the inner cosine, to account for overtones. Adding a couple of extra constants, the resulting function is complex indeed.

For example, Fig. 6.8a shows the function $\cos(2\pi t)$, and Fig. 6.8b is another sinusoid at twice the frequency. A cosine of a cosine is the more interesting function Fig. 6.8c, and finally, with carrier (angular) frequency 2π and modulating (angular) frequency 4π, we have the much more interesting curve Fig. 6.8d. Obviously, once we consider a more complex signal, such as the following [3]:

$$x(t) = A(t) \cos[\omega_c t + I(t) \cos(\omega_m t + \phi_m) + \phi_c] \tag{6.12}$$

we can create a most complicated signal.

This FM synthesis equation states that we make a signal using a basic carrier frequency ω_c and also use an additional, modulating frequency ω_m. In Fig. 6.8d,

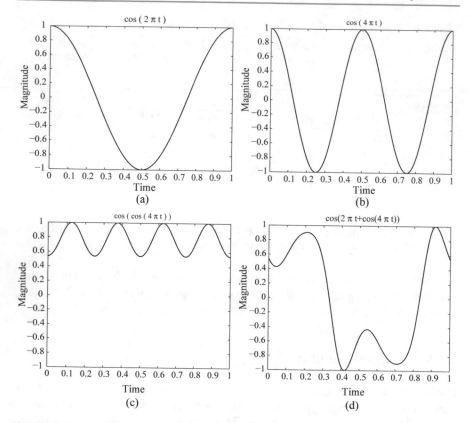

Fig. 6.8 Frequency modulation: **a** A single frequency; **b** twice the frequency; **c** usually, FM is carried out using a sinusoid argument to a sinusoid; **d** a more complex form arises from a carrier frequency 2π and a modulating frequency 4π cosine inside the sinusoid

these values are $\omega_c = 2\pi$ and $\omega_m = 4\pi$. The *phase* constants ϕ_m and ϕ_c create time-shifts for a more interesting sound. The time-dependent function $A(t)$ is called the *envelope*—it specifies overall loudness over time and is used to fade in and fade out the sound. A guitar string has an *attack* period, then a *decay* period, a *sustain* period, and finally a *release* period. This is shown in Fig. 6.10.

Finally, the time-dependent function $I(t)$ is used to produce a feeling of *harmonics* ("overtones") by changing the amount of modulation frequency heard. When $I(t)$ is small, we hear mainly low frequencies, and when $I(t)$ is larger, we hear higher frequencies as well. FM synthesis is used in low-end sound cards, but is also provided in many sound cards to provide backward compatibility.

A more accurate way of generating sounds from digital signals is called *wave-table synthesis*. In this technique, digital samples are stored sounds from real instruments. Since wave tables are stored in memory on the sound card, they can be manipulated by software so that sounds can be combined, edited, and enhanced. Sound reproduction is a good deal better with wave tables than with FM synthesis. To save memory space, a variety of special techniques, such as sample looping, pitch shifting, mathematical interpolation, and polyphonic digital filtering, can be applied [4,5].

For example, it is useful to be able to change the key—suppose a song is a bit too high for your voice. A wave table can be mathematically shifted so that it produces lower-pitched sounds. However, this kind of extrapolation can be used only just so far without sounding wrong. Wave tables often include sampling at various notes of the instrument, so that a key change need not be stretched too far. Wave-table synthesis is more expensive than FM synthesis, partly because the data storage needed is much larger. On the other hand, storage has become much less expensive, and it is possible to compress wave-table data, but, nonetheless, there are clearly simple tricks that one can accomplish easily using the compact formulation of FM synthesis, whereas making changes from a particular wave table is a good deal more complex. Nevertheless, with the advent of cheap storage, wave data has become generally used, including in ring tones.

6.2 MIDI: Musical Instrument Digital Interface

Wave-table files provide an accurate rendering of real instrument sounds but are quite large. For simple music, we might be satisfied with FM synthesis versions of audio signals that could easily be generated by a sound card. Essentially every computer is equipped with a sound card; a sound card is capable of manipulating and outputting sounds through speakers connected to the board, recording sound input from a microphone or line-in connection to the computer, and manipulating sound stored in memory.

If we are willing to be satisfied with the sound card's defaults for many of the sounds we wish to include in a multimedia project, we can use a simple scripting language and hardware setup called MIDI.

6.2.1 MIDI Overview

MIDI, which dates from the early 1980s, is an acronym that stands for *musical instrument digital interface*. It forms a protocol adopted by the electronic music industry that enables computers, synthesizers, keyboards, and other musical devices to communicate with each other. A synthesizer produces synthetic music and is included on sound cards, using one of the two methods discussed above. The MIDI standard is supported by most synthesizers, so sounds created on one can be played and manipulated on another and sound reasonably close. Computers must have a special MIDI interface, but this is incorporated into most sound cards. The sound card must also have both DA and AD converters.

MIDI is a scripting language—it codes "events" that stand for the production of certain sounds. Therefore, MIDI files are generally very small. For example, a MIDI event might include values for the pitch of a single note, its volume, and what instrument sound to play.

Role of MIDI. MIDI makes music notes (among other capabilities), so is useful for inventing, editing, and exchanging musical ideas that can be encapsulated as notes. This is quite a different idea than sampling, where the specifics of actual sounds are captured. Instead, MIDI is aimed at music, which can then be altered as the "user" wishes. Since MIDI is intimately related to music composition (music notation) programs, MIDI is a very useful vehicle for music education.

One strong capability of MIDI-based musical communication is the availability of a single MIDI instrument to control other MIDI instruments, allowing a master-slave relationship: the other MIDI instruments must play the same music, in part, as the master instrument, thus allowing interesting music. MIDI instruments may include excellent, or poor, musical capabilities. For example, suppose a keyboard mimics a traditional instrument well, but generates poor actual sound with a built-in synthesizer. Then "daisy-chaining" to a different synthesizer, one that generates excellent sound, may make an overall good combination. Since MIDI comes with a built-in timecode, the master's clock can be used to synchronize all the slave timecodes, making for more exact synchronization.

A so-called "sequencer-sampler" can be used to reorder and manipulate sets of digital-audio samples *and/or* sequences of MIDI. In a Digital Audio Workstation, running ProTools, for example, multitrack recording is possible, either sequentially or concurrently. For example, one could simultaneously record eight vocal tracks and eight instrument tracks.

MIDI Concepts

- Music is organized into *tracks* in a sequencer. Each track can be turned on or off on recording or playing back. Usually, a particular instrument is associated with a MIDI *channel*. MIDI channels are used to separate messages. There are 16 channels, numbered from 1 to 16. The idea is that each channel is associated with a particular instrument—for example, Channel 1 is the piano, Channel 10 is the drums. Nevertheless, you can switch instruments midstream, if desired, and associate another instrument with any channel.

- Along with *channel messages* (which include a channel number), several other types of messages are sent, such as a general message for all instruments indicating a change in tuning or timing; these are called *system messages*. It is also possible to send a special message to an instrument's channel that allows sending many notes without a channel specified. We will describe these messages in detail later.

- The way a synthetic musical instrument responds to a MIDI message is usually by simply ignoring any "play sound" message that is not for its channel. If several messages are for its channel, say several simultaneous notes being played on a piano, then the instrument responds, provided it is *multi-voice*—that is, can play more than a single note at once.

MIDI Terminology

- A *synthesizer* was, and still can be, a stand-alone sound generator that can vary pitch, loudness, and tone color. (The pitch is the musical note the instrument plays—a C, as opposed to a G, say. Whereas frequency in Hz is an absolute musical sound, pitch is relative, e.g., tuning your guitar to itself may sound fine but not have the same absolute notes as another guitar.) It can also change additional music characteristics, such as attack and delay time. A good (musician's) synthesizer often has a microprocessor, keyboard, control panels, memory, and so on. However, inexpensive synthesizers are also included on PC sound cards. Units that generate sound are referred to as *tone modules* or sound modules.
- A *sequencer* started off as a special hardware device for storing and editing a *sequence* of musical events, in the form of MIDI data. Now it is more often a software *music editor* on the computer.
- A *MIDI keyboard* produces no sound, instead generating sequences of MIDI instructions, called *MIDI messages* (but can also include a synthesizer for generating sound). MIDI messages are rather like assembler code and usually consist of just a few bytes. Stored as a sequence of MIDI messages, you might have 3 min of music, say, stored in only 3 kB. In comparison, a wave-table file (WAV) stores 1 minute of music in about 10 MB. In MIDI parlance, the keyboard is referred to as a *keyboard controller*.
- It is easy to confuse the term *voice* with the term *timbre*. The latter is MIDI terminology for just what instrument we are trying to emulate—for example, a piano as opposed to a violin. It is the quality of the sound. An instrument (or sound card) that is *multi-timbral* is capable of playing many different sounds at the same time, (e.g., piano, brass, drums).
- On the other hand, the term "voice," while sometimes used by musicians to mean the same thing as timbre, is used in MIDI to mean every different timbre and pitch that the tone module can produce at the same time. Synthesizers can have many (typically 16, 32, 64, 256, etc.) voices. Each voice works independently and simultaneously to produce sounds of different timbre and pitch.
- The term *polyphony* refers to the number of voices that can be produced at the same time. So a typical tone module may be able to produce "64 voices of polyphony" (64 different notes at once) and be "16-part multi-timbral" (can produce sounds like 16 different instruments at once).

MIDI Specifics

How different timbres are produced digitally is by using a *patch*, which is the set of control settings that define a particular timbre. Patches are often organized into databases, called *banks*. For true aficionados, software patch editors are available.

A standard mapping specifying just what instruments (patches) will be associated with what channels has been agreed on and is called *General MIDI*. In General

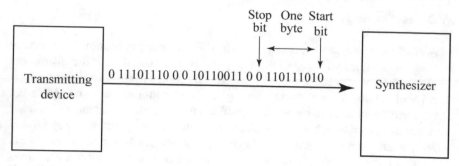

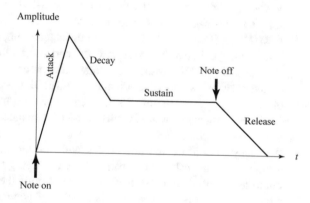

Fig. 6.9 Stream of 10-bit bytes; for typical MIDI messages, these consist of {status byte, data byte, data byte} = {Note On, Note Number, Note Velocity}

Fig. 6.10 Stages of amplitude versus time for a music note

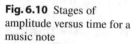

MIDI, there are 128 patches associated with standard instruments, and Channel 10 is reserved for percussion instruments.

For most instruments, a typical message might be Note On (meaning, e.g., a keypress), consisting of what channel, what pitch, and what *velocity* (i.e., volume). For percussion instruments, the pitch data means which kind of drum. A Note On message thus consists of a *status* byte—which channel, what pitch—followed by two data bytes. It is followed by a Note Off message (key release), which also has a pitch (which note to turn off) and —for consistency, one supposes—a velocity (often set to zero and ignored).

The data in a MIDI status byte is between 128 and 255; each of the data bytes is between 0 and 127. Actual MIDI bytes are 8 bit, plus a 0 start and stop bit, making them 10-bit "bytes." Figure 6.9 shows the MIDI data stream.

A MIDI device often is capable of *programmability*, which means it has filters available for changing the bass and treble response and can also change the "envelope" describing how the amplitude of a sound changes over time. Figure 6.10 shows a model of a digital instrument's response to Note On/Note Off messages.

MIDI sequencers (editors) allow you to work with standard music notation or get right into the data, if desired. MIDI files can also store wave-table data. The advantage of wave-table data (WAV files) is that it much more precisely stores the exact sound of an instrument. A sampler is used to sample the audio data—for example, a "drum machine" always stores wave-table data of real drums. So one could have a music editor using MIDI on one track plus digital audio such as vocals, say, on another track.

Of interest to computer science or engineering students is the fact that MIDI provides a full messaging protocol that can be used for whatever one likes, for example, for controlling lights in a theater. We shall see below that there is a MIDI message for sending any kind and any number of bytes, which can then be used as the programmer desires.

Sequencers employ several techniques for producing more music from what is actually available. For example, looping over (repeating) a few bars can be more or less convincing. Volume can be easily controlled over time—this is called *time-varying amplitude modulation*. More interestingly, sequencers can also accomplish time compression or expansion with no pitch change.

While it is possible to change the pitch of a sampled instrument, if the key change is large, the resulting sound begins to sound displeasing. For this reason, samplers employ *multisampling*. A sound is recorded using several band-pass filters, and the resulting recordings are assigned to different keyboard keys. This makes frequency shifting for a change of key more reliable, since less shift is involved for each note.

MIDI machine control (MMC) is a subset of the MIDI specification that can be used for controlling recording equipment, e.g., multitrack recorders. The most common use of this facility is to effectively press "Play" on a remote device. An example usage would be if a MIDI device has a poor timer, then the master could activate it at just the right time. In general, MIDI can be used for controlling and synchronizing musical instrument synthesizers and recording equipment, and even control lighting.

6.2.2   Hardware Aspects of MIDI

The MIDI hardware setup consists of a 31.25 kbps (kilobits per second) serial connection, with the 10-bit bytes including a 0 start and stop bit. Usually, MIDI-capable units are either input devices or output devices, not both.

Figure 6.11 shows a traditional synthesizer. The modulation wheel adds vibrato. Pitch bend alters the frequency, much like pulling a guitar string over slightly. There are often other controls, such as foots pedals, sliders, and so on.

The physical MIDI ports consist of 5-pin connectors labeled IN and OUT and there can also be a third connector, THRU. This last data channel simply copies data entering the IN channel. MIDI communication is half-duplex. MIDI IN is the connector via which the device receives all MIDI data. MIDI OUT is the connector through which the device transmits all the MIDI data it generates itself. MIDI THRU is the connector by which the device echoes the data it receives from MIDI IN (and only that—all the data generated by the device itself is sent via MIDI OUT). These

Keyboard

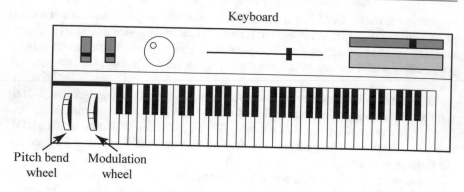

Pitch bend   Modulation
wheel   wheel

Fig. 6.11 A MIDI synthesizer

Fig. 6.12 A typical MIDI setup

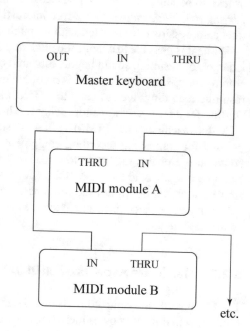

ports are on the sound card or interface externally, either on a separate card or using a special interface to a serial port.

Figure 6.12 shows a typical MIDI sequencer setup. Here, the MIDI OUT of the keyboard is connected to the MIDI IN of a synthesizer and then THRU to each of the additional sound modules. During recording, a keyboard-equipped synthesizer sends MIDI messages to a sequencer, which records them. During playback, messages are sent from the sequencer to all the sound modules and the synthesizer, which play the music.

MIDI Message Transmission

The 31.25 kbps data rate is actually quite restrictive. To initiate playing a note, a 3-byte message is sent (with bytes equal to 10 bits). If my hands are playing chords with all ten fingers, then a single crashing chord will take ten notes at 30 bits each, requiring transmission of 300 bits. At 31.25 kbps transmission of this chord will take about 0.01 s, at a speed of about 0.001 s per note—and all this not counting the additional messages we'll have to send to turn *off* these ten notes. Moreover, using the pitch bend and modulation wheels in a synthesizer could generate many messages as well, all taking time to transmit. Hence, there could be an audible time lag generated by the slow bit transmission rate.

A trick used to tackle this problem is called *Running Status*: MIDI allows sending just the data, provided the command from the previous message has not changed. So instead of 3 bytes—command, data, data—MIDI would use just 2 bytes for the next message having the same command.

6.2.3   Structure of MIDI Messages

MIDI messages can be classified into two types, as in Fig. 6.13—channel messages and system messages—and further classified as shown. Each type of message will be examined below.

Channel Messages. A channel message can have up to 3 bytes; the first is the status byte (the opcode, as it were), and has its most significant bit set to 1. The four least significant bits (LSBs) identify which of the 16 possible channels this message belongs to,[2] with the three remaining bits holding the message. For a data byte, the most significant bit is set to zero.

Voice Messages. This type of channel message controls a voice—that is, sends information specifying which note to play or to turn off—and encodes key pressure. Voice messages are also used to specify controller effects, such as sustain, vibrato, tremolo, and the pitch wheel. Table 6.3 lists these operations.

For *Note On* and *Note Off* messages, the *velocity* is how quickly the key is played. Typically, a synthesizer responds to a higher velocity by making the note louder or brighter. Note On makes a note occur, and the synthesizer also attempts to make the note sound like the real instrument while the note is playing. *Pressure* messages can be used to alter the sound of notes while they are playing. The *Channel Pressure* message is a force measure for the keys on a specific channel (instrument) and has an identical effect on all notes playing on that channel. The other pressure message, *Polyphonic Key Pressure* (also called *Key Pressure*), specifies how much volume keys

---

[2]The hexadecimal numbers derived from the four LSBs range from 0 to 15. In practice, musicians and software refer to the MIDI channels from 1 to 16, so there is a difference of 1 when coding them in hexadecimal, e.g., Channel 1 is coded "0," and Channel 16 is coded "F."

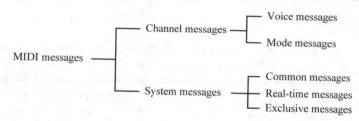

Fig. 6.13 MIDI message taxonomy

Table 6.3 MIDI voice messages

Voice message	Status byte	Data byte1	Data byte2
Note off	&H8n	Key number	Note off velocity
Note on	&H9n	Key number	Note on velocity
Polyphonic key pressure	&HAn	Key number	Amount
Control change	&HBn	Controller number	Controller value
Program change	&HCn	Program number	None
Channel pressure	&HDn	Pressure value	None
Pitch Bend	&HEn	MSB	LSB

&H indicates hexadecimal, and *n* in the Status byte hex value stands for a channel number. All values are in 0..127 except controller number, which is in 0..120

played together are to have and can be different for each note in a chord. Pressure is also called *aftertouch*.

The control change instruction sets various controllers (faders, vibrato, etc.). Each manufacturer may make use of different controller numbers for different tasks. However, controller 1 is likely the modulation wheel (for vibrato).

For example, a Note On message is followed by 2 bytes, one to identify the note and one to specify the velocity. Therefore, to play note number 80 with maximum velocity on Channel 13, the MIDI device would send the following three hex byte values: &H9C &H50 &H7F. (As explained earlier, "&HC" refers to Channel 13, not 12.) Notes are numbered such that middle C has number 60.

To play two notes simultaneously (effectively), first, we would send a Program Change message for each of two channels. Recall that Program Change means to load a particular patch for that channel. So far, we have attached two timbres to two different channels. Then sending two Note On messages (in serial) would turn on both channels. Alternatively, we could also send a Note On message for a particular channel and then another Note On message, with another pitch, before sending the Note Off message for the first note. Then we would be playing two notes effectively at the same time on the same instrument.

Recall that the Running Status method allows one to send one status byte, e.g., a Note On message, followed by a stream of data bytes that all are associated with the same status byte. For instance, a Note On message on Channel 1, "&H90," could

Table 6.4 MIDI mode messages

First data byte	Description	Meaning of second data byte
&H79	Reset all controllers	None; set to 0
&H7A	Local control	0 = off; 127 = on
&H7B	All notes off	None; set to 0
&H7C	Omni mode off	None; set to 0
&H7D	Omni mode on	None; set to 0
&H7E	Mono mode on (Poly mode off)	Controller number
&H7F	Poly mode on (Mono mode off)	None; set to 0

be followed by two data bytes as indicated in Table 6.3. But with Running Status we need not send another Note On but instead simply keep sending data byte pairs for the next stream of Note On data. As well, in fact, Running Status has another trick: if the velocity data byte is 0 then that Note On message is interpreted as a Note Off. Hence, one can send a single "&H90" followed by numerous Note On and Note Off datasets. So, for example, a Note On, Note Off pair for playing middle C on Channel 1 could be sent as &H90 &H3C &H7F; &H3C &H00 (middle C is note number 60 = "&H3C").

Polyphonic Pressure refers to how much force simultaneous notes have on several instruments. Channel Pressure refers to how much force a single note has on one instrument.

Channel Mode Messages. Channel mode messages form a special case of the Control Change message, and therefore all mode messages have opcode B (so the message is "&HBn," or 1011nnnn). However, a Channel Mode message has its first data byte in 121 through 127 (&H79–7F).

Channel mode messages determine how an instrument processes MIDI voice messages. Some examples include respond to all messages, respond just to the correct channel, don't respond at all, or go over to local control of the instrument.

Recall that the status byte is "&HBn," where *n* is the channel. The data bytes have meanings as shown in Table 6.4. *Local Control Off* means that the keyboard should be disconnected from the synthesizer (and another, external, device will be used to control the sound). *All Notes Off* is a handy command, especially if, as sometimes happens, a bug arises such that a note is left playing inadvertently. *Omni* means that devices respond to messages from all channels. The usual mode is OMNI OFF—pay attention to your own messages only, and do not respond to every message regardless of what channel it is on. *Poly* means a device will play back several notes at once if requested to do so. The usual mode is POLY ON.

In POLY OFF—monophonic mode—the argument that represents the number of monophonic channels can have a value of zero, in which case it defaults to the number of voices the receiver can play; or it may set to a specific number of channels. However, the exact meaning of the combination of OMNI ON/OFF and *Mono/Poly*

Table 6.5 MIDI system common messages

System common message	Status byte	Number of data bytes
MIDI timing code	&HF1	1
Song position pointer	&HF2	2
Song select	&HF3	1
Tune request	&HF6	None
EOX (terminator)	&HF7	None

Table 6.6 MIDI system real-time messages

System real-time message	Status byte
Timing clock	&HF8
Start sequence	&HFA
Continue sequence	&HFB
Stop sequence	&HFC
Active sensing	&HFE
System reset	&HFF

depends on the specific combination, with four possibilities. Suffice it to say that the usual combination is OMNI OFF, POLY ON.

System Messages. System messages have no channel number and are meant for commands that are not channel specific, such as timing signals for synchronization, positioning information in prerecorded MIDI sequences, and detailed setup information for the destination device. Opcodes for all system messages start with "&HF." System messages are divided into three classifications, according to their use.

System Common Messages. Table 6.5 sets out these messages, which relate to timing or positioning. Song position is measured in beats. The messages determine what is to be played upon receipt of a "start" real-time message (see below).

System Real-Time Messages. Table 6.6 sets out system real-time messages, which are related to synchronization.

System Exclusive Message. The final type of system message, *System Exclusive* messages, is included so that manufacturers can extend the MIDI standard. After the initial code "&HF0," they can insert a stream of any specific messages that apply to their own product. A System Exclusive message is supposed to be terminated by a terminator byte "&HF7," as specified in Table 6.5. However, the terminator is optional, and the data stream may simply be ended by sending the status byte of the next message.

6.2.4 MIDI-to-WAV Conversion

Some programs, such as early versions of Adobe Premiere, cannot include MIDI files—instead, they insist on WAV format files. Various shareware programs can approximate a reasonable conversion between these formats. The programs essentially consist of large lookup files that try to do a reasonable job of substituting predefined or shifted WAV output for some MIDI messages, with inconsistent success.

6.2.5 General MIDI

For MIDI music to sound more or less the same on every machine, we would at least like to have the same patch numbers associated with the same instruments, for example, patch 1 should always be a piano, not a flugelhorn. To this end, General MIDI [5] is a scheme for assigning instruments to patch numbers. A standard percussion map also specifies 47 percussion sounds, where a "note" appearing on a musical score determines just what percussion element is being struck. This book's website includes both the General MIDI Instrument Path Map and the Percussion Key map.

Other requirements for General MIDI compatibility are that a MIDI device must support all 16 channels; must be multi-timbral (i.e., each channel can play a different instrument/program); must be polyphonic (i.e., each channel is able to play many voices); and must have a minimum of 24 dynamically allocated voices.

General MIDI Level2

An extended General MIDI, GM-2, was defined in 1999 and updated in 2003, with a standard SMF *Standard MIDI File* format defined. A nice extension is the inclusion of extra character information, such as karaoke lyrics, which can be displayed on a good sequencer.

6.2.6 MIDI 2.0

After 35 years of MIDI 1.0, a significant update in the MIDI standard has evolved from General Level2, entitled MIDI 2.0. The principles of this new standard were set out in January of 2020 as a collaboration of Google, Microsoft, Yamaha, and others, under the auspices of The MIDI Association [6].

Some of the proposed changes are as follows:

- The old and slow 5-pin MIDI connector/cable will be replaced by USB.
- The total number of channels will be increased from 16 to 256, organized in 16 groups of 16 channels.

- While the number of pitch values remains at 128, the possible velocity values will be increased from 128 to 65,536, i.e., from using 7 to 16 bits.
- MIDI 2.0 messages are no longer restricted to the old 2- or 3-byte messages with status and data bytes. They will be sent in Universal MIDI Packets, with a length of 32, 64, 96, or 128 bits depending on the type of the message.
- MIDI-CI (MIDI Capability Inquiry) is created to take care of protocol negotiation and backward compatibility with MIDI 1.0.

6.3 Quantization and Transmission of Audio

To be transmitted, sampled audio information must be digitized, and here we look at some of the details of this process. Once the information has been quantized, it can then be transmitted or stored. We go through a few examples in complete detail, which helps in understanding what is being discussed.

6.3.1 Coding of Audio

Quantization and transformation of data are collectively known as *coding* of the data. For audio, the μ-law technique for companding audio signals is usually combined with a simple algorithm that exploits the temporal redundancy present in audio signals. Differences in signals between the present and a previous time can effectively reduce the size of signal values and, most important, concentrate the histogram of pixel values (differences, now) into a much smaller range. The result of reducing the variance of values is that lossless compression methods that produce a bitstream with shorter bit lengths for more likely values, introduced in Chap. 7, fare much better and produce a greatly compressed bitstream.

In general, producing quantized sampled output for audio is called *pulse code modulation* or *PCM*. The differences version is called *DPCM* (and a crude but efficient variant is called *DM*). The adaptive version is called *ADPCM*, and variants that take into account speech properties follow from these. More complex models for audio are outlined in Chap. 13.

6.3.2 Pulse Code Modulation

PCM in General

Audio is analog—the waves we hear travel through the air to reach our eardrums. We know that the basic techniques for creating digital signals from analog ones consist of *sampling* and *quantization*. Sampling is invariably done uniformly—we select a sampling rate and produce one value for each sampling time.

In the magnitude direction, we digitize by quantization, selecting breakpoints in magnitude and remapping any value within an interval to one representative output level. The set of interval boundaries is sometimes called *decision boundaries*, and the representative values are called *reconstruction levels*.

We say that the boundaries for quantizer input intervals that will all be mapped into the same output level form a *coder mapping*, and the representative values that are the output values from a quantizer are a *decoder mapping*. Since we quantize, we may choose to create either an accurate or less accurate representation of sound magnitude values. Finally, we may wish to *compress* the data, by assigning a bitstream that uses fewer bits for the most prevalent signal values.

Every compression scheme has three stages:

1. **Transformation**. The input data is *transformed* to a new representation that is easier or more efficient to compress. For example, in predictive coding (discussed later in the chapter), we predict the next signal from previous ones and transmit the prediction error.
2. **Loss**. We may introduce *loss* of information. Quantization is the main lossy step. Here we use a limited number of reconstruction levels, fewer than in the original signal. Therefore, quantization necessitates some loss of information.
3. **Coding**. Here, we assign a *codeword* (thus forming a binary bitstream) to each output level or symbol. This could be a fixed-length code or a variable-length code, such as Huffman coding (discussed in Chap. 7).

For audio signals, we first consider PCM, the digitization method that enables us to consider lossless predictive coding as well as the DPCM scheme; these methods use *differential coding*. We also look at the adaptive version, ADPCM, which is meant to provide better compression.

Pulse code modulation is a formal term for the sampling and quantization we have already been using. *Pulse* comes from an engineer's point of view that the resulting digital signals can be thought of as infinitely narrow vertical "pulses." As an example of PCM, audio samples on a CD are sampled at a rate of 44.1 kHz, with 16 bits per sample. For stereo sound, with two channels, this amounts to a data rate of about 1,400 kbps.

PCM in Speech Compression

Recall that in Sect. 6.1.6 we considered *companding* the so-called compressor and expander stages for speech signal processing, for telephony. For this application, signals are first transformed using the μ-law (or A-law for Europe) rule into what is essentially a logarithmic scale. Only then is PCM, using uniform quantization, applied. The result is that finer increments in sound volume are used at the low-volume end of speech rather than at the high-volume end, where we can't discern small changes in any event.

Assuming a bandwidth for speech from about 50 Hz to about 10 kHz, the Nyquist rate would dictate a sampling rate of 20 kHz. Using uniform quantization without companding, the minimum sample size we could get away with would likely be about 12 bits. Hence, for mono speech transmission the bitrate would be 240 kbps. With companding, we can safely reduce the sample size to 8 bits with the same perceived level of quality and thus reduce the bitrate to 160 kbps. However, the standard approach to telephony assumes that the highest frequency audio signal we want to reproduce is about 4 kHz. Therefore, the sampling rate is only 8 kHz, and the companded bitrate thus reduces to only 64 kbps.

We must also address two small wrinkles to get this comparatively simple form of speech compression right. First because only sounds up to 4 kHz are to be considered, all other frequency contents must be noise. Therefore, we should remove this high-frequency content from the analog input signal. This is done using a band-limiting filter that blocks out high frequencies as well as very low ones. The "band" of not-removed ("passed") frequencies are what we wish to keep. This type of filter is therefore also called a band-pass filter.

Second, once we arrive at a pulse signal, such as the one in Fig. 6.14a, we must still perform digital-to-analog conversion and then construct an output analog signal. But the signal we arrive at is effectively the staircase shown in Fig. 6.14b. This type of discontinuous signal contains not just frequency components due to the original signal but, because of the sharp corners, also a theoretically infinite set of higher frequency components (from the theory of Fourier analysis, in signal processing). We know these higher frequencies are extraneous. Therefore, the output of the digital-to-analog converter is in turn passed to a *low-pass filter*, which allows only frequencies up to the original maximum to be retained. Figure 6.15 shows the complete scheme for encoding and decoding telephony signals as a schematic. As a result of the low-pass filtering, the output becomes smoothed, as Fig. 6.14c shows. For simplicity, Fig. 6.14 does not show the effect of companding.

A-law or μ-law PCM coding is used in the older International Telegraph and Telephone Consultative Committee (CCITT) standard G.711, for digital telephony. This CCITT standard is now subsumed into standards promulgated by a newer organization, the International Telecommunication Union (ITU).

6.3.3   Differential Coding of Audio

Audio is often stored not in simple PCM but in a form that exploits differences. For a start, differences will generally be smaller numbers and hence offer the possibility of using fewer bits to store.

An advantage of forming differences is that the histogram of a difference signal is usually considerably more peaked than the histogram for the original signal. For example, as an extreme case, the histogram for a linear ramp signal that has constant slope is uniform, whereas the histogram for the derivative of the signal (i.e., the

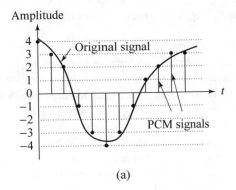

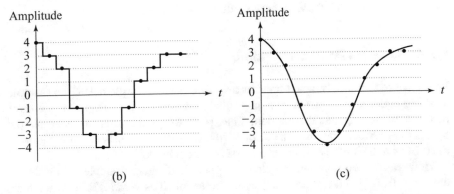

Fig. 6.14 Pulse code modulation (PCM): **a** Original analog signal and its corresponding PCM signals; **b** decoded staircase signal; **c** reconstructed signal after low-pass filtering

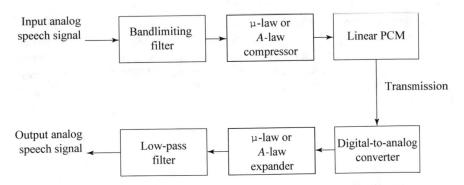

Fig. 6.15 PCM signal encoding and decoding

differences, from sampling point to sampling point) consists of a spike at the slope value.

Generally, if a time-dependent signal has some consistency over time (*temporal redundancy*), the difference signal—subtracting the current sample from the previous one—will have a more peaked histogram, with a maximum around zero. Consequently, if we then go on to assign bitstring codewords to differences, we can assign short codes to prevalent values and long codewords to rarely occurring ones.

To begin with, consider a lossless version of this scheme. Loss arises when we quantize. If we apply no quantization, we can still have compression—via the decrease in the variance of values that occurs in differences, compared to the original signal. Chapter 7 introduces more sophisticated versions of lossless compression methods, but it helps to see a simple version here as well. With quantization, Predictive Coding becomes DPCM, a lossy method; we'll also try out that scheme.

6.3.4 Lossless Predictive Coding

Predictive coding simply means transmitting differences—we predict the next sample as being equal to the current sample and send not the sample itself but the error involved in making this assumption. That is, if we predict that the next sample equals the previous one, then the error is just the difference between previous and next. Our prediction scheme could also be more complex.

However, we do note one problem. Suppose our integer sample values are in the range $0..255$. Then differences could be as much as $-255..255$. So we have unfortunately increased our *dynamic range* (ratio of maximum to minimum) by a factor of two: we may well need more bits than we needed before to transmit some differences. Fortunately, we can use a trick to get around this problem, as we shall see.

So, basically, predictive coding consists of finding differences and transmitting them, using a PCM system such as the one introduced in Sect. 6.3.2. First, note that differences of integers will at least be integers. Let's formalize our statement of what we are doing by defining the integer signal as the set of values f_n. Then we *predict* values $\hat{f}_n$ as simply the previous value, and we define the error e_n as the difference between the actual and predicted signals:

$$\hat{f}_n = f_{n-1}$$
$$e_n = f_n - \hat{f}_n \qquad (6.13)$$

We certainly would like our error value e_n to be as small as possible. Therefore, we would wish our prediction $\hat{f}_n$ to be as close as possible to the actual signal f_n. But for a particular sequence of signal values, some *function* of a few of the previous values, f_{n-1}, f_{n-2}, f_{n-3}, etc., may provide a better prediction of f_n. Typically, a linear *predictor* function is used:

$$\hat{f}_n = \sum_{k=1}^{2 \text{ to } 4} a_{n-k} f_{n-k} \qquad (6.14)$$

Such a predictor can be followed by a truncating or rounding operation to result in integer values. In fact, since now we have such coefficients a_{n-k} available, we can even change them adaptively (see Sect. 6.3.7).

The idea of forming differences is to make the histogram of sample values more peaked. For example, Fig. 6.16a plots 1 s of sampled speech at 8 kHz, with magnitude resolution of 8 bits per sample.

A histogram of these values is centered around zero, as in Fig. 6.16b. Figure 6.16c shows the histogram for corresponding speech signal *differences*: difference values are much more clustered around zero than are sample values themselves. As a result, a method that assigns short codewords to frequently occurring symbols will assign a short code to *zero* and do rather well. Such a coding scheme will much more efficiently code sample differences than samples themselves, and a similar statement applies if we use a more sophisticated predictor than simply the previous signal value.

However, we are still left with the problem of what to do if, for some reason, a particular set of difference values does indeed consist of some exceptional large differences. A clever solution to this difficulty involves defining two new codes to add to our list of difference values, denoted SU and SD, standing for Shift-Up and Shift-Down. Some special values will be reserved for them.

Suppose samples are in the range 0..255, and differences are in −255..255. Define SU and SD as shifts by 32. Then we could, in fact, produce codewords for a limited set of signal differences, say only the range −15..16. Differences (that inherently are in the range −255..255) lying in the limited range can be coded as is, but if we add the extra two values for SU, SD, a value outside the range −15..16 can be transmitted as a series of shifts, followed by a value that is indeed inside the range −15..16. For example, 100 is transmitted as SU, SU, SU, 4, where (the codes for) SU and for 4 are what are sent.

Lossless predictive coding is ... lossless! That is, the decoder produces the same signals as the original. It is helpful to consider an explicit scheme for such coding considerations, so let's do that here (we won't use the most complicated scheme, but we'll try to carry out an entire calculation). As a simple example, suppose we devise a predictor for $\hat{f}_n$ as follows:

$$\hat{f}_n = \lfloor \frac{1}{2}(f_{n-1} + f_{n-2}) \rfloor$$
$$e_n = f_n - \hat{f}_n \qquad (6.15)$$

Then the error e_n (or a codeword for it) is what is actually transmitted.

Let's consider an explicit example. Suppose we wish to code the sequence $f_1, f_2, f_3, f_4, f_5 = 21, 22, 27, 25, 22$. For the purposes of the predictor, we'll invent an

Fig. 6.16 Differencing
concentrates the histogram:
a Digital speech signal;
b histogram of digital speech
signal values; **c** histogram of
digital speech signal
differences

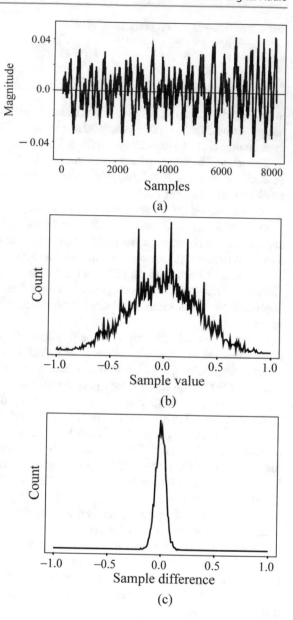

(a)

(b)

(c)

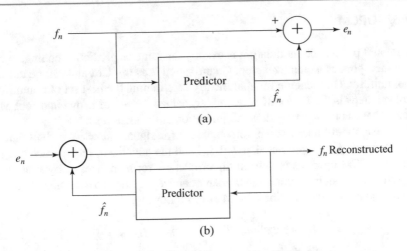

(a)

(b)

Fig. 6.17 Schematic diagram for predictive coding: **a** Encoder; **b** decoder

extra signal value f_0, equal to $f_1 = 21$, and first transmit this initial value, uncoded; after all, every coding scheme has the extra expense of some header information.

Then the first error, e_1, is zero, and subsequently

$$\hat{f}_2 = 21, \quad e_2 = 22 - 21 = 1$$

$$\hat{f}_3 = \lfloor \tfrac{1}{2}(f_2 + f_1) \rfloor = \lfloor \tfrac{1}{2}(22 + 21) \rfloor = 21$$

$$e_3 = 27 - 21 = 6$$

$$\hat{f}_4 = \lfloor \tfrac{1}{2}(f_3 + f_2) \rfloor = \lfloor \tfrac{1}{2}(27 + 22) \rfloor = 24$$

$$e_4 = 25 - 24 = 1$$

$$\hat{f}_5 = \lfloor \tfrac{1}{2}(f_4 + f_3) \rfloor = \lfloor \tfrac{1}{2}(25 + 27) \rfloor = 26$$

$$e_5 = 22 - 26 = -4 \tag{6.16}$$

The error does center around zero, we see, and coding (assigning bitstring codewords) will be efficient. Figure 6.17 shows a typical schematic diagram used to encapsulate this type of system. Notice that the predictor emits the predicted value $\hat{f}_n$. What is invariably (and annoyingly) left out of such schematics is the fact that the predictor is based on $f_{n-1}, f_{n-2}, \ldots$ Therefore, the predictor must involve a memory. At the least, the predictor includes a circuit for incorporating a delay in the signal, to store f_{n-1}.

6.3.5 DPCM

Differential pulse code modulation is exactly the same as predictive coding, except that it incorporates a quantizer step. Quantization is as in PCM and can be uniform or nonuniform. One scheme for analytically determining the best set of nonuniform quantizer steps is the *Lloyd–Max* quantizer, named for Stuart Lloyd and Joel Max, which is based on a least-squares minimization of the error term.

Here we should adopt some nomenclature for signal values. We shall call the original signal f_n, the predicted signal $\hat{f}_n$, and the quantized, reconstructed signal $\tilde{f}_n$. How DPCM operates is to form the prediction, form an error e_n by subtracting the prediction from the actual signal, then quantize the error to a quantized version, $\tilde{e}_n$. The equations that describe DPCM are as follows:

$$
\begin{aligned}
\hat{f}_n &= function\_of\ (\tilde{f}_{n-1}, \tilde{f}_{n-2}, \tilde{f}_{n-3}, \ldots) \\
e_n &= f_n - \hat{f}_n \\
\tilde{e}_n &= Q[e_n] \\
&\quad \text{transmit } codeword(\tilde{e}_n) \\
&\quad \text{reconstruct:} \tilde{f}_n = \hat{f}_n + \tilde{e}_n
\end{aligned}
\tag{6.17}
$$

Codewords for quantized error values $\tilde{e}_n$ are produced using entropy coding, such as Huffman coding (discussed in Chap. 7).

Notice that the predictor is always based on the reconstructed, quantized version of the signal: the reason for this is that then the encoder side is not using any information not available to the decoder side. Generally, if by mistake we made use of the *actual* signals f_n in the predictor instead of the reconstructed ones $\tilde{f}_n$, quantization error would tend to accumulate and could get worse rather than being centered on zero.

The main effect of the coder–decoder process is to produce reconstructed, quantized signal values $\tilde{f}_n = \hat{f}_n + \tilde{e}_n$. The "distortion" is the average squared error $[\sum_{n=1}^{N} (\tilde{f}_n - f_n)^2]/N$, and one often sees diagrams of distortion versus the number of bit levels used. A Lloyd–Max quantizer will do better (have less distortion) than a uniform quantizer.

For any signal, we want to choose the size of quantization steps so that they correspond to the range (the maximum and minimum) of the signal. Even using a uniform, equal-step quantization will naturally do better if we follow such a practice. For speech, we could modify quantization steps as we go, by estimating the mean and variance of a patch of signal values and shifting quantization steps accordingly, for every block of signal values. That is, starting at time i we could take a block of N values f_n and try to minimize the quantization error:

$$
\min \sum_{n=i}^{i+N-1} (f_n - Q[f_n])^2
\tag{6.18}
$$

Since signal *differences* are very peaked, we could model them using a Laplacian probability distribution function, which is also strongly peaked at zero [7]: it looks

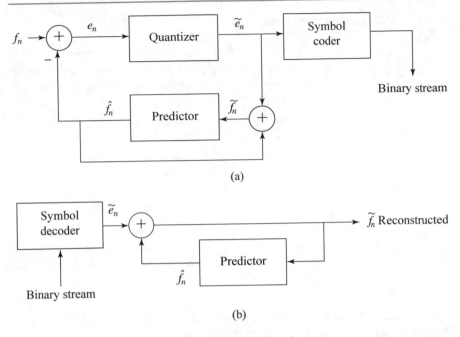

(a)

(b)

Fig. 6.18 Schematic diagram for DPCM: **a** Encoder; **b** decoder

like $l(x) = (1/\sqrt{2}\,\sigma)\, exp(-\sqrt{2}\,|x|/\sigma)$, for variance σ^2. So typically, we assign quantization steps for a quantizer with nonuniform steps by assuming that signal differences, d_n, say, are drawn from such a distribution and then choosing steps to minimize

$$\min \sum_{n=i}^{i+N-1} (d_n - Q[d_n])^2\, l(d_n) \tag{6.19}$$

This is a least-squares problem and can be solved iteratively using the Lloyd–Max quantizer.

Figure 6.18 shows a schematic diagram for the DPCM coder and decoder. As is common in such diagrams, several interesting features are more or less not indicated. First, we notice that the predictor makes use of the reconstructed, quantized signal values $\tilde{f}_n$, not actual signal values f_n, that is, the encoder simulates the decoder in the predictor path. The quantizer can be uniform or nonuniform.

The box labeled "Symbol coder" in the block diagram simply means a Huffman coder—the details of this step are set out in Chap. 7. The prediction value $\hat{f}_n$ is based on, however, much history the prediction scheme requires: we need to buffer previous values of $\tilde{f}$ to form the prediction. Notice that the quantization noise, $f_n - \tilde{f}_n$, is equal to the quantization effect on the error term, $e_n - \tilde{e}_n$.

Table 6.7 DPCM quantizer reconstruction levels

e_n in range	Quantized to value
$-255 .. -240$	-248
$-239 .. -224$	-232
$\vdots$	$\vdots$
$-31 .. -16$	-24
$-15 .. 0$	-8
$1 .. 16$	8
$17 .. 32$	24
$\vdots$	$\vdots$
$225 .. 240$	232
$241 .. 255$	248

It helps us explicitly understand the process of coding to look at actual numbers. Suppose we adopt a particular predictor as follows:

$$\hat{f}_n = \text{trunc}\left[\left(\tilde{f}_{n-1} + \tilde{f}_{n-2}\right)/2\right]$$

so that $e_n = f_n - \hat{f}_n$ is an integer   (6.20)

Let us use the particular quantization scheme

$$\tilde{e}_n = Q[e_n] = 16 * \text{trunc}\left[(255 + e_n)/16\right] - 256 + 8$$
$$\tilde{f}_n = \hat{f}_n + \tilde{e}_n$$

(6.21)

First, we note that the error is in the range $-255 .. 255$, that is, 511 levels are possible for the error term. The quantizer takes the simple course of dividing the error range into 32 patches of about 16 levels each. It also makes the representative reconstructed value for each patch equal to the midway point for each group of 16 levels.

Table 6.7 gives output values for any of the input codes: 4-bit codes are mapped to 32 reconstruction levels in a staircase fashion. (Notice that the final range includes only 15 levels, not 16.)

As an example stream of signal values, consider the set of values

$$f_1 \quad f_2 \quad f_3 \quad f_4 \quad f_5$$
$$130 \ \ 150 \ \ 140 \ \ 200 \ \ 230$$

We prepend extra values $f = 130$ in the data stream that replicate the first value, f_1, and initialize with quantized error $\tilde{e}_1 \equiv 0$, so that we ensure the first reconstructed value is exact: $\hat{f}_1 = 130$. Then subsequent values calculated are as follows (with prepended values in a box):

$$\hat{f} = \boxed{130}, \ 130, \ 142, \ 144, \ 167$$
$$e = \boxed{0}, \ \ 20, \ -2, \ \ 56, \ \ 63$$
$$\tilde{e} = \boxed{0}, \ \ 24, \ -8, \ \ 56, \ \ 56$$
$$\tilde{f} = \boxed{130}, \ 154, \ 134, \ 200, \ 223$$

On the decoder side, we again assume extra values $\tilde{f}$ equal to the correct value $\tilde{f}_1$, so that the first reconstructed value $\hat{f}_1$ is correct. What is received is $\tilde{e}_n$, and the reconstructed $\tilde{f}_n$ is identical to the one on the encoder side, provided we use exactly the same prediction rule.

6.3.6   DM

DM stands for *delta modulation*, a much simplified version of DPCM often used as a quick analog-to-digital converter. We include this scheme here for completeness.

Uniform-Delta DM

The idea in DM is to use only a *single* quantized error value, either positive or negative. Such a 1-bit coder thus produces coded output that follows the original signal in a staircase fashion. The relevant set of equations is as follows:

$$\hat{f}_n = \tilde{f}_{n-1}$$
$$e_n = f_n - \hat{f}_n = f_n - \tilde{f}_{n-1}$$
$$\tilde{e}_n = \begin{cases} +k \ \text{if } e_n > 0, \ \text{where } k \text{ is a constant} \\ -k \ otherwise, \end{cases} \tag{6.22}$$
$$\tilde{f}_n = \hat{f}_n + \tilde{e}_n$$

Note that the prediction simply involves a delay.

Again, let's consider actual numbers. Suppose signal values are as follows:

$$f_1 \ f_2 \ f_3 \ f_4$$
$$10 \ 11 \ 13 \ 15$$

We also define an exact reconstructed value $\tilde{f}_1 = f_1 = 10$.

Suppose we use a step value $k = 4$. Then we arrive at the following values:

$$\begin{aligned}
\hat{f}_2 &= 10, \; e_2 = 11 - 10 = 1, \quad \tilde{e}_2 = 4, \quad \tilde{f}_2 = 10 + 4 = 14 \\
\hat{f}_3 &= 14, \; e_3 = 13 - 14 = -1, \; \tilde{e}_3 = -4, \; \tilde{f}_3 = 14 - 4 = 10 \\
\hat{f}_4 &= 10, \; e_4 = 15 - 10 = 5, \quad \tilde{e}_4 = 4, \quad \tilde{f}_4 = 10 + 4 = 14
\end{aligned}$$

We see that the reconstructed set of values 10, 14, 10, 14 never strays far from the correct set 10, 11, 13, 15.

Nevertheless, it is not difficult to discover that DM copes well with more or less constant signals, but not as well with rapidly changing signals. One approach to mitigating this problem is to simply increase the sampling, perhaps to many times the Nyquist rate. This scheme can work well and makes DM a very simple yet effective analog-to-digital converter.

Adaptive DM

However, if the slope of the actual signal curve is high, the staircase approximation cannot keep up. A straightforward approach to dealing with a steep curve is to simply change the step size k *adaptively*, that is, in response to the signal's current properties.

6.3.7 ADPCM

Adaptive DPCM takes the idea of adapting the coder to suit the input much further. Basically, two pieces make up a DPCM coder: the quantizer and the predictor. Above, in adaptive DM, we adapted the quantizer step size to suit the input. In DPCM, we can *adaptively modify the quantizer*, by changing the step size as well as decision boundaries in a nonuniform quantizer.

We can carry this out in two ways: using the properties of the input signal (called *forward adaptive quantization*) or the properties of the quantized output. If quantized errors become too large, we should change the nonuniform Lloyd–Max quantizer (this is called *backward adaptive quantization*).

We can also *adapt the predictor*, again using forward or backward adaptation. Generally, making the predictor coefficients adaptive is called *adaptive predictive coding* (APC). It is interesting to see how this is done. Recall that the predictor is usually taken to be a linear function of previously reconstructed quantized values, $\tilde{f}_n$. The number of previous values used is called the *order* of the predictor. For example, if we use M previous values, we need M coefficients $a_i, \; i = 1 .. M$ in a predictor

$$\hat{f}_n = \sum_{i=1}^{M} a_i \tilde{f}_{n-i} \tag{6.23}$$

However, we can get into a difficult situation if we try to *change* the prediction coefficients that multiply previous quantized values, because that makes a complicated set of equations to solve for these coefficients. Suppose we decide to use a least-squares approach to solving a minimization, trying to find the best values of the a_i:

$$\min \sum_{n=1}^{N} (f_n - \hat{f}_n)^2 \tag{6.24}$$

where here we would sum over a large number of samples f_n for the current patch of speech, say. But because $\hat{f}_n$ depends on the quantization, we have a difficult problem to solve. Also, we should really be changing the fineness of the quantization at the same time, to suit the signal's changing nature; this makes things problematical.

Instead, we usually resort to solving the simpler problem that results from using not $\tilde{f}_n$ in the prediction but simply the signal f_n itself. This is indeed simply solved, since, explicitly writing in terms of the coefficients a_i, we wish to solve

$$\min \sum_{n=1}^{N} (f_n - \sum_{i=1}^{M} a_i f_{n-i})^2 \tag{6.25}$$

Differentiation with respect to each of the a_i and setting to zero produces a linear system of M equations that is easy to solve. (The set of equations is called the Wiener–Hopf equations.).

Thus, we indeed find a simple way to adaptively change the predictor as we go. For speech signals, it is common to consider *blocks* of signal values, just as for image coding, and adaptively change the predictor, quantizer, or both. If we sample at 8 kHz, a common block size is 128 samples—16 ms of speech. Figure 6.19 shows a schematic diagram for the ADPCM coder and decoder [8].

6.4 Exercises

1. We wish to develop a new Internet service, for doctors. Medical ultrasound is in the range 2-10 MHz; what should our sampling rate be chosen as?
2. My old Soundblaster card is an 8-bit card.

   (a) What is it 8 bits of?
   (b) What is the best SQNR (signal-to-quantization-noise ratio) it can achieve?

3. If a tuba is 20 dB louder than a singer's voice, what is the ratio of intensities (power) of the tuba to the voice?
4. If a set of ear protectors reduces the noise level by 30 dB, how much do they reduce the intensity (the power)?
5. It is known that a loss of audio output at both ends of the audible frequency range is inevitable due to the frequency response function of audio amplifier.

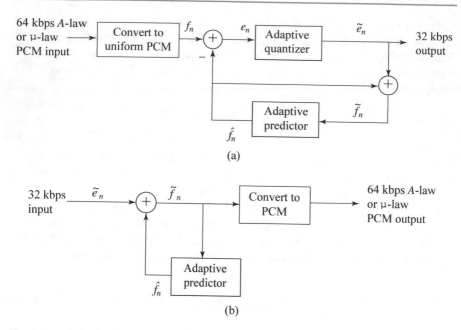

Fig. 6.19 Schematic diagram for: **a** ADPCM encoder; **b** decoder

(a) If the output was 1 volt for frequencies at mid-range, after a loss of -3 dB at 18 KHz, what is the output voltage at this frequency?

(b) To compensate the loss, a listener can adjust the gain (and hence the output) at different frequencies from an equalizer. If the loss remains -3 dB and a gain through the equalizer is 6 dB at 18 KHz, what is the output voltage now?

[Hint: Assume $\log_{10} 2 = 0.3$.]

6. Suppose the sampling frequency is 1.5 times the true frequency. What is the alias frequency?

7. In a crowded room, we can still pick out and understand a nearby speaker's voice notwithstanding the fact that general noise levels may be high. This is what is known as the "cocktail-party effect"; how it operates is that our hearing can localize a sound source by taking advantage of the difference in phase between the two signals entering our left and right ears ("binaural auditory perception"). In mono, we could not hear our neighbor's conversation very well if the noise level were at all high.

State how you think a karaoke machine works.

Hint: the mix for commercial music recordings is such that the "pan" parameter is different going to the left and right channels for each instrument. That is, for an instrument, the left, or the right, channel is emphasized. How would the singer's track timing have to be recorded in order to make it easy to subtract out the sound of the singer? (And this is typically done.)

8. The *dynamic range* of a signal V is the ratio of the maximum to the minimum, expressed in decibels. The dynamic range expected in a signal is to some extent an expression of the signal quality. It also dictates the number of bits per sample needed in order to reduce the quantization noise down to an acceptable level, e.g., we may like to reduce the noise to at least an order of magnitude below V_{min}.

   Suppose the dynamic range for a signal is 60 dB. Can we use 10 bits for this signal? Can we use 16 bits?

9. Suppose the dynamic range of speech in telephony implies a ratio V_{max}/V_{min} of about 256. Using uniform quantization, how many bits should we use to encode speech, so as to make the quantization noise at least an order of magnitude less than the smallest detectable telephonic sound?

10. *Perceptual nonuniformity* is a general term for describing the nonlinearity of human perception, e.g., when a certain parameter of an audio signal varies, humans do not necessarily perceive the difference in proportion to the amount of change.

    (a) Briefly describe at least two types of perceptual nonuniformities in human auditory perception.
    (b) Which one of them does A-law (or μ-law) attempt to approximate? Why could it improve the quantization?

11. Suppose we mistakenly always use the 0.75 point instead of the 0.50 point in a quantization interval as the decision point, in deciding to which quantization level an analog value should be mapped. Above, we have a rough calculation of SQNR. What effect does this mistake have on the SQNR?

12. State the Nyquist frequency for the following digital sample intervals. Express the result in Hertz in each case.

    (a) 1 millisecond,
    (b) 0.005 s, and
    (c) 1 h.

13. Draw a diagram showing a sinusoid at 5.5 kHz, and sampling at 8 kHz (just show eight intervals between samples in your plot). Draw the alias at 2.5 kHz and show that in the eight sample intervals, exactly 5.5 cycles of the true signal fit into 2.5 cycles of the alias signal.

14. In an old Western movie, we notice that a stagecoach wheel appears to be moving backward at 5° per frame, even though the stagecoach is moving forward. To what is this effect due? What is the true situation?

15. Suppose a signal contains tones at 1, 10, and 21 kHz, and is sampled at the rate 12 kHz (and then processed with an anti-aliasing filter limiting output to 6 kHz).

What tones are included in the output?

Hint: Most of the output consists of aliasing.

16. The Pitch Bend opcode in MIDI is followed by two data bytes specifying how the control is to be altered. How many bits of accuracy does this amount of data correspond to? Why?

17. (a) Can a single MIDI message produce more than one note sounding?

(b) Is it possible that more than one note can be sounding on a particular instrument at once? How is that done in MIDI?

(c) Is the Program Change MIDI message a Channel Message? What does this message accomplish? Based on the Program Change message, how many different instruments are there in General MIDI? Why?

(d) In general, what are the two main kinds of MIDI messages? In terms of data, what is the main difference between the two types of messages? Within those two categories, please list the different sub-types.

18. The note "A above Middle C" (with frequency 440 Hz) is note 69 in General MIDI. What MIDI bytes (in hex) should be sent to play a note twice the frequency of (i.e., one octave above) "A above Middle C" at maximum volume on Channel 1? (Don't include start/stop bits.)

Information: An octave is 12 steps on a piano, i.e., 12 notes up.

19. Give an example (in English, not hex) of a MIDI voice message. Describe the parts of the "assembler" statement for the message you suggested above.

What does a "program change" message do? Suppose "Program change" is hex &HC1 . What does the instruction &HC103 do?

20. We have suddenly invented a new kind of music: "18-tone music," that requires a keyboard with 180 keys. How would we have to change the MIDI standard to be able to play this music?

21. In PCM, what is the *delay*, assuming 8 kHz sampling? Generally, delay is the penalty associated with any algorithm due to sampling, processing, and analysis.

22. (a) Suppose we use a predictor as follows:

$$\hat{f}_n = \text{trunc}\left(\tfrac{1}{2}(\tilde{f}_{n-1} + \tilde{f}_{n-2})\right),$$
$$e_n = f_n - \hat{f}_n. \tag{6.26}$$

Also, suppose we adopt the quantizer Eq. (6.21). If the input signal has values as follows:

20 38 56 74 92 110 128 146 164 182 200 218 236 254

then show that the output from a DPCM coder (without entropy coding) is as follows:

20 44 56 74 89 105 121 153 161 181 195 212 243 251

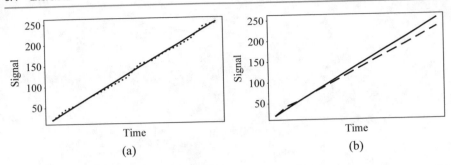

Fig. 6.20 a DPCM reconstructed signal (dotted line) tracks the input signal (solid line). **b** DPCM reconstructed signal (dashed line) steers farther and farther from the input signal (solid line)

Figure 6.20a shows how the quantized reconstructed signal tracks the input signal. As a programming project, write a small piece of code to verify your results.

(b) Now, suppose by mistake on the coder side we inadvertently use the predictor for *lossless coding*, Eq. (6.15), using original values f_n instead of quantized ones $\tilde{f}_n$. Show that on the decoder side we end up with reconstructed signal values as follows:

$$20 \quad 44 \quad 56 \quad 74 \quad 89 \quad 105 \quad 121 \quad 137 \quad 153 \quad 169 \quad 185 \quad 201 \quad 217 \quad 233$$

so that the error gets progressively worse.

Figure 6.20b shows how this appears: the reconstructed signal gets progressively worse. Modify your code from above to verify this statement.

References

1. B. Truax, *Handbook for Acoustic Ecology*, 2nd edn. (Cambridge Street Publishing, 1999)
2. K.C. Pohlmann, *Principles of Digital Audio*, 6th edn. (McGraw-Hill, 2010)
3. J.H. McClellan, R.W. Schafer, M.A. Yoder, *DSP First: a Multimedia Approach*. (Prentice-Hall PTR, 1998)
4. J. Heckroth, Tutorial on MIDI and music synthesis. The MIDI Manufacturers Association, POB 3173, La Habra CA 90632-3173 (1995). http://www.harmony-central.com/MIDI/Doc/tutorial.html
5. P.K. Andleigh, K. Thakrar, *Multimedia Systems Design*. (Prentice-Hall PTR, 1995)
6. The MIDI Association. Strategic overview and introduction to midi 2.0. in *The National Association of Music Merchants (NAMM) Show* (2020)
7. K. Sayood, *Introduction to Data Compression*, 5th edn. (Morgan Kaufmann, San Francisco, 2017)
8. R.L. Freeman, *Reference Manual for Telecommunications Engineering*, 3rd edn. (Wiley, 2001)

Part II
Multimedia Data Compression

In this part, we examine the role played in multimedia by data compression, perhaps the most important enabling technology that makes modern multimedia systems possible. So much data exists, in archives, via streaming, and elsewhere, that it has become critical to compress this information.

We start off in Chap. 7 looking at lossless data compression, i.e., involving no distortion of the original signal once it is decompressed or reconstituted. A good example is archival storage of precious artworks. Here, we may go to the trouble of imaging an Old Master's painting using a high-powered camera mounted on a dolly to avoid parallax. Certainly we do not wish to lose any of this hard-won information, so we'd best use lossless compression. WinZip and WinRAR, for example, are ubiquitous tools that utilize lossless compression.

On the other hand, when it comes to my home movies I am more willing to lose some information. If there is a choice between losing some information anyway because my computer, tablet, or smartphone cannot handle all the data I want to push through it, or else losing some information on purpose using a "lossy" compression method, I'll choose the latter. Nowadays, almost all videos you see are compressed in some way, and the compression used is mostly lossy. As well, most images on the web are in the standard JPEG format. And this is almost always a lossy compression format. It is known that lossy compression methods achieve a much higher level of compression than lossless ones.

So in Chap. 8 we go on to look at the fundamentals of lossy methods of compression, mainly focusing on the discrete cosine transform and the discrete wavelet transform. The major applications of these important methods are in the set of JPEG still-image compression standards, including JPEG 2000. These are examined in Chap. 9. We then go on to look at how data compression methods can be applied to moving images-videos. We start with basic video compression techniques in Chap. 10. We examine the ideas behind the MPEG standards, starting with MPEG-1, MPEG-2, and then MPEG-4, and MPEG-7 in Chap. 11. In Chap. 12, we introduce the modern video compression standards H.264, H.265, and H.266.

Audio compression in a sense stands by itself, and we consider some basic audio compression techniques in Chap. 13, while in Chap. 14 we look at MPEG audio, including MP3 and AAC.

Lossless Compression Algorithms

7

7.1 Introduction

The emergence of multimedia technologies has made *digital libraries* a reality. Nowadays, libraries, museums, film studios, and governments are converting more and more data and archives into digital form. Some of the data (e.g., precious books and paintings) indeed need to be stored without any loss.

As a start, suppose we want to encode the call numbers of the 120 million or so items in the Library of Congress (a mere 20 million, if we consider just books). Why don't we just transmit each item as a 27-bit number, giving each item a unique binary code (since $2^{27} > 120,000,000$)?

The main problem is that this "great idea" requires too many bits. And in fact there exist many coding techniques that will effectively reduce the total number of bits needed to represent the above information. The process involved is generally referred to as *compression* [1,2].

In Chap. 6, we had a beginning look at compression schemes aimed at audio. Therefore, we had to first consider the complexity of transforming analog signals to digital ones, whereas here we shall consider that we at least start with digital signals. For example, even though we know an image is captured using analog signals, the file produced by a digital camera is indeed digital. The more general problem of coding (compressing) a set of any symbols, not just byte values, say, has been studied for a long time.

Getting back to our Library of Congress problem, it is well known that certain parts of call numbers appear more frequently than others, so it would be more economic to assign fewer bits as their codes. This is known as *variable-length coding (VLC)*—the more frequently appearing symbols are coded with fewer bits per symbol, and vice versa. As a result, fewer bits are usually needed to represent the whole collection.

In this chapter, we study the basics of information theory and several popular lossless compression techniques. Figure 7.1 depicts a general data compression scheme,

© Springer Nature Switzerland AG 2021
Z.-N. Li et al., *Fundamentals of Multimedia*, Texts in Computer Science,
https://doi.org/10.1007/978-3-030-62124-7_7

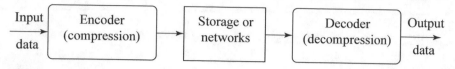

Fig. 7.1 A general data compression scheme

in which compression is performed by an encoder and decompression is performed by a decoder.

We call the output of the encoder *codes* or *codewords*. The intermediate medium could either be data storage or a communication/computer network. If the compression and decompression processes induce no information loss, the compression scheme is *lossless*; otherwise, it is *lossy*. The next several chapters deal with lossy compression algorithms as they are commonly used for image, video, and audio compression. Here, we concentrate on lossless compression.

If the total number of bits required to represent the data before compression is B_0 and the total number of bits required to represent the data after compression is B_1, then we define the *compression ratio* as

$$compression\ ratio = \frac{B_0}{B_1} \tag{7.1}$$

In general, we would desire any *codec* (encoder/decoder scheme) to have a *compression ratio* much larger than 1.0. The higher the *compression ratio*, the better the lossless compression scheme, as long as it is computationally feasible.

7.2 Basics of Information Theory

According to the famous scientist Claude E. Shannon, of Bell Labs [3,4], the *entropy* η of an information *source* with alphabet $S = \{s_1, s_2, \ldots, s_n\}$ is defined as

$$\eta = H(S) = \sum_{i=1}^{n} p_i \log_2 \frac{1}{p_i} \tag{7.2}$$

$$= -\sum_{i=1}^{n} p_i \log_2 p_i \tag{7.3}$$

where p_i is the probability that symbol s_i in S will occur.

The term $\log_2 \frac{1}{p_i}$ indicates the amount of information (the so-called *self-information* defined by Shannon [3]) contained in s_i, which corresponds to the number of bits[1] needed to encode s_i. For example, if the probability of having the

---

[1] Since we have chosen 2 as the base for logarithms in the above definition, the unit of information is *bit*—naturally also most appropriate for the binary code representation used in digital computers. If the log base is 10, the unit is *hartley*; if the base is e, the unit is *nat*.

character n in a manuscript is $1/32$, the amount of information associated with receiving this character is 5 bits. In other words, a character string nnn will require 15 bits to code. This is the basis for possible data reduction in text compression, since it will lead to character coding schemes different from the ASCII representation, in which each character requires at least 7 bits.

What is the entropy? In science, entropy is a measure of the *disorder* of a system—the more entropy, the more disorder. Typically, we add *negative* entropy to a system when we impart more order to it. For example, suppose we sort a deck of cards. (Think of a bubble sort for the deck—perhaps this is not the usual way you actually sort cards, though.) For every decision to swap or not, we impart 1 bit of information to the card system and transfer 1 bit of negative entropy to the card deck.

Now suppose we wish to *communicate* those swapping decisions, via a network, say. If we had to make two consecutive swap decisions, the possible number of outcomes will be 4. If all outcomes have an equal probability of $1/4$, then the number of bits to send is on average $4 \times (1/4) \times \log_2(1/(1/4)) = 2$ bits—no surprise here. To communicate (transmit) the results of our two decisions, we would need to transmit 2 bits.

But if the probability for one of the outcomes were higher than the others, the average number of bits we'd send would be different. (This situation might occur if the deck were already partially ordered, so that the probability of a not-swap were higher than for a swap.) Suppose the probabilities of one of our four states were $1/2$, and the other three states each had probability $1/6$ of occurring. To extend our modeling of how many bits to send on average, we need to go to noninteger powers of 2 for probabilities. Then we can use a logarithm to ask how many (float) bits of information must be sent to transmit the information content. Equation (7.3) says that in this case, we'd have to send just $(1/2) \times \log_2(2) + 3 \times (1/6) \times \log_2(6) = 1.7925$ bits, a value less than 2. This reflects the idea that if we could somehow *encode* our four states, such that the most occurring one means fewer bits to send, we'd do better (fewer bits) on average.

The definition of entropy is aimed at identifying often occurring symbols in the datastream as good candidates for *short* codewords in the compressed bitstream. As described earlier, we use a *variable-length coding* scheme for entropy coding—frequently occurring symbols are given codes that are quickly transmitted, while infrequently occurring ones are given longer codes. For example, E occurs frequently in English, so we should give it a shorter code than Q, for example.

This aspect of "surprise" in receiving an infrequent symbol in the data stream is reflected in Eq. (7.3). For if a symbol occurs rarely, its probability p_i is low (e.g., $1/100$), and thus its self-information $\log_2 \frac{1}{p_i} = \log_2 100$ is a relatively large number. This reflects the fact that it takes a longer bitstring to encode it. The probabilities p_i sitting outside the logarithm in Eq. (7.3) say that over a long stream, the symbols come by with an average frequency equal to the probability of their occurrence. This weighting should multiply the long or short information content given by the element of "surprise" in seeing a particular symbol.

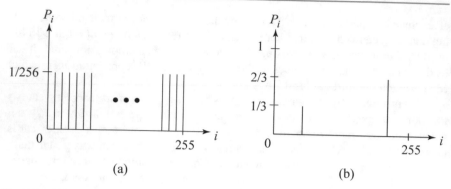

Fig. 7.2 Histograms for two gray-level images

One wrinkle in the algorithm implied by Eq. (7.3) is that if a symbol occurs with zero frequency, we simply don't count it into the entropy: we cannot take a log of zero.

As another concrete example, if the information source S is a gray-level digital image, each s_i is a gray-level intensity ranging from 0 to $(2^k - 1)$, where k is the number of bits used to represent each pixel in an uncompressed image. The range is often [0, 255], since 8 bits are typically used: this makes a convenient 1 byte per pixel. The image histogram (as discussed in Chap. 3) is a way of calculating the probability p_i of having pixels with gray-level intensity i in the image.

Figure 7.2a shows the histogram of an image with *uniform* distribution of gray-level intensities, that is, $\forall i\ p_i = 1/256$. Hence, the entropy of this image is

$$\eta = \sum_{i=0}^{255} \frac{1}{256} \cdot \log_2 256 = 256 \cdot \frac{1}{256} \cdot \log_2 256 = 8 \qquad (7.4)$$

As can be seen in Eq. (7.3), the entropy η is a weighted sum of terms $\log_2 \frac{1}{p_i}$; hence, it represents the *average* amount of information contained per symbol in the source S. For a memoryless source[2] S, the entropy η represents the minimum average number of bits required to represent each symbol in S. In other words, it specifies the lower bound for the average number of bits to code each symbol in S.

If we use $\bar{l}$ to denote the average length (measured in bits) of the codewords produced by the encoder, the Shannon coding theorem states that the entropy is the *best* we can do (under certain conditions):

$$\eta \leq \bar{l} \qquad (7.5)$$

Coding schemes aim to get as close as possible to this theoretical lower bound.

---

[2]An information source that is independently distributed, meaning that the value of the current symbol does not depend on the values of the previously appeared symbols.

It is interesting to observe that in the above uniform-distribution example we found that $\eta = 8$—the minimum average number of bits to represent each gray-level intensity is at least 8. No compression is possible for this image! In the context of imaging, this will correspond to the "worst case," where neighboring pixel values have no similarity.

Figure 7.2b shows the histogram of another image, in which one-third of the pixels are rather dark and two-third of them are rather bright. The entropy of this image is

$$\eta = \frac{1}{3} \cdot \log_2 3 + \frac{2}{3} \cdot \log_2 \frac{3}{2}$$
$$= 0.33 \times 1.59 + 0.67 \times 0.59 = 0.52 + 0.40 = 0.92$$

In general, the entropy is greater when the probability distribution is flat and smaller when it is more peaked.

7.3 Run-Length Coding

Instead of assuming a memoryless source, *run-length coding* (RLC) exploits memory present in the information source. It is one of the simplest forms of data compression. The basic idea is that if the information source we wish to compress has the property that symbols tend to form continuous groups, instead of coding each symbol in the group individually, we can code one such symbol and the length of the group.

As an example, consider a bi-level image (one with only 1-bit black and white pixels) with monotone regions—like a fax. This information source can be efficiently coded using run-length coding. In fact, since there are only two symbols, we do not even need to code any symbol at the start of each run. Instead, we can assume that the starting run is always of a particular color (either black or white) and simply code the length of each run.

The above description is the one-dimensional run-length coding algorithm. A two-dimensional variant of it is usually used to code bi-level images. This algorithm uses the coded run information in the previous row of the image to code the run in the current row. A full description of this algorithm can be found in [5].

7.4 Variable-Length Coding (VLC)

Since the entropy indicates the information content in an information source S, it leads to a family of coding methods commonly known as *entropy coding* methods. As described earlier, *variable-length coding* (VLC) is one of the best known such methods. Here, we will study the Shannon–Fano algorithm, Huffman coding, and adaptive Huffman coding.

7.4.1   Shannon–Fano Algorithm

The Shannon–Fano algorithm was independently developed by Shannon at Bell Labs and Robert Fano at MIT [6]. To illustrate the algorithm, let's suppose the symbols to be coded are the characters in the word HELLO. The frequency count of the symbols is

Symbol	H	E	L	O
Count	1	1	2	1

The encoding steps of the Shannon–Fano algorithm can be presented in the following *top-down* manner:

1. Sort the symbols according to the frequency count of their occurrences.
2. Recursively divide the symbols into two parts, each with approximately the same number of counts, until all parts contain only one symbol.

A natural way of implementing the above procedure is to build a binary tree. As a convention, let's assign bit 0 to its left branches and 1 to the right branches.

Initially, the symbols are sorted as LHEO. As Fig. 7.3 shows, the first division yields two parts: (a) L with a count of 2, denoted as L:(2) and (b) H, E, and O with a total count of 3, denoted as H,E,O:(3). The second division yields H:(1) and E,O:(2). The last division is E:(1) and O:(1).

Table 7.1 summarizes the result, showing each symbol, its frequency count, information content $\left(\log_2 \frac{1}{p_i} \right)$, resulting codeword, and the number of bits needed to encode each symbol in the word HELLO. The total number of bits used is shown at the bottom.

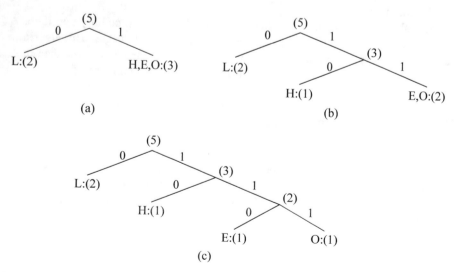

Fig. 7.3 Coding tree for HELLO by the Shannon–Fano algorithm

Table 7.1 One result of performing the Shannon–Fano algorithm on HELLO

Symbol	Count	$\log_2 \frac{1}{p_i}$	Code	Number of bits used
L	2	1.32	0	2
H	1	2.32	10	2
E	1	2.32	110	3
O	1	2.32	111	3
Total number of bits:				10

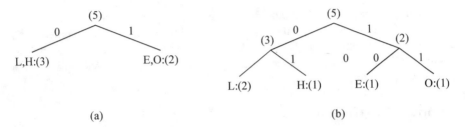

(a)   (b)

Fig. 7.4 Another coding tree for HELLO by the Shannon–Fano algorithm

To revisit the previous discussion on entropy, in this case

$$\eta = p_L \cdot \log_2 \frac{1}{p_L} + p_H \cdot \log_2 \frac{1}{p_H} + p_E \cdot \log_2 \frac{1}{p_E} + p_O \cdot \log_2 \frac{1}{p_O}$$
$$= 0.4 \times 1.32 + 0.2 \times 2.32 + 0.2 \times 2.32 + 0.2 \times 2.32 = 1.92$$

This suggests that the minimum average number of bits to code each character in the word HELLO would be at least 1.92. In this example, the Shannon–Fano algorithm uses an average of $10/5 = 2$ bits to code each symbol, which is fairly close to the lower bound of 1.92. Apparently, the result is satisfactory.

It should be pointed out that the outcome of the Shannon–Fano algorithm is not necessarily unique. For instance, at the first division in the above example, it would be equally valid to divide into the two parts L,H:(3) and E,O:(2). This would result in the coding in Fig. 7.4. Table 7.2 shows the codewords are different now. Also, these two sets of codewords may behave differently when errors are present. Coincidentally, the total number of bits required to encode the world HELLO remains at 10.

The Shannon–Fano algorithm delivers satisfactory coding results for data compression, but it was soon outperformed and overtaken by the Huffman coding method.

7.4.2 Huffman Coding

First presented by David A. Huffman in a 1952 paper [7], this method attracted an overwhelming amount of research and has been adopted in many important and/or commercial applications, such as fax machines, JPEG, and MPEG.

Table 7.2 Another result of performing the Shannon–Fano algorithm on HELLO

Symbol	Count	$\log_2 \frac{1}{p_i}$	Code	Number of bits used
L	2	1.32	00	4
H	1	2.32	01	2
E	1	2.32	10	2
O	1	2.32	11	2
Total number of bits:				10

In contradistinction to Shannon–Fano, which is top-down, the encoding steps of the Huffman algorithm are described in the following *bottom-up* manner. Let's use the same example word, HELLO. A similar binary coding tree will be used as above, in which the left branches are coded 0 and right branches 1. A simple list data structure is also used.

Algorithm 7.1 (Huffman Coding).

1. Initialization: put all symbols on the list sorted according to their frequency counts.
2. Repeat until the list has only one symbol left.

   (a) From the list, pick two symbols with the lowest frequency counts. Form a Huffman subtree that has these two symbols as child nodes and create a parent node for them.
   (b) Assign the sum of the children's frequency counts to the parent and insert it into the list, such that the order is maintained.
   (c) Delete the children from the list.

3. Assign a codeword for each leaf based on the path from the root.

In the above figure, new symbols P1, P2, P3 are created to refer to the parent nodes in the Huffman coding tree. The contents in the list are illustrated below:

After initialization :	L H E O
After iteration (a) :	L P1 H
After iteration (b) :	L P2
After iteration (c) :	P3

For this simple example, the Huffman algorithm apparently generated the same coding result as one of the Shannon–Fano results shown in Fig. 7.3, although the results are usually better. The average number of bits used to code each character is also 2, (i.e., $(1 + 1 + 2 + 3 + 3)/5 = 2$). As another simple example, consider a text string containing a set of characters and their frequency counts as follows: A:(15), B:(7), C:(6), D:(6) and E:(5). It is easy to show that the Shannon–Fano algorithm needs a total of 89 bits to encode this string, whereas the Huffman algorithm needs only 87.

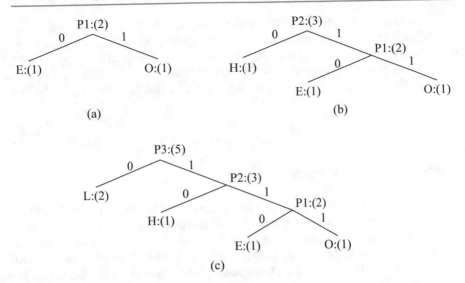

Fig. 7.5 Coding tree for HELLO using the Huffman algorithm

As shown above, if correct probabilities ("prior statistics") are available and accurate, the Huffman coding method produces good compression results. Decoding for the Huffman coding is trivial as long as the statistics and/or coding tree are sent before the data to be compressed (in the file header, say). This overhead becomes negligible if the data file is sufficiently large.

The following are important properties of Huffman coding:

- **Unique prefix property.**
  No Huffman code is a prefix of any other Huffman code. For instance, the code 0 assigned to L in Fig. 7.5c is not a prefix of the code 10 for H or 110 for E or 111 for O; nor is the code 10 for H a prefix of the code 110 for E or 111 for O. It turns out that the unique prefix property is guaranteed by the above Huffman algorithm, since it always places all input symbols at the leaf nodes of the Huffman tree. The Huffman code is one of the *prefix codes* for which the unique prefix property holds. The code generated by the Shannon–Fano algorithm is another such example.
  This property is essential and also makes for an efficient decoder, since it precludes any ambiguity in decoding. In the above example, if a bit 0 is received, the decoder can immediately produce a symbol L without waiting for any more bits to be transmitted.
- **Optimality.** The Huffman code is a *minimum-redundancy code*, as shown in Huffman's 1952 paper [7]. It has been proven [2,8] that the Huffman code is *optimal* for a given data model (i.e., a given, accurate, probability distribution):
  - The two least frequent symbols will have the same length for their Huffman codes, differing only at the last bit. This should be obvious from the above algorithm.

- Symbols that occur more frequently will have shorter Huffman codes than symbols that occur less frequently. Namely, for symbols s_i and s_j, if $p_i \geq p_j$ then $l_i \leq l_j$, where l_i is the number of bits in the codeword for s_i.
- It has been shown (see Sayood [2]) that the average code length for an information source S is strictly less than $\eta + 1$. Combined with Eq. (7.5), we have

$$\eta \leq \bar{l} < \eta + 1 \tag{7.6}$$

Extended Huffman Coding

The discussion of Huffman coding so far assigns each symbol a codeword that has an *integer* bit length. As stated earlier, $\log_2 \frac{1}{p_i}$ indicates the amount of information contained in the information source s_i, which corresponds to the number of bits needed to represent it. When a particular symbol s_i has a large probability (close to 1.0), $\log_2 \frac{1}{p_i}$ will be close to 0, and assigning 1 bit to represent that symbol will be costly. Only when the probabilities of all symbols can be expressed as 2^{-k}, where k is a positive integer, would the average length of codewords be truly optimal—that is, $\bar{l} \equiv \eta$. Clearly, $\bar{l} > \eta$ in most cases.

One way to address the problem of integral codeword length is to group several symbols and assign a single codeword to the group. Huffman coding of this type is called *Extended Huffman Coding* [2]. Assume an information source has alphabet $S = \{s_1, s_2, \ldots, s_n\}$. If k symbols are grouped together, then the *extended alphabet* is

$$S^{(k)} = \{\overbrace{s_1 s_1 \ldots s_1}^{k\ symbols}, s_1 s_1 \ldots s_2, \ldots, s_1 s_1 \ldots s_n, s_1 s_1 \ldots s_2 s_1, \ldots, s_n s_n \ldots s_n\}$$

Note that the size of the new alphabet $S^{(k)}$ is n^k. If k is relatively large (e.g., $k \geq 3$), then for most practical applications where $n \gg 1$, n^k would be a very large number, implying a huge symbol table. This overhead makes extended Huffman coding impractical.

As shown in [2], if the entropy of S is η, then the average number of bits needed for each symbol in S is now

$$\eta \leq \bar{l} < \eta + \frac{1}{k} \tag{7.7}$$

so we have shaved quite a bit from the coding schemes' bracketing of the theoretical best limit. Nevertheless, this is not as much of an improvement over the original Huffman coding (where group size is 1) as one might have hoped for.

Extended Huffman Example

As an explicit example of the power of extended Huffman coding, let's construct a binary Huffman code for a source S with just three symbols A, B, and C, having probabilities 0.6, 0.3, and 0.1, respectively. That is, here we have $n = 3$.

We're interested in what is the average codeword length, in bits per symbol. To start, in comparison let's first look at the value of the entropy of this source, and the bitrate given by Huffman coding for single symbols, not blocks.

Generating a Huffman tree gives

$$A: 0; \quad B: 10; \quad C: 11;$$

$$\text{Average} = 0.6 \times 1 + 0.3 \times 2 + 0.1 \times 2 = 1.4 \text{ bits/symbol.}$$

Below we calculate the entropy and get

$$\eta = -\sum_i p_i \log_2 p_i = -0.6 \times \log_2 0.6 - 0.3 \times \log_2 0.3 - 0.1 \times \log_2 0.1 \approx 1.2955.$$

Now let's *extend* this code by grouping symbols into two-character groups—i.e., we use blocks of $k = 2$ characters. We wish to compare the performance now, in bits per original source symbol, with the best possible.

The extended Huffman tree is shown in Fig. 7.6:

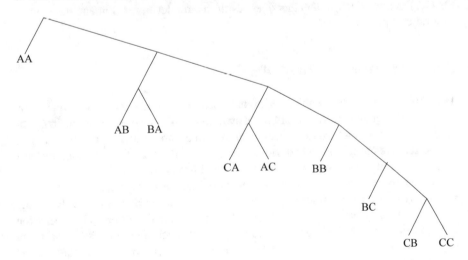

Fig. 7.6 Huffman tree for extended alphabet

Codeword bitlengths are then as follows:

Symbol group	Probability	Codeword	Bitlength
AA	0.36	0	1
AB	0.18	100	3
BA	0.18	101	3
CA	0.06	1100	4
AC	0.06	1101	4
BB	0.09	1110	4
BC	0.03	11110	5
CB	0.03	111110	6
CC	0.01	111111	6

Consequently, the average bitrate per symbol is

$$\text{Average} = 0.5 \times (0.36 + 3 \times 0.18 + 3 \times 0.18 + 4 \times 0.06 + 4 \times 0.06 + 4 \times 0.09$$
$$+5 \times 0.03 + 6 \times 0.03 + 6 \times 0.01) = 1.3350$$

(The reason for the factor 0.5 is that each leaf gives us a length-2 ($k = 2$) block of symbols, whereas we want to compare to the bitrate per single symbol.)

Now recall that the average bitrate per symbol was 1.4 for length-1 symbols, and the best possible is the entropy: $\eta \approx 1.2955$.

So we found that, indeed, the extended Huffman bitrate does fit into the bound Eq. (7.7) as advertised, and in fact does result in a modest improvement in compression ratio in practice.

7.4.3 Adaptive Huffman Coding

The Huffman algorithm requires prior statistical knowledge about the information source, and such information is often not available. This is particularly true in multimedia applications, where future data is unknown before its arrival, as, for example, in live (or streaming) audio and video. Even when the statistics are available, the transmission of the symbol table could represent heavy overhead.

For the non-extended version of Huffman coding, the above discussion assumes a so-called *order-0* model, that is, symbols/characters were treated singly, without any context or history maintained. One possible way to include contextual information is to examine k preceding (or succeeding) symbols each time; this is known as an *order-k* model. For example, an order-1 model can incorporate such statistics as the probability of "qu" in addition to the individual probabilities of "q" and "u." Nevertheless, this again implies that much more statistical data has to be stored and sent for the order-k model when $k \geq 1$.

The solution is to use *adaptive* compression algorithms, in which statistics are gathered and updated dynamically as the data stream arrives. The probabilities are no longer based on prior knowledge but on the actual data received so far. The new coding methods are "adaptive" because, as the probability distribution of the received symbols changes, symbols will be given new (longer or shorter) codes. This is especially desirable for multimedia data, when the content (the music or the color of the scene) and hence the statistics can change rapidly.

As an example, we introduce the *adaptive Huffman coding* algorithm in this section. Many ideas, however, are also applicable to other adaptive compression algorithms.

Procedure 7.1 (Procedures for Adaptive Huffman Coding)

```
ENCODER                          DECODER
-------                          -------

Initial_code();                  Initial_code();
while not EOF                     while not EOF
   {                                {
     get(c);                          decode(c);
     encode(c);                       output(c);
     update_tree(c)                   update_tree(c);
   }                                }
```

- `Initial_code` assigns symbols with some initially agreed-upon codes, without any prior knowledge of the frequency counts for them. For example, some conventional codes such as ASCII may be used for coding character symbols.
- `update_tree` is a procedure for constructing an adaptive Huffman tree. It basically does two things: it increments the frequency counts for the symbols (including any new ones) and updates the configuration of the tree.
  - The Huffman tree must always maintain its *sibling property*, that is, all nodes (internal and leaf) are arranged in the order of increasing counts. Nodes are numbered in order from left to right, bottom to top. (See Fig. 7.7, in which the first node is 1.A:(1), the second node is 2.B:(1), and so on, where the numbers in parentheses indicate the count.) If the sibling property is about to be violated, a *swap* procedure is invoked to update the tree by rearranging the nodes.
  - When a swap is necessary, the farthest node with count N is swapped with the node whose count has just been increased to $N + 1$. Note that if the node with count N is not a leaf node—it is the root of a subtree—the entire subtree will go with it during the swap.
- The encoder and decoder must use exactly the same `Initial_code` and `update_tree` routines.

Figure 7.7a depicts a Huffman tree with some symbols already received. Figure 7.7b shows the updated tree after an additional A (i.e., the second A) was received. This increased the count of As to $N + 1 = 2$ and triggered a swap. In this case, the farthest node with count $N = 1$ was D:(1). Hence, A:(2) and D:(1) were swapped.

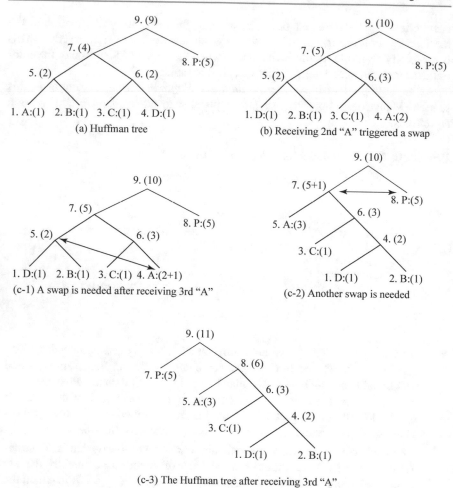

Fig. 7.7 Node swapping for updating an adaptive Huffman tree: **a** A Huffman tree; **b** receiving 2nd "A" triggered a swap; (c-1) a swap is needed after receiving 3rd "A"; (c-2) another swap is needed; (c-3) the Huffman tree after receiving 3rd "A"

Apparently, the same result could also be obtained by first swapping A:(2) with B:(1), then with C:(1), and finally with D:(1). The problem is that such a procedure would take three swaps; the rule of swapping with "the farthest node with count N" helps avoid such unnecessary swaps.

The update of the Huffman tree after receiving the third A is more involved and is illustrated in the three steps shown in Fig. 7.7(c-1) to (c-3). Since A:(2) will become A:(3) (temporarily denoted as A:(2+1)), it is now necessary to swap A:(2+1) with the fifth node. This is illustrated with an arrow in Fig. 7.7(c-1).

Since the fifth node is a non-leaf node, the subtree with nodes 1. D:(1), 2. B:(1), and 5. (2) is swapped as a whole with A:(3). Figure 7.7(c-2) shows the tree after this first swap. Now the seventh node will become (5+1), which triggers another swap with the eighth node. Figure 7.7(c-3) shows the Huffman tree after this second swap.

Table 7.3 Initial code assignment for AADCCDD using adaptive Huffman coding

Symbol	Initial code
NEW	0
A	00001
B	00010
C	00011
D	00100
⋮	⋮

The above example shows an update process that aims to maintain the sibling property of the adaptive Huffman tree—the update of the tree sometimes requires more than one swap. When this occurs, the swaps should be executed in multiple steps in a "bottom-up" manner, starting from the lowest level where a swap is needed. In other words, the update is carried out sequentially: tree nodes are examined in order, and swaps are made whenever necessary.

To clearly illustrate more implementation details, let's examine another example. Here, we show exactly what *bits* are sent, as opposed to simply stating how the tree is updated.

Example 7.1 (Adaptive Huffman Coding for Symbol String AADCCDD).

Let's assume that the initial code assignment for both the encoder and decoder simply follows the ASCII order for the 26 symbols in an alphabet, A through Z, as Table 7.3 shows. To improve the implementation of the algorithm, we adopt an additional rule: if any character/symbol is to be sent the first time, it must be preceded by a special symbol, NEW. The initial code for NEW is 0. The *count* for NEW is always kept as 0 (the count is never increased); hence, it is always denoted as NEW:(0) in Fig. 7.8.

Figure 7.8 shows the Huffman tree after each step. Initially, there is no tree. For the first A, 0 for NEW and the initial code 00001 for A are sent. Afterward, the tree is built and shown as the first one, labeled A. Now both the encoder and decoder have constructed the same first tree, from which it can be seen that the code for the second A is 1. The code sent is thus 1.

After the second A, the tree is updated, shown labeled as AA. The updates after receiving D and C are similar. More subtrees are spawned, and the code for NEW is getting longer—from 0 to 00 to 000.

From AADC to AADCC takes two swaps. To illustrate the update process clearly, this is shown in three steps, with the required swaps again indicated by arrows.

- **AADCC Step 1**. The frequency count for C is increased from 1 to $1 + 1 = 2$; this necessitates its swap with D:(1).

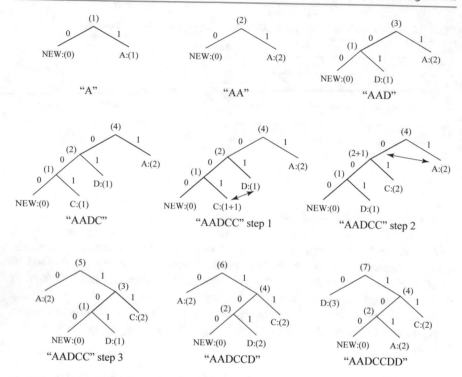

Fig. 7.8 Adaptive Huffman tree for AADCCDD

Table 7.4 Sequence of symbols and codes sent to the decoder

Symbol	New	A	A	New	D	New	C	C	D	D
Code	0	00001	1	0	00100	00	00011	001	101	101

- **AADCC Step 2**. After the swap between C and D, the count of the parent node of C:(2) will be increased from 2 to $2 + 1 = 3$; this requires its swap with A:(2).
- **AADCC Step 3**. The swap between A and the parent of C is completed.

Table 7.4 summarizes the sequence of symbols and code (zeros and ones) being sent to the decoder.

It is important to emphasize that the code for a particular symbol often changes during the adaptive Huffman coding process. The more frequent the symbol up to the moment, the shorter the code. For example, after AADCCDD, when the character D overtakes A as the most frequent symbol, its code changes from 101 to 0. This is of course fundamental for the adaptive algorithm—codes are reassigned dynamically according to the new probability distribution of the symbols.

The "demo" for adaptive Huffman coding on this book's website should aid you in learning this algorithm.

7.5 Dictionary-Based Coding

The Lempel–Ziv–Welch (LZW) algorithm employs an adaptive, dictionary-based compression technique. Unlike variable-length coding, in which the lengths of the codewords are different, LZW uses fixed-length codewords to represent variable-length strings of symbols/characters that commonly occur together, such as words in English text.

As in the other adaptive compression techniques, the LZW encoder and decoder build up the same dictionary dynamically while receiving the data—the encoder and the decoder both develop the same dictionary. Since a single code can now represent more than one symbol/character, data compression is realized.

LZW proceeds by placing longer and longer repeated entries into a dictionary, then emitting the *code* for an element rather than the string itself, if the element has already been placed in the dictionary. The predecessors of LZW are LZ77 [9] and LZ78 [10], due to Jacob Ziv and Abraham Lempel in 1977 and 1978. Terry Welch [11] improved the technique in 1984. LZW is used in many applications, such as UNIX compress, GIF for images, WinZip, and others.

Algorithm 7.2 (LZW Compression).

```
BEGIN
    s = next input character;
    while not EOF
        {
           c = next input character;

           if s + c exists in the dictionary
              s = s + c;
           else
              {
                 output the code for s;
                 add string s + c to the dictionary with a new code;
                 s = c;
              }
        }
     output the code for s;
END
```

Example 7.2 (LZW Compression for String ABABBABCABABBA).

Let's start with a very simple dictionary (also referred to as a *string table*), initially containing only three characters, with codes as follows:

```
          code       string
          ---------------
            1          A
            2          B
            3          C
```

Now if the input string is ABABBABCABABBA, the LZW compression algorithm works as follows:

s	c	output	code	string
			1	A
			2	B
			3	C
A	B	1	4	AB
B	A	2	5	BA
A	B			
AB	B	4	6	ABB
B	A			
BA	B	5	7	BAB
B	C	2	8	BC
C	A	3	9	CA
A	B			
AB	A	4	10	ABA
A	B			
AB	B			
ABB	A	6	11	ABBA
A	EOF	1		

The output codes are 1 2 4 5 2 3 4 6 1. Instead of 14 characters, only 9 codes need to be sent. If we assume each character or code is transmitted as a byte that is quite a saving (the compression ratio would be $14/9 = 1.56$). (Remember, the LZW is an adaptive algorithm, in which the encoder and decoder independently build their own string tables. Hence, there is no overhead involving transmitting the string table.)

Obviously, for our illustration, the above example is replete with a great deal of redundancy in the input string, which is why it achieves compression so quickly. In general, savings for LZW would not come until the text is more than a few hundred bytes long.

The above LZW algorithm is simple, and it makes no effort in selecting optimal new strings to enter into its dictionary. As a result, its string table grows rapidly, as illustrated above. A typical LZW implementation for textual data uses a 12-bit code length. Hence, its dictionary can contain up to 4,096 entries, with the first 256 (0–255) entries being ASCII codes. If we take this into account, the above compression ratio is reduced to $(14 \times 8)/(9 \times 12) = 1.04$.

Algorithm 7.3 (LZW Decompression (Simple Version)).

```
BEGIN
    s = NIL;
    while not EOF
    {
        k = next input code;
        entry = dictionary entry for k;
        output entry;
```

```
        if (s != NIL)
            add string s + entry[0] to dictionary
            with a new code;
        s = entry;
    }
END
```

Example 7.3 (LZW decompression for string ABABBABCABABBA).

Input codes to the decoder are 1 2 4 5 2 3 4 6 1. The initial string table is identical to what is used by the encoder.

The LZW decompression algorithm then works as follows:

s	k	entry/output	code	string
			1	A
			2	B
			3	C
NIL	1	A		
A	2	B	4	AB
B	4	AB	5	BA
AB	5	BA	6	ABB
BA	2	B	7	BAB
B	3	C	8	BC
C	4	AB	9	CA
AB	6	ABB	10	ABA
ABB	1	A	11	ABBA
A	EOF			

Apparently, the output string is ABABBABCABABBA—a truly lossless result!

LZW Algorithm Details

A more careful examination of the above simple version of the LZW decompression algorithm will reveal a potential problem. In adaptively updating the dictionaries, the encoder is sometimes ahead of the decoder. For example, after the sequence ABABB, the encoder will output code 4 and create a dictionary entry with code 6 for the new string ABB.

On the decoder side, after receiving the code 4, the output will be AB, and the dictionary is updated with code 5 for a new string, BA. This occurs several times in the above example, such as after the encoder outputs another code 4, code 6. In a way, this is anticipated—after all, it is a sequential process, and the encoder had to be ahead. In this example, this did not cause a problem.

Welch [11] points out that the simple version of the LZW decompression algorithm will break down when the following scenario occurs. Assume that the input string is ABABBABCABBABBAX....

The LZW encoder:

```
s      c    output    code   string
----------------------------------------
                        1      A
                        2      B
                        3      C
----------------------------------------
A      B      1         4      AB
B      A      2         5      BA
A      B
AB     B      4         6      ABB
B      A
BA     B      5         7      BAB
B      C      2         8      BC
C      A      3         9      CA
A      B
AB     B
ABB    A      6        10      ABBA
A      B
AB     B
ABB    A
ABBA   X     10        11      ABBAX
                .
                .
                .
```

The sequence of output codes from the encoder (and hence the input codes for the decoder) is 1 2 4 5 2 3 6 10 ...

The simple LZW decoder:

```
s      k    entry/output    code   string
--------------------------------------------
                             1      A
                             2      B
                             3      C
--------------------------------------------
NIL    1         A
A      2         B           4      AB
B      4         AB          5      BA
AB     5         BA          6      ABB
BA     2         B           7      BAB
B      3         C           8      BC
C      6         ABB         9      CA
ABB    10        ???
```

Here "???" indicates that the decoder has encountered a difficulty: no dictionary entry exists for the last input code, 10. A closer examination reveals that code 10 was most recently created at the encoder side, formed by a concatenation of character, string, character. In this case, the character is A, and string is BB, that is, A + BB +

A. Meanwhile, the sequence of the output symbols from the encoder is A, BB, A, BB, A.

This example illustrates that whenever the sequence of symbols to be coded is character, string, character, string, character, and so on, the encoder will create a new code to represent Character + String + Character and use it right away, before the decoder has had a chance to create it!

Fortunately, this is the only case in which the above simple LZW decompression algorithm will fail. Also, when this occurs, the variable s = Character + String. A modified version of the algorithm can handle this exceptional case by checking whether the input code has been defined in the decoder's dictionary. If not, it will simply assume that the code represents the symbols $s + s[0]$, that is, Character + String + Character.

Algorithm 7.4 (LZW Decompression (Modified)).

```
BEGIN
    s = NIL;
    while not EOF
        {
            k = next input code;
            entry = dictionary entry for k;

            /* exception handler */
            if (entry == NULL)
                entry = s + s[0];

            output entry;
            if (s != NIL)
                add string s + entry[0] to dictionary
                with a new code;
            s = entry;
        }
END
```

Actual implementations of the LZW algorithm require some practical limit for the dictionary size, for example, a maximum of 4,096 entries for GIF. Nevertheless, this still yields a 12-bit or 11-bit code length for LZW codes, which is longer than the word length for the original data—8 bit for ASCII.

In real applications, the code length l is kept in the range of $[l_0, l_{max}]$. For the UNIX compress command, $l_0 = 9$ and l_{max} is by default 16. The dictionary initially has a size of 2^{l_0}. When it is filled up, the code length will be increased by 1; this is allowed to repeat until $l = l_{max}$.

If the data to be compressed lacks any repetitive structure, the chance of using the new codes in the dictionary entries could be low. Sometimes, this will lead to *data expansion* instead of data reduction, since the code length is often longer than the word length of the original data. To deal with this, the algorithm can build in two modes: *compressed* and *transparent*. The latter turns off compression and is invoked when data expansion is detected.

Since the dictionary has a maximum size, once it reaches $2^{l_{max}}$ entries, LZW loses its adaptive power and becomes a static, dictionary-based technique. UNIX compress, for example, will monitor its own performance at this point. It will simply flush and reinitialize the dictionary when the compression ratio falls below a threshold. A better dictionary management is perhaps to remove the LRU (least recently used) entries. The algorithm will look for any entry that is not a prefix to any other dictionary entry, because this indicates that the code has not been used since its creation.

7.6 Arithmetic Coding

Arithmetic coding is a more advanced coding method that usually outperforms Huffman coding in practice. It was initially developed in the late 1970s and 1980s [12–14]. The initial idea of arithmetic coding was introduced in Shannon's 1948 work [3]. Peter Elias developed its first recursive implementation (which was not published but was mentioned in Abramson's 1963 book [15]). The method was further developed and described in Jelinek's 1968 book [16]. Some better known improved arithmetic coding methods can be attributed to Pasco [17] and Rissanen and Langdon [12].

Various modern versions of arithmetic coding have been developed for newer multimedia standards, for example, Fast Binary Arithmetic Coding in JBIG, JBIG2, and JPEG 2000 and Context-Adaptive Binary Arithmetic Coding (CABAC) in H.264, H.265, and H.266. We will introduce some of the fundamentals in this section and later in Chap. 12.

Normally (in its non-extended mode), Huffman coding assigns each symbol a codeword that has an integral bit length. As stated earlier, $\log_2 \frac{1}{p_i}$ indicates the amount of information contained in the information source s_i, which corresponds to the number of bits needed to represent it. For example, when a particular symbol s_i has a large probability (close to 1.0), $\log_2 \frac{1}{p_i}$ will be close to 0, and even assigning only 1 bit to represent that symbol will be very costly if we have to transmit that 1 bit many times.

Although it is possible to group symbols into metasymbols for codeword assignment (as in extended Huffman coding) to overcome the limitation of integral number of bits per symbol, the increase in the resultant symbol table required by the Huffman encoder and decoder would be formidable.

Arithmetic coding can treat the whole message as one unit and achieve fractional number of bits for each input symbol. In practice, the input data is usually broken up into chunks to avoid error propagation. In our presentation below, we will start with a simplistic approach and include a terminator symbol. Then we will introduce some improved methods for practical implementations.

7.6.1   Basic Arithmetic Coding Algorithm

A message is represented by a half-open interval $[a, b)$ where a and b are real numbers between 0 and 1. Initially, the interval is $[0, 1)$. When the message becomes longer, the length of the interval shortens, and the number of bits needed to represent

Symbol	Probability	Range	Range_low	Range_high
A	0.2	[0, 0.2)	0	0.2
B	0.1	[0.2, 0.3)	0.2	0.3
C	0.2	[0.3, 0.5)	0.3	0.5
D	0.05	[0.5, 0.55)	0.5	0.55
E	0.3	[0.55, 0.85)	0.55	0.85
F	0.05	[0.85, 0.9)	0.85	0.9
$	0.1	[0.9, 1.0)	0.9	1.0

(a)

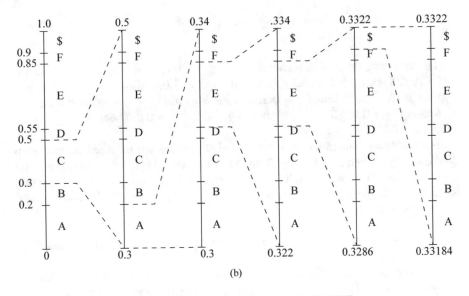

(b)

Symbol	low	high	range
	0	1.0	1.0
C	0.3	0.5	0.2
A	0.30	0.34	0.04
E	0.322	0.334	0.012
E	0.3286	0.3322	0.0036
$	0.33184	0.33220	0.00036

(c)

Fig. 7.9 Arithmetic coding: encode symbols CAEE$: **a** Probability distribution of symbols; **b** graphical display of shrinking ranges; **c** new *low*, *high*, and *range* generated

the interval increases. Suppose the alphabet is [A, B, C, D, E, F, $], in which $ is a special symbol used to terminate the message, and the known probability distribution is shown in Fig. 7.9a.

Algorithm 7.5 (Arithmetic Coding Encoder).

```
BEGIN
    low = 0.0;    high = 1.0;    range = 1.0;
    initialize symbol;               // so symbol != terminator

    while (symbol != terminator)
        {
            get (symbol);
            high = low + range * Range_high(symbol);
            low =  low + range * Range_low(symbol);
            range = high - low;
        }

    output a code so that low <= code < high;
END
```

The encoding process is illustrated in Fig. 7.9b, c, in which a string of symbols CAEE$ is encoded. Initially, $low = 0$, $high = 1.0$, and $range = 1.0$. The first symbol is C, $Range\_low(C) = 0.3$, $Range\_high(C) = 0.5$, so after the symbol C, $low = 0 + 1.0 \times 0.3 = 0.3$, $high = 0 + 1.0 \times 0.5 = 0.5$. The new $range$ is now reduced to 0.2.

For clarity of illustration, the ever-shrinking ranges are enlarged in each step (indicated by dashed lines) in Fig. 7.9b. After the second symbol A, low, $high$, and $range$ are 0.30, 0.34, and 0.04. The process repeats itself until after the terminating symbol $ is received. By then low and $high$ are 0.33184 and 0.33220, respectively. It is apparent that finally we have

$$range = P_C \times P_A \times P_E \times P_E \times P_\$ = 0.2 \times 0.2 \times 0.3 \times 0.3 \times 0.1 = 0.00036$$

The final step in encoding calls for generation of a number that falls within the range [low, high). This number is referred to as a *tag*, i.e., a unique identifier for the interval that represents the sequence of symbols. Although it is trivial to pick such a number in decimal, such as 0.33184, 0.33185, or 0.332 in the above example, it is less obvious how to do it with a binary fractional number. The following algorithm will ensure that the shortest binary codeword is found if low and $high$ are the two ends of the range and $low < high$.

Procedure 7.2 (Generating Codeword for Encoder).

```
BEGIN
    code = 0;
    k = 1;
    while (value(code) < low)
```

```
{
    assign 1 to the kth binary fraction bit;
    if (value(code) > high)
        replace the kth bit by 0;
    k = k + 1;
}
END
```

For the above example, $low = 0.33184$, $high = 0.3322$. If we assign 1 to the first binary fraction bit, it would be 0.1 in binary, and its decimal $value(code) = value(0.1) = 0.5 > high$. Hence, we assign 0 to the first bit. Since $value(0.0) = 0 < low$, the while loop continues.

Assigning 1 to the second bit makes a binary $code$ 0.01 and $value(0.01) = 0.25$, which is less than $high$, so it is accepted. Since it is still true that $value(0.01) < low$, the iteration continues. Eventually, the binary codeword generated is 0.01010101, which is $2^{-2} + 2^{-4} + 2^{-6} + 2^{-8} = 0.33203125$.

It must be pointed out that we were lucky to have found a codeword of only 8 bits to represent this sequence of symbols CAEE\$. In this case, $\log_2 \frac{1}{P_C} + \log_2 \frac{1}{P_A} + \log_2 \frac{1}{P_E} + \log_2 \frac{1}{P_E} + \log_2 \frac{1}{P_\$} = \log_2 \frac{1}{range} = \log_2 \frac{1}{0.00036} \approx 11.44$, which would suggest that it could take 12 bits to encode a string of symbols like this.

It can be proven [2] that $\lceil \log_2(1/\prod_i P_i) \rceil$ is the upper bound. Namely, in the worst case, the shortest codeword in arithmetic coding will require k bits to encode a sequence of symbols, and

$$k = \lceil \log_2 \frac{1}{range} \rceil = \lceil \log_2 \frac{1}{\prod_i P_i} \rceil \qquad (7.8)$$

where P_i is the probability for each symbol i in the sequence, and $range$ is the final range generated by the encoder.

Apparently, when the length of the message is long, its $range$ quickly becomes very small, and hence $\log_2 \frac{1}{range}$ becomes very large; the difference between $\log_2 \frac{1}{range}$ and $\lceil \log_2 \frac{1}{range} \rceil$ is negligible.

Generally, arithmetic coding achieves better performance than Huffman coding, because the former treats an entire sequence of symbols as one unit, whereas the latter has the restriction of assigning an integral number of bits to each symbol. For example, Huffman coding would require 12 bits for CAEE\$, equaling the worst-case performance of arithmetic coding.

Moreover, Huffman coding cannot always attain the upper bound given in Eq. (7.8). It can be shown (see Exercise 7.) that if the alphabet is $[A, B, C]$ and the known probability distribution is $P_A = 0.5$, $P_B = 0.4$, $P_C = 0.1$, then for sending BBB, Huffman coding will require 6 bits, which is more than $\lceil \log_2 \frac{1}{0.4 \times 0.4 \times 0.4} \rceil = 4$, whereas arithmetic coding will need only 4 bits.

Table 7.5 Arithmetic coding: decode symbols CAEE$

Value	Output symbol	Range_low	Range_high	Range
0.33203125	C	0.3	0.5	0.2
0.16015625	A	0.0	0.2	0.2
0.80078125	E	0.55	0.85	0.3
0.8359375	E	0.55	0.85	0.3
0.953125	$	0.9	1.0	0.1

Algorithm 7.6 (Arithmetic Coding Decoder).

```
BEGIN
    get binary code and convert to decimal value = value(code);
    Do
        {
            find a symbol s so that
                Range_low(s) <= value < Range_high(s);
            output s;
            low = Rang_low(s);
            high = Range_high(s);
            range = high - low;
            value = [value - low] / range;
        }
    Until symbol s is a terminator
END
```

Table 7.5 illustrates the decoding process for the above example. Initially, $value = 0.33203125$. Since $Range\_low(C) = 0.3 \le 0.33203125 < 0.5 = Range\_high(C)$, the first output symbol is C. This yields $value = [0.33203125 - 0.3]/0.2 = 0.16015625$, which in turn determines that the second symbol is A. Eventually, $value$ is 0.953125, which falls in the range [0.9, 1.0) of the terminator $.

In the above discussion, a special symbol, $, is used as a terminator of the string of symbols. This is analogous to sending *end-of-line* (EOL) in image transmission. However, the coding of the EOL symbol itself is an interesting problem. Usually, EOL ends up being relatively long. Lei et al. [18] address some of these issues and propose an algorithm that controls the length of the EOL codeword it generates. Also, if the transmission channel/network is noisy (lossy), the protection of having a terminator (or EOL) symbol is crucial for the decoder to regain synchronization with the encoder.

Naturally, if both the encoder and decoder know the number of symbols in the sequence, no terminator symbol is needed.

7.6.2 Scaling and Incremental Coding

The basic algorithm described in the last section has the following limitations that make its practical implementation infeasible:

- When it is used to code long sequences of symbols, the tag intervals shrink to a very small range. Representing these small intervals requires very high-precision numbers (i.e., even 32-bit or 64-bit floating point may not be enough) when more than a few symbols are coded.
- The encoder will not produce any output codeword until the entire sequence is entered. Likewise, the decoder needs to have the codeword for the entire sequence of the input symbols before decoding.

Some key observations [2, 13]:

1. Although the binary representations for the *low*, *high*, and any number within the small interval usually require many bits, they always have the same MSBs (most significant bits). For example, 0.1000110 for 0.5469 (low), 0.1000111 for 0.5547 (high).
2. Subsequent intervals will always stay within the current interval. Hence, we can output the common MSBs and remove them from subsequent considerations.

These are the bases for *scaling and incremental coding*. They provide a solution to the above problems.

Scaling

There are three types of scaling methods. They are shown in Fig. 7.10. Dark segments indicate the tag intervals $[low, high)$, with $low < high$.

- **E1 scaling**: This applies when the tag interval is entirely in the first half of the unit interval, i.e., $high \leq 0.5$ (Fig. 7.10a). A "0" is sent to the decoder and the resulting interval is rescaled as double what it was, i.e., $low = 2 \times low, high = 2 \times high$. The multiplication can be realized by $\ll 1$, i.e., left-shift by 1 bit.
- **E2 scaling**: This applies when the tag interval is entirely in the second half of the unit interval, i.e., $low \geq 0.5$ (Fig. 7.10b). A "1" is sent to the decoder and $low = 2 \times (low - 0.5), high = 2 \times (high - 0.5)$. This can also be realized by $\ll 1$, i.e., left-shift by 1 bit because the bit "1" will simply be moved out.
- **E3 scaling**: The tag interval straddles the midpoint 0.5, $low \geq 0.25$, and $high \leq 0.75$ (Fig. 7.10c). This can be handled by $low = 2 \times (low - 0.25), high = 2 \times (high - 0.25)$. The signaling of the E3 scaling is a bit more complicated [2]. We will address it later after we show the E1 and E2 scalings.

Below we show a procedure and an example that involves E1 and E2 scalings only.

Fig. 7.10 Scaling in
arithmetic coding

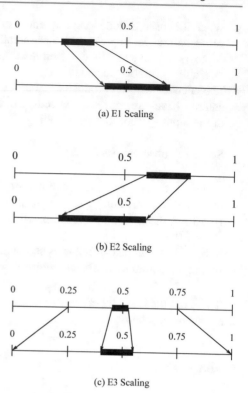

(a) E1 Scaling

(b) E2 Scaling

(c) E3 Scaling

Procedure 7.3 (E1 and E2 Scalings in Arithmetic Coding).

```
BEGIN
    while (high <= 0.5) OR (low >= 0.5)
      { if (high <= 0.5)          // E1 scaling
          { output '0';
            low = 2 * low;
            high = 2 * high;
          }
        else                      // E2 scaling
          { output '1';
            low = 2 * (low - 0.5);
            high = 2 * (high - 0.5);
          }
      }
END
```

Example 7.4 (Arithmetic Coding with Scaling and Incremental Coding).

Assume we only have three symbols A, B, C, and their probabilities are: A: 0.7,
B: 0.2, C: 0.1. Suppose the input sequence for this example is ACB, and both the
encoder and decoder know that the length of the sequence is 3.

The encoder works as (see Fig. 7.11) follows:

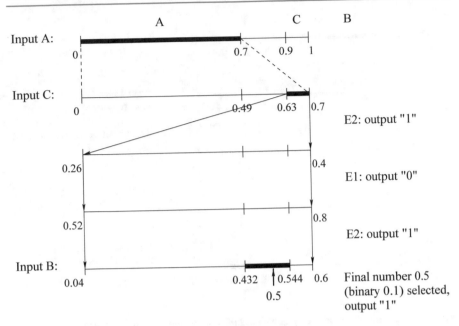

Fig. 7.11 Example: Arithmetic coding with scaling and incremental coding—encoder

- After receiving "A," the first interval [0, 0.7) is selected and further examined (shown as expanded).
- After receiving "C," the last interval [0.63, 0.7) is selected.
- Because [0.63, 0.7) is entirely in the second half of the unit interval, an E2 scaling is triggered and results in output "1" and the new interval [0.26, 0.4), because $2 \times (0.63 - 0.5) = 0.26$ and $2 \times (0.7 - 0.5) = 0.4$.
- Because [0.26, 0.4) is entirely in the first half of the unit interval, an additional E1 scaling is now necessary and results in output "0" and the new interval [0.52, 0.8), because $2 \times 0.26 = 0.52$ and $2 \times 0.4 = 0.8$.
- Again, an E2 scaling is triggered and results in output "1" and the new interval [0.04, 0.6).
  At this point, no more scaling is needed. As the third (and last) input symbol is "B," the interval [0.432, 0.544) is selected. We then generate and output the shortest codeword "1" (according to the procedure discussed in the last section) for this tag interval.

In summary, the encoder has produced 1011 which is equivalent to 0.6875.

It is important to point out that without scaling, the interval [0.63, 0.7) would have been chosen after the input symbol C, and the same code 1011, i.e., 0.6875 would have been chosen as the shortest codeword to represent the sequence ACB. In other words, we have demonstrated that the scaling procedure does produce a correct result!

The decoder works as (see Fig. 7.12) follows.

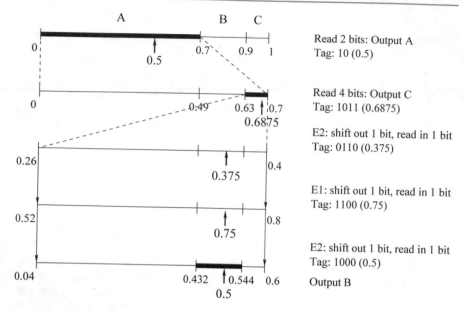

Fig. 7.12 Example: Arithmetic coding with scaling and incremental coding—decoder

In general, the length k of the codeword produced by the encoder can be relatively large. The decoder can choose to use a maximum of $l \leq k$ bits to work with at any time during the decoding process. The value l is determined by the size of the smallest interval. In this example, the smallest interval is $[0.9, 1.0)$, so $l = \lceil \log_2 \frac{1}{0.1} \rceil = 4$. When needed, the old bits will be shifted out and new bits gradually read in. This, of course, is the essence of the incremental coding—both encoder and decoder can start working with a limited number of bits, and working continuously without seeing the entire sequence of symbols and the final code.

- After reading the first 2 bits "10," the symbol A can already be unambiguously decoded. Output A. Tag: 10, i.e., 0.5.
- After reading 4 bits "1011," the symbol C can be unambiguously decoded. Output C. Tag: 1011, i.e., 0.6875.
- Because $[0.63, 0.7)$ is entirely in the second half of the unit interval, an E2 scaling is triggered: shift out the first bit, read in 1 bit (by default, it will be "0"). Tag: 0110, i.e., 0.375.
- Because $[0.26, 0.4)$ is entirely in the first half of the unit interval, an E1 scaling is triggered: shift out 1 bit, read in 1 bit. Tag: 1100, i.e., 0.75.
- Again, an E2 scaling is triggered: shift out the first bit, read in 1 bit. Tag: 1000, i.e., 0.5.
  At this point, no more scaling is needed. The last symbol B can be decoded. Output B.

E3 Scaling

As an example, let's first see how E3 scaling works. Given an interval [0.48, 0.51), an E3 scaling will yield [0.46, 0.52), another E3 will yield [0.42, 0.54), and so on. Apparently, the intervals are gradually expanding. This addresses the original concern that representing tiny intervals could require very high precision numbers. In the subsequent steps of the incremental coding, it is very likely that a chosen range will fall in the first half ($high \leq 0.5$) or the second half ($low \geq 0.5$) of the unit interval, so an E1 or E2 will be triggered. The following can be shown [2]:

- N E3 scaling steps followed by an E1 is equivalent to an E1 followed by N E2 steps.
- N E3 scaling steps followed by an E2 is equivalent to an E2 followed by N E1 steps.

Therefore, a good way to handle the signaling of the E3 scaling is to postpone until there is an E1 or E2. If there is an E1 after N E3 operations, send "0" followed by N "1's" after the E1; if there is an E2 after N E3 operations, send "1" followed by N "0's" after the E2.

7.6.3 Integer Implementation

The algorithm described in the last section also has an integer implementation that uscs only integer arithmetic [2,13,19]. It is quite common in modern multimedia applications. It is a fairly straightforward extension of the original implementation. Basically, the unit interval is replaced by a range [0, N), where N is an integer, e.g., 255. Because the integer range could be so small, e.g., [0, 255), applying the scaling techniques similar to what was discussed above, now in integer arithmetic, is a necessity.

The main motivation of the integer implementation is, of course, to avoid any floating number operations.

7.6.4 Binary Arithmetic Coding

As described, the implementation of arithmetic coding involves continuous generation (calculation) of new intervals, and checking against delimiters of the intervals. When the number of symbols is large, this involves many calculations (integer or floating number) so it can be slow.

As the name suggests, *binary arithmetic coding* deals with two symbols only, i.e., 0 and 1. Figure 7.13 illustrates a simple scenario. It is obvious that only one new value inside the tag interval is generated at each step, i.e., 0.7, 0.49, 0.637, and 0.5929. The decision of which interval to take (first or second) is also simpler.

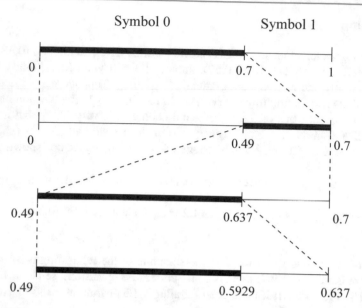

Fig. 7.13 Binary Arithmetic coding

The encoder and decoder including the scaling and possible integer implementation work the same way as for non-binary symbols.

Non-binary symbols can be converted to binary for binary arithmetic coding through *binarization*. Many coding schemes can be used for the binarization. We will introduce one of them, the Exp-Golomb code, in Chap. 12.

Fast binary arithmetic coding (Q-coder, MQ-coder) was developed in multimedia standards such as JBIG, JBIG2, and JPEG-2000. The more advanced version, context-adaptive binary arithmetic coding (CABAC), is used in H.264 (M-coder), H.265, and H.266.

7.6.5   Adaptive Arithmetic Coding

We now know that arithmetic coding can be performed incrementally. Hence, there is no need to know the probability distribution of all symbols in advance. This makes the codec process especially adaptive—we can record the current counts of the symbols received so far, and update the probability distribution after each symbol. The updated probability distribution will be used for dividing up the interval in the next step.

As in the adaptive Huffman coding, as long as the encoder and decoder are synchronized (i.e., using the same update rules), the adaptive process will work flawlessly. Nevertheless, adaptive arithmetic coding has a major advantage over adaptive Huffman coding: there is now no need to keep a (potentially) large and dynamic symbol table and constantly update the adaptive Huffman tree.

Below we outline the procedures for adaptive arithmetic coding, and give an example to illustrate how it also works for *binary* arithmetic coding.

Procedure 7.4 (Procedures for Adaptive Arithmetic Coding).

```
ENCODER                              DECODER
-------                              -------

Initialization (reset counters)      Initialization (reset counters)
while (symbol != terminator)         while (symbol != terminator)
    {                                    {
      get(symbol);                         decode(symbol);
      encode(symbol);                      output(symbol);
      update stats and interval;           update stats and interval;
    }                                    }
```

Example 7.5 (Adaptive Binary Arithmetic Coding).

Figure 7.14 illustrates the encoder for adaptive binary arithmetic coding. We assume the input symbol to the encoder is 10001. Again, for this simple example, we assume that both the encoder and decoder know the length of the sequence. [For clarity and simplicity, we will not invoke the scaling procedure (for E1, E2, etc.) introduced earlier.]

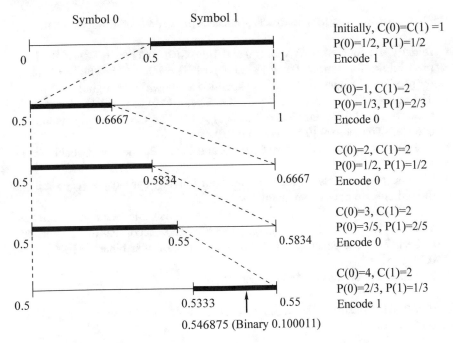

Fig. 7.14 Adaptive binary arithmetic coding—encoder. [Input symbols:10001]

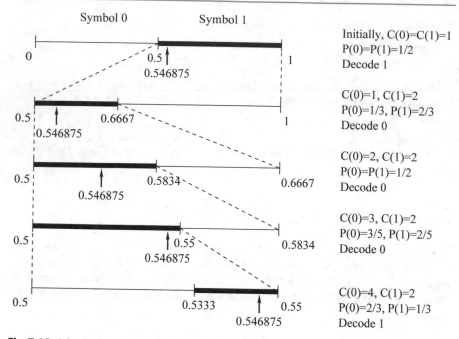

Fig. 7.15 Adaptive binary arithmetic coding—decoder. [Input: 0.546875 (Binary 0.100011)]

- Initially, the counters for symbols 0 and 1 are $C(0) = C(1) = 1$. Hence, the initial probabilities for symbols 0 and 1 are $P(0) = P(1) = 1/2$. The first binary symbol 1 is encoded. The interval $[0.5, 1)$ is expanded as shown in the figure.
- The counter $C(1)$ is updated, so $C(1) = 2$ (and $C(0)$ remains 1). Accordingly, the new probabilities are $P(0) = 1/3$, $P(1) = 2/3$. The second binary symbol 0 is encoded. The interval $[0.5, 0.6667)$ is expanded.
- The counter $C(0)$ is updated, so $C(0) = 2$ and $C(1) = 2$. The new probabilities are $P(0) = 1/2$, $P(1) = 1/2$.
- The encoding process for the third and fourth binary symbols is very similar to the last one. So now we are at the last line.
- We will encode the fifth (last) symbol 1, and finish this example by selecting a number that is inside the interval $[0.5333, 0.55)$. According to the procedure in Sect. 7.6.1 for generating the shortest codeword, it is binary 0.100011, i.e., 0.546875.
  The encoder outputs 0.546875 (binary 0.100011).

Figure 7.15 illustrates the decoder. It receives the input code 0.546875 (binary 0.100011) from the encoder.

- As in the encoder, initially, the counters for symbols 0 and 1 are $C(0) = C(1) = 1$ in the decoder. Hence, the initial probabilities for symbols 0 and 1 are $P(0) = P(1) = 1/2$. The first binary symbol 1 is decoded. The interval $[0.5, 1)$ is expanded.

- The counter C(1) is updated, so C(1) = 2 (and C(0) remains 1). The new probabilities are P(0) = 1/3, P(1) = 2/3. The second binary symbol 0 is decoded. The interval [0.5, 0.6667) is expanded.
- The counter C(0) is updated, so C(0) = 2 and C(1) = 2. The new probabilities are P(0) = 1/2, P(1) = 1/2.
- The decoding process for the third and fourth binary symbols is very similar to the last one. So now we are again at the last line.
- The fifth (last) symbol 1 is decoded in this small example, because 0.546875 is inside the interval [0.5333, 0.55).

Note, normally, the encoder and decoder work incrementally, and they are capable of continuously handling a long sequence of symbols until a terminator is finally reached.

7.7   Lossless Image Compression

One of the most commonly used compression techniques in multimedia data compression is *differential coding*. The basis of data reduction in differential coding is the redundancy in consecutive symbols in a data stream. Recall that we considered lossless differential coding in Chap. 6, when we examined how audio must be dealt with via subtraction from predicted values. Audio is a signal indexed by one dimension, time. Here we consider how to apply the lessons learned from audio to the context of digital image signals that are indexed by two, spatial, dimensions (x, y).

7.7.1   Differential Coding of Images

Let's consider differential coding in the context of digital images. In a sense, we move from signals with domain in one dimension to signals indexed by numbers in two dimensions (x, y)—the rows and columns of an image. Later, we'll look at video signals. These are even more complex, in that they are indexed by space and time (x, y, t).

Because of the continuity of the physical world, the gray-level intensities (or color) of background and foreground objects in images tend to change relatively slowly across the image frame. Since we were dealing with signals in the time domain for audio, practitioners generally refer to images as signals in the *spatial domain*. The generally slowly changing nature of imagery spatially produces a high likelihood that neighboring pixels will have similar intensity values. Given an original image $I(x, y)$, using a simple difference operator, we can define a difference image $d(x, y)$ as follows:

$$d(x, y) = I(x, y) - I(x - 1, y) \tag{7.9}$$

This is a simple approximation of a partial differential operator $\partial/\partial x$ applied to an image defined in terms of integer values of x and y.

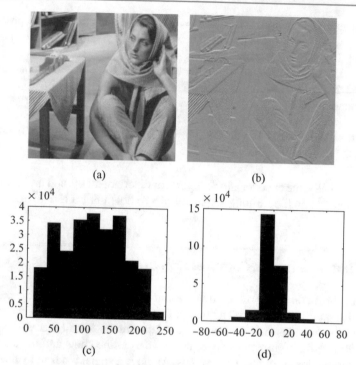

Fig. 7.16 Distributions for original versus derivative images. **a, b** Original gray-level image and its partial derivative image; **c, d** histograms for original and derivative images. This figure uses a commonly employed image called Barb

Another approach is to use the discrete version of the 2D Laplacian operator to define a difference image $d(x, y)$ as

$$d(x, y) = 4\,I(x, y) - I(x, y - 1) - I(x, y + 1) - I(x + 1, y) - I(x - 1, y)$$

$$(7.10)$$

In both cases, the difference image will have a histogram as in Fig. 7.16d, derived from the $d(x, y)$ partial derivative image in Fig. 7.16b for the original image I in Fig. 7.16a. Notice that the histogram for the unsubtracted I itself is much broader, as in Fig. 7.16c. It can be shown that image I has larger entropy than the difference image d, since it has a more even distribution in its intensity values. Consequently, Huffman coding or some other variable-length coding scheme will produce shorter bit-length codewords for the difference image. Compression will work better on a difference image.

7.7.2 Lossless JPEG

Lossless JPEG is a special case of JPEG image compression. It differs drastically from other JPEG modes in that the algorithm has no lossy steps. Thus, we treat it here

Fig. 7.17 Neighboring pixels for predictors in lossless JPEG. Note that any of A, B, or C has already been decoded before it is used in the predictor, on the decoder side of an encode/decode cycle

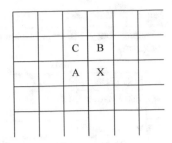

and consider the more used lossy JPEG methods in Chap. 9. Lossless JPEG is invoked when the user selects a 100% *quality factor* in an image tool. Essentially, lossless JPEG is included in the JPEG compression standard simply for completeness.

The following predictive method is applied on the unprocessed original image (or each color band of the original color image). It essentially involves two steps: forming a differential prediction and encoding.

1. A predictor combines the values of up to three neighboring pixels as the predicted value for the current pixel, indicated by X in Fig. 7.17. The predictor can use any one of the seven schemes listed in Table 7.6. If predictor P1 is used, the neighboring intensity value A will be adopted as the predicted intensity of the current pixel; if predictor P4 is used, the current pixel value is derived from the three neighboring pixels as $A + B - C$; and so on.
2. The encoder compares the prediction with the actual pixel value at position X and encodes the difference using one of the lossless compression techniques we have discussed, such as the Huffman coding scheme.

Since prediction must be based on previously encoded neighbors, the very first pixel in the image $I(0, 0)$ will have to simply use its own value. The pixels in the first row always use predictor P1, and those in the first column always use P2.

Lossless JPEG usually yields a relatively low compression ratio, which renders it impractical for most multimedia applications. An empirical comparison using some 20 images indicates that the compression ratio for lossless JPEG with any one of the seven predictors ranges from 1.0 to 3.0, with an average of around 2.0. Predictors

Table 7.6 Predictors for lossless JPEG

Predictor	Prediction
P1	A
P2	B
P3	C
P4	$A + B - C$
P5	$A + (B - C) / 2$
P6	$B + (A - C) / 2$
P7	$(A + B) / 2$

Table 7.7 Comparison of lossless JPEG with other lossless compression programs

Compression program	Compression ratio			
	Lena	Football	F-18	Flowers
Lossless JPEG	1.45	1.54	2.29	1.26
Optimal lossless JPEG	1.49	1.67	2.71	1.33
Compress (LZW)	0.86	1.24	2.21	0.87
Gzip (LZ77)	1.08	1.36	3.10	1.05
Gzip-9 (optimal LZ77)	1.08	1.36	3.13	1.05
Pack (Huffman coding)	1.02	1.12	1.19	1.00

4–7 that consider neighboring nodes in both horizontal and vertical dimensions offer slightly better compression (approximately 0.2–0.5 higher) than predictors 1–3.

Table 7.7 shows a comparison of the compression ratio for several lossless compression techniques using test images Lena, football, F-18, and flowers. These standard images used for many purposes in imaging work are shown on the textbook website for this chapter.

This chapter has been devoted to the discussion of lossless compression algorithms. It should be apparent that their *compression ratio* is generally limited (with a maximum at about 2–3). However, many of the multimedia applications we will address in the next several chapters require a much higher compression ratio. This is accomplished by *lossy* compression schemes.

7.8   Exercises

1. Calculate the *entropy* of a "checkerboard" image in which half of the pixels are BLACK and half of them are WHITE.
2. Suppose eight characters have a distribution A:(1), B:(1), C:(1), D:(2), E:(3), F:(5), G:(5), H:(10). Draw a Huffman tree for this distribution. (Because the algorithm may group subtrees with equal probability in a different order, your answer is not strictly unique.)
3. (a) What is the entropy η of the image below, where numbers (0, 20, 50, 99) denote the gray-level intensities?

$$
\begin{array}{cccccccc}
99 & 99 & 99 & 99 & 99 & 99 & 99 & 99 \\
20 & 20 & 20 & 20 & 20 & 20 & 20 & 20 \\
0 & 0 & 0 & 0 & 0 & 0 & 0 & 0 \\
0 & 0 & 50 & 50 & 50 & 50 & 0 & 0 \\
0 & 0 & 50 & 50 & 50 & 50 & 0 & 0 \\
0 & 0 & 50 & 50 & 50 & 50 & 0 & 0 \\
0 & 0 & 50 & 50 & 50 & 50 & 0 & 0 \\
0 & 0 & 0 & 0 & 0 & 0 & 0 & 0 \\
\end{array}
$$

(b) Show step by step how to construct the Huffman tree to encode the above four intensity values in this image. Show the resulting code for each intensity value.

(c) What is the average number of bits needed for each pixel, using your Huffman code? How does it compare to η?

4. Consider an alphabet with two symbols A, B, with probability $P(A) = x$ and $P(B) = 1 - x$.

   (a) Plot the entropy as a function of x. You might want to use $\log_2 3 = 1.6$ and $\log_2 7 = 2.8$.

   (b) Discuss why it must be the case that if the probability of the two symbols is $1/2 + \epsilon$ and $1/2 - \epsilon$, with small ϵ, the entropy is less than the maximum.

   (c) Generalize the above result by showing that, for a source generating N symbols, the entropy is maximum when the symbols are all equiprobable.

   (d) As a small programming project, write code to verify the conclusions above.

5. Extended Huffman coding assigns one codeword to each group of k symbols. Why is $average(l)$ (the average number of bits for each symbol) still no less than the entropy η as indicated in Eq. (7.7)?

6. (a) Suppose we are coding a *binary* source, i.e., the alphabet consists of 0 or 1. For example, a fax is like this.
   Suppose the probability of a 0 is 7/8, and that for a 1 is 1/8.
   What is the entropy? (Note: In case you don't have a calculator, $\log_2 7 \approx 2.81$.)
   What is the set of Huffman codes? And what is the average bitrate?

   (b) Now code the problem in terms of extended Huffman compression, using $n = 2$ and groups of $k = 2$ pairs of symbols. What is the average bitrate now? Show your work (you can just use fractions, not decimal numbers, if you like).

7. Arithmetic coding and Huffman coding are two popular lossless compression methods.

   (a) What are the advantages and disadvantages of arithmetic coding as compared to Huffman coding?

   (b) Suppose the alphabet is $[A, B, C]$, and the known probability distribution is $P_A = 0.5$, $P_B = 0.4$, $P_C = 0.1$. For simplicity, let's also assume that both encoder and decoder know that the length of the messages is always 3, so there is no need for a terminator.

      i. How many bits are needed to encode the message BBB by Huffman coding?

      ii. How many bits are needed to encode the message BBB by arithmetic coding?

Fig. 7.18 Adaptive Huffman
tree

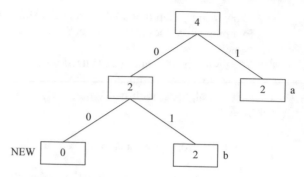

8. (a) What are the advantages of adaptive Huffman coding compared to the original Huffman coding algorithm?
   (b) Assume that adaptive Huffman coding is used to code an information source
   S with a vocabulary of four letters (a, b, c, d). Before any transmission, the
   initial coding is a = 00, b = 01, c = 10, d = 11. As in the example illustrated
   in Fig. 7.8, a special symbol NEW will be sent before any letter if it is to be
   sent the first time.

   Figure 7.18 is the adaptive Huffman tree after sending letters **aabb**. After
   that, the additional bitstream received by the decoder for the next few letters
   is 01010010101.

   i. What are the additional letters received?
   ii. Draw the adaptive Huffman trees after each of the additional letters is
   received.

9. Work out the details of scaling and incremental coding in arithmetic coding when
the probabilities for the three symbols are A: 0.8, B: 0.02, C: 0.18, and the input
sequence is ACBA.

10. We will use arithmetic coding to encode images.

   (a) At most, how many bits does it take to encode the first row of the image
   below, where all numbers are gray-level intensity values?

200 200 80 40 20 80 200 200

200 200 80 40 20 80 200 200

200 200 80 40 20 80 200 200

200 200 80 40 20 80 200 200

200 200 80 40 20 80 200 200

200 200 80 40 20 80 200 200

200 200 80 40 20 80 200 200

200 200 80 40 20 80 200 200

   (b) (i) Show in detail how to encode the first four numbers (200, 200, 80, 40)
   in the image above using *arithmetic coding*. (Note: You must use E1 or E2

scaling when applicable.)

(ii) What is the code (sequence of bits) that the encoder generates for the four numbers? Explain how you get the code.

11. Work out the details of the encoder and decoder for adaptive binary arithmetic coding when the input symbols are 01111.
12. Compare the rate of adaptation of adaptive Huffman coding and adaptive arithmetic coding. What prevents each method from adapting to quick changes in source statistics?
13. Consider the dictionary-based LZW compression algorithm. Suppose the alphabet is the set of symbols { 0 , 1 }. Show the dictionary (symbol sets plus associated codes) and output for LZW compression of the input

$$0 \quad 1 \quad 1 \quad 0 \quad 0 \quad 1 \quad 1$$

14. Implement Huffman coding, LZW coding, and arithmetic coding algorithms using your favorite programming language. Generate at least three types of statistically different artificial data sources to test your implementation of these algorithms. Compare and comment on each algorithm's performance in terms of compression ratio for each type of data source.

Optionally, implement adaptive Huffman and adaptive arithmetic coding algorithms.

References

1. M. Nelson, J.L. Gailly, *The Data Compression Book*, 2nd edn. (M&T Books, New York, 1995)
2. K. Sayood, *Introduction to Data Compression*, 5th edn. (Morgan Kaufmann, San Francisco, 2017)
3. C.E. Shannon, A mathematical theory of communication. Bell Syst. Tech. J. **27**, 379–423, 623–656 (1948)
4. C.E. Shannon, W. Weaver, *The Mathematical Theory of Communication* (University of Illinois Press, 1971)
5. R.C. Gonzalez, R.E. Woods, *Digital Image Processing*, 3rd edn. (Prentice-Hall, 2007)
6. R. Fano, *Transmission of Information*. (MIT Press, 1961)
7. D.A. Huffman, A method for the construction of minimum-redundancy codes. Proc. IRE **40**(9), 1098–1101 (1952)
8. T.H. Cormen, C.E. Leiserson, R.L. Rivest, *Introduction to Algorithms*, 3rd edn. (The MIT Press, Cambridge, Massachusetts, 2009)
9. J. Ziv, A. Lempel, A universal algorithm for sequential data compression. IEEE Trans. Inf. Theory **23**(3), 337–343 (1977)
10. J. Ziv, A. Lempel, Compression of individual sequences via variable-rate coding. IEEE Trans. Inf. Theory **24**(5), 530–536 (1978)
11. T.A. Welch, A technique for high performance data compression. IEEE Comput. **17**(6), 8–19 (1984)

12. J. Rissanen, G.G. Langdon, Arithmetic coding. IBM J. Res. Dev. **23**(2), 149–162 (1979)
13. I.H. Witten, R.M. Neal, J.G. Cleary, Arithmetic coding for data compression. Commun. ACM **30**(6), 520–540 (1987)
14. T.C. Bell, J.G. Cleary, I.H. Witten, *Text Compression* (Prentice Hall, Englewood Cliffs, New Jersey, 1990)
15. N. Abramson, *Information Theory and Coding* (McGraw-Hill, New York, 1963)
16. F. Jelinek, *Probabilistic Information Theory* (McGraw-Hill, New York, 1968)
17. R. Pasco, *Source Coding Algorithms for Data Compression*. Ph.D. thesis, Department of Electrical Engineering. (Stanford University, 1976)
18. S.M. Lei, M.T. Sun, An entropy coding system for digital HDTV applications. IEEE Trans. Circuits Syst. Video Technol. **1**(1), 147–154 (1991)
19. P.G. Howard, J.S. Vitter, Practical implementation of arithmetic coding, in *Image and Text Compression*, ed. by J.A. Storer, pp. 85–112. (Kluwer Academic Publishers, 1992)

Lossy Compression Algorithms

8

In this chapter, we consider *lossy* compression methods. Since information loss implies some trade-off between error and bitrate, we first consider measures of *distortion*—e.g., squared error. Different quantizers are introduced, each of which has a different distortion behavior. A discussion of transform coding leads into an introduction to the Discrete Cosine Transform used in JPEG compression (see Chap. 9) and the Karhunen Loève transform. Another transform scheme, wavelet-based coding, is then set out.

Sayood [1] deals extensively with the subject of lossy data compression in a well-organized and easy-to-understand manner. The mathematical foundation for the development of many lossy data compression algorithms is the study of *stochastic processes*. Stark and Woods [2] is an excellent textbook on this subject.

8.1 Introduction

As discussed in Chap. 7, the *compression ratio* for image data using lossless compression techniques (e.g., Huffman Coding, Arithmetic Coding, and LZW) is low when the image histogram is relatively flat. For image compression in multimedia applications, where a higher compression ratio is required, lossy methods are usually adopted. In lossy compression, the compressed image is not the same as the original image but is meant to form a close approximation to the original image *perceptually*. To quantitatively describe how close the approximation is to the original data, some form of distortion measure is required.

© Springer Nature Switzerland AG 2021
Z.-N. Li et al., *Fundamentals of Multimedia*, Texts in Computer Science,
https://doi.org/10.1007/978-3-030-62124-7_8

8.2   Distortion Measures

A *distortion measure* is a mathematical quantity that specifies how close an approximation is to its original, using some distortion criteria. When looking at compressed data, it is natural to think of the distortion in terms of the numerical difference between the original data and the reconstructed data. However, when the data to be compressed is an image, such a measure may not yield the intended result.

For example, if the reconstructed image is the same as the original image except that it is shifted to the right by one vertical scan line, an average human observer would have a hard time distinguishing it from the original and would therefore conclude that the distortion is small. However, when the calculation is carried out numerically, we find a large distortion, because of the large changes in individual pixels of the reconstructed image. The problem is that we need a measure of *perceptual distortion*, not a more naive numerical approach. However, the study of perceptual distortions is beyond the scope of this book.

Of the many numerical distortion measures that have been defined, we present the three most commonly used in image compression. If we are interested in the average pixel difference, the *Mean Squared Error (MSE)* σ_d^2 is often used. It is defined as

$$\sigma_d^2 = \frac{1}{N} \sum_{n=1}^{N} (x_n - y_n)^2 \tag{8.1}$$

where x_n, y_n, and N are the input data sequence, reconstructed data sequence, and length of the data sequence, respectively.

If we are interested in the size of the error relative to the signal, we can measure the *signal-to-noise ratio (SNR)* by taking the ratio of the average square of the original data sequence and the Mean Squared Error (MSE), as discussed in Chap. 6. In decibel units (dB), it is defined as

$$SNR = 10 \log_{10} \frac{\sigma_x^2}{\sigma_d^2} \tag{8.2}$$

where σ_x^2 is the average squared value of the original data sequence and σ_d^2 is the MSE. Another commonly used measure for distortion is the *peak-signal-to-noise ratio (PSNR)*, which measures the size of the error relative to the peak value of the signal x_{peak}. It is given by

$$PSNR = 10 \log_{10} \frac{x_{\text{peak}}^2}{\sigma_d^2} \tag{8.3}$$

8.3   The Rate-Distortion Theory

Lossy compression always involves a trade-off between rate and distortion. Rate is the average number of bits required to represent each source symbol. Within this framework, the trade-off between rate and distortion is represented in the form of a rate-distortion function $R(D)$.

Intuitively, for a given source and a given distortion measure, if D is a tolerable amount of distortion, $R(D)$ specifies the lowest rate at which the source data can be encoded while keeping the distortion bounded above by D. It is easy to see that when $D = 0$, we have a lossless compression of the source. The rate-distortion function is meant to describe a fundamental limit for the performance of a coding algorithm and so can be used to evaluate the performance of different algorithms.

Figure 8.1 shows a typical rate-distortion function. Notice that the minimum possible rate at $D = 0$, no loss, is the entropy of the source data. The distortion corresponding to a rate $R(D) \equiv 0$ is the maximum amount of distortion incurred when "nothing" is coded.

Finding a closed-form analytic description of the rate-distortion function for a given source is difficult, if not impossible. Gyorgy [3] presents analytic expressions of the rate-distortion function for various sources. For sources for which an analytic solution cannot be readily obtained, the rate-distortion function can be calculated numerically, using algorithms developed by Arimoto [4] and Blahut [5].

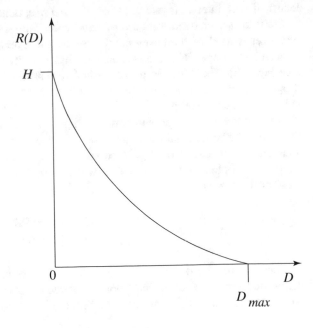

Fig. 8.1 Typical rate-distortion function

8.4 Quantization

Quantization in some form is the heart of any lossy scheme. Without quantization, we would indeed be losing little information. Here, we embark on a more detailed discussion of quantization than in Sect. 6.3.2.

The source we are interested in compressing may contain a large number of distinct output values (or even infinite, if analog). To efficiently represent the source output, we have to reduce the number of distinct values to a much smaller set, via quantization.

Each algorithm (each *quantizer*) can be uniquely determined by its partition of the input range, on the encoder side, and the set of output values, on the decoder side. The input and output of each quantizer can be either scalar values or vector values, thus leading to *scalar quantizers* and *vector quantizers*. In this section, we examine the design of both uniform and nonuniform scalar quantizers and briefly introduce the topic of *vector quantization* (VQ).

8.4.1 Uniform Scalar Quantization

A uniform scalar quantizer partitions the domain of input values into equally spaced intervals, except possibly at the two outer intervals. The end points of partition intervals are called the quantizer's *decision boundaries*. The output or reconstruction value corresponding to each interval is taken to be the midpoint of the interval. The length of each interval is referred to as the *step size*, denoted by the symbol Δ.

Uniform scalar quantizers are of two types: *midrise* and *midtread*. A midrise quantizer is used with an even number of output levels, and a midtread quantizer with an odd number. The midrise quantizer has a partition interval that brackets zero (see Fig. 8.2). The midtread quantizer has zero as one of its output values, hence, it is also known as *dead-zone* quantizer, because it turns a range of nonzero input values into the zero output.

The midtread quantizer is important when source data represents the zero value by fluctuating between small positive and negative numbers. Applying the midtread quantizer in this case would produce an accurate and steady representation of the value zero. For the special case $\Delta = 1$, we can simply compute the output values for these quantizers as

$$Q_{midrise}(x) = \lceil x \rceil - 0.5 \tag{8.4}$$
$$Q_{midtread}(x) = \lfloor x + 0.5 \rfloor \tag{8.5}$$

The goal for the design of a successful uniform quantizer is to minimize the distortion for a given source input with a desired number of output values. This can be done by adjusting the step size Δ to match the input statistics.

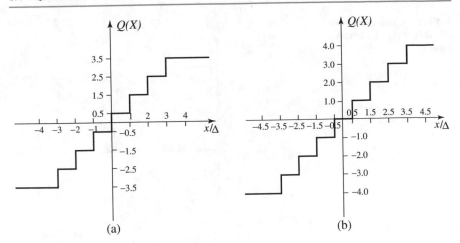

Fig. 8.2 Uniform scalar quantizers: **a** Midrise; **b** midtread

Let's examine the performance of an M level quantizer. Let $B = \{b_0, b_1, \ldots, b_M\}$ be the set of decision boundaries and $Y = \{y_1, y_2, \ldots, y_M\}$ be the set of reconstruction or output values. Suppose the input is uniformly distributed in the interval $[-X_{max}, X_{max}]$. The rate of the quantizer is

$$R = \lceil \log_2 M \rceil \tag{8.6}$$

That is, R is the number of bits required to code M things—in this case, the M output levels.

The step size Δ is given by

$$\Delta = \frac{2X_{max}}{M} \tag{8.7}$$

since the entire range of input values is from $-X_{max}$ to X_{max}. For bounded input, the quantization error caused by the quantizer is referred to as *granular distortion*. If the quantizer replaces a whole range of values, from a maximum value to ∞, and similarly for negative values, that part of the distortion is called the *overload* distortion.

To get an overall figure for granular distortion, notice that decision boundaries b_i for a midrise quantizer are $[(i-1)\Delta, i\Delta]$, $i = 1 .. M/2$, covering positive data X (and another half for negative X values). Output values y_i are the midpoints $i\Delta - \Delta/2$, $i = 1 .. M/2$, again just considering positive data. The total distortion is twice the sum over the positive data, or

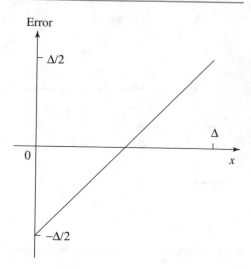

Fig. 8.3 Quantization error of a uniformly distributed source

$$D_{gran} = 2 \sum_{i=1}^{\frac{M}{2}} \int_{(i-1)\Delta}^{i\Delta} \left(x - \frac{2i-1}{2}\Delta \right)^2 \frac{1}{2X_{max}} \, dx \tag{8.8}$$

where we divide by the range of X to normalize to a value of at most 1.

Since the reconstruction values y_i are the midpoints of each interval, the quantization error must lie within the values $[-\frac{\Delta}{2}, \frac{\Delta}{2}]$. Figure 8.3 is a graph of quantization error for a uniformly distributed source. The quantization error in this case is also uniformly distributed. Therefore, the average squared error is the same as the variance σ_d^2 of the quantization error calculated from just the interval $[0, \Delta]$ with error values in $[-\frac{\Delta}{2}, \frac{\Delta}{2}]$. The error value at x is $e(x) = x - \Delta/2$, so the variance of errors is given by

$$
\begin{aligned}
\sigma_d^2 &= \frac{1}{\Delta} \int_0^\Delta (e(x) - \bar{e})^2 \, dx \\
&= \frac{1}{\Delta} \int_0^\Delta \left(x - \frac{\Delta}{2} - 0 \right)^2 \, dx \\
&= \frac{\Delta^2}{12}
\end{aligned} \tag{8.9}
$$

Similarly, the *signal* variance is $\sigma_x^2 = (2X_{max})^2/12$ for a random signal, so if the quantizer is n bits, $M = 2^n$, then from Eq. (8.2) we have

$$
\begin{aligned}
SQNR &= 10\log_{10}\left(\frac{\sigma_x^2}{\sigma_d^2}\right) \\
&= 10\log_{10}\left(\frac{(2X_{max})^2}{12} \cdot \frac{12}{\Delta^2}\right) \\
&= 10\log_{10}\left(\frac{(2X_{max})^2}{12} \cdot \frac{12}{\left(\frac{2X_{max}}{M}\right)^2}\right) \\
&= 10\log_{10} M^2 = 20\, n\, \log_{10} 2 \qquad\qquad (8.10) \\
&= 6.02\, n\ (dB) \qquad\qquad (8.11)
\end{aligned}
$$

Hence, we have rederived the formula (6.3) derived more simply in Sect. 6.1. From Eq. (8.11), we have the important result that increasing one bit in the quantizer increases the signal-to-quantization noise ratio by 6.02 dB. In Sect. 6.1.5, we also showed that if we know the signal probability density function we can get a more accurate figure for the SQNR: there we assumed a sinusoidal signal and derived a more exact SQNR Eq. (6.4). As well, more sophisticated estimates of D result from more sophisticated models of the probability distribution of errors.

8.4.2 Nonuniform Scalar Quantization

If the input source is not uniformly distributed, a uniform quantizer may be inefficient. Increasing the number of decision levels within the region where the source is densely distributed can effectively lower granular distortion. In addition, without having to increase the total number of decision levels, we can enlarge the region in which the source is sparsely distributed. Such *nonuniform quantizers* thus have nonuniformly defined decision boundaries.

There are two common approaches for nonuniform quantization: the Lloyd–Max quantizer and the companded quantizer, both introduced in Chap. 6.

Lloyd–Max Quantizer*

For a uniform quantizer, the total distortion is equal to the granular distortion, as in Eq. (8.8). If the source distribution is not uniform, we must explicitly consider its probability distribution (*probability density function*) $f_X(x)$. Now we need the correct decision boundaries b_i and reconstruction values y_i, by solving for both simultaneously. To do so, we plug variables b_i, y_i into a total distortion measure

$$D_{gran} = \sum_{j=1}^{M} \int_{b_{j-1}}^{b_j} (x - y_j)^2 \frac{1}{X_{max}} f_X(x)\, dx \tag{8.12}$$

Then we can minimize the total distortion by setting the derivative of Eq. (8.12) to zero. Differentiating with respect to y_j yields the set of reconstruction values

$$y_j = \frac{\int_{b_{j-1}}^{b_j} x f_X(x)\, dx}{\int_{b_{j-1}}^{b_j} f_X(x)\, dx} \tag{8.13}$$

This says that the optimal reconstruction value is the weighted centroid of the x interval. Differentiating with respect to b_j and setting the result to zero yields

$$b_j = \frac{y_{j+1} + y_j}{2} \tag{8.14}$$

This gives a decision boundary b_j at the midpoint of two adjacent reconstruction values. Solving these two equations simultaneously is carried out by iteration. The result is termed the Lloyd–Max quantizer.

Algorithm 8.1 (Lloyd–Max Quantization)

```
BEGIN
   Choose initial level set y₀
   i = 0
   Repeat
      Compute bᵢ using Equation 8.14
      i = i + 1
      Compute yᵢ using Equation 8.13
   Until |yᵢ − yᵢ₋₁| < ε
END
```

Starting with an initial guess of the optimal reconstruction levels, the algorithm above iteratively estimates the optimal boundaries, based on the current estimate of the reconstruction levels. It then updates the current estimate of the reconstruction levels, using the newly computed boundary information. The process is repeated until the reconstruction levels converge. For an example of the algorithm in operation, see Exercise 3.

Companded Quantizer

In companded quantization, the input is mapped by a *compressor function G* and then quantized using a uniform quantizer. After transmission, the quantized values are mapped back using an *expander function* G^{-1}. The block diagram for the companding process is shown in Fig. 8.4, where $\hat{X}$ is the quantized version of X. If the input source is bounded by x_{max}, then any nonuniform quantizer can be represented as a companded quantizer. The two commonly used companders are the μ-law and A-law companders (Sect. 6.1).

8.4.3   Vector Quantization

One of the fundamental ideas in Shannon's original work on information theory is that any compression system performs better if it operates on vectors or groups of samples rather than on individual symbols or samples. We can form vectors of input samples by concatenating a number of consecutive samples into a single vector. For example, an input vector might be a segment of a speech sample, a group of consecutive pixels in an image, or a chunk of data in any other format.

The idea behind *vector quantization* (VQ) is similar to that of scalar quantization but extended into multiple dimensions. Instead of representing values within an interval in 1D space by a reconstruction value, as in scalar quantization, in VQ an n-component *code vector* represents vectors that lie within a region in n-dimensional space. A collection of these code vectors forms the *codebook* for the vector quantizer.

Since there is no implicit ordering of code vectors, as there is in the 1D case, an index set is also needed to index into the codebook. Figure 8.5 shows the basic vector quantization procedure. In the diagram, the encoder finds the closest code vector to the input vector and outputs the associated index. On the decoder side, exactly the same codebook is used. When the coded index of the input vector is received, a simple table lookup is performed to determine the reconstruction vector.

Finding the appropriate codebook and searching for the closest code vector at the encoder end may require considerable computational resources. However, the decoder can execute quickly, since only a constant time operation is needed to obtain the reconstruction. Because of this property, VQ is attractive for systems with a lot of resources at the encoder end while the decoder has only limited resources, and the need is for quick execution time. Most multimedia applications fall into this category.

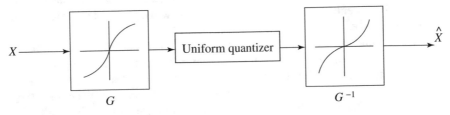

Fig. 8.4 Companded quantization

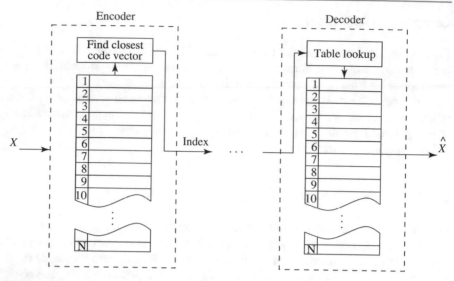

Fig. 8.5 Basic vector quantization procedure

Gersho and Gray [6] cover quantization, especially vector quantization, comprehensively. In addition to the basic theory, this book provides a nearly exhaustive description of available VQ methods.

8.5   Transform Coding

From basic principles of information theory, we know that coding vectors are more efficient than coding scalars (see Sect. 7.4.2). To carry out such an intention, we need to group blocks of consecutive samples from the source input into vectors.

Let $\mathbf{X} = \{x_1, x_2, \ldots, x_k\}^T$ be a vector of samples. Whether our input data is an image, a piece of music, an audio or video clip, or even a piece of text, there is a good chance that a substantial amount of correlation is inherent among neighboring samples x_i. The rationale behind transform coding is that if $\mathbf{Y}$ is the result of a linear transform $\mathbf{T}$ of the input vector $\mathbf{X}$ in such a way that the components of $\mathbf{Y}$ are much less correlated, then $\mathbf{Y}$ can be coded more efficiently than $\mathbf{X}$.

For example, if most information in an RGB image is contained in the main axis, rotating so that this direction is the first component means that luminance can be compressed differently from color information. This will approximate the luminance channel in the eye.

In higher dimensions than three, if most information is accurately described by the first few components of a transformed vector, the remaining components can be coarsely quantized, or even set to zero, with little signal distortion. The more *decorrelated*—that is, the less effect one dimension has on another (the more orthogonal the axes)—the more chance we have of dealing differently with the axes that store relatively minor amounts of information without affecting reasonably accurate reconstruction of the signal from its quantized or truncated transform coefficients.

Generally, the transform **T** itself does not compress any data. The compression comes from the processing and *quantization* of the components of **Y**. In this section, we will study the Discrete Cosine Transform (DCT) as a tool to decorrelate the input signal. We will also examine the Karhunen–Loève transform (KLT), which *optimally* decorrelates the components of the input **X**.

8.5.1 Discrete Cosine Transform (DCT)

The discrete cosine transform (DCT), a widely used transform coding technique, is able to perform decorrelation of the input signal in a data-independent manner [7,8]. Because of this, it has gained tremendous popularity. We will examine the definition of the DCT and discuss some of its properties, in particular the relationship between it and the more familiar discrete Fourier transform (DFT).

Definition of DCT

Let's start with the 2D DCT. Given a function $f(i, j)$ over two integer variables i and j (a piece of an image), the 2D DCT transforms it into a new function $F(u, v)$, with integers u and v running over the same range as i and j. The general definition of the transform is

$$F(u, v) = \frac{2\, C(u)\, C(v)}{\sqrt{MN}} \sum_{i=0}^{M-1} \sum_{j=0}^{N-1} \cos \frac{(2i + 1)u\pi}{2M} \cos \frac{(2j + 1)v\pi}{2N} f(i, j) \quad (8.15)$$

where $i, u = 0, 1, \ldots, M - 1$, $j, v = 0, 1, \ldots, N - 1$, and the constants $C(u)$ and $C(v)$ are determined by

$$C(\xi) = \begin{cases} \frac{\sqrt{2}}{2} & \text{if } \xi = 0, \\ 1 & \text{otherwise.} \end{cases} \quad (8.16)$$

In the JPEG image compression standard (see Chap. 9), an image block is defined to have dimension $M = N = 8$. Therefore, the definitions for the 2D DCT and its inverse (IDCT) in this case are as follows.

2D Discrete Cosine Transform (2D DCT)

$$F(u, v) = \frac{C(u)\, C(v)}{4} \sum_{i=0}^{7} \sum_{j=0}^{7} \cos \frac{(2i + 1)u\pi}{16} \cos \frac{(2j + 1)v\pi}{16} f(i, j), \quad (8.17)$$

where $i, j, u, v = 0, 1, \ldots, 7$, and the constants $C(u)$ and $C(v)$ are determined by Eq. (8.16).

2D Inverse Discrete Cosine Transform (2D IDCT)

The inverse function is almost the same, with the roles of $f(i, j)$ and $F(u, v)$ reversed, except that now $C(u)C(v)$ must stand inside the sums:

$$\tilde{f}(i, j) = \sum_{u=0}^{7} \sum_{v=0}^{7} \frac{C(u)\,C(v)}{4} \cos \frac{(2i + 1)u\pi}{16} \cos \frac{(2j + 1)v\pi}{16} F(u, v) \quad (8.18)$$

where $i, j, u, v = 0, 1, \ldots, 7$, and the constants $C(u)$ and $C(v)$ are determined by Eq. (8.16).

The 2D transforms are applicable to 2D signals, such as digital images. As shown below, the 1D version of the DCT and IDCT is similar to the 2D version.

1D Discrete Cosine Transform (1D DCT)

$$F(u) = \frac{C(u)}{2} \sum_{i=0}^{7} \cos \frac{(2i + 1)u\pi}{16} f(i), \quad (8.19)$$

where $i = 0, 1, \ldots, 7$, $u = 0, 1, \ldots, 7$, and the constant $C(u)$ is the same as in Eq. (8.16).

1D Inverse Discrete Cosine Transform (1D-IDCT)

$$\tilde{f}(i) = \sum_{u=0}^{7} \frac{C(u)}{2} \cos \frac{(2i + 1)u\pi}{16} F(u), \quad (8.20)$$

where $i = 0, 1, \ldots, 7$, $u = 0, 1, \ldots, 7$, and the constant $C(u)$ is the same as in Eq. (8.16).

A Closer Look at 1D DCT

Let's examine the DCT for a 1D signal; almost all concepts are readily extensible to the 2D DCT.

An electrical signal with constant magnitude is known as a DC (direct current) signal. A common example is a battery that carries 1.5 or 9 volts DC. An electrical signal that changes its magnitude periodically at a certain frequency is known as an AC (alternating current) signal. A good example is the household electric power circuit, which carries electricity with sinusoidal waveform at 110 V AC, 60 Hz (or 220 V, 50 Hz in many other countries).

Most real signals are more complex. Speech signals or a row of gray-level intensities in a digital image are examples of such 1D signals. However, any signal can be expressed as a sum of multiple signals that are sine or cosine waveforms at various amplitudes and frequencies. This is known as Fourier analysis. The terms DC and AC, originating in electrical engineering, are carried over to describe these components of a signal (usually) composed of one DC and several AC components.

If a cosine function is used, the process of determining the amplitudes of the AC and DC components of the signal is called a *Cosine Transform*, and the integer indices make it a *Discrete Cosine Transform*. When $u = 0$, Eq. (8.19) yields the DC coefficient; when $u = 1$, or 2, ..., up to 7, it yields the first or second, etc., up to the seventh AC coefficient.

Equation (8.20) shows the *Inverse Discrete Cosine Transform*. This uses a sum of the products of the DC or AC coefficients and the cosine functions to reconstruct (recompose) the function $f(i)$. Since computing the DCT and IDCT involves some loss, $f(i)$ is now denoted by $\tilde{f}(i)$.

In short, the role of the DCT is to decompose the original signal into its DC and AC components; the role of the IDCT is to reconstruct (recompose) the signal. The DCT and IDCT use the same set of cosine functions; they are known as *basis functions*. Figure 8.6 shows the family of eight 1D DCT basis functions: $u = 0..7$.

The DCT enables a new means of signal processing and analysis in the *frequency domain*. In the original *Signal Processing* that deals with electrical and electronic signals (e.g., electricity and speech), $f(i)$ usually represents a signal that changes with time i (we will not be bothered here by the convention that time is usually denoted as t). The 1D DCT transforms $f(i)$, which is in the *time domain*, to $F(u)$, which is in the *frequency domain*. The coefficients $F(u)$ are known as the *frequency responses* and form the *frequency spectrum* of $f(i)$. In *Image Processing*, the image content $f(i.j)$ changes with the spatial indices i and j in the *space domain*. The 2D DCT now transforms $f(i, j)$ to $F(u, v)$, which is in the *spatial frequency domain*. For the convenience of discussion, we sometimes use 1D images and 1D DCT as examples.

Let's use some examples to illustrate frequency responses.

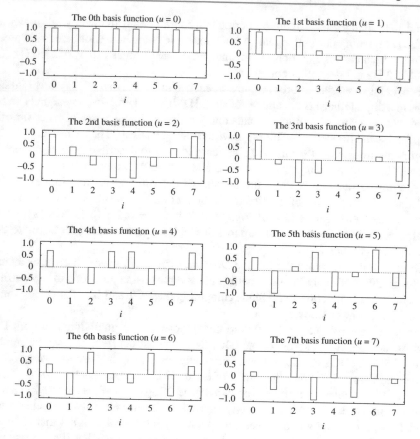

Fig. 8.6 The 1D DCT basis functions

Example 8.1

The left side of Fig. 8.7a shows a DC signal with a magnitude of 100, i.e., $f_1(i) = 100$. Since we are examining the *Discrete* Cosine Transform, the input signal is discrete, and its domain is [0, 7].

When $u = 0$, regardless of the i value, all the cosine terms in Eq. (8.19) become $\cos 0$, which equals 1. Taking into account that $C(0) = \sqrt{2}/2$, $F_1(0)$ is given by

$$
\begin{aligned}
F_1(0) = {} & \frac{\sqrt{2}}{2 \cdot 2} \cdot (1 \cdot 100 + 1 \cdot 100 + 1 \cdot 100 + 1 \cdot 100 \\
& + 1 \cdot 100 + 1 \cdot 100 + 1 \cdot 100 + 1 \cdot 100) \\
\approx {} & 283
\end{aligned}
$$

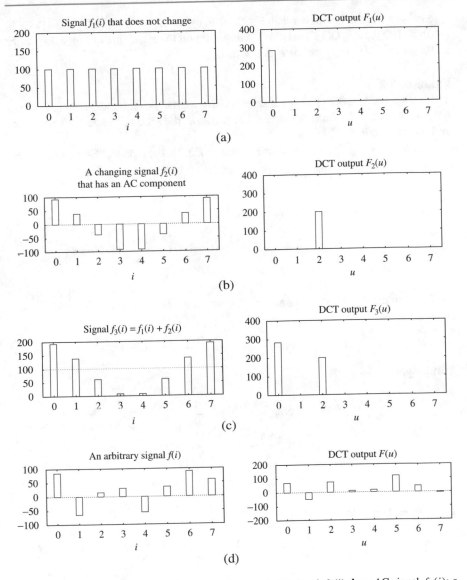

Fig. 8.7 Examples of 1D discrete cosine transform: **a** A DC signal $f_1(i)$; **b** an AC signal $f_2(i)$; **c** $f_3(i) = f_1(i) + f_2(i)$; and **d** an arbitrary signal $f(i)$

When $u = 1$, $F_1(u)$ is as below. Because $\cos \frac{\pi}{16} = -\cos \frac{15\pi}{16}$, $\cos \frac{3\pi}{16} = -\cos \frac{13\pi}{16}$, etc., and $C(1) = 1$, we have

$$
\begin{aligned}
F_1(1) = \frac{1}{2} \cdot (\cos \frac{\pi}{16} \cdot 100 + \cos \frac{3\pi}{16} \cdot 100 + \cos \frac{5\pi}{16} \cdot 100 + \cos \frac{7\pi}{16} \cdot 100 \\
+ \cos \frac{9\pi}{16} \cdot 100 + \cos \frac{11\pi}{16} \cdot 100 + \cos \frac{13\pi}{16} \cdot 100 + \cos \frac{15\pi}{16} \cdot 100) \\
= 0
\end{aligned}
$$

Similarly, it can be shown that $F_1(2) = F_1(3) = \ldots = F_1(7) = 0$. The 1D DCT result $F_1(u)$ for this DC signal $f_1(i)$ is depicted on the right side of Fig. 8.7a—only a DC (i.e., first) component of F is nonzero.

Example 8.2

The left side of Fig. 8.7b shows a discrete cosine signal $f_2(i)$. Incidentally (or, rather, purposely), it has the same frequency and phase as the second cosine basis function, and its amplitude is 100.

When $u = 0$, again, all the cosine terms in Eq. (8.19) equal 1. Because $\cos\frac{\pi}{8} = -\cos\frac{7\pi}{8}$, $\cos\frac{3\pi}{8} = -\cos\frac{5\pi}{8}$, and so on, we have

$$
\begin{aligned}
F_2(0) = {} & \frac{\sqrt{2}}{2 \cdot 2} \cdot 1 \cdot (100\cos\frac{\pi}{8} + 100\cos\frac{3\pi}{8} + 100\cos\frac{5\pi}{8} + 100\cos\frac{7\pi}{8} \\
& + 100\cos\frac{9\pi}{8} + 100\cos\frac{11\pi}{8} + 100\cos\frac{13\pi}{8} + 100\cos\frac{15\pi}{8}) \\
= {} & 0
\end{aligned}
$$

To calculate $F_2(u)$, we first note that when $u = 2$, because $\cos\frac{3\pi}{8} = \sin\frac{\pi}{8}$, we have

$$
\cos^2\frac{\pi}{8} + \cos^2\frac{3\pi}{8} = \cos^2\frac{\pi}{8} + \sin^2\frac{\pi}{8} = 1
$$

Similarly,

$$
\cos^2\frac{5\pi}{8} + \cos^2\frac{7\pi}{8} = 1
$$
$$
\cos^2\frac{9\pi}{8} + \cos^2\frac{11\pi}{8} = 1
$$
$$
\cos^2\frac{13\pi}{8} + \cos^2\frac{15\pi}{8} = 1
$$

Then we end up with

$$
\begin{aligned}
F_2(2) = {} & \frac{1}{2} \cdot (\cos\frac{\pi}{8} \cdot \cos\frac{\pi}{8} + \cos\frac{3\pi}{8} \cdot \cos\frac{3\pi}{8} + \cos\frac{5\pi}{8} \cdot \cos\frac{5\pi}{8} \\
& + \cos\frac{7\pi}{8} \cdot \cos\frac{7\pi}{8} + \cos\frac{9\pi}{8} \cdot \cos\frac{9\pi}{8} + \cos\frac{11\pi}{8} \cdot \cos\frac{11\pi}{8} \\
& + \cos\frac{13\pi}{8} \cdot \cos\frac{13\pi}{8} + \cos\frac{15\pi}{8} \cdot \cos\frac{15\pi}{8}) \cdot 100 \\
= {} & \frac{1}{2} \cdot (1 + 1 + 1 + 1) \cdot 100 = 200
\end{aligned}
$$

We will not show the other derivations in detail. It turns out that $F_2(1) = F_2(3) = F_2(4) = \cdots = F_2(7) = 0$.

Example 8.3

In the third row of Fig. 8.7, the input signal to the DCT is now the sum of the previous two signals—that is, $f_3(i) = f_1(i) + f_2(i)$. The output $F(u)$ values are

$$F_3(0) = 283,$$
$$F_3(2) = 200,$$
$$F_3(1) = F_3(3) = F_3(4) = \cdots = F_3(7) = 0$$

Thus, we discover that $F_3(u) = F_1(u) + F_2(u)$.

Example 8.4

The fourth row of the figure shows an arbitrary (or at least relatively complex) input signal $f(i)$ and its DCT output $F(u)$:

$$f(i)(i = 0..7): \quad 85 \quad -65 \quad 15 \quad 30 \quad -56 \quad 35 \quad 90 \quad 60$$
$$F(u)(u = 0..7): \quad 69 \quad -49 \quad 74 \quad 11 \quad 16 \quad 117 \quad 44 \quad -5$$

Note that in this more general case, all the DCT coefficients $F(u)$ are nonzero and some are negative.

From the above examples, the characteristics of the DCT can be summarized as follows:

1. The DCT produces the spatial frequency spectrum $F(u)$ corresponding to the spatial signal $f(i)$.

   In particular, the 0th DCT coefficient $F(0)$ is the DC component of the signal $f(i)$. Up to a constant factor (i.e., $\frac{1}{2} \cdot \frac{\sqrt{2}}{2} \cdot 8 = 2 \cdot \sqrt{2}$ in the 1D DCT and $\frac{1}{4} \cdot \frac{\sqrt{2}}{2} \cdot \frac{\sqrt{2}}{2} \cdot 64 = 8$ in the 2D DCT), $F(0)$ equals the average magnitude of the signal. In Fig. 8.7a, the average magnitude of the DC signal is obviously 100, and $F(0) = 2\sqrt{2} \times 100$; in Fig. 8.7b, the average magnitude of the AC signal is 0, and so is $F(0)$; in Fig. 8.7c, the average magnitude of $f_3(i)$ is apparently 100, and again we have $F(0) = 2\sqrt{2} \times 100$.

   The other seven DCT coefficients reflect the various changing (i.e., AC) components of the signal $f(i)$ at different frequencies. If we denote $F(1)$ as AC1, $F(2)$ as AC2, ..., $F(7)$ as AC7, then AC1 is the first AC component, which completes half a cycle as a cosine function over $[0, 7]$; AC2 completes a full cycle; AC3 completes one and a half cycles; ...; and AC7, three and a half cycles. All these are, of course, due to the cosine basis functions, which are arranged in exactly this manner. In other words, the second basis function corresponds to AC1, the third corresponds to AC2, and so on. In the example in Fig. 8.7b, since the signal $f_2(i)$ and the third basis function have exactly the same cosine waveform, with identical frequency and phase, they will reach the maximum (positive) and minimum (negative) values synchronously. As a result, their products are always positive, and the sum of their products ($F_2(2)$ or AC2) is large. It turns out that all the other AC coefficients are zero, since $f_2(i)$ and all the other basis functions happen to be orthogonal. (We will discuss orthogonality later in this chapter.)

It should be pointed out that the DCT coefficients can easily take on negative values. For DC, this occurs when the average of $f(i)$ is less than zero. (For an image, this never happens so the DC is nonnegative.) For AC, a special case occurs when $f(i)$ and some basis function have the same frequency but one of them happens to be half a cycle behind—this yields a negative coefficient, possibly with a large magnitude.

In general, signals will look more like the ones in Fig. 8.7d. Then $f(i)$ will produce many nonzero AC components, with the ones toward AC7 indicating higher frequency content. A signal will have large (positive or negative) response in its high-frequency components only when it alternates rapidly within the small range [0, 7].

As an example, if AC7 is a large positive number, this indicates that the signal $f(i)$ has a component that alternates synchronously with the eighth basis function—three and half cycles. According to the Nyquist theorem, this is the highest frequency in the signal that can be sampled with eight discrete values without significant loss and aliasing.

2. DCT is a *linear transform*.

In general, a transform T (or function) is *linear*, iff

$$T(\alpha p + \beta q) = \alpha T(p) + \beta T(q), \tag{8.21}$$

where α and β are constants, and p and q are any functions, variables, or constants. From the definition in Eq. (8.19), this property can readily be proven for the DCT, because it uses only simple arithmetic operations.

1D Inverse DCT

Let's finish the example in Fig. 8.7d by showing its inverse DCT (IDCT). Recall that $F(u)$ contains the following:

$$F(u)(u = 0..7): 69 \quad -49 \quad 74 \quad 11 \quad 16 \quad 117 \quad 44 \quad -5$$

The 1D IDCT, as indicated in Eq. (8.20), can readily be implemented as a loop with eight iterations, as illustrated in Fig. 8.8.

Iteration 0: $\tilde{f}(i) = \frac{C(0)}{2} \cdot \cos 0 \cdot F(0) = \frac{\sqrt{2}}{2 \cdot 2} \cdot 1 \cdot 69 \approx 24.3$.

Iteration 1: $\tilde{f}(i) = \frac{C(0)}{2} \cdot \cos 0 \cdot F(0) + \frac{C(1)}{2} \cdot \cos \frac{(2i+1)\pi}{16} \cdot F(1)$
$\approx 24.3 + \frac{1}{2} \cdot (-49) \cdot \cos \frac{(2i+1)\pi}{16} \approx 24.3 - 24.5 \cdot \cos \frac{(2i+1)\pi}{16}$.

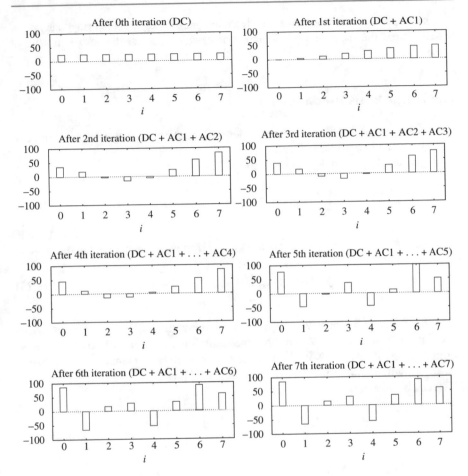

Fig. 8.8 An example of 1D IDCT

Iteration 2: $\tilde{f}(i) = \frac{C(0)}{2} \cdot \cos 0 \cdot F(0) + \frac{C(1)}{2} \cdot \cos \frac{(2i+1)\pi}{16} \cdot F(1) + \frac{C(2)}{2} \cdot$

$\quad\quad\quad\quad \cos \frac{(2i+1)\pi}{8} \cdot F(2)$

$\quad\quad\quad \approx 24.3 - 24.5 \cdot \cos \frac{(2i+1)\pi}{16} + 37 \cdot \cos \frac{(2i+1)\pi}{8}.$

$\vdots$

After iteration 0, $\tilde{f}(i)$ has a constant value of approximately 24.3, which is the recovery of the DC component in $f(i)$; after iteration 1, $\tilde{f}(i) \approx 24.3 - 24.5 \cdot \cos \frac{(2i+1)\pi}{16}$, which is the sum of the DC and first AC component; after iteration 2, $\tilde{f}(i)$ reflects the sum of DC and AC1 and AC2; and so on. As shown, the process of the sum-of-product in IDCT eventually reconstructs (recomposes) the function $f(i)$, which is approximately

$$\tilde{f}(i)(i = 0..7) : 85 \quad -65 \quad 15 \quad 30 \quad -56 \quad 35 \quad 90 \quad 60$$

As it happens, even though we went from integer to integer via intermediate *floats*, we recovered the signal exactly. This is not always true, but the answer is always close.

The Cosine Basis Functions

For a better decomposition, the basis functions should be *orthogonal*, so as to have the least redundancy among them.

Functions $B_p(i)$ and $B_q(i)$ are orthogonal if

$$\sum_i [B_p(i) \cdot B_q(i)] = 0 \quad if \quad p \neq q \tag{8.22}$$

Functions $B_p(i)$ and $B_q(i)$ are *orthonormal* if they are orthogonal and

$$\sum_i [B_p(i) \cdot B_q(i)] = 1 \quad if \quad p = q \tag{8.23}$$

The orthonormal property is desirable. With this property, the signal is not amplified during the transform. When the same basis function is used in both the transformation and its inverse (sometimes called *forward transform* and *backward transform*), we will get (approximately) the same signal back.

It can be shown that

$$\sum_{i=0}^{7} \left[\cos \frac{(2i+1) \cdot p\pi}{16} \cdot \cos \frac{(2i+1) \cdot q\pi}{16} \right] = 0 \quad if \quad p \neq q$$

$$\sum_{i=0}^{7} \left[\frac{C(p)}{2} \cos \frac{(2i+1) \cdot p\pi}{16} \cdot \frac{C(q)}{2} \cos \frac{(2i+1) \cdot q\pi}{16} \right] = 1 \quad if \quad p = q$$

The cosine basis functions in the DCT are indeed orthogonal. With the help of constants $C(p)$ and $C(q)$, they are also orthonormal. (Now we understand why constants $C(u)$ and $C(v)$ in the definitions of DCT and IDCT seemed to have taken some arbitrary values.)

Recall that because of the orthogonality, for $f_2(i)$ in Fig. 8.7b, only $F_2(2)$ (for $u = 2$) has a nonzero output whereas all other DCT coefficients are zero. This is desirable for some signal processing and analysis in the frequency domain, since we are now able to precisely identify the frequency components in the original signal.

The cosine basis functions are analogous to the basis vectors **x**, **y**, and **z** for the 3D Cartesian space, or the so-called *3D vector space*. The vectors are orthonormal, because

$$\mathbf{x} \cdot \mathbf{y} = (1, 0, 0) \cdot (0, 1, 0) = 0$$
$$\mathbf{x} \cdot \mathbf{z} = (1, 0, 0) \cdot (0, 0, 1) = 0$$
$$\mathbf{y} \cdot \mathbf{z} = (0, 1, 0) \cdot (0, 0, 1) = 0$$
$$\mathbf{x} \cdot \mathbf{x} = (1, 0, 0) \cdot (1, 0, 0) = 1$$
$$\mathbf{y} \cdot \mathbf{y} = (0, 1, 0) \cdot (0, 1, 0) = 1$$
$$\mathbf{z} \cdot \mathbf{z} = (0, 0, 1) \cdot (0, 0, 1) = 1$$

Any point $\mathbf{P} = (x_p, y_p, z_p)$ can be represented by a vector $\mathbf{OP} = (x_p, y_p, z_p)$, where O is the origin, which can in turn be decomposed into $x_p \cdot \mathbf{x} + y_p \cdot \mathbf{y} + z_p \cdot \mathbf{z}$.

If we view the sum-of-products operation in Eq. (8.19) as the dot product of one of the discrete cosine basis functions (for a specified u) and the signal $f(i)$, then the analogy between the DCT and the Cartesian projection is remarkable. Namely, to get the x-coordinate of point $\mathbf{P}$, we simply project $\mathbf{P}$ onto the x-axis. Mathematically, this is equivalent to a dot product $\mathbf{x} \cdot \mathbf{OP} = x_p$. Obviously, the same goes for obtaining y_p and z_p.

Now, compare this to the example in Fig. 8.7(b), for a point $\mathbf{P} = (0, 5, 0)$ in the Cartesian space. Only its projection onto the y-axis is $y_p = 5$, and its projections onto the x and z axes are both 0.

Finally, for reconstruction of $\mathbf{P}$, use the dot product of (x_p, y_p, z_p) and $(\mathbf{x}, \mathbf{y}, \mathbf{z})$ to obtain $x_p \cdot \mathbf{x} + y_p \cdot \mathbf{y} + z_p \cdot \mathbf{z}$.

2D Basis Functions

For 2D DCT functions, we use the basis depicted as 8×8 *images* in Fig. 8.9, where u and v indicate their spatial frequencies, white indicates positive values, and black indicates negative. For a particular pair of u and v, the respective basis function is

$$\cos \frac{(2i + 1) \cdot u\pi}{16} \cdot \cos \frac{(2j + 1) \cdot v\pi}{16}, \tag{8.24}$$

where i and j are their row and column indices.

For example, for the enlarged block shown in Fig. 8.9, where $u = 1$ and $v = 2$, it is

$$\cos \frac{(2i + 1) \cdot 1\pi}{16} \cdot \cos \frac{(2j + 1) \cdot 2\pi}{16}.$$

To obtain DCT coefficients, we essentially just form the inner product of each of these 64 basis functions with an 8×8 block from an original image. Again, we are talking about an original signal indexed by space, not time. The 64 products we calculate make up an 8×8 *spatial frequency* response $F(u, v)$. We do this for each 8×8 image block.

Fig. 8.9 Graphical
illustration of 8 × 8 2D DCT
basis

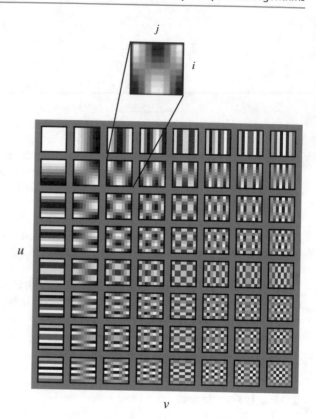

2D Separable Basis Functions

Of course, for speed, most software implementations use fixed point arithmetic to calculate the DCT transform. Just as there is a mathematically derived Fast Fourier Transform, there is also a Fast DCT. Some fast implementations approximate coefficients so that all multiplies are shifts and adds. Moreover, a much simpler mechanism is used to produce 2D DCT coefficients—*factorization* into two 1D DCT transforms.

The 2D DCT can be *separated* into a sequence of two 1D DCT steps. First, we calculate an intermediate function $G(u, j)$ by performing a 1D DCT in each column—in this way, we have taken care of the 1D transform vertically, i.e., replacing the row index by its frequency counterpart u. When the block size is 8 × 8,

$$G(u, j) = \frac{1}{2} C(u) \sum_{i=0}^{7} \cos \frac{(2i + 1)u\pi}{16} f(i, j) \qquad (8.25)$$

Then we calculate another 1D DCT horizontally in each row, this time replacing the column index j by its frequency counterpart v:

$$F(u, v) = \frac{1}{2}C(v) \sum_{j=0}^{7} \cos \frac{(2j+1)v\pi}{16} G(u, j) \qquad (8.26)$$

This is possible because the 2D DCT basis functions are *separable* (multiply separate functions of i and j). It is straightforward to see that this simple change saves many arithmetic steps. The number of iterations required is reduced from 8×8 to $8 + 8$.

2D DCT Matrix Implementation

The above factorization of a 2D DCT into two 1D DCTs can be implemented by two consecutive matrix multiplications, i.e.,

$$F(u, v) = \mathbf{T} \cdot f(i, j) \cdot \mathbf{T}^T, \qquad (8.27)$$

where superscript T means transpose. We will name $\mathbf{T}$ the *DCT-matrix*.

$$\mathbf{T}[i, j] = \begin{cases} \frac{1}{\sqrt{N}}, & \text{if } i = 0 \\ \sqrt{\frac{2}{N}} \cdot \cos \frac{(2j+1) \cdot i\pi}{2N}, & \text{if } i > 0 \end{cases} \qquad (8.28)$$

where $i = 0, ..., N - 1$ and $j = 0, ..., N - 1$ are the row and column indices, and the block size is $N \times N$.

When $N = 8$, we have

$$\mathbf{T_8}[i, j] = \begin{cases} \frac{1}{2\sqrt{2}}, & \text{if } i = 0 \\ \frac{1}{2} \cdot \cos \frac{(2j+1) \cdot i\pi}{16}, & \text{if } i > 0 \end{cases} \qquad (8.29)$$

Hence,

$$\mathbf{T_8} = \begin{bmatrix} \frac{1}{2\sqrt{2}} & \frac{1}{2\sqrt{2}} & \frac{1}{2\sqrt{2}} & \cdots & \frac{1}{2\sqrt{2}} \\ \frac{1}{2} \cdot \cos \frac{\pi}{16} & \frac{1}{2} \cdot \cos \frac{3\pi}{16} & \frac{1}{2} \cdot \cos \frac{5\pi}{16} & \cdots & \frac{1}{2} \cdot \cos \frac{15\pi}{16} \\ \frac{1}{2} \cdot \cos \frac{\pi}{8} & \frac{1}{2} \cdot \cos \frac{3\pi}{8} & \frac{1}{2} \cdot \cos \frac{5\pi}{8} & \cdots & \frac{1}{2} \cdot \cos \frac{15\pi}{8} \\ \frac{1}{2} \cdot \cos \frac{3\pi}{16} & \frac{1}{2} \cdot \cos \frac{9\pi}{16} & \frac{1}{2} \cdot \cos \frac{15\pi}{16} & \cdots & \frac{1}{2} \cdot \cos \frac{45\pi}{16} \\ \vdots & \vdots & \vdots & \ddots & \vdots \\ \frac{1}{2} \cdot \cos \frac{7\pi}{16} & \frac{1}{2} \cdot \cos \frac{21\pi}{16} & \frac{1}{2} \cdot \cos \frac{35\pi}{16} & \cdots & \frac{1}{2} \cdot \cos \frac{105\pi}{16} \end{bmatrix} \qquad (8.30)$$

A closer look at the DCT-matrix will reveal that each row of the matrix is basically a 1D DCT basis function, ranging from DC to AC1, AC2, ..., AC7. Compared to the functions in Fig. 8.6, the only difference is that we have added some constants and taken care of the orthonormal property of the DCT basis functions. Indeed, the constants and basis functions in Eqs. 8.19 and 8.29 are exactly the same. We will leave it as an exercise (see Exercise 8.) to verify that the rows and columns of $\mathbf{T_8}$ are orthonormal vectors, i.e., $\mathbf{T_8}$ is an Orthogonal Matrix.

In summary, the implementation of the 2D DCT is now a simple matter of applying two matrix multiplications as in Eq.8.27. The first multiplication applies 1D DCT vertically (for each column), and the second applies 1D DCT horizontally (for each row). What has been achieved is exactly the two steps as indicated in Eqs. 8.25 and 8.26.

2D IDCT Matrix Implementation

In this section, we will show how to reconstruct $f(i, j)$ from $F(u, v)$ losslessly by matrix multiplications. In the next several chapters, when we discuss lossy compression of images and videos, quantization steps will usually be applied to the DCT coefficients $F(u, v)$ before the IDCT.

It turns out that the 2D IDCT matrix implementation is simply

$$f(i, j) = \mathbf{T}^T \cdot F(u, v) \cdot \mathbf{T}. \tag{8.31}$$

Its derivation is as follows:

First, because $\mathbf{T} \cdot \mathbf{T}^{-1} = \mathbf{T}^{-1} \cdot \mathbf{T} = \mathbf{I}$, where $\mathbf{I}$ is the identity matrix, we can simply rewrite $f(i, j)$ as

$$f(i, j) = \mathbf{T}^{-1} \cdot \mathbf{T} \cdot f(i, j) \cdot \mathbf{T}^T \cdot (\mathbf{T}^T)^{-1}.$$

According to Eq. 8.27,

$$F(u, v) = \mathbf{T} \cdot f(i, j) \cdot \mathbf{T}^T.$$

Hence,

$$f(i, j) = \mathbf{T}^{-1} \cdot F(u, v) \cdot (\mathbf{T}^T)^{-1}.$$

As stated above, the DCT-matrix $\mathbf{T}$ is orthogonal, therefore,

$$\mathbf{T}^T = \mathbf{T}^{-1}.$$

It follows:

$$f(i, j) = \mathbf{T}^T \cdot F(u, v) \cdot \mathbf{T}.$$

Comparison of DCT and DFT

The discrete cosine transform is a close counterpart to the *Discrete Fourier Transform (DFT)* [9], and in the world of signal processing, the latter is likely more common. We have started off with the DCT instead because it is simpler and is also much used in multimedia. Nevertheless, we should not entirely ignore DFT.

For a continuous signal, we define the continuous Fourier transform $\mathcal{F}$ as follows:

$$\mathcal{F}(\omega) = \int_{-\infty}^{\infty} f(t) e^{-i\omega t}\, dt \qquad (8.32)$$

Using Euler's formula, we have

$$e^{ix} = \cos(x) + i\sin(x) \qquad (8.33)$$

Thus, the continuous Fourier transform is composed of an infinite sum of sine and cosine terms. Because digital computers require us to discretize the input signal, we define a DFT that operates on eight samples of the input signal $\{f_0, f_1, \ldots, f_7\}$ as

$$F_\omega = \sum_{x=0}^{7} f_x \cdot e^{-\frac{2\pi i\omega x}{8}} \qquad (8.34)$$

Writing the sine and cosine terms explicitly, we have

$$F_\omega = \sum_{x=0}^{7} f_x \cos\left(\frac{2\pi\omega x}{8}\right) - i \sum_{x=0}^{7} f_x \sin\left(\frac{2\pi\omega x}{8}\right) \qquad (8.35)$$

Even without giving an explicit definition of the DCT, we can guess that the DCT is likely a transform that involves only the real part of the DFT. The intuition behind the formulation of the DCT that allows it to use only the cosine basis functions of the DFT is that we can cancel out the imaginary part of the DFT by making a symmetric copy of the original input signal.

This works because sine is an odd function; thus, the contributions from the sine terms cancel each other out when the signal is symmetrically extended. Therefore, the DCT of eight input samples corresponds to the DFT of 16 samples made up of the original eight input samples and a symmetric copy of these, as in Fig. 8.10.

With the symmetric extension, the DCT is now working on a triangular wave, whereas the DFT tries to code the repeated ramp. Because the DFT is trying to model the artificial discontinuity created between each copy of the samples of the ramp function, a lot of high-frequency components are needed. (Refer to [9] for a thorough discussion and comparison of DCT and DFT.)

Table 8.1 shows the calculated DCT and DFT coefficients. We can see that more energy is concentrated in the first few coefficients in the DCT than in the DFT. If

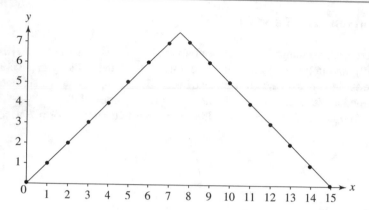

Fig. 8.10 Symmetric extension of the ramp function

Table 8.1 DCT and DFT coefficients of the ramp function

Ramp	DCT	DFT
0	9.90	28.00
1	−6.44	−4.00
2	0.00	9.66
3	−0.67	−4.00
4	0.00	4.00
5	−0.20	−4.00
6	0.00	1.66
7	−0.51	−4.00

we try to approximate the original ramp function using only three terms of both the DCT and DFT, we notice that the DCT approximation is much closer. Figure 8.11 shows the comparison.

8.5.2   Karhunen–Loève Transform*

The Karhunen–Loève transform (KLT) is a reversible linear transform that exploits the statistical properties of the vector representation. Its primary property is that it optimally decorrelates the input. To do so, it fits an n-dimensional ellipsoid around the (mean-subtracted) data. The main ellipsoid axis is the major direction of change in the data.

Think of a cigar that has unfortunately been stepped on. Cigar data consists of a cloud of points in 3-space giving the coordinates of positions of measured points in the cigar. The long axis of the cigar will be identified by a statistical program as the first KLT axis. The second most important axis is the horizontal axis across the squashed cigar, perpendicular to the first axis. The third axis is orthogonal to both

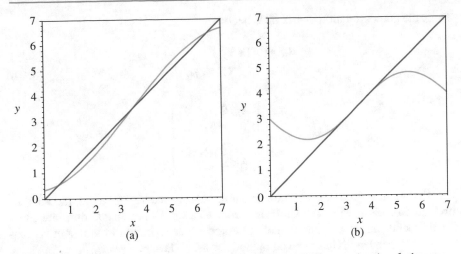

Fig. 8.11 Approximation of the ramp function: **a** Three-term DCT approximation; **b** three-term DFT approximation

and is in the vertical, thin direction. A KLT component program carries out just this analysis.

To understand the optimality of the KLT, consider the autocorrelation matrix $\mathbf{R_X}$ of the set of k input vectors $\mathbf{X}$, defined in terms of the expectation value $E(\cdot)$ as

$$\mathbf{R_X} = E[\mathbf{XX}^T] \tag{8.36}$$

$$= \begin{bmatrix} R_X(1,1) & R_X(1,2) & \cdots & R_X(1,k) \\ R_X(2,1) & R_X(2,2) & \cdots & R_X(2,k) \\ \vdots & \vdots & \ddots & \vdots \\ R_X(k,1) & R_X(k,2) & \cdots & R_X(k,k) \end{bmatrix} \tag{8.37}$$

where $R_X(t,s) = E[X_t X_s]$ is the autocorrelation function. Our goal is to find a transform $\mathbf{T}$ such that the components of the output $\mathbf{Y}$ are uncorrelated—that is, $E[Y_t Y_s] = 0$, if $t \neq s$. Thus, the autocorrelation matrix of $\mathbf{Y}$ takes on the form of a positive diagonal matrix.

Since any autocorrelation matrix is symmetric and nonnegative definite, there are k orthogonal eigenvectors $\mathbf{u}_1, \mathbf{u}_2, \ldots, \mathbf{u}_k$, and k corresponding real and nonnegative eigenvalues $\lambda_1 \geq \lambda_2 \geq \cdots \geq \lambda_k \geq 0$. We define the Karhunen-Loève transform as

$$\mathbf{T} = [\mathbf{u}_1, \mathbf{u}_2, \cdots, \mathbf{u}_k]^T \tag{8.38}$$

Then, the autocorrelation matrix of $\mathbf{Y}$ becomes

$$\mathbf{R}_\mathbf{Y} = E[\mathbf{Y}\mathbf{Y}^T] \tag{8.39}$$

$$= E[\mathbf{T}\mathbf{X}\mathbf{X}^T\mathbf{T}] \tag{8.40}$$

$$= \mathbf{T}\mathbf{R}_\mathbf{X}\mathbf{T}^T \tag{8.41}$$

$$= \begin{bmatrix} \lambda_1 & 0 & \cdots & 0 \\ 0 & \lambda_2 & \cdots & 0 \\ 0 & \vdots & \ddots & 0 \\ 0 & 0 & \cdots & \lambda_k \end{bmatrix} \tag{8.42}$$

Clearly, we have the required autocorrelation matrix for $\mathbf{Y}$. Therefore, the KLT is optimal, in the sense that it completely decorrelates the input. In addition, since the KLT depends on the computation of the autocorrelation matrix of the input vector, it is data dependent: it has to be computed for every dataset.

Example 8.5

To illustrate the mechanics of the KLT, consider the four 3D input vectors $\mathbf{x}_1 = (4, 4, 5), \mathbf{x}_2 = (3, 2, 5), \mathbf{x}_3 = (5, 7, 6)$, and $\mathbf{x}_4 = (6, 7, 7)$. To find the required transform, we must first estimate the autocorrelation matrix of the input. The mean of the four input vectors is

$$\mathbf{m}_x = \frac{1}{4} \begin{bmatrix} 18 \\ 20 \\ 23 \end{bmatrix}$$

We can estimate the autocorrelation matrix using the formula

$$\mathbf{R}_\mathbf{X} = \frac{1}{n} \sum_{i=1}^{n} \mathbf{x}_i \mathbf{x}_i^T - \mathbf{m}_x \mathbf{m}_x^T \tag{8.43}$$

where n is the number of input vectors, which is 4. From this equation, we obtain

$$\mathbf{R}_\mathbf{X} = \begin{bmatrix} 1.25 & 2.25 & 0.88 \\ 2.25 & 4.50 & 1.50 \\ 0.88 & 1.50 & 0.69 \end{bmatrix}$$

We are trying to diagonalize matrix $\mathbf{R}_\mathbf{X}$, which is the same as forming an eigenvector–eigenvalue decomposition (from linear algebra). That is, we want to rewrite $\mathbf{R}_\mathbf{X}$ as $\mathbf{R}_\mathbf{X} = \mathbf{T}\mathbf{D}\mathbf{T}^{-1}$, where the matrix $\mathbf{D}$ is diagonal, with off-diagonal values equaling zero: $\mathbf{D} = diag(\lambda_1, \lambda_2, \lambda_3)$, with the λ values called the eigenvalues and the columns of matrix $\mathbf{T}$ called the eigenvectors. These are easy to calculate using various math libraries; in MATLAB, function `eig` will do the job.

Here, the eigenvalues of $\mathbf{R_X}$ are $\lambda_1 = 6.1963$, $\lambda_2 = 0.2147$, and $\lambda_3 = 0.0264$. Clearly, the first component is by far the most important. The corresponding eigenvectors are

$$\mathbf{u}_1 = \begin{bmatrix} 0.4385 \\ 0.8471 \\ 0.3003 \end{bmatrix} \quad \mathbf{u}_2 = \begin{bmatrix} 0.4460 \\ -0.4952 \\ 0.7456 \end{bmatrix} \quad \mathbf{u}_3 = \begin{bmatrix} -0.7803 \\ 0.1929 \\ 0.5949 \end{bmatrix}$$

Therefore, the KLT is given by the matrix

$$\mathbf{T} = \begin{bmatrix} 0.4385 & 0.8471 & 0.3003 \\ 0.4460 & -0.4952 & 0.7456 \\ -0.7803 & 0.1929 & 0.5949 \end{bmatrix}$$

Subtracting the mean vector from each input vector and applying the KLT, we have

$$\mathbf{y}_1 = \begin{bmatrix} -1.2916 \\ -0.2870 \\ -0.2490 \end{bmatrix} \quad \mathbf{y}_2 = \begin{bmatrix} -3.4242 \\ 0.2573 \\ 0.1453 \end{bmatrix}$$

$$\mathbf{y}_3 = \begin{bmatrix} 1.9885 \\ -0.5809 \\ 0.1445 \end{bmatrix} \quad \mathbf{y}_4 = \begin{bmatrix} 2.7273 \\ 0.6107 \\ -0.0408 \end{bmatrix}$$

Since the rows of $\mathbf{T}$ are orthonormal vectors, the inverse transform is just the transpose: $\mathbf{T}^{-1} = \mathbf{T}^T$. We can obtain the original vectors from the transform coefficients using the inverse relation

$$\mathbf{x} = \mathbf{T}^T \mathbf{y} + \mathbf{m}_x \tag{8.44}$$

In terms of the transform coefficients $\mathbf{y}_i$, the magnitude of the first few components is usually considerably larger than that of the other components. In general, after the KLT, most of the "energy" of the transform coefficients is concentrated within the first few components. This is the *energy compaction* property of the KLT.

For an input vector $\mathbf{x}$ with n components, if we coarsely quantize the output vector $\mathbf{y}$ by setting its last k components to zero, calling the resulting vector $\hat{\mathbf{y}}$, the KLT minimizes the mean squared error between the original vector and its reconstruction.

8.6 Wavelet-Based Coding

8.6.1 Introduction

Decomposing the input signal into its constituents allows us to apply coding techniques suitable for each constituent, to improve compression performance. Consider again a time-dependent signal $f(t)$ (it is best to base the discussion on continuous functions to start with). The traditional method of signal decomposition is the

Fourier transform. Above, in our discussion of the DCT, we considered a special cosine-based transform. If we carry out analysis based on both sine and cosine, then a concise notation assembles the results into a function $\mathcal{F}(\omega)$, a complex-valued function of real-valued frequency ω given in Eq. (8.32). Such decomposition results in very fine resolution in the *frequency* domain. However, since a sinusoid is theoretically infinite in extent in time, such a decomposition gives no *temporal* resolution.

Another method of decomposition that has gained a great deal of popularity in recent years is the *wavelet transform*. It seeks to represent a signal with good resolution in *both* time and frequency, by using a set of basis functions called wavelets.

There are two types of wavelet transforms: the *Continuous Wavelet Transform* (CWT) and the *Discrete Wavelet Transform* (DWT). We assume that the CWT is applied to the large class of functions $f(x)$ that are square integrable on the real line—that is, $\int [f(x)]^2 \, dx < \infty$. In mathematics, this is written as $f(x) \in \mathbf{L}^2(R)$.

The other kind of wavelet transform, the DWT, operates on discrete samples of the input signal. The DWT resembles other discrete linear transforms, such as the DFT and the DCT, and is very useful for image processing and compression.

Before we begin a discussion of the theory of wavelets, let's develop an intuition about this approach by going through an example using the simplest wavelet transform, the so-called *Haar Wavelet Transform*, to form averages and differences of a sequence of `float` values.

If we repeatedly take averages and differences and keep results for every step, we effectively create a *multiresolution analysis* of the sequence. For images, this would be equivalent to creating smaller and smaller summary images, one-quarter the size for each step, and keeping track of differences from the average as well. Mentally stacking the full-size image, the quarter-size image, the sixteenth-size image, and so on creates a *pyramid*. The full set, along with difference images, is the multiresolution decomposition.

Example 8.6 (A Simple Wavelet Transform).
The objective of the wavelet transform is to decompose the input signal, for compression purposes, into components that are easier to deal with, have special interpretations, or have some components that can be thresholded away. Furthermore, we want to be able to at least approximately reconstruct the original signal, given these components. Suppose we are given the following input sequence:

$$\{x_{n,i}\} = \{10, 13, 25, 26, 29, 21, 7, 15\} \tag{8.45}$$

Here, $i \in [0 .. 7]$ indexes "pixels," and n stands for the level of a *pyramid* we are on. At the top, $n = 3$ for this sequence, and we shall form three more sequences, for $n = 2, 1$, and 0. At each level, less information will be retained in the beginning elements of the transformed signal sequence. When we reach pyramid level $n = 0$, we end up with the sequence average stored in the first element. The remaining elements store detail information.

Consider the transform that replaces the original sequence with its pairwise *average* $x_{n-1,i}$ and *difference* $d_{n-1,i}$, defined as follows:

$$x_{n-1,i} = \frac{x_{n,2i} + x_{n,2i+1}}{2} \tag{8.46}$$

$$d_{n-1,i} = \frac{x_{n,2i} - x_{n,2i+1}}{2} \tag{8.47}$$

where now $i \in [0..3]$. Notice that the averages and differences are applied only on consecutive *pairs* of input sequences whose first element has an even index. Therefore, the number of elements in each set $\{x_{n-1,i}\}$ and $\{d_{n-1,i}\}$ is exactly half the number of elements in the original sequence. We can form a new sequence having length equal to that of the original sequence by concatenating the two sequences $\{x_{n-1,i}\}$ and $\{d_{n-1,i}\}$. The resulting sequence is thus

$$\{x_{n-1,i}, d_{n-1,i}\} = \{11.5, 25.5, 25, 11, -1.5, -0.5, 4, -4\} \tag{8.48}$$

where we are now at level $n - 1 = 2$. This sequence has exactly the same number of elements as the input sequence—the transform did not increase the amount of data. Since the first half of the above sequence contains averages from the original sequence, we can view it as a coarser approximation to the original signal.

The second half of this sequence can be viewed as the details or approximation errors of the first half. Most of the values in the detail sequence are much smaller than those of the original sequence. Thus, most of the energy is effectively concentrated in the first half. Therefore, we can potentially store $\{d_{n-1,i}\}$ using fewer bits.

It is easily verified that the original sequence can be reconstructed from the transformed sequence, using the relations

$$x_{n,2i} = x_{n-1,i} + d_{n-1,i}$$
$$x_{n,2i+1} = x_{n-1,i} - d_{n-1,i} \tag{8.49}$$

This transform is the discrete Haar wavelet transform. Averaging and differencing can be carried out by applying a so-called *scaling function* and *wavelet function* along the signal. Figure 8.12 shows the Haar version of these functions.

We can further apply the same transform to $\{x_{n-1,i}\}$, to obtain another level of approximation $x_{n-2,i}$ and detail $d_{n-2,i}$:

$$\{x_{n-2,i}, d_{n-2,i}, d_{n-1,i}\} = \{18.5, 18, -7, 7, -1.5, -0.5, 4, -4\} \tag{8.50}$$

This is the essential idea of multiresolution analysis. We can now study the input signal in three different scales, along with the details needed to go from one scale to another. This process can continue n times, until only one element is left in the approximation sequence. In this case, $n = 3$, and the final sequence is given below:

$$\{x_{n-3,i}, d_{n-3,i}, d_{n-2,i}, d_{n-1,i}\} = \{18.25, 0.25, -7, 7, -1.5, -0.5, 4, -4\} \tag{8.51}$$

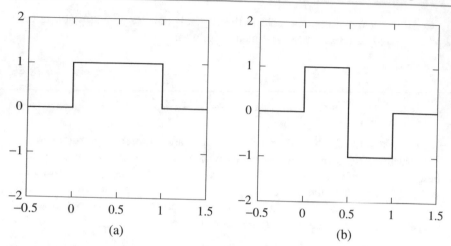

Fig. 8.12 Haar wavelet transform: **a** Scaling function; **b** wavelet function

0	0	0	0	0	0	0	0
0	0	0	0	0	0	0	0
0	0	63	127	127	63	0	0
0	0	127	255	255	127	0	0
0	0	127	255	255	127	0	0
0	0	63	127	127	63	0	0
0	0	0	0	0	0	0	0
0	0	0	0	0	0	0	0

(a)

(b)

Fig. 8.13 Input image for the 2D Haar wavelet transform: **a** Pixel values; **b** an 8×8 image

Now we realize that n was 3 because only three resolution changes were available until we reached the final form.

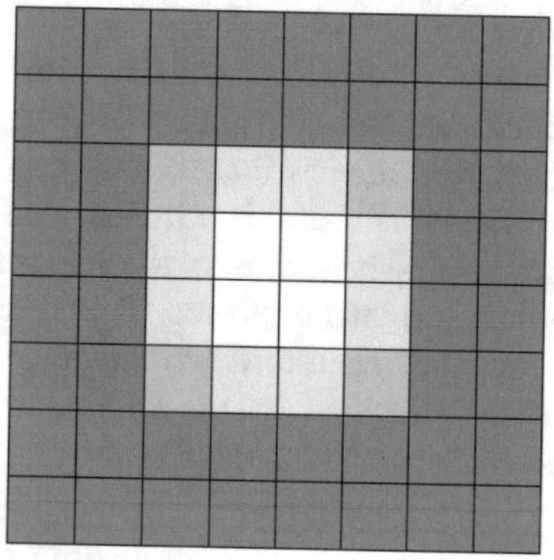

The value 18.25, corresponding to the coarsest approximation to the original signal, is the average of all the elements in the original sequence. From this example, it is easy to see that the cost of computing this transform is proportional to the number of elements N in the input sequence—that is, $O(N)$.

Extending the 1D Haar wavelet transform into two dimensions is relatively easy: we simply apply the 1D transform to the rows and columns of the 2D input separately. We will demonstrate the 2D Haar transform applied to the 8×8 input image shown in Fig. 8.13.

Example 8.7 (2D Haar Transform).
This example of the 2D Haar transform not only serves to illustrate how the wavelet transform is applied to 2D inputs but also points out useful interpretations of the transformed coefficients. However, it is intended only to provide the reader with an intuitive feeling of the kinds of operations involved in performing a general 2D wavelet transform. Subsequent sections provide more detailed description of the forward and inverse 2D wavelet transform algorithms, as well as a more elaborate example using a more complex wavelet.

2D Haar Wavelet Transform

We begin by applying a 1D Haar wavelet transform to each row of the input. The first and last two rows of the input are trivial. After performing the averaging and differencing operations on the remaining rows, we obtain the intermediate output shown in Fig. 8.14.

We continue by applying the same 1D Haar transform to each column of the intermediate output. This step completes one level of the 2D Haar transform. Figure 8.15 gives the resulting coefficients.

We can naturally divide the result into four quadrants. The upper left quadrant contains the averaged coefficients from both the horizontal and vertical passes. Therefore, it can be viewed as a low-pass-filtered version of the original image, in the sense that higher frequency edge information is lost, while low-spatial frequency smooth information is retained.

The upper right quadrant contains the vertical averages of the horizontal differences and can be interpreted as information about the *vertical edges* within the original image. Similarly, the lower left quadrant contains the vertical differences of the horizontal averages and represents the *horizontal edges* in the original image. The

Fig. 8.14 Intermediate output of the 2D Haar wavelet transform

0	0	0	0	0	0	0	0
0	0	0	0	0	0	0	0
0	95	95	0	0	-32	32	0
0	191	191	0	0	-64	64	0
0	191	191	0	0	-64	64	0
0	95	95	0	0	-32	32	0
0	0	0	0	0	0	0	0
0	0	0	0	0	0	0	0

Fig. 8.15 Output of the first
level of the 2D Haar wavelet
transform

0	0	0	0	0	0	0	0
0	143	143	0	0	−48	48	0
0	143	143	0	0	−48	48	0
0	0	0	0	0	0	0	0
0	0	0	0	0	0	0	0
0	−48	−48	0	0	16	−16	0
0	48	48	0	0	−16	16	0
0	0	0	0	0	0	0	0

Fig. 8.16 A simple
graphical illustration of the
wavelet transform

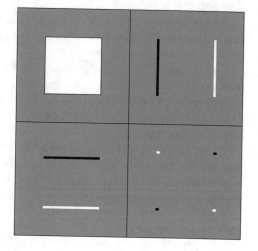

lower right quadrant contains the differences from both the horizontal and vertical
passes. The coefficients in this quadrant represent *diagonal edges*.

These interpretations are shown more clearly as images in Fig. 8.16, where bright
pixels code positive and dark pixels code negative image values.

The inverse of the 2D Haar transform can be calculated by first inverting the
columns using Eq. (8.49), and then inverting the resulting rows.

8.6.2   Continuous Wavelet Transform*

We noted that the motivation for the use of wavelets is to provide a set of basis
functions that decompose a signal in time over parameters in the frequency domain
and the time domain simultaneously. A Fourier transform aims to pin down only the
frequency content of a signal, in terms of spatially varying rather than time-varying

signals. What wavelets aim to do is pin down the frequency content at different parts of the image.

For example, one part of the image may be "busy" with texture and thus high-frequency content, while another part may be smooth, with little high-frequency content. Naturally, one can think of obvious ways to consider frequencies for localized areas of an image: divide an image into parts and fire away with Fourier analysis. The time-sequence version of that idea is called the *Short-Term* (or *Windowed*) *Fourier Transform*. And other ideas have also arisen. However, it turns out that wavelets, a much newer development, have neater characteristics.

To further motivate the subject, we should consider the *Heisenberg uncertainty principle*, from physics. In the context of signal processing, this says that there is a trade-off between accuracy in pinning down a function's frequency and its extent in time. We cannot do both accurately, in general, and still have a useful basis function. For example, a sine wave is exact in terms of its frequency but infinite in extent.

As an example of a function that dies away quickly and also has limited frequency content, suppose we start with a Gaussian function,

$$f(t) = \frac{1}{\sigma\sqrt{2\pi}} e^{\frac{-t^2}{2\sigma^2}} \tag{8.52}$$

The parameter σ expresses the *scale* of the Gaussian (bell-shaped) function.

The second derivative of this function, called $\psi(t)$, looks like a Mexican hat, as in Fig. 8.17a. Clearly, the function $\psi(t)$ is limited in time. Its equation is as follows:

$$\psi(t) = \frac{1}{\sigma^3\sqrt{2\pi}} \left[e^{\frac{-t^2}{2\sigma^2}} \left(\frac{t^2}{\sigma^2} - 1 \right) \right] \tag{8.53}$$

We can explore the frequency content of function $\psi(t)$ by taking its Fourier transform. This turns out to be given by

$$\mathcal{F}(\omega) = \omega^2 e^{-\frac{\sigma^2\omega^2}{2}} \tag{8.54}$$

Figure 8.17b displays this function: the candidate wavelet (8.53) is indeed limited in frequency as well.

In general, a wavelet is a function $\psi \in \mathbf{L}^2(R)$ with a zero average,

$$\int\limits_{-\infty}^{+\infty} \psi(t)\, dt = 0 \tag{8.55}$$

that satisfies some conditions that ensure it can be utilized in a multiresolution decomposition. The conditions ensure that we can use the decomposition for zooming in locally in some part of an image, much as we might be interested in closer or farther views of some neighborhood in a map.

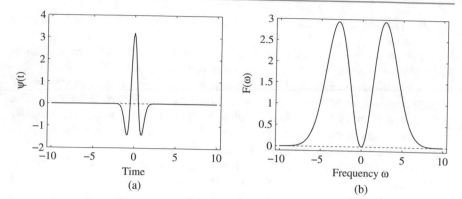

Fig. 8.17 A Mexican-hat Wavelet: **a** $\sigma = 0.5$; **b** its Fourier transform

The constraint (8.55) is called the *admissibility condition* for wavelets. A function that sums to zero must oscillate around zero. Also, from (8.32), we see that the DC value, the Fourier transform of $\psi(t)$ for $\omega = 0$, is zero. Another way to state this is that the 0th moment M_0 of $\psi(t)$ is zero. The pth moment is defined as

$$M_p = \int\limits_{-\infty}^{\infty} t^p \psi(t) \, dt \qquad (8.56)$$

The function ψ is normalized with $\|\psi\| = 1$ and centered in the neighborhood of $t = 0$. We can obtain a *family* of wavelet functions by scaling and translating the *mother wavelet* ψ as follows:

$$\psi_{s,u}(t) = \frac{1}{\sqrt{s}} \psi\left(\frac{t-u}{s}\right) \qquad (8.57)$$

If $\psi(t)$ is normalized, so is $\psi_{s,u}(t)$.

The *Continuous Wavelet Transform* (CWT) of $f \in \mathbf{L}^2(R)$ at time u and scale s is defined as

$$\mathcal{W}(f, s, u) = \int\limits_{-\infty}^{+\infty} f(t)\psi_{s,u}(t) \, dt \qquad (8.58)$$

The CWT of a 1D signal is a 2D function—a function of both *scale s* and *shift u*.

A very important issue is that, in contradistinction to (8.32), where the Fourier analysis function is stipulated to be the sinusoid, here (8.58) does not state what $\psi(t)$ actually is! Instead, we create a set of rules such functions must obey and then invent useful functions that obey these rules—different functions for different uses.

Just as we defined the DCT in terms of products of a function with a set of basis functions, here the transform $\mathcal{W}$ is written in terms of inner products with basis functions that are a scaled and shifted version of the mother wavelet $\psi(t)$.

The mother wavelet $\psi(t)$ is a *wave*, since it must be an oscillatory function. Why is it wave*let*? The spatial frequency analyzer parameter in (8.58) is s, the scale. We choose some scale s and see how much content the signal has around that scale. To make the function decay rapidly, away from the chosen s, we have to choose a mother wavelet $\psi(t)$ that decays as fast as some power of s.

It is actually easy to show, from (8.58), that if all moments of $\psi(t)$ up to the nth are zero (or quite small, practically speaking), then the CWT coefficient $\mathcal{W}(f, s, u)$ has a Taylor expansion around $u = 0$ that is of order s^{n+2} (see Exercise 13). This is the localization in frequency we desire in a good mother wavelet.

We derive wavelet coefficients by applying wavelets at different scales over many locations of the signal. Excitingly, if we shrink the wavelets down small enough that they cover a part of the function $f(t)$ that is a polynomial of degree n or less, the coefficient for that wavelet and all smaller ones will be zero. The condition that the wavelet should have vanishing moments up to some order is one way of characterizing mathematical *regularity conditions* on the mother wavelet.

The inverse of the continuous wavelet transform is

$$f(t) = \frac{1}{C_\psi} \int\limits_0^{+\infty} \int\limits_{-\infty}^{+\infty} \mathcal{W}(f, s, u) \frac{1}{\sqrt{s}} \, \psi\left(\frac{t-u}{s}\right) \frac{1}{s^2} \, du \, ds \tag{8.59}$$

where

$$C_\psi = \int\limits_0^{+\infty} \frac{|\Psi(\omega)|^2}{\omega} \, d\omega < +\infty \tag{8.60}$$

and $\Psi(\omega)$ is the Fourier transform of $\psi(t)$. Equation (8.60) is another phrasing of the admissibility condition.

The trouble with CWT is that (8.58) is nasty: most wavelets are not analytic but result simply from numerical calculations. The resulting infinite set of scaled and shifted functions is not necessary for the analysis of *sampled* functions, such as the ones that arise in image processing. For this reason, we apply the ideas that pertain to CWT to the discrete domain.

8.6.3 Discrete Wavelet Transform*

Discrete wavelets are again formed from a mother wavelet, but with scale and shift in discrete steps.

Multiresolution Analysis and the Discrete Wavelet Transform

The connection between wavelets in the continuous time domain and *filter banks* in the discrete time domain is multiresolution analysis; we discuss the DWT within this framework. Mallat [10] showed that it is possible to construct wavelets ψ such that the dilated and translated family

$$\left\{\psi_{j,n}(t) = \frac{1}{\sqrt{2^j}}\,\psi\left(\frac{t - 2^j n}{2^j}\right)\right\}_{(j,n)\in\mathbf{Z}^2} \tag{8.61}$$

is an *orthonormal basis* of $\mathbf{L}^2(R)$, where $\mathbf{Z}$ represents the set of integers. This is known as "dyadic" scaling and translation and corresponds to the notion of zooming out in a map by factors of 2. (If we draw a cosine function $\cos(t)$ from time 0 to 2π and then draw $\cos(t/2)$, we see that while $\cos(t)$ goes over a whole cycle, $\cos(t/2)$ has only a half cycle: the function $\cos(2^{-1}t)$ is a *wider* function and thus is at a broader scale.)

Note that we change the scale of translations along with the overall scale 2^j, so as to keep the movement in the lower resolution image in proportion. Notice also that the notation used says that a larger index j corresponds to a coarser version of the image.

Multiresolution analysis provides the tool to *adapt signal resolution to only relevant details* for a particular task. The *octave decomposition* introduced by Mallat [11] initially decomposes a signal into an approximation component and a detail component. The approximation component is then recursively decomposed into approximation and detail at successively coarser scales. Wavelets are set up such that the approximation at resolution 2^{-j} contains all the necessary information to compute an approximation at the coarser resolution $2^{-(j+1)}$.

Wavelets are used to characterize detail information. The averaging information is formally determined by a kind of dual to the mother wavelet, called the *scaling function* $\phi(t)$.

The main idea in the theory of wavelets is that at a particular level of resolution j, the set of *translates* indexed by n form a basis at that level. Interestingly, the set of translates forming the basis at the $j + 1$ next level, a coarser level, can all be written as a sum of weights times the level-j basis. The scaling function is chosen such that the coefficients of its translates are all necessarily bounded (less than infinite).

The scaling function, along with its translates, forms a basis at the coarser level $j + 1$ (say 3, or the 1/8 level) but not at level j (say 2, or the 1/4 level). Instead, at level j the set of translates of the scaling function ϕ *along with the set of translates of the mother wavelet* ϕ do form a basis. We are left with the situation that the scaling function describes smooth, or approximation, information and the wavelet describes what is leftover—detail information.

Since the set of translates of the scaling function ϕ at a coarser level can be written exactly as a weighted sum of the translates at a finer level, the scaling function must satisfy the so-called *dilation equation* [12]:

$$\phi(t) = \sum_{n\in\mathbf{Z}} \sqrt{2}h_0[n]\phi(2t - n) \tag{8.62}$$

The square brackets come from the theory of *filters*, and their use is carried over here. The dilation equation is a recipe for finding a function that can be built from a sum of copies of itself that are first scaled, translated, and dilated. Equation (8.62) expresses a condition that a function must satisfy to be a scaling function and at the same time forms a definition of the *scaling vector h_0*.

Not only is the scaling function expressible as a sum of translates, but as well the *wavelet* at the coarser level is also expressible as such

$$\psi(t) = \sum_{n \in \mathbf{Z}} \sqrt{2} h_1[n]\phi(2t - n) \tag{8.63}$$

Below, we'll show that the set of coefficients h_1 for the wavelet can in fact be derived from the scaling function ones h_0 (Eq. (8.65)), so we also have that the wavelet can be derived from the scaling function, once we have one. The equation reads

$$\psi(t) = \sum_{n \in \mathbf{Z}} (-1)^n h_0[1 - n]\phi(2t - n) \tag{8.64}$$

So the condition on a wavelet is similar to that on the scaling function, Eq. (8.62), and in fact uses the same coefficients, only in the opposite order and with alternating signs.

Clearly, for efficiency, we would like the sums in (8.62) and (8.63) to be as few as possible, so we choose wavelets that have as few vector entries h_0 and h_1 as possible. The effect of the scaling function is a kind of smoothing, or filtering, operation on a signal. Therefore it acts as a low-pass filter, screening out high-frequency content. The vector values $h_0[n]$ are called the low-pass filter *impulse response* coefficients, since they describe the effect of the filtering operation on a signal consisting of a single spike with magnitude unity (an impulse) at time $t = 0$. A complete discrete signal is made of a set of such spikes, shifted in time from 0 and weighted by the magnitudes of the discrete samples.

Hence, to specify a DWT, only the discrete low-pass filter impulse response $h_0[n]$ is needed. These specify the approximation filtering, given by the scaling function. The discrete *high-pass* impulse response $h_1[n]$, describing the details using the wavelet function, can be derived from $h_0[n]$ using the following equation:

$$h_1[n] = (-1)^n h_0[1 - n] \tag{8.65}$$

The number of coefficients in the impulse response is called the number of *taps* in the filter. If $h_0[n]$ has only a finite number of nonzero entries, the resulting wavelet is said to have *compact support*. Additional constraints, such as orthonormality and regularity, can be imposed on the coefficients $h_0[n]$. The vectors $h_0[n]$ and $h_1[n]$ are called the low-pass and high-pass *analysis* filters.

To *reconstruct* the original input, an inverse operation is needed. The inverse filters are called *synthesis* filters. For orthonormal wavelets, the forward transform and its inverse are transposes of each other, and the analysis filters are identical to the synthesis filters.

Without orthogonality, the wavelets for analysis and synthesis are called *biorthogonal*, a weaker condition. In this case, the synthesis filters are not identical to the

Table 8.2 Orthogonal wavelet filters

Wavelet	Number of taps	Start index	Coefficients
Haar	2	0	$[0.707, 0.707]$
Daubechies 4	4	0	$[0.483, 0.837, 0.224, -0.129]$
Daubechies 6	6	0	$[0.332, 0.807, 0.460, -0.135,$ $-0.085, 0.0352]$
Daubechies 8	8	0	$[0.230, 0.715, 0.631, -0.028,$ $-0.187, 0.031, 0.033, -0.011]$

analysis filters. We denote them as $\tilde{h}_0[n]$ and $\tilde{h}_1[n]$. To specify a biorthogonal wavelet transform, we require both $h_0[n]$ and $\tilde{h}_0[n]$. As before, we can compute the discrete high-pass filters in terms of sums of the low-pass ones:

$$h_1[n] = (-1)^n \tilde{h}_0[1 - n] \qquad (8.66)$$
$$\tilde{h}_1[n] = (-1)^n h_0[1 - n] \qquad (8.67)$$

Tables 8.2 and 8.3 (cf. [13]) give some commonly used orthogonal and biorthogonal wavelet filters. The "start index" columns in these tables refer to the starting value of the index n used in Eqs. (8.66) and (8.67).

Figure 8.18 shows a block diagram for the 1D dyadic wavelet transform. Here, $x[n]$ is the discrete sampled signal. The box $\boxed{\downarrow 2}$ means subsampling by taking every second element, and the box $\boxed{\uparrow 2}$ means upsampling by replication. The reconstruction phase yields series $y[n]$.

For analysis, at each level we transform the series $x[n]$ into another series of the same length, in which the first half of the elements is approximation information and the second half consists of detail information. For an N-tap filter, this is simply the series

$$\{x[n]\} \rightarrow y[n] = \left\{ \sum_j x[j]h_0[n - j] \ ; \ \sum_j x[j]h_1[n - j] \right\} \qquad (8.68)$$

where for each half, the odd-numbered results are discarded. The summation over shifted coefficients in (8.68) is referred to as a *convolution*.

2D Discrete Wavelet Transform

The extension of the wavelet transform to two dimensions is quite straightforward. A 2D scaling function is said to be *separable* if it can be factored into a product of two 1D scaling functions. That is,

$$\phi(x, y) = \phi(x)\phi(y) \qquad (8.69)$$

Table 8.3 Biorthogonal wavelet filters

Wavelet	Filter	Number of taps	Start index	Coefficients
Antonini 9/7	$h_0[n]$	9	−4	[0.038, −0.024, −0.111, 0.377, 0.853, 0.377, −0.111, −0.024, 0.038]
	$\tilde{h}_0[n]$	7	−3	[−0.065, −0.041, 0.418, 0.788, 0.418, −0.041, −0.065]
Villa 10/18	$h_0[n]$	10	−4	[0.029, 0.0000824, −0.158, 0.077, 0.759, 0.759, 0.077, −0.158, 0.0000824, 0.029]
	$\tilde{h}_0[n]$	18	−8	[0.000954, −0.00000273, −0.009, −0.003, 0.031, −0.014, −0.086, 0.163, 0.623, 0.623, 0.163, −0.086, −0.014, 0.031, −0.003, −0.009, −0.00000273, 0.000954]
Brislawn	$h_0[n]$	10	−4	[0.027, −0.032, −0.241, 0.054, 0.900, 0.900, 0.054, −0.241, −0.032, 0.027]
	$\tilde{h}_0[n]$	10	−4	[0.020, 0.024, −0.023, 0.146, 0.541, 0.541, 0.146, −0.023, 0.024, 0.020]

For simplicity, only separable wavelets are considered in this section. Furthermore, let's assume that the width and height of the input image are powers of 2.

For an N by N input image, the 2D DWT proceeds as follows:

1. Convolve each row of the image with $h_0[n]$ and $h_1[n]$, discard the odd-numbered columns of the resulting arrays, and concatenate them to form a transformed row.
2. After all rows have been transformed, convolve each column of the result with $h_0[n]$ and $h_1[n]$. Again discard the odd-numbered rows and concatenate the result.

After the above two steps, one stage of the DWT is complete. The transformed image now contains four subbands LL, HL, LH, and HH, standing for low-low, high-low, and so on, as Fig. 8.19a shows. As in the 1D transform, the LL subband can be further decomposed to yield yet another level of decomposition. This process can be continued until the desired number of decomposition levels is reached or the LL component only has a single element left. A two-level decomposition is shown in Fig. 8.19b.

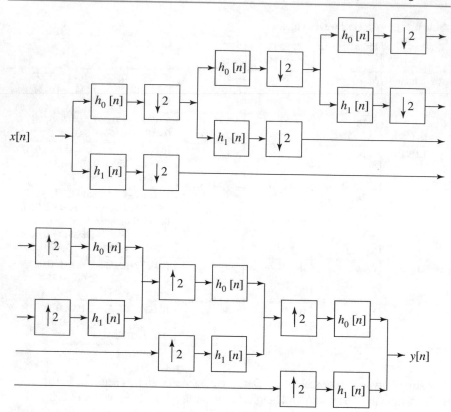

Fig. 8.18 Block diagram of the 1D dyadic wavelet transform

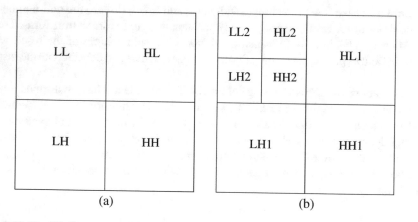

Fig. 8.19 The 2D discrete wavelet transform: **a** One-level transform; **b** two-level transform

The inverse transform simply reverses the steps of the forward transform.

1. For each stage of the transformed image, starting with the last, separate each column into low-pass and high-pass coefficients. Upsample each of the low-pass and high-pass arrays by inserting a zero after each coefficient.
2. Convolve the low-pass coefficients with $h_0[n]$ and high-pass coefficients with $h_1[n]$ and add the two resulting arrays.
3. After all columns have been processed, separate each row into low-pass and high-pass coefficients and upsample each of the two arrays by inserting a zero after each coefficient.
4. Convolve the low-pass coefficients with $h_0[n]$ and high-pass coefficients with $h_1[n]$ and add the two resulting arrays.

If biorthogonal filters are used for the forward transform, we must replace the $h_0[n]$ and $h_1[n]$ above with $\tilde{h}_0[n]$ and $\tilde{h}_1[n]$ in the inverse transform.

Example 8.8

The input image is a subsampled version of the image Lena, as shown in Fig. 8.20. The size of the input is 16×16. The filter used in the example is the Antonini 9/7 filter set given in Table 8.3.

Before we begin, we need to compute the analysis and synthesis high-pass filters using Eqs. (8.66) and (8.67). The resulting filter coefficients are

$$h_1[n] = [-0.065, 0.041, 0.418, -0.788, 0.418, 0.041, -0.065]$$
$$\tilde{h}_1[n] = [-0.038, -0.024, 0.111, 0.377, -0.853, 0.377, 0.111, -0.024, -0.038]$$

$$(8.70)$$

(a)

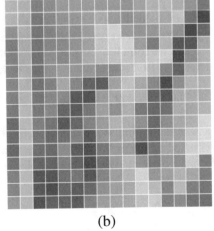

(b)

Fig. 8.20 Lena: **a** Original 128×128 image; **b** 16×16 subsampled image

The input image in numerical form is

$$I_{00}(x, y) =$$

$$\begin{bmatrix}
158 & 170 & 97 & 104 & 123 & 130 & 133 & 125 & 132 & 127 & 112 & 158 & 159 & 144 & 116 & 91 \\
164 & 153 & 91 & 99 & 124 & 152 & 131 & 160 & 189 & 116 & 106 & 145 & 140 & 143 & 227 & 53 \\
116 & 149 & 90 & 101 & 118 & 118 & 131 & 152 & 202 & 211 & 84 & 154 & 127 & 146 & 58 & 58 \\
95 & 145 & 88 & 105 & 188 & 123 & 117 & 182 & 185 & 204 & 203 & 154 & 153 & 229 & 46 & 147 \\
101 & 156 & 89 & 100 & 165 & 113 & 148 & 170 & 163 & 186 & 144 & 194 & 208 & 39 & 113 & 159 \\
103 & 153 & 94 & 103 & 203 & 136 & 146 & 92 & 66 & 192 & 188 & 103 & 178 & 47 & 167 & 159 \\
102 & 146 & 106 & 99 & 99 & 121 & 39 & 60 & 164 & 175 & 198 & 46 & 56 & 56 & 156 & 156 \\
99 & 146 & 95 & 97 & 144 & 61 & 103 & 107 & 108 & 111 & 192 & 62 & 65 & 128 & 153 & 154 \\
99 & 140 & 103 & 109 & 103 & 124 & 54 & 81 & 172 & 137 & 178 & 54 & 43 & 159 & 149 & 174 \\
84 & 133 & 107 & 84 & 149 & 43 & 158 & 95 & 151 & 120 & 183 & 46 & 30 & 147 & 142 & 201 \\
58 & 153 & 110 & 41 & 94 & 213 & 71 & 73 & 140 & 103 & 138 & 83 & 152 & 143 & 128 & 207 \\
56 & 141 & 108 & 58 & 92 & 51 & 55 & 61 & 88 & 166 & 58 & 103 & 146 & 150 & 116 & 211 \\
89 & 115 & 188 & 47 & 113 & 104 & 56 & 67 & 128 & 155 & 187 & 71 & 153 & 134 & 203 & 95 \\
35 & 99 & 151 & 67 & 35 & 88 & 88 & 128 & 140 & 142 & 176 & 213 & 144 & 128 & 214 & 100 \\
89 & 98 & 97 & 51 & 49 & 101 & 47 & 90 & 136 & 136 & 157 & 205 & 106 & 43 & 54 & 76 \\
44 & 105 & 69 & 69 & 68 & 53 & 110 & 127 & 134 & 146 & 159 & 184 & 109 & 121 & 72 & 113
\end{bmatrix}$$

I represents the pixel values. The first subscript of I indicates the current stage of the transform, while the second subscript indicates the current step within a stage. We start by convolving the first row with both $h_0[n]$ and $h_1[n]$ and discarding the values with odd-numbered index. The results of these two operations are

$$(I_{00}(:, 0) * h_0[n]) \downarrow 2 = [245, 156, 171, 183, 184, 173, 228, 160]$$
$$(I_{00}(:, 0) * h_1[n]) \downarrow 2 = [-30, 3, 0, 7, -5, -16, -3, 16]$$

where the colon in the first index position indicates that we are showing a whole row. If you like, you can verify these operations using MATLAB's conv function.

Next, we form the transformed output row by concatenating the resulting coefficients. The first row of the transformed image is then

$$[245, 156, 171, 183, 184, 173, 228, 160, -30, 3, 0, 7, -5, -16, -3, 16]$$

Similar to the simple 1D Haar transform examples, most of the energy is now concentrated on the first half of the transformed image. We continue the same process for the remaining rows and obtain the following result:

$I_{11}(x,y) =$

$$
\begin{bmatrix}
245 & 156 & 171 & 183 & 184 & 173 & 228 & 160 & -30 & 3 & 0 & 7 & -5 & -16 & -3 & 16 \\
239 & 141 & 181 & 197 & 242 & 158 & 202 & 229 & -17 & 5 & -20 & 3 & 26 & -27 & 27 & 141 \\
195 & 147 & 163 & 177 & 288 & 173 & 209 & 106 & -34 & 2 & 2 & 19 & -50 & -35 & -38 & -1 \\
180 & 139 & 226 & 177 & 274 & 267 & 247 & 163 & -45 & 29 & 24 & -29 & -2 & 30 & -101 & -78 \\
191 & 145 & 197 & 198 & 247 & 230 & 239 & 143 & -49 & 22 & 36 & -11 & -26 & -14 & 101 & -54 \\
192 & 145 & 237 & 184 & 135 & 253 & 169 & 192 & -47 & 38 & 36 & 4 & -58 & 66 & 94 & -4 \\
176 & 159 & 156 & 77 & 204 & 232 & 51 & 196 & -31 & 9 & -48 & 30 & 11 & 58 & 29 & 4 \\
179 & 148 & 162 & 129 & 146 & 213 & 92 & 217 & -39 & 18 & 50 & -10 & 33 & 51 & -23 & 8 \\
169 & 159 & 163 & 97 & 204 & 202 & 85 & 234 & -29 & 1 & -42 & 23 & 37 & 41 & -56 & -5 \\
155 & 153 & 149 & 159 & 176 & 204 & 65 & 236 & -32 & 32 & 85 & 39 & 38 & 44 & -54 & -31 \\
145 & 148 & 158 & 148 & 164 & 157 & 188 & 215 & -55 & 59 & -110 & 28 & 26 & 48 & -1 & -64 \\
134 & 152 & 102 & 70 & 153 & 126 & 199 & 207 & -47 & 38 & 13 & 10 & -76 & 3 & -7 & -76 \\
127 & 203 & 130 & 94 & 171 & 218 & 171 & 228 & 12 & 88 & -27 & 15 & 1 & 76 & 24 & 85 \\
70 & 188 & 63 & 144 & 191 & 257 & 215 & 232 & -5 & 24 & -28 & -9 & 19 & -46 & 36 & 91 \\
129 & 124 & 87 & 96 & 177 & 236 & 162 & 77 & -2 & 20 & -48 & 1 & 17 & -56 & 30 & -24 \\
103 & 115 & 85 & 142 & 188 & 234 & 184 & 132 & -37 & 0 & 27 & -4 & 5 & -35 & -22 & -33
\end{bmatrix}
$$

We now go on and apply the filters to the columns of the above resulting image. As before, we apply both $h_0[n]$ and $h_1[n]$ to each column and discard the odd indexed results:

$$(I_{11}(0, :) * h_0[n]) \downarrow 2 = [353, 280, 269, 256, 240, 206, 160, 153]^T$$
$$(I_{11}(0, :) * h_1[n]) \downarrow 2 = [-12, 10, -7, -4, 2, -1, 43, 16]^T$$

Concatenating the above results into a single column and applying the same procedure to each of the remaining columns, we arrive at the final transformed image:

$I_{12}(x,y) =$

$$
\begin{bmatrix}
353 & 212 & 251 & 272 & 281 & 234 & 308 & 289 & -33 & 6 & -15 & 5 & 24 & -29 & 38 & 120 \\
280 & 203 & 254 & 250 & 402 & 269 & 297 & 207 & -45 & 11 & -2 & 9 & 31 & -26 & -74 & 23 \\
269 & 202 & 312 & 280 & 316 & 353 & 337 & 227 & -70 & 43 & 56 & -23 & -41 & 21 & 82 & -81 \\
256 & 217 & 247 & 155 & 236 & 328 & 114 & 283 & -52 & 27 & -14 & 23 & -2 & 90 & 49 & 12 \\
240 & 221 & 226 & 172 & 264 & 294 & 113 & 330 & -41 & 14 & 31 & 23 & 57 & 60 & -78 & -3 \\
206 & 204 & 201 & 192 & 230 & 219 & 232 & 300 & -76 & 67 & -53 & 40 & 4 & 46 & -18 & -107 \\
160 & 275 & 150 & 135 & 244 & 294 & 267 & 331 & -2 & 90 & -17 & 10 & -24 & 49 & 29 & 89 \\
153 & 189 & 113 & 173 & 260 & 342 & 256 & 176 & -20 & 18 & -38 & -4 & 24 & -75 & 25 & -5 \\
-12 & 7 & -9 & -13 & -6 & 11 & 12 & -69 & -10 & -1 & 14 & 6 & -38 & 3 & -45 & -99 \\
10 & 3 & -31 & 16 & -1 & -51 & -10 & -30 & 2 & -12 & 0 & 24 & -32 & -45 & 109 & 42 \\
-7 & 5 & -44 & -35 & 67 & -10 & -17 & -15 & 3 & -15 & -28 & 0 & 41 & -30 & -18 & -19 \\
-4 & 9 & -1 & -37 & 41 & 6 & -33 & 2 & 9 & -12 & -67 & 31 & -7 & 3 & 2 & 0 \\
2 & -3 & 9 & -25 & 2 & -25 & 60 & -8 & -11 & -4 & -123 & -12 & -6 & -4 & 14 & -12 \\
-1 & 22 & 32 & 46 & 10 & 48 & -11 & 20 & 19 & 32 & -59 & 9 & 70 & 50 & 16 & 73 \\
43 & -18 & 32 & -40 & -13 & -23 & -37 & -61 & 8 & 22 & 2 & 13 & -12 & 43 & -8 & -45 \\
16 & 2 & -6 & -32 & -7 & 5 & -13 & -50 & 24 & 7 & -61 & 2 & 11 & -33 & 43 & 1
\end{bmatrix}
$$

This completes one stage of the Discrete Wavelet Transform. We can perform another stage by applying the same transform procedure to the upper left 8×8 DC image of $I_{12}(x, y)$. The resulting two-stage transformed image is

$I_{22}(x, y) =$

$$
\begin{bmatrix}
558 & 451 & 608 & 532 & 75 & 26 & 94 & 25 & -33 & 6 & -15 & 5 & 24 & -29 & 38 & 120 \\
463 & 511 & 627 & 566 & 66 & 68 & -43 & 68 & -45 & 11 & -2 & 9 & -31 & -26 & -74 & 23 \\
464 & 401 & 478 & 416 & 14 & 84 & -97 & -229 & -70 & 43 & 56 & -23 & -41 & 21 & 82 & -81 \\
422 & 335 & 477 & 553 & -88 & 46 & -31 & -6 & -52 & 27 & -14 & 23 & -2 & 90 & 49 & 12 \\
14 & 33 & -56 & 42 & 22 & -43 & -36 & 1 & -41 & 14 & 31 & 23 & 57 & 60 & -78 & -3 \\
-13 & 36 & 54 & 52 & 12 & -21 & 51 & 70 & -76 & 67 & -53 & 40 & 4 & 46 & -18 & -107 \\
25 & -20 & 25 & -7 & -35 & 35 & -56 & -55 & -2 & 90 & -17 & 10 & -24 & 49 & 29 & 89 \\
46 & 37 & -51 & 51 & -44 & 26 & 39 & -74 & -20 & 18 & -38 & -4 & 24 & -75 & 25 & -5 \\
-12 & 7 & -9 & -13 & -6 & 11 & 12 & -69 & -10 & -1 & 14 & 6 & -38 & 3 & -45 & -99 \\
10 & 3 & -31 & 16 & -1 & -51 & -10 & -30 & 2 & -12 & 0 & 24 & -32 & -45 & 109 & 42 \\
-7 & 5 & -44 & -35 & 67 & -10 & -17 & -15 & 3 & -15 & -28 & 0 & 41 & -30 & -18 & -19 \\
-4 & 9 & -1 & -37 & 41 & 6 & -33 & 2 & 9 & -12 & -67 & 31 & -7 & 3 & 2 & 0 \\
2 & -3 & 9 & -25 & 2 & -25 & 60 & -8 & -11 & -4 & -123 & -12 & -6 & -4 & 14 & -12 \\
-1 & 22 & 32 & 46 & 10 & 48 & -11 & 20 & 19 & 32 & -59 & 9 & 70 & 50 & 16 & 73 \\
43 & -18 & 32 & -40 & -13 & -23 & -37 & -61 & 8 & 22 & 2 & 13 & -12 & 43 & -8 & -45 \\
16 & 2 & -6 & -32 & -7 & 5 & -13 & -50 & 24 & 7 & -61 & 2 & 11 & -33 & 43 & 1
\end{bmatrix}
$$

Notice that I_{12} corresponds to the subband diagram shown in Fig. 8.19a, and I_{22} corresponds to Fig. 8.19b. At this point, we may apply *different levels of quantization* to each subband according to some preferred bit allocation algorithm, given a desired bitrate. *This is the basis for a simple wavelet-based compression algorithm.* However, since in this example we are illustrating the mechanics of the DWT, here we will simply bypass the quantization step and perform an inverse transform to reconstruct the input image.

We refer to the top left 8×8 block of values as the innermost stage in correspondence with Fig. 8.19. Starting with the innermost stage, we extract the first column and separate the low-pass and high-pass coefficients. The low-pass coefficient is simply the first half of the column, and the high-pass coefficients are the second half. Then we upsample them by appending a zero after each coefficient. The two resulting arrays are

$$\mathbf{a} = [558, 0, 463, 0, 464, 0, 422, 0]^T$$
$$\mathbf{b} = [14, 0, -13, 0, 25, 0, 46, 0]^T$$

Since we are using biorthogonal filters, we convolve $\mathbf{a}$ and $\mathbf{b}$ with $\tilde{h}_0[n]$ and $\tilde{h}_1[n]$, respectively. The results of the two convolutions are then added to form a single 8×1 array. The resulting column is

$$[414, 354, 323, 338, 333, 294, 324, 260]^T$$

All columns in the innermost stage are processed in this manner. The resulting image is

$$I'_{21}(x, y) =$$

$$\begin{bmatrix}
414 & 337 & 382 & 403 & 70 & -16 & 48 & 12 & -33 & 6 & -15 & 5 & 24 & -29 & 38 & 120 \\
354 & 322 & 490 & 368 & 39 & 59 & 63 & 55 & -45 & 11 & -2 & 9 & -31 & -26 & -74 & 23 \\
323 & 395 & 450 & 442 & 62 & 25 & -26 & 90 & -70 & 43 & 56 & -23 & -41 & 21 & 82 & -81 \\
338 & 298 & 346 & 296 & 23 & 77 & -117 & -131 & -52 & 27 & -14 & 23 & -2 & 90 & 49 & 12 \\
333 & 286 & 364 & 298 & 4 & 67 & -75 & -176 & -41 & 14 & 31 & 23 & 57 & 60 & -78 & -3 \\
294 & 279 & 308 & 350 & -2 & 17 & 12 & -53 & -76 & 67 & -53 & 40 & 4 & 46 & -18 & -107 \\
324 & 240 & 326 & 412 & -96 & 54 & -25 & -45 & -2 & 90 & -17 & 10 & -24 & 49 & 29 & 89 \\
260 & 189 & 382 & 359 & -47 & 14 & -63 & 69 & -20 & 18 & -38 & -4 & 24 & -75 & 25 & -5 \\
-12 & 7 & -9 & -13 & -6 & 11 & 12 & -69 & -10 & -1 & 14 & 6 & -38 & 3 & -45 & -99 \\
10 & 3 & -31 & 16 & -1 & -51 & -10 & -30 & 2 & -12 & 0 & 24 & -32 & -45 & 109 & 42 \\
-7 & 5 & -44 & -35 & 67 & -10 & -17 & -15 & 3 & -15 & -28 & 0 & 41 & -30 & -18 & -19 \\
-4 & 9 & -1 & -37 & 41 & 6 & -33 & 2 & 9 & -12 & -67 & 31 & -7 & 3 & 2 & 0 \\
2 & -3 & 9 & -25 & 2 & -25 & 60 & -8 & -11 & -4 & -123 & -12 & -6 & -4 & 14 & -12 \\
-1 & 22 & 32 & 46 & 10 & 48 & -11 & 20 & 19 & 32 & -59 & 9 & 70 & 50 & 16 & 73 \\
43 & -18 & 32 & -40 & -13 & -23 & -37 & -61 & 8 & 22 & 2 & 13 & -12 & 43 & -8 & -45 \\
16 & 2 & -6 & -32 & -7 & 5 & -13 & -50 & 24 & 7 & -61 & 2 & 11 & -33 & 43 & 1
\end{bmatrix}$$

We are now ready to process the rows. For each row of the upper left 8 × 8 subimage, we again separate them into low-pass and high-pass coefficients. Then we upsample both by adding a zero after each coefficient. The results are convolved with the appropriate $\tilde{h}_0[n]$ and $\tilde{h}_1[n]$ filters. After these steps are completed for all rows, we have

$$I'_{12}(x, y) =$$

$$\begin{bmatrix}
353 & 212 & 251 & 272 & 281 & 234 & 308 & 289 & -33 & 6 & -15 & 5 & 24 & -29 & 38 & 120 \\
280 & 203 & 254 & 250 & 402 & 269 & 297 & 207 & -45 & 11 & -2 & 9 & -31 & -26 & -74 & 23 \\
269 & 202 & 312 & 280 & 316 & 353 & 337 & 227 & -70 & 43 & 56 & -23 & -41 & 21 & 82 & -81 \\
256 & 217 & 247 & 155 & 236 & 328 & 114 & 283 & -52 & 27 & -14 & 23 & -2 & 90 & 49 & 12 \\
240 & 221 & 226 & 172 & 264 & 294 & 113 & 330 & -41 & 14 & 31 & 23 & 57 & 60 & -78 & -3 \\
206 & 204 & 201 & 192 & 230 & 219 & 232 & 300 & -76 & 67 & -53 & 40 & 4 & 46 & -18 & -107 \\
160 & 275 & 150 & 135 & 244 & 294 & 267 & 331 & -2 & 90 & -17 & 10 & -24 & 49 & 29 & 89 \\
153 & 189 & 113 & 173 & 260 & 342 & 256 & 176 & -20 & 18 & -38 & -4 & 24 & -75 & 25 & -5 \\
-12 & 7 & -9 & -13 & -6 & 11 & 12 & -69 & -10 & -1 & 14 & 6 & -38 & 3 & -45 & -99 \\
10 & 3 & -31 & 16 & -1 & -51 & -10 & -30 & 2 & -12 & 0 & 24 & -32 & -45 & 109 & 42 \\
-7 & 5 & -44 & -35 & 67 & -10 & -17 & -15 & 3 & -15 & -28 & 0 & 41 & -30 & -18 & -19 \\
-4 & 9 & -1 & -37 & 41 & 6 & -33 & 2 & 9 & -12 & -67 & 31 & -7 & 3 & 2 & 0 \\
2 & -3 & 9 & -25 & 2 & -25 & 60 & -8 & -11 & -4 & -123 & -12 & -6 & -4 & 14 & -12 \\
-1 & 22 & 32 & 46 & 10 & 48 & -11 & 20 & 19 & 32 & -59 & 9 & 70 & 50 & 16 & 73 \\
43 & -18 & 32 & -40 & -13 & -23 & -37 & -61 & 8 & 22 & 2 & 13 & -12 & 43 & -8 & -45 \\
16 & 2 & -6 & -32 & -7 & 5 & -13 & -50 & 24 & 7 & -61 & 2 & 11 & -33 & 43 & 1
\end{bmatrix}$$

We then repeat the same inverse transform procedure on $I'_{12}(x, y)$, to obtain $I'_{00}(x, y)$. Notice that $I'_{00}(x, y)$ is not exactly the same as $I_{00}(x, y)$, but the difference is small. These small differences are caused by round-off errors during the forward and inverse transforms, and truncation errors when converting from floating-point numbers to integer grayscale values.

$I'_{00}(x, y) =$

$$
\begin{bmatrix}
158 & 170 & 97 & 103 & 122 & 129 & 132 & 125 & 132 & 126 & 111 & 157 & 159 & 144 & 116 & 91 \\
164 & 152 & 90 & 98 & 123 & 151 & 131 & 159 & 188 & 115 & 106 & 145 & 140 & 143 & 227 & 52 \\
115 & 148 & 89 & 100 & 117 & 118 & 131 & 151 & 201 & 210 & 84 & 154 & 127 & 146 & 58 & 58 \\
94 & 144 & 88 & 104 & 187 & 123 & 117 & 181 & 184 & 203 & 202 & 153 & 152 & 228 & 45 & 146 \\
100 & 155 & 88 & 99 & 164 & 112 & 147 & 169 & 163 & 186 & 143 & 193 & 207 & 38 & 112 & 158 \\
103 & 153 & 93 & 102 & 203 & 135 & 145 & 91 & 66 & 192 & 188 & 103 & 177 & 46 & 166 & 158 \\
102 & 146 & 106 & 99 & 99 & 121 & 39 & 60 & 164 & 175 & 198 & 46 & 56 & 56 & 156 & 156 \\
99 & 146 & 95 & 97 & 143 & 60 & 102 & 106 & 107 & 110 & 191 & 61 & 65 & 128 & 153 & 154 \\
98 & 139 & 102 & 109 & 103 & 123 & 53 & 80 & 171 & 136 & 177 & 53 & 43 & 158 & 148 & 173 \\
84 & 133 & 107 & 84 & 148 & 42 & 157 & 94 & 150 & 119 & 182 & 45 & 29 & 146 & 141 & 200 \\
57 & 152 & 109 & 41 & 93 & 213 & 70 & 72 & 139 & 102 & 137 & 82 & 151 & 143 & 128 & 207 \\
56 & 141 & 108 & 58 & 91 & 50 & 54 & 60 & 87 & 165 & 57 & 102 & 146 & 149 & 116 & 211 \\
89 & 114 & 187 & 46 & 113 & 104 & 55 & 66 & 127 & 154 & 186 & 71 & 153 & 134 & 203 & 94 \\
35 & 99 & 150 & 66 & 34 & 88 & 88 & 127 & 140 & 141 & 175 & 212 & 144 & 128 & 213 & 100 \\
88 & 97 & 96 & 50 & 49 & 101 & 47 & 90 & 136 & 136 & 156 & 204 & 105 & 43 & 54 & 76 \\
43 & 104 & 69 & 69 & 68 & 53 & 110 & 127 & 134 & 145 & 158 & 183 & 109 & 121 & 72 & 113
\end{bmatrix}
$$

Figure 8.21 shows a three-level image decomposition using the Haar wavelet.

8.7   Wavelet Packets

Wavelet packets can be viewed as a generalization of wavelets. They were first introduced by Coifman, Meyer, Quake, and Wickerhauser [14] as a family of orthonormal bases for discrete functions of $\mathbf{R}^N$. A complete subband decomposition can be viewed as a decomposition of the input signal, using an analysis tree of depth $\log N$.

In the usual dyadic wavelet decomposition, only the low-pass-filtered subband is recursively decomposed and thus can be represented by a logarithmic tree structure. However, a wavelet packet decomposition allows the decomposition to be represented by any pruned subtree of the full tree topology. Therefore, this representation of the decomposition topology is isomorphic to all permissible subband topologies [15]. The leaf nodes of each pruned subtree represent one permissible orthonormal basis.

The wavelet packet decomposition offers a number of attractive properties, including

- Flexibility, since the best wavelet basis in the sense of some cost metric can be found within a large library of permissible bases.
- Favorable localization of wavelet packets in both frequency and space.
- Low computational requirement for wavelet packet decomposition, because each decomposition can be computed in the order of $N \log N$ using fast filter banks.

Wavelet packets are currently being applied to solve various practical problems such as image compression, signal de-noising, fingerprint identification, and so on.

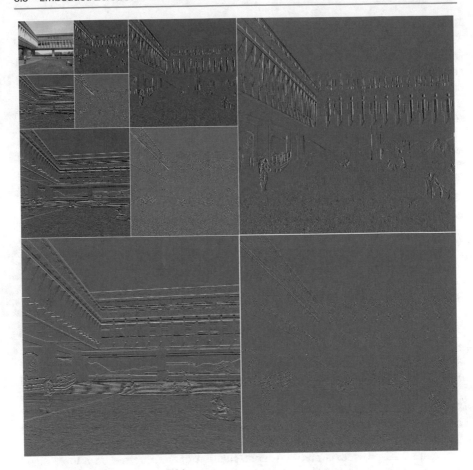

Fig. 8.21 Haar wavelet decomposition

8.8   Embedded Zerotree of Wavelet Coefficients

So far, we have described a wavelet-based scheme for image decomposition. However, aside from referring to the idea of quantizing away small coefficients, we have not really addressed how to code the wavelet transform values—how to form a bitstream. This problem is precisely what is dealt with in terms of a new data structure, the Embedded Zerotree.

The *Embedded Zerotree Wavelet* (EZW) algorithm introduced by Shapiro [16] is an effective and computationally efficient technique in image coding. This work has inspired a number of refinements to the initial EZW algorithm, the most notable being Said and Pearlman's *Set Partitioning in Hierarchical Trees* (SPIHT) algorithm [17] and Taubman's *Embedded Block Coding with Optimized Truncation* (EBCOT) algorithm [18], which is adopted into the JPEG2000 standard.

The EZW algorithm addresses two problems: obtaining the best image quality for a given bitrate and accomplishing this task in an embedded fashion. An *embedded* code is one that contains all lower rate codes "embedded" at the beginning of the bitstream. The bits are effectively ordered by importance in the bitstream. An embedded code allows the encoder to terminate the encoding at any point and thus meet any target bitrate exactly. Similarly, a decoder can cease to decode at any point and produce reconstructions corresponding to all lower rate encodings.

To achieve this goal, the EZW algorithm takes advantage of an important aspect of low-bitrate image coding. When conventional coding methods are used to achieve low bitrates, using scalar quantization followed by entropy coding, say, the most likely symbol, after quantization, is zero. It turns out that a large fraction of the bit budget is spent encoding the *significance map*, which flags whether input samples (in the case of the 2D discrete wavelet transform, the transform coefficients) have a zero or nonzero quantized value. The EZW algorithm exploits this observation to turn any significant improvement in encoding the significance map into a corresponding gain in compression efficiency. The EZW algorithm consists of two central components: the zerotree data structure and the method of successive approximation quantization.

8.8.1   The Zerotree Data Structure

The coding of the significance map is achieved using a new data structure called the *zerotree*. A wavelet coefficient x is said to be *insignificant* with respect to a given threshold T if $|x| < T$. The zerotree operates under the hypothesis that if a wavelet coefficient at a coarse scale is insignificant with respect to a given threshold T, all wavelet coefficients of the same orientation in the same spatial location at finer scales are likely to be insignificant with respect to T. Using the hierarchical wavelet decomposition presented in Chap. 8, we can relate every coefficient at a given scale to a set of coefficients at the next finer scale of similar orientation.

Figure 8.22 provides a pictorial representation of the zerotree on a three-stage wavelet decomposition. The coefficient at the coarse scale is called the *parent* while all corresponding coefficients are the next finer scale of the same spatial location and similar orientation are called *children*. For a given parent, the set of all coefficients at all finer scales are called *descendants*. Similarly, for a given child, the set of all coefficients at all coarser scales are called *ancestors*.

The scanning of the coefficients is performed in such a way that no child node is scanned before its parent. Figure 8.23 depicts the scanning pattern for a three-level wavelet decomposition.

Given a threshold T, a coefficient x is an element of the zerotree if it is insignificant and all its descendants are insignificant as well. An element of a zerotree is a *zerotree root* if it is not the descendant of a previously found zerotree root. The significance map is coded using the zerotree with a four-symbol alphabet. The four symbols are

- **The zerotree root**. The root of the zerotree is encoded with a special symbol indicating that the insignificance of the coefficients at finer scales is completely predictable.

Fig. 8.22 Parent–child
relationship in a zerotree

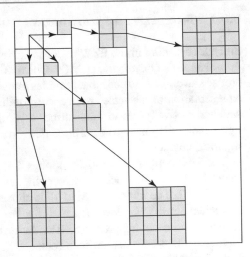

Fig. 8.23 EZW scanning
order

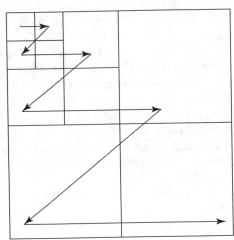

- **Isolated zero**. The coefficient is insignificant but has some significant descendants.
- **Positive significance**. The coefficient is significant with a positive value.
- **Negative significance**. The coefficient is significant with a negative value.

The cost of encoding the significance map is substantially reduced by employing the zerotree. The zerotree works by exploiting self-similarity on the transform coefficients. The underlying justification for the success of the zerotree is that even though the image has been transformed using a decorrelating transform, the occurrences of insignificant coefficients are not independent events.

In addition, the zerotree coding technique is based on the observation that it is much easier to predict insignificance than to predict significant details across scales. This technique focuses on reducing the cost of encoding the significance map so that more bits will be available to encode the expensive significant coefficients.

8.8.2 Successive Approximation Quantization

Embedded coding in the EZW coder is achieved using a method called *Successive Approximation Quantization* (SAQ). One motivation for developing this method is to produce an embedded code that provides a coarse-to-fine, multiprecision logarithmic representation of the scale space corresponding to the wavelet-transformed image. Another motivation is to take further advantage of the efficient encoding of the significance map using the zerotree data structure, by allowing it to encode more significance maps.

The SAQ method sequentially applies a sequence of thresholds $T_0, \ldots, T_{N-1}$ to determine the significance of each coefficient. The thresholds are chosen such that $T_i = T_{i-1}/2$. The initial threshold T_0 is chosen so that $|x_j| < 2T_0$ for all transform coefficients x_j. A *dominant list* and a *subordinate list* are maintained during the encoding and decoding processes. The dominant list contains the coordinates of the coefficients that have not yet been found to be significant in the same relative order as the initial scan.

Using the scan ordering shown in Fig. 8.23, all coefficients in a given subband appear on the initial dominant list prior to coefficients in the next subband. The subordinate list contains the magnitudes of the coefficients that have been found to be significant. Each list is scanned only once for each threshold.

During a dominant pass, coefficients having their coordinates on the dominant list imply that they are not yet significant. These coefficients are compared to the threshold T_i to determine their significance. If a coefficient is found to be significant, its magnitude is appended to the subordinate list, and the coefficient in the wavelet transform array is set to zero to enable the possibility of a zerotree occurring on future dominant passes at smaller thresholds. The resulting significance map is zerotree-coded.

The dominant pass is followed by a subordinate pass. All coefficients on the subordinate list are scanned, and their magnitude, as it is made available to the decoder, is refined to an additional bit of precision. Effectively, the width of the uncertainty interval for the true magnitude of the coefficients is cut in half. For each magnitude on the subordinate list, the refinement can be encoded using a binary alphabet with a 1 indicating that the true value falls in the upper half of the uncertainty interval and a 0 indicating that it falls in the lower half. The string of symbols from this binary alphabet is then entropy-coded. After the subordinate pass, the magnitudes on the subordinate list are sorted in decreasing order to the extent that the decoder can perform the same sort.

The process continues to alternate between the two passes, with the threshold halved before each dominant pass. The encoding stops when some target stopping criterion has been met.

8.8.3   EZW Example

The following example demonstrates the concept of zerotree coding and successive approximation quantization. Shapiro [16] presents an example of EZW coding in his paper for an 8×8 three-level wavelet transform. However, unlike the example given by Shapiro, we will complete the encoding and decoding process and show the output bitstream up to the point just before entropy coding.

Figure 8.24 shows the coefficients of a three-stage wavelet transform that we attempt to code using the EZW algorithm. We will use the symbols p, n, t, and z to denote positive significance, negative significance, zerotree root, and isolated zero, respectively.

Since the largest coefficient is 57, we will choose the initial threshold T_0 to be 32. In the beginning, the dominant list contains the coordinates of all the coefficients. We begin scanning in the order shown in Fig. 8.23 and determine the significance of the coefficients. The following is the list of coefficients visited, in the order of the scan:

$$\{57, -37, -29, 30, 39, -20, 17, 33, 14, 6, 10, 19, 3, 7, 8, 2, 2, 3, 12, -9, 33, 20, 2, 4\}$$

With respect to the threshold $T_0 = 32$, it is easy to see that the coefficients 57 and -37 are significant. Thus, we output a p and an n to represent them. The coefficient -29 is insignificant but contains a significant descendant, 33, in LH1. Therefore, it is coded as z. The coefficient 30 is also insignificant, and all its descendants are insignificant with respect to the current threshold, so it is coded as t.

Since we have already determined the insignificance of 30 and all its descendants, the scan will bypass them, and no additional symbols will be generated. Continuing

Fig. 8.24 Coefficients of a three-stage wavelet transform used as input to the EZW algorithm

57	−37	39	−20	3	7	9	10
−29	30	17	33	8	2	1	6
14	6	15	13	9	−4	2	3
10	19	−7	9	−7	14	12	−9
12	15	33	20	−2	3	1	0
0	7	2	4	4	−1	1	1
4	1	10	3	2	0	1	0
5	6	0	0	3	1	2	1

in this manner, the dominant pass outputs the following symbols:

$$D_0 : pnztpttptzttpttt$$

Five coefficients are found to be significant: 57, −37, 39, 33, and another 33. Since we know that no coefficients are greater than $2T_0 = 64$, and the threshold used in the first dominant pass is 32, the uncertainty interval is thus [32, 64). Therefore, we know that the value of significant coefficients lies somewhere inside this uncertainty interval.

The subordinate pass following the dominant pass refines the magnitude of these coefficients by indicating whether they lie in the first half or the second half of the uncertainty interval. The output is 0 if the values lie in [32, 48) and 1 for values within [48, 64). According to the order of the scan, the subordinate pass outputs the following bits:

$$S_0 : 10000$$

Now the dominant list contains the coordinates of all the coefficients except those found to be significant, and the subordinate list contains the values {57, 37, 39, 33, 33}. After the subordinate pass is completed, we attempt to rearrange the values in the subordinate list such that larger coefficients appear before smaller ones, with the constraint that the decoder is able do exactly the same.

Since the subordinate pass halves the uncertainty interval, the decoder is able to distinguish values between [32, 48) and [48, 64). Since 39 and 37 are not distinguishable in the decoder, their order will not be changed. Therefore, the subordinate list remains the same after the reordering operation.

Before we move on to the second round of dominant and subordinate passes, we need to set the values of the significant coefficients to 0 in the wavelet transform array so that they do not prevent the emergence of a new zerotree.

The new threshold for a second dominant pass is $T_1 = 16$. Using the same procedure as above, the dominant pass outputs the following symbols. Note that the coefficients in the dominant list will not be scanned.

$$D_1 : zznptnpttztptttttttttttttptttttt \tag{8.71}$$

The subordinate list is now {57, 37, 39, 33, 33, 29, 30, 20, 17, 19, 20}. The subordinate pass that follows will halve each of the three current uncertainty intervals [48, 64), [32, 48), and [16, 32). The subordinate pass outputs the following bits:

$$S_1 : 10000110000$$

Now we set the value of the coefficients found to be significant to 0 in the wavelet transform array.

The output of the subsequent dominant and subordinate passes is shown below:

D_2 :zzzzzzzzptpzpptnttptppttpttpttpnpptttttttpttttttttttttttt

S_2 :011001110011011000000110110

D_3 :zzzzzzztzpztztnttptttttptnnttttptttpptppttpttttt

S_3 :0010001000111010011000100111110 1100010

D_4 :zzzzzttztztztzztzzpttppptttttpttpttnpttptptttpt

S_4 :11111010011010110000010111011011000100100101010101 0

D_5 :zzzztztttttztzzzzttpttpttttttnptppttttppttp

Since the length of the uncertainty interval in the last pass is 1, the last subordinate
pass is unnecessary.

On the decoder side, suppose we received information only from the first dominant
and subordinate passes. We can reconstruct a lossy version of the transform coefficients by reversing the encoding process. From the symbols in D_0, we can obtain the
position of the significant coefficients. Then, using the bits decoded from S_0, we can
reconstruct the value of these coefficients using the center of the uncertainty interval.
Figure 8.25 shows the resulting reconstruction.

It is evident that we can stop the decoding process at any point to reconstruct
a coarser representation of the original input coefficients. Figure 8.26 shows the
reconstruction if the decoder received only D_0, S_0, D_1, S_1, D_2, and only the first
10 bits of S_2. The coefficients that were not refined during the last subordinate pass
appear as if they were quantized using a coarser quantizer than those that were.

In fact, the reconstruction value used for these coefficients is the center of the
uncertainty interval from the previous pass. The heavily shaded coefficients in the
figure are those that were refined, while the lightly shaded coefficients are those that
were not refined. As a result, it is not easy to see where the decoding process ended,
and this eliminates much of the visual artifact contained in the reconstruction.

Fig. 8.25 Reconstructed transform coefficients from the first dominant and subordinate passes

56	−40	40	0	0	0	0	0
0	0	0	40	0	0	0	0
0	0	0	0	0	0	0	0
0	0	0	0	0	0	0	0
0	0	40	0	0	0	0	0
0	0	0	0	0	0	0	0
0	0	0	0	0	0	0	0
0	0	0	0	0	0	0	0

Fig. 8.26 Reconstructed transform coefficients from D_0, S_0, D_1, S_1, D_2, and the first 10 bits of S_2

58	−38	38	−22	0	0	12	12
−30	30	18	34	12	0	0	0
12	0	12	12	12	0	0	0
12	20	0	12	0	12	12	−12
12	12	34	22	0	0	0	0
0	0	0	0	0	0	0	0
0	0	12	0	0	0	0	0
0	0	0	0	0	0	0	0

8.9   Set Partitioning in Hierarchical Trees (SPIHT)

SPIHT is a revolutionary extension of the EZW algorithm. Based on EZW's underlying principles of the partial ordering of transformed coefficients, ordered bitplane transmission of refinement bits, and the exploitation of self-similarity in the transformed wavelet image, the SPIHT algorithm significantly improves the performance of its predecessor by changing the ways subsets of coefficients are partitioned and refinement information is conveyed.

A unique property of the SPIHT bitstream is its compactness. The resulting bitstream from the SPIHT algorithm is so compact that passing it through an entropy coder would produce only marginal gain in compression at the expense of much more computation. Therefore, a fast SPIHT coder can be implemented without any entropy coder or possibly just a simple patent-free Huffman coder.

Another signature of the SPIHT algorithm is that no ordering information is explicitly transmitted to the decoder. Instead, the decoder reproduces the execution path of the encoder and recovers the ordering information. A desirable side effect of this is that the encoder and decoder have similar execution times, which is rarely the case for other coding methods. Said and Pearlman [17] give a full description of this algorithm.

8.10   Exercises

1. Assume we have an unbounded source we wish to quantize using an M-bit midtread uniform quantizer. Derive an expression for the total distortion if the step size is 1.

2. Suppose the domain of a uniform quantizer is $[-b_M, b_M]$. We define the loading fraction as

$$\gamma = \frac{b_M}{\sigma}$$

where σ is the standard deviation of the source. Write a simple program to quantize a Gaussian distributed source having zero-mean and unit variance using a 4-bit uniform quantizer. Plot the SNR against the loading fraction and estimate the optimal step size that incurs the least amount of distortion from the graph.

3. *Suppose the input source is Gaussian-distributed with zero-mean and unit variance—that is, the probability density function is defined as

$$f_X(x) = \frac{1}{\sqrt{2\pi}} e^{-\frac{x^2}{2}} \tag{8.72}$$

We wish to find a four-level Lloyd–Max quantizer. Let $y_i = [y_i^0, \ldots, y_i^3]$ and $b_i = [b_i^0, \ldots, b_i^3]$. The initial reconstruction levels are set to $y_0 = [-2, -1, 1, 2]$. This source is unbounded, so the outer two boundaries are $+\infty$ and $-\infty$.

Follow the Lloyd–Max algorithm in this chapter: the other boundary values are calculated as the midpoints of the reconstruction values. We now have $b_0 = [-\infty, -1.5, 0, 1.5, \infty]$. Continue one more iteration for $i = 1$, using Eq. 8.13 and find y_0^1, y_1^1, y_2^1, and y_3^1, using numerical integration. Also calculate the squared error of the difference between y_1 and y_0.

Iteration is repeated until the squared error between successive estimates of the reconstruction levels is below some predefined threshold ϵ. Write a small program to implement the Lloyd–Max quantizer described above.

4. If the block size for a 2D DCT transform is 8×8, and we use only the DC components to create a thumbnail image, what fraction of the original pixels would we be using?

5. When the block size is 8×8, the definition of the DCT is given in Eq. 8.17.

(a) If an 8×8 grayscale image is in the range $0 .. 255$, what is the largest value a DCT coefficient could be, and for what input image? (Also, state *all* the DCT coefficient values for that image.)

(b) If we first subtract the value 128 from the whole image and then carry out the DCT, what is the exact effect on the DCT value $F[2, 3]$?

(c) Why would we carry out that subtraction? Does the subtraction affect the number of bits we need to code the image?

(d) Would it be possible to invert that subtraction, in the IDCT? If so, how?

6. Write a simple program or refer to the sample DCT program dct_1D.c in the book's website to verify the results in Example 8.2 of the 1D DCT example in this chapter.

7. As Eq. 8.27 indicates, the 2D DCT can be implemented by the following matrix multiplications:

$$F(u, v) = \mathbf{T} \cdot f(i, j) \cdot \mathbf{T}^T.$$

(a) What is in each row of $\mathbf{T}$?
(b) Explain what has been achieved in each of the two matrix multiplications.

8. Write a program to verify that the DCT-matrix $\mathbf{T_8}$ as defined in Eqs. 8.29 and 8.30 is an Orthogonal Matrix, i.e., all its rows and columns are orthogonal unit vectors (orthonormal vectors).

9. Write a program to verify that the 2D DCT and IDCT matrix implementations as defined in Eqs. 8.27 and 8.31 are lossless, i.e., they can transform any 8×8 values $f(i, j)$ to $F(u, v)$ and back to $f(i.j)$. (Here, we are not concerned with possible/tiny floating point calculation errors.)

10. We could use a similar DCT scheme for *video streams* by using a 3D version of DCT. Suppose one color component of a video has pixels f_{ijk} at position (i, j) and time k. How could we define its 3D DCT transform?

11. Suppose a uniformly colored sphere is illuminated and has shading varying smoothly across its surface, as in Fig. 8.27.

(a) What would you expect the DCT coefficients for its image to look like?
(b) What would be the effect on the DCT coefficients of having a checkerboard of colors on the surface of the sphere?
(c) For the uniformly colored sphere again, describe the DCT values for a block that straddles the top edge of the sphere, where it meets the black background.
(d) Describe the DCT values for a block that straddles the left edge of the sphere.

12. The Haar wavelet has a scaling function which is defined as follows:

$$\phi(t) = \begin{cases} 1 & 0 \leq t \leq 1 \\ 0 & \text{otherwise} \end{cases} \qquad (8.73)$$

and its scaling vector is $h_0[0] = h_0[1] = 1/\sqrt{2}$.

(a) Draw the scaling function, then verify that its dilated translates $\phi(2t)$ and $\phi(2t - 1)$ satisfy the dilation Eq. (8.62). Draw the combination of these functions that makes up the full function $\phi(t)$.
(b) Derive the wavelet vector $h_1[0], h_1[1]$ from Eq. 8.65 and then derive and draw the Haar wavelet function $\psi(t)$ from Eq. 8.63.

13. Suppose the mother wavelet $\psi(t)$ has vanishing moments M_p up to and including M_n. Expand $f(t)$ in a Taylor series around $t = 0$, up to the nth derivative of f [i.e., up to leftover error of order $O(n + 1)$]. Evaluate the summation of integrals

Fig. 8.27 Sphere shaded by a light

produced by substituting the Taylor series into (8.58) and show that the result is of order $O(s^{n+2})$.

14. The program `wavelet_compression.c` on this book's website is in fact simple to implement as a MATLAB function (or similar fourth-generation language). The advantage in doing so is that the `imread` function can input image formats of a great many types, and `imwrite` can output as desired. Using the given program as a template, construct a MATLAB program for wavelet-based image reduction, with perhaps the number of wavelet levels being a function parameter.

15. It is interesting to find the Fourier transform of functions, and this is easy if you have available a symbolic manipulation system such as MAPLE. In that language, you can just invoke the `fourier` function and view the answer directly! As an example, try the following code fragment:

```
with('inttrans');
f := 1;
F := fourier(f,t,w);
```

The answer should be $2\pi\delta(w)$. Let's try a Gaussian:

```
f := exp(-t^2);
F := fourier(f,t,w);
```

Now the answer should be $\sqrt{\pi}e^{(-w^2/4)}$: the Fourier transform of a Gaussian is simply another Gaussian.

16. Suppose we define the wavelet function

$$\psi(t) = exp(-t^{1/4})\sin(t^4), \quad t \geq 0 \tag{8.74}$$

This function oscillates about the value 0. Use a plotting package to convince yourself that the function has a zero moment M_p for any value of p.

17. Implement both a DCT-based and a wavelet-based image coder. Design your user interface so that the compressed results from both coders can be seen side by side for visual comparison. The PSNR for each coded image should also be shown, for quantitative comparisons.

Include a slider bar that controls the target bitrate for both coders. As you change the target bitrate, each coder should compress the input image in real time and show the compressed results immediately on your user interface.

Discuss both qualitative and quantitative compression results observed from your program at target bitrates of 4, 1, and 0.25 bpp.

References

1. K. Sayood, *Introduction to Data Compression*, 5th edn. (Morgan Kaufmann, San Francisco, 2017)
2. H. Stark, J.W. Woods, *Probability, Statistics, and Random Processes for Engineers*, 4th edn. (Pearson, 2012)
3. A. György, On the theoretical limits of lossy source coding, 1998. Tudományos Diákkör (TDK) Conference (Hungarian Scientific Student's Conference) at Technical University of Budapest. http://www.szit.bme.hu/gya/mixed.ps
4. S. Arimoto, An algorithm for calculating the capacity of an arbitrary discrete memoryless channel. IEEE Trans. Inf. Theory **18**, 14–20 (1972)
5. R. Blahut, Computation of channel capacity and rate-distortion functions. IEEE Trans. Inform. Theory **18**, 460–473 (1972)
6. A. Gersho, R.M. Gray, *Vector Quantization and Signal Compression*. (Springer, 1991)
7. A.K. Jain, *Fundamentals of Digital Image Processing* (Prentice-Hall, Englewood Cliffs, NJ, 1988)
8. K.R. Rao, P. Yip, *Discrete Cosine Transform: algorithms, Advantages Applications* (Academic Press, Boston, MA, 1990)
9. J.F. Blinn, What's the deal with the DCT? IEEE Comput. Graph. Appl. **13**(4), 78–83 (1993)
10. S. Mallat, *A Wavelet Tour of Signal Processing*, 3rd edn. (Academic Press, San Diego, 2008)
11. S. Mallat, A theory for multiresolution signal decomposition: the wavelet representation. IEEE Trans. Pattern Anal. Mach. Intell. **11**, 674–693 (1989)
12. R.C. Gonzalez, R.E. Woods, *Digital Image Processing*, 3rd edn. (Prentice-Hall, 2007)
13. B.E. Usevitch, A tutorial on modern lossy wavelet image compression: foundations of JPEG 2000. IEEE Signal Process. Mag. **18**(5), 22–35 (2001)
14. R. Coifman, Y. Meyer, S. Quake, V. Wickerhauser, *Signal Processing and Compression with Wavelet Packets* (Yale University, Numerical Algorithms Research Group, 1990)
15. K. Ramachandran, M. Vetterli, Best wavelet packet basis in a rate-distortion sense. IEEE Trans. Image Process. **2**, 160–173 (1993)
16. J. Shapiro, Embedded image coding using zerotrees of wavelet coefficients. IEEE Trans. Signal Process. **41**(12) (1993)
17. A. Said, W.A. Pearlman, A new, fast, and efficient image codec based on set partitioning in hierarchical trees. IEEE Trans. CSVT **6**(3), 243–249 (1996)
18. D. Taubman, High performance scalable image compression with EBCOT. IEEE Trans. Image Process. **9**(7), 1158–1170 (2000)

Image Compression Standards 9

Recent years have seen an explosion in the availability of digital images, because of the increase in numbers of digital imaging devices such as smartphones, webcams, digital cameras, and scanners. The need to efficiently process and store images in digital form has motivated the development of many image compression *standards* for various applications and needs. In general, standards have greater longevity than particular programs or devices and therefore warrant careful study. In this chapter, we examine some image compression standards and demonstrate how topics presented in Chaps. 7 and 8 are applied in practice.

We first explore the standard JPEG definition, then go on to look at the wavelet-based JPEG 2000 standard. Two other standards, JPEG-LS—aimed particularly at a lossless JPEG, outside the main JPEG standard—and JBIG, for bi-level image compression, are included for completeness.

9.1 The JPEG Standard

JPEG is an image compression standard developed by the *Joint Photographic Experts Group*. It was formally accepted as an international standard in 1992 [1]. Remarkably, almost 30 years later, it is still going strong—used in most images. At its 25th anniversary, Hudson et al., four of the original JPEG development team members, presented an informative historical review of the development of the standard [2].

JPEG consists of a number of steps, each of which contributes to compression. We'll look at the motivation behind these steps, then take apart the algorithm piece by piece.

© Springer Nature Switzerland AG 2021
Z.-N. Li et al., *Fundamentals of Multimedia*, Texts in Computer Science,
https://doi.org/10.1007/978-3-030-62124-7_9

9.1.1   Main Steps in JPEG Image Compression

As we know, unlike 1D audio signals, a digital image $f(i, j)$ is not defined over the time domain. Instead, it is defined over a *spatial domain*—that is, an image is a function of the two dimensions i and j (or, conventionally, x and y). The 2D DCT is used as one step in JPEG, to yield a frequency response that is a function $F(u, v)$ in the *spatial frequency domain*, indexed by two integers u and v.

JPEG is a lossy image compression method. The effectiveness of the DCT transform coding method in JPEG relies on three major observations:

Observation 1. Useful image contents change relatively slowly across the image—that is, it is unusual for intensity values to vary widely several times in a small area—for example, in an 8×8 image block. Spatial frequency indicates how many times pixel values change across an image block. The DCT formalizes this notion with a measure of how much the image contents change in relation to the number of cycles of a cosine wave per block.

Observation 2. Psychophysical experiments suggest that humans are much less likely to notice the loss of very high-spatial frequency components than lower frequency components.

JPEG's approach to the use of DCT is basically to reduce high-frequency contents and then efficiently code the result into a bitstring. The term *spatial redundancy* indicates that much of the information in an image is repeated: if a pixel is red, then its neighbor is likely red also. Because of Observation 2 above, the DCT coefficients for the lowest frequencies are most important. Therefore, as the frequency gets higher, it becomes less important to represent the DCT coefficient accurately. It may even be safely set to zero without losing much perceivable image information.

Clearly, a string of zeros can be represented efficiently as the length of such a run of zeros, and compression of bits required is possible. Since we end up using fewer numbers to represent the pixels in blocks, by removing some location-dependent information, we have effectively removed spatial redundancy.

JPEG works for both color and grayscale images. In the case of color images, such as YCbCr, the encoder works on each component separately, using the same routines. If the source image is in a different color format, the encoder performs a color-space conversion to YCbCr. As discussed in Chap. 5, the chrominance images Cr and Cb are subsampled: JPEG uses the 4:2:0 scheme, making use of another observation about vision:

Observation 3. Visual acuity (accuracy in distinguishing closely spaced lines) is much greater for gray ("black and white") than for color. We simply cannot see much change in color if it occurs in close proximity—think of the blobby ink used in comic books. This works simply because our eye sees the black lines best, and our brain just pushes the color into place. In fact, traditionally, broadcast TV makes use of this phenomenon to transmit much less color information than gray information.

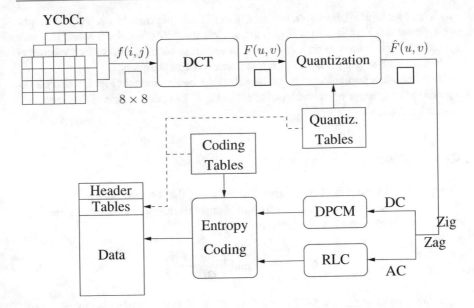

Fig. 9.1 Block diagram for JPEG encoder

When the JPEG image is needed for viewing, the three compressed component images can be decoded independently and eventually combined. For the color channels, each pixel must be first enlarged to cover a 2 × 2 block. Without loss of generality, we will simply use one of them—for example, the Y image, in the description of the compression algorithm below.

Figure 9.1 shows a block diagram for a JPEG encoder. If we reverse the arrows in the figure, we basically obtain a JPEG decoder. The JPEG encoder consists of the following main steps:

- Transform RGB to YCbCr and subsample color
- Perform DCT on image blocks
- Apply Quantization
- Perform Zigzag ordering and run-length encoding
- Perform Entropy coding.

DCT on Image Blocks

Each image is divided into 8 × 8 blocks. The 2D DCT (Eq. 8.17) is applied to each block image $f(i, j)$, with output being the DCT coefficients $F(u, v)$ for each block. The choice of small block size in JPEG is a compromise reached by the committee: a number larger than 8 would have made accuracy at low frequencies better, but using 8 makes the DCT (and IDCT) computation very fast.

Using blocks at all, however, has the effect of isolating each block from its neighboring context. This is why JPEG images look choppy ("blocky") when the user specifies a high *compression ratio*—we can see these blocks. And in fact, removing such "blocking artifacts" is an important concern of researchers.

To calculate a particular $F(u, v)$, we select the basis image in Fig. 8.9 that corresponds to the appropriate u and v and use it in Eq. 8.17 to derive one of the frequency responses $F(u, v)$.

Quantization

The quantization step in JPEG is aimed at reducing the total number of bits needed for a compressed image [3]. It consists of simply dividing each entry in the DCT coefficient block by an integer, then rounding

$$\hat{F}(u, v) = \text{round}\left(\frac{F(u, v)}{Q(u, v)}\right) \tag{9.1}$$

Here, $F(u, v)$ represents a DCT coefficient, $Q(u, v)$ is a *quantization matrix* entry, and $\hat{F}(u, v)$ represents the *quantized DCT coefficient* that JPEG will use in the succeeding entropy coding.

The default values in the 8×8 quantization matrix $Q(u, v)$ are listed in Tables 9.1 and 9.2 for luminance and chrominance images, respectively. These numbers resulted from psychophysical studies, with the goal of maximizing the compression ratio while minimizing perceptual losses in JPEG images. The following should be apparent:

- Since the numbers in $Q(u, v)$ are relatively large, the magnitude and variance of $\hat{F}(u, v)$ are significantly smaller than those of $F(u, v)$. We'll see later that $\hat{F}(u, v)$ can be coded with much fewer bits. *The quantization step is the main source for loss in JPEG compression.*
- The entries of $Q(u, v)$ tend to have larger values toward the lower right corner. This aims to introduce more loss at the higher spatial frequencies—a practice supported by Observations 1 and 2.

We can handily change the compression ratio simply by multiplicatively *scaling* the numbers in the $Q(u, v)$ matrix. In fact, the *quality factor* (qf), a user choice offered in every JPEG implementation, can be specified. It is usually in the range of $1 .. 100$, where $qf = 100$ corresponds to the highest quality compressed images and $qf = 1$ the lowest quality. The relationship between qf and the *scaling_factor* is as below:

Table 9.1 The luminance quantization table

16	11	10	16	24	40	51	61
12	12	14	19	26	58	60	55
14	13	16	24	40	57	69	56
14	17	22	29	51	87	80	62
18	22	37	56	68	109	103	77
24	35	55	64	81	104	113	92
49	64	78	87	103	121	120	101
72	92	95	98	112	100	103	99

Table 9.2 The chrominance quantization table

17	18	24	47	99	99	99	99
18	21	26	66	99	99	99	99
24	26	56	99	99	99	99	99
47	66	99	99	99	99	99	99
99	99	99	99	99	99	99	99
99	99	99	99	99	99	99	99
99	99	99	99	99	99	99	99
99	99	99	99	99	99	99	99

```
// qf is the user-selected compression quality
// Q is the default Quantization Matrix
// Qx is the scaled Quantization Matrix
// Q1 is a Quantization Matrix which is all 1's

if qf >= 50
    scaling_factor = (100-qf)/50;
else
    scaling_factor = (50/qf);
end
if scaling_factor != 0    // if qf is not 100
    Qx = round( Q*scaling_factor );
else
    Qx = Q1;              // no quantization
end
Qx = uint8(Qx);          // max is clamped to 255 for qf=1
```

As an example, when $qf = 50$, $scaling\_factor$ will be 1. The resulting Q values will be equal to the table entries. When $qf = 10$, the $scaling\_factor$ will be 5. The resulting Q values will be 5 times the table entry values. For $qf = 100$, the table entries simply become all 1 values, meaning no quantization from this source. Very low-quality factors, like $qf = 1$, are a special case: if indeed $qf = 1$ then the $scaling\_factor$ will be 50, and the quantization matrix will contain very large numbers. However, because of the type-cast to `uint8` in a typical implementation, the table entries go to their effective maximum value of 255. Realistically, almost for all applications, qf should not be less than 10.

JPEG also allows custom quantization tables to be specified and put in the header; it is interesting to use low-constant or high-constant values such as $Q \equiv 2$ or $Q \equiv 128$ to observe the basic effects of Q on visual artifacts.

Figures 9.2 and 9.3 show some results of JPEG image coding and decoding on the test image *Lena*. Only the luminance image (Y) is shown. Also, the lossless coding steps after quantization are not shown, since they do not affect the quality/loss of the JPEG images. These results show the effect of compression and decompression applied to a relatively smooth block in the image and a more textured (higher frequencycontent) block, respectively.

Suppose $f(i, j)$ represents one of the 8×8 blocks extracted from the image, $F(u, v)$ the DCT coefficients, and $\hat{F}(u, v)$ the quantized DCT coefficients. Let $\tilde{F}(u, v)$ denote the de-quantized DCT coefficients, determined by simply multiplying by $Q(u, v)$, and let $\tilde{f}(i, j)$ be the reconstructed image block. To illustrate the quality of the JPEG compression, especially the loss, the error $\epsilon(i, j) = f(i, j) - \tilde{f}(i, j)$ is shown in the last row in Figs. 9.2 and 9.3.

In Fig. 9.2, an image block (indicated by a black box in the image) is chosen at the area where the luminance values change smoothly. Actually, the left side of the block is brighter, and the right side is slightly darker. As expected, except for the DC and the first few AC components, representing low spatial frequencies, most of the DCT coefficients $F(u, v)$ have small magnitudes. This is because the pixel values in this block contain few high-spatial frequency changes.

An explanation of a small implementation detail is in order. The range of 8-bit luminance values $f(i, j)$ is [0, 255]. In the JPEG implementation, each Y value is first reduced by 128 by simply subtracting 128 before encoding. The idea here is to turn the Y component into a zero-mean image, the same as the chrominance images. As a result, we do not waste any bits coding the mean value. (Think of an 8×8 block with intensity values ranging from 120 to 135.) Using $f(i, j) - 128$ in place of $f(i, j)$ will not affect the output of the AC coefficients—it alters only the DC coefficient. After decoding, the subtracted 128 will be added back onto the Y values.

In Fig. 9.3, the image block chosen has rapidly changing luminance. Hence, many more AC components have large magnitudes (including those toward the lower right corner, where u and v are large). Notice that the error $\epsilon(i, j)$ is also larger now than in Fig. 9.2—JPEG does introduce more loss if the image has quickly changing details.

An easy-to-use JPEG demo written in Java is available in the Demos section of the text website for you to try.

Preparation for Entropy Coding

We have so far seen two of the main steps in JPEG compression: DCT and quantization. The remaining small steps shown in the block diagram in Fig. 9.1 all lead up to *entropy coding* of the quantized DCT coefficients. These additional data compres-

An 8×8 block from the Y image of 'Lena'

200	202	189	188	189	175	175	175
200	203	198	188	189	182	178	175
203	200	200	195	200	187	185	175
200	200	200	200	197	187	187	187
200	205	200	200	195	188	187	175
200	200	200	200	200	190	187	175
205	200	199	200	191	187	187	175
210	200	200	200	188	185	187	186

$$f(i, j)$$

515	65	-12	4	1	2	-8	5
-16	3	2	0	0	-11	-2	3
-12	6	11	-1	3	0	1	-2
-8	3	-4	2	-2	-3	-5	-2
0	-2	7	-5	4	0	-1	-4
0	-3	-1	0	4	1	-1	0
3	-2	-3	3	3	-1	-1	3
-2	5	-2	4	-2	2	-3	0

$$F(u, v)$$

32	6	-1	0	0	0	0	0
-1	0	0	0	0	0	0	0
-1	0	1	0	0	0	0	0
-1	0	0	0	0	0	0	0
0	0	0	0	0	0	0	0
0	0	0	0	0	0	0	0
0	0	0	0	0	0	0	0
0	0	0	0	0	0	0	0

$$\hat{F}(u, v)$$

512	66	-10	0	0	0	0	0
-12	0	0	0	0	0	0	0
-14	0	16	0	0	0	0	0
-14	0	0	0	0	0	0	0
0	0	0	0	0	0	0	0
0	0	0	0	0	0	0	0
0	0	0	0	0	0	0	0
0	0	0	0	0	0	0	0

$$\tilde{F}(u, v)$$

199	196	191	186	182	178	177	176
201	199	196	192	188	183	180	178
203	203	202	200	195	189	183	180
202	203	204	203	198	191	183	179
200	201	202	201	196	189	182	177
200	200	199	197	192	186	181	177
204	202	199	195	190	186	183	181
207	204	200	194	190	187	185	184

$$\tilde{f}(i, j)$$

1	6	-2	2	7	-3	-2	-1
-1	4	2	-4	1	-1	-2	-3
0	-3	-2	-5	5	-2	2	-5
-2	-3	-4	-3	-1	-4	4	8
0	4	-2	-1	-1	-1	5	-2
0	0	1	3	8	4	6	-2
1	-2	0	5	1	1	4	-6
3	-4	0	6	-2	-2	2	2

$$\epsilon(i, j) = f(i, j) - \tilde{f}(i, j)$$

Fig. 9.2 JPEG compression for a smooth image block

Another 8×8 block from the Y image of 'Lena'

70	70	100	70	87	87	150	187
85	100	96	79	87	154	87	113
100	85	116	79	70	87	86	196
136	69	87	200	79	71	117	96
161	70	87	200	103	71	96	113
161	123	147	133	113	113	85	161
146	147	175	100	103	103	163	187
156	146	189	70	113	161	163	197

$$f(i,j)$$

-80	-40	89	-73	44	32	53	-3
-135	-59	-26	6	14	-3	-13	-28
47	-76	66	-3	-108	-78	33	59
-2	10	-18	0	33	11	-21	1
-1	-9	-22	8	32	65	-36	-1
5	-20	28	-46	3	24	-30	24
6	-20	37	-28	12	-35	33	17
-5	-23	33	-30	17	-5	-4	20

$$F(u,v)$$

-5	-4	9	-5	2	1	1	0
-11	-5	-2	0	1	0	0	-1
3	-6	4	0	-3	-1	0	1
0	1	-1	0	1	0	0	0
0	0	-1	0	0	1	0	0
0	-1	1	-1	0	0	0	0
0	0	0	0	0	0	0	0
0	0	0	0	0	0	0	0

$$\hat{F}(u,v)$$

-80	-44	90	-80	48	40	51	0
-132	-60	-28	0	26	0	0	-55
42	-78	64	0	-120	-57	0	56
0	17	-22	0	51	0	0	0
0	0	-37	0	0	109	0	0
0	-35	55	-64	0	0	0	0
0	0	0	0	0	0	0	0
0	0	0	0	0	0	0	0

$$\tilde{F}(u,v)$$

70	60	106	94	62	103	146	176
85	101	85	75	102	127	93	144
98	99	92	102	74	98	89	167
132	53	111	180	55	70	106	145
173	57	114	207	111	89	84	90
164	123	131	135	133	92	85	162
141	159	169	73	106	101	149	224
150	141	195	79	107	147	210	153

$$\tilde{f}(i,j)$$

0	10	-6	-24	25	-16	4	11
0	-1	11	4	-15	27	-6	-31
2	-14	24	-23	-4	-11	-3	29
4	16	-24	20	24	1	11	-49
-12	13	-27	-7	-8	-18	12	23
-3	0	16	-2	-20	21	0	-1
5	-12	6	27	-3	2	14	-37
6	5	-6	-9	6	14	-47	44

$$\epsilon(i,j) = f(i,j) - \tilde{f}(i,j)$$

Fig. 9.3 JPEG compression for a textured image block

Fig. 9.4 Zigzag scan in
JPEG

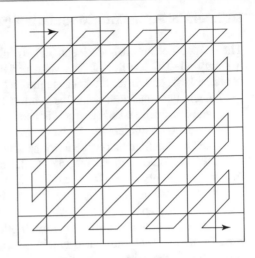

sion steps are lossless. Interestingly, the DC and AC coefficients are treated quite differently before entropy coding: run-length encoding on ACs versus DPCM on DCs.

Run-Length Coding (RLC) on AC Coefficients

Notice in Fig. 9.2 the many zeros in $\hat{F}(u, v)$ after quantization is applied. *Run-length Coding (RLC)* (or *Run-length Encoding, RLE*) is therefore useful in turning the $\hat{F}(u, v)$ values into sets {*#-zeros-to-skip, next nonzero value*}. RLC is even more effective when we use an addressing scheme, making it most likely to hit a long run of zeros: a *zigzag scan* turns the 8×8 matrix $\hat{F}(u, v)$ into a *64-vector*, as Fig. 9.4 illustrates. After all, most image blocks tend to have small high-spatial-frequency components, which are zeroed out by quantization. Hence, the zigzag-scan order has a good chance of concatenating long runs of zeros. For example, $\hat{F}(u, v)$ in Fig. 9.2 will be turned into

$$(32, 6, -1, -1, 0, -1, 0, 0, 0, -1, 0, 0, 1, 0, 0, \ldots, 0)$$

with three runs of zeros in the middle and a run of 51 zeros at the end.

The RLC step replaces values by a pair (RUNLENGTH, VALUE) for each run of zeros in the AC coefficients of $\hat{F}$, where RUNLENGTH is the number of zeros in the run and VALUE is the next nonzero coefficient. To further save bits, a special pair (0,0) indicates the end of block after the last nonzero AC coefficient is reached. In the above example, not considering the first (DC) component, we will thus have

$$(0, 6)(0, -1)(0, -1)(1, -1)(3, -1)(2, 1)(0, 0)$$

Differential Pulse Code Modulation (DPCM) on DC Coefficients

The DC coefficients are coded separately from the AC ones. Each 8×8 image block has only one DC coefficient. The values of the DC coefficients for various blocks could be large and different, because the DC value reflects the average intensity of each block, but consistent with Observation 1 above, the DC coefficient is unlikely to change drastically within a short distance. This makes DPCM an ideal scheme for coding the DC coefficients.

If the DC coefficients for the first five image blocks are 150, 155, 149, 152, and 144, DPCM would produce 150, 5, -6, 3, and -8, assuming the predictor for the ith block is simply $d_i = DC_i - DC_{i-1}$, and $d_0 = DC_0$. We expect DPCM codes to generally have smaller magnitude and variance, which is beneficial for the next *entropy coding* step.

It is worth noting that unlike the run-length coding of the AC coefficients, which is performed on each individual block, DPCM for the DC coefficients in JPEG is carried out on the entire image at once.

Entropy Coding

The DC and AC coefficients finally undergo an entropy coding step. Below, we will discuss only the basic (or *baseline*[1]) entropy coding method, which uses Huffman coding and supports only 8-bit pixels in the original images (or color image components).

Let's examine the two entropy coding schemes, using a variant of Huffman coding for DCs and a slightly different scheme for ACs.

Huffman Coding of DC Coefficients. Each DPCM-coded DC coefficient is represented by a pair of symbols (SIZE, AMPLITUDE), where SIZE indicates how many bits are needed for representing the coefficient and AMPLITUDE contains the actual bits.

Table 9.3 illustrates the size category for the different possible amplitudes. Notice that DPCM values could require more than 8 bits and could be negative values. The one's-complement scheme is used for negative numbers—that is, binary code 10 for 2, 01 for -2, 11 for 3, 00 for -3, and so on. In the example we are using, codes 150, 5, -6, 3, and -8 will be turned into

$$(8, 10010110), (3, 101), (3, 001), (2, 11), (4, 0111)$$

In the JPEG implementation, SIZE is Huffman coded and is hence a variable-length code. In other words, SIZE 2 might be represented as a single bit (0 or 1) if it appeared most frequently. In general, smaller SIZEs occur much more often—the

---

[1]The JPEG standard allows both Huffman coding and Arithmetic coding; both are entropy coding methods. It also supports both 8-bit and 12-bit pixel sizes.

Table 9.3 Baseline entropy coding details—size category

Size	Amplitude
1	$-1, 1$
2	$-3, -2, 2, 3$
3	$-7..-4, 4..7$
4	$-15..-8, 8..15$
$\vdots$	$\vdots$
10	$-1023..-512, 512..1023$

entropy of SIZE is low. Hence, deployment of Huffman coding brings additional compression. After encoding, a custom Huffman table can be stored in the JPEG image header; otherwise, a default Huffman table is used.

On the other hand, AMPLITUDE is not Huffman coded. Since its value can change widely, Huffman coding has no appreciable benefit.

Huffman Coding of AC Coefficients. Recall we said that the AC coefficients are run-length coded and are represented by pairs of numbers (RUNLENGTH, VALUE). However, in an actual JPEG implementation, VALUE is further represented by SIZE and AMPLITUDE, as for the DCs. To save bits, RUNLENGTH and SIZE are allocated only 4 bits each and squeezed into a single byte—let's call this *Symbol 1*. *Symbol 2* is the AMPLITUDE value; its number of bits is indicated by SIZE:

Symbol 1: (RUNLENGTH, SIZE)
Symbol 2: (AMPLITUDE)

The 4-bit RUNLENGTH can represent only zero-runs of length 0 to 15. Occasionally, the zero-run length exceeds 15; then a special extension code, (15, 0), is used for Symbol 1. In the worst case, three consecutive (15, 0) extensions are needed before a normal terminating Symbol 1, whose RUNLENGTH will then complete the actual run length. As in DC, Symbol 1 is Huffman coded, whereas Symbol 2 is not.

9.1.2 JPEG Modes

The JPEG standard supports numerous modes (variations). Some of the commonly used ones are

1. Sequential Mode
2. Progressive Mode
3. Hierarchical Mode
4. Lossless Mode.

1. Sequential Mode

This is the default JPEG mode. Each gray-level image or color image component is encoded in a single left-to-right, top-to-bottom scan. We implicitly assumed this mode in the discussions so far. The "Motion JPEG" video codec uses Baseline Sequential JPEG, applied to each image frame in the video.

2. Progressive Mode

Progressive JPEG delivers low-quality versions of the image quickly, followed by higher quality passes, and has become widely supported in web browsers. Such multiple scans of images are of course most useful when the speed of the communication line is low. In Progressive Mode, the first few scans carry only a few bits and deliver a rough picture of what is to follow. After each additional scan, more data is received, and image quality is gradually enhanced. The advantage is that the user end has a choice whether to continue receiving image data after the first scan(s).

Progressive JPEG can be realized in one of the following two ways. The main steps (DCT, quantization, etc.) are identical to those in Sequential Mode.

Spectral selection: This scheme takes advantage of the *spectral* (spatial frequency spectrum) characteristics of the DCT coefficients: the higher AC components provide only detail information.

Scan 1: Encode DC and the first few AC components, e.g., AC1 and AC2.
Scan 2: Encode a few more AC components, e.g., AC3, AC4, and AC5.
$\vdots$
Scan k: Encode the last few ACs, e.g., AC61, AC62, and AC63.

Successive approximation: Instead of gradually encoding spectral bands, all DCT coefficients are encoded simultaneously, but with their most significant bits (MSBs) first.

Scan 1: Encode the first few MSBs, e.g., Bits 7, 6, 5, and 4.
Scan 2: Encode a few more less-significant bits, e.g., Bit 3.
$\vdots$
Scan m: Encode the least significant bit (LSB), Bit 0.

3. Hierarchical Mode

As its name suggests, Hierarchical JPEG encodes the image in a hierarchy of several different resolutions. The encoded image at the lowest resolution is basically a compressed low-pass-filtered image, whereas the images at successively higher resolutions provide additional details (differences from the lower resolution images).

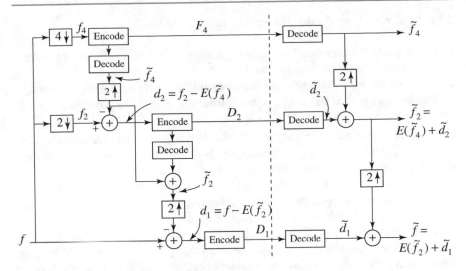

Fig. 9.5 Block diagram for Hierarchical JPEG

Similar to Progressive JPEG, Hierarchical JPEG images can be transmitted in multiple passes with progressively improving quality.

Figure 9.5 illustrates a three-level hierarchical JPEG encoder and decoder (separated by the dashed line in the figure).

Algorithm 9.1 (Three-Level Hierarchical JPEG Encoder)

1. **Reduction of image resolution**. Reduce resolution of the input image f (e.g., 512×512) by a factor of 2 in each dimension to obtain f_2 (e.g., 256×256). Repeat this to obtain f_4 (e.g., 128×128).
2. **Compress low-resolution image f_4**. Encode f_4 using any other JPEG method (e.g., Sequential and Progressive) to obtain F_4.
3. **Compress difference image d_2**.
   (a) Decode F_4 to obtain $\tilde{f}_4$. Use any interpolation method to expand $\tilde{f}_4$ to be of the same resolution as f_2 and call it $E(\tilde{f}_4)$.
   (b) Encode difference $d_2 = f_2 - E(\tilde{f}_4)$ using any other JPEG method (e.g., Sequential and Progressive) to generate D_2.

4. **Compress difference image d_1**.
   (a) Decode D_2 to obtain $\tilde{d}_2$; add it to $E(\tilde{f}_4)$ to get $\tilde{f}_2 = E(\tilde{f}_4) + \tilde{d}_2$, which is a version of f_2 after compression and decompression.
   (b) Encode difference $d_1 = f - E(\tilde{f}_2)$ using any other JPEG method (e.g., Sequential and Progressive) to generate D_1.

Algorithm 9.2 (Three-Level Hierarchical JPEG Decoder)

1. **Decompress the encoded low-resolution image** F_4. Decode F_4 using the same JPEG method as in the encoder, to obtain $\tilde{f}_4$.
2. **Restore image** $\tilde{f}_2$ at the **intermediate resolution**. Use $E(\tilde{f}_4) + \tilde{d}_2$ to obtain $\tilde{f}_2$.
3. **Restore image** $\tilde{f}$ at the **original resolution**. Use $E(\tilde{f}_2) + \tilde{d}_1$ to obtain $\tilde{f}$.

It should be pointed out that at Step 3 in the encoder, the difference d_2 is not taken as $f_2 - E(f_4)$ but as $f_2 - E(\tilde{f}_4)$. Employing $\tilde{f}_4$ has its overhead, since an additional decoding step must be introduced on the encoder side, as shown in the figure.

So, is it necessary? It is, because the *decoder* never has a chance to see the original f_4. The restoration step in the decoder uses $\tilde{f}_4$ to obtain $\tilde{f}_2 = E(\tilde{f}_4) + \tilde{d}_2$. Since $\tilde{f}_4 \neq f_4$ when a lossy JPEG method is used in compressing f_4, the encoder must use $\tilde{f}_4$ in $d_2 = f_2 - E(\tilde{f}_4)$ to avoid unnecessary error at decoding time. This kind of decoder–encoder step is typical in many compression schemes. In fact, we have seen it in Sect. 6.3.5. It is present simply because the decoder has access only to encoded, not original, values.

Similarly, at Step 4 in the encoder, d_1 uses the difference between f and $E(\tilde{f}_2)$, not $E(f_2)$.

4. **Lossless Mode**

Lossless JPEG is a very special case of JPEG which indeed has no loss in its image quality. As discussed in Chap. 7, however, it employs only a simple differential coding method, involving no transform coding. It is rarely used, since its compression ratio is very low compared to other lossy modes. On the other hand, it meets a special need, and the subsequently developed JPEG-LS standard is specifically aimed at lossless image compression (see Sect. 9.3).

9.1.3 A Glance at the JPEG Bitstream

Figure 9.6 provides a hierarchical view of the organization of the bitstream for JPEG images. Here, a *frame* is a picture, a *scan* is a pass through the pixels (e.g., the red component), a *segment* is a group of blocks, and a *block* consists of 8×8 pixels. Examples of some header information are

- **Frame header**
  - Bits per pixel
  - (Width, height) of image
  - Number of components
  - Unique ID (for each component)
  - Horizontal/vertical sampling factors (for each component)
  - Quantization table to use (for each component).

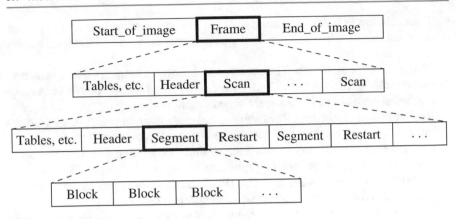

Fig. 9.6 JPEG bitstream

- **Scan header**
  - Number of components in scan
  - Component ID (for each component)
  - Huffman/Arithmetic coding table (for each component).

9.2 The JPEG 2000 Standard

The JPEG standard is no doubt the most successful and popular image format to date. The main reason for its success is the quality of its output for a relatively good compression ratio. However, in anticipating the needs and requirements of next-generation imagery applications, the JPEG committee also defined a new standard: JPEG 2000. The main part, the core coding system, is specified in ISO/IEC 15444-1 [4].

The JPEG 2000 standard [5–7] aims to provide not only a better rate-distortion trade-off and improved subjective image quality but also additional functionalities that the JPEG standard lacks. In particular, the JPEG 2000 standard addresses the following problems:

- **Low-bitrate compression**. The JPEG standard offers excellent rate-distortion performance at medium and high bitrates. However, at bitrates below 0.25 bpp, subjective distortion becomes unacceptable. This is important if we hope to receive images on our web-enabled ubiquitous devices, such as web-aware wristwatches, and so on.
- **Lossless and lossy compression**. Before JPEG 2000, no standard could provide superior lossless compression and lossy compression in a single bitstream.
- **Large images**. The new standard will allow image resolutions greater than 64 k × 64 k without tiling. It can handle image sizes up to $2^{32} - 1$.
- **Single decompression architecture**. The JPEG standard had 44 modes, many of them were application-specific and not used by the majority of JPEG decoders.

- **Transmission in noisy environments**. The new standard will provide improved error resilience for transmission in noisy environments such as wireless networks and the Internet.
- **Progressive transmission**. The new standard provides seamless quality and resolution scalability from low to high bitrates. The target bitrate and reconstruction resolution need not be known at the time of compression.
- **Region-of-interest coding**. The new standard permits specifying *Regions of Interest (ROI)*, which can be coded with better quality than the rest of the image. We may, for example, like to code the face of someone making a presentation with more quality than the surrounding furniture.
- **Computer-generated imagery**. The JPEG standard was optimized for natural imagery and did not perform well on computer-generated imagery.
- **Compound documents**. The new standard offers metadata mechanisms for incorporating additional non-image data as part of the file. This may be useful for including text along with imagery, as one important example.

In addition, JPEG 2000 is able to handle up to 256 channels of information, whereas the JPEG standard was only able to handle three color channels. Such huge quantities of data are routinely produced in satellite imagery.

Consequently, JPEG 2000 is designed to address a variety of applications, such as the Internet, color facsimile, printing, scanning, digital photography, remote sensing, mobile applications, medical imagery, digital library, e-commerce, and so on. The method looks ahead and provides the power to carry out remote browsing of large compressed images.

The JPEG 2000 standard operates in two coding modes: DCT based and wavelet based. The DCT-based coding mode is offered for backward compatibility with the JPEG standard and implements baseline JPEG. All the new functionalities and improved performance reside in the wavelet-based mode.

To date, JPEG 2000 is mostly used in high-end applications such as remote sensing satellite images, medical imaging, and digital cinema, where very high-resolution and high-quality images are required.

9.2.1 Main Steps of JPEG 2000 Image Compression*

The main compression method used in JPEG 2000 is the *Embedded Block Coding with Optimized Truncation algorithm (EBCOT)*, designed by Taubman [8]. In addition to providing excellent compression efficiency, EBCOT produces a bitstream with a number of desirable features, including quality, resolution scalability, and *random access*.

The basic idea of EBCOT is the partition of each subband LL, LH, HL, and HH produced by the wavelet transform into small blocks called *code blocks*. Each code block is coded independently, in such a way that no information for any other block is used.

A separate, scalable bitstream is generated for each code block. With its block-based coding scheme, the EBCOT algorithm has improved error resilience. The EBCOT algorithm consists of three steps:

1. Block coding and bitstream generation
2. Post compression rate-distortion (PCRD) optimization
3. Layer formation and representation.

1. Block Coding and Bitstream Generation

Each subband generated for the 2D discrete wavelet transform is first partitioned into small code blocks, typically 64×64, or other size no less than 32×32. Then the EBCOT algorithm generates a highly scalable bitstream for each code block B_i. The bitstream associated with B_i may be independently truncated to any member of a predetermined collection of different lengths R_i^n, with associated distortion D_i^n.

For each code block B_i (see Fig. 9.7), let $s_i[\mathbf{k}] = s_i[k_1, k_2]$ be the 2D sequence of small code blocks of subband samples, with k_1 and k_2 the row and column indices. (With this definition, the horizontal high-pass subband HL must be transposed so that k_1 and k_2 will have a meaning consistent with the other subbands. This transposition means that the HL subband can be treated in the same way as the LH, HH, and LL subbands and use the same context model.)

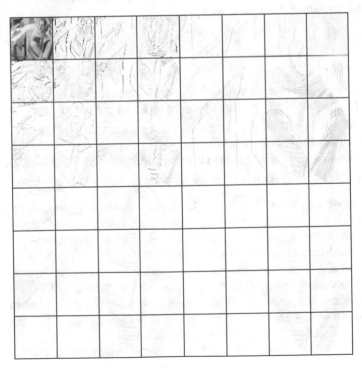

Fig. 9.7 Code block structure of EBCOT

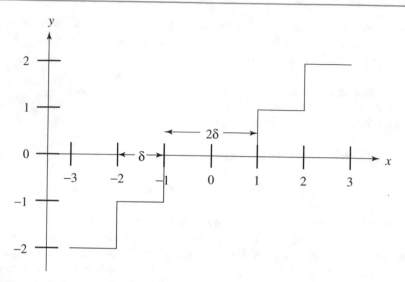

Fig. 9.8 Dead-zone quantizer. The length of the dead zone is 2δ. Values inside the dead zone are quantized to 0

The algorithm uses a *dead-zone* quantizer shown in Fig. 9.8—a double-length region straddling 0. Let $\chi_i[\mathbf{k}] \in \{-1, 1\}$ be the sign of $s_i[\mathbf{k}]$ and let $\nu_i[\mathbf{k}]$ be the quantized magnitude. Explicitly, we have

$$\nu_i[\mathbf{k}] = \frac{\|s_i[\mathbf{k}]\|}{\delta_{\beta_i}} \qquad (9.2)$$

where δ_{β_i} is the step size for subband β_i, which contains code block B_i. Let $\nu_i^p[\mathbf{k}]$ be the pth bit in the binary representation of $\nu_i[\mathbf{k}]$, where $p = 0$ corresponds to the least significant bit, and let p_i^{max} be the maximum value of p such that $\nu_i^{p_i^{max}}[\mathbf{k}] \neq 0$ for at least one sample in the code block.

The encoding process is similar to that of a bitplane coder, in which the most significant bit $\nu_i^{p_i^{max}}[\mathbf{k}]$ is coded first for all samples in the code block, followed by the next most significant bit $\nu_i^{p_i^{(max-1)}}[\mathbf{k}]$, and so on, until all bitplanes have been coded. In this way, if the bitstream is truncated, then some samples in the code block may be missing one or more least-significant bits. This is equivalent to having used a coarser dead-zone quantizer for these samples.

In addition, it is important to exploit the previously encoded information about a particular sample and its neighboring samples. This is done in EBCOT by defining a binary-valued state variable $\sigma_i[\mathbf{k}]$, which is initially 0 but changes to 1 when the relevant sample's first nonzero bitplane $v_i^p[\mathbf{k}] = 1$ is encoded. This binary state variable is referred to as the *significance* of a sample.

Section 8.8 introduces the zerotree data structure as a way of efficiently coding the bitstream for wavelet coefficients. The underlying observation behind the zerotree data structure is that significant samples tend to be clustered, so that it is often possible to dispose of a large number of samples by coding a single binary symbol.

EBCOT takes advantage of this observation; however, with efficiency in mind, it exploits the clustering assumption only down to relatively large sub-blocks of size 16×16. As a result, each code block is further partitioned into a 2D sequence of sub-blocks $B_i[\mathbf{j}]$. For each bitplane, explicit information is first encoded that identifies sub-blocks containing one or more significant samples. The other sub-blocks are bypassed in the remaining coding phases for that bitplane.

Let $\sigma^p(B_i[\mathbf{j}])$ be the significance of sub-block $B_i[\mathbf{j}]$ in bitplane p. The significance map is coded using a quad-tree. The tree is constructed by identifying the sub-blocks with leaf nodes—that is, $B_i^0[\mathbf{j}] = B_i[\mathbf{j}]$. The higher levels are built using recursion: $B_i^t[\mathbf{j}] = \cup_{\mathbf{z} \in \{0,1\}^2} B_i^{t-1}[2\mathbf{j} + \mathbf{z}], 0 \leq t \leq T$. The root of the tree represents the entire code block: $B_i^T[\mathbf{0}] = \cup_{\mathbf{j}} B_i[\mathbf{j}]$.

The significance of the code block is identified one quad level at a time, starting from the root at $t = T$ and working toward the leaves at $t = 0$. The significance values are then sent to an arithmetic coder for entropy coding. Significance values that are *redundant* are skipped. A value is taken as redundant if any of the following conditions is met:

- The parent is insignificant.
- The current quad was already significant in the previous bitplane.
- This is the last quad visited among those that share the same significant parent, and the other siblings are insignificant.

EBCOT uses four different coding primitives to code new information for a single sample in a bitplane p, as follows:

- **Zero coding**. This is used to code $\nu_i^p[\mathbf{k}]$, given that the quantized sample satisfies $\nu_i[\mathbf{k}] < 2^{p+1}$. Because the sample statistics are measured to be approximately Markovian, the significance of the current sample depends on the values of its eight immediate neighbors. The significance of these neighbors can be classified into three categories:
  - **Horizontal**. $h_i[\mathbf{k}] = \sum_{z \in \{1,-1\}} \sigma_i[k_1 + z, k_2]$, with $0 \leq h_i[\mathbf{k}] \leq 2$
  - **Vertical**. $v_i[\mathbf{k}] = \sum_{z \in \{1,-1\}} \sigma_i[k_1, k_2 + z]$, with $0 \leq v_i[\mathbf{k}] \leq 2$
  - **Diagonal**. $d_i[\mathbf{k}] = \sum_{z_1, z_2 \in \{1,-1\}} \sigma_i[k_1 + z_1, k_2 + z_2]$, with $0 \leq d_i[\mathbf{k}] \leq 4$.
  The neighbors outside the code block are considered to be insignificant, but note that sub-blocks are not at all independent. The 256 possible neighborhood configurations are reduced to the nine distinct context assignments listed in Table 9.4.

- **Run-length coding**. The run-length coding primitive is aimed at producing runs of the 1-bit significance values, as a prelude for the arithmetic coding engine. When a horizontal run of insignificant samples having insignificant neighbors is

Table 9.4 Context assignment for the zero coding primitive

Label	LL, LH, and HL subbands			HH subband	
	$h_i[\mathbf{k}]$	$v_i[\mathbf{k}]$	$d_i[\mathbf{k}]$	$d_i[\mathbf{k}]$	$h_i[\mathbf{k}] + v_i[\mathbf{k}]$
0	0	0	0	0	0
1	0	0	1	0	1
2	0	0	>1	0	>1
3	0	1	x	1	0
4	0	2	x	1	1
5	1	0	0	1	>1
6	1	0	>0	2	0
7	1	>0	x	2	>0
8	2	x	x	>2	x

found, it is invoked instead of the zero coding primitive. Each of the following four conditions must be met for the run-length coding primitive to be invoked:

- Four consecutive samples must be insignificant.
- The samples must have insignificant neighbors.
- The samples must be within the same sub-block.
- The horizontal index k_1 of the first sample must be even.

The last two conditions are simply for efficiency. When four symbols satisfy these conditions, one special bit is encoded instead, to identify whether any sample in the group is significant in the current bitplane (using a separate context model). If any of the four samples becomes significant, the index of the first such sample is sent as a 2-bit quantity.

- **Sign coding.** The sign coding primitive is invoked at most once for each sample, immediately after the sample makes a transition from being insignificant to significant during a zero coding or run-length coding operation. Since it has four horizontal and vertical neighbors, each of which may be insignificant, positive, or negative, there are $3^4 = 81$ different context configurations. However, exploiting both horizontal and vertical symmetries and assuming that the conditional distribution of $\chi_i[\mathbf{k}]$, given any neighborhood configuration, is the same as that of $-\chi_i[\mathbf{k}]$, the number of contexts is reduced to 5.

  Let $\bar{h}_i[\mathbf{k}]$ be 0 if both horizontal neighbors are insignificant, 1 if at least one horizontal neighbor is positive, or -1 if at least one horizontal neighbor is negative (and $\bar{v}_i[\mathbf{k}]$ is defined similarly). Let $\hat{\chi}_i[\mathbf{k}]$ be the sign prediction. The binary symbol coded using the relevant context is $\chi_i[\mathbf{k}] \cdot \hat{\chi}_i[\mathbf{k}]$. Table 9.5 lists these context assignments.

- **Magnitude refinement.** This primitive is used to code the value of $v_i^p[\mathbf{k}]$, given that $v_i[\mathbf{k}] \geq 2^{p+1}$. Only three context models are used for the magnitude refinement primitive. A second state variable $\tilde{\sigma}_i[\mathbf{k}]$ is introduced that changes from 0 to 1 after the magnitude refinement primitive is first applied to $s_i[\mathbf{k}]$. The context models depend on the value of this state variable: $v_i^p[\mathbf{k}]$ is coded with context 0 if $\tilde{\sigma}[\mathbf{k}] = h_i[\mathbf{k}] = v_i[\mathbf{k}] = 0$, with context 1 if $\tilde{\sigma}_i[\mathbf{k}] = 0$ and $h_i[\mathbf{k}] + v_i[\mathbf{k}] \neq 0$,

Table 9.5 Context assignments for the sign coding primitive

Label	$\hat{\chi}_i[\mathbf{k}]$	$\bar{h}_i[\mathbf{k}]$	$\bar{v}_i[\mathbf{k}]$
4	1	1	1
3	1	0	1
2	1	−1	1
1	−1	1	0
0	1	0	0
1	1	−1	0
2	−1	1	−1
3	−1	0	−1
4	−1	−1	−1

and with context 2 if $\tilde{\sigma}_i[\mathbf{k}] = 1$.

To ensure that each code block has a finely embedded bitstream, the coding of each bitplane p proceeds in four distinct passes, $(\mathcal{P}_1^p)$ to $(\mathcal{P}_4^p)$:

- **Forward-significance-propagation pass** $(\mathcal{P}_1^{\mathbf{p}})$. The sub-block samples are visited in scan line order. Insignificant samples and samples that do not satisfy the neighborhood requirement are skipped. For the LH, HL, and LL subbands, the neighborhood requirement is that at least one of the horizontal neighbors has to be significant. For the HH subband, the neighborhood requirement is that at least one of the four diagonal neighbors must be significant.

  For significant samples that pass the neighborhood requirement, the zero coding and run-length coding primitives are invoked as appropriate, to determine whether the sample first becomes significant in bitplane p. If so, the sign coding primitive is invoked to encode the sign. This is called the forward-significance-propagation pass, because a sample that has been found to be significant helps in the new significance determination steps that propagate in the direction of the scan.

- **Reverse-significance-propagation pass** $(\mathcal{P}_2^{\mathbf{p}})$. This pass is identical to $\mathcal{P}_1^p$, except that it proceeds in the reverse order. The neighborhood requirement is relaxed to include samples that have at least one significant neighbor in any direction.

- **Magnitude refinement pass** $(\mathcal{P}_3^{\mathbf{p}})$. This pass encodes samples that are already significant but that have not been coded in the previous two passes. Such samples are processed with the magnitude refinement primitive.

- **Normalization pass** $(\mathcal{P}_4^{\mathbf{p}})$. The value $v_i^p[\mathbf{k}]$ of all samples not considered in the previous three coding passes is coded using the sign coding and run-length coding primitives, as appropriate. If a sample is found to be significant, its sign is immediately coded using the sign coding primitive.

| $S^{p_i^{max}}$ | $P_4^{p_i^{max}}$ | $P_1^{p_i^{max}-1}$ | $P_2^{p_i^{max}-1}$ | $P_3^{p_i^{max}-1}$ | $S^{p_i^{max}-1}$ | $P_4^{p_i^{max}-1}$ | $\cdots$ | P_1^0 | P_2^0 | P_3^0 | S^0 | P_4^0 |

Fig. 9.9 Appearance of coding passes and quad-tree codes in each block's embedded bitstream

Figure 9.9 shows the layout of coding passes and quad-tree codes in each block's embedded bitstream. S^p denotes the quad-tree code identifying the significant sub-blocks in bitplane p. Notice that for any bitplane p, S^p appears just before the final coding pass $\mathcal{P}_4^p$, not the initial coding pass $\mathcal{P}_1^p$. This implies that sub-blocks that become significant for the first time in bitplane p are ignored until the final pass.

2. Post Compression Rate-Distortion Optimization

After all the subband samples have been compressed, a *post compression rate distortion (PCRD)* step is performed. The goal of PCRD is to produce an optimal truncation of each code block's independent bitstream such that distortion is minimized, subject to the bitrate constraint. For each truncated embedded bitstream of code block B_i having rate $R_i^{n_i}$, the overall distortion of the reconstructed image is (assuming distortion is additive)

$$D = \sum_i D_i^{n_i} \tag{9.3}$$

where $D_i^{n_i}$ is the distortion from code block B_i having truncation point n_i. For each code block B_i, distortion is computed by

$$D_i^n = w_{b_i}^2 \sum_{\mathbf{k} \in B_i} (\hat{s}_i^n[\mathbf{k}] - s_i[\mathbf{k}])^2 \tag{9.4}$$

where $s_i[\mathbf{k}]$ is the 2D sequence of subband samples in code block B_i and $\hat{s}_i^n[\mathbf{k}]$ is the quantized representation of these samples associated with truncation point n. The value $w_{b_i}^2$ is the L_2 norm of the wavelet basis function for the subband b_i that contains code block B_i.

The optimal selection of truncation points n_i can be formulated into a minimization problem subject to the following constraint:

$$R = \sum_i R_i^{n_i} \leq R^{max} \tag{9.5}$$

where R^{max} is the available bitrate. For some λ, any set of truncation points $\{n_i^\lambda\}$ that minimizes

$$(D(\lambda) + \lambda R(\lambda)) = \sum_i \left(D_i^{n_i^\lambda} + \lambda R_i^{n_i^\lambda} \right) \tag{9.6}$$

is optimal in the rate-distortion sense. Thus, finding the set of truncation points that minimizes Eq. (9.6) with total rate $R(\lambda) = R^{max}$ would yield the solution to the entire optimization problem.

Since the set of truncation points is discrete, it is generally not possible to find a value of λ for which $R(\lambda)$ is exactly equal to $R^{\max}$. However, since the EBCOT algorithm uses relatively small code blocks, each of which has many truncation points, it is sufficient to find the smallest value of λ such that $R(\lambda) \leq R^{\max}$.

It is easy to see that each code block B_i can be minimized independently. Let $\mathcal{N}_i$ be the set of feasible truncation points and let $j_1 < j_2 < \cdots$ be an enumeration of these feasible truncation points having corresponding distortion-rate slopes given by the ratios

$$S_i^{j_k} = \frac{\Delta D_i^{j_k}}{\Delta R_i^{j_k}} \tag{9.7}$$

where $\Delta R_i^{j_k} = R_i^{j_k} - R_i^{j_{k-1}}$ and $\Delta D_i^{j_k} = D_i^{j_k} - D_i^{j_{k-1}}$. It is evident that the slopes are strictly decreasing, since the operational distortion-rate curve is convex and strictly decreasing. The minimization problem for a fixed value of λ is simply the trivial selection

$$n_i^\lambda = \max \left\{ j_k \in \mathcal{N}_i | S_i^{j_k} > \lambda \right\} \tag{9.8}$$

The optimal value λ^* can be found using a simple bisection method operating on the distortion-rate curve. A detailed description of this method can be found in [9].

3. Layer Formation and Representation

The EBCOT algorithm offers both resolution and quality scalability, as opposed to other well-known scalable image compression algorithms such as EZW and SPIHT, which offer only quality scalability. This functionality is achieved using a layered bitstream organization and a two-tiered coding strategy.

The final bitstream EBCOT produces is composed of a collection of quality layers. The quality layer $\mathcal{Q}_1$ contains the initial $R_i^{n_i^1}$ bytes of each code block B_i and the other layers $\mathcal{Q}_q$ contain the incremental contribution $L_i^q = R_i^{n_i^q} - R_i^{n_i^{q-1}} \geq 0$ from code block B_i. The quantity n_i^q is the truncation point corresponding to the rate-distortion threshold λ_q selected for the qth quality layer. Figure 9.10 illustrates the layered bitstream (after [8]).

Along with these incremental contributions, auxiliary information such as the length L_i^q, the number of new coding passes $N_i^q = n_i^q - n_i^{q-1}$, the value $p_i^{\max}$ when B_i makes its first nonempty contribution to the quality layer $\mathcal{Q}_q$, and the index q_i of the quality layer to which B_i first makes a nonempty contribution must be explicitly stored. This auxiliary information is compressed in the second-tier coding engine. Hence, in this two-tiered architecture, the first tier produces the embedded block bitstreams, while the second encodes the block contributions to each quality layer.

The focus of this subsection is the second-tier processing of the auxiliary information accompanying each quality layer. The second-tier coding engine handles carefully the two quantities that exhibit substantial interblock redundancy. These two quantities are $p_i^{\max}$ and the index q_i of the quality layer to which B_i first makes a nonempty contribution.

Fig. 9.10 Three quality layers with eight blocks each

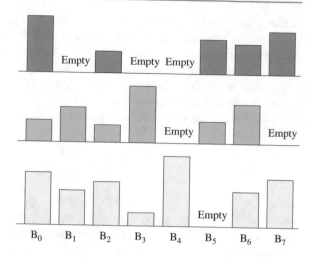

The quantity q_i is coded using a separate embedded quad-tree code within each subband. Let $B_i^0 = B_i$ be the leaves and B_i^T be the root of the tree that represents the entire subband. Let $q_i^t = \min\{q_j | B_j \subset B_i^t\}$ be the index of the first layer in which any code block in quad B_i^t makes a nonempty contribution. A single bit identifies whether $q_i^t > q$ for each quad at each level t, with redundant quads omitted. A quad is redundant if either $q_i^t < q - 1$ or $q_j^{t+1} > q$ for some parent quad B_j^{t+1}.

The other redundant quantity to consider is $p_i^{\max}$. It is clear that $p_i^{\max}$ is irrelevant until the coding of the quality layer $\mathcal{Q}_q$. Thus, any unnecessary information concerning $p_i^{\max}$ need not be sent until we are ready to encode $\mathcal{Q}_q$. EBCOT does this using a modified embedded quad-tree driven from the leaves rather than from the root.

Let B_i^t be the elements of the quad-tree structure built on top of the code blocks B_i from any subband, and let $p_i^{\max,t} = \max\{p_j^{\max} | B_j \subset B_i^t\}$. In addition, let $B_{i_t}^t$ be the ancestor of quads from which B_i descends and let P be a value guaranteed to be larger than $p_i^{\max}$ for any code block B_i. When code block B_i first contributes to the bitstream in the quality layer $\mathcal{Q}_q$, the value of $p_i^{\max} = p_{i_0}^{\max,0}$ is coded using the following algorithm:

- For $p = P - 1, P - 2, \dots, 0$
  - Send binary digits to identify whether $p_{i_t}^{\max,t} < p$. The redundant bits are skipped.
  - If $p_i^{\max} = p$, then stop.

The redundant bits are those corresponding to the condition $p_{i_t}^{\max,t} < p$ that can be inferred either from ancestors such that $p_{i_{t+1}}^{\max,t+1} < p$ or from the partial quad-tree code used to identify $p_j^{\max}$ for a different code block B_j.

9.2.2   Adapting EBCOT to JPEG 2000

JPEG 2000 uses the EBCOT algorithm as its primary coding method. However, the algorithm is slightly modified to enhance compression efficiency and reduce computational complexity.

To further enhance compression efficiency, as opposed to initializing the entropy coder using equiprobable states for all contexts, the JPEG 2000 standard makes an assumption of highly skewed distributions for some contexts, to reduce the model adaptation cost for typical images. Several small adjustments are made to the original algorithm to further reduce its execution time.

First, a low-complexity arithmetic coder that avoids multiplications and divisions, known as the MQ coder [10], replaces the usual arithmetic coder used in the original algorithm. Furthermore, JPEG 2000 does not transpose the HL subband's code blocks. Instead, the corresponding entries in the zero coding context assignment map are transposed.

To ensure a consistent scan direction, JPEG 2000 combines the forward- and reverse-significance-propagation passes into a single forward-significance-propagation pass with a neighborhood requirement equal to that of the original reverse pass. In addition, reducing the sub-block size to 4×4 from the original 16×16 eliminates the need to explicitly code sub-block significance. The resulting probability distribution for these small sub-blocks is highly skewed, so the coder behaves as if all sub-blocks are significant.

The cumulative effect of these modifications is an increase of about 40% in software execution speed, with an average loss of about 0.15dB relative to the original algorithm.

9.2.3   Region-of-Interest Coding

A significant feature of the JPEG 2000 standard is the ability to perform region-of-interest (ROI) coding. Here, particular regions of the image may be coded with better quality than the rest of the image or the background. The method is called MAXSHIFT, a scaling-based method that scales up the coefficients in the ROI so that they are placed into higher bitplanes. During the embedded coding process, the resulting bits are placed in front of the non-ROI part of the image. Therefore, given a reduced bitrate, the ROI will be decoded and refined before the rest of the image. As a result of these mechanisms, the ROI will have much better quality than the background.

One thing to note is that regardless of scaling, full decoding of the bitstream will result in the reconstruction of the entire image with the highest fidelity available. Figure 9.11 demonstrates the effect of region-of-interest coding as the target bitrate of the sample image is increased.

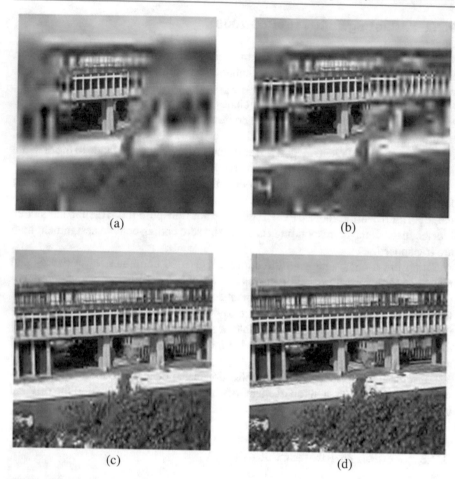

Fig. 9.11 Region of interest (ROI) coding of an image with increasing bitrate using a circularly shaped ROI: **a** 0.4 bpp; **b** 0.5 bpp; **c** 0.6 bpp; **d** 0.7 bpp

9.2.4 Comparison of JPEG and JPEG 2000 Performance

After studying the internals of the JPEG 2000 compression algorithm, a natural question that comes to mind is: how well does JPEG 2000 perform compared to other well-known standards, in particular JPEG? Many comparisons have been made between JPEG and other well-known standards, so here we compare JPEG 2000 only to the popular JPEG.

Various criteria, such as computational complexity, error resilience, compression efficiency, and so on, have been used to evaluate the performance of systems. Since our main focus is on the compression aspect of the JPEG 2000 standard, here we simply compare compression efficiency. (Interested readers can refer to [7, 11] for comparisons using other criteria.)

Given a fixed bitrate, let's compare the quality of compressed images quantitatively by the PSNR: for color images, the PSNR is calculated based on the average of the mean square error of all the RGB components. Also, we visually show results for both JPEG 2000 and JPEG compressed images, so that you can make your own qualitative assessment. We perform a comparison for three categories of images: natural, computer-generated, and medical, using three images from each category. The test images used are shown on the textbook website.

For each image, we compress using JPEG and JPEG 2000, at four bitrates: 0.25, 0.5, 0.75, and 1.0 bpp. Figure 9.12 shows plots of the average PSNR of the images in each category against bitrate. We see that JPEG 2000 substantially outperforms JPEG in all categories.

For a qualitative comparison of the compression results, let's choose a single image and show decompressed output for the two algorithms using a low bitrate (0.75 bpp) and the lowest bitrate (0.25 bpp). From the results in Fig. 9.13, it should be obvious that images compressed using JPEG 2000 show significantly fewer visual artifacts.

9.3   The JPEG-LS Standard

Generally, we would likely apply a lossless compression scheme to images that are critical in some sense, say medical images of a brain, or perhaps images that are difficult or costly to acquire. A scheme in competition with the lossless mode provided in JPEG 2000 is the JPEG-LS standard, specifically aimed at lossless encoding [12]. The main advantage of JPEG-LS over JPEG 2000 is that JPEG-LS is based on a low-complexity algorithm. JPEG-LS is part of a larger ISO effort aimed at better compression of medical images.

JPEG-LS is in fact the current ISO/ITU standard for lossless or "near lossless" compression of continuous-tone images. The core algorithm in JPEG-LS is called *LOw COmplexity LOssless COmpression for Images (LOCO-I)*, proposed by Hewlett Packard [12]. The design of this algorithm is motivated by the observation that complexity reduction is often more important overall than any small increase in compression offered by more complex algorithms. JPEG-LS was standardized in 1999 with Huffman coding. It was extended in 2003 to include Arithmetic coding. Typically, JPEG-LS can achieve a compression ratio of about 4 or better.

LOCO-I exploits a concept called *context modeling*. The idea of context modeling is to take advantage of the structure in the input source—conditional probabilities of what pixel values follow from each other in the image. This extra knowledge is called the *context*. If the input source contains substantial structure, as is usually the case, we could potentially compress it using fewer bits than the order-0 entropy.

As a simple example, suppose we have a binary source with $P(0) = 0.4$ and $P(1) = 0.6$. Then the order-0 entropy $H(S) = -0.4\log_2(0.4) - 0.6\log_2(0.6) = 0.97$. Now suppose we also know that this source has the property that if the previous symbol is 0, the probability of the current symbol being 0 is 0.8, and if the previous symbol is 1, the probability of the current symbol being 0 is 0.1.

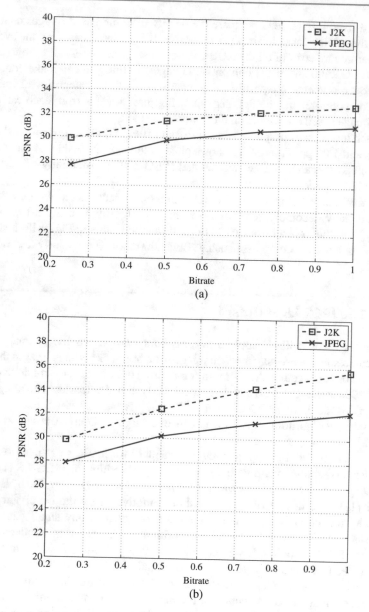

Fig. 9.12 Performance comparison for JPEG and JPEG 2000 on different image types: **a** Natural images; **b** computer generated images; **c** medical images

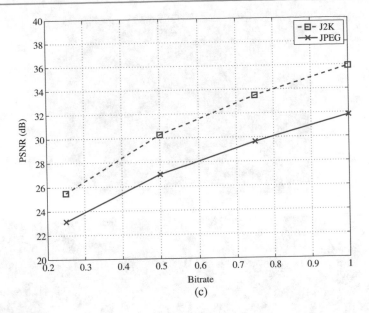

Fig. 9.12 (continued)

If we use the previous symbol as our *context*, we can divide the input symbols into two sets, corresponding to context 0 and context 1, respectively. Then the entropy of each of the two sets is

$$H(S_1) = -0.8 \log_2(0.8) - 0.2 \log_2(0.2) = 0.72$$
$$H(S_2) = -0.1 \log_2(0.1) - 0.9 \log_2(0.9) = 0.47$$

The average bitrate for the entire source would be $0.4 \times 0.72 + 0.6 \times 0.47 = 0.57$, which is substantially less than the order-0 entropy of the entire source in this case.

LOCO-I uses a context model shown in Fig. 9.14. In raster scan order, the context pixels a, b, c, and d all appear before the current pixel x. Thus, this is called a *causal context*.

LOCO-I can be broken down into three components:

- **Prediction**. Predicting the value of the next sample x' using a causal template
- **Context determination**. Determining the context in which x' occurs
- **Residual coding**. Entropy coding of the prediction residual conditioned by the context of x'.

9.3.1 Prediction

A better version of prediction can use an adaptive model based on a calculation of the local edge direction. However, because JPEG-LS is aimed at low complexity, the LOCO-I algorithm instead uses a fixed predictor that performs primitive tests

(a)

(b)

(c)

Fig. 9.13 Comparison of JPEG and JPEG 2000: **a** Original image; **b** JPEG (*left*) and JPEG 2000 (*right*) images compressed at 0.75 bpp; **c** JPEG (*left*) and JPEG 2000 (*right*) images compressed at 0.25 bpp

Fig. 9.14 JPEG-LS context model

c	a	d
b	x	

to detect vertical and horizontal edges. The fixed predictor used by the algorithm is given as follows:

$$\hat{x}' = \begin{cases} \min(a, b) & c \geq \max(a, b) \\ \max(a, b) & c \leq \min(a, b) \\ a + b - c & \text{otherwise} \end{cases} \tag{9.9}$$

It is easy to see that this predictor switches between three simple predictors. It outputs a when there is a vertical edge to the left of the current location; it outputs b when there is a horizontal edge above the current location; and finally it outputs $a + b - c$ when the neighboring samples are relatively smooth.

9.3.2 Context Determination

The context model that conditions the current prediction error (the *residual*) is indexed using a three-component context vector $\mathbf{Q} = (q_1, q_2, q_3)$, whose components are

$$\begin{aligned} q_1 &= d - b \\ q_2 &= b - c \\ q_3 &= c - a \end{aligned} \tag{9.10}$$

These differences represent the local gradient that captures the local smoothness or edge contents that surround the current sample. Because these differences can potentially take on a wide range of values, the underlying context model is huge, making the context-modeling approach impractical. To solve this problem, parameter reduction methods are needed.

An effective method is to quantize these differences so that they can be represented by a limited number of values. The components of $\mathbf{Q}$ are quantized using a quantizer with decision boundaries $-T, \cdots, -1, 0, 1, \cdots, T$. In JPEG-LS, $T = 4$. The context size is further reduced by replacing any context vector $\mathbf{Q}$ whose first element is negative by $-\mathbf{Q}$. Therefore, the number of different context states is $\frac{(2T+1)^3+1}{2} = 365$ in total. The vector $\mathbf{Q}$ is then mapped into an integer in $[0, 364]$.

9.3.3 Residual Coding

For any image, the prediction residual has a finite size, α. For a given prediction $\hat{x}$, the residual ε is in the range $-\hat{x} \leq \varepsilon < \alpha - \hat{x}$. Since the value $\hat{x}$ can be generated by the decoder, the dynamic range of the residual ε can be reduced modulo α and mapped into a value between $-\lfloor \frac{\alpha}{2} \rfloor$ and $\lceil \frac{\alpha}{2} \rceil - 1$.

It can be shown that the error residuals follow a *two-sided geometric distribution* (TSGD). As a result, they are coded using adaptively selected codes based on *Golomb codes*, which are optimal for sequences with geometric distributions [13].

9.3.4   Near-Lossless Mode

The JPEG-LS standard also offers a near-lossless mode, in which the reconstructed samples deviate from the original by no more than an amount δ. The main lossless JPEG-LS mode can be considered a special case of the near-lossless mode with $\delta = 0$. Near-lossless compression is achieved using quantization: residuals are quantized using a uniform quantizer having intervals of length $2\delta + 1$. The quantized values of ε are given by

$$Q(\varepsilon) = \text{sign}(\varepsilon) \left\lfloor \frac{|\varepsilon| + \delta}{2\delta + 1} \right\rfloor \tag{9.11}$$

Since δ can take on only a small number of integer values, the division operation can be implemented efficiently using lookup tables. In near-lossless mode, the prediction and context determination step described previously are based on the quantized values only.

9.4   Bi-Level Image Compression Standards

As more and more documents are handled in electronic form, efficient methods for compressing bi-level images (those with only 1-bit, black-and-white pixels) are sometimes called for. A familiar example is fax images. Algorithms that take advantage of the binary nature of the image data often perform better than generic image compression algorithms. Earlier facsimile standards, such as G3 and G4, use simple *models* of the structure of bi-level images. Each scan line in the image is treated as a run of black-and-white pixels. However, considering the neighboring pixels and the nature of data to be coded allows much more efficient algorithms to be constructed. This section examines the JBIG standard and its successor, JBIG2, as well as the underlying motivations and principles for these two standards.

9.4.1   The JBIG Standard

JBIG is the coding standard recommended by Joint Bi-level Image Processing Group for binary images. This lossless compression standard is used primarily to code scanned images of printed or handwritten text, computer-generated text, and facsimile transmissions. It offers progressive encoding and decoding capability, in the sense that the resulting bitstream contains a set of progressively higher resolution images. This standard can also be used to code grayscale and color images by coding each bitplane independently, but this is not the main objective.

The JBIG compression standard has three separate modes of operation: *progressive, progressive-compatible sequential*, and *single-progression sequential*. The progressive-compatible sequential mode uses a bitstream compatible with the progressive mode. The only difference is that the data is divided into *strips* in this mode.

The single-progression sequential mode has only a single lowest resolution layer. Therefore, an entire image can be coded without any reference to other higher resolution layers. Both these modes can be viewed as special cases of the progressive mode. Therefore, our discussion covers only the progressive mode.

The JBIG encoder can be decomposed into two components:

- Resolution-reduction and differential-layer encoder
- Lowest resolution-layer encoder.

The input image goes through a sequence of resolution-reduction and differential-layer encoders. Each is equivalent in functionality, except that their input images have different resolutions. Some implementations of the JBIG standard may choose to recursively use one such physical encoder. The lowest resolution image is coded using the lowest resolution-layer encoder. The design of this encoder is somewhat simpler than that of the resolution-reduction and differential-layer encoders, since the resolution-reduction and deterministic-prediction operations are not needed.

9.4.2 The JBIG2 Standard

While the JBIG standard offers both lossless and progressive (lossy to lossless) coding abilities, the lossy image produced by this standard has significantly lower quality than the original, because the lossy image contains at most only one-quarter of the number of pixels in the original image. By contrast, the JBIG2 standard is explicitly designed for lossy, lossless, and lossy to lossless image compression. The design goal for JBIG2 aims not only at providing superior lossless compression performance over existing standards but also at incorporating lossy compression at a much higher compression ratio, with as little visible degradation as possible.

A unique feature of JBIG2 is that it is both *quality progressive* and *content progressive*. By quality progressive, we mean that the bitstream behaves similar to that of the JBIG standard, in which the image quality progresses from lower to higher (or possibly lossless) quality. On the other hand, content progressive allows different types of image data to be added progressively. The JBIG2 encoder decomposes the input bi-level image into regions of different attributes and codes each separately, using different coding methods.

As in other image compression standards, only the JBIG2 bitstream, and thus the decoder, is explicitly defined. As a result, any encoder that produces the correct bitstream is "compliant", regardless of the actions it actually takes. Another feature of JBIG2 that sets it apart from other image compression standards is that it is able to represent multiple pages of a document in a single file, enabling it to exploit interpage similarities.

For example, if a character appears on one page, it is likely to appear on other pages as well. Thus, using a dictionary-based technique, this character is coded only once instead of multiple times for every page on which it appears. This compression technique is somewhat analogous to video coding, which exploits inter-frame redundancy to increase compression efficiency.

JBIG2 offers content-progressive coding and superior compression performance through *model-based coding*, in which different models are constructed for different data types in an image, realizing an additional coding gain.

Model-Based Coding

The idea behind model-based coding is essentially the same as that of context-based coding. From the study of the latter, we know we can realize better compression performance by carefully designing a context template and accurately estimating the probability distribution for each context. Similarly, if we can separate the image content into different categories and derive a model specifically for each, we are much more likely to accurately model the behavior of the data and thus achieve higher compression ratio.

In the JBIG style of coding, adaptive and model templates capture the structure within the image. This model is general, in the sense that it applies to all kinds of data. However, being general implies that it does not explicitly deal with the structural differences between text and halftone data that comprise nearly all the contents of bi-level images. JBIG2 takes advantage of this by designing custom models for these data types.

The JBIG2 specification expects the encoder to first segment the input image into regions of different data types, in particular, text and halftone regions. Each region is then coded independently, according to its characteristics.

Text-Region Coding

Each text region is further segmented into pixel blocks containing connected black pixels. These blocks correspond to characters that make up the content of this region. Then, instead of coding all pixels of each character, the bitmap of one representative instance of this character is coded and placed into a *dictionary*. For any character to be coded, the algorithm first tries to find a match with the characters in the dictionary. If one is found, then both a pointer to the corresponding entry in the dictionary and the position of the character on the page are coded. Otherwise, the pixel block is coded directly and added to the dictionary. This technique is referred to as *pattern matching and substitution* in the JBIG2 specification.

However, for *scanned* documents, it is unlikely that two instances of the same character will match pixel by pixel. In this case, JBIG2 allows the option of including refinement data to reproduce the original character on the page. The refinement data codes the current character using the pixels in the matching character *in the dictionary*. The encoder has the freedom to choose the refinement to be exact or lossy. This method is called *soft pattern matching*.

The numeric data, such as the index of a matched character in the dictionary and the position of the characters on the page, are either bitwise or Huffman encoded. Each bitmap for the characters in the dictionary is coded using JBIG-based techniques.

Halftone-Region Coding

The JBIG2 standard suggests two methods for halftone image coding. The first is similar to the context-based arithmetic coding used in JBIG. The only difference is that the new standard allows the context template to include as many as 16 template pixels, four of which may be adaptive.

The second method is called *descreening*. This involves converting back to grayscale and coding the grayscale values. In this method, the bi-level region is divided into blocks of size $m_b \times n_b$. For an $m \times n$ bi-level region, the resulting grayscale image has dimension $m_g = \lfloor (m + (m_b - 1))/m_b \rfloor$ by $n_g = \lfloor (n + (n_b - 1))/n_b \rfloor$. The grayscale value is then computed to be the sum of the binary pixel values in the corresponding $m_b \times n_b$ block. The bitplanes of the grayscale image are coded using context-based arithmetic coding. The grayscale values are used as indices into a dictionary of halftone bitmap patterns. The decoder can use this value to index into this dictionary, to reconstruct the original halftone image.

Preprocessing and Post-processing

JBIG2 allows the use of lossy compression but does not specify a method for doing so. From the decoder point of view, the decoded bitstream is lossless with respect to the image encoded by the encoder, although not necessarily with respect to the original image. The encoder may modify the input image in a preprocessing step, to increase coding efficiency. The preprocessor usually tries to change the original image to lower the code length in a way that does not generally affect the image's appearance. Typically, it tries to remove noisy pixels and smooth out pixel blocks.

Post-processing, another issue not addressed by the specification, can be especially useful for halftones, potentially producing more visually pleasing images. It is also helpful to tune the decoded image to a particular output device, such as a laser printer.

9.5 Exercises

1. You are given a computer cartoon picture and a photograph. If you have a choice of using either JPEG compression or GIF, which compression would you apply for these two images? Justify your answer.

2. Suppose we view a decompressed 512×512 JPEG image but use only the *color* part of the stored image information, not the luminance part, to decompress. What does the 512×512 color image look like? Assume JPEG is compressed using a 4:2:0 scheme.

3. An X-ray photo is usually a gray-level image. It often has low contrast and low gray-level intensities, i.e., all intensity values are in the range of $[a, b]$, where a and b are positive integers, much less than the maximum intensity value 255 if it is an 8-bit image. In order to enhance the appearance of this image, a simple "stretch" operation can be used to convert all original intensity values f_0 to f:

$$f = \frac{255}{b - a} \cdot (f_0 - a).$$

For simplicity, assuming $f_0(i, j)$ and $f(i, j)$ are 8×8 images:

(a) If the DC value for the original image f_0 is m, what is the DC value for the stretched image f?

(b) If one of the AC values $F_0(2, 1)$ for the original image f_0 is n, what is the $F(2, 1)$ value for the stretched image f?

4. (a) JPEG uses the discrete cosine transform (DCT) for image compression.

 (i) What is the value of F(0, 0) if the image $f(i, j)$ is as below?
 (ii) Which AC coefficient $|F(u, v)|$ is the largest for this $f(i, j)$? Why? Is this $F(u, v)$ positive or negative? Why?

$$
\begin{array}{cccccccc}
20 & 20 & 20 & 20 & 20 & 20 & 20 & 20 \\
20 & 20 & 20 & 20 & 20 & 20 & 20 & 20 \\
80 & 80 & 80 & 80 & 80 & 80 & 80 & 80 \\
80 & 80 & 80 & 80 & 80 & 80 & 80 & 80 \\
140 & 140 & 140 & 140 & 140 & 140 & 140 & 140 \\
140 & 140 & 140 & 140 & 140 & 140 & 140 & 140 \\
200 & 200 & 200 & 200 & 200 & 200 & 200 & 200 \\
200 & 200 & 200 & 200 & 200 & 200 & 200 & 200
\end{array}
$$

(b) Show in detail how a three-level hierarchical JPEG will encode the image above, assuming that

 (i) The encoder and decoder at all three levels use Lossless JPEG.
 (ii) *Reduction* simply averages each 2×2 block into a single pixel value.
 (iii) *Expansion* duplicates the single pixel value four times.

5. JPEG employs the block-based coding method.

   (a) Why was 8×8 chosen as the block size in JPEG?
   (b) Given the ever-increasing image resolution, what would be a better block size now?

6. In JPEG, the Discrete Cosine Transform is applied to 8×8 blocks in the image. For now, let's call it DCT-8. Generally, we can define a DCT-N to be applied to $N \times N$ blocks in the image. DCT-N is defined as

$$F_N(u, v) = \frac{2C(u)C(v)}{N} \sum_{i=0}^{N-1} \sum_{j=0}^{N-1} \cos \frac{(2i+1)u\pi}{2N} \cos \frac{(2j+1)v\pi}{2N} f(i, j)$$

$$C(\xi) = \begin{cases} \frac{\sqrt{2}}{2} & \text{for } \xi = 0 \\ 1 & \text{otherwise} \end{cases}$$

   Given $f(i, j)$ as below, show your work for deriving all pixel values of $F_2(u, v)$. (That is, show the result of applying DCT-2 to the image below.)

$$
\begin{array}{rrrrrrrr}
100 & -100 & 100 & -100 & 100 & -100 & 100 & -100 \\
100 & -100 & 100 & -100 & 100 & -100 & 100 & -100 \\
100 & -100 & 100 & -100 & 100 & -100 & 100 & -100 \\
100 & -100 & 100 & -100 & 100 & -100 & 100 & -100 \\
100 & -100 & 100 & -100 & 100 & -100 & 100 & -100 \\
100 & -100 & 100 & -100 & 100 & -100 & 100 & -100 \\
100 & -100 & 100 & -100 & 100 & -100 & 100 & -100 \\
100 & -100 & 100 & -100 & 100 & -100 & 100 & -100 \\
\end{array}
$$

7. According to the DCT-N definition above, $F_N(1)$ and $F_N(N-1)$ are the AC coefficients representing the lowest and highest spatial frequencies, respectively.

   (a) It is known that $F_{16}(1)$ and $F_8(1)$ *do not* capture the same (lowest) frequency response in image filtering. Explain why.
   (b) Do $F_{16}(15)$ and $F_8(7)$ capture the same (highest) frequency response?

8. (a) How many principal modes does JPEG have? What are their names?
   (b) In the hierarchical model, explain briefly why we must include an encode/decode cycle on the coder side before transmitting difference images to the decode side.
   (c) What are the two methods used to decode only part of the information in a JPEG file, so that the image can be coarsely displayed quickly and iteratively increased in quality?

9. Could we use wavelet-based compression in ordinary JPEG? How?

10. We decide to create a new image compression standard based on JPEG, for use with images that will be viewed by an alien species. What part of the JPEG workflow would we likely have to change?

11. Unlike EZW, EBCOT does not explicitly take advantage of the spatial relationships of wavelet coefficients. Instead, it uses the PCRD optimization approach. Discuss the rationale behind this approach.

12. Is the JPEG 2000 bitstream SNR scalable? If so, explain how it is achieved using the EBCOT algorithm.

13. Implement transform coding, quantization, and hierarchical coding for the encoder and decoder of a three-level Hierarchical JPEG. Your code should include a (minimal) graphical user interface for the purpose of demonstrating your results. You do not need to implement the entropy (lossless) coding part; optionally, you may include any publicly available code for it.

References

1. W.B. Pennebaker, J.L. Mitchell, *The JPEG Still Image Data Compression Standard* (Van Nostrand Reinhold, New York, 1992)
2. G. Hudson, A. Léger, B. Niss, I. Sebestyén, JPEG at 25: still going strong. IEEE Multimed. **24**(2), 96–103 (2017)
3. V. Bhaskaran, K. Konstantinides, *Image and Video Compression Standards: algorithms and Architectures*, 2nd edn. (Kluwer Academic Publishers, Boston, 1997)
4. ISO/IEC 15444-1, *Information Technology—JPEG 2000 Image Coding System: core Coding System.* (ISO/IEC, 2004)
5. D.S. Taubman, M.W. Marcellin, *JPEG2000: image Compression Fundamentals* (Springer, Standards and Practice, 2013)
6. M. Rabbani, R. Joshi, An overview of the JPEG 2000 still image compression Standard. Signal Process. Image Commun. **17**, 3–48 (2002)
7. P. Schelkens, A. Skodras, T. Ebrahimi (eds.), *The JPEG 2000 Suite*. (Wiley, 2009)
8. D. Taubman, High performance scalable image compression with EBCOT. IEEE Trans. Image Process. **9**(7), 1158–1170 (2000)
9. K. Ramachandran, M. Vetterli, Best wavelet packet basis in a rate-distortion sense. IEEE Trans. Image Process. **2**, 160–173 (1993)
10. I. Ueno, F. Ono, T. Yanagiya, T. Kimura, M. Yoshida. *Proposal of the Arithmetic Coder for JPEG2000.* (ISO/IEC JTC1/SC29/WG1 N1143, 1999)
11. D. Santa-Cruz, R. Grosbois, T. Ebrahimi, JPEG 2000 performance evaluation and assessment. Signal Proces. Image Commun. **17**, 113–130 (2002)

12. M. Weinberger, G. Seroussi, G. Sapiro, The LOCO-I lossless image compression algorithm: Principles and standardization into JPEG-LS. Technical Report HPL-98-193R1, HP Technical Report (1998)
13. N. Merhav, G. Seroussi, M.J. Weinberger, Optimal prefix codes for sources with two-sided geometric distributions. IEEE Trans. Inf. Theory **46**(1), 121–135 (2000)

Basic Video Compression Techniques

10

As discussed in Chap. 7, the volume of uncompressed video data could be extremely large. Even a modest CIF video with a picture resolution of only 352×288, if uncompressed, would carry more than 35 Mbps. In HDTV and UHDTV, the bitrate far exceeds 1 Gbps. This poses challenges and problems for storage and network communications.

This chapter introduces some basic video compression techniques and illustrates them in standards H.261 and H.263—two video compression standards aimed mostly at videoconferencing. The next two chapters further introduce several MPEG video compression standards and the latest, H.264, H.265, and H.266.

Tekalp [1] and Poynton [2] set out the fundamentals of digital video processing. They provide a good overview of the mathematical foundations of the problems to be addressed in a video. The books by Bhaskaran and Konstantinides [3] and Wang et al. [4] include good descriptions of early video compression algorithms.

10.1 Introduction to Video Compression

A video consists of a time-ordered sequence of frames—images. An obvious solution to video compression would be predictive coding based on previous frames. For example, suppose we simply created a predictor such that the prediction equals the previous frame. Then compression proceeds by subtracting images: instead of subtracting the image from itself (i.e., use a spatial-domain derivative), we subtract in time order and code the residual error.

And this works. Suppose most of the video is unchanging in time, then we get a nice histogram peaked sharply at zero—a great reduction in terms of the entropy of the original video, just what we wish for.

© Springer Nature Switzerland AG 2021
Z.-N. Li et al., *Fundamentals of Multimedia*, Texts in Computer Science,
https://doi.org/10.1007/978-3-030-62124-7_10

However, it turns out that at an acceptable cost, we can do even better by searching for just the right parts of the image to subtract from the previous frame. After all, our naive subtraction scheme will likely work well for a background of office furniture and sedentary university types. But wouldn't a football game have players zooming around the frame, producing large values when subtracted from the previously static green playing field?

So in the next section, we examine how to do better. The idea of looking for the football player in the next frame is called *motion estimation*, and the concept of shifting pieces of the frame around so as to best subtract away the player is called *motion compensation*.

10.2   Video Compression Based on Motion Compensation

The image compression techniques discussed in the previous chapters (e.g., JPEG and JPEG 2000) exploit *spatial redundancy*, the phenomenon that picture contents often change relatively slowly across images, making a large suppression of higher spatial frequency components viable.

A video can be viewed as a sequence of images stacked in the *temporal* dimension. Since the frame rate of the video is often relatively high (e.g., > 15 frames per second) and the camera parameters (focal length, position, viewing angle, etc.) usually do not change rapidly between frames, the contents of consecutive frames are usually similar, unless certain objects in the scene move extremely fast or the scene changes. In other words, the video has *temporal redundancy*.

Temporal redundancy is often significant and it is indeed exploited, so that not every frame of the video needs to be coded independently as a new image. Instead, the difference between the current frame and other frame(s) in the sequence is coded. If redundancy between them is great enough, the difference images could consist mainly of small values and low entropy, which is good for compression .

All modern digital video compression algorithms (including H.264, H.265, and H.266) adopt this *Hybrid coding* approach, i.e., predicting and compensating for the differences between video frames to remove the temporal redundancy, and then transform coding on the residual errors (the differences).

As we mentioned, although a simplistic way of deriving the difference image is to subtract one image from the other (pixel by pixel), such an approach is ineffective in yielding a high compression ratio. Since the main cause of the difference between frames is a camera and/or object motion, these motion generators can be "compensated" by detecting the displacement of corresponding pixels or regions in these frames and measuring their differences. Video compression algorithms that adopt this approach are said to be based on motion compensation (MC). The three main steps of these algorithms are

1. Motion estimation (motion vector search)
2. Motion-compensation-based prediction
3. Derivation of the prediction error—the difference.

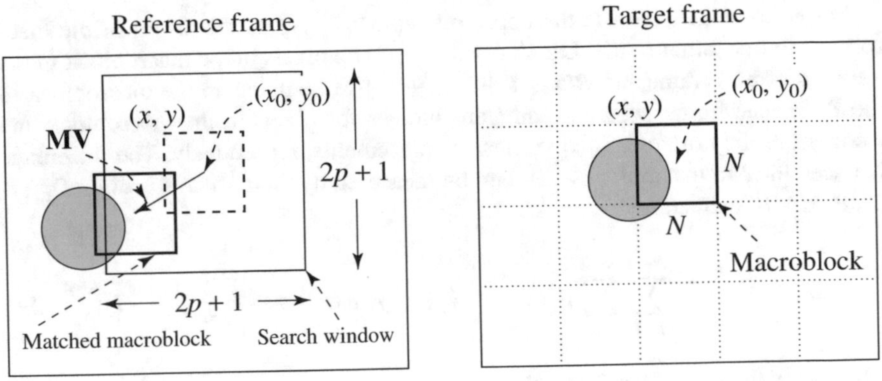

Fig. 10.1 Macroblocks and motion vector in video compression: **a** Reference frame; **b** target frame

For efficiency, each image is divided into *macroblocks* of size $N \times N$; by default, $N = 16$ for luminance images. For chrominance images, $N = 8$ if 4:2:0 chroma subsampling is adopted. Motion compensation is not performed at the pixel level, nor at the level of *video object*, as in later video standards (such as MPEG-4). Instead, it is at the macroblock level.

The current image frame is referred to as the *Target frame*. A match is sought between the macroblock under consideration in the Target frame and the most similar macroblock in previous and/or future frame(s) [referred to as *Reference frame(s)*]. In that sense, the Target macroblock is predicted from the Reference macroblock(s).

The displacement of the reference macroblock to the target macroblock is called a *motion vector* **MV**. Figure 10.1 shows the case of *forward prediction*, in which the Reference frame is taken to be a previous frame. If the Reference frame is a future frame, it is referred to as *backward prediction*. The *difference* of the two corresponding macroblocks is the prediction error.

For video compression based on motion compensation, after the first frame, only the motion vectors and difference macroblocks need be coded, since they are sufficient for the decoder to regenerate all macroblocks in subsequent frames.

We will return to the discussion of some common video compression standards after the following section, in which we discuss search algorithms for motion vectors.

10.3   Search for Motion Vectors

The search for motion vectors $\mathbf{MV}(u, v)$ as defined above is a matching problem, also called a *correspondence* problem [5]. Since the MV search is computationally expensive, it is usually limited to a small immediate neighborhood. Horizontal and vertical displacements i and j are in the range $[-p, p]$, where p is a positive integer with a relatively small value. This makes a search window of size $(2p + 1) \times (2p + 1)$, as Fig. 10.1 shows. The center of the macroblock (x_0, y_0) can be placed at each of the grid positions in the window.

For convenience, we use the upper left corner (x, y) as the origin of the macroblock in the Target frame. Let $C(x + k, y + l)$ be pixels in the macroblock in the Target (current) frame and $R(x + i + k, y + j + l)$ be pixels in the macroblock in the Reference frame, where k and l are indices for pixels in the macroblock and i and j are the horizontal and vertical displacements, respectively. The difference between the two macroblocks can then be measured by their *Mean Absolute Difference (MAD)*, defined as

$$MAD(i, j) = \frac{1}{N^2} \sum_{k=0}^{N-1} \sum_{l=0}^{N-1} |C(x + k, y + l) - R(x + i + k, y + j + l)|, \quad (10.1)$$

where N is the size of the macroblock.

The goal of the search is to find a vector (i, j) as the motion vector $\mathbf{MV} = (\mathbf{u}, \mathbf{v})$, such that $MAD(i, j)$ is minimum:

$$(u, v) = [(i, j) \mid MAD(i, j) \text{ is minimum}, \ i \in [-p, p], j \in [-p, p]] \quad (10.2)$$

We used the mean absolute difference in the above discussion. However, this measure is by no means the only possible choice. In fact, some encoders (e.g., H.263) will simply use the *Sum of Absolute Difference (SAD)*. Some other common error measures, such as the *Mean Squared Error (MSE)*, would also be appropriate.

10.3.1 Sequential Search

The simplest method for finding motion vectors is to sequentially search the whole $(2p + 1) \times (2p + 1)$ window in the Reference frame (also referred to as *full search*). A macroblock centered at each of the positions within the window is compared to the macroblock in the Target frame, pixel by pixel, and their respective MAD is then derived using Eq. (10.1). The vector (i, j) that offers the least MAD is designated the $\mathbf{MV}$ (u, v) for the macroblock in the Target frame.

Procedure 10.1 (Motion-vector: Sequential search)

```
BEGIN
    min_MAD = LARGE_NUMBER;        /* Initialization */
    for i = −p to p
        for j = −p to p
            {
                cur_MAD = MAD(i, j);
                if cur_MAD < min_MAD
                    {
                        min_MAD = cur_MAD;
                        u = i;            /* Get the coordinates for MV. */
                        v = j;
```

```
                    }
            }
END
```

Clearly, the sequential search method is very costly. From Eq. (10.1), each pixel comparison requires three operations (subtraction, absolute value, and addition). Thus the cost for obtaining a motion vector for a single macroblock is $(2p + 1) \cdot (2p + 1) \cdot N^2 \cdot 3 \Rightarrow O(p^2 N^2)$.

As an example, let's assume the video has a resolution of 720×480 and a frame rate of 30 fps; also, assume $p = 15$ and $N = 16$. The number of operations needed for each motion vector search is thus

$$(2p + 1)^2 \cdot N^2 \cdot 3 = 31^2 \times 16^2 \times 3.$$

Considering that a single image frame has $\frac{720 \times 480}{N \cdot N}$ macroblocks, and 30 frames each second, the total operations needed per second is

$$
\begin{aligned}
\text{OPS\_per\_second} &= (2p + 1)^2 \cdot N^2 \cdot 3 \cdot \frac{720 \times 480}{N \cdot N} \cdot 30 \\
&= 31^2 \times 16^2 \times 3 \times \frac{720 \times 480}{16 \times 16} \times 30 \approx 29.89 \times 10^9.
\end{aligned}
$$

This would certainly make real-time encoding of this video difficult.

10.3.2 2D Logarithmic Search

A more efficient version, suboptimal but still usually effective, is called *Logarithmic Search*. The procedure for a 2D Logarithmic Search of motion vectors takes several iterations and is akin to a binary search. As Fig. 10.2 illustrates, only nine locations in the search window, marked "1," are initially used as seeds for a MAD-based search. After the one that yields the minimum MAD is located, the center of the new search region is moved to it, and the step size (*offset*) is reduced to half. In the next iteration, the nine new locations are marked "2," and so on.[1] For the macroblock centered at (x_0, y_0) in the Target frame, the procedure is as follows:

---

[1] The procedure is heuristic. It assumes a general continuity (monotonicity) of image contents—that they do not change randomly within the search window. Otherwise, the procedure might not find the best match.

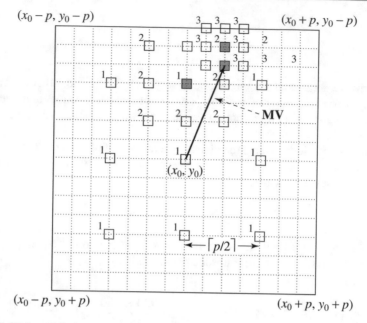

Fig. 10.2 2D Logarithmic search for motion vectors

Procedure 10.2 (Motion-vector: 2D-Logarithmic-search)

BEGIN
    offset $= \lceil \frac{p}{2} \rceil$;
    Specify nine macroblocks within the search window in the Reference frame,
    they are centered at (x_0, y_0) and separated by offset horizontally and/or vertically;
    WHILE last $\neq$ TRUE
        {
            Find one of the nine specified macroblocks that yields the minimum
            MAD;
            if offset $= 1$ then last $=$ TRUE;
            offset $= \lceil$offset/2$\rceil$;
            Form a search region with the new offset and new center found;
        }
END

Instead of sequentially comparing with $(2p + 1)^2$ macroblocks from the Reference frame, the 2D Logarithmic Search will compare with only $9 \cdot (\lceil \log_2 p \rceil + 1)$ macroblocks. In fact, it would be $8 \cdot (\lceil \log_2 p \rceil + 1) + 1$, since the comparison that yielded the least *MAD* from the last iteration can be reused. Therefore, the complexity is dropped to $O(\log p \cdot N^2)$. Since p is usually of the same order of magnitude as N, the saving is substantial compared to $O(p^2 N^2)$.

Using the same example as in the previous subsection, the total operations per second drop to

$$\text{OPS\_per\_second} = \left(8 \cdot (\lceil \log_2 p \rceil + 1) + 1\right) \cdot N^2 \cdot 3 \cdot \frac{720 \times 480}{N \cdot N} \cdot 30$$
$$= \left(8 \cdot \lceil \log_2 15 \rceil + 9\right) \times 16^2 \times 3 \times \frac{720 \times 480}{16 \times 16} \times 30$$
$$\approx 1.25 \times 10^9.$$

10.3.3 Hierarchical Search

The search for motion vectors can benefit from a hierarchical (multiresolution) approach in which the initial estimation of the motion vector can be obtained from images with a significantly reduced resolution. Figure 10.3 depicts a three-level hierarchical search in which the original image is at level 0, images at levels 1 and 2 are obtained by downsampling from the previous levels by a factor of 2, and the initial search is conducted at level 2. Since the size of the macroblock is smaller and p can also be proportionally reduced at this level, the number of operations required is greatly reduced (by a factor of 16 at this level).

The initial estimation of the motion vector is coarse because of the lack of image detail and resolution. It is then refined level by level toward level 0. Given the estimated motion vector (u^k, v^k) at level k, a 3×3 neighborhood centered at $(2 \cdot u^k, 2 \cdot v^k)$ at level $k - 1$ is searched for the refined motion vector. In other words, the refinement is such that at level $k - 1$, the motion vector (u^{k-1}, v^{k-1}) satisfies

$$(2u^k - 1 \leq u^{k-1} \leq 2u^k + 1, \quad 2v^k - 1 \leq v^{k-1} \leq 2v^k + 1),$$

and yields minimum MAD for the macroblock under examination.

Let (x_0^k, y_0^k) denote the center of the macroblock at level k in the Target frame. The procedure for hierarchical motion vector search for the macroblock centered at (x_0^0, y_0^0) in the Target frame can be outlined as follows:

Procedure 10.3 (Motion-vector: Hierarchical-search)

BEGIN
    // Get macroblock center position at the lowest resolution level k, e.g., level 2.
    $x_0^k = x_0^0/2^k; \quad y_0^k = y_0^0/2^k;$
    Use Sequential (or 2D Logarithmic) search method to get initial estimated
    $\mathbf{MV}(u^k, v^k)$ at level k;
    WHILE last $\neq$ TRUE

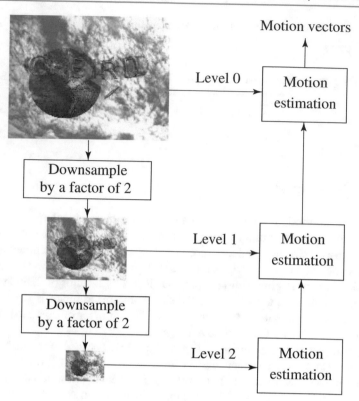

Fig. 10.3 A three-level hierarchical search for motion vectors

{
    Find one of the nine macroblocks that yields minimum MAD
    at level $k - 1$ centered at
    $(2(x_0^k + u^k) - 1 \leq x \leq 2(x_0^k + u^k) + 1, \quad 2(y_0^k + v^k) - 1 \leq y \leq 2(y_0^k + v^k) + 1)$;
    if $k = 1$ then last = TRUE;
    $k = k - 1$;
    Assign (x_0^k, y_0^k) and (u^k, v^k) with the new center location and motion
    vector;
}
END

We will use the same example as in the previous sections to estimate the total operations needed each second for a three-level hierarchical search. For simplicity, the overhead for initially generating multiresolution target and reference frames will not be included, and it will be assumed that Sequential search is used at each level.

Table 10.1 Comparison of computational cost of motion vector search methods according to the examples

Search method	OPS_per_second for 720 × 480 at 30 fps	
	$p = 15$	$p = 7$
Sequential search	29.89×10^9	7.00×10^9
2D Logarithmic search	1.25×10^9	0.78×10^9
Three-level Hierarchical search	0.51×10^9	0.40×10^9

The total number of macroblocks processed each second is still $\frac{720 \times 480}{N \cdot N} \times 30$. However, the operations needed for each macroblock are reduced to

$$\left[\left(2 \left\lceil \frac{p}{4} \right\rceil + 1\right)^2 \left(\frac{N}{4}\right)^2 + 9 \left(\frac{N}{2}\right)^2 + 9N^2 \right] \times 3.$$

Hence,

$$\text{OPS\_per\_second} = \left[\left(2 \left\lceil \frac{p}{4} \right\rceil + 1\right)^2 \left(\frac{N}{4}\right)^2 + 9 \left(\frac{N}{2}\right)^2 + 9N^2 \right]$$

$$\times 3 \times \frac{720 \times 480}{N \cdot N} \times 30$$

$$= \left[\left(\frac{9}{4}\right)^2 + \frac{9}{4} + 9 \right] \times 16^2 \times 3 \times \frac{720 \times 480}{16 \times 16} \times 30$$

$$\approx 0.51 \times 10^9.$$

Table 10.1 summarizes the comparison of the three motion vector search methods for a 720 × 480, 30 fps video when $p = 15$ and 7, respectively.

10.4 H.261

H.261 is an earlier digital video compression standard. Because its principle of motion-compensation-based compression is very much retained in all later video compression standards, we will start with a detailed discussion of H.261.

The International Telegraph and Telephone Consultative Committee (CCITT) initiated the development of H.261 in 1988. The final recommendation was adopted by the ITU (International Telecommunication Union)—Telecommunication standardization sector (ITU-T), formerly CCITT, in 1990 [6].

The standard was designed for videophone, videoconferencing, and other audio-visual services over ISDN telephone lines. Initially, it was intended to support multiples (from 1 to 5) of 384 kbps channels. In the end, however, the video codec supports bitrates of $p \times 64$ kbps, where p ranges from 1 to 30. Hence, the standard was once known as $p * 64$, pronounced "p star 64". The standard requires video encoder delay to be less than 150 ms, so that the video can be used for real-time, bidirectional video conferencing.

H.261 belongs to the following set of ITU recommendations for visual telephony systems:

- **H.221**. Frame structure for an audiovisual channel supporting 64–1,920 kbps
- **H.230**. Frame control signals for audiovisual systems
- **H.242**. Audiovisual communication protocols
- **H.261**. Video encoder/decoder for audiovisual services at $p \times 64$ kbps
- **H.320**. Narrowband audiovisual terminal equipment for $p \times 64$ kbps transmission.

Table 10.2 lists the video formats supported by H.261. Chroma subsampling in H.261 is 4:2:0. Considering the relatively low bitrate in network communications at the time, support for CCIR 601 QCIF is specified as required, whereas support for CIF is optional.

Figure 10.4 illustrates a typical H.261 frame sequence. Two types of image frames are defined: intra-frames (*I-frames*) and inter-frames (*P-frames*).

I-frames are treated as independent images. Basically, a transform coding method similar to JPEG is applied within each I-frame, hence the name "intra".

P-frames are not independent. They are coded by a forward predictive coding method in which current macroblocks are predicted from similar macroblocks in the preceding I- or P-frame, and *differences* between the macroblocks are coded. *Temporal redundancy removal* is hence included in P-frame coding, whereas I-frame

Table 10.2 Video formats supported by H.261

Video format	Luminance image resolution	Chrominance image resolution	Bitrate (Mbps) (if 30 fps and uncompressed)	H.261 support
QCIF	176 × 144	88 × 72	9.1	Required
CIF	352 × 288	176 × 144	36.5	Optional

Fig. 10.4 H.261 frame sequence

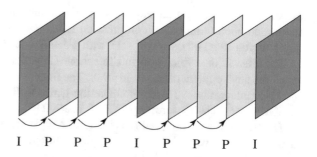

I   P   P   P   I   P   P   P   I

coding performs only *spatial redundancy removal*. It is important to remember that the prediction from a previous P-frame is allowed (not just from a previous I-frame).

The interval between pairs of I-frames is a variable and is determined by the encoder. Usually, an ordinary digital video has a couple of I-frames per second. Motion vectors in H.261 are always measured in units of full pixels and have a limited range of ± 15 pixels—that is, $p = 15$.

10.4.1 Intra-Frame (I-Frame) Coding

Macroblocks are of size 16×16 pixels for the Y frame of the original image. For Cb and Cr frames, they correspond to areas of 8×8, since 4:2:0 chroma subsampling is employed. Hence, a macroblock consists of four Y blocks, one Cb, and one Cr, 8×8 blocks.

For each 8×8 block, a DCT transform is applied. As in JPEG (discussed in detail in Chap. 9), the DCT coefficients go through a quantization stage. Afterwards, they are zigzag-scanned and eventually entropy-coded (as shown in Fig. 10.5).

10.4.2 Inter-Frame (P-Frame) Predictive Coding

Figure 10.6 shows the H.261 P-frame coding scheme based on motion compensation. For each macroblock in the Target frame, a motion vector is allocated by one of the search methods discussed earlier. After the prediction, a *difference macroblock* is derived to measure the *prediction error*. It is also carried in the form of four Y blocks, one Cb, and one Cr block. Each of these 8×8 blocks goes through DCT, quantization, zigzag scan, and entropy coding. The motion vector is also coded.

Sometimes, a good match cannot be found—the prediction error exceeds a certain acceptable level. The macroblock itself is then encoded (treated as an intra macroblock) and in this case it is termed a *non-motion-compensated macroblock*.

P-frame coding encodes the difference macroblock (not the Target macroblock itself). Since the difference macroblock usually has a much smaller entropy than the Target macroblock, a large *compression ratio* is attainable.

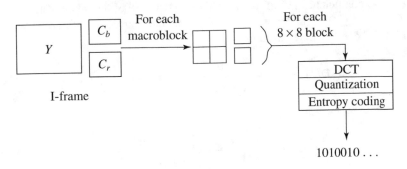

Fig. 10.5 I-frame coding

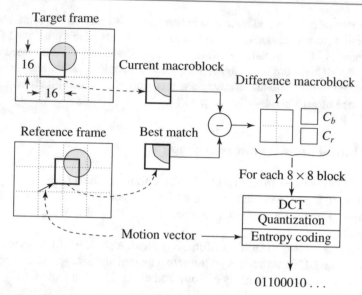

Fig. 10.6 H.261 P-frame coding based on motion compensation

In fact, even the motion vector is not directly coded. Instead, the difference, **MVD**, between the motion vectors of the preceding macroblock and current macroblock is sent for entropy coding:

$$\mathbf{MVD} = \mathbf{MV}_{\text{Preceding}} - \mathbf{MV}_{\text{Current}} \tag{10.3}$$

10.4.3 Quantization in H.261

The quantization in H.261 does not use 8×8 quantization matrices, as in JPEG and MPEG. Instead, it uses a constant, called *step_size*, for all DCT coefficients within a macroblock. According to the need (e.g., bitrate control of the video), *step_size* can take on any one of the 31 even values from 2 to 62. One exception, however, is made for the DC coefficient in intra mode, where a step_size of 8 is always used. If we use DCT and $QDCT$ to denote the DCT coefficients before and after quantization, then for DC coefficients in intra mode,

$$QDCT = \text{round}\left(\frac{DCT}{\text{step\_size}}\right) = \text{round}\left(\frac{DCT}{8}\right). \tag{10.4}$$

For all other coefficients:

$$QDCT = \left\lfloor \frac{DCT}{\text{step\_size}} \right\rfloor = \left\lfloor \frac{DCT}{2 \times \text{scale}} \right\rfloor, \tag{10.5}$$

where *scale* is an integer in the range of $[1,31]$.

The midtread quantizer, discussed in Sect. 8.4.1 typically uses a `round` operator. Equation (10.4) uses this type of quantizer. However, Eq. (10.5) uses a `floor` operator and, as a result, leaves a center dead zone (as Fig. 9.8 shows) in its quantization space, with a larger input range mapped to zero.

10.4.4 H.261 Encoder and Decoder

Figure 10.7 shows a relatively complete picture of how the H.261 encoder and decoder work. Here, Q and Q^{-1} stand for quantization and its inverse, respectively. Switching of the intra- and inter-frame modes can be readily implemented by a multiplexer. To avoid the propagation of coding errors,

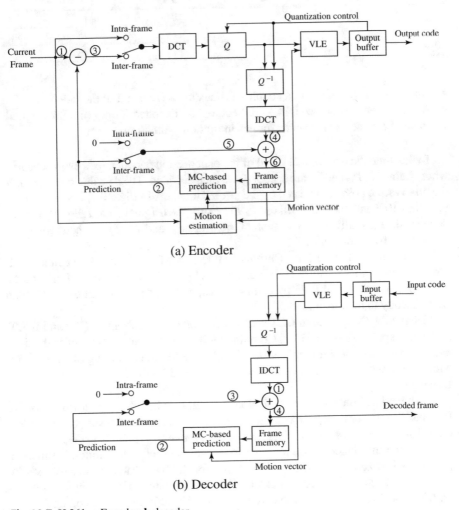

(a) Encoder

(b) Decoder

Fig. 10.7 H.261: **a** Encoder; **b** decoder

Table 10.3 Data flow at the observation points in H.261 encoder

Current frame	Observation point					
	1	2	3	4	5	6
I	I			$\tilde{I}$	0	$\tilde{I}$
P_1	P_1	P_1'	D_1	$\tilde{D}_1$	P_1'	$\tilde{P}_1$
P_2	P_2	P_2'	D_2	$\tilde{D}_2$	P_2'	$\tilde{P}_2$

Table 10.4 Data flow at the observation points in H.261 decoder

Current frame	Observation point			
	1	2	3	4
I	$\tilde{I}$		0	$\tilde{I}$
P_1	$\tilde{D}_1$	P_1'	P_1'	$\tilde{P}_1$
P_2	$\tilde{D}_2$	P_2'	P_2'	$\tilde{P}_2$

- An I-frame is usually sent a couple of times in each second of the video.
- As discussed earlier (see DPCM in Sect. 6.3.5), decoded frames (not the original frames) are used as reference frames in motion estimation.

To illustrate the operational detail of the encoder and decoder, let's use a scenario where frames I, P_1, and P_2 are encoded and then decoded. The data that goes through the observation points, indicated by the circled numbers in Fig. 10.7, is summarized in Tables 10.3 and 10.4. We will use I, P_1, P_2 for the original data, $\tilde{I}$, $\tilde{P}_1$, $\tilde{P}_2$ for the decoded data (usually a lossy version of the original), and P_1', P_2' for the predictions in the Inter-frame mode.

For the encoder, when the Current Frame is an Intra-frame, Point number 1 receives macroblocks from the I-frame, denoted by I in Table 10.3. Each I undergoes DCT, Quantization, and Entropy Coding steps, and the result is sent to the Output Buffer, ready to be transmitted.

Meanwhile, the quantized DCT coefficients for I are also sent to Q^{-1} and IDCT and hence appear at Point 4 as $\tilde{I}$. Combined with a zero input from Point 5, the data at Point 6 remains as $\tilde{I}$ and this is stored in Frame Memory, waiting to be used for Motion Estimation and Motion-Compensation-based Prediction for the subsequent frame P_1.

Quantization Control serves as feedback—that is, when the Output Buffer is too full, the quantization step_size is increased, so as to reduce the size of the coded data. This is known as an *encoding rate control process*.

When the subsequent Current Frame P_1 arrives at Point 1, the Motion Estimation process is invoked to find the motion vector for the best matching macroblock in frame $\tilde{I}$ for each of the macroblocks in P_1. The estimated motion vector is sent to both Motion-Compensation-based Prediction and Variable-Length Encoding (VLE).

The MC-based Prediction yields the best matching macroblock in P_1. This is denoted as P_1' appearing at Point 2.

At Point 3, the "prediction error" is obtained, which is $D_1 = P_1 - P_1'$. Now D_1 undergoes DCT, Quantization, and Entropy Coding, and the result is sent to the Output Buffer. As before, the DCT coefficients for D_1 are also sent to Q^{-1} and IDCT and appear at Point 4 as $\tilde{D}_1$.

Added to P_1' at Point 5, we have $\tilde{P}_1 = P_1' + \tilde{D}_1$ at Point 6. This is stored in Frame Memory, waiting to be used for Motion Estimation and Motion-Compensation-based Prediction for the subsequent frame P_2. The steps for encoding P_2 are similar to those for P_1, except that P_2 will be the Current Frame and P_1 becomes the Reference Frame.

For the decoder, the input code for frames will be decoded first by Entropy Decoding, Q^{-1}, and IDCT. For Intra-frame mode, the first decoded frame appears at Point 1 and then Point 4 as $\tilde{I}$. It is sent as the first output and at the same time stored in the Frame Memory.

Subsequently, the input code for Inter-frame P_1 is decoded, and prediction error $\tilde{D}_1$ is received at Point 1. Since the motion vector for the current macroblock is also entropy-decoded and sent to Motion-Compensation-based Prediction, the corresponding predicted macroblock P_1' can be located in frame $\tilde{I}$ and will appear at Points 2 and 3. Combined with $\tilde{D}_1$, we have $\tilde{P}_1 = P_1' + \tilde{D}_1$ at Point 4, and it is sent out as the decoded frame and also stored in the Frame Memory. Again, the steps for decoding P_2 are similar to those for P_1.

10.4.5 A Glance at the H.261 Video Bitstream Syntax

Let's take a brief look at the H.261 video bitstream syntax (see Fig. 10.8). This consists of a hierarchy of four layers: *Picture*, *Group of Blocks (GOB)*, *Macroblock*, and *Block*.

1. **Picture layer**. *Picture Start Code (PSC)* delineates boundaries between pictures. *Temporal Reference (TR)* provides a timestamp for the picture. Since temporal subsampling can sometimes be invoked such that some pictures will not be transmitted, it is important to have TR, to maintain synchronization with audio. *Picture Type (PType)* specifies, for example, whether it is a CIF or QCIF picture.

2. **GOB layer**. H.261 pictures are divided into regions of 11×3 macroblocks (i.e., regions of 176×48 pixels in luminance images), each of which is called a *Group of Blocks (GOB)*. Figure 10.9 depicts the arrangement of GOBs in a CIF or QCIF luminance image. For instance, the CIF image has 2×6 GOBs, corresponding to its image resolution of 352×288 pixels.

   Each GOB has its *Start Code (GBSC)* and *Group number (GN)*. The GBSC is unique and can be identified without decoding the entire variable-length code in the bitstream. In case a network error causes a bit error or the loss of some bits, the H.261 video can be recovered and resynchronized at the next identifiable GOB, preventing the possible propagation of errors.

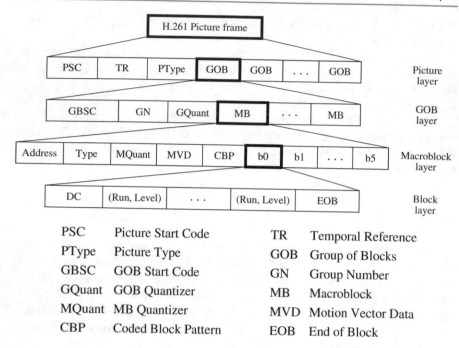

Fig. 10.8 Syntax of H.261 video bitstream

Fig. 10.9 Arrangement of GOBs in H.261 luminance images

GOB 0	GOB 1
GOB 2	GOB 3
GOB 4	GOB 5
GOB 6	GOB 7
GOB 8	GOB 9
GOB 10	GOB 11

CIF

GOB 0
GOB 1
GOB 2

QCIF

GQuant indicates the quantizer to be used in the GOB, unless it is overridden by any subsequent *Macroblock Quantizer* (*MQuant*). GQuant and MQuant are referred to as *scale* in Eq. (10.5).

3. **Macroblock layer**. Each *macroblock* (*MB*) has its own *Address*, indicating its position within the GOB, quantizer (MQuant), and six 8×8 image blocks (4 Y, 1 Cb, 1 Cr). *Type* denotes whether it is an Intra- or Inter-, motion-compensated or non-motion-compensated macroblock. *Motion Vector Data* (*MVD*) is obtained by taking the difference between the motion vectors of the preceding and current macroblocks. Moreover, since some blocks in the macroblocks match well and some match poorly in Motion Estimation, a bitmask *Coded Block Pattern* (*CBP*) is used to indicate this information. Only well-matched blocks will have their coefficients transmitted.

4. **Block layer**. For each 8×8 block, the bitstream starts with *DC value*, followed by pairs of length of zero-run (*Run*) and the subsequent nonzero value (*Level*) for ACs, and finally the *End of Block* (*EOB*) code. The range of "Run" is [0, 63]. "Level" reflects quantized values—its range is $[-127, 127]$, and Level $\neq 0$.

10.5   H.263

H.263 is an improved video coding standard [7] for videoconferencing and other audiovisual services transmitted on Public Switched Telephone Networks (PSTN). It aims at low bitrate communications at bitrates of less than 64 kbps. It was adopted by the ITU-T Study Group 15 in 1995. Similar to H.261, it uses predictive coding for inter-frames, to reduce temporal redundancy, and transform coding for the remaining signal, to reduce spatial redundancy (for both intra-frames and difference macroblocks from inter-frame prediction) [7].

In addition to CIF and QCIF, H.263 supports sub-QCIF, 4CIF, and 16CIF. Table 10.5 summarizes video formats supported by H.263. If not compressed and assuming 30 fps, the bitrate for high-resolution videos (e.g., 16CIF) could be very high (>500 Mbps). For compressed video, the standard defines maximum bitrate per picture (BPPmaxKb), measured in units of 1,024 bits. In practice, a lower bitrate for compressed H.263 video can be achieved.

As in H.261, the H.263 standard also supports the notion of group of blocks. The difference is that GOBs in H.263 do not have a fixed size, and they always start and end at the left and right borders of the picture. As Fig. 10.10 shows, each QCIF luminance image consists of 9 GOBs and each GOB has 11×1 MBs (176×16 pixels), whereas each 4CIF luminance image consists of 18 GOBs and each GOB has 44×2 MBs (704×32 pixels).

10.5.1   Motion Compensation in H.263

The process of motion compensation in H.263 is similar to that of H.261. The motion vector (**MV**) is, however, not simply derived from the current macroblock. The

Table 10.5 Video formats supported by H.263

Video format	Luminance image resolution	Chrominance image resolution	Bitrate (Mbps) (if 30 fps and uncompressed)	Bitrate (kbps) BPPmaxKb (compressed)
Sub-QCIF	128×96	64×48	4.4	64
QCIF	176×144	88×72	9.1	64
CIF	352×288	176×144	36.5	256
4CIF	704×576	352×288	146.0	512
16CIF	1408×1152	704×576	583.9	1024

| GOB 0 |
| GOB 1 |
| GOB 2 |
| GOB 3 |
| GOB 4 |
| GOB 5 |

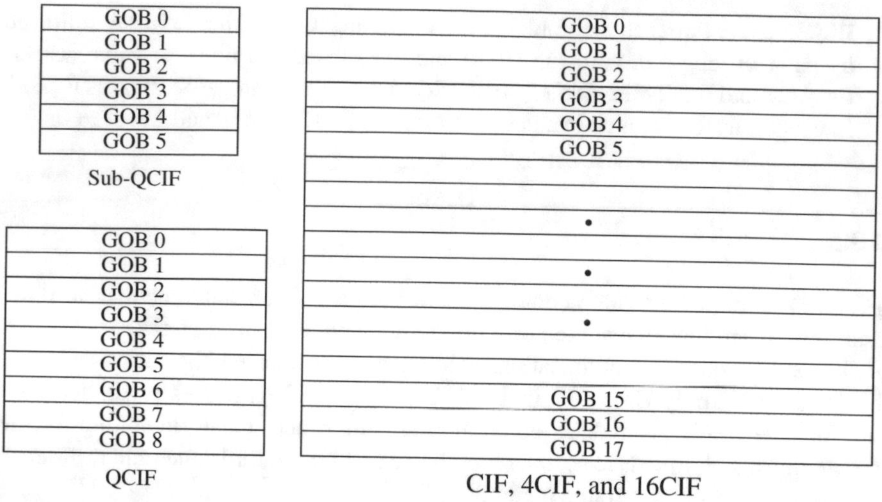

Sub-QCIF

| GOB 0 |
| GOB 1 |
| GOB 2 |
| GOB 3 |
| GOB 4 |
| GOB 5 |
| GOB 6 |
| GOB 7 |
| GOB 8 |

QCIF

| GOB 0 |
| GOB 1 |
| GOB 2 |
| GOB 3 |
| GOB 4 |
| GOB 5 |
| • |
| • |
| • |
| GOB 15 |
| GOB 16 |
| GOB 17 |

CIF, 4CIF, and 16CIF

Fig. 10.10 Arrangement of GOBs in H.263 luminance images

horizontal and vertical components of the **MV** are predicted from the median values of the horizontal and vertical components, respectively, of **MV1**, **MV2**, and **MV3** from the "previous," "above," and "above and right" macroblocks (see Fig. 10.11a). Namely, for the macroblock with **MV**(u, v),

$$u_p = median(u_1, u_2, u_3),$$
$$v_p = median(v_1, v_2, v_3). \tag{10.6}$$

Instead of coding the **MV**(u, v) itself, the error vector $(\delta u, \delta v)$ is coded, where $\delta u = u - u_p$ and $\delta v = v - v_p$. As shown in Fig. 10.11b, when the current MB is at the border of the picture or GOB, either $(0, 0)$ or **MV1** is used as the motion vector for the out-of-bound MB(s).

To improve the quality of motion compensation—that is, to reduce the prediction error—H.263 supports *half-pixel precision* as opposed to full-pixel precision only in H.261. The default range for both the horizontal and vertical components u and v of **MV**(u, v) is now $[-16, 15.5]$.

The pixel values needed at half-pixel positions are generated by a simple *bilinear interpolation* method, as shown in Fig. 10.12, where A, B, C, D and a, b, c, d are pixel values at full-pixel positions and half-pixel positions, respectively, and "/" indicates division by truncation (also known as integer division).

10.5.2 Optional H.263 Coding Modes

Besides its core coding algorithm, H.263 specifies many negotiable coding options in its various Annexes. Four of the common options are as follows:

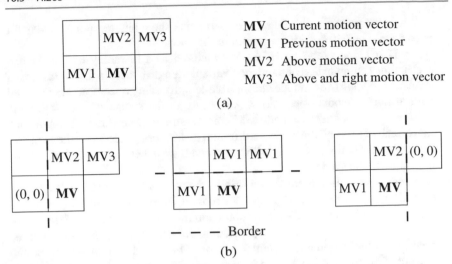

Fig. 10.11 Prediction of motion vector in H.263: **a** Predicted **MV** of the current macroblock is the median of (**MV1**, **MV2**, **MV3**); **b** special treatment of MVs when the current macroblock is at border of picture or GOB

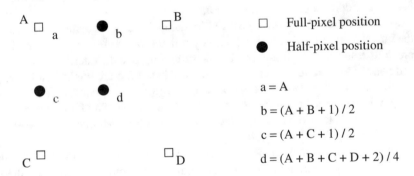

Fig. 10.12 Half-pixel prediction by bilinear interpolation in H.263

- **Unrestricted motion vector mode**. The pixels referenced are no longer restricted to within the boundary of the image. When the motion vector points outside the image boundary, the value of the boundary pixel geometrically closest to the referenced pixel is used. This is beneficial when image content is moving across the edge of the image, often caused by object and/or camera movements. This mode also allows an extension of the range of motion vectors. The maximum range of motion vectors is $[-31.5, 31.5]$, which enables efficient coding of fast-moving objects in videos.
- **Syntax-based arithmetic coding mode**. Like H.261, H.263 uses variable-length coding (VLC) as a default coding method for the DCT coefficients. Variable-length coding implies that each symbol must be coded into a fixed, integral number of bits. By employing arithmetic coding, this restriction is removed, and a higher

compression ratio can be achieved. Experiments show bitrate savings of 4% for inter-frames and 10% for intra-frames in this mode.

As in H.261, the syntax of H.263 is structured as a hierarchy of four layers, each using a combination of fixed- and variable-length codes. In the *syntax-based arithmetic coding* (SAC) mode, all variable-length coding operations are replaced with arithmetic coding operations. According to the syntax of each layer, the arithmetic encoder needs to code a different bitstream from various components. Since each of these bitstreams has a different distribution, H.263 specifies a model for each distribution, and the arithmetic coder switches the model on the fly, according to the syntax.

- **Advanced prediction mode**. In this mode, the macroblock size for motion compensation is reduced from 16 to 8. Four motion vectors (from each of the 8×8 blocks) are generated for each macroblock in the luminance image. Afterwards, each pixel in the 8×8 luminance prediction block takes a weighted sum of three predicted values based on the motion vector of the current luminance block and two out of the four motion vectors from the neighboring blocks—that is, one from the block at the left or right side of the current luminance block and one from the block above or below. Although sending four motion vectors incurs some additional overhead, the use of this mode generally yields better prediction and hence considerable gain in compression.

- **PB-frames mode**. As shown by MPEG (detailed discussions in the next chapter), the introduction of a B-frame, which is predicted bidirectionally from both the previous frame and the future frame, can often improve the quality of prediction and hence the compression ratio without sacrificing picture quality. In H.263, a PB-frame consists of two pictures coded as one unit: one P-frame, predicted from the previous decoded I-frame or P-frame (or P-frame part of a PB-frame), and one B-frame, predicted from both the previous decoded I- or P-frame and the P-frame currently being decoded (Fig. 10.13).

The use of the PB-frames mode is indicated in *PTYPE*. Since the P- and B-frames are closely coupled in the PB-frame, the bidirectional motion vectors for the B-frame need not be independently generated. Instead, they can be temporally scaled and further enhanced from the forward motion vector of the P-frame [8] so as to reduce the bitrate overhead for the B-frame. PB-frames mode yields satisfactory results for videos with moderate motion. Under large motions, PB-frames do not compress so do B-frames. An improved mode has been developed in H.263 version 2.

10.5.3 H.263+ and H.263++

The second version of H.263, also known as H.263+, was approved in January 1998 by ITU-T Study Group 16. It is fully backward-compatible with the design of H.263 version 1.

Fig. 10.13 A PB-frame in H.263

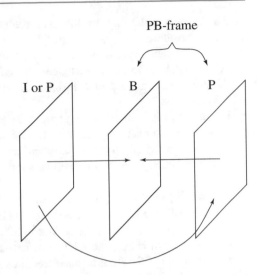

The aim of H.263+ is to broaden the potential applications and offer additional flexibility in terms of custom source formats, different pixel aspect ratios, and clock frequencies. H.263+ includes numerous recommendations to improve code efficiency and error resilience [9]. It also provides 12 new negotiable modes, in addition to the four optional modes in H.263.

Since its development came after the standardization of MPEG-1 and 2, it is not surprising that it also adopts many aspects of the MPEG standards. Below, we mention only briefly some of these features and leave their detailed discussion to the next chapter, where we study the MPEG standards.

- The unrestricted motion vector mode is redefined under H.263+. It uses *Reversible Variable Length Coding (RVLC)* to encode the difference motion vectors. The RVLC encoder is able to minimize the impact of transmission error by allowing the decoder to decode from both forward and reverse directions. The range of motion vectors is extended again to [−256, 256]. Refer to [10,11] for more detailed discussions on the construction of RVLC.
- A *slice* structure is used to replace GOB for additional flexibility. A slice can contain a variable number of macroblocks. The transmission order can be either sequential or arbitrary, and the shape of a slice is not required to be rectangular.
- H.263+ implements *Temporal, SNR,* and *Spatial scalabilities*. Scalability refers to the ability to handle various constraints, such as display resolution, bandwidth, and hardware capabilities. The enhancement layer for Temporal scalability increases perceptual quality by inserting B-frames between two P-frames.

  SNR scalability is achieved by using various quantizers of smaller-and-smaller step_size to encode additional enhancement layers into the bitstream. Thus, the decoder can decide how many enhancement layers to decode according to computational or network constraints. The concept of Spatial scalability is similar to that of SNR scalability. In this case, the enhancement layers provide increased spatial resolution.

- H.263+ supports improved PB-frames mode, in which the two motion vectors of the B-frame do not have to be derived from the forward motion vector of the P-frame, as in version 1. Instead, they can be generated independently, as in MPEG-1 and 2.
- Deblocking filters in the coding loop reduce blocking effects. The filter is applied to the edge boundaries of the four luminance and two chrominance blocks. The coefficient weights depend on the quantizer step_size for the block. This technique results in better prediction as well as a reduction in blocking artifacts.

The development of H.263 continued beyond its second version. The third version, H.263 v3, also known as H.263++, was initially approved in the year 2000. A further developed version was approved in 2005 [7]. H.263++ includes the baseline coding methods of H.263 and additional recommendations for *enhanced reference picture selection* (ERPS), *data partition slice* (DPS), and additional supplemental enhancement information.

The ERPS mode operates by managing a multiframe buffer for stored frames, enhancing coding efficiency and error resilience. The DPS mode provides additional enhancement to error resilience by separating header and motion-vector data from the DCT coefficient data in the bitstream and protects the motion-vector data by using a reversible code. The additional supplemental enhancement information provides the ability to add backward-compatible enhancements to an H.263 bitstream.

Since we will describe in detail the newer standards H.264, H.265, and H.266 in Chap. 12, and many fundamental ideas are quite similar to the latest H.263, we will leave the rest of the discussions to that chapter.

10.6 Exercises

1. Describe how H.261 deals with *temporal* and *spatial* redundancies in a video.
2. An H.261 video has the three color channels Y, C_r, and C_b. Should **MV**s be computed for each channel and then transmitted? Justify your answer. If not, which channel should be used for motion compensation?
3. Thinking about my large collection of JPEG images (of my family taken in various locales), I decide to unify them and make them more accessible by simply combining them into a big H.261-compressed file. My reasoning is that I can simply use a viewer to step through the file, making a cohesive whole out of my collection. Comment on the utility of this idea, in terms of the compression ratio achievable for the set of images.
4. In block-based video coding, what takes more effort: compression or decompression? Briefly explain why.
5. Work out the following problem of 2D Logarithmic Search for motion vectors in detail (see Fig. 10.14).

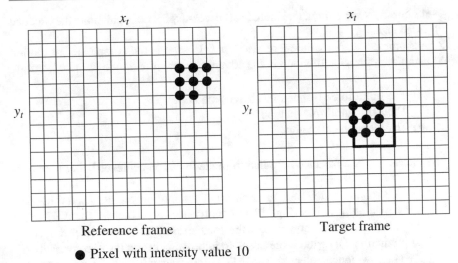

Reference frame   Target frame

● Pixel with intensity value 10

Other background (unmarked) pixels all have intensity value 100

Fig. 10.14 2D Logarithmic search for motion vectors

The target (current) frame is a P-frame. The size of macroblocks is 4×4. The motion vector is $\mathbf{MV}(\Delta x, \Delta y)$, in which $\Delta x \in [-p, p]$, $\Delta y \in [-p, p]$. In this question, assume $p \equiv 5$.

The macroblock in question (darkened) in the frame has its upper left corner at (x_t, y_t). It contains 9 dark pixels, each with intensity value 10; the other 7 pixels are part of the background, which has a uniform intensity value of 100. The reference (previous) frame has 8 dark pixels.

(a) What is the best Δx, Δy, and mean absolute error (MAE) for this macroblock?
(b) Show step by step how the 2D Logarithmic Search is performed, include the locations and passes of the search and all intermediate Δx, Δy, and MAEs.

6. The logarithmic **MV** search method is suboptimal, in that it relies on continuity in the residual frame.

(a) Explain why that assumption is necessary, and offer a justification for it.
(b) Give an example where this assumption fails.
(c) Does the hierarchical search method suffer from suboptimality too?

7. A video sequence is to be encoded using H.263 in PB-mode, having a frame size of 4CIF, frame rate of 30 fps, and video length of 90 min. The following is known about the compression parameters: on average, two I-frames are encoded per second. The video at the required quality has an I-frame average compression ratio of 10:1, an average P-frame compression ratio twice as good as I-frame, and an average B-frame compression ratio twice as good as P-frame. Assuming

the compression parameters include all necessary headers, calculate the encoded video size.

8. Assuming a search window of size $2p + 1$, what is the complexity of motion estimation for a QCIF video in the advanced prediction mode of H.263, using

   (a) The brute-force (sequential search) method?
   (b) The 2D logarithmic method?
   (c) The hierarchical method?

9. Discuss how the advanced prediction mode in H.263 achieves better compression.

10. In H.263 motion estimation, the *median* of the motion vectors from three preceding macroblocks (see Fig. 10.11a) is used as a prediction for the current macroblock. It can be argued that the median may not necessarily reflect the best prediction. Describe some possible improvements on the current method.

11. H.263+ allows independent forward **MV**s for B-frames in a PB-frame. Compared to H.263 in PB-mode, what are the trade-offs? What is the point in having PB joint coding if B-frames have independent motion vectors?

References

1. A.M. Tekalp, *Digital Video Processing*, 2nd edn. (Prentice Hall, 2015)
2. C.A. Poynton, *Digital Video and HD: Algorithms and Interfaces*, 2nd edn. (Elsevier, 2012)
3. V. Bhaskaran, K. Konstantinides, *Image and Video Compression Standards: algorithms and Architectures*, 2nd edn. (Kluwer Academic Publishers, Boston, 1997)
4. Y. Wang, J. Ostermann, Y.Q. Zhang, *Video Processing and Communications*. (Prentice Hall, 2002)
5. D. Marr, *Vision*. (The MIT Press, San Francisco, 2010)
6. *Video Codec for Audiovisual Services at $p \times 64$ kbit/s*. ITU-T Recommendation H.261, Version 1, Dec. 1990, Version 2, (1993)
7. *Video Coding for Low Bit Rate Communication*. ITU-T Recommendation H.263, Version 1, 1995, Version 2, 1998, Version 3, 2000 (2005)
8. B.G. Haskell, A. Puri, A. Netravali, *Digital Video: an Introduction to MPEG-2* (Chapman & Hall, New York, 1996)
9. G. Cote, B. Erol, M. Gallant, H.263+: video coding at low bit rates. IEEE Trans. Circuits Syst. Video Technol. **8**(7), 849–866 (1998)
10. Y. Takishima, M. Wada, H. Murakami, Reversible variable length codes. IEEE Trans. Commun. **43**(2–4), 158–162 (1995)
11. C.W. Tsai, J.L. Wu, On constructing the Huffman-code-based reversible variable-length codes. IEEE Trans. Commun. **49**(9), 1506–1509 (2001)

MPEG Video Coding: MPEG-1, 2, 4, and 7

11.1 Overview

Moving Picture Experts Group (MPEG) was established in 1988 to create a standard for the delivery of digital video and audio. Membership grew rapidly from about 25 experts in 1988 to a community of hundreds of companies and organizations [1]. It is appropriately recognized that proprietary interests need to be maintained within the family of MPEG standards. This is accomplished by defining only an MPEG-compliant compressed bitstream that implicitly defines the decoder. The continuous improvement and optimization of the compression algorithms on the encoder side are completely up to the manufacturers.

In this chapter, we will study some of the most important design issues of MPEG-1 and 2, followed by some basics of the later standards, MPEG-4 and 7, which have very different objectives.

With the modern video compression standards such as H.264, H.265, and H.266 (to be discussed in the next chapter), one might view these MPEG standards as *old*, i.e., outdated. This is simply not a concern because (a) The fundamental technology of hybrid coding and most important concepts that we will introduce here, such as motion compensation, DCT-based transform coding, and scalabilities, are used in all old and new standards. (b) Although the visual object-based video representation and compression approach developed in MPEG-4 and 7 has not been commonly used in current popular standards, it has a great potential to be adopted in the future when the necessary Computer Vision technology for automatic object detection and recognition becomes more readily available.

© Springer Nature Switzerland AG 2021
Z.-N. Li et al., *Fundamentals of Multimedia*, Texts in Computer Science,
https://doi.org/10.1007/978-3-030-62124-7_11

11.2   MPEG-1

The MPEG-1 audio/video digital compression standard [2,3] was approved by the International Organization for Standardization/International Electrotechnical Commission (ISO/IEC) MPEG group in November 1991 for *Coding of Moving Pictures and Associated Audio for Digital Storage Media at up to about* 1.5 Mbit/s [4]. Common digital storage media include compact discs (CDs) and video compact discs (VCDs). Out of the specified 1.5, 1.2 Mbps is intended for coded video, and 256 kbps can be used for stereo audio. This yields a picture quality comparable to VHS cassettes and a sound quality equal to CD audio.

In general, MPEG-1 adopts the CCIR601 digital TV format, also known as *Source Input Format* (SIF). MPEG-1 supports only noninterlaced video. Normally, its picture resolution is 352×240 for NTSC video at 30 fps, or 352×288 for PAL video at 25 fps. It uses 4:2:0 chroma subsampling.

The MPEG-1 standard, also referred to as ISO/IEC 11172 [4], has five parts: 11172-1 Systems, 11172-2 Video, 11172-3 Audio, 11172-4 Conformance, and 11172-5 Software. Briefly, Systems takes care of, among many things, dividing output into packets of bitstreams, multiplexing, and synchronization of the video and audio streams. Conformance (or compliance) specifies the design of tests for verifying whether a bitstream or decoder complies with the standard. Software includes a complete software implementation of the MPEG-1 standard decoder and a sample software implementation of an encoder.

As in H.261 and H.263, MPEG-1 employs *Hybrid Coding*, i.e., a combination of motion compensation and transform coding on prediction residual errors. We will examine the main features of MPEG-1 video coding and leave discussions of MPEG audio coding to Chap. 14.

11.2.1 Motion Compensation in MPEG-1

As discussed in the last chapter, motion-compensation-based video encoding in H.261 works as follows: In motion estimation, each macroblock of the target P-frame is assigned the best matching macroblock from the previously coded I- or P-frame. This is called a *prediction*. The difference between the macroblock and its matching macroblock is the *prediction error*, which is sent to DCT and its subsequent encoding steps.

Since the prediction is from a previous frame, it is called *forward prediction*. Due to unexpected movements and occlusions in real scenes, the target macroblock may not have a good matching entity in the previous frame. Figure 11.1 illustrates that the macroblock containing part of a ball in the target frame cannot find a good matching macroblock in the previous frame, because half of the ball was occluded by another object. However, a match can readily be obtained from the next frame.

MPEG introduces a third frame type—*B-frames*—and their accompanying bidirectional motion compensation. Figure 11.2 illustrates the motion-compensation-based B-frame coding idea. In addition to the forward prediction, a backward pre-

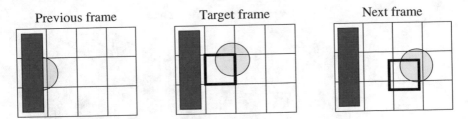

Fig. 11.1 The need for bidirectional search

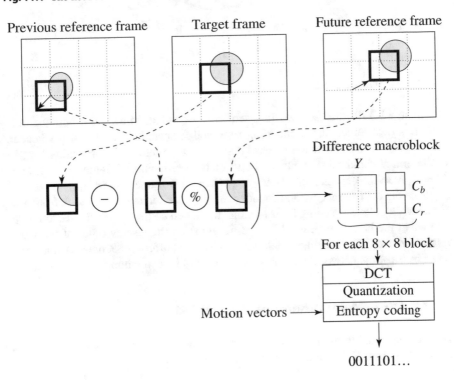

Fig. 11.2 B-frame coding based on bidirectional motion compensation

diction is also performed, in which the matching macroblock is obtained from a future I- or P-frame in the video sequence. Consequently, each macroblock from a B-frame will specify up to *two* motion vectors, one from the forward and one from the backward prediction.

If matching in both directions is successful, two motion vectors will be sent, and the two corresponding matching macroblocks are averaged (indicated by "%" in the figure) before comparing to the target macroblock for generating the prediction error. If an acceptable match can be found in only one of the reference frames, only one motion vector and its corresponding macroblock will be used from either the forward or backward prediction.

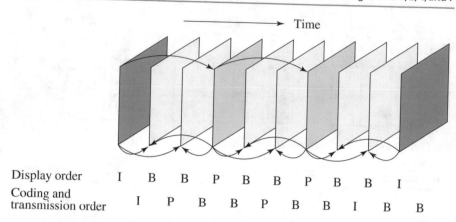

Time

Display order   I   B   B   P   B   B   P   B   B   I
Coding and
transmission order   I   P   B   B   P   B   B   I   B   B

Fig. 11.3 MPEG frame sequence

Figure 11.3 illustrates a possible sequence of video frames. The actual frame pattern is determined at the encoding time and is specified in the video's header. MPEG uses M to indicate the interval between a P-frame and its preceding I- or P-frame, and N to indicate the interval between two consecutive I-frames. In Fig. 11.3, $M = 3$, $N = 9$. A special case is $M = 1$, when no B-frame is used.

Since the MPEG encoder and decoder cannot work for any macroblock from a B-frame without its succeeding P- or I-frame, the actual coding and transmission order (shown at the bottom of Fig. 11.3) is different from the display order of the video (shown above). The inevitable delay and need for buffering become an important issue in real-time network transmission, especially in a streaming MPEG video.

11.2.2 Other Major Differences from H.261

Besides introducing bidirectional motion compensation (the B-frames), MPEG-1 also differs from H.261 in the following aspects:

- **Source formats**. H.261 supports only CIF (352×288) and QCIF (176×144) source formats. MPEG-1 supports SIF (352×240 for NTSC and 352×288 for PAL). It also allows the specification of other formats, as long as the *constrained parameter set* (*CPS*), shown in Table 11.1, is satisfied.
- **Slices**. Instead of GOBs, as in H.261, an MPEG-1 picture can be divided into one or more *slices* (Fig. 11.4), which are more flexible than GOBs. They may contain variable numbers of macroblocks in a single picture and may also start and end anywhere, as long as they fill the whole picture. Each slice is coded independently. For example, the slices can have different scale factors in the quantizer. This provides additional flexibility in bitrate control.

  Moreover, the slice concept is important for error recovery, because each slice has a unique *slice_start_code*. A slice in MPEG is similar to the GOB in H.261

Table 11.1 The MPEG-1 constrained parameter set

Parameter	Value
Horizontal size of picture	≤768
Vertical size of picture	≤ 576
Number of macroblocks/picture	≤396
Number of macroblocks/s	≤9,900
Frame rate	≤30 fps
Bitrate	≤1,856 kbps

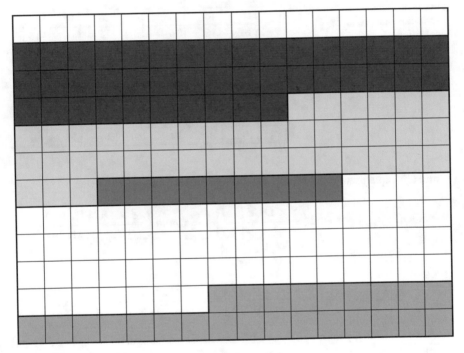

Fig. 11.4 Slices in an MPEG-1 picture

(and H.263): it is the lowest level in the MPEG layer hierarchy that can be fully recovered without decoding the entire set of variable-length codes in the bitstream.

- **Quantization**. MPEG-1 quantization uses different quantization tables for its intra- and inter-coding (Tables 11.2 and 11.3). The quantizer numbers for intra-coding (Table 11.2) vary within a macroblock. This is different from H.261, where all quantizer numbers for AC coefficients are constant within a macroblock. The $step\_size[i, j]$ value is now determined by the product of $Q[i, j]$ and $scale$, where Q_1 or Q_2 is one of the above quantization tables and $scale$ is an integer in the range [1, 31]. Using DCT and $QDCT$ to denote the DCT coefficients before

Table 11.2 Default
quantization table (Q_1) for
intra-coding

8	16	19	22	26	27	29	34
16	16	22	24	27	29	34	37
19	22	26	27	29	34	34	38
22	22	26	27	29	34	37	40
22	26	27	29	32	35	40	48
26	27	29	32	35	40	48	58
26	27	29	34	38	46	56	69
27	29	35	38	46	56	69	83

Table 11.3 Default
quantization table (Q_2) for
inter-coding

16	16	16	16	16	16	16	16
16	16	16	16	16	16	16	16
16	16	16	16	16	16	16	16
16	16	16	16	16	16	16	16
16	16	16	16	16	16	16	16
16	16	16	16	16	16	16	16
16	16	16	16	16	16	16	16
16	16	16	16	16	16	16	16

and after quantization, for DCT coefficients in intra mode,

$$\text{QDCT}[i, j] = \text{round}\left(\frac{8 \times DCT[i, j]}{step\_size[i, j]}\right) = \text{round}\left(\frac{8 \times DCT[i, j]}{Q_1[i, j] \times scale}\right), \tag{11.1}$$

and for DCT coefficients in inter mode,

$$\text{QDCT}[i, j] = \left\lfloor\frac{8 \times DCT[i, j]}{step\_size[i, j]}\right\rfloor = \left\lfloor\frac{8 \times DCT[i, j]}{Q_2[i, j] \times scale}\right\rfloor, \tag{11.2}$$

where Q_1 and Q_2 refer to Tables 11.2 and 11.3, respectively.

Again, a round operator is typically used in Eq. (11.1) and hence leaves no dead zone, whereas a floor operator is used in Eq. (11.2), leaving a center dead zone in its quantization space.

- To increase the precision of the motion-compensation-based predictions and hence reduce prediction errors, MPEG-1 allows motion vectors to be of sub-pixel precision (1/2 pixel). The technique of bilinear interpolation discussed in Sect. 10.5.1 for H.263 can be used to generate the needed values at half-pixel locations.

- MPEG-1 supports larger gaps between I- and P-frames and consequently a much larger motion-vector search range. Compared to the maximum range of ±15 pixels for motion vectors in H.261, MPEG-1 supports a range of $[-512, 511.5]$ for half-pixel precision and $[-1,024, 1,023]$ for full-pixel precision motion vectors. However, due to the practical limitation in its picture resolution, such a large maximum range might never be used.

Table 11.4 Typical compression performance of MPEG-1 frames

Type	Size (kB)	Compression
I	18	7:1
P	6	20:1
B	2.5	50:1
Average	4.8	27:1

- The MPEG-1 bitstream allows random access. This is accomplished by the *Group of Pictures (GOP)* layer, in which each GOP is time-coded. In addition, the first frame in any GOP is an I-frame, which eliminates the need to reference other frames. Thus, the GOP layer allows the decoder to seek a particular position within the bitstream and start decoding from there.

Table 11.4 lists typical sizes (in kilobytes) for all types of MPEG-1 frames. It can be seen that the typical size of compressed P-frames is significantly smaller than that of I-frames, because inter-frame compression exploits temporal redundancy. Notably, B-frames are even smaller than P-frames, due partially to the advantage of bidirectional prediction. It is also because B-frames are often given the lowest priority in terms of preservation of quality; hence, a higher compression ratio can be assigned.

11.2.3 MPEG-1 Video Bitstream

Figure 11.5 depicts the six hierarchical layers for the bitstream of an MPEG-1 video.

1. **Sequence layer**. A video sequence consists of one or more groups of pictures (GOPs). It always starts with a sequence header. The header contains information about the picture, such as *horizontal_size* and *vertical_size*, *pixel_aspect_ratio*, *frame_rate*, *bit_rate*, *buffer_size*, *quantization_matrix*, and so on. Optional sequence headers between GOPs can indicate parameter changes.
2. **Group of Pictures (GOPs) layer**. A GOP contains one or more pictures, one of which must be an I-picture. The GOP header contains information such as *time_code* to indicate hour-minute-second-frame from the start of the sequence.
3. **Picture layer**. The three common MPEG-1 picture types are *I-picture* (intra-coding), *P-picture* (predictive coding), and *B-picture* (bidirectional predictive coding), as discussed above. There is also an uncommon type, *D-picture* (DC coded), in which only DC coefficients are retained. MPEG-1 does not allow mixing D-pictures with other types, which makes D-pictures impractical.
4. **Slice layer**. As mentioned earlier, MPEG-1 introduced the slice notion for bitrate control and for recovery and synchronization after lost or corrupted bits. Slices may have variable numbers of macroblocks in a single picture. The length and position of each slice are specified in the header.

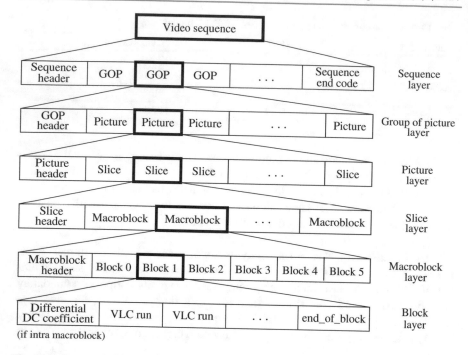

Fig. 11.5 Layers of MPEG-1 video bitstream

5. **Macroblock layer**. Each macroblock consists of four Y blocks, one C_b block, and one C_r block. All blocks are 8×8.
6. **Block layer**. If the blocks are intra-coded, the differential DC coefficient (DPCM of DCs, as in JPEG) is sent first, followed by variable-length codes (VLC), for AC coefficients. Otherwise, DC and AC coefficients are both coded using the variable-length codes.

Mitchell et al. [5] provide detailed information on the headers in various MPEG-1 layers.

11.3   MPEG-2

The development of the MPEG-2 standard started in 1990. Unlike MPEG-1, which is basically a standard for storing and playing videos on the CD of a single computer at a low bitrate (1.5 Mbps), MPEG-2 [6] is for higher quality video at a bitrate of more than 4 Mbps. It was initially developed as a standard for digital broadcast TV.

In the late 1980s, *Advanced TV (ATV)* was envisioned, to broadcast HDTV via terrestrial networks. During the development of MPEG-2, digital ATV finally took precedence over various early attempts at analog solutions to HDTV. MPEG-2 has managed to meet the compression and bitrate requirements of digital TV/HDTV

Table 11.5 Profiles and levels in MPEG-2

Level	Simple profile	Main profile	SNR scalable profile	Spatially scalable profile	High profile	4:2:2 profile	Multiview profile
High		*			*	*	
High 1440		*		*	*	*	
Main	*	*	*		*	*	*
Low	*	*	*				

and in fact supersedes a separate standard, MPEG-3, initially thought necessary for HDTV.

The MPEG-2 audio/video compression standard, also referred to as ISO/IEC 13818 [7], was approved by ISO/IEC Moving Picture Experts Group in November 1994. Similar to MPEG-1, it has Parts for Systems, Video, Audio, Conformance, and Software, plus other aspects. Part 2, the video compression part of the standard ISO/IEC 13818-2 is also known as H.262 in ITU-T (International Telecommunication Union—Telecommunications). MPEG-2 has gained wide acceptance beyond broadcasting digital TV over terrestrial, satellite, or cable networks. Among various applications such as Interactive TV, it was also adopted for *digital video discs* or *digital versatile discs* (DVDs).

MPEG-2 defined seven *profiles* aimed at different applications (e.g., low-delay videoconferencing, scalable video, and HDTV). The profiles are *Simple, Main, SNR scalable, Spatially scalable, High, 4:2:2,* and *Multiview* (where two views would refer to a stereoscopic video). Within each profile, up to four *levels* are defined. As Table 11.5 shows, not all profiles have four levels. For example, the Simple profile only has the Main and Low levels, whereas the High profile does not have the Low level.

Table 11.6 lists the four levels in the Main profile, with the maximum amount of data and targeted applications. For example, the High level supports a high picture resolution of $1,920 \times 1,152$, a maximum frame rate of 60 fps, maximum pixel rate of 62.7×10^6 per second, and a maximum data rate after coding of 80 Mbps. The Low level is targeted at SIF video; hence, it provides backward compatibility with MPEG-1. The Main level is for CCIR601 video, whereas High 1440 and High levels are aimed at European HDTV and North American HDTV, respectively.

The DVD video specification allows only four display resolutions: for example, at 29.97 frames per second, interlaced, 720×480, 704×480, 352×480, and 352×240. Hence, the DVD video standard uses only a restricted form of the MPEG-2 Main profile at the Main and Low levels.

Table 11.6 Four levels in the main profile of MPEG-2

Level	Maximum resolution	Maximum fps	Maximum pixels/s	Maximum coded data rate (Mbps)	Application
High	$1,920 \times 1,152$	60	62.7×10^6	80	Film production
High 1440	$1,440 \times 1,152$	60	47.0×10^6	60	Consumer HDTV
Main	720×576	30	10.4×10^6	15	Studio TV
Low	352×288	30	3.0×10^6	4	Consumer tape equivalent

11.3.1 Supporting Interlaced Video

MPEG-1 supports only noninterlaced (progressive) video. Since MPEG-2 is adopted by digital broadcast TV, it must also support interlaced video, because this is one of the options for digital broadcast TV and HDTV.

As mentioned earlier, in interlaced video each frame consists of two fields, referred to as the *top-field* and the *bottom-field*. In a *frame-picture*, all scan lines from both fields are interleaved to form a single frame. This is then divided into 16×16 macroblocks and coded using motion compensation. On the other hand, if each field is treated as a separate picture, then it is called *field-picture*. As Fig. 11.6a shows, each frame-picture can be split into two field-pictures. The figure shows 16 scan lines from a frame-picture on the left, as opposed to 8 scan lines in each of the two field portions of a field-picture on the right.

We see that, in terms of the display area on the monitor/TV, each 16-column $\times$ 16-row macroblock in the field-picture corresponds to a 16×32 block area in the frame-picture, whereas each 16×16 macroblock in the frame-picture corresponds to a 16×8 block area in the field-picture. As shown below, this observation will become an important factor in developing different modes of predictions for motion-compensation-based video coding.

Five Modes of Predictions

MPEG-2 defines *frame prediction* and *field prediction* as well as five different prediction modes, suitable for a wide range of applications where the requirement for the accuracy and speed of motion compensation varies.

1. **Frame prediction for frame-pictures**. This is identical to MPEG-1 motion-compensation-based prediction methods in both P-frames and B-frames. Frame prediction works well for videos containing only slow and moderate object and camera motions.

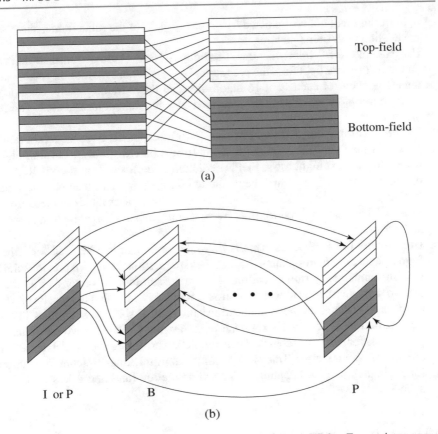

Fig. 11.6 Field-pictures and field prediction for field-pictures in MPEG-2: **a** Frame-picture versus field-pictures; **b** field prediction for field-pictures

2. **Field prediction for field-pictures**. (See Fig. 11.6b). This mode uses a macroblock size of 16×16 from field-pictures. For P-field-pictures (the rightmost ones shown in the figure), predictions are made from the two most recently encoded fields. Macroblocks in the top-field picture are forward-predicted from the top-field or bottom-field pictures of the preceding I- or P-frame. Macroblocks in the bottom-field picture are predicted from the top-field picture of the same frame or the bottom-field picture of the preceding I- or P-frame.

   For B-field-pictures, both forward and backward predictions are made from field-pictures of preceding and succeeding I- or P-frames. No regulation requires that field "parity" be maintained—that is, the top-field and bottom-field pictures can be predicted from either the top or bottom fields of the reference pictures.

3. **Field prediction for frame-pictures**. This mode treats the top-field and bottom-field of a frame-picture separately. Accordingly, each 16×16 macroblock from the target frame-picture is split into two 16×8 parts, each coming from one field. Field prediction is carried out for these 16×8 parts in a manner similar to that shown in Fig. 11.6b. Besides the smaller block size, the only difference is that the

bottom-field will not be predicted from the top-field of the same frame, since we are dealing with frame-pictures now.

For example, for P-frame-pictures, the bottom 16×8 part will instead be predicted from either field from the preceding I- or P-frame. Two motion vectors are thus generated for each 16×16 macroblock in the P-frame-picture. Similarly, up to four motion vectors can be generated for each macroblock in the B-frame-picture.

4. **16×8 MC for field-pictures**. Each 16×16 macroblock from the target field-picture is now split into top and bottom 16×8 halves—that is, the first eight rows and the next eight rows. Field prediction is performed on each half. As a result, two motion vectors will be generated for each 16×16 macroblock in the P-field-picture and up to four motion vectors for each macroblock in the B-field-picture. This mode is good for finer motion compensation when motion is rapid and irregular.

5. **Dual-prime for P-pictures**. This is the only mode that can be used for either frame-pictures or field-pictures. At first, field prediction from each previous field with the same parity (top or bottom) is made. Each motion vector **MV** is then used to derive a calculated motion vector **CV** in the field with the opposite parity, taking into account the temporal scaling and vertical shift between lines in the top and bottom fields. In this way, the pair **MV** and **CV** yields two preliminary predictions for each macroblock. Their prediction errors are averaged and used as the final prediction error. This mode is aimed at mimicking B-picture prediction for P-pictures without adopting backward prediction (and hence less encoding delay).

Alternate Scan and Field_DCT

Alternate Scan and *Field_DCT* are techniques aimed at improving the effectiveness of DCT on prediction errors. They are applicable only to frame-pictures in interlaced videos.

After frame prediction in frame-pictures, the prediction error is sent to DCT, where each block is of size 8×8. Due to the nature of the interlaced video, the consecutive rows in these blocks are from different fields; hence, there is less correlation between them than between the alternate rows. This suggests that the DCT coefficients at low vertical spatial frequencies tend to have reduced magnitudes, compared to the ones in the noninterlaced video.

Based on the above analysis, an alternate scan is introduced. It may be applied on a picture-by-picture basis in MPEG-2 as an alternative to a zigzag scan. As Fig. 11.7a indicates, zigzag scan assumes that in a noninterlaced video, the DCT coefficients at the upper left corner of the block often have larger magnitudes. Alternate scan (Fig. 11.7b) recognizes that in interlaced video, the vertically higher spatial frequency components may have larger magnitudes and thus allows them to be scanned earlier in the sequence. Experiments have shown [6] that alternate scan can improve the

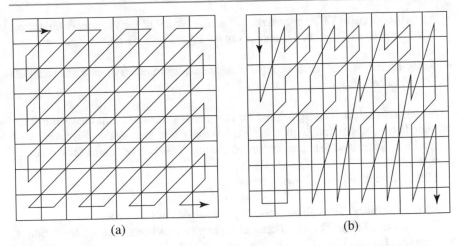

Fig. 11.7 a Zigzag (progressive) and **b** alternate (interlaced) scans of DCT coefficients for videos in MPEG-2

PSNR by up to 0.3 dB over zigzag scan and is most effective for videos with fast motion.

In MPEG-2, Field_DCT can address the same issue. Before applying DCT, rows in the macroblock of frame-pictures can be reordered, so that the first eight rows are from the top-field and the last eight are from the bottom-field. This restores the higher spatial redundancy (and correlation) between consecutive rows. The reordering will be reversed after the IDCT. Field_DCT is not applicable to chrominance images, where each macroblock has only 8×8 pixels.

11.3.2 MPEG-2 Scalabilities

As in JPEG 2000, *scalability* is also an important issue for MPEG-2. Since MPEG-2 is designed for a variety of applications, including digital TV and HDTV, the video will often be transmitted over networks with very different characteristics. Therefore, it is necessary to have a single coded bitstream that is *scalable* to various bitrates.

MPEG-2 *scalable coding* is also known as *layered coding*, in which a base layer and one or more enhancement layers can be defined. The base layer can be independently encoded, transmitted, and decoded, to obtain basic video quality. The encoding and decoding of the enhancement layer, however, depend on the base layer or the previous enhancement layer. Often, only one enhancement layer is employed, which is called two-layer scalable coding.

Scalable coding is suitable for MPEG-2 video transmitted over networks with the following characteristics.

- **Very different bitrates**. If the link speed is slow, only the bitstream from the base layer will be sent. Otherwise, bitstreams from one or more enhancement layers will also be sent, to achieve improved video quality.

- **Variable-bitrate (VBR) channels**. When the bitrate of the channel deteriorates, bitstreams from fewer or no enhancement layers will be transmitted, and vice versa.
- **Noisy connections**. The base layer can be better protected or sent via channels known to be less noisy.

Moreover, scalable coding is ideal for progressive transmission: bitstreams from the base layer are sent first, to give users a fast and basic view of the video, followed by gradually increased data and improved quality. This can be useful for delivering compatible digital TV (ATV) and HDTV.

MPEG-2 supports the following scalabilities:

- **SNR scalability**. The enhancement layer provides higher SNR.
- **Spatial scalability**. The enhancement layer provides higher spatial resolution.
- **Temporal scalability**. The enhancement layer facilitates higher frame rate.
- **Hybrid scalability**. This combines any two of the above three scalabilities.
- **Data partitioning**. Quantized DCT coefficients are split into partitions.

SNR Scalability

Figure 11.8 illustrates how SNR scalability works in the MPEG-2 encoder and decoder.

The MPEG-2 SNR scalable encoder generates output bitstreams `Bits_base` and `Bits_enhance` at two layers. At the base layer, a coarse quantization of the DCT coefficients is employed, which results in fewer bits and a relatively low-quality video. After variable-length coding, the bitstream is called `Bits_base`.

The coarsely quantized DCT coefficients are then inversely quantized (Q^{-1}) and fed to the enhancement layer, to be compared with the original DCT coefficient. Their difference is finely quantized to generate a *DCT coefficient refinement*, which, after variable-length coding, becomes the bitstream called `Bits_enhance`. The inversely quantized coarse and refined DCT coefficients are added back, and after inverse DCT (IDCT), they are used for motion-compensated prediction for the next frame. Since the enhancement/refinement over the base layer improves the signal-to-noise ratio, this type of scalability is called *SNR scalability*.

If, for some reason (e.g., the breakdown of some network channel), `Bits_enhance` from the enhancement layer cannot be obtained, the above scalable scheme can still work using `Bits_base` only. In that case, the input from the inverse quantizer (Q^{-1}) of the enhancement layer simply has to be treated as zero.

The decoder (Fig. 11.8b) operates in reverse order to the encoder. Both `Bits_base` and `Bits_enhance` are variable-length decoded (VLD) and inversely quantized (Q^{-1}) before they are added together to restore the DCT coefficients. The remaining steps are the same as in any motion-compensation-based video decoder. If both bitstreams (`Bits_base` and `Bits_enhance`) are used, the output video

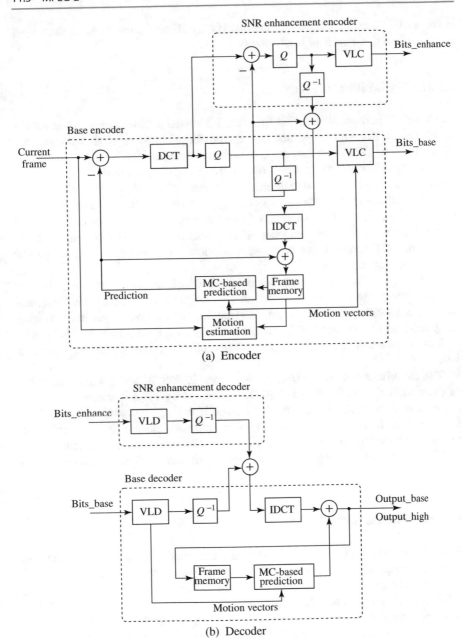

Fig. 11.8 MPEG-2 SNR scalability: **a** Encoder; **b** decoder

is `Output_high` with enhanced quality. If only `Bits_base` is used, the output video `Output_base` is of basic quality.

Spatial Scalability

The base and enhancement layers for MPEG-2 spatial scalability are not as tightly coupled as in SNR scalability; hence, this type of scalability is somewhat less complicated. We will not show the details of both encoder and decoder, as we did above, but will explain only the encoding process, using high-level diagrams.

The base layer is designed to generate a bitstream of reduced-resolution pictures. Combining them with the enhancement layer produces pictures at the original resolution. As Fig. 11.9a shows, the original video data is spatially decimated by a factor of 2 and sent to the base layer encoder. After the normal coding steps of motion compensation, DCT on prediction errors, quantization, and entropy coding, the output bitstream is `Bits_base`.

As Fig. 11.9b indicates, the predicted macroblock from the base layer is now spatially interpolated to get to resolution 16×16. This is then combined with the normal, temporally predicted macroblock from the enhancement layer itself, to form the prediction macroblock for the purpose of motion compensation in this layered coding. The spatial interpolation here adopts *bilinear interpolation*, as discussed before.

The combination of macroblocks uses a simple weight table, where the value of the weight w is in the range of [0, 1.0]. If $w = 0$, no consideration is given to the predicted macroblock from the base layer. If $w = 1$, the prediction is entirely from the base layer. Normally, both predicted macroblocks are linearly combined, using the weights w and $1 - w$, respectively. To achieve minimum prediction errors, MPEG-2 encoders have an analyzer to choose different w values from the weight table on a macroblock basis.

Temporal Scalability

Temporally scalable coding has both the base and enhancement layers of a video at a reduced temporal rate (frame rate). The reduced frame rates for the layers are often the same; however, they could also be different. Pictures from the base layer and enhancement layer(s) have the same spatial resolution as in the input video. When combined, they restore the video to its original temporal rate.

Figure 11.10 illustrates the MPEG-2 implementation of temporal scalability. The input video is temporally demultiplexed into two pieces, each carrying half the original frame rate. As before, the base layer encoder carries out the normal single-layer coding procedures for its own input video and yields the output bitstream `Bits_base`.

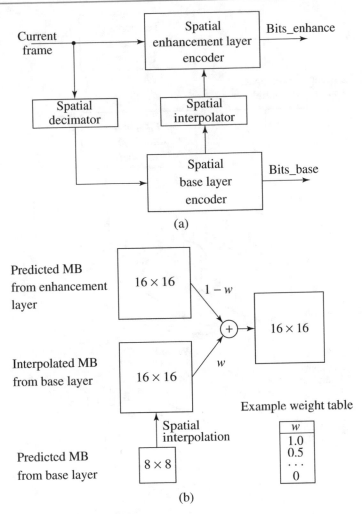

Fig. 11.9 Encoder for MPEG-2 Spatial scalability: **a** Block diagram; **b** combining temporal and spatial predictions for encoding at enhancement layer

The prediction of matching macroblocks at the enhancement layer can be obtained in two ways [6]: *Interlayer motion-compensated prediction* or *combined motion-compensated prediction and interlayer motion-compensated prediction*.

- **Interlayer motion-compensated prediction** (Fig. 11.10b). The macroblocks of B-frames for motion compensation at the enhancement layer are predicted from the preceding and succeeding frames (either I-, P-, or B-) at the base layer, so as to exploit the possible interlayer redundancy in motion compensation.
- **Combined motion-compensation prediction and interlayer motion-compensation prediction** (Fig. 11.10c). This further combines the advantages of the ordinary forward prediction and the above interlayer prediction. Macroblocks

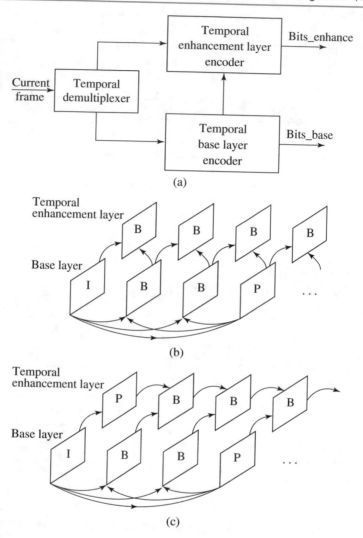

Fig. 11.10 Encoder for MPEG-2 temporal scalability: **a** Block diagram; **b** interlayer motion-compensated prediction; **c** combined motion-compensated prediction and interlayer motion-compensated prediction

of B-frames at the enhancement layer are forward-predicted from the preceding frame at its own layer and "backward"-predicted from the preceding (or, alternatively, succeeding) frame at the base layer. At the first frame, the P-frame at the enhancement layer adopts only forward prediction from the I-frame at the base layer.

Hybrid Scalability

Any two of the above three scalabilities can be combined to form hybrid scalability. These combinations are

- Spatial and temporal hybrid scalability
- SNR and spatial hybrid scalability
- SNR and temporal hybrid scalability.

Usually, a three-layer hybrid coder will be adopted, consisting of base layer, enhancement layer 1, and enhancement layer 2.

For example, for Spatial and temporal hybrid scalability, the base layer and enhancement layer 1 will provide spatial scalability, and enhancement layers 1 and 2 will provide temporal scalability, in which enhancement layer 1 is effectively serving as a base layer.

For the encoder, the incoming video data is first temporally demultiplexed into two streams: one to enhancement layer 2; the other to enhancement layer 1 and the base layer (after further spatial decimation for the base layer).

The encoder generates three output bitstreams: (a) `Bits_base` from the base layer, (b) spatially enhanced `Bits_enhance1` from enhancement layer 1, and (c) spatially and temporally enhanced `Bits_enhance2` from enhancement layer 2.

The implementations of the other two hybrid scalabilities are similar and are left as exercises.

Data Partitioning

The compressed video stream is divided into two partitions. The base partition contains lower frequency DCT coefficients, and the enhancement partition contains high-frequency DCT coefficients. Although the partitions are sometimes also referred to as layers (base layer and enhancement layer), strictly speaking, data partitioning does not conduct the same type of layered coding, since a single stream of video data is simply divided up and does not depend further on the base partition in generating the enhancement partition. Nevertheless, data partitioning can be useful for transmission over noisy channels and for progressive transmission.

11.3.3 Other Major Differences from MPEG-1

- **Better resilience to bit errors**. Since MPEG-2 video will often be transmitted on various networks, some of them noisy and unreliable, bit errors are inevitable. To cope with this, MPEG-2 systems have two types of streams: *Program* and *Transport*. The Program stream is similar to the Systems stream in MPEG-1; hence, it also facilitates backward compatibility with MPEG-1.

Table 11.7 Possible nonlinear scale in MPEG-2

i	1	2	3	4	5	6	7	8	9	10	11	12	13	14	15	16
$scale_i$	1	2	3	4	5	6	7	8	10	12	14	16	18	20	22	24
i	17	18	19	20	21	22	23	24	25	26	27	28	29	30	31	
$scale_i$	28	32	36	40	44	48	52	56	64	72	80	88	96	104	112	

The Transport stream aims at providing error resilience and the ability to include multiple programs with independent time bases in a single stream, for asynchronous multiplexing and network transmission. Instead of using long, variable-length packets, as in MPEG-1 and in the MPEG-2 Program stream, it uses fixed-length (188-byte) packets. It also has a new header syntax, for better error checking and correction.

- **Support of 4:2:2 and 4:4:4 chroma subsampling**. In addition to 4:2:0 chroma subsampling, as in H.261 and MPEG-1, MPEG-2 also allows 4:2:2 and 4:4:4 to increase color quality. As discussed in Chap. 5, each chrominance picture in 4:2:2 is horizontally subsampled by a factor of 2, whereas 4:4:4 is a special case, where no chroma subsampling actually takes place.

- **Nonlinear quantization**. Quantization in MPEG-2 is similar to that in MPEG-1. Its *step_size* is also determined by the product of $Q[i, j]$ and *scale*, where Q is one of the default quantization tables for intra- or inter- coding. Two types of scales are allowed. For the first, *scale* is the same as in MPEG-1, in which it is an integer in the range of [1, 31] and $scale_i = i$. For the second type, however, a nonlinear relationship exists—that is, $scale_i \neq i$. The ith scale value can be looked up in Table 11.7.

- **More restricted slice structure**. MPEG-1 allows slices to cross macroblock row boundaries. As a result, an entire picture can be a single slice. MPEG-2 slices must start and end in the same macroblock row. In other words, the left edge of a picture always starts a new slice, and the longest slice in MPEG-2 can have only one row of macroblocks.

- **More flexible video formats**. According to the standard, MPEG-2 picture sizes can be as large as 16 k × 16 k pixels. In reality, MPEG-2 is used mainly to support various picture resolutions as defined by DVD, ATV, and HDTV.

Similar to H.261, H.263, and MPEG-1, MPEG-2 specifies only its bitstream syntax and the decoder. This leaves much room for future improvement, especially on the encoder side. The MPEG-2 video-stream syntax is more complex than that of MPEG-1, and good references for it can be found in [6,7].

11.4 MPEG-4

11.4.1 Overview of MPEG-4

MPEG-1 and 2 employ *frame-based* coding techniques, in which each rectangular video frame is divided into macroblocks and then blocks for compression. This is also known as *block-based* coding. Their main concern is the high compression ratio and satisfactory quality of video under such compression techniques. MPEG-4 has a very different emphasis [8]. Besides compression, it pays great attention to user interactivity. This allows a larger number of users to create and communicate their multimedia presentations and applications on new infrastructures, such as the Internet, the World Wide Web (WWW), and mobile/wireless networks. MPEG-4 departs from its predecessors in adopting a new *object-based coding* approach—*media objects* are now entities for MPEG-4 coding. Media objects (also known as *audio and visual objects*) can be either *natural* or *synthetic*; that is to say, they may be captured by a video camera or created by computer programs.

Object-based coding not only has the potential of offering higher compression ratio but is also beneficial for digital video composition, manipulation, indexing, and retrieval. Figure 11.11 illustrates how MPEG-4 videos can be composed and manipulated by simple operations such as insertion/deletion, translation/rotation, scaling, and so on, on the visual objects.

MPEG-4 (version 1) was finalized in October 1998 and became an international standard in early 1999, referred to as ISO/IEC 14496 [9]. An improved version (version 2) was finalized in December 1999 and acquired International Standard status in 2000. Similar to the previous MPEG standards, its first five parts are Systems, Video,

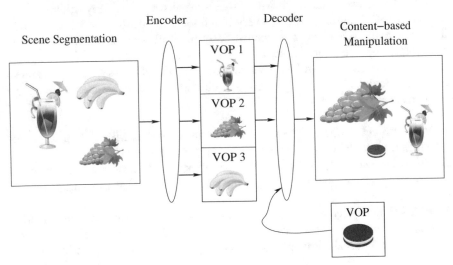

Fig. 11.11 Composition and manipulation of MPEG-4 videos (VOP = Video Object Plane)

Audio, Conformance, and Software. This chapter will discuss the video compression issues in MPEG-4 Part 2 (formally ISO/IEC 14496-2).[1]

Originally targeted at low-bitrate communication (4.8–64 kbps for mobile applications and up to 2 Mbps for other applications), the bitrate for MPEG-4 video now covers a large range, between 5 kbps and 10 Mbps.

As the *Reference Models* in Fig. 11.12a show, an MPEG-1 system simply delivers audio and video data from its storage and does not allow any user interactivity. MPEG-2 added an Interaction component (indicated by dashed lines in Fig. 11.12a) and thus permits limited user interactions in applications such as networked video and Interactive TV. MPEG-4 (Fig. 11.12b) is a standard that emphasizes (a) composing media objects to create desirable audiovisual scenes, (b) multiplexing and synchronizing the bitstreams for these media data entities so that they can be transmitted with guaranteed Quality of Service (QoS), and (c) interacting with the audiovisual scene at the receiving end.

MPEG-4 provides a toolbox of advanced coding modules and algorithms for audio and video compression.

MPEG-4 defines *BInary Format for Scenes* (BIFS) [10] that facilitates the composition of media objects into a scene. BIFS is often represented by a scene graph, in which the nodes describe audiovisual primitives and their attributes, and the graph structure enables a description of spatial and temporal relationships of objects in the scene. BIFS is an enhancement of virtual reality modeling language (VRML). In particular, it emphasizes timing and synchronization of objects, which were lacking in the original VRML design. In addition to BIFS, MPEG-4 (version 2) provides a programming environment, *MPEG-J* [11], in which Java applications (called *MPE-Glets*) can access Java packages and APIs so as to enhance end users' interactivity.

The hierarchical structure of MPEG-4 visual bitstreams is very different from that of MPEG-1 and 2 in that it is very much video-object-oriented. Figure 11.13 illustrates five levels of the hierarchical description of a scene in MPEG-4 visual bitstreams. In general, each *Video-object Sequence* (VS) will have one or more *Video Objects* (VOs), each VO will have one or more *Video Object Layers* (VOLs), and so on. Syntactically, all five levels have a unique start code in the bitstream, to enable random access.

1. **Video-object Sequence (VS)**. VS delivers the complete MPEG-4 visual scene, which may contain 2D or 3D natural or synthetic objects.
2. **Video Object (VO)**. VO is a particular object in the scene, which can be of arbitrary (nonrectangular) shape, corresponding to an object or background of the scene.
3. **Video Object Layer (VOL)**. VOL facilitates a way to support (multilayered) scalable coding. A VO can have multiple VOLs under scalable coding or a single VOL under nonscalable coding. As a special case, MPEG-4 also supports a special

---

[1]MPEG-4 Part 10 which is identical to ITU-T H.264 AVC will be discussed in the next chapter.

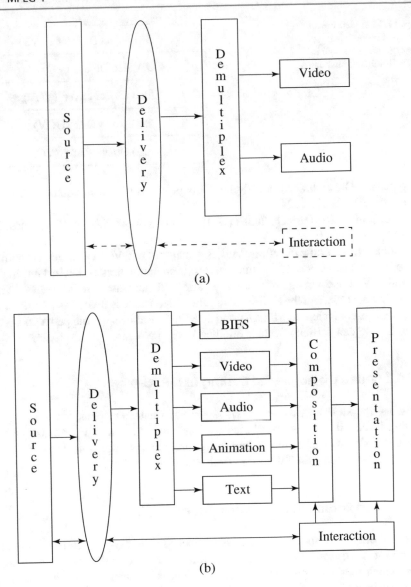

Fig. 11.12 Comparison of interactivity in MPEG standards: **a** Reference models in MPEG-1 and 2 (interaction in dashed lines supported only by MPEG-2); **b** MPEG-4 reference model

Fig. 11.13 Video-object-oriented hierarchical description of a scene in MPEG-4 visual bitstreams

| Video-object Sequence (VS) |
| Video Object (VO) |
| Video Object Layer (VOL) |
| Group of VOPs (GOV) |
| Video Object Plane (VOP) |

type of VOL with a shorter header. This provides bitstream compatibility with the baseline H.263 [12].

4. **Group of Video Object Planes (GOV)**. GOV groups video object planes. It is an optional level.

5. **Video Object Plane (VOP)**. A VOP is a snapshot of a VO at a particular moment, reflecting the VO's shape, texture, and motion parameters at that instant. In general, a VOP is an image of arbitrary shape. A degenerate case in MPEG-4 video coding occurs when the entire rectangular video frame is treated as a VOP. In this case, it is equivalent to MPEG-1 and 2. MPEG-4 allows overlapped VOPs—that is, a VOP can partially occlude another VOP in a scene.

11.4.2 Video Object-Based Coding in MPEG-4

MPEG-4 encodes/decodes each VOP separately (instead of considering the whole frame). Hence, its video-object-based coding is also known as *VOP-based coding*. Our discussion will start with coding for natural objects (more details can be found in [13, 14]). Section 11.4.3 describes synthetic object coding.

VOP-Based Coding Versus Frame-Based Coding

MPEG-1 and 2 do not support the VOP concept; hence, their coding method is referred to as *frame-based*. Since each frame is divided into many macroblocks from which motion-compensation-based coding is conducted, it is also known as *block-based coding*. Figure 11.14a shows three frames from a video sequence with a vehicle moving toward the left and a pedestrian walking in the opposite direction. Figure 11.14b shows the typical block-based coding in which the motion vector (**MV**) is obtained for one of the macroblocks.

MPEG-1 and 2 video coding are concerned only with *compression ratio* and do not consider the existence of visual objects. Therefore, the motion vectors generated may be inconsistent with object-level motion and would not be useful for object-based video analysis and indexing.

Figure 11.14c illustrates a possible example in which both potential matches yield small prediction errors. If Potential Match 2 yields a (slightly) smaller prediction

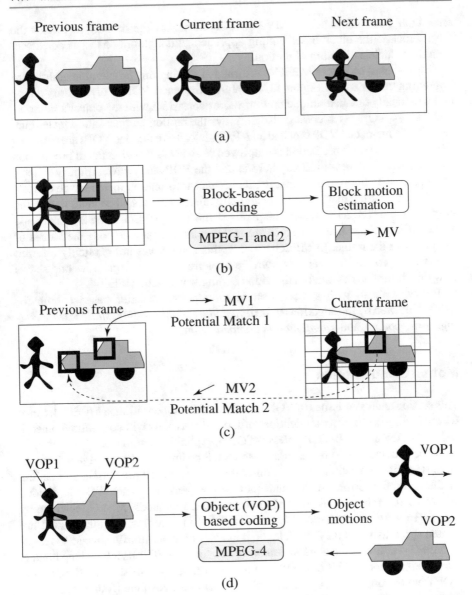

Fig. 11.14 Comparison between block-based coding and object-based coding: **a** A video sequence; **b** MPEG-1 and 2 block-based coding; **c** two potential matches in MPEG-1 and 2; **d** object-based coding in MPEG-4

error than Potential Match 1, **MV**2 will be chosen as the motion vector for the macroblock in the block-based coding approach, although only **MV**1 is consistent with the vehicle's direction of motion.

Object-based coding in MPEG-4 is aimed at solving this problem, in addition to improving compression. Figure 11.14d shows that each VOP is of arbitrary shape and will ideally obtain a unique motion vector consistent with the object's motion.

MPEG-4 VOP-based coding also employs the motion compensation technique. An Intra-frame-coded VOP is called an *I-VOP*. Inter-frame-coded VOPs are called *P-VOP*s if only forward prediction is employed or *B-VOP*s if bidirectional predictions are employed. The new difficulty here is that the VOPs may have arbitrary shapes. Therefore, in addition to their texture, their shape information must now be coded.

It is worth noting that *texture* here actually refers to the visual content, that is, the gray-level and chroma values of the pixels in the VOP. MPEG-1 and 2 do not code shape information, since all frames are rectangular, but they do code the values of the pixels in the frame. In MPEG-1 and 2, this coding was not explicitly referred to as texture coding. The term "texture" comes from computer graphics and shows how this discipline has entered the video coding world with MPEG-4.

Below, we start with a discussion of motion-compensation-based coding for VOPs, followed by introductions to *texture coding*, *shape coding*, *static texture coding*, *sprite coding*, and *global motion compensation*.

Motion Compensation

This section addresses issues of VOP-based motion compensation in MPEG-4. Since I-VOP coding is relatively straightforward, our discussions will concentrate on coding for P-VOP and/or B-VOP unless I-VOP is explicitly mentioned.

As before, motion-compensation-based VOP coding in MPEG-4 again involves three steps: motion estimation, motion-compensation-based prediction, and coding of the prediction error. To facilitate motion compensation, each VOP is divided into many macroblocks, as in previous frame-based methods. Macroblocks are by default 16×16 in luminance images and 8×8 in chrominance images and are treated specially when they straddle the boundary of an arbitrarily shaped VOP.

MPEG-4 defines a rectangular *bounding box* for each VOP. Its left and top bounds are the left and top bounds of the VOP, which in turn specify the shifted origin for the VOP from the original $(0, 0)$ for the video frame in the absolute (frame) coordinate system (see Fig. 11.15). Both horizontal and vertical dimensions of the bounding box must be multiples of 16 in the luminance image. Therefore, the box is usually slightly larger than a conventional bounding box.

Macroblocks entirely within the VOP are referred to as *interior macroblocks*. As is apparent from Fig. 11.15, many of the macroblocks straddle the boundary of the VOP and are called *boundary macroblocks*.

Motion compensation for interior macroblocks is carried out in the same manner as in MPEG-1 and 2. However, boundary macroblocks could be difficult to match in motion estimation, since VOPs often have arbitrary (nonrectangular) shape, and

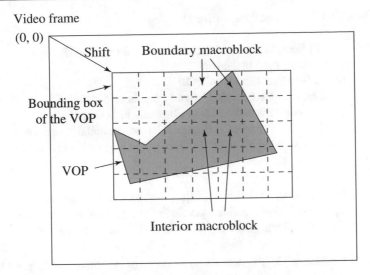

Fig. 11.15 Bounding box and boundary macroblocks of VOP

Fig. 11.16 A sequence of paddings for reference VOPs in MPEG-4

their shape may change from one instant in the video to another. To help match every pixel in the target VOP and meet the mandatory requirement of rectangular blocks in transform coding (e.g., DCT), a preprocessing step of *padding* is applied to the reference VOPs prior to motion estimation.

Only pixels within the VOP of the current (target) VOP are considered for matching in motion compensation, and padding takes place only in the reference VOPs.

For quality, some better extrapolation method than padding could have been developed. Padding was adopted in MPEG-4 largely due to its simplicity and speed.

The first two steps of motion compensation are padding and motion-vector coding.

Padding. For all boundary macroblocks in the reference VOP, *horizontal repetitive padding* is invoked first, followed by *vertical repetitive padding* (Fig. 11.16). Afterwards, for all *exterior macroblocks* that are outside of the VOP but adjacent to one or more boundary macroblocks, *extended padding* is applied.

The horizontal repetitive padding algorithm examines each row in the boundary macroblocks in the reference VOP. Each boundary pixel is replicated to the left and/or right to fill in the values for the interval of pixels outside the VOP in the macroblock. If the interval is bounded by two boundary pixels, their average is adopted.

Algorithm 11.1 (*Horizontal Repetitive Padding*)
begin

    for all rows in Boundary macroblocks in the Reference VOP
        if ∃ (boundary pixel) in the row
            for all *interval* outside of VOP
                if *interval* is bounded by only one boundary pixel b
                    assign the value of b to all pixels in *interval*
                else   *interval* is bounded by two boundary pixels b_1 and b_2
                    assign the value of $(b_1 + b_2)/2$ to all pixels in *interval*
end

    The subsequent vertical repetitive padding algorithm works similarly. It examines each column, and the newly padded pixels by the preceding horizontal padding process are treated as pixels inside the VOP for the purpose of this vertical padding.

Example 11.1 Figure 11.17 illustrates an example of repetitive padding in a boundary macroblock of a reference VOP. Figure 11.17a shows the luminance (or chrominance) intensity values of pixels in the VOP, with the VOP's boundary shown as darkened lines. For simplicity, the macroblock's resolution is reduced to 6×6 in this example, although its actual macroblock size is 16×16 in luminance images and 8×8 in chrominance images.

1. **Horizontal repetitive padding** (Fig. 11.17b)
   Row 0. The rightmost pixel of the VOP is the only boundary pixel. Its intensity value, 60, is used repetitively as the value of the pixels outside the VOP.
   Row 1. Similarly, the rightmost pixel of the VOP is the only boundary pixel. Its intensity value, 50, is used repetitively as the pixel value outside of the VOP.
   Rows 2 and 3. No horizontal padding, since no boundary pixels exist.
   Row 4. There exist two intervals outside the VOP, each bounded by a single boundary pixel. Their intensity values, 60 and 70, are used as the pixel values of the two intervals, respectively.

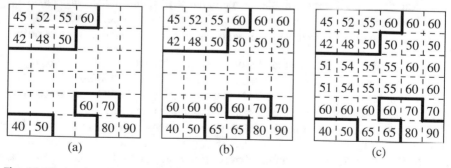

(a)   (b)   (c)

Fig. 11.17 An example of repetitive padding in a boundary macroblock of a reference VOP: **a** Original pixels within the VOP; **b** after horizontal repetitive padding; **c** followed by vertical repetitive padding

Row 5. A single interval outside the VOP is bounded by a pair of boundary pixels of the VOP. The average of their intensity values, $(50 + 80)/2 = 65$, is used repetitively as the value of the pixels between them.

2. **Vertical repetitive padding** (Fig. 11.17c)

Column 0. A single interval is bounded by a pair of boundary pixels of the VOP. One is 42 in the VOP; the other is 60, which just arose from horizontal padding. The average of their intensity values, $(42 + 60)/2 = 51$, is repetitively used as the value of the pixels between them.

Columns 1, 2, 3, 4, and 5. These columns are padded similar to Column 0.

Extended Padding. Macroblocks entirely outside the VOP are *exterior* macroblocks. Exterior macroblocks immediately next to boundary macroblocks are filled by replicating the values of the border pixels of the boundary macroblock. We note that boundary macroblocks are by now fully padded, so all their horizontal and vertical border pixels have defined values. If an exterior macroblock has more than one boundary macroblock as its immediate neighbor, the boundary macroblock to use for extended padding follows a priority list: left, top, right, and bottom.

Later versions of MPEG-4 allow some average values of these macroblocks to be used. This extended padding process can be repeated to fill in all exterior macroblocks within the rectangular bounding box of the VOP.

Motion-Vector Coding. Each macroblock from the target VOP will find the best matching macroblock from the reference VOP through the following motion estimation procedure.

Let $C(x + k, y + l)$ be pixels of the macroblock in the target VOP, and $R(x + i + k, y + j + l)$ be pixels of the macroblock in the reference VOP. Similar to MAD in Eq. (10.1), a *Sum of Absolute Difference* (SAD) for measuring the difference between the two macroblocks can be defined as

$$SAD(i, j) = \sum_{k=0}^{N-1} \sum_{l=0}^{N-1} |C(x + k, y + l) - R(x + i + k, y + j + l)| \\ \cdot Map(x + k, y + l)$$

where N is the size of the macroblock. $Map(p, q) = 1$ when $C(p, q)$ is a pixel within the target VOP; otherwise, $Map(p, q) = 0$. The vector (i, j) that yields the minimum SAD is adopted as the motion vector $\mathbf{MV}(u, v)$:

$$(u, v) = \{ (i, j) \mid SAD(i, j) \text{ is minimum, } i \in [-p, p], \ j \in [-p, p] \} \quad (11.3)$$

where p is the maximal allowable magnitude for u and v.

For motion compensation, the *motion vector* $\mathbf{MV}$ is coded. As in H.263 (see Fig. 10.11), the motion vector of the target macroblock is not simply taken as the $\mathbf{MV}$. Instead, $\mathbf{MV}$ is predicted from three neighboring macroblocks. The prediction error for the motion vector is then variable-length coded.

Following are some of the advanced motion compensation techniques adopted similar to the ones in H.263 (see Sect. 10.5).

- Four motion vectors (each from an 8 × 8 block) can be generated for each macroblock in the luminance component of a VOP.
- Motion vectors can have subpixel precision. At half-pixel precision, the range of motion vectors is [−2,048, 2,047]. MPEG-4 also allows quarter-pixel precision in the luminance component of a VOP.
- Unrestricted motion vectors are allowed: **MV** can point beyond the boundaries of the reference VOP. When a pixel outside the VOP is referenced, its value is still defined, due to padding.

Texture Coding

Texture refers to gray-level (or chroma) variations and/or patterns in the VOP. Texture coding in MPEG-4 can be based either on DCT or *shape-Adaptive DCT* (SA-DCT).

Texture Coding Based on DCT. In I-VOP, the gray (or chroma) values of the pixels in each macroblock of the VOP are directly coded, using the DCT followed by VLC, which is similar to what is done in JPEG for still pictures. P-VOP and B-VOP use motion-compensation-based coding; hence, it is the prediction error that is sent to DCT and VLC. The following discussion will be focused on motion-compensation-based texture coding for P-VOP and B-VOP.

Coding for Interior macroblocks, each 16 × 16 in the luminance VOP and 8 × 8 in the chrominance VOP, is similar to the conventional motion-compensation-based coding in H.261, H.263, and MPEG-1 and 2. Prediction errors from the six 8 × 8 blocks of each macroblock are obtained after the conventional motion estimation step. These are sent to a DCT routine to obtain six 8 × 8 blocks of DCT coefficients.

For boundary macroblocks, areas outside the VOP in the reference VOP are padded using repetitive padding, as described above. After motion compensation, texture prediction errors within the target VOP are obtained. For portions of the boundary macroblocks in the target VOP outside the VOP, zeros are padded to the block sent to DCT, since ideally, prediction errors would be near zero inside the VOP. While repetitive padding and extended padding were for better matching in motion compensation, this additional zero padding is for better DCT results in texture coding.

The quantization *step_size* for the DC component is 8. For the AC coefficients, one of the following two methods can be employed:

- The H.263 method, in which all coefficients receive the same quantizer controlled by a single parameter, and different macroblocks can have different quantizers.
- The MPEG-2 method, in which DCT coefficients in the same macroblock can have different quantizers and are further controlled by the *step_size* parameter.

Shape-Adaptive DCT (SA-DCT)-Based Coding for Boundary Macroblocks. SA-DCT [15] is another texture coding method for boundary macroblocks. Due to its effectiveness, SA-DCT has been adopted for coding boundary macroblocks in MPEG-4 version 2.

1D DCT-N is a variation of the 1D DCT described earlier (Eqs. (8.19) and (8.20)), in that N elements are used in the transform instead of a fixed $N = 8$. (For short, we will denote the 1D DCT-N transform by DCT-N in this section.)

Equations (11.4) and (11.5) describe the DCT-N transform and its inverse, IDCT-N.

1D Discrete Cosine Transform-N (DCT-N)

$$F(u) = \sqrt{\frac{2}{N}} C(u) \sum_{i=0}^{N-1} \cos \frac{(2i+1)u\pi}{2N} f(i) \tag{11.4}$$

1D Inverse Discrete Cosine Transform-N (IDCT-N)

$$\tilde{f}(i) = \sum_{u=0}^{N-1} \sqrt{\frac{2}{N}} C(u) \cos \frac{(2i+1)u\pi}{2N} F(u) \tag{11.5}$$

where $i = 0, 1, \ldots, N-1$, $u = 0, 1, \ldots, N-1$, and

$$C(u) = \begin{cases} \frac{\sqrt{2}}{2} & \text{if } u = 0, \\ 1 & \text{otherwise} \end{cases}$$

SA-DCT is a 2D DCT and is computed as a separable 2D transform in two iterations of DCT-N. Figure 11.18 illustrates the process of texture coding for boundary macroblocks using SA-DCT. The transform is applied to each of the 8×8 blocks in the boundary macroblock.

Figure 11.18a shows one of the 8×8 blocks of a boundary macroblock, where pixels inside the macroblock, denoted by $f(x, y)$, are shown in gray. The gray pixels are first shifted upward to obtain $f'(x, y)$, as Fig. 11.18b shows. In the first iteration, DCT-N is applied to each column of $f'(x, y)$, with N determined by the number of gray pixels in the column. Hence, we use DCT-2, DCT-3, DCT-5, and so on. The resulting DCT-N coefficients are denoted by $F'(x, v)$, as Fig. 11.18c shows, where the dark dots indicate the DC coefficients of the DCT-Ns. The elements of $F'(x, v)$ are then shifted to the left to obtain $F''(x, v)$ in Fig. 11.18d.

In the second iteration, DCT-N is applied to each row of $F''(x, v)$ to obtain $G(u, v)$ (Fig. 11.18e), in which the single dark dot indicates the DC coefficient $G(0, 0)$ of the 2D SA-DCT.

Some coding considerations:

- The total number of DCT coefficients in $G(u, v)$ is equal to the number of gray pixels inside the 8×8 block of the boundary macroblock, which is less than 8×8. Hence, the method is *shape adaptive* and is more efficient to compute.
- At decoding time, since the array elements must be shifted back properly after each iteration of IDCT-Ns, a binary mask of the original shape is required to decode the texture information coded by SA-DCT. The binary mask is the same as the *binary alpha map* described below.

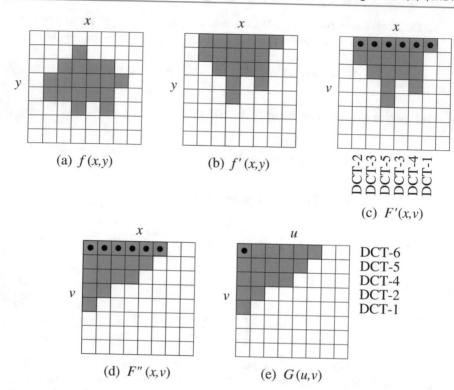

Fig. 11.18 Texture coding for boundary macroblocks using the shape-adaptive DCT (SA-DCT)

Shape Coding

Unlike in MPEG-1 and 2, MPEG-4 must code the shape of the VOP, since shape is one of the intrinsic features of visual objects.

MPEG-4 supports two types of shape information: *binary* and *grayscale*. Binary shape information can be in the form of a binary map (also known as a *binary alpha map*) that is of the same size as the VOP's rectangular bounding box. A value of 1 (opaque) or 0 (transparent) in the bitmap indicates whether the pixel is inside or outside the VOP. Alternatively, the grayscale shape information actually refers to the shape's *transparency*, with gray values ranging from 0 (transparent) to 255 (opaque).

Binary Shape Coding. To encode the binary alpha map more efficiently, the map is divided into 16 × 16 blocks, also known as *Binary Alpha Blocks* (BAB). If a BAB is entirely opaque or transparent, it is easy to code, and no special technique of shape coding is necessary. It is the boundary BABs that contain the contour and hence the shape information for the VOP. They are the subject of binary shape coding.

Various contour-based and bitmap-based (or area-based) algorithms have been studied and compared for coding boundary BABs. Two of the finalists were both bitmap-based. One was the *Modified Modified READ* (MMR) *algorithm*, which was also an optional enhancement in the fax Group 3 (G3) standard [16] and the mandatory compression method in the Group 4 (G4) standard [17]. The other finalist was *Context-based Arithmetic Encoding* (CAE), which was initially developed for JBIG [18]. CAE was finally chosen as the binary shape-coding method for MPEG-4 because of its simplicity and compression efficiency.

MMR is basically a series of simplifications of the *Relative Element Address Designate* (READ) algorithm. The basic idea behind the READ algorithm is to code the current line relative to the pixel locations in the previously coded line. The algorithm starts by identifying five pixel locations in the previous and current lines:

- a_0: the last pixel value known to both the encoder and decoder
- a_1: the transition pixel to the right of a_0
- a_2: the second transition pixel to the right of a_0
- b_1: the first transition pixel whose color is opposite to a_0 in the previously coded line
- b_2: the first transition pixel to the right of b_1 on the previously coded line.

READ works by examining the relative positions of these pixels. At any time, both the encoder and decoder know the position of a_0, b_1, and b_2, while the positions a_1 and a_2 are known only in the encoder.

Three coding modes are used:

- If the run lengths on the previous and the current lines are similar, the distance between a_1 and b_1 should be much smaller than the distance between a_0 and a_1. Thus, the *vertical mode* encodes the current run length as $a_1 - b_1$.
- If the previous line has no similar run length, the current run length is coded using 1D run-length coding. This is called the *horizontal mode*.
- If $a_0 \le b_1 < b_2 < a_1$, we can simply transmit a codeword indicating it is in *pass mode* and advance a_0 to the position under b_2, and continue the coding process.

Some simplifications can be made to the READ algorithm for practical implementation. For example, if $|a_1 - b_1| < 3$, then it is enough to indicate that we can apply the vertical mode. Also, to prevent error propagation, a k-factor is defined, such that every k line must contain at least one line coded using conventional run-length coding. These modifications constitute the *Modified READ* algorithm used in the G3 standard. The modified modified READ (MMR) algorithm simply removes the restrictions imposed by the k-factor.

For Context-based Arithmetic Encoding, Fig. 11.19 illustrates the "context" for a pixel in the boundary BAB. In intra-CAE mode, when only the target alpha map is involved (Fig. 11.19a), ten neighboring pixels (numbered from 0 to 9) in the same alpha map form the context. The ten binary numbers associated with these pixels can offer up to $2^{10} = 1,024$ possible contexts.

Now, it is apparent that certain contexts (e.g., all 1s or all 0s) appear more frequently than others. With some prior statistics, a probability table can be built to indicate the probability of occurrence for each of the 1,024 contexts.

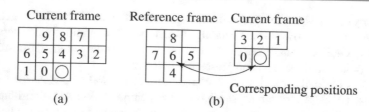

Fig. 11.19 Contexts in CAE for binary shape coding in MPEG-4. ○ indicates the current pixel, and digits indicate the other pixels in the neighborhood: **a** Intra-CAE; **b** inter-CAE

Recall that Arithmetic Coding (Chap. 7) is capable of encoding a sequence of probabilistic symbols with a single number. Now, each pixel can look up the table to find a probability value for its context. CAE simply scans the 16×16 pixels in each BAB sequentially and applies Arithmetic coding to eventually derive a single floating-point number for the BAB.

Inter-CAE mode is a natural extension of intra-CAE: it involves both the target and reference alpha maps. For each boundary macroblock in the target frame, a process of motion estimation (in integer precision) and compensation is invoked first to locate the matching macroblock in the reference frame. This establishes the corresponding positions for each pixel in the boundary BAB.

Figure 11.19b shows the context of each pixel includes four neighboring pixels from the target alpha map and five pixels from the reference alpha map. According to its context, each pixel in the boundary BAB is assigned one of the $2^9 = 512$ probabilities. Afterward, the CAE algorithm is applied.

The 16×16 binary map originally contains 256 bits of information. Compressing it to a single floating number achieves a substantial saving.

The above CAE method is *lossless*! The MPEG-4 group also examined some simple lossy versions of the above shape-coding method. For example, the binary alpha map can be simply subsampled by a factor of 2 or 4 before arithmetic coding. The trade-off is, of course, the deterioration of the shape.

Grayscale Shape Coding. The term *grayscale shape coding* in MPEG-4 could be misleading, because the true shape information is coded in the binary alpha map. Grayscale here is used to describe the *transparency* of the shape, not the texture!

In addition to the bitplanes for RGB frame buffers, raster graphics uses extra bitplanes for an *alpha map*, which can be used to describe the transparency of the graphical object. When the alpha map has more than one bitplane, multiple levels of transparency can be introduced—for example, 0 for transparent, 255 for opaque, and any number in between for various degrees of intermediate transparency. The term grayscale is used for transparency coding in MPEG-4 simply because the transparency number happens to be in the range of 0 to 255—the same as conventional 8-bit grayscale intensities.

Grayscale shape coding in MPEG-4 employs the same technique as in the texture coding described above. It uses the alpha map and block-based motion compensation and encodes prediction errors by DCT. The boundary macroblocks need padding, as before, since not all pixels are in the VOP.

Coding of the transparency information (grayscale shape coding) is lossy, as opposed to the coding of the binary shape information, which is by default lossless.

Static Texture Coding

MPEG-4 uses wavelet coding for the texture of static objects. This is particularly applicable when the texture is used for mapping onto 3D surfaces.

As introduced in Chap. 8, wavelet coding can recursively decompose an image into *subbands* of multiple frequencies. The embedded zerotree wavelet (EZW) algorithm [19] provides a compact representation by exploiting the potentially large number of insignificant coefficients in the subbands.

The coding of subbands in MPEG-4 static texture coding is conducted as follows:

- The subbands with the lowest frequency are coded using DPCM. The prediction of each coefficient is based on three neighbors.
- Coding of other subbands is based on a multiscale zerotree wavelet coding method.

The multiscale zerotree has a *parent–child relation* (PCR) *tree* for each coefficient in the lowest frequency subband. As a result, the location information of all coefficients is better tracked.

In addition to the original magnitude of the coefficients, the degree of quantization affects the data rate. If the magnitude of a coefficient is zero after quantization, it is considered insignificant. At first, a large quantizer is used; only the most significant coefficients are selected and subsequently coded using arithmetic coding. The difference between the quantized and the original coefficients is kept in residual subbands, which will be coded in the next iteration in which a smaller quantizer is employed. The process can continue for additional iterations; hence, it is very scalable.

Sprite Coding

Video photography often involves camera movements such as pan, tilt, zoom in/out, and so on. Often, the main objective is to track and examine foreground (moving) objects. Under these circumstances, the background can be treated as a static image. This creates a new VO type, the *sprite*—a graphic image that can freely move around within a larger graphic image or set of images.

To separate the foreground object from the background, we introduce the notion of a *sprite panorama*—a still image that describes the static background over a sequence of video frames. It can be generated using image "stitching" and warping techniques [20]. The large sprite panoramic image can be encoded and sent to the decoder only once, at the beginning of the video sequence. When the decoder receives separately

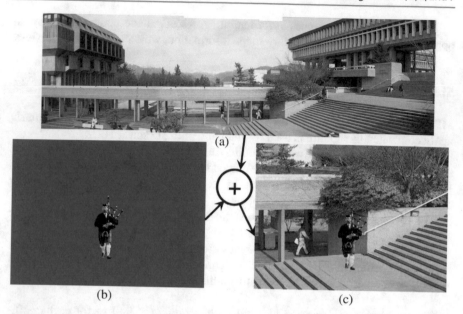

Fig. 11.20 Sprite coding: **a** The sprite panoramic image of the background; **b** the foreground object (in this case, a piper) in a bluescreen image; **c** the composed video scene. *Piper image courtesy of Simon Fraser University Pipe Band*

coded foreground objects and parameters describing the camera movements thus far, it can efficiently reconstruct the scene.

Figure 11.20a shows a sprite that is a panoramic image stitched from a sequence of video frames. By combining the sprite background with the piper in the bluescreen image (Fig. 11.20b), the new video scene (Fig. 11.20c) can readily be decoded with the aid of the sprite code and the additional pan/tilt and zoom parameters. Clearly, foreground objects can either be from the original video scene or newly created to realize flexible object-based composition of MPEG-4 videos.

Global Motion Compensation

Common camera motions, such as pan, tilt, rotation, and zoom (so-called *global* motions, since they apply to every block), often cause a rapid content change between successive video frames. Traditional block-based motion compensation would result in a large number of significant motion vectors. Also, these types of camera motions cannot all be described using the translational motion model employed by block-based motion compensation. *Global motion compensation (GMC)* is designed to solve this problem. There are four major components:

- **Global motion estimation**. Global motion estimation computes the motion of the current image with respect to the sprite. By "global," it means overall change due to camera change—zooming in, panning to the side, and so on. It is computed by minimizing the sum of square differences between the sprite S and the global motion-compensated image I'.

$$E = \sum_{i=1}^{N} (S(x_i, y_i) - I'(x_i', y_i'))^2. \tag{11.6}$$

The idea here is that if the background (possibly stitched) image is a sprite $S(x_i, y_i)$, we expect the new frame to consist mainly of the same background, altered by these global camera motions. To further constrain the global motion estimation problem, the motion over the whole image is parameterized by a perspective motion model using eight parameters, defined as

$$x_i' = \frac{a_0 + a_1 x_i + a_2 y_i}{a_6 x_i + a_7 y_i + 1},$$
$$y_i' = \frac{a_3 + a_4 x_i + a_5 y_i}{a_6 x_i + a_7 y_i + 1}. \tag{11.7}$$

This resulting constrained minimization problem can be solved using a gradient-descent-based method [21].

- **Warping and blending**. Once the motion parameters are computed, the background images are warped to align with respect to the sprite. The coordinates of the warped image are computed using Eq. (11.7). Afterwards, the warped image is blended into the current sprite to produce the new sprite. This can be done using simple averaging or some form of weighted averaging.

- **Motion trajectory coding**. Instead of directly transmitting the motion parameters, we encode only the displacements of reference points. This is called *trajectory coding* [21]. Points at the corners of the VOP bounding box are used as reference points, and their corresponding points in the sprite are calculated. The difference between these two entities is coded and transmitted as differential motion vectors.

- **Choice of local motion compensation (LMC) or GMC**. Finally, a decision has to be made whether to use GMC or LMC. For this purpose, we can apply GMC to the moving background and LMC to the foreground. Heuristically (and with much detail skipped), if $SAD_{GMC} < SAD_{LMC}$, then use GMC to generate the predicted reference VOP. Otherwise, use LMC as before.

11.4.3 Synthetic Object Coding in MPEG-4

The number of objects in videos that are created by computer graphics and animation software is increasing. These are denoted as *synthetic objects* and can often be presented together with natural objects and scenes in games, TV ads and programs, and animation or feature films.

In this section, we briefly discuss *2D mesh-based* and *3D model-based* coding and animation methods for synthetic objects. Beek, Petajan, and Ostermann [22] provide a more detailed survey of this subject.

2D Mesh Object Coding

A *2D mesh* is a tessellation (or partition) of a 2D planar region using polygonal patches. The vertices of the polygons are referred to as *nodes* of the mesh. The most popular meshes are *triangular meshes*, where all polygons are triangles. The MPEG-4 standard makes use of two types of 2D mesh: *uniform mesh* and *Delaunay mesh* [23]. Both are triangular meshes that can be used to model natural video objects as well as synthetic animated objects.

Since the triangulation structure (the edges between nodes) is known and can be readily regenerated by the decoder, it is not coded explicitly in the bitstream. Hence, 2D mesh object coding is compact. All coordinate values of the mesh are coded in half-pixel precision.

Each 2D mesh is treated as a *mesh object plane* (MOP). Figure 11.21 illustrates the encoding process for 2D MOPs. Coding can be divided into *geometry coding* and *motion coding*. As shown, the input data is the x- and y- coordinates of all the nodes and the triangles (t_m) in the mesh. The output data is the displacements (dx_n, dy_n) and the prediction errors of the motion (ex_n, ey_n), both of which are explained below.

2D Mesh Geometry Coding. MPEG-4 allows four types of uniform meshes with different triangulation structures. Figure 11.22 shows such meshes with 4×5 mesh nodes. Each uniform mesh can be specified by five parameters: the first two specify the number of nodes in each row and column, respectively; the next two specify the horizontal and vertical size of each rectangle (containing two triangles) respectively; and the last specifies the type of the uniform mesh.

Uniform meshes are simple and are especially good for representing 2D rectangular objects (e.g., the entire video frame). When used for objects of arbitrary shape,

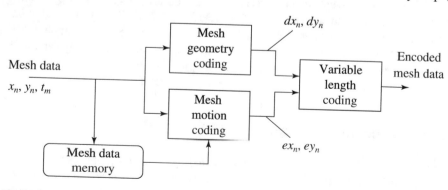

Fig. 11.21 2D mesh object plane (MOP) encoding process

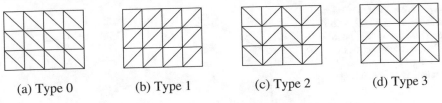

(a) Type 0   (b) Type 1   (c) Type 2   (d) Type 3

Fig. 11.22 Four types of uniform meshes: **a** type 0; **b** type 1; **c** type 2; **d** type3

they are applied to (overlaid on) the bounding boxes of the VOPs, which incurs some inefficiency.

A Delaunay mesh is a better object-based mesh representation for arbitrary-shaped 2D objects.

Definition 1 If $\mathcal{D}$ is a Delaunay triangulation, then any of its triangles $t_n = (P_i, P_j, P_k) \in \mathcal{D}$ satisfies the property that the circumcircle of t_n does not contain in its interior any other node point P_l.

A Delaunay mesh for a video object can be obtained in the following steps:

1. **Select boundary nodes of the mesh.** A polygon is used to approximate the boundary of the object. The polygon vertices are the *boundary nodes* of the Delaunay mesh. A possible heuristic is to select boundary points with high curvatures as boundary nodes.
2. **Choose interior nodes.** Feature points within the object's boundary such as edge points or corners can be chosen as interior nodes for the mesh.
3. **Perform Delaunay triangulation.** A *constrained Delaunay triangulation* is performed on the boundary and interior nodes, with the polygonal boundary used as a constraint. The triangulation will use line segments connecting consecutive boundary nodes as edges and form triangles only within the boundary.

Constrained Delaunay Triangulation. Interior edges are first added to form new triangles. The algorithm will examine each interior edge to make sure it is *locally Delaunay*. Given two triangles (P_i, P_j, P_k) and (P_j, P_k, P_l) sharing an edge $\overline{jk}$, if (P_i, P_j, P_k) contains P_l or (P_j, P_k, P_l) contains P_i in the interior of its circumcircle, then $\overline{jk}$ is not locally Delaunay and will be replaced by a new edge $\overline{il}$.

If P_l falls exactly on the circumcircle of (P_i, P_j, P_k) (and accordingly, P_i also falls exactly on the circumcircle of (P_j, P_k, P_l)), then $\overline{jk}$ will be viewed as locally Delaunay only if P_i or P_l has the largest x-coordinate among the four nodes.

Figure 11.23a, b show the set of Delaunay mesh nodes and the result of the constrained Delaunay triangulation. If the total number of nodes is N, and $N = N_b + N_i$ where N_b and N_i denote the number of boundary nodes and interior nodes, respectively, then the total number of triangles in the Delaunay mesh is $N_b + 2N_i - 2$. In the above figure, this sum is $8 + 2 \times 6 - 2 = 18$.

Fig. 11.23 Delaunay mesh:
a Boundary nodes (P_0 to P_7)
and interior nodes (P_8 to
P_{13}); **b** triangular mesh
obtained by constrained
Delaunay triangulation

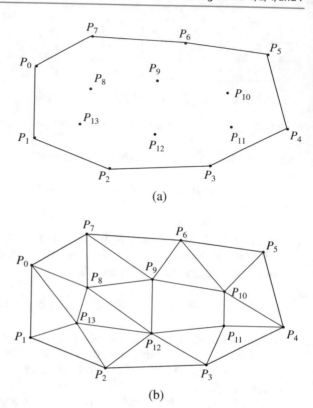

(a)

(b)

Unlike a uniform mesh, the node locations in a Delaunay mesh are irregular; hence, they must be coded. By convention in MPEG-4, the location (x_0, y_0) of the top left boundary node.[2] Is coded first, followed by the other boundary points counterclockwise (see Fig. 11.23a) or clockwise. Afterwards, the locations of the interior nodes are coded in any order.

Except for the first location (x_0, y_0), all subsequent coordinates are coded differentially—that is, for $n \geq 1$,

$$dx_n = x_n - x_{n-1}, \qquad dy_n = y_n - y_{n-1}, \qquad (11.8)$$

and afterwards, dx_n and dy_n are variable-length coded.

2D Mesh Motion Coding. The motion of each MOP triangle in either a uniform or Delaunay mesh is described by the motion vectors of its three vertex nodes. A new mesh structure can be created only in the intra-frame, and its triangular topology will not alter in the subsequent inter-frames. This enforces one-to-one mapping in 2D mesh motion estimation.

---

[2]The top left boundary node is defined as the one that has the minimum $x + y$ coordinate value. If more than one boundary node has the same $x + y$, the one with the minimum y is chosen.

For any MOP triangle (P_i, P_j, P_k), if the motion vectors for P_i and P_j are known to be $\mathbf{MV_i}$ and $\mathbf{MV_j}$, then a prediction $\mathbf{Pred_k}$ will be made for the motion vector of P_k, rounded to a half-pixel precision:

$$\mathbf{Pred_k} = 0.5 \cdot (\mathbf{MV_i} + \mathbf{MV_j}). \tag{11.9}$$

The prediction error $\mathbf{e_k}$ is coded as

$$\mathbf{e_k} = \mathbf{MV_k} - \mathbf{Pred_k}. \tag{11.10}$$

Once the three motion vectors of the first MOP triangle t_0 are coded, at least one neighboring MOP triangle will share an edge with t_0, and the motion vector for its third vertex node can be coded, and so on.

The estimation of motion vectors will start at the *initial triangle* t_0, which is the triangle that contains the top left boundary node and the boundary node next to it, clockwise. Motion vectors for all other nodes in the MOP are coded differentially, according to Eq. (11.10). A breadth-first order is established for traversing the MOP triangles in the 2D mesh motion coding process.

Figure 11.24 shows how a spanning tree can be generated to obtain the breadth-first order of the triangles. As shown, the initial triangle t_0 has two neighboring triangles t_1 and t_2, which are not visited yet. They become child nodes of t_0 in the spanning tree.

Triangles t_1 and t_2, in turn, have their unvisited neighboring triangles (and hence child nodes) t_3, t_4, and t_5, t_6, respectively. The traverse order so far is t_0, t_1, t_2, t_3, t_4, t_5, in a breadth-first fashion. One level down the spanning tree, t_3 has only one child node t_7, since the other neighbor t_1 is already visited; t_4 has only one child node t_8; and so on.

2D Object Animation. The above mesh motion coding established a one-to-one mapping between the mesh triangles in the reference MOP and the target MOP. It generated motion vectors for all node points in the 2D mesh. Mesh-based texture mapping is now used to generate the texture for the new animated surface by warping [20] the texture of each triangle in the reference MOP onto the corresponding triangle in the target MOP. This facilitates the animation of 2D synthetic video objects.

Fig. 11.24 A breadth-first order of MOP triangles for 2D mesh motion coding

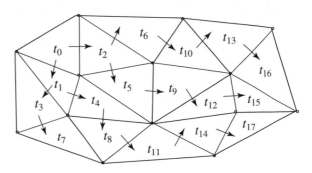

For triangular meshes, a common mapping function for the warping is the *affine transform*, since it maps a line to a line and can guarantee that a triangle is mapped to a triangle. It will be shown below that given the six vertices of the two matching triangles, the parameters for the affine transform can be obtained, so that the transform can be applied to all points within the target triangle for texture mapping.

Given a point $\mathbf{P} = (x, y)$ on a 2D plane, a *linear transform* can be specified, such that

$$[x'\ y'] = [x\ y] \begin{bmatrix} a_{11} & a_{12} \\ a_{21} & a_{22} \end{bmatrix} \tag{11.11}$$

A transform T is linear if $T(\alpha \mathbf{X} + \beta \mathbf{Y}) = \alpha T(\mathbf{X}) + \beta T(\mathbf{Y})$, where α and β are scalars. The above linear transform is suitable for geometric operations such as rotation and scaling but not for translation, since addition of a constant vector is not possible.

Definition 2 A transform A is an *affine transform* if and only if there exists a vector $\mathbf{C}$ and a linear transform T such that $A(\mathbf{X}) = T(\mathbf{X}) + \mathbf{C}$.

If the point (x, y) is represented as $[x, y, 1]$ in the homogeneous coordinate system commonly used in graphics [24], then an *affine transform* that transforms $[x, y, 1]$ to $[x', y', 1]$ is defined as

$$[x'\ y'\ 1] = [x\ y\ 1] \begin{bmatrix} a_{11} & a_{12} & 0 \\ a_{21} & a_{22} & 0 \\ a_{31} & a_{32} & 1 \end{bmatrix}. \tag{11.12}$$

It realizes the following mapping:

$$x' = a_{11}x + a_{21}y + a_{31} \tag{11.13}$$
$$y' = a_{12}x + a_{22}y + a_{32} \tag{11.14}$$

and has at most 6 degrees of freedom represented by the parameters $a_{11}, a_{21}, a_{31}, a_{12}, a_{22},$ and a_{32}.

The following 3×3 matrices are the affine transforms for translating by (T_x, T_y), rotating counterclockwise by θ, and scaling by factors S_x and S_y:

$$\begin{bmatrix} 1 & 0 & 0 \\ 0 & 1 & 0 \\ T_x & T_y & 1 \end{bmatrix}, \quad \begin{bmatrix} \cos\theta & \sin\theta & 0 \\ -\sin\theta & \cos\theta & 0 \\ 0 & 0 & 1 \end{bmatrix}, \quad \begin{bmatrix} S_x & 0 & 0 \\ 0 & S_y & 0 \\ 0 & 0 & 1 \end{bmatrix}$$

The following are the affine transforms for a shear along the x-axis and y-axis, respectively:

$$\begin{bmatrix} 1 & 0 & 0 \\ H_x & 1 & 0 \\ 0 & 0 & 1 \end{bmatrix}, \quad \begin{bmatrix} 1 & H_y & 0 \\ 0 & 1 & 0 \\ 0 & 0 & 1 \end{bmatrix}$$

where H_x and H_y are constants determining the degree of shear.

The above simple affine transforms can be combined (by matrix multiplications) to yield composite affine transforms—for example, for a translation followed by a rotation, or a shear followed by other transforms.

It can be proven (see Exercise 15) that any composite transform thus generated will have exactly the same matrix form and will have at most 6 degrees of freedom, specified by $a_{11}, a_{21}, a_{31}, a_{12}, a_{22}$, and a_{32}.

If the triangle in the target MOP is

$$(\mathbf{P}_0, \mathbf{P}_1, \mathbf{P}_2) = ((x_0, y_0), (x_1, y_1), (x_2, y_2))$$

and the matching triangle in the reference MOP is

$$(\mathbf{P}'_0, \mathbf{P}'_1, \mathbf{P}'_2) = ((x'_0, y'_0), (x'_1, y'_1), (x'_2, y'_2)),$$

then the mapping between the two triangles can be uniquely defined by the following:

$$\begin{bmatrix} x'_0 & y'_0 & 1 \\ x'_1 & y'_1 & 1 \\ x'_2 & y'_2 & 1 \end{bmatrix} = \begin{bmatrix} x_0 & y_0 & 1 \\ x_1 & y_1 & 1 \\ x_2 & y_2 & 1 \end{bmatrix} \begin{bmatrix} a_{11} & a_{12} & 0 \\ a_{21} & a_{22} & 0 \\ a_{31} & a_{32} & 1 \end{bmatrix} \qquad (11.15)$$

Equation (11.15) contains six linear equations (three for x's and three for y's) required to resolve the six unknown coefficients $a_{11}, a_{21}, a_{31}, a_{12}, a_{22}$, and a_{32}. Let Eq. (11.15) be stated as $\mathbf{X}' = \mathbf{X}A$. Then it is known that $A = \mathbf{X}^{-1}\mathbf{X}'$, with inverse matrix given by $\mathbf{X}^{-1} = adj(\mathbf{X})/det(\mathbf{X})$, where $adj(\mathbf{X})$ is the adjoint of $\mathbf{X}$ and $det(\mathbf{X})$ is the determinant. Therefore,

$$\begin{bmatrix} a_{11} & a_{12} & 0 \\ a_{21} & a_{22} & 0 \\ a_{31} & a_{32} & 1 \end{bmatrix} = \begin{bmatrix} x_0 & y_0 & 1 \\ x_1 & y_1 & 1 \\ x_2 & y_2 & 1 \end{bmatrix}^{-1} \begin{bmatrix} x'_0 & y'_0 & 1 \\ x'_1 & y'_1 & 1 \\ x'_2 & y'_2 & 1 \end{bmatrix}$$

$$= \frac{1}{det(\mathbf{X})} \begin{bmatrix} y_1 - y_2 & y_2 - y_0 & y_0 - y_1 \\ x_2 - x_1 & x_0 - x_2 & x_1 - x_0 \\ x_1 y_2 - x_2 y_1 & x_2 y_0 - x_0 y_2 & x_0 y_1 - x_1 y_0 \end{bmatrix} \begin{bmatrix} x'_0 & y'_0 & 1 \\ x'_1 & y'_1 & 1 \\ x'_2 & y'_2 & 1 \end{bmatrix} \qquad (11.16)$$

where $det(\mathbf{X}) = x_0(y_1 - y_2) - y_0(x_1 - x_2) + (x_1 y_2 - x_2 y_1)$.

Since the three vertices of the mesh triangle are never colinear points, it is ensured that $\mathbf{X}$ is not singular—that is, $det(\mathbf{X}) \neq 0$. Therefore Eq. (11.16) always has a unique solution.

The above affine transform is piecewise—that is, each triangle can have its own affine transform. It works well only when the object is mildly deformed during the animation sequence. Figure 11.25a shows a Delaunay mesh with a simple word mapped onto it. Figure 11.25b shows the warped word in a subsequent MOP in the animated sequence after an affine transform.

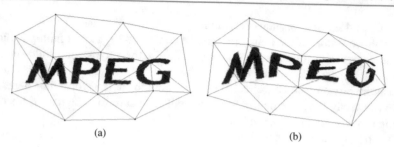

(a)   (b)

Fig. 11.25 Mesh-based texture mapping for 2D object animation

3D Model-Based Coding

Because of the frequent appearances of human faces and bodies in videos, MPEG-4 has defined special 3D models for *face objects* and *body objects*. Some of the potential applications for these new video objects include teleconferencing, human–computer interfaces, games, and e-commerce. In the past, 3D wireframe models and their animations were studied for 3D object animation [25]. MPEG-4 goes beyond wireframes, so that the surfaces of the face or body objects can be shaded or texture-mapped.

Face Object Coding and Animation. Face models for individual faces could either be created manually or generated automatically through computer vision and pattern recognition techniques. However, the former is cumbersome and nevertheless inadequate, and the latter has yet to be achieved reliably.

MPEG-4 has adopted a generic default face model, developed by the Virtual Reality Modeling Language (VRML) Consortium [26]. *Face Animation Parameters (FAPs)* can be specified to achieve desirable animations—deviations from the original "neutral" face. In addition, *Face Definition Parameters (FDPs)* can be specified to better describe individual faces. Figure 11.26 shows the feature points for FDPs. Feature points that can be affected by animation (FAPs) are shown as solid circles, and those that are not affected are shown as empty circles.

Sixty-eight FAPs are defined [22]: FAP 1 is for visemes and FAP 2 for facial expressions. Visemes code highly realistic lip motions by modeling the speaker's current mouth position. All other FAPs are for possible movements of head, jaw, lip, eyelid, eyeball, eyebrow, pupil, chin, cheek, tongue, nose, ear, and so on.

For example, expressions include *neutral, joy, sadness, anger, fear, disgust, and surprise*. Each is expressed by a set of features—sadness, for example, by slightly closed eyes, relaxed mouth, and upward-bent inner eyebrows. FAPs for movement include *head_pitch, head_yaw, head_roll, open_jaw, thrust_jaw, shift_jaw, push_bottom_lip, push_top_lip*, and so on.

For compression, the FAPs are coded using predictive coding. Predictions for FAPs in the target frame are made based on FAPs in the previous frame, and prediction errors are then coded using arithmetic coding. DCT can also be employed to improve the compression ratio, although it is considered more computationally expensive. FAPs are also quantized, with different quantization step sizes employed to explore

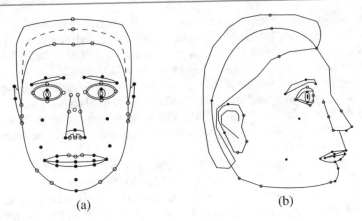

(a)   (b)

Fig. 11.26 Feature points for face definition parameters (FDPs). (Feature points for teeth and tongue are not shown.)

the fact that certain FAPs (e.g., open_jaw) need less precision than others (e.g., push_top_lip).

Body Object Coding and Animation. MPEG-4 Version 2 introduced *body objects*, which are a natural extension to face objects.

Working with the Humanoid Animation (H-Anim) Group in the VRML Consortium, MPEG adopted a generic virtual human body with default posture. The default is standing, with feet pointing to the front, arms at the sides, with palms facing inward. There are 296 *Body Animation Parameters* (BAPs). When applied to any MPEG-4-compliant generic body, they will produce the same animation.

A large number of BAPs describe joint angles connecting different body parts, including the spine, shoulder, clavicle, elbow, wrist, finger, hip, knee, ankle, and toe. This yields 186 degrees of freedom to the body, 25 to each hand alone. Furthermore, some body movements can be specified in multiple levels of detail. For example, five different levels, supporting 9, 24, 42, 60, and 72 degrees of freedom can be used for the spine, depending on the complexity of the animation.

For specific bodies, *Body Definition Parameters* (BDPs) can be specified for body dimensions, body surface geometry, and, optionally, texture. Body surface geometry uses a *3D polygon mesh* representation, consisting of a set of polygonal planar surfaces in 3D space [24]. The 3D mesh representation is popular in computer graphics for surface modeling. Coupled with texture mapping, it can deliver good (photorealistic) renderings.

The coding of BAPs is similar to that of FAPs: quantization and predictive coding are used, and the prediction errors are further compressed by arithmetic coding.

11.4.4 MPEG-4 Parts, Profiles, and Levels

So far, MPEG-4 has over 28 Parts [9], and more are still being developed. It not only specified Visual in Part 2 and Audio in Part 3, but also specialized subjects such as Graphics, Animation, Music, Scene description, Object descriptor, Delivery Multimedia Integration Framework (DMIF), Streaming, and Intellectual Property Management and Protection (IPMP) in various Parts. MPEG-4 Part 10 is about Advanced Video Coding (AVC) which is identical to ITU-T H.264 AVC.

MPEG-4 Part 2 defined more than 20 visual Profiles, like Simple, Advanced Simple, Core, Main, Simple Studio, etc. The commonly used ones are *Simple Profile (SP)* and *Advanced Simple Profile (ASP)*. The latter is adopted by several popular video coding software such as DivX, Nero Digital, and Quicktime 6. The open-source software Xvid supports both SP and ASP.

To target various applications, MPEG-4 Part 2 also defined multiple Levels in each profile, e.g., L0 to L3 for SP and L0 to L5 for ASP. In general, the lower Levels in these profiles support low-bitrate video formats (CIF, QCIF) and applications such as videoconferencing on the web, whereas the higher Levels support higher quality videos.

11.5   MPEG-7

As more and more multimedia content becomes an integral part of various applications, effective and efficient retrieval becomes a primary concern. In October 1996, the MPEG group therefore took on the development of another major standard, MPEG-7, following on MPEG-1, 2, and 4.

One common ground between MPEG-4 and MPEG-7 is the focus on audiovisual *objects*. The main objective of MPEG-7 [27–29] is to serve the need of audiovisual content-based retrieval (or audiovisual object retrieval) in applications such as digital libraries. Nevertheless, it is certainly not limited to retrieval—it is applicable to any multimedia applications involving the generation (*content creation*) and usage (*content consumption*) of multimedia data. Unlike MPEG-1, 2 and 4, it is not "yet another" standard for video coding.

MPEG-7 became an international standard in September 2001. Its formal name is *Multimedia Content Description Interface*, documented in ISO/IEC 15938 [30]. The standard's first seven Parts are Systems, Description Definition Language, Visual, Audio, Multimedia Description Schemes, Reference Software, and Conformance and Testing. Since 2002, there are further developments in Parts 8–13, mostly focusing on various profiles and query formats. For example, Part 13 released in 2015 was about Compact Descriptors for Visual Search.

MPEG-7 supports a variety of multimedia applications. Its data may include still pictures, graphics, 3D models, audio, speech, video, and composition information (how to combine these elements). These MPEG-7 data elements can be represented in textual or binary format, or both. Part 1 (Systems) specifies the syntax of *Binary format for MPEG-7* (BiM) data. Part 2 (Description Definition Language) specifies the syntax of the textual format which adopts XML Schema as its language of choice. A bidirectional lossless mapping is defined between the textual and binary representations.

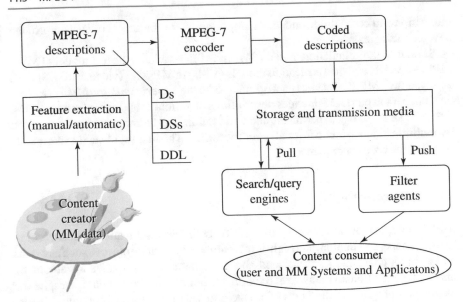

Fig. 11.27 Possible applications using MPEG-7

Figure 11.27 illustrates some possible applications that will benefit from MPEG-7. As shown, features are extracted and used to instantiate MPEG-7 *descriptions*. They are then coded by the MPEG-7 encoder and sent to the *storage and transmission media*. Various search and query engines issue search and browsing requests, which constitute the *pull* activities of the Internet, whereas the agents filter out numerous materials *pushed* onto the *terminal*—users and/or computer systems and applications that consume the data.

For multimedia content description, MPEG-7 has developed *Descriptors* (D), *Description Schemes* (DS), and a *Description Definition Language* (DDL). Following are some of the important terms:

- **Feature**. A characteristic of the data
- **Descriptor (D)**. A definition (syntax and semantics) of the feature
- **Description Scheme (DS)**. Specification of the structure and relationship between Ds and DSs (see Salembier and Smith [31])
- **Description**. A set of instantiated Ds and DSs that describes the structural and conceptual information of the content, storage and usage of the content, and so on
- **Description Definition Language (DDL)**. Syntactic rules to express and combine DSs and Ds (see Hunter and Nack [32]).

It is made clear [30] that the scope of MPEG-7 is to standardize the Ds, DSs, and DDL for descriptions. The mechanism and process of producing and consuming the descriptions are beyond the scope of MPEG-7. These are left open for industry innovation and competition and, more importantly, for the arrival of ever-improving new technologies.

Similar to the simulation model (SM) in MPEG-1 video, the test model (TM) in MPEG-2 video, and the verification models (VMs) in MPEG-4 (video, audio, SNHC, and systems), MPEG-7 names its working model the *Experimentation Model (XM)*— an alphabetical pun! XM provides descriptions of various tools for evaluating the Ds, DSs, and DDL, so that experiments and verifications can be conducted and compared by multiple independent parties all over the world. The first set of such experiments is called the *core experiments*.

11.5.1 Descriptor (D)

MPEG-7 descriptors are designed to describe both low-level features, such as color, texture, shape, and motion, and high-level features of semantic objects, such as events and abstract concepts. As mentioned above, methods and processes for automatic and even semiautomatic feature extraction are not part of the standard. Despite the efforts and progress in the fields of image and video processing, computer vision, and pattern recognition, automatic and reliable feature extraction is not expected in the near future, especially at the high level.

The descriptors are chosen based on a comparison of their performance, efficiency, and size. Low-level visual descriptors for basic visual features [33] include

- **Color**
  - **Color space**. (a) RGB, (b) YCbCr, (c) HSV (hue, saturation, value) [24], (d) HMMD (HueMaxMinDiff) [34], (e) 3D color space derivable by a 3×3 matrix from RGB, (f) monochrome.
  - **Color quantization**. (a) Linear, (b) nonlinear, (c) lookup tables.
  - **Dominant colors**. A small number of representative colors in each region or image. These are useful for image retrieval based on color similarity.
  - **Scalable color**. A color histogram in HSV color space. It is encoded by a Haar transform and hence is scalable.
  - **Color layout**. Spatial distribution of colors for color-layout-based retrieval.
  - **Color structure**. The frequency of a *color structuring element* describes both the color content and its structure in the image. The color structure element is composed of several image samples in a local neighborhood that have the same color.
  - **Group of Frames/Group of Pictures (GoF/GoP) color**. Similar to the scalable color, except this is applied to a video segment or a group of still images. An aggregated color histogram is obtained by the application of *average*, *median*, or *intersection* operations to the respective bins of all color histograms in the GoF/GoP and is then sent to the Haar transform.
- **Texture**
  - **Homogeneous texture**. Uses orientation and scale-tuned Gabor filters [35] that quantitatively represent regions of homogeneous texture. The advantage

of Gabor filters is that they provide simultaneous optimal resolution in both space and spatial frequency domains [36]. Also, they are bandpass filters that conform to the human visual profile. A filter bank consisting of 30 Gabor filters, at five different scales and six different directions for each scale, is used to extract the texture descriptor.

- **Texture browsing**. Describes the *regularity*, *coarseness*, and *directionality* of edges used to represent and browse homogeneous textures. Again, Gabor filters are used.

- **Edge histogram**. Represents the spatial distribution of four directional (0°, 45°, 90°, 135°) edges and one nondirectional edge. Images are divided into small subimages, and an edge histogram with five bins is generated for each subimage.

- **Shape**
  - **Region-based shape**. A set of *Angular Radial Transform* (ART) [28] coefficients is used to describe an object's shape. An object can consist of one or more regions, with possibly some holes in the object. ART transform is a 2D complex transform defined in terms of polar coordinates on a unit disk. ART basis functions are separable along the angular and radial dimensions. Thirty-six basis functions, 12 angular and three radial, are used to extract the shape descriptor.
  - **Contour-based shape**. Uses a *curvature scale space* (CSS) representation [37] that is invariant to scale and rotation, and robust to nonrigid motion and partial occlusion of the shape.
  - **3D shape**. Describes 3D mesh models and *shape index* [38]. The histogram of the shape indices over the entire mesh is used as the descriptor.

- **Motion**
  - **Camera motion**. Fixed, pan, tilt, roll, dolly, track, and boom. (See Fig. 11.28 and [39].)
  - **Object motion trajectory**. A list of keypoints (x, y, z, t). Optional interpolation functions are used to specify the acceleration along the path. (See [39].)
  - **Parametric object motion**. The basic model is the 2D affine model for translation, rotation, scaling, shear, and the combination of these. A planar perspective model and quadratic model can be used for perspective distortion and more complex movements.
  - **Motion activity**. Provides descriptions such as the intensity, pace, mood, and so on, of the video—for example, "scoring in a hockey game" or "interviewing a person."

- **Localization**
  - **Region locator**. Specifies the localization of regions in images with a box or a polygon.
  - **Spatiotemporal locator**. Describes spatiotemporal regions in video sequences. Uses one or more sets of descriptors of regions and their motions.

- **Others**
  - **Face recognition**. A normalized face image is represented as a 1D vector, then projected onto a set of 49 basis vectors, representing all possible face vectors.

Fig. 11.28 Camera motions:
pan, tilt, roll, dolly, track,
and boom. (Camera has an
effective focal length of f. It
is shown initially at the
origin, pointing to the
direction of z-axis.)

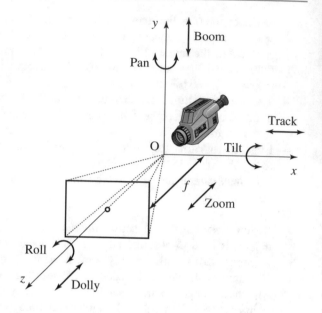

11.5.2 Description Scheme (DS)

This section provides a brief overview of MPEG-7 Description Schemes (DSs) in
the areas of Basic elements, Content management, Content description, Navigation
and access, Content organization, and User interaction.

- **Basic elements**
    - **Datatypes and mathematical structures**. Vectors, matrices, histograms, and
      so on.
    - **Constructs**. Links media files and localizing segments, regions, and so on.
    - **Schema tools**. Includes root elements (starting elements of MPEG-7 XML
      documents and descriptions), top-level elements (organizing DSs for specific
      content-oriented descriptions), and package tools (grouping related DS com-
      ponents of a description into packages).
- **Content Management**
    - **Media Description**. Involves a single DS, the *MediaInformation* DS, com-
      posed of a *MediaIdentification* D and one or more *Media Profile* Ds that
      contain information such as coding method, transcoding hints, storage and
      delivery formats, and so on.
    - **Creation and Production Description**. Includes information about creation
      (title, creators, creation location, date, etc.), classification (genre, language,
      parental guidance, etc.), and related materials.
    - **Content Usage Description**. Various DSs to provide information about usage
      rights, usage record, availability, and finance (cost of production and income
      from content use).

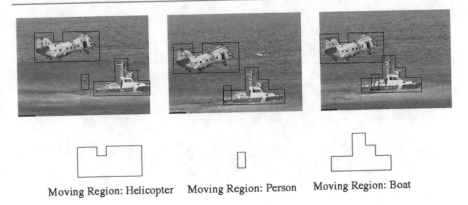

Moving Region: Helicopter   Moving Region: Person   Moving Region: Boat

Fig. 11.29 MPEG-7 video segment

- **Content Description**
  - **Structural Description**. A *Segment* DS describes structural aspects of the content. A *segment* is a section of an audiovisual object. The relationship among segments is often represented as a *segment tree*. When the relationship is not purely hierarchical, a *segment graph* is used.

    The *Segment* DS can be implemented as a class object. It has five subclasses: *Audiovisual segment DS*, *Audio segment DS*, *Still region DS*, *Moving region DS*, and *Video segment DS*. The subclass DSs can recursively have their own subclasses.

    A Still region DS, for example, can be used to describe an image in terms of its creation (title, creator, date), usage (copyright), media (file format), textual annotation, color histogram, and possibly texture descriptors, and so on. The initial region (image, in this case) can be further decomposed into several regions, which can in turn have their own DSs.

    Figure 11.29 shows a Video segment for a marine rescue mission, in which a person was lowered onto a boat from a helicopter. Three moving regions are inside the Video segment. A segment graph can be constructed to include such structural descriptions as composition of the video frame (helicopter, person, and boat) spatial relationship and motion (above, on, close-to, move-toward, etc.) of the regions.
  - **Conceptual Description**. This involves higher level (nonstructural) description of the content, such as *Event* DS for basketball game or Lakers ballgame, *Object* DS for John or person, *State* DS for semantic properties at a given time or location, and *Concept* DS for abstract notations such as "freedom" or "mystery." As for Segment DSs, the concept DSs can also be organized in a tree or graph.
- **Navigation and access**
  - **Summaries**. These provide a video summary for quick browsing and navigation of the content, usually by presenting only the keyframes. The following DSs are supported: *Summarization* DS, *HierarchicalSummary* DS, *HighlightLevel* DS, *SequentialSummary* DS. Hierarchical summaries provide

Fig. 11.30 A video summary

a keyframe hierarchy of multiple levels, whereas sequential summaries often provide a slide show or audiovisual skim, possibly with synchronized audio and text.

Figure 11.30 illustrates a summary for a video of a "dragon-boat" parade and race in a park. The summary is organized in a three-level hierarchy. Each video segment at each level is depicted by a keyframe of thumbnail size.

- **Partitions and Decompositions**. This refers to *view* partitions and decompositions. The *View partitions* (specified by *View* DSs) describe different space and frequency views of the audiovisual data, such as a spatial view (this could be a spatial segment of an image), temporal view (as in a temporal segment of a video), frequency view (as in a wavelet subband of an image), or resolution view (as in a thumbnail image), and so on. The *View decompositions* DSs specify different tree or graph decompositions for organizing the views of the audiovisual data, such as a SpaceTree DS (a quad-tree image decomposition).
- **Variations of the Content**. A *Variation* DS specifies a variation from the original data in image resolution, frame rate, color reduction, compression, and so on. It can be used by servers to adapt audiovisual data delivery to network and terminal characteristics for a given Quality of Service (QoS).
- **Content Organization**
  - **Collections**. The *CollectionStructure* DS groups audiovisual contents into clusters. It specifies common properties of the cluster elements and relationships among the clusters.
  - **Models**. *Model* DSs include a Probability model DS, Analytic model DS, and Classifier DS that extract the models and statistics of the attributes and features of the collections.
- **User Interaction**
  - **UserPreference**. DSs describe user preferences in the consumption of audiovisual contents, such as content types, browsing modes, privacy characteristics, and whether preferences can be altered by an agent that analyzes user behavior.

11.5.3 Description Definition Language (DDL)

MPEG-7 adopted the XML Schema Language initially developed by the WWW Consortium (W3C) as its Description Definition Language (DDL). Since XML Schema Language was not designed specifically for audiovisual contents, some extensions are made to it. Without the details, the MPEG-7 DDL has the following components:

- **XML Schema structure components**
  - The Schema—the wrapper around definitions and declarations
  - Primary structural components, such as simple and complex type definitions, and attribute and element declarations
  - Secondary structural components, such as attribute group definitions, identity-constraint definitions, group definitions, and notation declarations
  - "Helper" components, such as annotations, particles, and wildcards.
- **XML Schema datatype components**
  - Primitive and derived data types
  - Mechanisms for the user to derive new data types
  - Type checking better than XML 1.0.
- **MPEG-7 Extensions**
  - Array and matrix data types
  - Multiple media types, including audio, video, and audiovisual presentations
  - Enumerated data types for `MimeType`, `CountryCode`, `RegionCode`, `CurrencyCode`, and `CharacterSetCode`
  - *Intellectual Property Management and Protection* (IPMP) for Ds and DSs.

11.6   Exercises

1. As we know, MPEG video compression uses I-, P-, and B-frames. However, the earlier H.261 standard does not use B-frames. Describe a situation in which video compression would not be as effective without B-frames. (Your answer should be different from the one in Fig. 11.1.)
2. The MPEG-1 standard introduced B-frames, and the motion-vector search range has accordingly been increased from $[-15, 15]$ in H.261 to $[-512, 511.5]$. Why was this necessary? Calculate the number of B-frames between consecutive P-frames that would justify this increase.
3. B-frames provide obvious coding advantages, such as increase in SNR at low bitrates and bandwidth savings. What are some of the disadvantages of B-frames?
4. Redraw Fig. 11.8 of the MPEG-2 two-layer SNR scalability encoder and decoder to include a second enhancement layer.
5. Draw block diagrams for an MPEG-2 encoder and decoder for (a) SNR and spatial hybrid scalability, (b) SNR and temporal hybrid scalability.
6. Why are B-frames not used as reference frames for motion compensation? Suppose there is a mode where any frame type can be specified as a reference frame.

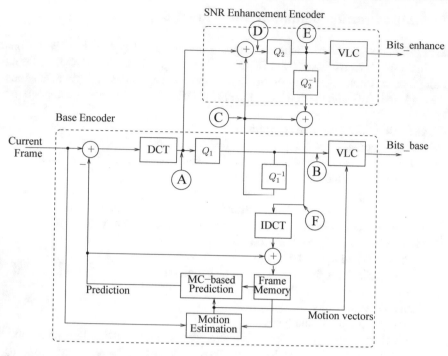

Fig. 11.31 MPEG-2 encoder for SNR scalability

Discuss the trade-offs of using reference B-frames instead of P-frames in a video
sequence (i.e., eliminating P-frames completely).

7. Scalability is an important issue in video compression.

(a) Describe how *SNR Scalability* in MPEG-2 works.
(b) In Fig. 11.31, six checkpoints (i.e., A, B, C, D, E, and F) are indicated.
    Assume the block size for transform coding is 4×4, and quantization
    is done by $F'(u, v) = round(\frac{F(u,v)}{Q(u,v)})$. The quantization matrices are

$$Q_1 = \begin{pmatrix} 10\ 10\ 10\ 10 \\ 10\ 10\ 10\ 10 \\ 10\ 10\ 10\ 10 \\ 10\ 10\ 10\ 10 \end{pmatrix}, \quad and \quad Q_2 = \begin{pmatrix} 2\ 2\ 2\ 2 \\ 2\ 2\ 2\ 2 \\ 2\ 2\ 2\ 2 \\ 2\ 2\ 2\ 2 \end{pmatrix}.$$

If we use **A** to indicate the 4×4 block at checkpoint A, and $\mathbf{A} =$

$$\begin{pmatrix} 3 & 25 & -12 & 0 \\ 7 & 0 & 0 & 0 \\ 0 & 0 & 0 & 0 \\ 0 & 0 & 0 & 0 \end{pmatrix},$$

show all the numbers observed at the other 5 checkpoints, i.e., **B, C, D,
E**, and **F**.

8. Write a program to implement the SNR scalability in MPEG-2. Your program should be able to work on any macroblock using any quantization *step_sizes* and should output both `Bits_base` and `Bits_enhance` bitstreams. The variable-length coding step can be omitted.

9. MPEG-4 motion compensation is supposed to be VOP-based. In the end, the VOP is still divided into macroblocks (interior macroblock, boundary macroblock, etc.) for motion compensation.

   (a) What are the potential problems of the current implementation? How can they be improved?
   (b) Can there be true VOP-based motion compensation? How would it compare to the current implementation?

10. MPEG-1, 2, and 4 are all known as decoder standards. The compression algorithms, hence the details of the encoder, are left open for future improvement and development. For MPEG-4, the major issue of *video object segmentation*—how to obtain the VOPs—is left unspecified.

    (a) Propose some of your own approaches to video object segmentation.
    (b) What are the potential problems of your approach?

11. Why was padding introduced in MPEG-4 VOP-based coding? Name some potential problems of padding.

12. Motion vectors can have subpixel precision. In particular, MPEG-4 allows quarter-pixel precision in the luminance VOPs. Describe an algorithm that will realize this precision.

13. As a programming project, compute the SA-DCT for the following 8×8 block:

$$
\begin{array}{cccccccc}
0 & 0 & 0 & 0 & 16 & 0 & 0 & 0 \\
4 & 0 & 8 & 16 & 32 & 16 & 8 & 0 \\
4 & 0 & 16 & 32 & 64 & 32 & 16 & 0 \\
0 & 0 & 32 & 64 & 128 & 64 & 32 & 0 \\
4 & 0 & 0 & 32 & 64 & 32 & 0 & 0 \\
0 & 16 & 0 & 0 & 32 & 0 & 0 & 0 \\
0 & 0 & 0 & 0 & 16 & 0 & 0 & 0 \\
0 & 0 & 0 & 0 & 0 & 0 & 0 & 0
\end{array}
$$

14. What is the computational cost of SA-DCT, compared to ordinary DCT? Assume the video object is a 4×4 square in the middle of an 8×8 block.

15. Affine transforms can be combined to yield a composite affine transform. Prove that the composite transform will have exactly the same form of matrix (with $[0\ 0\ 1]^T$ as the last column) and at most 6 degrees of freedom, specified by the parameters $a_{11}, a_{21}, a_{31}, a_{12}, a_{22}$, and a_{32}.

16. Mesh-based motion coding works relatively well for 2D animation and face animation. What are the main problems when it is applied to body animation?

17. What is the major motivation behind the development of MPEG-7? Give three examples of real-world applications that may benefit from MPEG-7.
18. Two of the main shape descriptors in MPEG-7 are "region-based" and "contour-based." There are, of course, numerous ways of describing the shape of regions and contours.

   (a) What would be your favorite shape descriptor?
   (b) How would it compare to ART and CSS in MPEG-7?

References

1. L. Chiariglione, The development of an integrated audiovisual coding standard: MPEG. Proc. IEEE **83**, 151–157 (1995)
2. D.J. Le Gall, MPEG: a video compression standard for multimedia applications. Commun. ACM **34**(4), 46–58 (1991)
3. R. Schafer, T. Sikora, Digital video coding standards and their role in video communications. Proc. IEEE **83**(6), 907–924 (1995)
4. *Information technology– Coding of moving pictures and associated audio for digital storage media at up to about 1.5 Mbit/s*. Int. Stand.: ISO/IEC 11172, Parts 1–5 (1992)
5. J.L. Mitchell, W.B. Pennebaker, C.E. Fogg, D.J. LeGall, *MPEG Video Compression Standard*. (Chapman & Hall, New York, 1996)
6. B.G. Haskell, A. Puri, A. Netravali, *Digital Video: an Introduction to MPEG-2*. (Chapman & Hall, New York, 1996)
7. *Information technology—Generic coding of moving pictures and associated audio information*. Int. Stand.: ISO/IEC 13818, Parts 1–11 (2004)
8. T. Sikora, The MPEG-4 video standard verification model. IEEE Trans. Circuits Syst. Video Technol. (Special issue on MPEG-4)
9. *Information technology—Generic coding of audio-visual objects*. International Standard: ISO/IEC 14496, Parts 1–28 (2012)
10. A. Puri, T. Chen (eds.), *Multimedia Systems, Standards, and Networks*. (Marcel Dekker, New York, 2000)
11. G. Fernando et al., Java in MPEG-4 (MPEG-J), in *Multimedia, Systems, Standards, and Networks*, ed. by A. Puri, T. Chen (Marcel Dekker, New York, 2000), pp. 449–460
12. *Video Coding for Low Bit Rate Communication*. ITU-T Recommendation H.263, Version 1, 1995, Version 2, 1998, Version 3, 2000 (2005)
13. A. Puri et al., MPEG-4 natural video coding—Part I, in *Multimedia, Systems, Standards, and Networks*, ed. by A. Puri, T. Chen (Marcel Dekker, New York, 2000), pp. 205–244
14. T. Ebrahimi, F. Dufaux, Y. Nakaya, MPEG-4 natural video coding—Part II, in *Multimedia, Systems, Standards, and Networks*, ed. by A. Puri, T. Chen (Marcel Dekker, New York, 2000), pp. 245–269
15. P. Kauff, et al., Functional coding of video using a shape-adaptive DCT algorithm and an object-based motion prediction toolbox. IEEE Trans. Circuits Syst. Video Technol. (Special issue on MPEG-4)
16. *Standardization of Group 3 facsimile apparatus for document transmission*. ITU-T Recommendation T.4 (1980)
17. *Facsimile coding schemes and coding control functions for Group 4 facsimile apparatus*. ITU-T Recommendation T.6 (1984)

18. *Information technology—Coded representation of picture and audio information—progressive bi-Level image compression.* International Standard: ISO/IEC 11544, also ITU-T Recommendation T.82 (1992)

19. J.M. Shapiro, Embedded image coding using zerotrees of wavelet coefficients. IEEE Trans. Signal Process. **41**(12), 3445–3462 (1993)

20. G. Wolberg, *Digital Image Warping.* (Computer Society Press, Los Alamitos, CA, 1990)

21. M.C. Lee, et al., A layered video object coding system using sprite and affine motion model. IEEE Trans. Circuits Syst. Video Technol. **7**(1), 130–145 (1997)

22. P. van Beek, MPEG-4 synthetic video, in *Multimedia, Systems, Standards, and Networks*, ed. by A. Puri, T. Chen, pp. 299–330. (Marcel Dekker, New York, 2000)

23. A.M. Tekalp, P. van Beek, C. Toklu, B. Gunsel, 2D mesh-based visual object representation for interactive synthetic/natural digital video. Proc. IEEE **86**, 1029–1051 (1998)

24. J.F. Hughes, A. van Dam, M. McGuire, D.F. Sklar, J.D. Foley, S.K. Feiner, K. Akeley, *Computer Graphics: Principles and Practice*, 3rd edn. (Addison-Wesley, 2013)

25. A. Watt, M. Watt, *Advanced Animation and Rendering Techniques.* (Addison-Wesley, 1992)

26. *Information technology—The Virtual Reality Modeling Language—Part 1: functional specification and UTF-8 encoding.* International Standard: ISO/IEC 14772-1 (1997)

27. S.F. Chang, T. Sikora, A. Puri, Overview of the MPEG-7 standard. *IEEE Trans. Circuits Syst Video Technol. (Special issue on MPEG-7)*, **11**(6), 688–695 (2001)

28. B.S. Manjunath, P. Salembier, T. Sikora (eds.), *Introduction to MPEG-7: Multimedia Content Description Interface.* (Wiley, 2002)

29. H.G. Kim, N. Moreau, T. Sikora, *MPEG-7 Audio and Beyond: Audio Content Indexing and Retrieval.* (Wiley, 2005)

30. *Information technology—Multimedia content description interface.* International Standard: ISO/IEC 15938, Parts 1–13 (2002–2015)

31. P. Salembier, J. R. Smith, MPEG-7 multimedia description schemes. IEEE Trans. Circuits Syst. Video Technol, **11**(6), 748–759 (2001)

32. J. Hunter, F. Nack, An overview of the MPEG-7 description definition language (DDL) proposals. Signal Process.: Image Commun. **16**(1–2), 271–293 (2001)

33. T. Sikora, The MPEG-7 visual standard for content description— An overview. IEEE Trans. Circuits Syst. Video Technol. (Special issue on MPEG-7). **11**(6), 696–702 (2001)

34. B.S. Manjunath, J.-R. Ohm, V.V. Vasudevan, A. Yamada, Color and texture descriptors. IEEE Trans. Cicuits Sys. Video Tech. **11**, 703–715 (2001)

35. B.S. Manjunath, G.M. Haley, W.Y. Ma. Multiband techniques for texture classification and segmentation, *Handbook of Image and Video Processing*, in ed. by A. Bovik, pp. 367–381. (Academic Press, 2000)

36. T.P. Weldon, W.E. Higgins, D.F. Dunn, Efficient Gabor filter design for texture segmentation. Pattern Recogn. (1996)

37. F. Mokhtarian, A.K. Mackworth, A theory of multiscale, curvature-based shape representation for planar curves. IEEE Trans Pattern Anal. Mach. Intell. **14**(8), 789–805 (1992)

38. J.J. Koenderink, A.J. van Doorn, Surface shape and curvature scales. Image Vis. Comput. **10**, 557–565 (1992)

39. S. Jeannin et al., Motion descriptor for content-based video representation. Signal Process.: Image Commun. **16**(1–2), 59–85 (2000)

Modern Video Coding Standards: H.264, H.265, and H.266

12

12.1 Overview

The video coding standards H.264, H.265, and H.266 were jointly developed by ISO/IEC MPEG and ITU-T VCEG (Video Coding Experts Group). The main objective of H.264 (aka. AVC) and H.265 (aka. HEVC) was to improve the coding efficiency so as to deal with the ever-increasing amount of video data due to the increases in both video resolution and frame rate, initially with the introduction of high definition (HD) and later ultra-high definition (UHD) video/TV. Lately, H.266 (aka. VVC) was developed to further address the above issues, as well as emerging applications of high dynamic range (HDR) video, 360° video, and others (e.g., panoramic video). In this chapter, we will study the principle techniques essential for the superior performance of these modern video coding standards.

The last decade also witnessed the success of several open and royalty-free coding standards. For example, Google developed VP8 and then VP9 which was first employed in the compression of YouTube videos. Since 2018 its successor AV1 developed by the Alliance for Open Media (AOMedia) has been used by Netflix, initially on Android mobile apps, and by several other streaming video servers over the Internet. Another example is the development of Audio and Video Coding Standard (AVS) in China with its independent intellectual property rights, currently at Phase 2 of AVS3. By and large, these standards share the same video coding principles as H.264, H.265, and H.266.

12.2 H.264

Joint Video Team (JVT) of ISO/IEC MPEG and ITU-T VCEG (Video Coding Experts Group) developed the H.264 video compression standard. It was formerly known by its working title "H.26L." Version 1 of H.264 was approved in May 2003

© Springer Nature Switzerland AG 2021
Z.-N. Li et al., *Fundamentals of Multimedia*, Texts in Computer Science,
https://doi.org/10.1007/978-3-030-62124-7_12

[1]. H.264 is also known as MPEG-4 Part 10, Advanced Video Coding (AVC) [2–4]. It is often referred to as the H.264/AVC (or H.264/MPEG-4 AVC) video coding standard.

H.264/AVC provides a higher video coding efficiency, up to 50% better compression than MPEG-2 and up to 30% better than H.263+ and MPEG-4 Advanced Simple Profile, while maintaining the same quality of the compressed video. It covers a broad range of applications, from high bitrate to very low bitrate. The vastly improved H.264 core features, together with new coding tools offer a significant improvement in compression ratio, error resiliency, and subjective quality over previous ITU-T and MPEG standards. It has since become the default standard for various applications, e.g., the Blu-ray disks, HDTV broadcasts, streaming video on the Internet, web software such as Flash and Silverlight, and apps on mobile and portable devices.

Similar to previous video compression standards, H.264 specifies a block-based hybrid coding scheme that supports a combination of inter-picture motion predictions and intra-picture spatial prediction, and transform coding on prediction residual errors. Again, each picture can be separated into macroblocks (16×16 blocks), and arbitrary-sized slices can group multiple macroblocks into self-contained units. The basic H.264/AVC encoder is shown in Fig. 12.1.

Main Features of H.264/AVC are

- Integer transform in 4×4 blocks. Low complexity, no drifting.
- Variable block-size motion compensation, from 16×16 to 4×4 in luma images.
- Quarter-pixel accuracy in motion vectors, accomplished by interpolations.

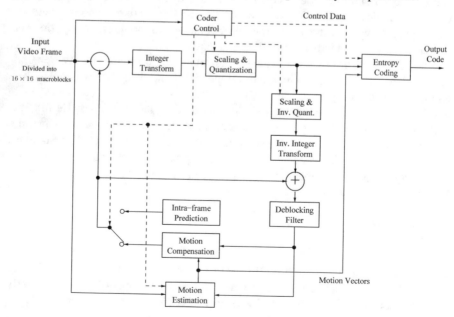

Fig. 12.1 Basic encoder for H.264/AVC

- Multiple reference picture motion compensation. More than just P or B frames for motion estimation.
- Directional spatial prediction for intra-frames.
- In-loop deblocking filtering.
- Context-Adaptive Variable Length Coding (CAVLC) and Context-Adaptive Binary Arithmetic Coding (CABAC).
- More robust to data errors and data losses, more flexible in synchronization and switching of video streams produced by different decoders.

The decoder has the following five major blocks:

- Entropy decoding
- Inverse quantization and transform of residual pixels
- Motion compensation or intra prediction
- Reconstruction
- In-loop deblocking filter on reconstructed pixels.

12.2.1 Motion Compensation

Variable Block-Size Motion Compensation

As before, Inter-frame motion estimation in H.264 is also block-based. By default, the macroblocks are of the size 16×16.

A macroblock can also be divided into four 8×8 *partitions*. While conducting motion estimation, each macroblock or each of the partitions can be further segmented into smaller partitions as shown in Fig. 12.2. The top four options are from the 16×16 macroblock (the so-called M Types), and the bottom four options are from each of the 8×8 partitions (the so-called 8×8 Types).

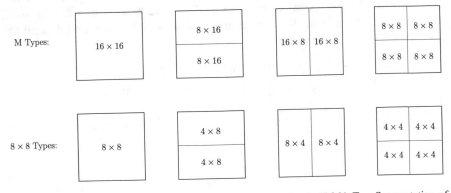

Fig. 12.2 Segmentation of the macroblock for motion estimation in H.264. Top: Segmentation of the macroblock. Bottom: Segmentation of the 8×8 partition

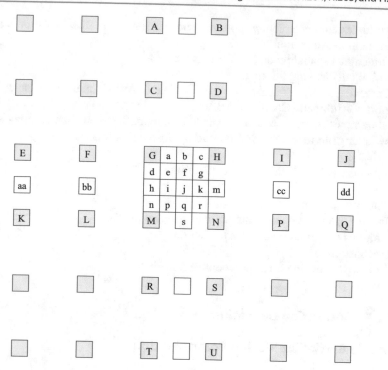

Fig. 12.3 Interpolation for fractional samples in H.264. Upper-case letters indicate pixels on the image grid. Lower-case letters indicate pixels at half-pixel and quarter-pixel positions

Quarter-Pixel Precision

The accuracy of motion compensation is of quarter-pixel precision in luma images. Figure 12.3 illustrates how the pixel values at half-pixel and quarter-pixel positions can be derived by interpolation. In order to derive the pixel values at half-pixel positions labeled b and h, the intermediate values b_1 and h_1 are first derived by applying the 6-tap filters as below.

$$b_1 = E - 5F + 20G + 20H - 5I + J$$
$$h_1 = A - 5C + 20G + 20M - 5R + T.$$

The values of b and h are then obtained by the following, clipped to the range 0–255.

$$b = (b_1 + 16) \gg 5$$
$$h = (h_1 + 16) \gg 5.$$

The special symbol "$\gg$" in the above formula indicates a right shift. For example, $b = (b_1 + 16) \gg 5$ is equivalent to $b = \text{round}(b_1/2^5) = \text{round}(b_1/32)$. However, the shift operation is much more efficient compared to a *round* function call.

The middle pixel "j" is obtained by

$$j_1 = aa_1 - 5bb_1 + 20h_1 + 20m_1 - 5cc_1 + dd_1,$$

where the intermediate values aa_1, bb_1, m_1, cc_1, and dd_1 are derived in a similar manner as h_1. Then, the value of j is obtained by the following, clipped to the range 0–255.

$$j = (j_1 + 512) \gg 10.$$

The pixel values at quarter-pixel positions labeled a, c, d, n, f, i, k, and q are obtained by averaging the values of the two nearest pixels at integer and half-pixel positions. For example,

$$a = (G + b + 1) \gg 1.$$

Finally, the pixel values at quarter-pixel positions labeled e, g, p, and r are obtained by averaging the values of the two nearest pixels at half-pixel positions in the diagonal direction. For example,

$$e = (b + h + 1) \gg 1.$$

Additional Options in Group of Pictures (GOP)

As shown in Fig. 11.3, in previous MPEG standards, the *Group of Pictures (GOP)* starts and ends with I-frames. In between, it has P-frames and B-frames. Either an I-frame or P-frame can be used as a *Reference frame*. Macroblocks in P-frames are predicted by forward prediction, and macroblocks in B-frames are predicted by a combination of forward prediction and backward prediction. H.264 will continue supporting this "classic" GOP structure. In addition, it will support the following GOP structures:

No B-Frames

The prediction of macroblocks in B-frames incurs more delay and requires more storage for the necessary I- and P-frames because of the bidirectional prediction. In this option, only I- and P-frames are allowed. Although the compression efficiency is relatively low, it is more suitable for certain applications, e.g., videoconferencing where a minimal delay is more desirable. This is compatible with the goals of the Baseline Profile or Constrained Baseline Profile of H.264.

Multiple Reference Frames

In order to find the best match for each macroblock in P-frames, H.264 allows up to N reference frames. Figure 12.4 illustrates an example where $N = 4$. The reference frame for P_1 is I_0, the reference frames for P_2 are I_0 and P_1, ... For P_4, the reference frames are I_0, P_1, P_2, and P_3. While this improves the compression efficiency, it requires much more computation in motion estimation at the encoder. Moreover, it requires a larger buffer to store up to N frames at the encoder and decoder.

Fig. 12.4 An illustration of
Multi-reference frames

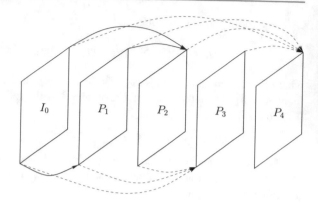

Hierarchical Prediction Structure

Hierarchical prediction structures are also allowed under H.264's flexible prediction options. For example, we can have a GOP that starts and ends with I_0 and I_{12}. In between, there are 11 consecutive B-frames, B_1 to B_{11}. First, B_6 is predicted using I_0 and I_{12} as references. Next, B_3 is predicted using I_0 and B_6, and B_9 is predicted using B_6 and I_{12} as references. At last, B_1 and B_2 are predicted using I_0 and B_3, B_4 and B_5 are predicted using B_3 and B_6, and so on. This hierarchical structure can be seen as having layers. For this example: Layer 0: I_0, I_{12}; Layer 1: B_6; Layer 2: B_3, B_9; Layer 3: B_1, B_2, B_4, B_5, B_7, B_8, B_{10}, B_{11}. Usually, increasingly larger quantization parameters will be associated with higher layers to control the compression efficiency. This is shown to be more efficient than the IBBP... structure used in previous video coding standards for temporal prediction.

12.2.2 Integer Transform

As in previous video coding standards, H.264 employs Transform coding after difference macroblocks are obtained. One of the most important features in H.264/AVC is the use of an *Integer Transform*.

The discrete cosine transform (DCT) in previous video coding standards is known to cause *prediction shift* because of floating point calculation and rounding errors in the transform and inverse transform. It is also slow due to many floating point multiplications. H.264 allows 4×4 blocks and various predictions; even intra-coding relies on spatial prediction followed by transform coding. Hence, it is very sensitive to the prediction shift. For example, a 4×4 block may be predicted from a neighboring intrablock, and the neighboring block may itself be predicted from another neighboring block, and so on. As a result, the prediction shift could be accumulated, causing a large error.

Given the powerful and accurate P- and I-prediction schemes in H.264, it is recognized that the spatial correlation in residual pixels is typically very low. Hence, a simple integer-precision 4×4 DCT is sufficient to compact the energy. The integer arithmetic allows exact inverse transform on all processors and eliminates encoder/decoder mismatch problems in previous transform-based codecs. H.264 also provides a quantization scheme with nonlinear *step sizes* to obtain accurate rate control at both the high and low ends of the quantization scale.

The 4×4 transforms in H.264 approximate the DCT and IDCT. They involve only integer, 16-bit arithmetic operations. They can be implemented very efficiently.

As discussed in Chap. 8, the 2D DCT is separable: it can be realized by two consecutive 1D transforms, i.e., in the vertical direction first and then the horizontal direction. This can be implemented by two matrix multiplications: $\mathbf{F} = \mathbf{T} \times \mathbf{f} \times \mathbf{T}^T$, where $\mathbf{f}$ is the input data, $\mathbf{F}$ is the transformed data, and $\mathbf{T}$ is the so-called *DCT-matrix*. The DCT-matrix is orthonormal, i.e., all the rows are orthogonal, and they all have norm 1.

The 4×4 DCT-matrix T_4 can be written as

$$\mathbf{T_4} = \begin{bmatrix} a & a & a & a \\ b & c & -c & -b \\ a & -a & -a & a \\ c & -b & b & -c \end{bmatrix}$$

where $a = 1/2$, $b = \frac{\sqrt{2}}{2} \cos \frac{\pi}{8}$, and $c = \frac{\sqrt{2}}{2} \cos \frac{3\pi}{8}$.

To derive a scaled 4×4 Integer Transform that approximates $\mathbf{T_4}$, we can simply scale the entries of $\mathbf{T_4}$ up and round them to the nearest integers [5]:

$$\mathbf{H} = \text{round}(\alpha \cdot \mathbf{T_4}). \tag{12.1}$$

When $\alpha = 26$, we have

$$\mathbf{H} = \begin{bmatrix} 13 & 13 & 13 & 13 \\ 17 & 7 & -7 & -17 \\ 13 & -13 & -13 & 13 \\ 7 & -17 & 17 & -7 \end{bmatrix}$$

Similar to T_4, this matrix has some nice properties: all of its rows are orthogonal; they also have the same norm, because $4 \times 13^2 = 2 \times (17^2 + 7^2)$. However, this matrix has a dynamic range gain of 52 (i.e., 4×13). Since it is used twice in $\mathbf{F} = \mathbf{H} \times \mathbf{f} \times \mathbf{H}^T$ in order to transform the columns and then the rows of $\mathbf{f}$, the total gain is $52^2 = 2704$. Because $\log_2 2704 \approx 11.4$, it would require 12 more bits for coefficients in $\mathbf{F}$ than the number of bits required for data in the original $\mathbf{f}$. This would make the 16-bit arithmetic insufficient and would require 32-bit arithmetic.

Hence, it is proposed in [5] that $\alpha = 2.5$ in Eq. 12.1. This yields

$$\mathbf{H} = \begin{bmatrix} 1 & 1 & 1 & 1 \\ 2 & 1 & -1 & -2 \\ 1 & -1 & -1 & 1 \\ 1 & -2 & 2 & -1 \end{bmatrix} \tag{12.2}$$

This new matrix $\mathbf{H}$ is still orthogonal, although its rows no longer have the same norm. To restore the orthonormal property, we could simply derive the following matrix $\bar{\mathbf{H}}$ by dividing all row entries in $\mathbf{H}$ by $\sqrt{\sum_j H_{ij}^2}$, where H_{ij} is the jth entry of the ith row in $\mathbf{H}$. However, this would no longer be an integer transform.

$$\bar{\mathbf{H}} = \begin{bmatrix} 1/2 & 1/2 & 1/2 & 1/2 \\ 2/\sqrt{10} & 1/\sqrt{10} & -1/\sqrt{10} & -2/\sqrt{10} \\ 1/2 & -1/2 & -1/2 & 1/2 \\ 1/\sqrt{10} & -2/\sqrt{10} & 2/\sqrt{10} & -1/\sqrt{10} \end{bmatrix}$$

In the H.264 implementation, this normalization issue is postponed. It is merged into the quantization process in the next step, since we can simply adjust the values in the quantization matrix to achieve both the objectives of quantization and normalization.

Because $\mathbf{H}$ is orthogonal, we could have used $\mathbf{H}^T$ as the inverse transform $\mathbf{H}^{-1}$, as long as the normalization issue is taken care of. Again, since we can also resolve this issue later at the de-quantization step, we simply introduce an ad hoc inverse transform $\mathbf{H}_{inv}$ to use

$$\mathbf{H}_{inv} = \begin{bmatrix} 1 & 1 & 1 & 1/2 \\ 1 & 1/2 & -1 & -1 \\ 1 & -1/2 & -1 & 1 \\ 1 & -1 & 1 & -1/2 \end{bmatrix} \tag{12.3}$$

$\mathbf{H}_{inv}$ is basically $\mathbf{H}^T$, but with the second and fourth columns scaled down by $1/2$. This is because the dynamic range of the input data to $\mathbf{H}_{inv}$ is larger than that to $\mathbf{H}$. Hence, a further scaling down is applied to the columns that would otherwise have a higher dynamic range gain.

H.264 also supports the 8×8 integer transform $\mathbf{H}_{8\times8}$. It is as in Eq. 12.4. We will use $\mathbf{H}$, the 4×4 version, in our discussions unless otherwise noted.

$$\mathbf{H}_{8\times8} = \begin{bmatrix} 8 & 8 & 8 & 8 & 8 & 8 & 8 & 8 \\ 12 & 10 & 6 & 3 & -3 & -6 & -10 & -12 \\ 8 & 4 & -4 & -8 & -8 & -4 & 4 & 8 \\ 10 & -3 & -12 & -6 & 6 & 12 & 3 & -10 \\ 8 & -8 & -8 & 8 & 8 & -8 & -8 & 8 \\ 6 & -12 & 3 & 10 & -10 & -3 & 12 & -6 \\ 4 & -8 & 8 & -4 & -4 & 8 & -8 & 4 \\ 3 & -6 & 10 & -12 & 12 & -10 & 6 & -3 \end{bmatrix} \tag{12.4}$$

12.2.3 Quantization and Scaling

As in previous video compression standards, *quantization* is used after the transform. Instead of designing simple quantization matrices, H.264 has a more sophisticated design [5] that accomplishes both tasks of the quantization and scaling (normalization) of $\mathbf{H}$.

Integer Transform and Quantization

Let $\mathbf{f}$ be the 4×4 input matrix, and $\hat{\mathbf{F}}$ the transformed and then quantized output. The forward integer transform, scaling, and quantization are implemented according to

$$\hat{\mathbf{F}} = \text{round}\left[(\mathbf{H} \times \mathbf{f} \times \mathbf{H}^T) \cdot \mathbf{M_f}/2^{15}\right]. \tag{12.5}$$

Here, "$\times$" denotes matrix multiplication, while "$\cdot$" denotes element-by-element multiplication. $\mathbf{H}$ is the same as in Eq. 12.2. $\mathbf{M_f}$ is the 4×4 quantization matrix derived from $\mathbf{m}$ which is a 6×3 matrix (see Table 12.1). QP is the *quantizaton parameter*.

For $0 \leq QP < 6$, we have

$$\mathbf{M_f} = \begin{bmatrix} \mathbf{m}(QP,0) & \mathbf{m}(QP,2) & \mathbf{m}(QP,0) & \mathbf{m}(QP,2) \\ \mathbf{m}(QP,2) & \mathbf{m}(QP,1) & \mathbf{m}(QP,2) & \mathbf{m}(QP,1) \\ \mathbf{m}(QP,0) & \mathbf{m}(QP,2) & \mathbf{m}(QP,0) & \mathbf{m}(QP,2) \\ \mathbf{m}(QP,2) & \mathbf{m}(QP,1) & \mathbf{m}(QP,2) & \mathbf{m}(QP,1) \end{bmatrix} \tag{12.6}$$

For $QP \geq 6$, each element $\mathbf{m}(QP, k)$ is replaced by $\mathbf{m}(QP\%6, k)/2^{\lfloor QP/6 \rfloor}$.

The quantization is followed by another scaling step, which can be implemented by a right shift "$\gg 15$."

Inverse Integer Transform and De-Quantization

Let $\tilde{\mathbf{f}}$ be the de-quantized and then inversely transformed result. The scaling, de-quantization, and inverse integer transform are implemented according to

$$\tilde{\mathbf{f}} = \text{round}\left[(\mathbf{H_{inv}} \times (\hat{\mathbf{F}} \cdot \mathbf{V_i}) \times \mathbf{H_{inv}}^T)/2^6\right]. \tag{12.7}$$

Table 12.1 The matrix $\mathbf{m}$—used to generate $\mathbf{M_f}$

QP	Positions in $\mathbf{M_f}$ (0, 0), (0, 2) (2, 0), (2, 2)	Positions in $\mathbf{M_f}$ (1, 1), (1, 3) (3, 1), (3, 3)	Remaining $\mathbf{M_f}$ positions
0	13107	5243	8066
1	11916	4660	7490
2	10082	4194	6554
3	9362	3647	5825
4	8192	3355	5243
5	7282	2893	4559

Table 12.2 The matrix $\mathbf{v}$—used to generate $\mathbf{V_i}$

QP	Positions in $\mathbf{V_i}$ $(0, 0), (0, 2)$ $(2, 0), (2, 2)$	Positions in $\mathbf{V_i}$ $(1, 1), (1, 3)$ $(3, 1), (3, 3)$	Remaining $\mathbf{V_i}$ positions
0	10	16	13
1	11	18	14
2	13	20	16
3	14	23	18
4	16	25	20
5	18	29	23

$\mathbf{H_{inv}}$ is the same as in Eq. 12.3. $\mathbf{V_i}$ is the 4×4 de-quantization matrix derived from $\mathbf{v}$ which is a 6×3 matrix (see Table 12.2). For $0 \le QP < 6$, we have

$$\mathbf{V_i} = \begin{bmatrix} \mathbf{v}(QP, 0) & \mathbf{v}(QP, 2) & \mathbf{v}(QP, 0) & \mathbf{v}(QP, 2) \\ \mathbf{v}(QP, 2) & \mathbf{v}(QP, 1) & \mathbf{v}(QP, 2) & \mathbf{v}(QP, 1) \\ \mathbf{v}(QP, 0) & \mathbf{v}(QP, 2) & \mathbf{v}(QP, 0) & \mathbf{v}(QP, 2) \\ \mathbf{v}(QP, 2) & \mathbf{v}(QP, 1) & \mathbf{v}(QP, 2) & \mathbf{v}(QP, 1) \end{bmatrix} \quad (12.8)$$

For $QP \ge 6$, each element $\mathbf{v}(QP, k)$ is replaced by $\mathbf{v}(QP\%6, k) \cdot 2^{\lfloor QP/6 \rfloor}$.

The de-quantization is also followed by another scaling step, which can be implemented by a right shift "$\gg 6$."

12.2.4 Examples of H.264 Integer Transform and Quantization

This section shows some examples of the H.264 integer transform and quantization and their inverse using various quantization parameters QPs. The input data is a 4×4 matrix $\mathbf{f}$ with arbitrary values; the transformed and then quantized coefficients are in $\hat{\mathbf{F}}$. $\mathbf{M_f}$ and $\mathbf{V_i}$ are the quantization and de-quantization matrices, and $\tilde{f}$ is the de-quantized and then inversely transformed output. For comparison, we will also show the compression loss $\epsilon = \mathbf{f} - \tilde{\mathbf{f}}$.

In order to improve the rate-distortion performance, H.264 adopts the dead-zone quantization (also known as *midtread* as discussed in Chap. 8). It can be described as a function that turns a real number x to an integer Z, as follows:

$$Z = \lfloor x + b \rfloor,$$

where x is the scaled value as discussed in the last section. By default, $b = 0.5$, the above function is then equivalent to the *round* function as specified in Eq. 12.5 or

$$
\mathbf{f} = \begin{matrix} 72 & 82 & 85 & 79 \\ 74 & 75 & 86 & 82 \\ 84 & 73 & 78 & 80 \\ 77 & 81 & 76 & 84 \end{matrix}
\qquad
\mathbf{M_f} = \begin{matrix} 13107 & 8066 & 13107 & 8066 \\ 8066 & 5243 & 8066 & 5243 \\ 13107 & 8066 & 13107 & 8066 \\ 8066 & 5243 & 8066 & 5243 \end{matrix}
\qquad
\hat{\mathbf{F}} = \begin{matrix} 507 & -12 & -2 & 2 \\ 0 & -7 & -14 & 5 \\ 2 & 0 & -8 & -11 \\ -1 & 8 & 4 & 3 \end{matrix}
$$

$$
\mathbf{V_i} = \begin{matrix} 10 & 13 & 10 & 13 \\ 13 & 16 & 13 & 16 \\ 10 & 13 & 10 & 13 \\ 13 & 16 & 13 & 16 \end{matrix}
\qquad
\tilde{\mathbf{f}} = \begin{matrix} 72 & 82 & 85 & 79 \\ 74 & 75 & 86 & 82 \\ 84 & 73 & 78 & 80 \\ 77 & 81 & 76 & 84 \end{matrix}
\qquad
\boldsymbol{\epsilon} = \mathbf{f} - \tilde{\mathbf{f}} = \begin{matrix} 0 & 0 & 0 & 0 \\ 0 & 0 & 0 & 0 \\ 0 & 0 & 0 & 0 \\ 0 & 0 & 0 & 0 \end{matrix}
$$

(a) $QP = 0$

$$
\mathbf{f} = \begin{matrix} 72 & 82 & 85 & 79 \\ 74 & 75 & 86 & 82 \\ 84 & 73 & 78 & 80 \\ 77 & 81 & 76 & 84 \end{matrix}
\qquad
\mathbf{M_f} = \begin{matrix} 6554 & 4033 & 6554 & 4033 \\ 4033 & 2622 & 4033 & 2622 \\ 6554 & 4033 & 6554 & 4033 \\ 4033 & 2622 & 4033 & 2622 \end{matrix}
\qquad
\hat{\mathbf{F}} = \begin{matrix} 254 & -6 & -1 & 1 \\ 0 & -4 & -7 & 3 \\ 1 & 0 & -4 & -6 \\ 0 & 4 & 2 & 1 \end{matrix}
$$

$$
\mathbf{V_i} = \begin{matrix} 20 & 26 & 20 & 26 \\ 26 & 32 & 26 & 32 \\ 20 & 26 & 20 & 26 \\ 26 & 32 & 26 & 32 \end{matrix}
\qquad
\tilde{\mathbf{f}} = \begin{matrix} 72 & 82 & 85 & 79 \\ 74 & 75 & 86 & 82 \\ 84 & 74 & 78 & 80 \\ 77 & 82 & 76 & 84 \end{matrix}
\qquad
\boldsymbol{\epsilon} = \mathbf{f} - \tilde{\mathbf{f}} = \begin{matrix} 0 & 0 & 0 & 0 \\ 0 & 0 & 0 & 0 \\ 0 & -1 & 0 & 0 \\ 0 & -1 & 0 & 0 \end{matrix}
$$

(b) $QP = 6$

$$
\mathbf{f} = \begin{matrix} 72 & 82 & 85 & 79 \\ 74 & 75 & 86 & 82 \\ 84 & 73 & 78 & 80 \\ 77 & 81 & 76 & 84 \end{matrix}
\qquad
\mathbf{M_f} = \begin{matrix} 1638 & 1008 & 1638 & 1008 \\ 1008 & 655 & 1008 & 655 \\ 1638 & 1008 & 1638 & 1008 \\ 1008 & 655 & 1008 & 655 \end{matrix}
\qquad
\hat{\mathbf{F}} = \begin{matrix} 63 & -2 & 0 & 0 \\ 0 & -1 & -2 & 1 \\ 0 & 0 & -1 & -1 \\ 0 & 1 & 0 & 0 \end{matrix}
$$

$$
\mathbf{V_i} = \begin{matrix} 80 & 104 & 80 & 104 \\ 104 & 128 & 104 & 128 \\ 80 & 104 & 80 & 104 \\ 104 & 128 & 104 & 128 \end{matrix}
\qquad
\tilde{\mathbf{f}} = \begin{matrix} 70 & 81 & 86 & 78 \\ 73 & 73 & 85 & 83 \\ 82 & 75 & 77 & 82 \\ 77 & 79 & 74 & 85 \end{matrix}
\qquad
\boldsymbol{\epsilon} = \mathbf{f} - \tilde{\mathbf{f}} = \begin{matrix} 2 & 1 & -1 & 1 \\ 1 & 2 & 1 & -1 \\ 2 & -2 & 1 & -2 \\ 0 & 2 & 2 & -1 \end{matrix}
$$

(c) $QP = 18$

$$
\mathbf{f} = \begin{matrix} 72 & 82 & 85 & 79 \\ 74 & 75 & 86 & 82 \\ 84 & 73 & 78 & 80 \\ 77 & 81 & 76 & 84 \end{matrix}
\qquad
\mathbf{M_f} = \begin{matrix} 410 & 252 & 410 & 252 \\ 252 & 164 & 252 & 164 \\ 410 & 252 & 410 & 252 \\ 252 & 164 & 252 & 164 \end{matrix}
\qquad
\hat{\mathbf{F}} = \begin{matrix} 16 & 0 & 0 & 0 \\ 0 & 0 & 0 & 0 \\ 0 & 0 & 0 & 0 \\ 0 & 0 & 0 & 0 \end{matrix}
$$

$$
\mathbf{V_i} = \begin{matrix} 320 & 416 & 320 & 416 \\ 416 & 512 & 416 & 512 \\ 320 & 416 & 320 & 416 \\ 416 & 512 & 416 & 512 \end{matrix}
\qquad
\tilde{\mathbf{f}} = \begin{matrix} 80 & 80 & 80 & 80 \\ 80 & 80 & 80 & 80 \\ 80 & 80 & 80 & 80 \\ 80 & 80 & 80 & 80 \end{matrix}
\qquad
\boldsymbol{\epsilon} = \mathbf{f} - \tilde{\mathbf{f}} = \begin{matrix} -8 & 2 & 5 & -1 \\ -6 & -5 & 6 & 2 \\ 4 & -7 & -2 & 0 \\ -3 & 1 & -4 & 4 \end{matrix}
$$

(d) $QP = 30$

Fig. 12.5 Examples of H.264 Integer transform and quantization with various QPs

Eq. 12.7. To minimize the quantization error, H.264 actually adopts adaptive quantization in which the width of the dead zone can be controlled by b. For example, $b = 1/3$ for intra-coding and $b = 1/6$ for inter-coding. For simplicity, in the following examples, we just use $b = 0.5$.

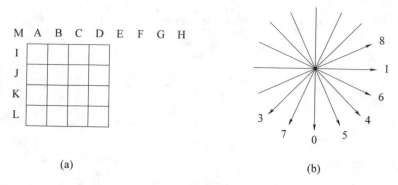

Fig. 12.6 H.264 Intra-frame prediction. **a** Intra_4 × 4 prediction using neighboring samples A to M. **b** Eight directions for Intra_4 × 4 predictions

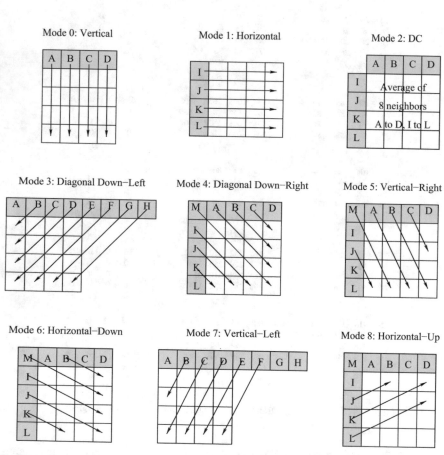

Fig. 12.7 The nine Intra_4 × 4 prediction modes in H.264 Intra-frame prediction

Figure 12.5a shows the result when $QP = 0$. This is the option that offers the least possible compression loss. The values of $\mathbf{M_f}$ and $\mathbf{V_i}$ are determined according to Eqs. 12.6 and 12.8. Since there is no scale-down of $\mathbf{F}$ values in the quantization step, the reconstructed $\tilde{\mathbf{f}}$ is exactly the same as $\mathbf{f}$, i.e., $\tilde{\mathbf{f}} = \mathbf{f}$.

Figure 12.5b shows the result when $QP = 6$. As expected, compared to corresponding matrix entries for $QP = 0$, the values in $\mathbf{M_f}$ are reduced in half, and the values in $\mathbf{V_i}$ are about twice. The quantization factor is now approximately 1.25 (i.e., equivalent to $qstep \approx 1.25$). As a result, $\tilde{\mathbf{f}} \neq \mathbf{f}$. The slight loss can be observed in ϵ.

Similarly, when $QP = 18$, the result is in Fig. 12.5c. The values in $\mathbf{M_f}$ and the values in $\mathbf{V_i}$ are further decreased or increased, respectively, by a factor of 4 compared to those for $QP = 6$. The quantization factor is approximately 5 (i.e., equivalent to $qstep \approx 5$). The loss shown in ϵ is more significant now.

Perhaps a more interesting result is shown in Fig. 12.5d when $QP = 30$. The quantization factor is now approximately 20 (i.e., equivalent to $qstep \approx 20$). Except for the larger quantized DC value in $\hat{F}$ that remains nonzero, all the AC coefficients become zero. As a result, the reconstructed $\tilde{f}$ has 80 in all entries. The compression loss in ϵ is no longer acceptable.

12.2.5 Intra-Coding

H.264 exploits much more *spatial prediction* than previous video standards such as MPEG-2 and H.263+. Intra-coded macroblocks are predicted using some of the neighboring reconstructed pixels (using either intra- or inter-coded reconstructed pixels).

Similar to the variable block sizes in inter-picture motion compensation, different intra prediction block sizes (4×4 or 16×16) can be chosen for each intra-coded macroblock. As shown in Fig. 12.6, there are nine prediction modes for the 4×4 blocks in Intra_4 $\times$ 4. Figure 12.7 illustrates their details.

Prediction modes 0, 1, 3, and 4 are fairly straightforward. For example, in Mode 0, the value of pixel A will be used as the predicted value for all pixels in the first column (i.e., the column below A), the value of pixel B will be used as the predicted value for all pixels in the second column, etc. Mode 2 (DC) is a special mode in which the average of the eight previously coded neighbors (A to D and I to L) is used as the predicted value for all pixels in the 4×4 block.

Mode 5 (Vertical-Right), Mode 6 (Horizontal-Down), Mode 7 (Vertical-Left), and Mode 8 (Horizontal-Up) are similar. As shown in Fig. 12.7, the direction of the prediction in Mode 5 is down and to the right at the ratio of 2:1 (i.e., 2 pixels down and 1 pixel to the right, or approximately 26.6° to the right). This works well for the pixels at the second and fourth rows. For example, if the row and column indices of the 4×4 block are in the range 0 .. 3, then the prediction value for pixels at [1, 1] and [3, 2] will be the value of A. However, the pixels at the first and third rows will not be able to use a single value from any of the previously coded neighbors. Instead, it must be extrapolated from two of them. For example, the prediction value for pixels at [0, 0] and [2, 1] will be a proportional combination of the values of M and A.

For each prediction mode, the predicted value and the actual value will be compared to produce the prediction error. The mode that produces the least prediction error will be chosen as the prediction mode for the block. The prediction errors (residuals) are then sent to transform coding where the 4×4 integer transform is employed. Each 4×4 block in a macroblock may have a different prediction mode. The sophisticated intra prediction is powerful as it drastically reduces the amount of data to be transmitted when temporal prediction fails.

There are only four prediction modes for the 16×16 blocks in Intra_16 $\times$ 16. Mode 0 (Vertical), Mode 1 (Horizontal), and Mode 2 (DC) are very similar to Intra_4 $\times$ 4 above except the larger block size. For Mode 3 (Plane) which is unique to 16×16 blocks, a plane (linear) function is fitted to the upper and left samples in the 16×16 block as the prediction.

In summary, the following four modes are specified for Intra-Coding:

- Intra_4 $\times$ 4 for luma macroblocks
- Intra_16 $\times$ 16 for luma macroblocks
- Intra-coding for chroma macroblocks—It uses the same four prediction modes as in Intra_16 $\times$ 16 luma. The prediction block size is 8×8 for 4:2:0, 8×16 for 4:2:2, and 16×16 for 4:4:4 chroma sampling.
- I_PCM (Pulse Code Modulation)—Bypass the spatial prediction and transform coding, and directly send the PCM coded (fixed-length) luma and chroma pixel values. It is invoked in rare cases when other prediction modes failed to produce any data compression/reduction.

12.2.6 In-loop Deblocking Filtering

One of the prominent deficiencies of the block-based coding methods is the generation of unwanted visible block structures. Pixels at the block boundaries are usually reconstructed less accurately: they tend to look like the interior pixels in the same block, hence the artificial appearance of blocks.

H.264 specifies a signal-adaptive deblocking filter in which a set of filters is applied on the 4×4 block edges. Filter length, strength, and type (deblocking/smoothing) vary, depending on the macroblock coding parameters (intra- or inter-coded, reference-frame differences, coefficients coded, etc.) and spatial activity (edge detection), so that blocking artifacts are eliminated without distorting visual features. The H.264 deblocking filter is important in increasing the subjective quality of the videos.

As shown in Fig. 12.1, in H.264 the deblocking filtering takes place in the loop, after the inverse transform in the encoder, before the decoded block data are fed to Motion Estimation.

Figure 12.8 depicts a simplified 1D edge, where the height of the pixels p_0, q_0, etc., indicates their value. The function of the *deblocking filtering* is basically smoothing of the block edges. For example, a "4-tap filtering" will take some weighted average of the values of p_1, p_0, q_0, and q_1 to generate new p_0 or q_0.

Fig. 12.8 Deblocking of a
1D edge on the block
boundary

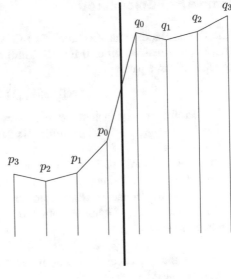

Block Boundary

Apparently, real edges across the block boundary will need to be protected from the deblocking filtering. The deblocking filtering on p_0 and q_0 will be applied only if all the following criteria are met:

$$|p_0 - q_0| < \alpha(QP),$$
$$|p_0 - p_1| < \beta(QP),$$
$$|q_0 - q_1| < \beta(QP),$$

where α and β are thresholds, and they are functions of the quantization parameter QP as defined in the standard. They are lower when QP is smaller. This is because when QP is small, a relatively significant difference, e.g., $|p_0 - q_0|$ is likely caused by a real edge.

In addition to p_0 and q_0, the deblocking filtering on p_1 or q_1 will be applied if

$$|p_0 - p_2| < \beta(QP) \quad \text{or} \quad |q_0 - q_2| < \beta(QP)$$

12.2.7 Entropy Coding

H.264 has developed a set of sophisticated entropy coding methods. When entropy_coding_mode = 0, a simple Exponential-Golomb (Exp-Golomb) code is used for header data, motion vectors, and other nonresidual data, while the more complex *Context-Adaptive Variable Length Coding (CAVLC)* is used for quantized residual coefficients. When entropy_coding_mode = 1, *Context-Adaptive Binary Arithmetic Coding (CABAC)* is used (see Sect. 12.2.9).

Simple Exp-Golomb code

The simple Exponential-Golomb (Exp-Golomb) code that is used for header data, etc., is the so-called order-0 Exp-Golomb code (EG_0). It is a binary code, and it consists of three parts:

$$[\text{Prefix}] \ [1] \ [\text{Suffix}]$$

The Prefix is a sequence of l zeros. Given an unsigned (positive) number N to be coded, $l = \lfloor \log_2(N+1) \rfloor$. The Suffix S is the binary number $N + 1 - 2^l$ represented in l bits.

As shown in Table 12.3, if an unsigned $N = 4$, then $l = \lfloor \log_2(4+1) \rfloor = 2$, the Prefix is 00, and the Suffix S is the binary number $S = 4 + 1 - 2^2 = 1$ represented in 2 bits, i.e., 01. Hence, the Exp-Golomb code for $N = 4$ is 00101.

To decode the Exp-Golomb codeword EG_0 for unsigned N, the following steps can be followed:

1. Read in the sequence of consecutive zeros, $l = $ number_of_zeros.
2. Skip the next "1."
3. Read in the next l bits and assign to S.
4. $N = S - 1 + 2^l$.

The unsigned numbers are used to indicate, e.g., macroblock type, reference frame index, etc. For signed numbers, e.g., motion vector difference, they will simply be squeezed in to produce a new set of table entries as listed in the second column (Table 12.3).

Table 12.3 The order-0 Exp-Golomb codewords (EG_0)

Unsigned N	Signed N	Codeword
0	0	1
1	1	010
2	−1	011
3	2	00100
4	−2	00101
5	3	00110
6	−3	00111
7	4	0001000
8	−4	0001001
⋮	⋮	⋮

Order-k Exp-Golomb code (EG_k)

In general, the Exp-Golomb code can have a higher order, i.e., the order-k EG_k. Similarly, it is a binary code and consists of three parts: [Prefix] [1] [Suffix]. The Prefix is a sequence of l zeros. Given an unsigned (positive) number N to be coded, $l = \lfloor \log_2(N/2^k + 1) \rfloor$. The Suffix S is the binary number $N + 2^k(1 - 2^l)$ represented in $l + k$ bits.

For example, the EG_1 code for $N = 4$ is 0110. It is because $l = \lfloor \log_2(4/2^1 + 1) \rfloor = 1$, and the Prefix is 0; the Suffix is the binary representation of $4 + 2^1(1 - 2^1) = 2$, and in $l + k = 1 + 1 = 2$ bits, it is 10. Table 12.4 provides some examples of the first- and second-order Exp-Golomb codes for nonnegative numbers.

To decode the order-k Exp-Golomb codeword EG_k for unsigned N, the following steps can be followed:

1. Read in the sequence of consecutive zeros, l = number_of_zeros.
2. Skip the next "1."
3. Read in the next $l + k$ bits and assign to S.
4. $N = S - 2^k(1 - 2^l)$.

Table 12.4 First- and second-order Exp-Golomb Codewords (EG_1 and EG_2)

Unsigned N	EG_1 Codeword	EG_2 Codeword
0	10	100
1	11	101
2	0100	110
3	0101	111
4	0110	01000
5	0111	01001
6	001000	01010
7	001001	01011
8	001010	01100
9	001011	01101
10	001100	01110
11	001101	01111
12	001110	0010000
13	001111	0010001
14	00010000	0010010
15	00010001	0010011
⋮	⋮	⋮

12.2.8 Context-Adaptive Variable Length Coding (CAVLC)

Previous video coding standards such as MPEG-2 and H.263 use fixed VLC. In CAVLC [6,7], multiple VLC tables are predefined for each data type (zero-runs, levels, etc.), and predefined rules predict the optimal VLC table based on the context, e.g., previously decoded neighboring blocks.

It is known that the matrices (by default 4 × 4) that contain the quantized frequency coefficients of the residual data are typically sparse, i.e., contain many zeros. Even when they are nonzero, the quantized coefficients for higher frequencies are often +1 or −1 (the so-called "trailing_1s"). CAVLC exploits these characteristics by carefully extracting the following parameters from the current block data:

- Total number of nonzero coefficients (TotalCoeff) and the number of trailing ±1s (Trailing_1s).
- Signs of the Trailing_1s.
- Level (sign and magnitude) of the other nonzero coefficients (not Trailing_1s).
- Total number of zeros before the last nonzero coefficient.
- Run of zeros (zeros_left and run_before) before each nonzero coefficient.

Figure 12.9 shows an example of a 4 × 4 block of the residual data after transform and quantization. After the zigzag scan, the 1D sequence is 0 3 0 −2 1 0 −1 1 0 0 0 0 0 0 0 0. Table 12.5 gives details of the CAVLC code generated. The nonzero coefficients are processed in reverse order, i.e., the last one indexed by "4" (Trailing_1[4]) is examined first, and so forth.

We will briefly explain the code generation process for the above example.

- A code 0000100 is generated for TotalCoeffs = 5, Trailing_1s = 3. These are looked up from "Table 9.5—coeff_token mapping to TotalCoeff and TrailingOnes" in the H.264 standard (2003). It is observed that, in general, the numbers of nonzero coefficients in the neighboring blocks are similar, hence the code for coeff_token is context-adaptive. For each pair of TotalCoeffs and Trailing_1s, its code can be assigned one of the four possible values depending on the numbers of nonzero coefficients in the blocks above and to the left of the current block. If the neigh-

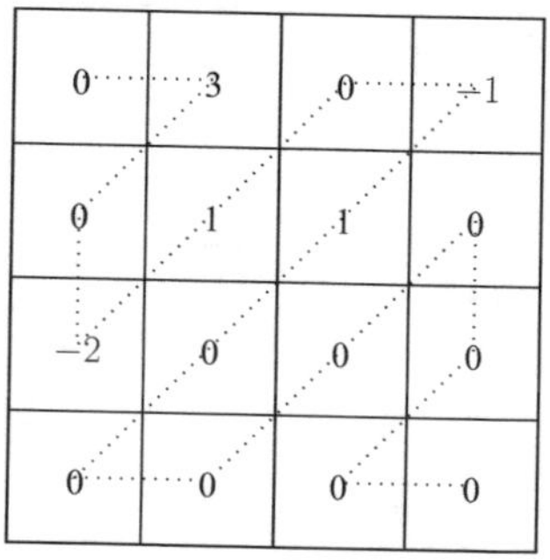

Fig. 12.9 Example: A 4 × 4 block of data for CAVLC encoder

Table 12.5 CAVLC code generation for data from Fig. 12.9

Data	Value	Code
coeff_token	TotalCoeffs = 5, Trailing_1s = 3	0000100
Trailing_1[4] Sign	+	0
Trailing_1[3] Sign	−	1
Trailing_1[2] Sign	+	0
Level [1]	−2 (SuffixLength = 0)	0001 (Prefix)
Level [0]	3 (SuffixLength = 1)	001 (Prefix) 0 (Suffix)
Total zeros	3	111
run_before[4]	zeros_left = 3, run_before = 0	11
run_before[3]	zeros_left = 3, run_before = 1	10
run_before[2]	zeros_left = 2, run_before = 0	1
run_before[1]	zeros_left = 2, run_before = 1	01
run_before[0]	zeros_left = 1, run_before = 1	No code required

boring blocks have a small number of nonzeros, then a code assignment favoring the small TotalCoeffs in the current block will be used (i.e., small TotalCoeffs are assigned very short codes and large TotalCoeffs are assigned particularly long codes), and vice versa. In this example, it is assumed that the number of nonzero coefficients in the two neighboring blocks is less than 2.

- Sign "+" is assigned the code 0, and "−" the code 1.
- The choice of the VLC code for Level is again context-adaptive; it depends on the magnitudes of the recently coded Levels. In reverse order, the first nonzero coefficient is −2. Initially, SuffixLength = 0, so the code for −2 is 0001 (prefix). Afterwards, SuffixLength is increased by 1, hence the next nonzero 3 gets the code 001 (Prefix) 0 (Suffix). The magnitude of Levels tends to increase (when examined in reverse order), so the SuffixLength is increased adaptively to accommodate larger magnitudes. For further details, the readers are referred to [4,6,7].
- The total number of zeros is 3. It gets the code 111.
- The last five rows in Table 12.5 record the information for the runs of zeros in the current block. For example, for the last nonzero coefficient "1," 3 zeros are in front of it, and no zero is immediately in front of it. The code 11 is looked up from Table 9.10 in the H.264 standard. To illustrate, we extracted part of it below as Table 12.6. This should also explain the codes in the next three rows, 10 1 and 01. Come down to the last row, only one zero is left for the only (last) nonzero coefficient, the encoder and decoder can unambiguously determine it, so no code for run_before is needed.

For this example, the resulting sequence of the code is 0000100 0 1 0 0001 001 0 111 11 10 1 01. Based on it, the decoder is able to reproduce the block data.

Table 12.6 Code for various run_before

Run_before	Zeros_left						
	1	2	3	4	5	6	>6
0	1	1	11	11	11	11	111
1	0	01	10	10	10	000	110
2	–	00	01	01	011	001	101
3	–	–	00	001	010	011	100
4	–	–	–	000	001	010	011
5	–	–	–	–	000	101	010
⋮							

12.2.9 Context-Adaptive Binary Arithmetic Coding (CABAC)

The VLC-based entropy coding methods (including CAVLC) are inefficient in deal-ing with symbols with a probability greater than 0.5, because usually a minimum of 1 bit has to be assigned to each symbol, which could be much larger than its *self-information* measured by $\log_2 \frac{1}{p_i}$, where p_i is the symbol's probability. For better coding efficiency in H.264 Main and High profiles, context-adaptive binary arith-metic coding (CABAC) [8] is used for some data and quantized residual coefficients when entropy_coding_mode = 1.

As shown in Fig. 12.10, CABAC has three major components:

- **Binarization**. All non-binary data are converted to binary bit (*bin*) strings first since CABAC uses Binary Arithmetic Coding. Five schemes can be used for the binarization: (a) Unary (U)—for $N \geq 0$, it is N 1s followed by a terminating 0. For example, it is 111110 for 5. (b) Truncated Unary (TU)—similar to U, but without the terminating 0. (c) kth order Exp-Golomb code. (d) The concatenation of (a) and (c). (e) Fixed-Length binary scheme.
- **Context Modeling**. This step deals with the context model selection and access. The *Regular Coding Mode* is for most symbols, e.g., macroblock type, mvd, infor-mation about prediction modes, information about slice and macroblock control, and residual data. Various "context models" are built to store the conditional prob-abilities for the bins of the binarized symbol to be 1 or 0. The probability models are derived from the statistics of the context, i.e., recently coded symbols and bins. The *Bypass Coding Mode* uses no context model; it is used to speed up the coding process.
- **Binary Arithmetic Coding**. For efficiency, a binary arithmetic coding method is developed [8]. Below is a brief description of *Binary Arithmetic Coding* in H.264. As shown in Chap. 7, Arithmetic Coding involves recursive subdivisions of the current range. The numerous multiplications involved there is its main disadvan-tage in terms of the computational cost when compared with other entropy coding

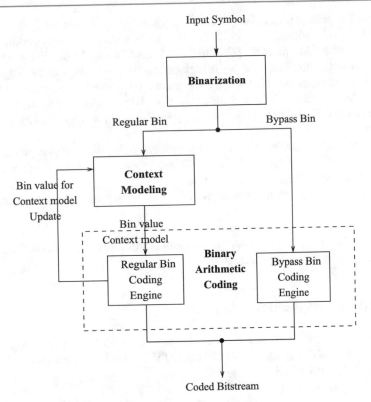

Fig. 12.10 Block diagram of CABAC in H.264

methods. Binary arithmetic coding only involves two symbols, so the number of multiplications is greatly reduced.

In the binary case, we name the two symbols *LPS (Least Probable Symbol)* and *MPS (Most Probable Symbol)*, the range as R, and the lower bound of R as L. (Note: LPS is the upper interval and MPS is the lower interval in R.) If the probability of the LPS is P_{LPS}, then $P_{MPS} = 1 - P_{LPS}$, and the following procedure can be employed for the generation of the next range given the new symbol S:

PROCEDURE   Calculating Ranges in Binary Arithmetic Coding

BEGIN
    If S is MPS
        $R = R \times (1 - P_{LPS})$;
    Else   // S is LPS.
        $L = L + R \times (1 - P_{LPS})$;
        $R = R \times P_{LPS}$;
END

However, the multiplication in $R \times P_{LPS}$ would be computationally expensive. Various "multiplication-free" binary arithmetic coding schemes have been developed, for example, the Q-coder for binary images, its improved QM-coder, and the MQ-coder adopted in JPEG 2000. The H.264 binary arithmetic coding method developed by Marpe et al. [8] is the so-called *M-coder (Modulo Coder)*. Here, the multiplication in $R \times P_{LPS}$ is replaced by a table lookup. In the Regular Bin Coding mode, the table has 4×64 pre-calculated product values to allow 4 different values for R and 64 values for p_{LPS}. (The Bypass Bin Coding mode assumes a uniform probability model, i.e., $P_{MPS} \approx P_{LPS}$ in order to simplify and speed up the process.)

Apparently, due to the limited size of the lookup table, the precision of the product values is limited. These multiplication-free methods are hence known as performing *Reduced-precision Arithmetic Coding*.

Previous study shows that the impact of the reduced precision on the code length is minimal.

The implementation of CABAC has a tremendous amount of details. Readers are referred to [8] for detailed discussions.

12.2.10   H.264 Profiles

As before, a number of profiles are provided to suit the needs of various applications ranging from mobile devices to broadcast HDTV. Figure 12.11 provides an overview of the H.264 profiles [4,9].

Baseline Profile

The Baseline profile of H.264 is intended for real-time conversational applications, such as videoconferencing. It contains all the core coding tools of H.264 discussed above and the following additional error-resilience tools, to allow for error-prone carriers such as IP and wireless networks:

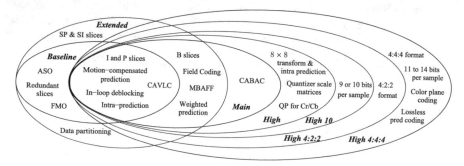

Fig. 12.11 H.264 profiles

- **Arbitrary slice order (ASO)**. The decoding order of slices within a picture may not follow monotonic increasing order. This allows decoding of out-of-order packets in a packet-switched network, thus reducing latency.
- **Flexible macroblock order (FMO)**. Macroblocks can be decoded in any order, such as checkerboard patterns, not just raster scan order. This is useful on error-prone networks, so that loss of a slice results in loss of macroblocks scattered in the picture, which can easily be masked from human eyes. This feature can also help reduce jitter and latency, as the decoder may decide not to wait for late slices and still be able to produce acceptable pictures.
- **Redundant slices**. Redundant copies of the slices can be decoded, to further improve error resilience.

Main Profile

The Main profile defined by H.264 represents non-low-delay applications such as standard definition (SD) digital broadcast TV and stored medium. The Main profile contains all Baseline profile features (except ASO, FMO, and redundant slices) plus the following non-low-delay and higher complexity features, for maximum compression efficiency:

- **B slices**. The bi-prediction mode in H.264 has been made more flexible than in existing standards. Bi-predicted pictures can also be used as reference frames. Two reference frames for each macroblock can be in any temporal direction, as long as they are available in the reference frame buffer. Hence, in addition to the normal forward + backward bi-prediction, it is legal to have backward + backward or forward + forward prediction as well.
- **Context-Adaptive Binary Arithmetic Coding (CABAC)**. This coding mode replaces VLC-based entropy coding with binary arithmetic coding that uses a different adaptive statistics model for different data types and contexts.
- **Weighted Prediction**. Global weights (multiplier and an offset) for modifying the motion-compensated prediction samples can be specified for each slice, to predict lighting changes and other global effects, such as fading.

Extended Profile

The eXtended profile (or profile X) is designed for the new video streaming applications. This profile allows non-low-delay features, bitstream switching features, and also more error-resilience tools. It includes all Baseline profile features plus the following:

- B slices
- Weighted prediction.

- **Slice data partitioning**. This partitions slice data with different importance into separate sequences (header information, residual information) so that more important data can be transmitted on more reliable channels.
- **SP (Switching P) and SI (Switching I) slice types**. These are slices that contain special temporal prediction modes, to allow efficient switching of bitstreams produced by different decoders. They also facilitate fast forward/backward and random access.

High Profiles

H.264/AVC also has four High profiles for applications that demand higher video qualities, i.e., high definition (HD).

- **High Profile**—This profile is adopted by the Blu-ray Disk format and DVB HDTV broadcast. It supports 8×8 integer transform for the parts of the pictures that do not have much detail, and 4×4 integer transform for the parts that do have details. It also allows 8×8 Intra prediction for better coding efficiency especially for higher resolution videos. It provides adjustable quantizer scale matrices, and separate quantizer parameters for Cb and Cr. It also supports monochrome video (4:0:0).
- **High 10 Profile**—Supports 9 or 10 bits per sample.
- **High 4:2:2 Profile**—Supports 4:2:2 chroma subsampling.
- **High 4:4:4 Predictive Profile**—It supports up to 4:4:4 chroma sampling, up to 14 bits per sample, coding of separate color planes, and efficient lossless predictive coding.

12.2.11   H.264 Scalable Video Coding (SVC)

The *scalable video coding (SVC)* extension of the H.264/AVC standard was approved in 2007 [10]. It provides bitstream scalability which is especially important for multimedia data transmission through various networks that may have very different bandwidths.

Similar to MPEG-2 and MPEG-4, H.264/AVC SVC provides *temporal scalability*, *spatial scalability*, *quality scalability*, and their possible combinations. Compared to previous standards, the coding efficiency is greatly improved. Other functions such as bitrate and power adaptation, and graceful degradation in lossy network transmissions are also provided.

We covered the issues of temporal scalability, spatial scalability, quality (SNR) scalability, and their possible combinations in sufficient detail under MPEG-2 in Chap. 11. Since the fundamental concepts and approaches are very similar, we will not discuss this topic in detail in this chapter. For more information about H264/AVC SVC, readers are referred to [10] and Annex G of the H.264/AVC standard.

12.2.12 H.264 Multiview Video Coding (MVC)

Multiview Video Coding (MVC) is an emerging issue. It has potential applications in some new areas such as *Free Viewpoint Video (FVV)* where users can specify their preferred views. Merkle et al. [11] described some possible MVC prediction structures. Figure 12.12 shows a small example in which there are only 4 views. The two most important features are

- **Interview Prediction**—Since there is apparent redundancy among the multiple views, the IPPP structure, for example, can be employed for the so-called Key Pictures (the first and ninth pictures in each view in the figure). This Interview Prediction structure can of course be extended to other structures, e.g., IBBP. This would be even more beneficial when many more views are involved.
- **Hierarchical B Pictures**—For temporal prediction in each view, a hierarchy of B picture, e.g., B_1, B_2, and B_3, similar to the ones discussed in Sect. 12.2.1

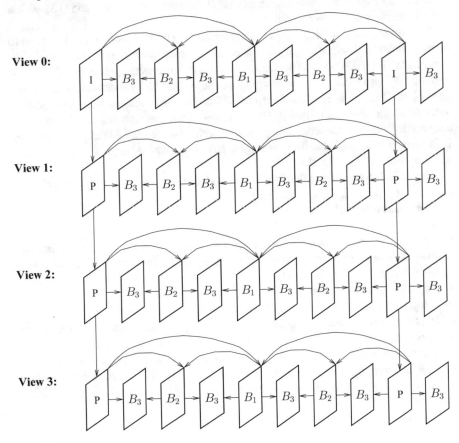

Fig. 12.12 H.264 MVC prediction structure

can be employed. It is made feasible because H.264/AVC is more flexible in supporting various prediction schemes at the picture/sequence level. As discussed earlier, increasingly larger quantization parameters can usually be applied down the hierarchy to control the compression efficiency.

For more information about H264/AVC MVC, readers are referred to [12] and Annex H of the H.264/AVC standard.

12.3   H.265

High efficiency video coding (HEVC) [13,14] was the standard jointly developed by Joint Collaborative Team on Video Coding (JCT-VC) from the groups of ITU-T VCEG (Video Coding Experts Group) and ISO/IEC MPEG. Version 1 of the standard was approved in April 2013. In ISO/IEC, HEVC became MPEG-H Part 2 (ISO/IEC 23008-2). It is also known as ITU-T Recommendation H.265 [15], which is the term we will use in this book.

The development of H.265 was largely motivated by two factors: (a) The need to further improve coding efficiency due to ever-increasing video resolution (e.g., 7,680 × 4,320 in 8K UHDTV). (b) The need to speed up the more complex coding/decoding methods by exploiting the increasingly available parallel processing devices and algorithms. The initial goal was a further 50% reduction of the size of the compressed video (with the same visual quality) from H.264, and it was reported that this goal was exceeded. With their superior compression performance over MPEG-2, H.264 and H.265 are currently the leading candidates to carry a whole range of video contents on many potential applications.

The default format for color video in H.265 is YCbCr. In Main profiles, chroma subsampling is 4:2:0.

Main Features of H.265 are

- Variable block-size motion compensation, from 4 × 4 up to 64 × 64 in luma images. The macroblock structure is replaced by a quad-tree structure of coding blocks at various levels and sizes.
- Exploration of parallel processing.
- Integer transform in various sizes, from 4 × 4, 8 × 8, 16 × 16 to 32 × 32.
- Improved interpolation methods for the quarter-pixel accuracy in motion vectors.
- Expanded directional spatial prediction (33 angular directions) for intra-coding.
- The potential use of discrete sine transform (DST) in luma intra-coding.
- In-loop filters including deblocking-filtering and sample adaptive offset (SAO).
- Only context-adaptive binary arithmetic coding (CABAC) will be used, i.e., no more CAVLC.

12.3.1 Motion Compensation

As in previous video coding standards, H.265 still uses the technology of *hybrid coding*, i.e., a combination of inter/intra predictions and transform coding on prediction residual errors.

Variable block sizes are used in inter/intra predictions as in H.264. However, more partitions of the prediction and transform blocks are encouraged in order to reduce the prediction errors. Unlike previous video coding standards, H.265 does not use the simple and fixed structure of **macroblocks**. Instead, a quad-tree hierarchy of various blocks is introduced as below for its efficiency.

- **CTB** and **CTU** (Coding Tree Block and Coding Tree Unit): CTB is the largest block: the root in the quad-tree hierarchy. The size of the luma CTB is $N \times N$, where N can be 16, 32, or 64. The chroma CTB is half-size, i.e., $N/2 \times N/2$. A CTU consists of 1 luma CTB and 2 chroma CTBs.
- **CB** and **CU** (Coding Block and Coding Unit): The CTB consists of CBs organized in the quad-tree structure. CB is a square block that can be as small as 8×8 in luma and 4×4 in chroma images. The CBs in a CTB are traversed and coded in Z-order. One luma CB and two chroma CBs form a CU.
- **PB** and **PU** (Prediction Block and Prediction Unit): A CB can be further split into PBs for the purpose of prediction. The prediction mode for a CU can be intra-picture (spatial) or inter-picture (temporal). For the intra prediction, the sizes of CB and PB are usually the same; except when the CB is 8×8, a split into 4 PBs is allowed so that each PB may have a different prediction mode. For the inter prediction, a luma or chroma CB can be split into one, two, or four PBs, i.e., the PBs may not be square albeit always rectangular. The PU contains the luma and chroma PBs and their prediction syntax.
- **TB** and **TU** (Transform Block and Transform Unit): A CB can be further split into TBs for the purpose of transform coding of residual errors. This is represented in the same quad-tree structure, hence it is very efficient. The range of the TB size is 32×32 down to 4×4. In H.265, the TBs are allowed to span across PB boundaries in inter-predicted CUs in order to gain higher coding efficiency. A TU consists of TBs from luma and chroma images.

Figure 12.13 illustrates an example in which a CTB is partitioned into CBs and then further into TBs in a quad-tree structure. In this example, the original CTB is 64×64 and the smallest TB is 4×4.

Slices and Tiles

As in H.264, H.265 supports Slices of any length consisting of a sequence of CTUs (Fig. 12.14a). They can be I-slices, P-slices, or B-slices.

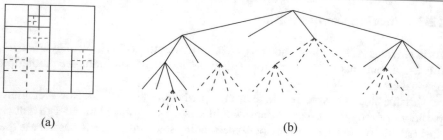

(a)   (b)

Fig. 12.13 Partitioning of a CTB. **a** The CTB and its partitioning (solid lines—CB boundaries, dotted line—TB boundaries). **b** The corresponding quad-tree

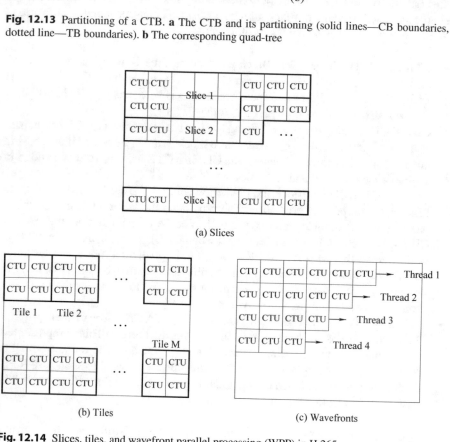

(a) Slices

(b) Tiles   (c) Wavefronts

Fig. 12.14 Slices, tiles, and wavefront parallel processing (WPP) in H.265

In addition to the Slices, the concept of *Tile* is introduced to facilitate parallel processing among multiple Tiles. A Tile is a rectangular structure consisting of CTUs (Fig. 12.14b); it may also contain multiple Slices.

An additional feature is the inclusion of the *Wavefront Parallel Processing (WPP)* technology. Basically, rows of CTUs can be processed in parallel, in multiple threads in the wavefront manner shown in (Fig. 12.14c).

For the time being, the standard does not allow the mixed use of both Tiles and Wavefronts.

Quarter-Pixel Precision in Luma Images

In inter-picture prediction for luma images, the precision for motion vectors is again at quarter-pixel as in H.264. The values at subpixel positions are derived through interpolations. As shown in Table 12.7, an 8-tap filter **hfilter** is used for half-pixel positions (e.g., position b in Fig. 12.15), and a 7-tap filter **qfilter** is used for quarter-pixel positions (e.g., positions a and c in Fig. 12.15).

As shown below, all values at subpixel positions are derived through separable filtering steps vertically and horizontally. This is different from H.264 which uses 6-tap filtering to get the values at half-pixel positions, and then averaging to obtain the values at quarter-pixel positions.

The values at positions a, b, and c can be derived using the following:

$$a_{i,j} = \sum_{t=-3}^{3} A_{i,j+t} \cdot \text{qfilter}[t], \tag{12.9}$$

$$b_{i,j} = \sum_{t=-3}^{4} A_{i,j+t} \cdot \text{hfilter}[t], \tag{12.10}$$

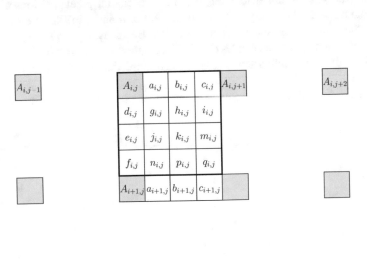

Fig. 12.15 Interpolation for fractional samples in H.265. Lower-case letters ($a_{i,j}, b_{i,j}$, etc.) indicate pixels at quarter-pixel and half-pixel positions. $A_{i,j}, A_{i,j+1}$, etc., indicate pixels on the image grid. (To save space, $A_{i,j-2}, A_{i,j+3}$, etc., are not drawn in the figure, although they will be used in the calculation.)

Table 12.7 Filters for sample interpolations in Luma images in H.265

Filter	Number of taps	Array index							
		−3	−2	−1	0	1	2	3	4
Hfilter	8	−1	4	−11	40	40	−11	4	−1
Qfilter	7	−1	4	−10	58	17	−5	1	

$$c_{i,j} = \sum_{t=-2}^{4} A_{i,j+t} \cdot \text{qfilter}[1-t]. \tag{12.11}$$

Numerically, according to Table 12.7,

$$a_{i,j} = -A_{i,j-3} + 4 \cdot A_{i,j-2} - 10 \cdot A_{i,j-1} + 58 \cdot A_{i,j} + 17 \cdot A_{i,j+1} - 5 \cdot A_{i,j+2} + A_{i,j+3},$$

$$b_{i,j} = -A_{i,j-3} + 4 \cdot A_{i,j-2} - 11 \cdot A_{i,j-1} + 40 \cdot A_{i,j} + 40 \cdot A_{i,j+1} - 11 \cdot A_{i,j+2} + 4 \cdot A_{i,j+3} - A_{i,j+4},$$

$$c_{i,j} = A_{i,j-2} - 5 \cdot A_{i,j-1} + 17 \cdot A_{i,j} + 58 \cdot A_{i,j+1} - 10 \cdot A_{i,j+2} + 4 \cdot A_{i,j+3} - A_{i,j+4}.$$

The actual implementation involves a right shift by $(B - 8)$ bits after the above calculations, where $B \geq 8$ is the number of bits per image sample.

It should be obvious that the 8-tap hfilter is symmetric, so it works well for the half-pixel positions which are in the middle of pixels that are on the image grid. The 7-tap qfilter is asymmetric, well-suited for the quarter-pixel positions which are not in the middle. The subtly different treatment of a and c in Eqs. 12.9 and 12.11 reflects the nature of this asymmetric operation. Basically, qfilter$[1 - t]$ is a flipped version of qfilter$[t]$. For example, $a_{i,j}$ is closest to $A_{i,j}$, and it will draw the most from $A_{i,j}$ with the weight 58, whereas $a_{i,j}$ will draw the most from $A_{i,j+1}$ with the weight 58.

Similarly, the values at positions d, e, and f can be derived using the following:

$$d_{i,j} = \sum_{t=-3}^{3} A_{i+t,j} \cdot \text{qfilter}[t], \tag{12.12}$$

$$e_{i,j} = \sum_{t=-3}^{4} A_{i+t,j} \cdot \text{hfilter}[t], \tag{12.13}$$

$$f_{i,j} = \sum_{t=-2}^{4} A_{i+t,j} \cdot \text{qfilter}[1-t]. \tag{12.14}$$

The other subpixel samples can be obtained from the vertically nearby a, b, or c pixels as below. To enable 16-bit operations, a right shift of 6 bits is introduced.

$$g_{i,j} = \left(\sum_{t=-3}^{3} a_{i+t,j} \cdot \text{qfilter}[t] \right) \gg 6,$$

$$j_{i,j} = \left(\sum_{t=-3}^{4} a_{i+t,j} \cdot \text{hfilter}[t] \right) \gg 6,$$

$$n_{i,j} = \left(\sum_{t=-2}^{4} a_{i+t,j} \cdot \text{qfilter}[1-t] \right) \gg 6,$$

$$h_{i,j} = \left(\sum_{t=-3}^{3} b_{i+t,j} \cdot \text{qfilter}[t] \right) \gg 6,$$

$$k_{i,j} = \left(\sum_{t=-3}^{4} b_{i+t,j} \cdot \text{hfilter}[t] \right) \gg 6,$$

$$p_{i,j} = \left(\sum_{t=-2}^{4} b_{i+t,j} \cdot \text{qfilter}[1-t] \right) \gg 6,$$

$$i_{i,j} = \left(\sum_{t=-3}^{3} c_{i+t,j} \cdot \text{qfilter}[t] \right) \gg 6,$$

$$m_{i,j} = \left(\sum_{t=-3}^{4} c_{i+t,j} \cdot \text{hfilter}[t] \right) \gg 6,$$

$$q_{i,j} = \left(\sum_{t=-2}^{4} c_{i+t,j} \cdot \text{qfilter}[1-t] \right) \gg 6.$$

12.3.2 Integer Transform

As in H.264, transform coding is applied to the prediction error residuals. The 2D transform is accomplished by applying a 1D transform in the vertical and then horizontal direction. This is implemented by two matrix multiplications: $\mathbf{F} = \mathbf{H} \times \mathbf{f} \times \mathbf{H}^T$, where $\mathbf{f}$ is the input residual data and $\mathbf{F}$ is the transformed data. $\mathbf{H}$ is the Integer Transform matrix that approximates the DCT-matrix.

Transform block sizes of 4×4, 8×8, 16×16, and 32×32 are supported. Only one Integer Transform Matrix, i.e., $\mathbf{H}_{32 \times 32}$ is specified in H.265. The other matrices for smaller TBs are subsampled versions of $\mathbf{H}_{32 \times 32}$. For example, $\mathbf{H}_{16 \times 16}$ shown below is for the 16×16 TBs.

$$
\mathbf{H}_{16 \times 16} =
\begin{bmatrix}
64 & 64 & 64 & 64 & 64 & 64 & 64 & 64 & 64 & 64 & 64 & 64 & 64 & 64 & 64 & 64 \\
90 & 87 & 80 & 70 & 57 & 43 & 25 & 9 & -9 & -25 & -43 & -57 & -70 & -80 & -87 & -90 \\
89 & 75 & 50 & 18 & -18 & -50 & -75 & -89 & -89 & -75 & -50 & -18 & 18 & 50 & 75 & 89 \\
87 & 57 & 9 & -43 & -80 & -90 & -70 & -25 & 25 & 70 & 90 & 80 & 43 & -9 & -57 & -87 \\
83 & 36 & -36 & -83 & -83 & -36 & 36 & 83 & 83 & 36 & -36 & -83 & -83 & -36 & 36 & 83 \\
80 & 9 & -70 & -87 & -25 & 57 & 90 & 43 & -43 & -90 & -57 & 25 & 87 & 70 & -9 & -80 \\
75 & -18 & -89 & -50 & 50 & 89 & 18 & -75 & -75 & 18 & 89 & 50 & -50 & -89 & -18 & 75 \\
70 & -43 & -87 & 9 & 90 & 25 & -80 & -57 & 57 & 80 & -25 & -90 & -9 & 87 & 43 & -70 \\
64 & -64 & -64 & 64 & 64 & -64 & -64 & 64 & 64 & -64 & -64 & 64 & 64 & -64 & -64 & 64 \\
57 & -80 & -25 & 90 & -9 & -87 & 43 & 70 & -70 & -43 & 87 & 9 & -90 & 25 & 80 & -57 \\
50 & -89 & 18 & 75 & -75 & -18 & 89 & -50 & -50 & 89 & -18 & -75 & 75 & 18 & -89 & 50 \\
43 & -90 & 57 & 25 & -87 & 70 & 9 & -80 & 80 & -9 & -70 & 87 & -25 & -57 & 90 & -43 \\
36 & -83 & 83 & -36 & -36 & 83 & -83 & 36 & 36 & -83 & 83 & -36 & -36 & 83 & -83 & 36 \\
25 & -70 & 90 & -80 & 43 & 9 & -57 & 87 & -87 & 57 & -9 & -43 & 80 & -90 & 70 & -25 \\
18 & -50 & 75 & -89 & 89 & -75 & 50 & -18 & -18 & 50 & -75 & 89 & -89 & 75 & -50 & 18 \\
9 & -25 & 43 & -57 & 70 & -80 & 87 & -90 & 90 & -87 & 80 & -70 & 57 & -43 & 25 & -9
\end{bmatrix}
\tag{12.15}
$$

$\mathbf{H}_{8 \times 8}$ can be obtained by using the first 8 entries of Rows 0, 2, 4, 6, ... of $\mathbf{H}_{16 \times 16}$. For $\mathbf{H}_{4 \times 4}$, use the first 4 entries of Rows 0, 4, 8, and 12.

$$
\mathbf{H}_{4 \times 4} =
\begin{bmatrix}
64 & 64 & 64 & 64 \\
83 & 36 & -36 & -83 \\
64 & -64 & -64 & 64 \\
36 & -83 & 83 & -36
\end{bmatrix}
\tag{12.16}
$$

Compared to Eq. 12.2, the entries in $\mathbf{H}_{4 \times 4}$ clearly have much larger magnitudes. In order to use 16-bit arithmetic and 16-bit memory, the dynamic range of the intermediate results from the first matrix multiplication must be reduced by introducing a 7-bit right shift and 16-bit clipping operation.

12.3.3 Quantization and Scaling

Unlike the $\mathbf{H}$ matrix in H.264 (Eq. 12.2), the numbers in the H.265 integer transform matrices, e.g., Eq. 12.15, are proportionally very close to the actual values of the DCT basis functions. Hence, the ad hoc scaling factors as built in Tables 12.1 and 12.2 are no longer needed.

For quantization, the quantization matrix and the same parameter QP as in H.264 are employed. The range of QP is $[0, 51]$. Similarly, the quantization step size doubles when the QP value is increased by 6.

12.3.4 Intra-Coding

As in H.264, spatial predictions are used in Intra-coding in H.265. The neighboring boundary samples from the blocks at the top and/or left of the current block are used for the predictions. The prediction errors are then sent for transform coding. The transform block (TB) size ranges from 4×4 to 32×32 in Intra-coding in H.265. Because of (a) the potentially much larger TB size, and (b) the effort to reduce prediction errors, the possible number of prediction modes is increased from 9 in H.264 to 35 in H.265. As shown in Fig. 12.16a, Mode 2 to Mode 34 are Intra_angular prediction modes. Note, the angle difference between modes is deliberately made uneven, e.g., to make it denser near horizontal or vertical directions. Most of the samples that are needed for angular predictions will be at subpixel positions. Bilinear interpolation of the two nearest pixels at integer positions is employed, and the precision is up to 1/32 pixel.

The two special prediction modes are Mode 0: Intra_Planar and Mode 1: Intra_DC. They are similar as in H.264. In Intra_DC, the average of the reference samples is used as the prediction. In Intra_planar, different from H.264, all four corners are used for the planar prediction, i.e., two plane predictions will be made and the average of their values will be adopted.

12.3.5 Discrete Sine Transform (DST)

In Intra_4×4, for luma residual blocks, HEVC introduced an alternative transform based on one of the variants of the *Discrete Sine Transform (DST)* (the so-called DST-VII) [16]. It is because the intra predictions are based on the neighboring boundary samples on the top or at the left of the block. The prediction error tends to increase for

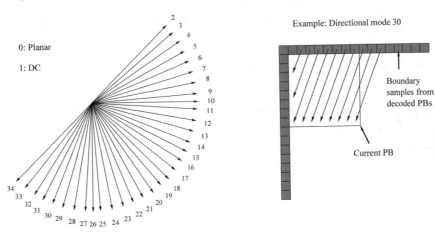

(a) Modes and Intra prediction directions

(b) Intra prediction for an 8×8 block

Fig. 12.16 H.265 Intra prediction

the nodes in the block that are farther away from the top or left neighboring samples. In general, DST is found to cope with this situation better than DCT at the transform coding step.

The integer matrix for DST can be described by

$$\mathbf{H}_{\text{DST}}[i, j] = \text{round} \left(128 \times \frac{2}{\sqrt{2N+1}} \sin \frac{(2i+1)(j+1)\pi}{2N+1} \right), \qquad (12.17)$$

where $i = 0, ..., N-1$ and $j = 0, ..., N-1$ are the row and column indices, and the block size is $N \times N$.

When $N = 4$, the following $\mathbf{H}_{\text{DST}}$ is obtained:

$$\mathbf{H}_{\text{DST}} = \begin{bmatrix} 29 & 55 & 74 & 84 \\ 74 & 74 & 0 & -74 \\ 84 & -29 & -74 & 55 \\ 55 & -84 & 74 & -29 \end{bmatrix} \qquad (12.18)$$

Saxena and Fernandes [17, 18] further studied the benefit of combining DCT and DST, i.e., allowing either DCT or DST in one of the two 1D transforms, because DST and DCT are shown to win in either the vertical and/or horizontal direction(s) for certain prediction modes. Although there are over 30 different intra prediction directions in H.265, they classify the prediction modes into

- Category 1—the samples for prediction are either all from the left neighbors of the current block (Fig. 12.17a), or all from the top neighbors of the current block (Fig. 12.17b).
- Category 2—the samples for prediction are from both the top and left neighbors of the current block (Fig. 12.17c, d).
- DC—a special prediction mode in which the average of a fixed set of neighboring samples is used.

Table 12.8 shows some of their recommendations.

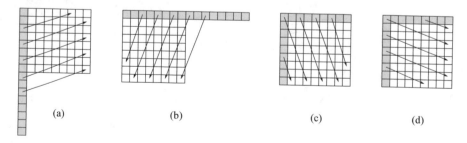

(a)   (b)   (c)   (d)

Fig. 12.17 Intra prediction directions in H.265. **a** Category 1, predictions from left neighbors only, **b** Category 1, predictions from top neighbors only, **c** and **d** Category 2, predictions from both top and left neighbors

Table 12.8 Combining DCT and DST for Intra-coding

Intra prediction category	Neighboring samples used	Vertical (Column) transform	Horizontal (Row) transform
Category 1	From left only	DCT	DST
Category 1	From top only	DST	DCT
Category 2	From both top and left	DST	DST
DC	Special (from a fixed set)	DCT	DCT

12.3.6 In-Loop Filtering

Similar to H.264, in-loop filtering processes are applied in order to remove some blocky and other artifacts. In addition to Deblocking Filtering, H.265 also introduces a *Sample Adaptive Offset (SAO)* process.

Deblocking Filtering

Instead of applying deblocking filtering to 4×4 blocks as in H.264, it is applied only to edges that are on the 8×8 image grid. This reduces the computation complexity, and it is especially good for parallel processing since the chance of cascading changes at nearby samples is greatly reduced. The visual quality is still good, partly due to the SAO process described below.

The deblocking filtering is applied first to the vertical edges, then to the horizontal edges in the picture, thus enabling parallel processing. Alternatively, it can be applied CTB by CTB.

Sample Adaptive Offset (SAO)

The SAO process can be invoked optionally after the Deblocking Filtering. Basically, an offset value is added to each sample based on certain conditions described below.

Two modes are defined for applying the SAO: *Band offset mode* and *Edge offset mode*.

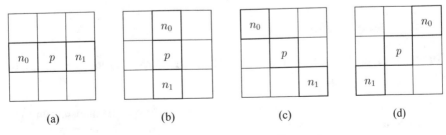

(a)   (b)   (c)   (d)

Fig. 12.18 Neighboring samples considered in SAO edge offset mode

In the Band offset mode, the range of the sample amplitudes is split into 32 bands. A band offset can be added to the sample values in four of the consecutive bands simultaneously. This helps to reduce the "banding artifacts" in smooth areas.

In the Edge Offset Mode, the gradient (edge) information is analyzed first. Figure 12.18 depicts the four possible gradient (edge) directions: (a) horizontal, (b) vertical, and (c, d) diagonals. A positive or negative offset, or zero offset, can be added to the sample p based on the following:

- Positive: p is a local minimum ($p < n_0$ & $p < n_1$), or p is an edge pixel ($p < n_0$ & $p = n_1$ or $p = n_0$ & $p < n_1$).
- Negative: p is a local maximum ($p > n_0$ & $p > n_1$), or p is an edge pixel ($p > n_0$ & $p = n_1$ or $p = n_0$ & $p > n_1$).
- Zero: None of the above.

12.3.7 Entropy Coding

H.265 only uses CABAC in entropy coding, i.e., CAVLC is no longer used. Because of the newly introduced coding tree and transform tree structure, the tree depth now becomes an important part of the *context modeling* in addition to the spatially neighboring context in H.264/AVC. As a result, the number of contexts is reduced, and the entropy coding efficiency is further improved.

Unlike previous video standards, three simple scanning methods are defined to read in the transform coefficients, i.e., *Diagonal up-right, Horizontal*, and *Vertical*. The goal is still to maximize the length of zero-runs. The scanning always takes place in 4×4 sub-blocks regardless of the TB size. The Diagonal up-right scan is used for all inter-predicted blocks and for intra-predicted blocks that are 16×16 or 32×32. For intra-predicted blocks that are 4×4 or 8×8, the following are used: Horizontal—for prediction directions close to vertical; Vertical—for prediction directions close to horizontal; Diagonal up-right—for the other prediction directions.

There are many improvements as to how to code the nonzero transform coefficients efficiently [13, 15]. Also, one of the goals of the new implementation of CABAC in H.265 is to simplify its context representations so its throughput can be increased. For details, readers are referred to [19] which provides an excellent reference.

12.3.8 Special Coding Modes

Three special coding modes are defined in H.265. They can be applied at the CU or TU level.

- **I_PCM**: As in H.264, the prediction, transform coding, quantization, and entropy coding steps are bypassed. The PCM coded (fixed-length) samples are sent directly. It is invoked when other prediction modes fail to produce any data reduction.

- **Lossless**: The residual errors from inter- or intra predictions are sent to entropy coding directly, thus to avoid any lossy steps, especially the quantization after transform coding.
- **Transform skipping**: Only the transform step is bypassed. This works for certain data (e.g., computer-generated images or graphics). It can only be applied to 4×4 TBs.

12.3.9 H.265 Profiles

In Version 1 of HEVC, only three profiles were defined: *Main profile*, *Main 10 profile*, and *Main Still Picture profile*. Version 2 of HEVC created 21 additional profiles [20], e.g., *Main 12, Main 4:2:2 10, Main 4:2:2 12, Main 4:4:4 12, Main 4:4:4 16 Intra, Main 4:4:4 16 Still Picture, Monochrome, Monochrome 12, Monochrome 16, ... profiles*.

The default format for color is YCbCr.

As an example, some of the video formats that are supported at various *levels* in the Main profile are listed in Table 12.9. As shown, the total number of levels proposed is 13. It covers very low-resolution videos, e.g., QCIF (176×144) at Level 1, as well as very high- resolution videos, e.g., UHDTV ($8, 192 \times 4, 320$) at Levels 6, 6.1, and 6.2.

In calculating the Max Luma Picture Size, the width and height are rounded up to the nearest multiples of 64 (as an implementation requirement). For example, 176×144 becomes $192 \times 192 = 36, 864$.

As shown, HDTV is at Levels 4 and 4.1 for Frame Rates of 30 and 60 fps. The maximum bitrates of the compressed video at the so-called Main Tier are 12 Mbits/sec and 20 Mbits/sec, respectively. At the High Tier, they can be as much as 2.5 times higher. The UHD videos at Level 5 and above demand much higher bitrates which remain as a challenge for all aspects of multimedia including storage, data transmission, and display devices.

Ohm et al. [21] reported many of their experimental results in comparing the coding efficiency between H.264 and H.265. As an example, Table 12.10 lists the average bitrate reductions when different video compression methods are compared at the same PSNR, in this case, in the range 32–42 dB. The test data are entertainment videos which are generally of higher quality and have higher resolutions. The notations are as follows: MP—Main Profile, HP—High Profile, and ASP—Advanced Simple Profile. For example, when H.265 MP is compared with H264/MPEG-4 AVC HP, a saving of 35.4% is realized.

The coding efficiency is also compared subjectively between H.264 and H.265. The *double stimulus* method as described in Sect. 5.6.2 is employed. It is reported [21] that when compared with H.264/MPEG-4 AVC HP, at approximately the same subjective quality, for the 9 test videos for entertainment applications, the average bitrate reductions by H.265 MP range from 29.8 to 66.6%, with an average of 49.3%. This is very close to the original goal of 50% reduction.

Table 12.9 Sample video formats supported in the H.265 main profile

Level(s)	Max Luma picture width × height	Max Luma picture size (samples)	Frame rate (fps)	Main Tier max Bitrate (Mb/s)
1	176 × 144	36,864	15	0.128
2	352 × 288	122,880	30	1.5
2.1	640 × 360	245,760	30	3.0
3	960 × 540	552,960	30	6.0
3.1	1280 × 720	983,040	30	10
4 / 4.1	2048 × 1080	2,228,224	30 / 60	12 / 20
5 / 5.1 / 5.2	4096 × 2160	8,912,896	30 / 60 / 120	25 / 40 / 60
6 / 6.1 / 6.2	8192 × 4320	35,651,584	30 / 60 / 120	60 / 120 / 240

Table 12.10 Average bitrate reductions under equal PSNR

Video compression method	H.264/MPEG-4 AVC HP	MPEG-4 ASP	MPEG-2/H.262 MP (%)
H.265 MP	35.4%	63.7%	70.8
H.264/MPEG-4 AVC HP	–	44.5%	55.4
MPEG-4 ASP	–	–	19.7

12.3.10   H.265 Extensions

Due to new developments in the video technology and a broader range of its applications, Joint Collaborative Team on Video Coding (JCT-VC) continued to work on the extensions of H.265 [22]. Initially, in Version 2 of HEVC, the extensions included Range Extensions (RExt), Scalable HEVC (SHVC), and Screen Content Coding (SCC). Later, Joint Collaborative Team on 3D Video Coding Extension Development (JCT-3V) was established to work on Multiview and 3D Video Coding Extensions (MV-HEVC and 3D-HEVC, respectively). Version 3 of HEVC including all the above extensions was approved in April 2015. More SCC extensions profiles, SHVC extensions profiles, and high throughput extensions profiles were included in Version 4 which was approved in December 2016.

Range Extensions

Version 1 of HEVC was mostly designed for applications with 4:2:0 chroma subsampling at 8 or 10 bits per sample. This became inadequate for various new applications. For example, many digital video broadcasting applications use 4:2:2 chroma format at 10 bits per sample, professional cameras use 4:4:4 and the R'G'B' color space, and high dynamic range (HDR) contents may need up to 16 bits per sample. Moreover, there is an increasing need for lossless compression (e.g., for content preservation and medical imaging), as well as coding of screen content where 4:4:4 with 8 or 10 bits per sample is commonly employed.

The main objective of *range extensions* described in HEVC RExt [20] is the support for 4:2:2 and 4:4:4 chroma formats and bit-depth of more than 10 bits (e.g., 12 or 16 bits) per sample. In addition, extended functionality and increased coding efficiency are supported, for example, coding of screen content, direct coding of R'G'B' source material, and coding of auxiliary pictures, e.g., alpha planes, depth maps, and lossless video coding.

The implementation of the range extensions is mostly realized by enhanced coding algorithms and tools, although in many cases, it also demands a higher dynamic range of the processing elements. Flynn et al. [20] provide a detailed description of RExt.

Scalability Extensions

The standard for scalable high efficiency video coding (SHVC) [23] was finalized in Version 2 of HEVC. In addition to the *temporal scalability* which was already supported by Version 1 of HEVC, SHVC supports *spatial scalability*, *SNR (signal-to-noise ratio) scalability*, as well as *bit-depth scalability* and *color gamut scalability* needed for UHD video where higher bit-depth and wider color gamut are employed. SHVC also supports combinations of any of the above.

Scalable coding usually involves multiple layers, i.e., base layer (BL) and enhancement layers (ELs). For scalable extensions to H.264, scalable video coding (SVC) was developed, which uses the so-called "single-loop decoding constraint," i.e., when decoding a bitstream containing multiple layers for temporal scalability, partial decoding of reference layers may be sufficient. As a result, it is more efficient, but could not handle more sophisticated extensions, e.g., the arbitrary switching in multiple layers for spatial scalability, etc. SHVC employs a multi-loop coding framework in which reference layers must be fully decoded first so that they can be used as prediction references. If there are more than two spatial or SNR layers, the intermediate ELs may also be used as reference layers. Moreover, SHVC uses multiple repurposed single-layer HEVC codec cores, with the addition of interlayer reference picture processing modules. The goal is to adopt a scalable coding architecture that relies on making high-level syntax only (HLS-only) changes to the underlying single-layer HEVC standard. Due to its success, the general multilayer high-level syntax design becomes common to all multilayer HEVC extensions, including SHVC, MV-HEVC, and 3D-HEVC.

Boyce et al. [23] provide a detailed description of SHVC.

Multiview and 3D Video Extensions

The multiview extension (MV-HEVC) [24] enables efficient coding of multiple camera views and associated auxiliary pictures. It can be implemented by reusing single-layer decoders without changing the block-level processing modules. Inter-view references in motion-compensated prediction are employed in order to realize some bitrate savings compared with HEVC simulcast.

In an autostereoscopic display, several images are emitted at the same time, and the user (with no need of glasses) only sees one stereo pair of images depending on his/her viewing position. The technique of depth-image-based rendering (DIBR) is often employed to generate these images. Naturally, it heavily relies on high-quality depth images. The 3D extension (3D-HEVC) [24] is designed for videos consisting of multiple views and associated depth maps. It also supports the generation of additional intermediate views in advanced 3D displays. New block-level video coding tools are specified to realize additional bitrate reduction compared with MV-HEVC especially in cases where depth maps must be coded. The tools explicitly exploit the unique characteristics of depth maps, and statistical dependencies between video texture and depth.

Both MV-HEVC and 3D-HEVC have been developed to support stereoscopic and autostereoscopic displays, hence so far they have limited ability in handling arrangements with a large number of views or arbitrary viewing positions.

Tech et al. [24] provide a detailed description of MV-HEVC and 3D-HEVC.

12.4 H.266

Joint Video Exploration Team (JVET) of the ITU-T Video Coding Experts Group (VCEG) was formed in 2015, 2 years after the H.265 (HEVC) Version 1 was finalized. Since then, the experts of VCEG and ISO/IEC MPEG have been collaboratively exploring the next major advances in coding efficiency for a new generation of the video coding standard. A Joint Exploration Model (JEM) and its software implementation were developed; its last version (JEM-7.0) achieved about 30% bitrate reduction compared to the HEVC test model [25]. This convinced VCEG and MPEG to formally call for proposals in October 2017 for a new video coding standard. The new standard will be suitable for standard dynamic range (SDR) video, high dynamic range (HDR) video, 360° video, and other new video formats (e.g., panoramic format), hence the name *Versatile Video Coding (VVC)*, also known as MPEG-I Part 3, or H.266. The H.266 standard was finalized in 2020.

H.266 supports up to 16K UHD videos, YCbCr 4:4:4, 4:2:2, 4:2:0 with 10 to 16 bits for each sample. It supports HDR of more than 16 stops (with peak brightness up to 10,000 nits[1]) and BT.2100 wide color gamut, auxiliary channels for depth, transparency, etc. It is aimed to achieve about 50% more compression than H.265 for the same perceptual quality.

Currently, compared to H.265, the proposed H.266 methods have much higher computational complexities on the encoder side. On the decoder side, they range from over 200% to over 300%. Major efforts are still being made to reduce the computational complexity on both the encoder and decoder sides.

---

[1]Nit is a measuring unit of luminance, which is equivalent to cd/m^2 (*candela per square meter*). $1\ nit = 1\ cd/m^2$.

Not surprisingly, there have been increasingly more studies on deep learning technologies for video coding [26,27]. Liu et al. [27] provided a survey of some of the recent developments. They are divided into two categories: (a) deep network-based coding tools that can be used in traditional coding schemes, (b) new coding schemes that are primarily built on deep networks.

The new techniques of the H.266 codec have a tremendous amount of details. In general, the developments are incremental from H.264 and H.265. No major deviation from the block-based hybrid coding was proposed. We will only discuss some of the essential new developments in this chapter. The papers [25,28–31] and others in the May 2020 Special issue of IEEE Transactions on Circuits and Systems for Video Technology provide much more details.

12.4.1 Motion Compensation

As in H.264 and H.265, H.266 employs the block-based hybrid coding scheme that supports a combination of inter-picture motion predictions and intra-picture spatial prediction, and transform coding on prediction residual errors.

One-Sixteenth-Pixel Precision

In H.264 and H.265, the precision for motion vectors (MVs) is one-quarter pixel for luma images (and one-eighth for chroma images in 4:2:0 videos). In JEM, the highest MV precision has been increased to 1/16 pixel for luma images (and 1/32 for chroma images in 4:2:0 videos). The 1/16 precision is not always used; it is used in inter prediction for the coding unit (CU) coded with the skip/merge mode, for example. In general, the search for the best MV will start with integer pixel precision. After the "optimal" MV is found at this stage, the search will continue progressively at 1/2, 1/4, ... pixel precisions.

The image samples at the fractional positions are derived by interpolation. Similar to H.265, separable 2D interpolation filters with a long tap, e.g., 8, are employed. The 1/32 fractional position samples are obtained by a simple average of the two neighboring 1/16 samples.

Block Partitioning

In H.266, the maximum size of the coding tree unit (CTU) is increased to 256×256 from 64×64 in H.265. As before, the CTU is further partitioned into coding units (CUs)—basic units for the transform and quantization of prediction residual errors. To improve the coding efficiency, in addition to the quad-tree (QT) in CTU as in H.265, binary tree (BT) partitioning was introduced in JEM, hence the term QTBT. In [28], it is further recommended that H.266 may offer the following options for block partitioning of the CTU as illustrated in Fig. 12.19:

- *Multi-Type Tree (MTT)*: Each QT leaf in the CTU can be further partitioned by *Binary Trees (BTs)* or *Ternary Trees (TTs)*. The BT and TT partitions can be

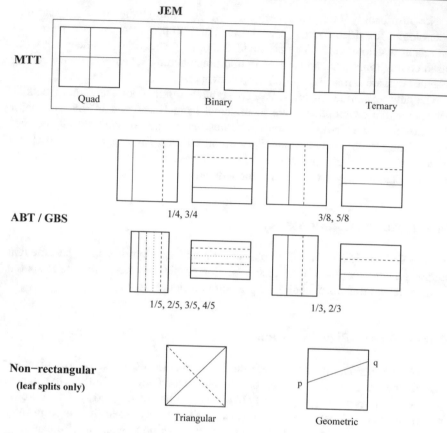

Fig. 12.19 H.266 block partitioning

applied vertically or horizontally. They can be applied recursively and interleaved. The TT uses a 1:2:1 ratio to make sure the height or width is of power of two. Once the BT/TT partitions are applied, no more QT partition is allowed. MTT was shown to be effective. Compared to JEM-7.0 where only QTBT was allowed, Luma BD-rates[2] [32] in the range of approximately -2 to -3% were reported [28].

- *Asymmetric Binary Tree (ABT)* and *Generalized Binary Tree with Shifts (GBS)*: As shown in Fig. 12.19, ABT uses a ratio of 1:3 (or 3:1). The combination of ABT and QTBT was reported to yield a Luma BD-rate of -3.2%. The GBS allows binary splits with ratios 1:2, 2:3, 1:4, and 3:5 (or the reverses) as long as the width and height of the resulting CUs are the multiples of 4 luma pixels.
- *Non-Rectangular Partitions*: For leaf nodes only, additional *triangular partitioning* and *Geometric Partitioning (GEO)* are also proposed. For GEO, the partitioning line is specified by points P and Q on the boundary of the block. When

---

[2]See Chap. 5 for the definition of BD-rate.

the luma block size is 16×16 and smaller, a lookup table can be employed for a total of 16 predefined GEO partitions. Otherwise, the partitioning line can be predicted from spatially or temporally neighboring blocks. Due to the block's nonrectangular shape, a Shape Adaptive DCT (SA-DCT) can also be employed.

12.4.2 Adaptive Multiple Transforms

As mentioned in Sect. 12.3.5, discrete sine transform (DST-VII) was introduced in H.265 for prediction residual errors in intra-coding, because the Sine basis function can fit the pattern of the errors better.

Further studies concluded that the following transforms can be beneficial for different modes in H.266 intra-coding [16, 25]. These are some of the 16 variants of DCT (DCT Type I–VIII) and DST (DST Type I–VIII) with different symmetry of their symmetric-periodic sequences (Table 12.11).

For intercoding in H.266, DCT-VIII and DST-VII are adopted. For intra-coding in H.266, three transform sets as listed in Table 12.12 are predefined.

H.265 uses 35 modes for intra prediction, i.e., Planar (Mode 0), DC (Mode 1), plus 33 directional angular modes (see Fig. 12.16a). In H.266, this has been further refined to a total of 67 modes, i.e., Planar (Mode 0), DC (Mode 1), plus 65 directional angular modes, basically doubling the number (hence the precision) of the angular modes in H.265. Table 12.13 lists the transform sets that should be considered for the two consecutive 1D transforms applied vertically (V) and horizontally (H). For each of the V and H transforms, the encoder should try both of the transforms in the set, and the one that is better in dealing with the residual errors will be adopted.

Because the intra prediction directions are symmetric, the selection of the transform set is also symmetric, i.e., except for the first two special modes (Planar and DC), the selected transform set for H of an angular mode i is the same as the one for V of mode $68 - i$, and vice versa.

12.4.3 Non-separable Secondary Transform

The DCT/DST transforms are known to be *separable*. A 2D separable transform can be implemented as two consecutive 1D transforms to yield much lower computational complexity. However, for complex texture patterns, arbitrary edge directions in the image, *Non-Separable Secondary Transforms (NSST)* may provide a better fit to their prediction residual errors. To limit the complexity, NSST will only be applied to the frequency coefficients after the primary separable transforms (DCT/DST), i.e., as a secondary transform to further improve the coding efficiency. NSST was studied at the time of the H.265 development, but was not included in the final version of H.265. In the JEM, a *Mode Dependent Non-Separable Secondary Transform (MD-NSST)* was introduced between the primary transform and quantizer at the encoder, and between the de-quantizer and inverse primary transform at the decoder.

Table 12.11 Basis functions for certain types of DCT and DST transforms

Transform type	Basis function $T_i(j)$, where $i, j = 0, 1, ..., N - 1$
DCT-II	$T_i(j) = C_0(i) \cdot \sqrt{\frac{2}{N}} \cdot \cos\left(\frac{i \cdot (2j+1) \cdot \pi}{2N}\right)$, where $C_0(i) = \begin{cases} \sqrt{\frac{2}{N}} & i = 0 \\ 1 & i \neq 0 \end{cases}$
DCT-V	$T_i(j) = C_0(i) \cdot C_1(j) \cdot \sqrt{\frac{2}{2N-1}} \cdot \cos\left(\frac{2i \cdot j \cdot \pi}{2N-1}\right)$, where $C_0(i) = \begin{cases} \sqrt{\frac{2}{N}} & i = 0 \\ 1 & i \neq 0 \end{cases}$, $C_1(j) = \begin{cases} \sqrt{\frac{2}{N}} & j = 0 \\ 1 & j \neq 0 \end{cases}$
DCT-VIII	$T_i(j) = \sqrt{\frac{4}{2N+1}} \cdot \cos\left(\frac{(2i+1) \cdot (2j+1) \cdot \pi}{4N+2}\right)$
DST-I	$T_i(j) = \sqrt{\frac{2}{N+1}} \cdot \sin\left(\frac{(i+1) \cdot (j+1) \cdot \pi}{N+1}\right)$
DST-VII	$T_i(j) = \sqrt{\frac{4}{2N+1}} \cdot \sin\left(\frac{(2i+1) \cdot (j+1) \cdot \pi}{2N+1}\right)$

Table 12.12 Predefined transform sets for H.266 intra-coding

Transform set	Transform candidates
0	DST-VII, DCT-VIII
1	DST-VII, DST-I
2	DST-VII, DCT-V

Table 12.13 Selected transform sets for each intra prediction mode in H.266

Intra mode	0	1	2	3	4	5	6	7	8	9	10	11	12	13	14	15	16	17
V	2	1	0	1	0	1	0	1	0	1	0	1	0	1	0	0	0	0
H	2	1	0	1	0	1	0	1	0	1	0	1	0	1	2	2	2	2

Intra mode	18	19	20	21	22	23	24	25	26	27	28	29	30	31	32	33	34
V	0	0	0	0	0	1	0	1	0	1	0	1	0	1	0	1	0
H	2	2	2	2	2	1	0	1	0	1	0	1	0	1	0	1	0

Intra mode	35	36	37	38	39	40	41	42	43	44	45	46	47	48	49	50	51
V	1	0	1	0	1	0	1	0	1	0	1	2	2	2	2	2	2
H	1	0	1	0	1	0	1	0	1	0	1	0	0	0	0	0	0

Intra mode	52	53	54	55	56	57	58	59	60	61	62	63	64	65	66
V	2	2	2	1	0	1	0	1	0	1	0	1	0	1	0
H	0	0	0	1	0	1	0	1	0	1	0	1	0	1	0

For lower complexity, MD-NSST will only be applied to the low-frequency coefficients after the primary transform. If both the width (W) and height (H) of the coefficient block are larger than or equal to 8, an 8×8 NSST will be applied to the top-left corner of the block. Otherwise, a 4×4 NSST will be applied. As an example, the implementation of NSST on a 4×4 block is described below.

Let $\mathbf{X}$ be the 4×4 coefficient block after the primary transform,

$$\mathbf{X} = \begin{bmatrix} X_{00} & X_{01} & X_{02} & X_{03} \\ X_{10} & X_{11} & X_{12} & X_{13} \\ X_{20} & X_{21} & X_{22} & X_{23} \\ X_{30} & X_{31} & X_{32} & X_{33} \end{bmatrix} \tag{12.19}$$

it is turned into a 1D vector $\overrightarrow{\mathbf{X}}$,

$$\overrightarrow{\mathbf{X}} = \begin{bmatrix} X_{00} \\ X_{01} \\ X_{02} \\ \cdot \\ \cdot \\ \cdot \\ X_{32} \\ X_{33} \end{bmatrix} \tag{12.20}$$

Then a 16×16 transform matrix $\mathbf{T}$ is used to calculate the NSST,

$$\overrightarrow{\mathbf{F}} = \mathbf{T} \cdot \overrightarrow{\mathbf{X}}. \tag{12.21}$$

The 1D NSST coefficient vector $\overrightarrow{\mathbf{F}}$ can be reorganized as another 4×4 block following the coefficient scanning order applied on the block for entropy coding.

In the JEM, 35×3 transforms $\mathbf{T}$s for MD-NSST are predefined for both 4×4 and 8×8 NSST, where 3 is the number of NSST candidates for each intra prediction mode. For mode with index number $i > 34$, the candidate transforms for mode $68 - i$ will be used. For example, the candidate transforms for Mode 35 are the same as the ones for Mode 33, etc.

In order to reduce the complexity of the NSST, instead of matrix multiplication (as indicated in Eq. 12.21), a hypercube-givens transform (HyGT) with butterfly implementation was recommended in the JEM [25]. The HyGT implementation required a number of iterations where the input of each subsequent iteration depended on the output of the previous iteration, and it was still deemed inefficient. Hence, a *Reduced Secondary Transform (RST)* is proposed.

A reduced transform (RT) can be realized by an $R \times N$ matrix, where $R < N$. As shown above, for a 4×4 block, the length N of the 1D vector $\overrightarrow{\mathbf{X}}$ is 16; for an 8×8 block, $N = 64$.

$$\mathbf{T_{RT}} = \begin{bmatrix} T_{00} & T_{01} & T_{02} & \cdots & T_{0(N-1)} \\ T_{10} & T_{11} & T_{12} & \cdots & T_{1(N-1)} \\ & & \vdots & & \\ T_{(R-1)0} & T_{(R-1)1} & T_{(R-1)2} & \cdots & T_{(R-1)(N-1)} \end{bmatrix} \tag{12.22}$$

Using $\mathbf{T_{RT}}$, the matrix multiplication below will yield a 1D vector $\overrightarrow{\mathbf{F_{RT}}}$ of length R.

$$\overrightarrow{\mathbf{F_{RT}}} = \mathbf{T_{RT}} \cdot \overrightarrow{\mathbf{X}}. \tag{12.23}$$

The ratio of N/R is the reduction factor. $\mathbf{T_{RT}}$ basically provides the first R bases of the N-dimensional space. When $N = 64$ and $R = 16$ for an 8×8 block, for example, only 16 nonzero RST coefficients will be generated, which will be used to fill the top-left 4×4 sub-block.

At the decoder, the transpose of matrix $\mathbf{T_{RT}}$ will be used.

12.4.4 In-Loop Filtering

The JEM proposes four in-loop filters and they are applied sequentially. They are (a) Bilateral filter, (b) Deblocking filter, (c) Sample Adaptive Offset (SAO) filter, and (d) Adaptive Loop Filter (ALF). Filters (b) and (c) are basically the same as in H.265, although stronger deblocking filters with longer taps are proposed as responses for Call for Proposal after the JEM [28].

Bilateral Filter

Bilateral filter is applied to all luma blocks with nonzero transform coefficients. It takes place right after the inverse transform in the loop in the encoder. Four neighboring (top, bottom, right, and left) samples are used to filter the current sample. The distance between the current and neighboring samples can be larger than 1, and it is controlled by a *spatial* parameter σ_d determined by the transform block size. In addition to the distance, the difference of the intensity values also has an impact on the filtering amount, and it is controlled by a *range* parameter σ_r determined by the QP used for the block. Given the current sample at location (i, j), the weight $w(i, j, k, l)$ that determines the amount of filtering by the neighboring sample at (k, l) is

$$w(i, j, k, l) = exp\left(-\frac{(i-k)^2 + (j-l)^2}{2\sigma_d^2} - \frac{(R(i, j) - R(k, l))^2}{2\sigma_r^2}\right), \tag{12.24}$$

where $R(i, j)$ and $R(k, j)$ are the original reconstructed intensity values at (i, j) and (k, l), respectively.

Adaptive Loop Filter (ALF)

Adaptive Loop Filter (ALF) is applied after SAO. The horizontal and vertical gradients of each block are derived first and used to determine the direction D of the block. The blocks are then classified into 15 categories based on D and the 2D-Laplacian activity so that different sets of filter coefficients can be applied for ALF filtering. Details of ALF can be found in [25].

Convolutional Neural Network (CNN) In-loop Filters

Zhou et al. [33] trained a CNN as the only in-loop filter before the ALF in intra-coding, thus replacing the bilateral filter, deblocking filter, and SAO. In addition to the reconstructed intensity values, the quantization parameter (QP) map is also fed to the CNN, which consists of 8 convolutional layers with rectified linear unit (ReLU).

In their deep learning for video coding (DLVC), Liu et al. [26] designed a deep CNN with 16 residual blocks (ResBlocks) and 2 convolutional layers. Each ResBlock consists of 2 convolutional layers separated by a ReLU and a skip connection. Hence, the network has a total of 34 layers. The proposed CNN-based in-loop filter is placed between the deblocking filter and SAO.

The above CNN-based filtering methods all reported good improvements, ranging from approximately -3 to -7% BD-rates over the JEM target for luma and chroma images.

The CNN-based methods for video coding appear promising. Nevertheless, their development is still in the early stage.

12.4.5 Tools for High Dynamic Range (HDR) Video

Xiu et al. [29] and François et al. [30] discussed various tools for HDR videos. The tools are software designed for conversions from and into different HDR video formats, analyzing and evaluating the quality of HDR videos. The coding tools are of three types:

1. **Encoder-only Adaptations**: The encoder can apply the following adaptations: (a) Local adaptation of the distortion metrics, (b) Local adaptation of the quantization parameters (see lumaDQP below), and (c) Adjustment of the luma chroma quality balance.
2. **Out-of-loop Sample Mapping**: These are some possible preprocessing steps before the signals are sent to the encoder, and post-processing steps after the signals are recovered by the decoder. They include (a) Color volume transform and (b) Out-of-loop luma mapping.
3. **In-loop Sample Mapping**: As the name suggests, the mapping is realized in the motion compensation loop in the encoder and decoder. They include (a) QP inference—to better control the quantization parameters, (b) In-loop reshaping (see In-loop Luma Reshaper below), (c) Luma-adaptive deblocking filter—to

adjust the strength of deblocking according to the average luma level at the block boundary, and (d) Internal bit-depth increase—so that the encoder and decoder will operate at higher bit-depth.

Luma-Dependent Quantization Scheme (lumaDQP)

In addition to the higher dynamic range and wider color gamut (BT.2100 color representation), HDR videos also adopt the perceptual quantization (PQ) in which many bits are redistributed from the bright areas of a video frame to the dark areas. Initial studies for H.266 found that PQ can introduce more undesirable artifacts in bright areas in the compressed video.

lumaDQP is a luma-dependent adaptive quantization scheme [29] that enables the adjustment of quantization steps depending on the luminance of the HDR content at the CTU level in order to shift the bits (back) from the dark areas to the bright areas.

$$dQP(Y) = \max(-3, \ \min(6, \ 0.015 \cdot Y - 1.5 - 6)), \qquad (12.25)$$

where Y is the average luma value of the CTU. The maximum value of Y, for example, for a 10-bit video is 1023.

dQP can be used to adaptively modify the QP, i.e., the original QP can be subtracted by the amount indicated by dQP.

In-Loop Luma Reshaper

In general, the quantizer (e.g., PQ for HDR) is designed to cover the full range of the video signals (e.g., from 0 to 10,000 nits). Due to large variations of actual scenes, the intent of the person who creates the video, and the limit of the display device, a large portion of the full range may be unused. For better coding efficiency, a *reshaper* can be introduced to modify or re-quantize the signal. Earlier work in MPEG/JCT-VC studied some out-of-loop reshapers. Basically, the video signals can be preprocessed before encoding to reduce the unused range and hence to reduce the excessive number of unused codewords. After decoding, the video will be post-processed before being displayed. It is usually applied to a group of pictures in the video, and cannot be readily adapted to the slice or block level.

In [29], an in-loop reshaper is introduced. In its implementation, a pair of lookup tables (LUTs) are employed, i.e., Forward LUT in encoding and Inverse LUT in decoding.

The Forward LUT (FwdLUT) can be obtained by the following three steps:

1. Derive the slope of the reshaper curve:

$$slope[Y] = 2^{dQP(Y)/6} \qquad (12.26)$$

2. Calculate the luma reshaping curve:

$$F[Y+1] = F[Y] + slope[Y], \qquad (12.27)$$

where Y is in the range of $[0, MaxY]$, e.g., $[0, 1023]$ for 10-bit HDR video.

3. Obtain FwdLUT[Y] by normalizing the values in $F[Y]$ to $[0, MaxY]$.

Accordingly, the Inverse LUT (InvLUT) can be derived by a simple reverse mapping of the FwdLUT.

See [29] for comparison of the in-loop and out-of-loop luma reshapers, and joint optimization of the luma QP adaptation and reshaper.

12.4.6 Tools for 360° Video

Xiu et al. [29] and Ye et al. [31] discussed several coding tools for 360° video. We will briefly describe some of them below.

Hybrid Angular Cubemap Projection (HAC)

In equirectangular projection (ERP), the 360° (spherical) video is projected onto a single 2D plane. It is most commonly used for viewing due to its convenience. However, ERP is known to produce large distortions (see Fig. 5.9), especially for the video content near the poles of the sphere. It will cause undesirable nonuniform sampling if it is used for 360° video representation and compression. The cubemap projection (CMP) uses the six faces of the cube as projection planes; its distortion is reduced compared to ERP. Therefore, it is recommended in H.266 for better 360° video compression. In order to further reduce the nonuniform sampling, especially toward the edges on each face of the CMP cube, the hybrid angular cubemap projection (HAC) is proposed [29]. It is derived from the equi-angular cubemap projection (EAC) as described in Sect. 5.4.2. The forward mapping functions from CMP to HAC are defined as

$$f_x(x) = \frac{4}{a \cdot \pi} \arctan\left[x \cdot \tan(a \cdot \frac{\pi}{4})\right], \qquad (12.28)$$

$$f_y(y) = \frac{4}{b \cdot \pi} \arctan\left[y \cdot \tan(b \cdot \frac{\pi}{4})\right], \qquad (12.29)$$

where a and b are adjustable parameters for the horizontal and vertical mapping functions, respectively; they have a range of $[0.3, 1.5]$.

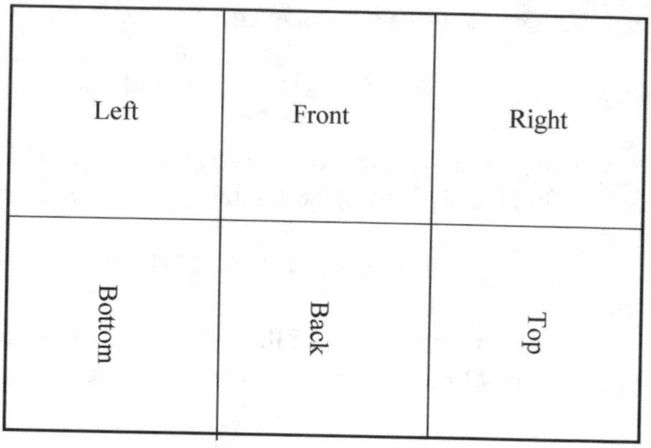

Fig. 12.20 3 × 2 frame packing for CMP-like projection formats

Left	Front	Right
Bottom	Back	Top

Adaptive Frame Packing

For coding efficiency, *frame packing* is adopted in which the six projection images from the six faces of the CMP cube are packed into a 2D rectangular picture, with either a 3 × 2 configuration (see Fig. 12.20), or 4 × 3 configuration. Apparently, no matter how one arranges the six images, there will be discontinuities at the vertical and/or horizontal boundaries between some faces, hence the "face seam" artifacts. This causes hardship in motion estimation, because the reference block that happens to be beyond the boundary of the current face and is (partially or entirely) in the neighboring face in the packed frame may not offer any continuous video content. To handle the face discontinuities, *Adaptive Frame Packing (AFP)* can be used in which the six face images can be "shuffled" and/or yaw-rotated by 0°, 90°, 180°, and 270° to dynamically create the needed continuity between frame-packed faces during the motion estimation.

Two types of neighbors can be defined in the block-based motion compensation: (a) *spherical neighbors*—neighboring blocks in the original 3D geometry, (b) *frame-packed neighbors*—neighboring blocks in the packed frame. It is proposed [29] that a frame-packed neighboring block is used for motion estimation of the current block only if it is also a spherical neighbor.

Geometry Padding

In motion estimation, when the needed video content goes beyond the reference image boundary, the technique of *padding* can be employed. Conventionally, *repetitive padding* is used, e.g., as described in Chap. 11. The neighboring face images in the packed frames are often discontinuous (with "face seams") as described above; they cannot readily be used for padding. *Geometry Padding* [29,31] is a technique where the content of the spherical neighbors is utilized for padding to provide more continuous and usable samples for areas outside of the face boundaries.

Table 12.14 H.266 BD-rates for the SDR-A (random access) category

	Over H.265 test model			Over JEM-7.0 test model		
	Y	U	V	Y	U	V
Average over 9 test videos (%)	−35.7	−44.7	−45.0	−4.0	−3.3	−3.2

Table 12.15 H.266 BD-rates for the SDR-B (low delay) category

	Over H.265 test model			Over JEM-7.0 test model		
	Y	U	V	Y	U	V
Average over 5 test videos (%)	−27.2	−43.8	−44.5	−3.6	−2.6	−4.5

12.4.7 H.266 Performance Report

BD-rate (short for *Bjontegaard's delta-rate*) [32] is again employed to measure the performance of the proposed H.266 compression methods. The BD-rate is the average difference of bitrates over a range of PSNRs (or quality levels), for example, between H.265 (HEVC) and H.266, or JEM and H.266 test models. A negative BD-rate indicates a saving of bitrate achieved by H.266.

When compared to H.265 (HEVC), Xiu et al. [29] reported luma BD-rates of −35.7 and −27.2% for their standard dynamic range (SDR) test videos for Random Access and Low Delay configurations, averaged over 9 and 5 test videos, respectively (see Tables 12.14 and 12.15). They are −4.0% and −3.6%, respectively, when compared to JEM-7.0.

Due to the substantial difference between SDR and HDR content (gray-level and color) distributions, several new evaluation metrics are introduced for the HDR performance measurement: (a) *wPSNR (weighted PSNR)* in which larger weights are given to brighter areas than darker areas in the image, (b) *PSNR-L100*, and (c) *DE100*. Both PSNR-L100 and DE100 are derived from the CIE DE2000 metric that is known to provide better perceptual uniformity in measuring color differences in the CIELAB color space. Table 12.16 shows the BD-rates for HDR (with common coding engine plus the default reshaper under random access, averaged over 5 test videos) of {−31.3, −31.7, −38.1%} for {wPSNR-Y, PSNR-L100, DE100} when compared to H.265 (HEVC). They are {−4.7, −5.2, −2.0%} when compared to JEM-7.0

For 360° videos encoded in the padded ERP format, the luma BD-rates (averaged over 5 test videos) are −33.9% and −13.5%, respectively, when compared to H.265 (HEVC) and JEM-7.0 (see Table 12.17).

Table 12.16 H.266 BD-rates for the HDR-B (PQ) category

	Over H.265 test model			Over JEM-7.0 test model		
	wPSNR-Y	PSNR-L100	DE100	wPSNR-Y	PSNR-L100	DE100
Average over 5 test videos (%)	−31.3	−31.7	−38.1	−4.7	−5.2	−2.0

Table 12.17 H.266 BD-rates for the 360° video category

	Over H.265 test model			Over JEM-7.0 test model		
	Y	U	V	Y	U	V
Average over 5 test videos (%)	−33.9	−54.0	−56.8	−13.5	−18.3	−21.0

12.5   Exercises

1. Integer Transforms are used in H.264, H.265, and H.266.

   (a) What is the relationship between the DCT and Integer Transform?
   (b) What are the main advantages of using Integer Transform instead of DCT?

2. H.264 and H.265 use quarter-pixel precision in motion compensation.

   (a) What is the main reason that subpixel (in this case, quarter-pixel) precision is advocated?
   (b) How do H.264 and H.265 differ in obtaining at quarter-pixel positions?

3. From Eq. 12.15, derive $\mathbf{H}_{8\times8}$ for the Integer Transform in H.265.
4. H.264, H.265, and H.266 support *in-loop deblocking filtering*.

   (a) Why is deblocking a good idea? What are its disadvantages?
   (b) What are the main differences in its H.264 and H.265 implementations?
   (c) Besides the deblocking filtering, what does H.265 do to improve the visual quality? How about H.266?

5. Name at least three features in H.265 that facilitate parallel processing.
6. P-frame coding in H.264 uses *Integer Transform*. For this exercise, assume

$$F(u, v) = H \cdot f(i, j) \cdot H^T, \text{ where } H = \begin{bmatrix} 1 & 1 & 1 & 1 \\ 2 & 1 & -1 & -2 \\ 1 & -1 & -1 & 1 \\ 1 & -2 & 2 & -1 \end{bmatrix}.$$

(a) What are the two advantages of using Integer Transform?

(b) Assume the Target Frame below is a P-frame. For simplicity, assume the size of the macroblock is 4×4. For the macroblock shown in the Target Frame:

(i) What should be the Motion Vector?

(ii) What are the values of $f(i, j)$ in this case?

(iii) Show all values of $F(u, v)$.

```
20   40   60   80  100 120 140 155        110 132 154 176 – – – –
30   50   70   90  110 130 150 165        120 142 164 186 – – – –
40   60   80  100  120 140 160 175        130 152 174 196 – – – –
50   70   90  110  130 150 170 185        140 162 184 206 – – – –
60   80  100  120  140 160 180 195         –   –   –   –  – – – –
70   90  110  130  150 170 190 205         –   –   –   –  – – – –
80  100  120  140  160 180 200 215         –   –   –   –  – – – –
85  105  125  145  165 185 205 220         –   –   –   –  – – – –
```

Reference Frame   Target Frame

7. You are to do a mini example of H.264 video compression. For simplicity, instead of 2D frames (images), your "video" consists of 1D frames as shown in Fig. 12.21. The Target Frame is a P-frame. The size of the macroblock is 1×4, and the darkened box indicates the current macroblock.

(a) Show all details for using *Logarithmic Search*, in which the range of the search $p = 7$, to find the Motion Vector for the current macroblock.

(b) Did you find the block [200, 200, 200, 200] as the best matching macroblock in your above search? Why?

(c) Complete the steps in the motion compensation (MC)-based compression before Quantization (i.e., you do not need to do Quantization, Scaling, etc.) in the encoder. Make sure you show the details of the Integer Transform.

Reference Frame: | 50 | 50 | 50 | 50 | 180 | 190 | 200 | 210 | 50 | 50 | 50 | 50 | 50 | 200 | 200 | 200 | 200 | 50 |

Target Frame: | 200 | 200 | 200 | 200 |

Current macroblock

Fig. 12.21 MC-based video compression in 1D frames

8. H.264, H.265, and H.266 all support Intra-coding.

   (a) What is the main objective of Intra-coding?
   (b) You are to try a simplified H.26* Intra_4 × 4 coding method in which only
       three modes (vertical, horizontal, and DC) are used. For the 8 × 8 image
       below, show the least prediction errors $d(i, j)$ by your method for the lower-
       right 4 × 4 block, i.e., the block that contains $\begin{matrix} 80 & 80 & 100 & 100 \\ 80 & 80 & 100 & 100 \\ 90 & 90 & 110 & 110 \\ 90 & 90 & 110 & 110 \end{matrix}$. Explain in detail
       how you derive the result.

$$
\begin{matrix}
20 & 20 & 40 & 40 & 60 & 60 & 80 & 80 \\
20 & 20 & 40 & 40 & 60 & 60 & 80 & 80 \\
30 & 30 & 50 & 50 & 70 & 70 & 90 & 90 \\
30 & 30 & 50 & 50 & 70 & 70 & 90 & 90 \\
40 & 40 & 60 & 60 & 80 & 80 & 100 & 100 \\
40 & 40 & 60 & 60 & 80 & 80 & 100 & 100 \\
50 & 50 & 70 & 80 & 90 & 90 & 110 & 110 \\
50 & 50 & 70 & 80 & 90 & 90 & 110 & 110
\end{matrix}
$$

   (c) Show the final result (all $D(u, v)$ values) produced by the encoder for this
       4 × 4 block after the integer transform.

9. Write a program for the Order-k Exp-Golomb encoder and decoder.

   (a) What is the EG_0 codeword for unsigned $N = 110$?
   (b) Given an EG_0 code 000000011010011, what is the decoded unsigned N?
   (c) What is the EG_3 codeword for unsigned $N = 110$?

10. Write a program to implement video compression with motion compensation,
    transform coding, and quantization for a simplified H.26* encoder and decoder.

    • Use 4:2:0 for chroma subsampling.
    • Choose a video frame sequence (I-, P-, B-frames) similar to MPEG-1, 2. No
      interlacing.
    • For I-frames, implement the H.264 Intra_4 × 4 predictive coding.
    • For P- and B-frames, use only 8 × 8 for motion estimation. Use logarithmic
      search for motion vectors. Afterwards, use the 4 × 4 Integer Transform as in
      H.264.
    • Use the quantization and scaling matrices as specified in Eqs. 12.5 and 12.7.
      Control and show the effect of various levels of compression and quantization
      losses.
    • Do not implement the entropy coding part. Optionally, you may include any
      publicly available code for this.

11. Write a program to verify the results in Table 12.8, for example, to show that DST will produce shorter code than DCT for Category 2 directional predictions.
12. For Intra-coding in H.266, some predefined transform sets are indicated in Table 12.12. As shown in Table 12.13, except for the first two special modes (Planar and DC), the selected transform set for H of an angular mode i is the same as the one for V of mode $68 - i$. Explain why.
13. Perceptual Quantization (PQ) is used for high dynamic range (HDR) video.

(a) What is PQ, why is it used for the HDR video?
(b) What is lumaDQP in H.266? Why is it adopted?

References

1. T. Wiegand, G.J. Sullivan, G. Bjøntegaard, A. Luthra, Overview of the H.264/AVC video coding standard. IEEE Trans. Circuits Syst. Video Technol. 13(7), 560–576 (2003)
2. ITU-T H.264 I ISO/IEC 14496-10, Advanced Video Coding for Generic Audio-Visual Services. (ITU-T and ISO/IEC, 2009)
3. ISO/IEC 14496, Part 10, Information Technology—Coding of Audio-Visual Objects—Part 10: advanced Video Coding. (ISO/IEC, 2012)
4. I.E. Richardson, The H.264 Advanced Video Compression Standard, 2nd ed. (Wiley, 2010)
5. H.S. Malvar, et al., Low-complexity transform and quantization in H.264/AVC. IEEE Trans. Circuits Syst. Video Technol. 13(7), 598–603 (2003)
6. G. Bjontegaard, K. Lillevold, Context-Adaptive VLC Coding of Coefficients. JVT document JVT-C028 (2002)
7. I.E. Richardson, H.264 and MPEG-4 Video Compression. (Wiley, 2003)
8. D. Marpe, H. Schwarz, T. Wiegand, Context-based adaptive binary arithmetic coding in the H.264/AVC video compression standard. IEEE Trans. Circuits Syst. Video Technol. 13(7), 620–636 (2003)
9. D. Marpe, T. Wiegand, The H.264/MPEG4 advanced video coding standard and its applications. IEEE Communi. Mag. 44(8), 134–143 (2006)
10. H. Schwarz, et al., Overview of scalable video coding extension of the H.264/AVC standard. IEEE Trans. Circuits Syst. Video Technol. 17(9), 1103–1120 (2007)
11. P. Merkle, et al., Efficient prediction structures for multiview video coding. IEEE Trans. Circuits Syst. Video Technol. 17(11), 1461–1473 (2007)
12. A. Vetro, T. Wiegand, G.J. Sullivan, Overview of the stereo and multiview video coding extensions of the H.264/MPEG-4 AVC standard. Proc. IEEE. 99(4), 626–642 (2011)
13. G.J. Sullivan, et al., Overview of the high efficiency video coding (HEVC) standard. IEEE Trans. Circuits Syst. Video Technol. 22(12), 1649–1668 (2012)
14. J.R. Ohm, G.J. Sullivan, High efficiency video coding: the next frontier in video compression. IEEE Signal Process. Mag. 30(1), 152–158 (2013)
15. ITU-T H.265 I ISO/IEC 23008-2, H.265: high Efficiency Video Coding. (ITU-T and ISO/IEC, 2013)
16. J. Wang et al., Joint separable and non-separable transforms for next-generation video coding. IEEE Trans. Image Process. 27(5), 2514–2525 (2018)
17. A. Saxena, F.C. Fernandes, Mode dependent DCT/DST for intra prediction in block-based image/video coding, in IEEE International Conference on Image Processing, pp. 1685–1688 (2011)
18. A. Saxena, F.C. Fernandes, DCT/DST based transform coding for intra prediction in image/video coding. IEEE Trans. Image Process. 22(10), 3974–3981 (2013)

19. V. Sze, M. Budagavi, High throughput CABAC entropy coding in HEVC. IEEE Trans. Circuits Syst. Video Technol. **22**(12), 1778–1791 (2012)
20. D. Flynn, D. Marpe, M. Naccari, T. Nguyen, C. Rosewarne, K. Sharman, J. Sole, J. Xu, Overview of the range extensions for the HEVC standard: tools, profiles, and performance. IEEE Trans. Circuits Syst. Video Technol. **26**(1), 4–19 (2016)
21. J.R. Ohm, et al., Comparison of the coding efficiency of video coding standards—Including high efficiency video coding (HEVC). IEEE Trans. Circuits Syst. Video Technol. **22**(12), 1669–1684 (2012)
22. G.J. Sullivan et al., Standardized extensions of high efficiency video coding (HEVC). IEEE J. Sel. Top. Signal Process. **7**(6), 1001–1016 (2013)
23. J.M. Boyce, Y. Ye, J. Chen, A.K. Ramasubramonian, Overview of SHVC: scalable extensions of the high efficiency video coding standard. IEEE Trans. Circuits Syst. Video Technol. **26**(1), 20–34 (2016)
24. G. Tech, Y. Chen, K. Müller, J.R. Ohm, A. Vetro, Y.K. Wang, Overview of the multiview and 3D extensions of high efficiency video coding. IEEE Trans. Circuits Syst. Video Technol. **26**(1), 35–49 (2016)
25. J. Chen, M. Karczewicz, Y.W. Huang, K. Choi, J.R. Ohm, G.J. Sullivan, The joint exploration model (JEM) for video compression with capability beyond HEVC. IEEE Trans. Circuits Syst. Video Technol. **30**(5), 1208–1225 (2020)
26. D. Liu, et al., Deep learning-based technology in responses to the joint call for proposals on video compression with capability beyond HEVC. IEEE Trans. Circuits Syst. Video Technol. **30**(5), 1267–1280 (2020)
27. D. Liu, et al., Deep learning-based video coding: a review and a case study. ACM Comput. Surv. **53**(1), 11, 1–11, 35 (2020)
28. B. Bross, et al., General video coding technology in responses to the joint call for proposals on video compression with capability beyond HEVC. IEEE Trans. Circuits Syst. Video Technol. **30**(5), 1226–1240 (2020)
29. X. Xiu, et al., A unified video codec for SDR, HDR, and 360° video applications. IEEE Trans. Circuits Syst. Video Technol. **30**(5), 1296–1310 (2020)
30. E. François, et al., High dynamic range video coding technology in responses to the joint call for proposals on video compression with capability beyond HEVC. IEEE Trans. Circuits Syst. Video Technol. **30**(5), 1253–1266 (2020)
31. Y. Ye, J.M. Boyce, P. Hanhart, Omnidirectional 360° video coding technology in responses to the joint call for proposals on video compression with capability beyond HEVC. IEEE Trans. Circuits Syst. Video Technol. **30**(5), 1241–1252 (2020)
32. G. Bjontegaard, Calculation of average PSNR differences between RD-curves. Technical Report VCEG-M33 (2001)
33. L. Zhou, et al., Convolutional neural network filter (CNNF) for intra frame. Technical Report JVET-10022 (2018)

Basic Audio Compression Techniques

13

Compression of audio information is somewhat special in multimedia systems. Some of the techniques used are familiar, while others are new. In this chapter, we take a look at basic audio compression techniques applied to speech compression, setting out a general introduction to a large topic with a long history. More extensive information can be found in the References section at the end of the chapter.

In the next chapter, we consider the set of tools developed for general audio compression under the aegis of the Motion Picture Experts Group (MPEG). Since this is generally of high interest to readers focusing on multimedia, we treat that subject in greater detail.

To begin with, let us recall some of the issues covered in Chap. 6 on digital audio in multimedia, such as the μ-law for companding audio signals. This is usually combined with a simple technique that exploits the temporal redundancy present in audio signals. We saw in Chap. 10, on video compression, that differences in signals between the present and a past time could very effectively reduce the size of signal values, and importantly, concentrate the histogram of pixel values (differences, now) into a much smaller range. The result of reducing the variance of values is that the entropy is greatly reduced, and subsequent Huffman coding can produce a greatly compressed bitstream.

The same applies here. Recall from Chap. 6 that quantized sampled output is called Pulse Code Modulation, or PCM. The differential version is called DPCM, and the adaptive version is called ADPCM. Variants that take into account the speech properties follow from these.

In this chapter, we look at ADPCM, Vocoders, and more general Speech Compression: LPC, CELP, MBE, and MELP. We also introduce the recent open source speech and audio codecs Speex and Opus.

© Springer Nature Switzerland AG 2021
Z.-N. Li et al., *Fundamentals of Multimedia*, Texts in Computer Science,
https://doi.org/10.1007/978-3-030-62124-7_13

13.1   ADPCM in Speech Coding

13.1.1 ADPCM

ADPCM forms the heart of the ITU's speech compression standards G.721, G.723, G.726, G.727, G.728, and G.729. The differences among these standards involve the bitrate and some details of the algorithm. The default input is μ-law-coded PCM 16-bit samples. Speech performance for ADPCM is such that the perceived quality of speech at 32 kbps is only slightly poorer than with the standard 64 kbps PCM transmission and is better than DPCM.

Figure 13.1 shows a one second speech sample of a voice speaking the word "audio." In Fig. 13.1a, the audio signal is stored as linear PCM (as opposed to the default μ-law PCM) recorded at 8,000 samples per second, with 16 bits per sample. After compression with ADPCM using ITU standard G.721, the signal appears as in Fig. 13.1b. Figure 13.1c shows the difference between the actual and reconstructed, compressed signals. Although differences are apparent electronically between the two, the compressed and original signals are *perceptually* very similar.

13.1.2 G.726 ADPCM, G.727-9

ITU G.726 provides another version of G.711, including companding, at a lower bitrate. G.726 can encode 13- or 14-bit PCM samples or 8-bit μ-law or A-law encoded data into 2-, 3-, 4-, or 5-bit codewords. It can be used in speech transmission over digital networks.

The G.726 standard works by adapting a *fixed* quantizer in a simple way. The different sizes of codewords used yield bitrates of 16, 24, 32, or 40 kbps. The standard defines a multiplier constant α that will change for every difference value e_n, depending on the current scale of signals. Define a scaled difference signal f_n as follows:

$$e_n = s_n - \hat{s}_n$$
$$f_n = e_n / \alpha \tag{13.1}$$

where $\hat{s}_n$ is the predicted signal value. f_n is then fed into the quantizer for quantization. The quantizer is as displayed in Fig. 13.2. Here, the input value is defined as a ratio of a difference with the factor α.

By changing the value of α, the quantizer can adapt to change in the range of the difference signal. The quantizer is a nonuniform midtread quantizer, so it includes the value zero. The quantizer is *backward adaptive*.

A backward-adaptive quantizer works in principle by noticing if too many values are quantized to values far from zero (which would happen if the quantizer step size in f was too small) or if too many values fell close to zero too much of the time (which would happen if the quantizer step size was too large).

Fig. 13.1 Waveform of the
word "audio:" **a** Speech
sample, linear PCM at 8 kHz
and 16 bits per sample;
b speech sample, restored
from G.721-compressed
audio at 4 bits per sample;
c difference signal between
(**a**) and (**b**)

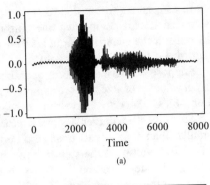

(a)

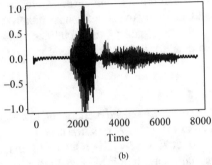

(b)

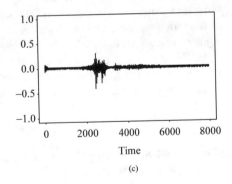

(c)

Fig. 13.2 G.726 quantizer

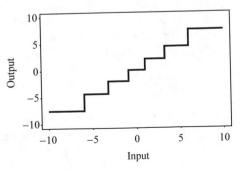

In fact, an algorithm due to Jayant [1] allows us to adapt a backward quantizer step size after receiving just one output! The Jayant quantizer simply expands the step size if the quantized input is in the outer levels of the quantizer and reduces the step size if the input is near zero.

Suppose we have a uniform quantizer, so that every range to which we compare input values is of size Δ. For example, for a 3-bit quantizer, there are $k = 0 .. 7$ levels. For 3-bit G.726, only 7 levels are used, grouped around zero.

The Jayant quantizer assigns *multiplier values* M_k to each level, with values smaller than 1 for levels near zero and values larger than 1 for outer levels. The multiplier multiplies the step size for the next signal value. That way, outer values enlarge the step size and are likely to bring the next quantized value back to the middle of the available levels. Quantized values near the middle reduce the step size and are likely to bring the next quantized value closer to the outer levels.

So, for signal f_n, the quantizer step size Δ is changed according to the quantized value k, for the previous signal value f_{n-1}, by the simple formula

$$\Delta \leftarrow M_k \Delta. \tag{13.2}$$

Since the *quantized* version of the signal is driving the change, this is indeed a backward-adaptive quantizer.

In G.726, how α is allowed to change depends on whether the audio signal is actually speech or is likely data that is simply using a voice band. In the former case, sample-to-sample differences can fluctuate a great deal, whereas, in the latter case of data transmission, this is less true. To adjust to either situation, the factor α is adjusted using a formula with two pieces.

G.726 works as a backward-adaptive Jayant quantizer by using fixed quantizer steps based on the logarithm of the input difference signal, e_n divided by α. The divisor α is written in terms of its logarithm:

$$\beta = \log_2 \alpha. \tag{13.3}$$

Since we wish to distinguish between situations when the difference values are usually small, and when they are large, α is divided into the so-called *locked* part, α_L, and an *unlocked* part, α_U. The idea is that the locked part is a scale factor for small difference values and changes slowly, whereas the unlocked part adapts quickly to larger differences. These correspond to log quantities β_L and β_U.

The logarithm value is written as a sum of two pieces,

$$\beta = A\beta_U + (1 - A)\beta_L \tag{13.4}$$

where A changes so that it is about 1 for speech, and about 0 for voice-band data. It is calculated based on the variance of the signal, keeping track of several past signal values.

The "unlocked" part adapts via the equation

$$\alpha_U \leftarrow M_k \alpha_U$$
$$\beta_U \leftarrow \log_2 M_k + \beta_U \qquad (13.5)$$

where M_k is a Jayant multiplier for the kth level. The locked part is slightly modified from the unlocked part, via

$$\beta_L \leftarrow (1 - B)\beta_L + B\beta_U \qquad (13.6)$$

where B is a small number, say 2^{-6}.

The G.726 predictor is complicated: it uses a linear combination of six quantized differences and two reconstructed signal values from the previous six signal values f_n.

ITU standards G.728 and G.729 use Code Excited Linear Prediction (CELP), discussed in Sect. 13.2.5.

13.2 Vocoders

The coders (encoding/decoding algorithms) we have studied so far could have been applied to any signals, not just speech. *Vocoders* are specifically voice coders.

Vocoders are concerned with modeling speech so that the salient features are captured in as few bits as possible. They use either a model of the speech waveform in time (*Linear Predictive Coding* (LPC) vocoding), or else break down the signal into frequency components and model these (channel vocoders and formant vocoders).

Incidentally, we all know that vocoder simulation of the voice is not wonderful yet—when the library calls you with your overdue notification, the automated voice is strangely lacking in zest.

13.2.1 Phase Insensitivity

Recall from Sect. 8.5 that we can break down a signal into its constituent frequencies by analyzing it using some variant of Fourier analysis. In principle, we can also reconstitute the signal from the frequency coefficients developed that way. But it turns out that a complete reconstituting of speech waveform is unnecessary, perceptually: all that is needed is for the amount of energy at any time to be about right, and the signal will sound about right.

"Phase" is a shift in the time argument, inside a function of time. Suppose we strike a piano key and generate a roughly sinusoidal sound $\cos(\omega t)$, with $\omega = 2\pi f$ where f is the frequency. If we wait for sufficient time to generate a phase shift $\pi/2$ and then strike another key, with sound $\cos(2\omega t + \pi/2)$, we generate a waveform like the solid line in Fig. 13.3. This waveform is the sum $\cos(\omega t) + \cos(2\omega t + \pi/2)$.

Fig. 13.3 The solid line
shows the superposition of
two cosines, with a phase
shift. The dashed line shows
the same with no phase shift.
The wave is very different,
yet the sound is the same,
perceptually

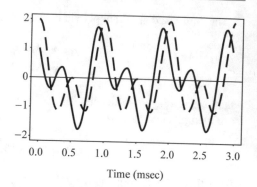

Time (msec)

If we did not wait before striking the second note (1/4 ms, in Fig. 13.3), our wave-form would be $\cos(\omega t) + \cos(2\omega t)$. But perceptually, the two notes would sound the same, even though in actuality they would be shifted in phase.

Hence, if we can get the energy spectrum right—where we hear loudness and quiet—then we don't really have to worry about the exact waveform.

13.2.2 Channel Vocoder

Subband filtering is the process of applying a bank of band-pass filters to the analog signal, thus actually carrying out the frequency decomposition indicated in a Fourier analysis. *Subband coding* is the process of making use of the information derived from this filtering to achieve better compression.

For example, an older ITU recommendation, G.722, uses subband filtering of analog signals into just two bands: voice frequencies in 50 Hz to 3.5 kHz and 3.5 to 7 kHz. Then the set of two signals is transmitted at 48 kbps for the low frequencies, where we can hear discrepancies well, and at only 16 kbps for the high frequencies.

Vocoders can operate at low bitrates, just 1–2 kbps. To do so, a *channel vocoder* first applies a filter bank to separate out the different frequency components, as in Fig. 13.4. However, as we saw above, only the energy is important, so first, the waveform is "rectified" to its absolute value. The filter bank derives relative power levels for each frequency range. A subband coder would not rectify the signal and would use wider frequency bands.

A channel vocoder also analyzes the signal to determine the general pitch of the speech—low (bass), or high (tenor)—and also the *excitation* of the speech. Speech excitation is mainly concerned with whether a sound is *voiced* or *unvoiced*. A sound is unvoiced if its signal simply looks like noise: the sounds *s* and *f* are unvoiced. Sounds such as the vowels *a*, *e*, and *o* are voiced, and their waveform looks periodic. The *o* at the end of the word "audio" in Fig. 13.1 is fairly periodic. During a vowel sound, air is forced through the vocal cords in a stream of regular, short puffs, occurring at the rate of 75–150 pulses per second for men and 150–250 per second for women.

Consonants can be voiced or unvoiced. For the nasal sounds of the letters *m* and *n*, the vocal cords vibrate, and air is exhaled through the nose rather than the mouth.

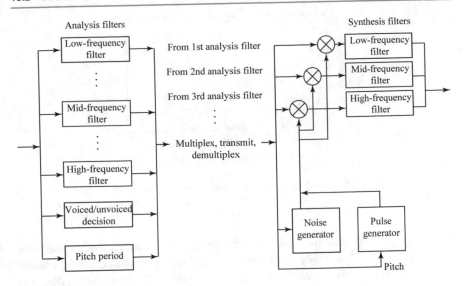

Fig. 13.4 Channel vocoder

These consonants are, therefore, voiced. The sounds b, d, and g, in which the mouth starts closed but then opens to the following vowel over a transition lasting a few milliseconds, are also voiced. The energy of voiced consonants is greater than that of unvoiced consonants but less than that of vowels. Examples of unvoiced consonants include the sounds sh, th, and h when used at the front of a word.

A channel vocoder applies a vocal-tract transfer model to generate a vector of excitation parameters that describe a model of the sound. The vocoder also guesses whether the sound is voiced or unvoiced and, for voiced sounds, estimates the period (i.e., the sound's pitch). Figure 13.4 shows that the decoder also applies a vocal-tract model.

Because voiced sounds can be approximated by sinusoids, a periodic pulse generator recreates voiced sounds. Since unvoiced sounds are noise-like, a pseudo-noise generator is applied, and all values are scaled by the energy estimates given by the band-pass filter set. A channel vocoder can achieve an intelligible but synthetic voice using 2,400 bps.

13.2.3 Formant Vocoder

It turns out that not all frequencies present in speech are equally represented. Instead, only certain frequencies show up strongly, and others are weak. This is a direct consequence of how speech sounds are formed, by resonance in only a few chambers of the mouth, throat, and nose. The important frequency peaks are called *formants* [2].

Figure 13.5 shows how this appears: only a few, usually just four or so, peaks of energy at certain frequencies are present. The peak locations, however, change

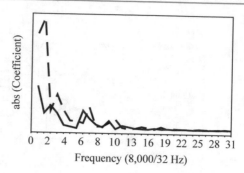

Fig. 13.5 Formants are the salient frequency components present in a sample of speech. Here, the solid line shows frequencies present in the first 40 ms of the speech sample in Fig. 6.16. The dashed line shows that while similar frequencies are still present, one second later, they have shifted

over time, as speech continues. For example, two different vowel sounds would activate different sets of formants—this reflects the different vocal-tract configurations necessary to form each vowel. Usually, a small segment of speech is analyzed, say 10–40 ms, and formants are found. A *Formant Vocoder* works by encoding only the most important frequencies. Formant vocoders can produce reasonably intelligible speech at only 1,000 bps.

13.2.4 Linear Predictive Coding

LPC vocoders extract salient features of speech directly from the waveform rather than transforming the signal to the frequency domain. LPC coding uses a time-varying model of vocal-tract sound generated from a given excitation. What is transmitted is a set of parameters modeling the shape and excitation of the vocal tract, not actual signals or differences.

Since what is sent is an analysis of the sound rather than the sound itself, the bitrate using LPC can be small. This is like using a simple descriptor such as MIDI to generate music: we send just the description parameters and let the sound generator do its best to create appropriate music. The difference is that as well as pitch, duration, and loudness variables, here we also send vocal-tract excitation parameters.

After a block of digitized samples, called a *segment* or *frame*, is analyzed, the speech signal generated by the output vocal-tract model is calculated as a function of the current speech output plus a second term linear in previous model coefficients. This is how "linear" in the coder's name arises. The model is adaptive—the encoder side sends a new set of coefficients for each new segment.

The typical number of sets of previous coefficients used is $N = 10$ (the "model order" is 10), and such an LPC-10 [3] system typically uses a rate of 2.4 kbps. The model coefficients a_i act as predictor coefficients, multiplying previous speech output sample values.

LPC starts by deciding whether the current segment is voiced or unvoiced. For unvoiced speech, a wide-band noise generator is used to create sample values $f(n)$

that act as input to the vocal-tract simulator. For voiced speech, a pulse-train generator creates values $f(n)$. Model parameters a_i are calculated by using a least-squares set of equations that minimize the difference between the actual speech and the speech generated by the vocal-tract model, excited by the noise or pulse-train generators that capture speech parameters.

If the output values generated are denoted $s(n)$, then for input values $f(n)$, the output depends on p previous *output* sample values, via

$$s(n) = \sum_{i=1}^{p} a_i s(n-i) + Gf(n). \tag{13.7}$$

Here, G is known as the *gain* factor. Note that the coefficients a_i act as values in a linear predictor model. The pseudo-noise generator and pulse generator are as discussed above and depicted in Fig. 13.4 in regard to the channel vocoder.

The speech encoder works in a blockwise fashion. The input digital speech signal is analyzed in some small, fixed-length segments, called speech frames. For the LPC speech coder, the frame length is usually selected as 22.5 ms, which corresponds to 180 samples for 8 kHz sampled digital speech. The speech encoder analyzes the speech frames to obtain the parameters such as LP coefficients a_i, $i = 1, \ldots, p$, gain G, pitch P, and voiced/unvoiced decision U/V.

To calculate LP coefficients, we can solve the following minimization problem for a_j:

$$\min E\{[s(n) - \sum_{j=1}^{p} a_j s(n-j)]^2\}. \tag{13.8}$$

By taking the derivative of a_i and setting it to zero, we get a set of p equations:

$$E\{[s(n) - \sum_{j=1}^{p} a_j s(n-j)]s(n-i)\} = 0, \qquad i = 1, \ldots, p \tag{13.9}$$

Letting $\phi(i, j) = E\{s(n-i)s(n-j)\}$, we have

$$\begin{bmatrix} \phi(1,1) & \phi(1,2) & \cdots & \phi(1,p) \\ \phi(2,1) & \phi(2,2) & \cdots & \phi(2,p) \\ \vdots & \vdots & \ddots & \vdots \\ \phi(p,1) & \phi(p,2) & \cdots & \phi(p,p) \end{bmatrix} \begin{bmatrix} a_1 \\ a_2 \\ \vdots \\ a_p \end{bmatrix} = \begin{bmatrix} \phi(0,1) \\ \phi(0,2) \\ \vdots \\ \phi(0,p) \end{bmatrix} \tag{13.10}$$

The *autocorrelation method* is often used to calculate LP coefficients, where

$$\phi(i,j) = \sum_{n=p}^{N-1} s_w(n-i)s_w(n-j) / \sum_{n=p}^{N-1} s_w^2(n) \qquad i = 1, \ldots, p, \ j = 1, \ldots, p \tag{13.11}$$

$s_w(n) = s(n + m)w(n)$ is the windowed speech frame starting from time m. Since $\phi(i, j)$ is determined only by $|i - j|$, we define $\phi(i, j) = R(|i - j|)$. Since we also have $R(0) \geq 0$, the matrix $\{\phi(i, j)\}$ is positive symmetric, and thus a fast scheme to calculate the LP coefficients is as follows:

Procedure 13.1 (LPC Coefficients).
$E(0) = R(0), i = 1$

while $i \leq p$
$\quad k_i = [R(i) - \sum_{j=1}^{i-1} a_j^{i-1} R(i - j)]/E(i - 1)$
$\quad a_i^i = k_i$
$\quad$ for $j = 1$ to $i - 1$
$\quad\quad a_j^i = a_j^{i-1} - k_i a_{i-j}^{i-1}$
$\quad E(i) = (1 - k_i^2)E(i - 1)$
$\quad i \leftarrow i + 1$
for $j = 1$ to p
$\quad a_j = a_j^p$

After getting the LP coefficients, gain G ca n be calculated as

$$G = E\{[s(n) - \sum_{j=1}^{p} a_j s(n - j)]^2\}$$

$$= E\{[s(n) - \sum_{j=1}^{p} a_j s(n - j)]s(n)\} \tag{13.12}$$

$$= \phi(0, 0) - \sum_{j=1}^{p} a_j \phi(0, j)$$

For the autocorrelation scheme, $G = R(0) - \sum_{j=1}^{p} a_j R(j)$. Order-10 LP analysis is found to be enough for speech coding applications.

The pitch P of the current speech frame can be extracted by the correlation method by finding the index of the peak of

$$v(i) = \sum_{n=m}^{N-1+m} s(n)s(n - i) \Bigg/ \left[\sum_{n=m}^{N-1+m} s^2(n) \cdot \sum_{n=m}^{N-1+m} s^2(n - i) \right]^{1/2},$$

$$i \in [P_{\min}, P_{\max}]. \tag{13.13}$$

The searching range $[P_{\min}, P_{\max}]$ is often selected as $[12, 140]$ for 8 kHz sampling speech. Denote P as the peak lag. If $v(P)$ is less than some given threshold, the current frame is classified as an unvoiced frame and will be reconstructed in the receiving end by stimulating with a white-noise sequence. Otherwise, the frame is determined as voiced and stimulated with a periodic waveform at the reconstruction stage. In practical LPC speech coders, the pitch estimation and U/V decision procedure are usually based on a dynamic programming scheme, so as to correct the often occurring errors of pitch doubling or halving in the single frame scheme.

In LPC-10, each segment has 180 samples, or with a length of 22.5 ms at 8 kHz. The speech parameters transmitted are the coefficients a_k; G, the gain factor; a voiced/unvoiced flag (1 bit); and the pitch period if the speech is voiced.

13.2.5 CELP

CELP, *Code Excited Linear Prediction* (sometimes *Codebook Excited*), is a more complex family of coders that attempts to mitigate the lack of quality of the simple LPC model by using a more complex description of the excitation. An entire set (a codebook) of excitation vectors is matched to the actual speech and the index of the best match is sent to the receiver. This complexity increases the bitrate to 4,800–9,600 bps, typically.

In CELP, since all speech segments make use of the same set of templates from the template codebook, the resulting speech is perceived as much more natural than the two-mode excitation scheme in the LPC-10 coder. The quality achieved is considered sufficient for audio conferencing.

In CELP coders, two kinds of prediction, Long Time Prediction (LTP) and Short Time Prediction (STP) is used to eliminate the redundancy in speech signals. STP is an analysis of *samples*—it attempts to predict the next sample from several previous ones. Here, redundancy is due to the fact that usually one sample will not change drastically from the next. LTP is based on the idea that in a *segment* of speech, or perhaps from segment to segment, especially for voiced sounds, a basic periodicity or pitch will cause a waveform that more or less repeats. We can reduce this redundancy by finding the pitch.

For concreteness, suppose we sample at 8,000 samples/s and use a 10 ms frame, containing 80 samples. Then we can roughly expect a pitch that corresponds to an approximately repeating pattern every 12–140 samples or so. (Notice that the pitch may actually be longer than the chosen frame size.)

STP is based on a short-time LPC analysis, discussed in the last section. It is "short-time" in that the prediction involves only a few samples, not a whole frame or several frames. STP is also based on minimizing the residue error over the whole speech frame, but it captures the correlation over just a short range of samples (10 for order-10 LPC).

After STP, we can subtract signal minus prediction to arrive at a differential coding situation. However, even in a set of errors $e(n)$, the basic pitch of the sequence may still remain. This is estimated by means of LTP. That is, LTP is used to further eliminate the periodic redundancy inherent in the voiced speech signals. Essentially, STP captures the formant structure of the short-term speech spectrum, while LTP recovers the long-term correlation in the speech signal that represents the periodicity in speech.

Thus, there are always two stages—and the order is in fact usually STP followed by LTP, since we always start off assuming zero error and then remove the pitch

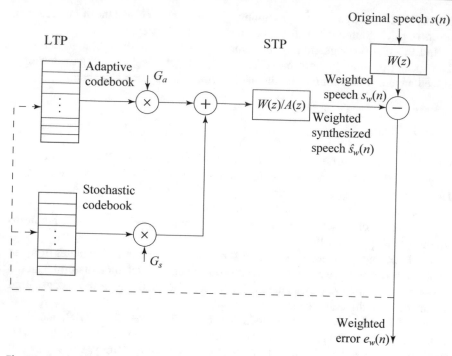

Fig. 13.6 CELP analysis model with adaptive and stochastic codebooks

component. (If we use a closed-loop scheme, STP usually is done first). LTP proceeds using whole frames—or, more often, subframes equal to one quarter of a frame. Figure 13.6 shows these two stages.

LTP is often implemented as *adaptive codebook searching*. The "codeword" in the adaptive codebook is a shifted speech residue segment indexed by the lag τ corresponding to the current speech frame or subframe. The idea is to look in a codebook of waveforms to find one that matches the current subframe. We generally look in the codebook using a *normalized* subframe of speech, so as well as a speech segment match, we also obtain a scaling value (the *gain*). The gain corresponding to the codeword is denoted as g_0.

There are two types of codeword searching: *open-loop* and *closed-loop*. Open-loop adaptive codebook searching tries to minimize the long-term prediction error but not the perceptual weighted reconstructed speech error

$$E(\tau) = \sum_{n=0}^{L-1} [s(n) - g_0 s(n - \tau)]^2 \qquad (13.14)$$

By setting the partial derivative of g_0 to zero, $\partial E(\tau)/\partial g_0 = 0$, we get

$$g_0 = \frac{\sum_{n=0}^{L-1} s(n)s(n - \tau)}{\sum_{n=0}^{L-1} s^2(n - \tau)} \qquad (13.15)$$

and hence a minimum summed-error value

$$E_{min}(\tau) = \sum_{n=0}^{L-1} s^2(n) - \frac{[\sum_{n=0}^{L-1} s(n)s(n-\tau)]^2}{\sum_{n=0}^{L-1} s^2(n-\tau)}. \qquad (13.16)$$

Notice that the sample $s(n-\tau)$ could be in the previous frame.

Now, to obtain the optimum adaptive codebook index τ, we can carry out a search exclusively in a small range determined by the pitch period. More often, CELP coders use a closed-loop search. Rather than simply considering sum-of-squares, speech is reconstructed, with perceptual error minimized via an adaptive codebook search. So in a closed-loop, adaptive codebook search, the best candidate in the adaptive codebook is selected to minimize the distortion of locally reconstructed speech. Parameters are found by minimizing a measure (usually the mean square) of the difference between the original and the reconstructed speech. Since this means that we are simultaneously incorporating synthesis, as well as analysis of the speech segment, this method is also called *analysis-by-synthesis*, or *A-B-S*.

The residue signal after STP based on LPC analysis and LTP based on adaptive codeword searching is like white noise and is encoded by codeword matching in the stochastic (random or probabilistic) codebook. This kind of sequential optimization of the adaptive codeword and stochastic codeword methods is used because jointly optimizing the adaptive and stochastic codewords is often too complex to meet real-time demands.

The decoding direction is just the reverse of the above process and works by combining the contribution from the two types of excitations.

DOD 4.8 KBPS CELP (FS1016)*

DOD 4.8 kbps CELP [4] is an early CELP coder adopted as a U.S. federal standard to update the 2.4 kbps LPC-10e (FS1015) vocoder. This vocoder is now a basic benchmark to test other low-bitrate vocoders. FS1016 uses an 8 kHz sampling rate and 30 ms frame size. Each frame is further split into four 7.5 ms subframes. In FS1016, STP is based on an open-loop order-10 LPC analysis.

To improve coding efficiency, a fairly sophisticated type of transform coding is carried out. Then, quantization and compression are done in terms of the transform coefficients.

First, in this field it is common to use the *z-transform*. Here, z is a complex number and represents a kind of complex "frequency." If $z = e^{-2\pi i/N}$, then the discrete z-transform reduces to a discrete Fourier transform. The z-transform makes Fourier transforms look like polynomials. Now we can write the error in a prediction equation

$$e(n) = s(n) - \sum_{i=1}^{p} a_i s(n-i) \qquad (13.17)$$

in the z domain as

$$E(z) = A(z)S(z) \tag{13.18}$$

where $E(z)$ is the z-transform of the error and $S(z)$ is the transform of the signal. The term $A(z)$ is the transfer function in the z domain, and equals

$$A(z) = 1 - \sum_{i=1}^{p} a_i z^{-i} \tag{13.19}$$

with the same coefficients a_i as appear in Eq. (13.7). How speech is reconstructed, then, is via

$$S(z) = E(z)/A(z) \tag{13.20}$$

with the estimated error. For this reason, $A(z)$ is usually stated in terms of $1/A(z)$.

The idea of going to the z-transform domain is to convert the LP coefficients to *Line Spectrum Pair* (*LSP*) coefficients, which are given in this domain. The reason is that the LSP space has several good properties with respect to quantization. LSP representation has become standard and has been applied to nearly all the recent LPC-based speech coders, such as G.723.1, G.729, and MELP. To get LSP coefficients, we construct two polynomials

$$P(z) = A(z) + z^{-(p+1)} A(z^{-1})$$
$$Q(z) = A(z) - z^{-(p+1)} A(z^{-1}) \tag{13.21}$$

where p is the order of the LPC analysis and $A(z)$ is the transform function of the LP filter, with z the transform domain variable. The z-transform is just like the Fourier transform but with a complex "frequency."

The roots of these two polynomials are spaced around the unit circle in the z plane and have mirror symmetry with respect to the x-axis. Assume p is even and denote the phase angles of the roots of $P(z)$ and $Q(z)$ above the x axis as $\theta_1 < \theta_2 < \ldots < \theta_{p/2}$ and $\varphi_1 < \varphi_2 < \ldots < \varphi_{p/2}$, respectively. Then the vector $\{\cos(\theta_1), \cos(\varphi_1), \cos(\theta_2), \cos(\varphi_1) \ldots \cos(\theta_{p/2}), \cos(\varphi_{p/2})\}$ is the LSP coefficient vector, and vector $\{\theta_1, \varphi_1, \theta_2, \varphi_1 \ldots, \theta_{p/2}, \varphi_{p/2}\}$ is usually called *Line Spectrum Frequency*, or *LSF*. Based on the relationship $A(z) = [P(z) + Q(z)]/2$, we can reconstruct the LP coefficients at the decoder end from the LSP or LSF coefficients.

Adaptive codebook searching in FS1016 is via a closed-loop search based on perceptually weighted errors. As opposed to considering just the mean squared error, here errors are weighted so as to take human perception into account. In terms of the z-transform, it is found that the following multiplier does a good job:

$$W(z) = \frac{A(z)}{A(z/\gamma)} = \frac{1 - \sum_{i=1}^{p} a_i z^{-i}}{1 - \sum_{i=1}^{p} a_i \gamma^i z^{-i}} \quad 0 < \gamma < 1 \tag{13.22}$$

with a constant parameter γ.

The adaptive codebook has 256 codewords for 128 integer delays and 128 non-integer delays (with half-sample interval, for better resolution), the former ranging from 20 to 147. To reduce searching complexity, even subframes are searched in an interval relative to the previous odd subframe, and the difference is coded with 6 bits. The gain is nonuniformly scalar coded between -1 and 2 with 5 bits.

Stochastic codebook search is applied for each of the four subframes. The stochastic codebook of FS1016 is generated by clipping a unit variance Gaussian distribution random sequence to within a threshold of absolute value 1.2 and quantizing to three values -1, 0, and 1. The stochastic codebook has 512 codewords. The codewords are overlapped, and each is shifted by 2 with respect to the previous codeword. This kind of stochastic design is called an *Algebraic Codebook*. It has many variations and is widely applied in recent CELP coders.

Denoting the excitation vector as $v^{(i)}$, the periodic component obtained in the first stage is $v^{(0)}$. $v^{(1)}$ is the stochastic component search result in the second stage. In closed-loop searching, the reconstructed speech can be represented as

$$\widehat{s} = \widehat{s_0} + (u + v^{(i)})H \tag{13.23}$$

where u is equal to zero at the first stage and $v^{(0)}$ at the second stage, and $\widehat{s_0}$ is the zero response of the LPC reconstructing filter. Matrix H is the truncated LPC reconstructing filter unit impulse response matrix

$$H = \begin{bmatrix} h_0 & h_1 & h_2 & \cdots & h_{L-1} \\ 0 & h_0 & h_1 & \cdots & h_{L-2} \\ 0 & 0 & h_0 & \cdots & h_{L-3} \\ \vdots & \vdots & \vdots & \ddots & \vdots \\ 0 & 0 & 0 & 0 & h_0 \end{bmatrix} \tag{13.24}$$

where L is the length of the subframe (this simply represents a convolution). Similarly, defining W as the unit response matrix of the perceptual weighting filter, the perceptually weighted error of reconstructed speech is

$$e = (s - \widehat{s})W = e_0 - v^{(i)}HW \tag{13.25}$$

where $e_0 = (s - \widehat{s_0})W - uHW$. The codebook searching process is to find a codeword $y^{(i)}$ in the codebook and corresponding $a^{(i)}$ such that $v^{(i)} = a^{(i)}y^{(i)}$ and ee^T is minimized. To make the problem tractable, adaptive and stochastic codebooks are searched sequentially. Denoting a quantized version by $\tilde{a}^{(i)} = Q[\hat{a}^{(i)}]$, then the criterion of codeword searching in the adaptive codebook or stochastic codebook is to minimize ee^T over all $y^{(i)}$ in terms of an expression in $\tilde{a}^{(i)}$, e_0, and $y^{(i)}$.

The decoder of the CELP codec is a reverse process of the encoder. Because of the unsymmetrical complexity property of vector quantization, the complexity in the decoder side is usually much lower.

G.723.1*

G723.1 [5] is an ITU standard aimed at multimedia communication. It has been incorporated into H.324 for audio encoding in videoconference applications. G.723.1 is a dual rate CELP-type speech coder that can work at bitrates of 5.3 and 6.3 kbps.

G.723.1 uses many techniques similar to FS1016, discussed in the last section. The input speech is again 8 kHz, sampled in 16-bit linear PCM format. The speech frame size is also 30 ms and is further divided into four equal-sized subframes. Order-10 LPC coefficients are estimated in each subframe. LP coefficients are further converted to LSP vectors and quantized by predictive splitting VQ. LP coefficients are also used to form the perceptually weighted filter.

G.723.1 first uses an open-loop pitch estimator to get a coarse pitch estimation in a time interval of every two subframes. Closed-loop pitch searching is done in every speech subframe by searching the data in a range of the open-loop pitch. After LP filtering and removing the harmonic components by LTP, the stochastic residue is quantized by *Multi-pulse Maximum Likelihood Quantization (MP-MLQ)* for the 5.3 kbps coder or *Algebraic-Code-Excited Linear Prediction (ACELP)* for the 6.3 kbps coder, which has a slightly higher speech quality. These two modes can be switched at any boundary of the 30 ms speech frames.

In MP-MLQ, the contribution of the stochastic component is represented as a sequence of pulses

$$v(n) = \sum_{i=1}^{M} g_i \delta(n - m_i) \tag{13.26}$$

where M is the number of pulses and g_i is the gain of pulse at position m_i. The closed-loop search is done by minimizing

$$e(n) = r(n) - \sum_{i=1}^{M} g_i h(n - m_i) \tag{13.27}$$

where $r(n)$ is the speech component after perceptual weighting and eliminating the zero response component and periodic component contributions. Based on methods similar to those presented in the last section, we can sequentially optimize the gain and position for each pulse. Say, we first assume there is only one pulse and find the best gain and position. After removing the contribution from this pulse, we can get the next optimal pulse based on the same method. This process is done recursively until we get all M pulses.

The stochastic codebook structure for the ACELP model is different from FS1016. The following table shows the ACELP excitation codebook:

$$
\begin{array}{ll}
Sign & Positions \\
\pm1 & 0, \ \ 8, 16, 24, 32, 40, 48, \ \ 56 \\
\pm1 & 2, 10, 18, 26, 34, 42, 50, \ \ 58 \\
\pm1 & 4, 12, 20, 28, 36, 44, 52, (60) \\
\pm1 & 6, 14, 22, 30, 38, 46, 54, (62)
\end{array}
\tag{13.28}
$$

There are only four pulses. Each can be in eight positions, coded by three bits each. Also, the sign of the pulse takes one bit, and another bit is to shift all possible positions to odd. Thus, the index of a codeword has 17 bits. Because of the special structure of the algebraic codebook, a fast algorithm exists for efficient codeword searching.

Besides the CELP coder we discussed above, there are many other CELP-type codecs, developed mainly for wireless communication systems. The basic concepts of these coders are similar, except for different implementation details on parameter analysis and codebook structuring.

Some examples include the 12.2 kbps GSM *Enhanced Full Rate* (*EFR*) algebraic CELP codec [6] and IS-641EFR [7], designed for the North American digital cellular IS-136 TDMA system. G.728 [8] is a low-delay CELP speech coder. G.729 [9] is another CELP based ITU standard aimed at toll-quality speech communications.

G.729 is a *Conjugate-Structure Algebraic-Code-Excited-Linear-Prediction* (*CS-ACELP*) codec. G.729 uses a 10 ms speech analysis frame, and thus has a lower delay than G.723.1, which uses a 30 ms speech frame. G.729 also has some inherent protection schemes to deal with packet loss in applications such as VoIP.

13.2.6 Hybrid Excitation Vocoders*

Hybrid Excitation Vocoders are another large class of speech coders. They are different from CELP, in which the excitation is represented as the contributions of the adaptive and stochastic codewords. Instead, hybrid excitation coders use model-based methods to introduce multi-model excitation.

MBE

The *Multi-Band Excitation* (*MBE*) [10] vocoder was developed by MIT's Lincoln Laboratory. The 4.15 kbps IMBE codec [11] has become the standard for IMMSAT. MBE is also a blockwise codec, in which a speech analysis is done in a speech frame unit of about 20–30 ms. In the analysis part of the MBE coder, a spectrum analysis such as FFT is first applied for the windowed speech in the current frame. The short-time speech spectrum is further divided into different spectrum bands. The bandwidth is usually an integer, times the basic frequency that equals the inverse of the pitch. Each band is described as "voiced" or "unvoiced".

The parameters of the MBE coder thus include the spectrum envelope, pitch, unvoiced/voiced (U/V) decisions for different bands. Based on different bitrate demands, the phase of the spectrum can be parameterized or discarded. In the speech decoding process, voiced bands and unvoiced bands are synthesized by different schemes and combined to generate the final output.

MBE utilizes the analysis-by-synthesis scheme in parameter estimation. Parameters such as basic frequency, spectrum envelope, and subband U/V decisions are all done via closed-loop searching. The criteria of the closed-loop optimization are based on minimizing the perceptually weighted reconstructed speech error, which can be represented in the frequency domain as

$$\varepsilon = \frac{1}{2\pi} \int_{-\pi}^{+\pi} G(\omega)|S_w(\omega) - S_{wr}(\omega)|d\omega \qquad (13.29)$$

where $S_w(\omega)$ and $S_{wr}(\omega)$ are the original speech short-time spectrum and reconstructed speech short-time spectrum, and $G(\omega)$ is the spectrum of the perceptual weighting filter.

Similar to the closed-loop searching scheme in CELP, a sequential optimization method is used to make the problem tractable. In the first step, all bands are assumed voiced bands, and the spectrum envelope and basic frequency are estimated. Rewriting the spectrum error with the all-voiced assumption, we have

$$\breve{\varepsilon} = \sum_{m=-M}^{M} [\frac{1}{2\pi} \int_{\alpha_m}^{\beta_m} G(\omega)|S_w(\omega) - A_m E_{wr}(\omega)|^2 d\omega] \qquad (13.30)$$

in which M is band number in $[0, \pi]$, A_m is the spectrum envelope of band m, $E_{wr}(\omega)$ is the short-time window spectrum, and $\alpha_m = (m - \frac{1}{2})\omega_0$, $\beta_m = (m + \frac{1}{2})\omega_0$. Setting $\partial \breve{\varepsilon}/\partial A_m = 0$, we get

$$A_m = \frac{\int_{\alpha_m}^{\beta_m} G(\omega)S_w(\omega)E_{wr}^*(\omega)d\omega}{\int_{\alpha_m}^{\beta_m} G(\omega)|E_{wr}(\omega)|^2 d\omega} \qquad (13.31)$$

The basic frequency is obtained at the same time by searching over a frequency interval to minimize $\breve{\varepsilon}$. Based on the estimated spectrum envelope, an adaptive thresholding scheme tests the matching degree for each band. We label a band as voiced if there is a good matching; otherwise, we declare the band as unvoiced and re-estimate the envelope for the unvoiced band as

$$A_m = \frac{\int_{\alpha_m}^{\beta_m} G(\omega)|S_w(\omega)|d\omega}{\int_{\alpha_m}^{\beta_m} G(\omega)d\omega}. \qquad (13.32)$$

The decoder uses separate methods to synthesize unvoiced and voiced speech, based on the unvoiced and voiced bands. The two types of reconstructed components are then combined to generate synthesized speech. The final step is overlapping the sum of the synthesized speech in each frame to get the final output.

Multiband Excitation Linear Predictive (MELP)

The MELP speech codec is a new U.S. federal standard to replace the old LPC-10 (FS1015) standard, with the application focus on low-bitrate safety communications. At 2.4 kbps, MELP [12] has comparable speech quality to the 4.8 kbps DOD-CELP (FS1016) and good robustness in a noisy environment.

MELP is also based on LPC analysis. Different from the hard-decision voiced/unvoiced model adopted in LPC-10, MELP uses a multiband soft-decision model for the excitation signal. The LP residue is band-passed, and a voicing strength parameter is estimated for each band. The decoder can reconstruct the excitation signal by combining the periodic pulses and white noises, based on the voicing strength in each band. Speech can be then reconstructed by passing the excitation through the LPC synthesis filter.

Different from MBE, MELP divides the excitation into five fixed bands of 0–500, 500–1000, 1000–2000, 2000–3000, and 3000–4000 Hz. It estimates a voice degree parameter in each band based on the normalized correlation function of the speech signal and the smoothed, rectified signal in the non-DC band. Let $s_k(n)$ denote the speech signal in band k, and $u_k(n)$ denote the DC-removed smoothed rectified signal of $s_k(n)$. The correlation function is defined as

$$R_x(P) = \frac{\sum_{n=0}^{N-1} x(n) \cdot x(n+P)}{\left[\sum_{n=0}^{N-1} x^2(n) \cdot \sum_{n=0}^{N-1} x^2(n+P)\right]^{1/2}} \tag{13.33}$$

where P is the pitch of the current frame, and N is the frame length. Then the voicing strength for band k is defined as $\max(R_{s_k}(P), R_{u_k}(P))$.

To further remove the buzziness of traditional LPC-10 speech coders for the voiced speech segment, MELP adopts a jittery voiced state to simulate the marginal voiced speech segments. The jittery state is indicated by an aperiodic flag. If the aperiodic flag is set in the analysis end, the receiver adds a random shifting component to the periodic pulse excitation. The shifting can be as big as $P/4$. The jittery state is determined by the peakiness of the full-wave rectified LP residue $e(n)$,

$$\text{peakiness} = \frac{\left[\frac{1}{N}\sum_{n=0}^{N-1} e(n)^2\right]^{1/2}}{\frac{1}{N}\sum_{n=0}^{N-1} |e(n)|}. \tag{13.34}$$

If peakiness is greater than some threshold, the speech frame is determined as jittered.

To better reconstruct the short-time spectrum of the speech signal, the spectrum of the residue signal is not assumed to be flat, as it is in the LPC-10 speech coder. After normalizing the LP residue signal, MELP preserves the magnitudes corresponding to the first $\min(10, P/4)$ basic frequency harmonics. Basic frequency is the inverse of the pitch period. The higher harmonics are discarded and assumed to be a unity spectrum.

The 10-d magnitude vector is quantized by 8-bit vector quantization, using a perceptually weighted distance measure. Similar to most modern LPC quantization schemes, MELP also converts LPC parameters to LSF and uses four-stage vector quantization. The bits allocated for the four stages are 7, 6, 6, and 6, respectively. Apart from integral pitch estimation similar to LPC-10, MELP applies a fractional pitch refinement procedure to improve the accuracy of pitch estimation.

In the speech reconstruction process, MELP does not use a periodic pulse to represent the periodic excitation signal but uses a dispersed waveform. To disperse the pulses, a *finite impulse response* (*FIR*) filter is applied to the pulses. MELP also applies a perceptual weighting filter to the reconstructed speech to suppress the quantization noise and improve the subject's speech quality.

Table 13.1 lists the speech codecs covered in the above sections. In Table 13.1, GSM at full rate 13 kbps uses a CELP-type speech codec RTE-LPC. The MOS scores of different codecs in the table indicate the quality of the encoded speech. A score of 4.2 and above is deemed as the toll quality. As the bit rate drops below 8kbps, the MOS score often drops rapidly and the encoding artifacts become audible. For very low bit rate speech codecs at 1.2 and 2.4 kbps, the encoding artifacts become easily noticeable. At such a low bit rate, the intelligibility of the decoded speech becomes more important.

As discussed in the previous sections, different speech coding schemes share many similar overall structures. Most of the speech coding standards were formed years ago. However, the research and engineering on speech coding have not stopped. As the Internet becomes almost everywhere, speech coding adapted to the network communication becomes more and more important. In the following, we discuss a few recent open source speech codecs targeted at voice over Internet applications.

13.3   Open Source Speech Codecs*

Speech coding has been evolving over the years. With the widespread of Internet, voice over network becomes more and more important. Different from traditional

Table 13.1 Comparison of different speech coding standards

Year	Bit rate (kbps)	Codec	MOS
1972	64	PCM	4.4
1976	2.4	LPC-10(FS-1015)	2.7
1984	16	G726 ADPCM	3.9
1990	4.15	MBE (Inmarsat)	3.2
1991	13	GSM (RTE-LPC)	3.6
1991	4.8	CELP (FS-1016)	3.2
1992	16	G.728 (LD-CELP)	4.0
1995	8	G.729	4.2
1995	6.3	G.723.1 (in H.323 and H.324)	3.4
1995	2.4	MELP	3.2

communication channels, Internet provides dramatically different bandwidths and thus speech and audio codecs should be able to support different bit rates. Variable bit rate encoding on the fly is also often preferred due to the constant bandwidth variations in the network. Many applications are also interactive, which require the codecs to have a short delay. Different speech and audio codecs have been developed to meet such a need. High bandwidth speech encoding is also desirable because sometimes the audio includes not only speech, but also generic audio signals such as music. Modern speech codecs thus often use hybrid structures which can handle both low bandwidth speech and high bandwidth audio simultaneously. Many of such codecs are open source and can be used freely with very few restrictions in commercial applications.

13.3.1 Speex

Speex [13] is an open source patent-free speech codec which provides a free alternative to more expensive proprietary speech codecs. Speex is designed for Voice over Internet. It is under the revised BSD license and can be used for free commercially.

Speex is a CELP type speech codec. It supports different input bandwidths: narrowband (8 kHz), wideband (16 kHz) and ultrawideband (32 kHz). Its target bitrate is from 2 to 44 kbps and it supports variable bit rate encoding. Similar to the standard CELP speech codec, Speex also uses the source filter model. The filter models the characteristics of the human vocal tract and the source stimulates the vocal track filter to generate output speech. Traditional linear prediction filter is used to model the vocal tract and used to flatten the spectrum of the residual signal. The excitation of Speex is generated from two types of sources: the periodic ones in an adaptive codebook and the noise-like ones in a fixed codebook. The adaptive codebook is constructed from the samples of previously decoded speech samples and models the periodic components in the source. For voiced speech segments, pitch is estimated using the correlation method and the long-term prediction approximates the speech excitation using the previous reconstructed ones shifted by the pitch and weighted by pitch gains. The innovation signal taken from the fixed codebook plus the long-term prediction forms the excitation. Speex uses the analysis-by-synthesis method to generate the excitation signal that optimally matches the original excitation in terms of the perceptually weighted spectrum.

For the narrowband mode (8 kHz), Speex is quite similar to a regular CELP. Speex uses a frame of 20 ms and 4 subframes, each of which has a length of 5 ms. It uses LSPs to represent the LPCs during quantization and encoding. Its analysis-by-synthesis encoder loop is similar to a standard CELP encoder. The long-term prediction uses a three-tap predictor with three gains but only one single pitch prediction. Speex encodes the gain in each subframe using a shared gain with a subframe gain correction. This is different from the standard CELP implementation and increases the robustness when packet loss happens. The innovation signal generation is different from a standard CELP. Speex does not use algebraic codebook because the technology has been patented. Instead, in Speex, each subframe is divided into sub-groups of length ranging from 5 to 20 samples and each sub-group is vector quantized using the bitrate dependent fixed codebook.

In the wideband mode (16 kHz), Speex uses a filter to split the input into two bands, the lower band is from 0 to 8 kHz and the upper band from 8 to 16 kHz. The lower band is processed the same as the previously mentioned narrowband mode encoder. The upper band operates similarly except that it does not estimate the pitch.

Since Speex is designed for Voice over Internet, it also includes the functions for packaging the encoded frames into RTP packets which can be streamed through the network. Each RTP packet may contain one encoded frame or multiple frames. In the Internet channel, the errors are mostly packet losses instead of bit corruptions. The design of Speex takes these factors into consideration.

13.3.2 Opus

Opus [14] is another open, royalty-free high quality audio codec. It is a standard of the Internet Engineering Task Force (IETF) known as RFC 6716. Different from Speex, which is a codec designed mainly for speech (Voice over Internet), Opus can be used to encode both speech and generic audio. Its speech codec is adapted from the Skype's SILK codec and audio codec is the CELT [15] codec. Opus has better speech compression quality than Speex. Due to its good quality and openness, it has been widely adopted in applications such as Voice over Internet, wideband video conference, music streaming, and low-delay broadcast. Opus supports bitrate from 6 to 510 kbps and can deal with input with bandwidth from 8 to 48 kHz. It supports both mono and stereo mode.

Opus audio codec's diagram is illustrated in Fig. 13.7. Opus can operate in three different modes: SILK-only mode for narrow, median, and wideband speech; hybrid of SILK and CELT mode for superband or fullband speech; and CELT-only mode for narrow or fullband music.

SILK is an LP type speech codec. It uses order-16 linear predictive coefficients (LPCs). The long-term prediction (LTP) uses a 5-tap filter. SILK encoder jointly optimizes the LPC and the LTP. SILK encoder is based on noise feedback coding [16] rather than the common analysis-by-synthesis scheme in CELP codecs. Noise feedback coding has also been used in the BroadVoice speech codecs (BV16 and BV32). Noise feedback coding has a lower computational cost and shorter delays than the analysis-by-synthesis method used in the traditional CELP based speech codecs. SILK's residue is encoded as a sequence of pulses with pulse dependent dither. SILK uses entropy coding to encode the LPC and the residue signal. SILK's decoder is a standard LPC type speech decoder.

CELT targets at wideband audio. However, its design borrows ideas from the CELP speech codecs. CELT is the short form of Constrained Energy Lapped Transform. The goal of CELT is to provide an audio codec that has a short delay so that it can be used in the interactive applications. Similar to other wideband generic audio codec, CELT uses MDCT to transform the input to the spectrum domain for encoding. Traditional MDCT based audio codecs use larger windows to obtain enough

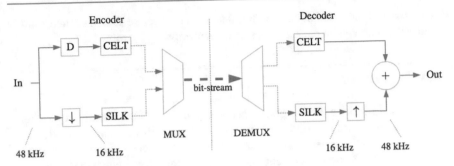

Fig. 13.7 Opus audio codec diagram

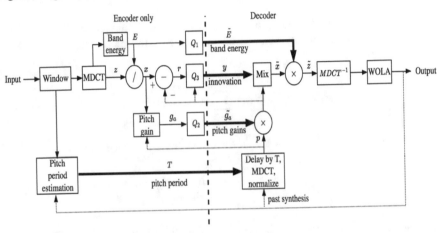

Fig. 13.8 CELT audio codec diagram

frequency resolution. However, larger window causes long delay. CELT uses a much shorter window to reduce the delay.

CELT borrows ideas from CELP: it flattens the "excitation" which is the summation of the adaptive (pitch) codebook and a fixed (innovation) codebook. The diagram of CELT is illustrated in Fig. 13.8. Different from CELP, CELT mainly operates in the spectrum domain. It flattens the MDCT spectrum by normalizing the energy in different frequency bands. Similar to CELP, CELT's adaptive codebook is based on pitch estimation and time shifted to the past with the associated gains, and the innovation part is predicted using the adaptive codebook. The long-term prediction is estimated using the decoded audio signal in a way similar to CELP's long-term prediction. The innovation estimation uses the algebraic codebook scheme similar to Algebraic CELP, which recursively constructs the optimal excitation.

Opus is designed for audio over Internet. It uses forward error correction, which lets each package include coarse encoding of previous ones when speech is active. This increases the robustness if there are packet losses. Opus also tries to encode packages so that they are independent of each other. Opus uses the so-called discontinuous transmission technique, which reduces the packet rate during silence. Opus's decoder also can handle packet losses to further improve the robustness.

13.4   Exercises

1. In Sect. 13.2.1, we discuss phase insensitivity. Explain the meaning of the term "phase" in regard to individual frequency components in a composite signal.
2. Input a speech segment, using C or MATLAB, and verify that formants indeed exist—that any speech segment has only a few important frequencies. Also, verify that formants change as the interval of speech being examined changes.

A simple approach to coding a frequency analyzer is to reuse the DCT coding ideas we have previously considered in Sect. 8.5. In one dimension, the DCT transform reads

$$F(u) = \sqrt{\frac{2}{N}} C(u) \sum_{i=0}^{N-1} \cos \frac{(2i+1)u\pi}{2N} f(i) \qquad (13.35)$$

where $i, u = 0, 1, \ldots, N-1$, and the constants $C(u)$ are given by

$$C(u) = \begin{cases} \frac{\sqrt{2}}{2} & \text{if } u = 0 \\ 1 & \text{otherwise} \end{cases} \qquad (13.36)$$

If we use the speech sample in Fig. 6.16, then taking the one-dimensional DCT of the first, or last, 40 ms (i.e., 32 samples), we arrive at the absolute frequency components as in Fig. 13.5.

3. Write code to read a WAV file. You will need the following set of definitions: a WAV file begins with a 44-byte header, in unsigned byte format. Some important parameter information is coded as follows:

Byte[22..23]   Number of channels
Byte[24..27]   Sampling rate
Byte[34..35]   Sampling bits
Byte[40..43]   Data length

4. Write a program to add fade-in and fade-out effects to sound clips (in WAV format). Specifications for the fades are as follows: The algorithm assumes a linear envelope; the fade-in duration is from 0 to 20% of the data samples; the fade-out duration is from 80 to 100% of the data samples.

If you like, you can make your code able to handle both mono and stereo WAV files. If necessary, impose a limit on the size of the input file, say 16 MB.

5. In the text, we study an adaptive quantization scheme for ADPCM. We can also use an adaptive prediction scheme. We consider the case of one tap prediction, $\hat{s}(n) = a \cdot s(n-1)$. Show how to estimate the parameter a in an open-loop method. Estimate the SNR gain you can get, compared to the direct PCM method based on a uniform quantization scheme.

6. Linear prediction analysis can be used to estimate the shape of the envelope of the short-time spectrum. Given ten LP coefficients $a_1, \ldots, a_{10}$, how do we get the formant position and bandwidth?

7. Download and implement a CELP coder (see the textbook web site). Try out this speech coder on your own recorded sounds.

8. In quantizing LSP vectors in G.723.1, splitting vector quantization is used: if the dimensionality of LSP is 10, we can split the vector into three subvectors of length 3, 3, and 4 each and use vector quantization for the subvectors separately. Compare the codebook space complexity with and without split vector quantization. Give the codebook searching time complexity improvement by using splitting vector quantization.

9. Discuss the advantage of using an algebraic codebook in CELP coding.

10. The LPC-10 speech coder's quality deteriorates rapidly with strong background noise. Discuss why MELP works better in the same noisy conditions.

11. Give a simple time-domain method for pitch estimation based on the autocorrelation function. What problem will this simple scheme have when based on one speech frame? If we have three speech frames, including a previous frame and a future frame, how can we improve the estimation result?

12. On the receiver side, speech is usually generated based on two frames' parameters instead of one, to avoid abrupt transitions. Give two possible methods to obtain smooth transitions. Use the LPC codec to illustrate your idea.

13. In real applications, speech signal usually has multiple channels. For instance, stereophonic sound uses two channels. Discuss how speech coding can take advantage of the similarity of the speech signal among different channels.

14. Implement a basic LPC speech codec in Matlab, Python or C++. The speech encoder should include functions to detect whether a speech segment is voiced or unvoiced, detect the pitch of the voiced segment, extract linear predictive coefficients, and the energy. The encoder quantizes these key parameters using scalar or vector quantization and encodes the parameters into a bitstream. The decoder decodes the parameters and reconstructs the speech. Experiment with different ways to quantize the linear predictive coefficients and other parameters and compare the speech quality at different bit rates.

   Encoded speech bitstreams may be corrupted when transmitted through a channel. Experiment with how the bit errors would affect the reconstructed speech and propose methods that can improve the bit error resistance of the speech codec.

References

1. N.S. Jayant, P. Noll, *Digital Coding of Waveforms*. (Prentice-Hall, 1984)
2. J.C. Bellamy, *Digital Telephony*. (Wiley, 2000)
3. T.E. Tremain, The government standard linear predictive coding algorithm: LPC-10, in *Speech Technology* (1982)
4. J.P. Campbell, Jr., T.E. Tremain, V.C. Welch, The DOD 4.8 kbps standard (Proposed Federal Standard 1016), in *Advances in Speech Coding*. (Kluwer Academic Publishers, 1991)
5. Dual rate speech coder for multimedia communications transmitting at 5.3 and 6.3 kbit/s. ITU-T Recommendation G.723.1 (1996)
6. GSM enhanced full rate speech transcoding (GSM 06.60). ETSI standards documentation, EN 301 245 (1998)
7. TDMA Cellular / PCS radio interface-enhanced full rate speech codec, TIA/EIA/IS-641 standard (1996)
8. Coding of speech at 16 kbit/s using low-delay code excited linear programming. ITU-T Recommendation G.728 (1992)
9. Coding of speech at 8 kbit/s using conjugate-structure algebraic-code-excited linear-prediction (CS-ACELP). ITU-T Recommendation G.729 (1996)
10. D.W. Griffin, J.S. Lim, Multi-band excitation vocoder. IEEE Trans. ASSP **36**(8), 1223–1235 (1988)
11. Inmarsat-m voice codec, v2. Inmarsat-M Specification, Inmarsat stands for Int. Mobile Satellite (1991)
12. T.P. III McCree, A.V. Barnwell, Mixed excitation LPC vocoder model for low bit rate speech coding. IEEE Trans. Speech Audio Process. **3**(4), 242–250 (1995)
13. Speex. https://github.com/xiph/speex
14. Opus. https://github.com/xiph/opus
15. J.-M. Valin, T.B. Terriberry, C. Montgomery, G. Maxwell, A high-quality speech and audio codec with less than 10 ms delay. IEEE Trans. Audio Speech Language Process. **18**(1) (2010)
16. J.-H. Chen, Novel codec structures for noise feedback coding of speech, in *Proceedings of the IEEE International Conference on Acoustics, Speech and Signal Processing* (2006)

MPEG Audio Compression

14

Have you ever attended a dance and found that for quite some time afterward you couldn't hear much? You were dealing with a type of *temporal masking*!

Have you ever noticed that the person on the soundboard at a dance basically cannot hear high frequencies anymore? Since many technicians have such hearing damage, some compensate by increasing the volume levels of the high frequencies, so they can hear them. If your hearing is not damaged, you experience this music mix as too piercing.

Moreover, if a very loud tone is produced, you also notice it is impossible to hear any sound nearby in the frequency spectrum—the band's singing may be drowned out by the lead guitar. If you've noticed this, you have experienced *frequency masking*!

MPEG audio uses this kind of perception phenomenon by simply giving up on the tones that can't be heard anyway. Using a curve of human hearing perceptual sensitivity, an MPEG audio codec makes decisions on when and to what degree frequency masking and temporal masking make some components of the music inaudible. It then controls the quantization process so that these components do not influence the output.

So far, in the previous chapter, we have concentrated on telephony applications—usually, LPC and CELP are tuned to speech parameters. In contrast, in this chapter, we consider compression methods applicable to general audio, such as music or perhaps broadcast digital TV@. Instead of modeling speech, the method used is a *waveform* coding approach—one that attempts to make the decompressed amplitude-versus-time waveform as much as possible like the input signal.

The main technique used in evaluating audio content for possible compression makes use of a *psychoacoustic model* of hearing. The kind of coding carried out, then, is generally referred to as *perceptual coding*.

In this chapter, we look at how such considerations impact MPEG audio compression standards and examine in some detail at the following topics:

© Springer Nature Switzerland AG 2021
Z.-N. Li et al., *Fundamentals of Multimedia*, Texts in Computer Science,
https://doi.org/10.1007/978-3-030-62124-7_14

- Psychoacoustics
- MPEG-1 Audio Compression
- Later MPEG audio developments: MPEG-2, 4, 7, and beyond.

14.1 Psychoacoustics

Recall that the range of human hearing is about 20 Hz to about 20 kHz (for people who have not gone to many dances). Sounds at higher frequencies are *ultrasonic*. However, the frequency range of the voice is typically only from about 500 Hz to 4 kHz. The dynamic range, the ratio of the maximum sound amplitude to the quietest sound humans can hear, is on the order of about 120 dB.

Recall that the decibel unit represents ratios of intensity on a logarithmic scale. The reference point for 0 dB is the threshold of human hearing—the quietest sound we can hear, measured at 1 kHz. Technically, this is a sound that creates a barely audible sound intensity of 10^{-12} Watt/m^2. Our range of magnitude perception is thus incredibly wide: the level at which the sensation of sound begins to give way to the sensation of pain is about 1 Watt/m^2, so we can perceive a ratio of 10^{12}!

The range of hearing actually depends on frequency. At a frequency of 2 kHz, the ear can readily respond to sound that is about 96 dB more powerful than the smallest perceivable sound at that frequency, or in other words a power ratio of 2^{32}. Table 6.1 lists some of the common sound levels in decibels.

14.1.1 Equal-Loudness Relations

Suppose we play two pure tones, sinusoidal sound waves, with the same amplitude but different frequencies. Typically, one may sound louder than the other. The reason is that the ear does not hear low or high frequencies, as well as it hears frequencies in the middle range. In particular, at normal sound volume levels, the ear is most sensitive to frequencies between 1 and 5 kHz.

Fletcher-Munson Curves

The Fletcher-Munson equal-loudness curves display the relationship between perceived loudness (*in phons*) for a given stimulus sound volume (*Sound Pressure Level*, in dB), as a function of frequency. Figure 14.1 shows the ear's perception of equal loudness. The abscissa—the x axis—(shown in a semi-log plot) is frequency, in kHz. The ordinate axis is sound pressure level—the *actual* intensity (i.e., loudness) of the tone generated in an experiment. The curves show the loudness with which such tones are *perceived* by humans. The bottom curve shows what level of pure tone stimulus is required to produce the perception of a 10 dB sound.

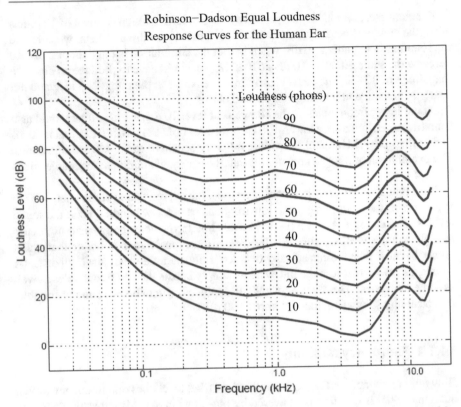

Fig. 14.1 Fletcher-Munson equal-loudness response curves for the human ear (remeasured by Robinson and Dadson)

All the curves are arranged so that the perceived loudness level gives the same loudness as for that loudness level of a pure tone at 1 kHz. Thus, the loudness level at the 1 kHz point is always equal to the dB level on the ordinate axis. The bottom curve, for example, is for 10 phons. All the tones on this curve will be perceived as loud as a 10 dB, 1000 Hz tone. The figure shows the equal-loudness curves remeasured by Robinson and Dadson in 1956 [1], which are more accurate at lower frequencies (<500 Hz) than the Fletcher and Munson originals [2]. The ISO standard (ISO 226:2003) further revised the curves in 2003, and recommended the term *equal-loudness contours* for these curves. The ISO 226:2003 contours actually show more agreement with the Fletcher and Munson curves at higher frequencies (>1000 Hz).

The idea is that a tone is produced at a certain frequency and *measured* loudness level, then a human rates the loudness as it is perceived. On the lowest curve shown, each pure tone between 20 Hz and 15 kHz would have to be produced at the volume level given by the ordinate for it to be perceived at a 10 dB loudness level [1]. The next curve shows what the magnitude would have to be for pure tones to each be perceived as being at 20 dB, and so on. The top curve is for perception at 90 dB.

For example, at 5000 Hz, we perceive a tone to have a loudness level of 10 phons when the source is actually only 5 dB. Notice that at the dip at 4 kHz, we perceive the sound as being about 10 dB, when in fact the stimulation is only about 2 dB. To perceive the same effective 10 dB at 10 kHz, we would have to produce an absolute magnitude of 20 dB. The ear is clearly more sensitive in the range 2–5 kHz and not nearly as sensitive in the range 6 kHz and above.

At the lower frequencies, if the source is at level 10 dB, a 1 kHz tone would also sound at 10 dB; however, a lower, 100 Hz tone must be at a level of 30 dB, that is 20 dB higher than the 1 kHz tone! So we are not very sensitive to the lower frequencies. The explanation of this phenomenon is that the ear canal amplifies frequencies from 2.5 to 4 kHz.

Note that as the overall loudness increases, the curves flatten somewhat. We are approximately equally sensitive to low frequencies of a few hundred Hz if the sound level is loud enough. And we perceive most low frequencies better than high ones at high volume levels. Hence, at the dance, loud music sounds better than quiet music, because then we can actually hear low frequencies and not just high ones. (A "loudness" switch on some sound systems simply boosts the low frequencies as well as some high ones.) However, above 90 dB, people begin to become uncomfortable. A typical city subway operates at about 100 dB.

14.1.2 Frequency Masking

How does one tone interfere with another? At what level does one frequency drown out another? This question is answered by masking curves. Also, masking answers the question of how much noise we can tolerate before we cannot hear the actual music. Lossy audio data compression methods, such as MPEG Audio or Dolby Digital (AC-3) encoding, which is popular in movies, remove some sounds that are masked anyway, thus reducing the total amount of information.

The general situation in regard to masking is as follows:

- A lower tone can effectively mask (make us unable to hear) a higher tone.
- The reverse is not true. A higher tone does not mask a lower tone well. Tones can in fact mask lower-frequency sounds, but not as effectively as they mask higher-frequency ones.
- The greater the power in the masking tone, the wider its influence—the broader the range of frequencies it can mask.
- As a consequence, if two tones are widely separated in frequency, little masking occurs.

Threshold of Hearing

Figure 14.2 shows a plot of the threshold of human hearing, for pure tones. To determine such a plot, a particular frequency tone is generated, say 1 kHz. Its volume

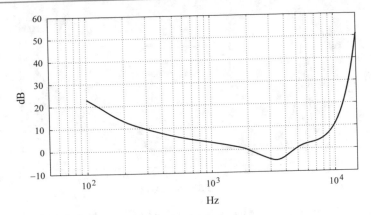

Fig. 14.2 Threshold of human hearing, for pure tones

is reduced to zero in a quiet room or using headphones, then turned up until the sound is just barely audible. Data points are generated for all audible frequencies in the same way.

The point of the threshold of hearing curve is that if a sound is above the dB level shown—say it is above 2 dB for a 6 kHz tone—then the sound is audible. Otherwise, we cannot hear it. Turning up the 6 kHz tone so that it equals or surpasses the curve means we can then distinguish the sound.

An approximate formula exists for this curve, as follows [3]:

$$Threshold(f) = 3.64(f/1000)^{-0.8} - 6.5e^{-0.6(f/1000-3.3)^2} + 10^{-3}(f/1000)^4 \tag{14.1}$$

The threshold units are dB. Since the dB unit is a ratio, we do have to choose which frequency will be pinned to the origin, $(0, 0)$. In Eq. (14.1), this frequency is 2,000 Hz: $Threshold(f) = 0$ at $f = 2$ kHz.

Frequency Masking Curves

Frequency masking is studied by playing a particular pure tone, say 1 kHz again, at a loud volume and determining how this tone affects our ability to hear tones at nearby frequencies. To do so, we would generate a 1 kHz *masking tone* at a fixed sound level of 60 dB, then raise the level of a nearby tone, say 1.1 kHz, until it is just audible. The threshold in Fig. 14.3 plots this audible level.

It is important to realize that this masking diagram holds only for a single masking tone: the plot changes if other masking tones are used. Figure 14.4 shows how this looks: the higher the frequency of the masking tone, the broader a range of influence it has.

If, for example, we play a 6 kHz tone in the presence of a 4 kHz masking tone, the masking tone has raised the threshold curve much higher. Therefore, at its neighbor frequency of 6 kHz, we must now surpass 30 dB to distinguish the 6 kHz tone.

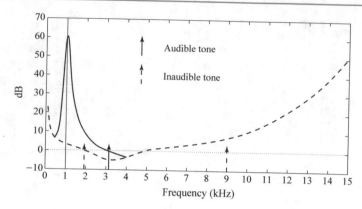

Fig. 14.3 Effect on the threshold of human hearing for a 1 kHz masking tone

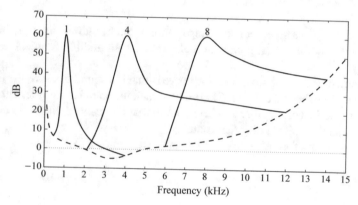

Fig. 14.4 Effect of masking tones at three different frequencies

The practical point is that if a signal can be decomposed into frequencies, then for frequencies that will be partially masked, only the audible part will be used to set quantization noise thresholds.

Critical Bands

The human hearing range naturally divides into *critical bands*, with the property that the human auditory system cannot resolve sounds better than within about one critical band when other sounds are present. Hearing has a limited, frequency-dependent resolution. According to [4], "In a complex tone, the critical bandwidth corresponds to the smallest frequency difference between two partials such that each can still be heard separately. ...the critical bandwidth represents the ear's resolving power for simultaneous tones or partials."

Table 14.1 Critical bands and their bandwidths

Band #	Lower bound (Hz)	Center (Hz)	Upper bound (Hz)	Bandwidth (Hz)
1	20	50	100	80
2	100	150	200	100
3	200	250	300	100
4	300	350	400	100
5	400	450	510	110
6	510	570	630	120
7	630	700	770	140
8	770	840	920	150
9	920	1000	1080	160
10	1080	1170	1270	190
11	1270	1370	1480	210
12	1480	1600	1720	240
13	1720	1850	2000	280
14	2000	2150	2320	320
15	2320	2500	2700	380
16	2700	2900	3150	450
17	3150	3400	3700	550
18	3700	4000	4400	700
19	4400	4800	5300	900
20	5300	5800	6400	1100
21	6400	7000	7700	1300
22	7700	8500	9500	1800
23	9500	10500	12000	2500
24	12000	13500	15500	3500
25	15500	18775	22050	6550

At the low-frequency end, a critical band is less than 100 Hz wide, while for high frequencies, the width can be greater than 4 kHz. This indeed is yet another kind of *perceptual nonuniformity.*

Experiments indicate that the critical bandwidth remains approximately constant in width for masking frequencies below about 500 Hz—this width is about 100 Hz. However, for frequencies above 500 Hz, the critical bandwidth increases approximately linearly with frequency.

Generally, the audio frequency range for hearing can be partitioned into about 24 critical bands (25 are typically used for coding applications), as Table 14.1 shows.

Notwithstanding the *general* definition of a critical band, it turns out that our hearing apparatus actually is somewhat tuned to *certain* critical bands. Since hearing depends on physical structures in the inner ear, the frequencies at which these struc-

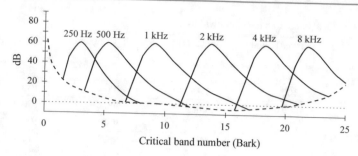

Fig. 14.5 Effect of masking tones, expressed in Bark units

tures best resonate is important. Frequency masking is a result of the ear structures becoming "saturated" at the masking frequency and nearby frequencies.

Hence, the ear operates something like a set of band-pass filters, where each allows a limited range of frequencies through and blocks all others. Experiments that illustrate this property are based on the observation that a constant-volume sound will seem louder if it spans the boundary between two critical bands than it would were it contained entirely within one critical band [5]. In effect, the ear is not very discriminating *within* a critical band, because of masking.

Bark Unit

Since the range of frequencies affected by masking is broader for higher frequencies, it is useful to define a new frequency unit such that, in terms of this new unit, each of the masking curves (the parts of Fig. 14.4 above the threshold in quiet) have about the same width.

The new unit defined is called the *Bark*, named after Heinrich Barkhausen (1881–1956), an early sound scientist. One Bark unit corresponds to the width of one critical band, for any masking frequency [6,7]. Figure 14.5 displays critical bands, with the frequency (the abscissa or *x*-axis) given in Bark units.

The conversion between a frequency f and its corresponding critical-band number b, expressed in Bark units, is as follows:

$$\text{Critical band number (Bark)} = \begin{cases} f/100, & \text{for } f < 500 \\ 9 + 4\log_2(f/1000), & \text{for } f \geq 500 \end{cases} \quad (14.2)$$

In terms of this new frequency measure, the critical-band number b equals 5 when $f = 500$ Hz. At double that frequency, for a masking frequency of 1 kHz, the Bark value goes up to 9. Another formula used for the Bark scale is as follows:

$$b = 13.0 \arctan(0.76f) + 3.5 \arctan(f^2/56.25) \quad (14.3)$$

where f is in kHz and b is in Barks. The inverse equation gives the frequency (in kHz) corresponding to a particular Bark value b

$$f = [(\exp(0.219 \times b)/352) + 0.1] \times b - 0.032 \times \exp[-0.15 \times (b - 5)^2]$$
(14.4)

Frequencies forming the boundaries between two critical bands are given by integer Bark values. The critical bandwidth (df) for a given center frequency f can also be approximated by the following [8]:

$$df = 25 + 75 \times [1 + 1.4 \times f^2]^{0.69}$$
(14.5)

where f is in kHz and df is in Hz.

The idea of the Bark unit is to define a more perceptually uniform unit of frequency, in that every critical band's width is roughly equal in terms of Barks.

14.1.3 Temporal Masking

Recall that after the dance it takes quite a while for our hearing to return to normal. Generally, any loud tone causes the hearing receptors in the inner ear (little hairlike structures called *cilia*) to become *saturated*, and they require time to recover. (Many other perceptual systems behave in this temporally slow fashion—for example, the receptors in the eye have this same kind of "capacitance" effect.)

To quantify this type of behavior, we can measure the time sensitivity of hearing by another masking experiment. Suppose we again play a masking tone at 1 kHz with a volume level of 60 dB, and a nearby tone at, say, 1.1 kHz with a volume level of 40 dB. Since the nearby test tone is masked, it cannot be heard. However, once the masking tone is turned off, we can again hear the 1.1 kHz tone, but only after a small amount of time. The experiment proceeds by stopping the test tone slightly after the masking tone is turned off, say 10 ms later.

The delay time is adjusted to the minimum amount of time such that the test tone can just be distinguished. In general, the louder the test tone, the less time it takes for our hearing to get over hearing the masking tone. Figure 14.6 shows this effect: it may take up to as much as 500 ms for us to discern a quiet test tone after a 60 dB masking tone has been played. Of course, this plot would change for different masking tone frequencies.

Test tones with frequencies near the masking tone are, of course, the most masked. Therefore, for a given masking tone, we have a two-dimensional temporal masking situation, as in Fig. 14.7. The closer the frequency to the masking tone and the closer in time to when the masking tone is stopped, the greater likelihood that a test tone cannot be heard. The figure shows the combined effect of both frequency and temporal masking.

The phenomenon of saturation also depends on just how long the masking tone has been applied. Figure 14.8 shows that for a masking tone played longer (200 ms) than another (100 ms), it takes longer before a test tone can be heard.

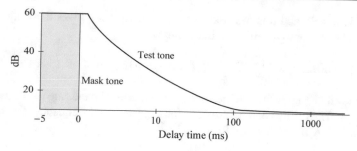

Fig. 14.6 The louder the test tone, the shorter the amount of time required before the test tone is audible once the masking tone is removed

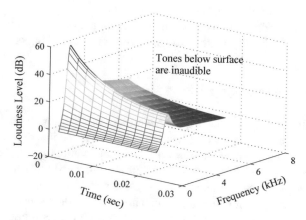

Fig. 14.7 Effect of temporal masking depends on both time and closeness in frequency

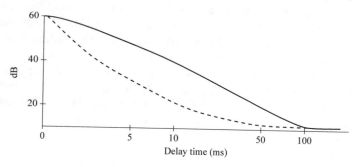

Fig. 14.8 Effect of temporal masking also depends on the length of time the masking tone is applied. Solid curve: masking tone played for 200 ms; dashed curve: masking tone played for 100 ms

As well as being able to mask other signals that occur just after it sounds (*post-masking*), a particular signal can even mask sounds played just before the stronger signal (*pre-masking*). Pre-masking has a much shorter effective interval (2–5 ms) in which it is operative than does post-masking (usually 50–200 ms).

MPEG audio compression takes advantage of these considerations in basically constructing a large, multi-dimensional lookup table. It uses this to transmit frequency components that are masked by frequency masking or temporal masking or both, using fewer bits.

14.2 MPEG Audio

MPEG Audio [9, 10] proceeds by first applying a filter bank to the input, to break the input into its frequency components. In parallel, it applies a psychoacoustic model to the data, and this model is used in a bit allocation block. Then the number of bits allocated is used to quantize the information from the filter bank. The overall result is that quantization provides the compression, and bits are allocated where they are most needed to lower the quantization noise below an audible level.

14.2.1 MPEG Layers

MP3 is a popular audio compression standard. The "3" stands for Layer 3, and "MP" stands for the MPEG-1 standard. Recall that we looked at MPEG video compression in Chap. 11. However, the MPEG standard actually delineates three different aspects of multimedia: audio, video, and systems. MP3 forms part of the audio component of this first phase of MPEG. It was released in 1992, and resulted in the international standard ISO/IEC 11172-3, published in 1993.

MPEG audio sets out three downward-compatible *layers* of audio compression, each able to understand the lower layers. Each offers more complexity in the psychoacoustic model applied and correspondingly better compression for a given level of audio quality. However, an increase in complexity, and concomitantly in compression effectiveness, is accompanied by extra delay.

Layers 1–3 in MPEG Audio are compatible, because all layers include the same file header information.

Layer 1 quality can be quite good, provided a comparatively high bitrate is available. While outdated by Layers 2 and 3, Layer 1 formed the basis for MPEG Audio. It is still largely supported, e.g., audio in packages in Ubuntu linux. Layer 2 has more complexity and was proposed for use in digital audio broadcasting. Layer 3 (MP3) is most complex and was originally aimed at audio transmission over ISDN lines. Each of the layers also uses a different frequency transform.

Most of the complexity increase is at the encoder rather than at the decoder side, and this accounts for the popularity of MP3 players. Layer 1 incorporates the simplest psychoacoustic model, and Layer 3 uses the most complex. The objective is a good trade-off between quality and bitrate. "Quality" is defined in terms of listening test scores (the psychologists hold sway here), where a quality measure is defined by

- 5.0 = "Transparent"—undetectable difference from original signal; equivalent to CD-quality audio at 14–16-bit PCM
- 4.0 = Perceptible difference, but not annoying
- 3.0 = Slightly annoying
- 2.0 = Annoying
- 1.0 = Very annoying.

(Now that's scientific!) At 64 kbps per channel, Layer 2 scores between 2.1 and 2.6, and Layer 3 scores between 3.6 and 3.8. So Layer 3 provides a substantial improvement but is still not perfect by any means.

14.2.2 MPEG Audio Strategy

Compression is certainly called for, since even audio can take fairly substantial bandwidth: CD audio is sampled at 44.1 kHz and 16 bits/channel, so for two channels needs a bitrate of about 1.4 Mbps. MPEG-1 aims at about 1.5 Mbps overall, with 1.2 Mbps for video and 256 kbps for audio. For the audio part, this amounts to a target reduction in the size of about 5.5 to 1 (6.0 to 1 for sampling at 48 kHz).

The MPEG approach to compression relies on quantization, of course, but also recognizes that the human auditory system is not accurate within the width of a critical band, both in terms of perceived loudness and audibility of a test frequency. The encoder employs a bank of filters that act to first analyze the frequency (*spectral*) components of the audio signal by calculating a frequency transform of a window of signal values. The bank of filters decomposes the signal into subbands. Layer 1 and Layer 2 codecs use a *quadrature-mirror filter* bank, while the Layer 3 codec adds a DCT. For the psychoacoustic model, a Fourier transform is used.

Then frequency masking can be brought to bear by using a psychoacoustic model to estimate the just noticeable noise level. In its quantization and coding stage, the encoder balances the masking behavior and the available number of bits by discarding inaudible frequencies and scaling quantization according to the sound level left over, above masking levels.

A sophisticated model would take into account the actual width of the critical bands centered at different frequencies. Within a critical band, our auditory system cannot finely resolve neighboring frequencies and instead tends to blur them. As mentioned earlier, audible frequencies are usually divided into 25 main critical bands, inspired by the auditory critical bands.

However, in keeping with design simplicity, the model adopts a *uniform width* for all frequency analysis filters, using 32 overlapping subbands [9, 11]. This means that at lower frequencies, each of the frequency analysis "subbands" covers the width of several critical bands of the auditory system, whereas at higher frequencies this is not so, since a critical band's width is less than 100 Hz at the low end and more than 4 kHz at the high end. For each frequency band, the sound level above the masking level dictates how many bits must be assigned to code signal values so that quantization noise is kept below the masking level, and hence cannot be heard.

In Layer 1, the psychoacoustic model uses only frequency masking. Bitrates range from 32 kbps (mono) to 448 kbps (stereo). Near-CD stereo quality is possible with a bitrate of 256–384 kbps. Layer 2 uses some temporal masking by accumulating more samples and examining temporal masking between the current block of samples and the ones just before and just after. Bitrates can be 32–192 kbps (mono) and 64–384 kbps (stereo). Stereo CD-audio quality requires a bitrate of about 192–256 kbps.

However, temporal masking is less important for compression than is frequency masking, which is why it is sometimes disregarded entirely in lower-complexity coders. Layer 3 is directed toward lower bitrate applications and uses a more sophisticated subband analysis, with nonuniform subband widths. It also adds nonuniform quantization and entropy coding. Bitrates are standardized at 32–320 kbps.

14.2.3 MPEG Audio Compression Algorithm

Basic Algorithm

Figure 14.9 shows the basic MPEG audio compression algorithm. It proceeds by dividing the input into 32 frequency subbands, via a filter bank. This is a linear operation that takes as its input a set of 32 PCM samples, sampled in time, and produces as its output 32 frequency coefficients. If the sampling rate is f_s, say $f_s = 48$ ksps (kilo samples per second; i.c., 48 kHz), then by the Nyquist theorem, the maximum

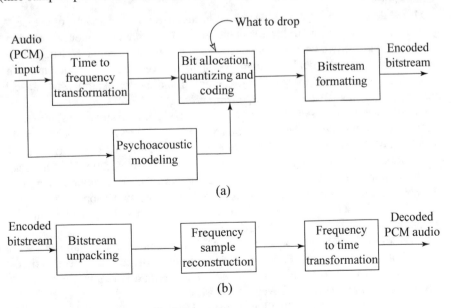

Fig. 14.9 a Basic MPEG audio encoder; and **b** decoder

| Header | SBS format | SBS | Ancillary data |

Fig. 14.10 Example MPEG audio frame

frequency mapped will be $f_s/2$. Thus, the mapped bandwidth is divided into 32 equal width segments, each of width $f_s/64$ (these segments overlap somewhat).

In the Layer 1 encoder, the sets of 32 PCM values are first assembled into a set of 12 groups of 32s. Hence, the coder has an inherent time lag, equal to the time to accumulate 384 (i.e., 12×32) samples. For example, if sampling proceeds at 32 kbps, then a time duration of 12 ms is required since each set of 32 samples is transmitted each millisecond. These sets of 12 samples, each of size 32, are called *segments*. The point of assembling them is to examine 12 sets of values at once in each of the 32 subbands after frequency analysis has been carried out, and then base quantization on just a summary figure for all 12 values.

The delay is actually somewhat longer than that required to accumulate 384 samples, since header information is also required. As well, *ancillary data*, such as multilingual data and surround sound data, is allowed. Higher layers also allow more than 384 samples to be analyzed, so the format of the subband samples (SBS) is also added, with a resulting *frame* of data, as in Fig. 14.10. The header contains a synchronization code (twelve 1s—111111111111), the sampling rate used, the bitrate, and stereo information. And as mentioned the frame format also contains room for so-called "ancillary" (extra) information. (In fact, an MPEG-1 audio decoder can at least partially decode an MPEG-2 audio bitstream, since the file header begins with an MPEG-1 header and places the MPEG-2 datastream into the MPEG-1 Ancillary Data location.)

MPEG Audio is set up to be able to handle stereo or mono channels, of course. A special *joint-stereo* mode produces a single stream by taking into account the redundancy between the two channels in stereo. This is the audio version of a composite video signal. It can also deal with *dual-monophonic*—two channels coded independently. This is useful for the parallel treatment of audio—for example, two speech streams, one in English and one in Spanish.

Consider the 32×12 segment as a 32×12 matrix. The next stage of the algorithm is concerned with scale, so that proper quantization levels can be set. For each of the 32 subbands, the maximum amplitude of the 12 samples in that row of the array is found, which is the *scaling factor* for that subband. This maximum is then passed to the bit allocation block of the algorithm, along with the SBS (subband samples). The key point of the bit allocation block is to determine how to apportion the total number of code bits available for the quantization of subband signals to minimize the audibility of the quantization noise.

As we know, the psychoacoustic model is fairly complex—more than just a set of simple lookup tables (and in fact this model is not standardized in the specification—it forms part of the "art" content of an audio encoder and is one major reason all encoders are not the same). In Layer 1, a decision step is included to decide whether each frequency band is basically like a tone or like noise. From that decision and the scaling factor, a masking threshold is calculated for each band and compared with the threshold of hearing.

The model's output consists of a set of what are known as *signal-to-mask ratios* (*SMRs*) that flag frequency components with an amplitude below the masking level. The SMR is the ratio of the short-term signal power within each frequency band to the minimum masking threshold for the subband. The SMR gives the amplitude resolution needed, and therefore, also controls the bit allocations that should be given to the subband. After the determination of the SMR, the scaling factors discussed above are used to set quantization levels such that the quantization error itself falls below the masking level. This ensures that more bits are used in regions where hearing is most sensitive. In sum, the coder uses fewer bits in critical bands when fewer can be used without making quantization noise audible.

The scaling factor is first quantized, using 6 bits. The 12 values in each subband are then quantized. Using 4 bits, the bit allocations for each subband are transmitted, after an iterative bit allocation scheme is used. Finally, the data is transmitted, with appropriate bit depths for each subband. Altogether, the data consisting of the quantized scaling factor and the 12 codewords are grouped into a collection known as the Subband-Sample format.

On the decoder side, the values are de-quantized, and the magnitudes of the 32 samples are reestablished. These are passed to a bank of *synthesis filters*, which reconstitute a set of 32 PCM samples. Note that the psychoacoustic model is not needed in the decoder.

Figure 14.11 shows how samples are organized. A Layer 2 or Layer 3 frame actually accumulates more than 12 samples for each subband: instead of 384 samples, a frame includes 1,152 samples.

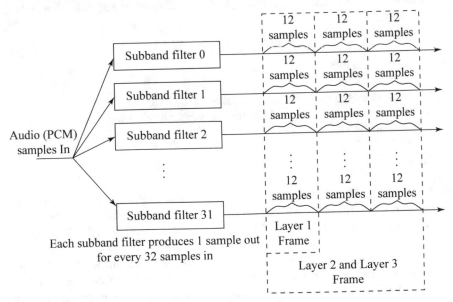

Fig. 14.11 MPEG audio frame sizes

Bit Allocation

The bit allocation algorithm (for Layer 1 and Layer 2) works in the following way. To reiterate, the aim is to ensure that all quantization noise values are below the masking thresholds. The psychoacoustic model is brought into play for such cases, to allocate more bits, from the number available, to the subbands where the increased resolution will be most beneficial.

Algorithm 14.1 (*Bit Allocation in MPEG Audio Compression (Layers 1 and 2)*)

1. From the psychoacoustic model, calculate the *Signal-to-Mask Ratio (SMR)* in decibels (dBs) for each subband:

$$SMR = 20 \log_{10} \frac{Signal}{Minimum\_masking\_threshold} \qquad (14.6)$$

- This determines the quantization, i.e., the minimum number of bits that is needed, if available. The amount of a signal above the threshold, i.e., SMR, s the amount that needs to be coded. Signals that are below the threshold do not.

2. Calculate *Signal-to-(quantization)-Noise Ratio (SNR)* for all signals.

   - A lookup table provides an estimate of SNR assuming a given number of quantizer levels.

3. *Mask-to-(quantization)-Noise Ratio (MNR)* is defined as the difference, in dB (See Fig. 14.12).

$$MNR = SNR - SMR \qquad (14.7)$$

4. Iterate until no bits left to allocate

   - Allocate bits to the subband with the lowest MNR
   - Look up new estimate of SNR for the subband allocated more bits, and re-calculate MNR.

Note:

- The masking effect means we can raise the quantization noise floor around a strong sound because the noise will be masked off anyway. As indicated in Fig. 14.12, adjusting the number of bits m allocated to a subband can move this floor up and down.
- To ensure that all the quantization noise values are inaudible, i.e., below the masking thresholds, so that all MNRs are ≥ 0, a minimum number of bits is needed. Otherwise, SNR could be too small, causing MNR to be <0, and the quality of the compressed audio could be significantly affected.

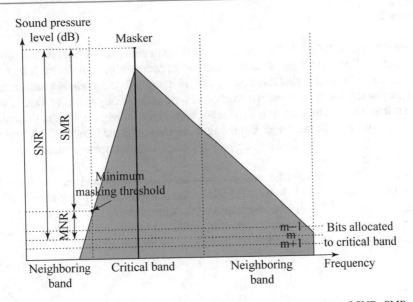

Fig. 14.12 Mask-to-noise ratio and signal-to-mask ratio. A qualitative view of SNR, SMR, and MNR, with one dominant masker and m bits allocated to a particular critical band

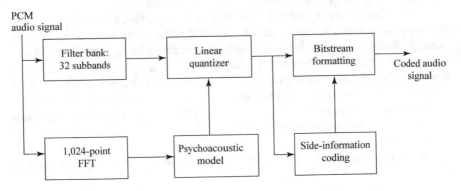

Fig. 14.13 MPEG-1 audio Layers 1 and 2

- If more bits than the minimum are allowed from the budget, allocate them anyway so as to further increase SNR. For each additional bit, we get 6 dB better SNR.

Mask calculations are performed in parallel with subband filtering, as in Fig. 14.13. The masking curve calculation requires an accurate frequency decomposition of the input signal, using a Discrete Fourier Transform (DFT). The frequency spectrum is usually calculated with a 1,024-point Fast Fourier Transform (FFT).

In Layer 1, 16 uniform quantizers are pre-calculated, and for each subband, the quantizer giving the lowest distortion is chosen. The index of the quantizer is sent as 4 bits of side information for each subband. The maximum resolution of each quantizer is 15 bits.

Layer 2

Layer 2 of the MPEG-1 Audio codec includes small changes to effect bitrate reduction and quality improvement, at the price of an increase in complexity. The main difference in Layer 2 is that three groups of 12 samples are encoded in each frame, and temporal masking is brought into play, as well as frequency masking. One advantage is that if the scaling factor is similar for each of the three groups, a single scaling factor can be used for all three. But using three frames in the filter (before, current, and next), for a total of 1,152 samples per channel, approximates taking temporal masking into account.

As well, the psychoacoustic model does better at modeling slowly-changing sound if the time window used is longer. Bit allocation is applied to window lengths of 36 samples instead of 12, and the resolution of the quantizers is increased from 15 bits to 16. To ensure that this greater accuracy does not mean poorer compression, the number of quantizers to choose from decreases for higher subbands.

Layer 3

Layer 3, or MP3, uses a bitrate similar to Layers 1 and 2 but produces substantially better audio quality, again at the price of increased complexity.

A filter bank similar to that used in Layer 2 is employed, except that now perceptual critical bands are more closely adhered to by using a set of filters with nonequal frequencies. This layer also takes into account stereo redundancy. It also uses a refinement of the Fourier transform: the *Modified Discrete Cosine Transform (MDCT)* addresses problems the DCT has at boundaries of the window used. The Discrete Fourier Transform can produce block edge effects. When such data is quantized and then transformed back to the time domain, the beginning and ending samples of a block may not be coordinated with the preceding and subsequent blocks, causing audible periodic noise.

The MDCT shown in Eq. (14.8), removes such effects by overlapping frames by 50%.

$$F(u) = \sum_{i=0}^{N-1} f(i) \cos \left[\frac{2\pi}{N} \left(i + \frac{N/2+1}{2} \right) (u + 1/2) \right], u = 0, \ldots, \frac{N}{2} - 1$$

(14.8)

Here the window length is $M = N/2$ and M is the number of transform coefficients.

The MDCT also gives better frequency resolution for the masking and bit allocation operations. Optionally, the window size can be reduced back to 12 samples from 36. Even so, since the window is 50% overlapped, a 12-sample window still includes an extra 6 samples. A size-36 window includes an extra 18 points. Since lower frequencies are more often tonelike rather than noiselike, they need not be analyzed as carefully, so a mixed mode is also available, with 36-point windows used for the lowest two frequency subbands and 12-point windows used for the rest.

PCM
audio signal

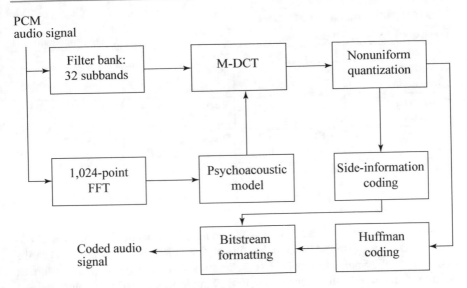

Fig. 14.14 MPEG-1 audio Layer 3

As well, instead of assigning scaling factors to uniform-width subbands, MDCT coefficients are grouped in terms of the auditory system's actual critical bands, and scaling factors, called *scale factor bands*, are calculated from these.

More bits are saved by carrying out entropy (Huffman) coding and making use of nonuniform quantizers. And, finally, a different bit allocation scheme is used, with two parts. Firstly, a nested loop is used, with an inner loop that adjusts the shape of the quantizer, and an outer loop that then evaluates the distortion from that bit configuration. If the error ("distortion") is too high, the scale factor band is amplified. Second, a *bit reservoir* banks bits from frames that don't need them and allocates them to frames that do. Figure 14.14 shows a summary of MPEG Audio Layer 3 coding.

Table 14.2 shows various achievable MP3 compression ratios. In particular, CD-quality audio is achieved with compression ratios in the range of 12:1 (i.e., bitrate of 112 kbps), assuming 16-bit samples at 44.1 kHz, times 2 for stereo. Table 14.2 shows typical performance data using MP3 compression.

14.2.4 MPEG-2 AAC (Advanced Audio Coding)

The MPEG-2 standard is the video compression standard for DVDs, and it, too, has an audio component. The *MPEG-2 Advanced Audio Coding (AAC)* standard [12] was originally aimed at transparent sound reproduction for theaters. It can deliver high-quality sound at 320 kbps for five channels, so that sound can be played from five directions: left, right, center, left surround, and right surround. So-called 5.1 channel systems also include a *low-frequency enhancement (LFE)* channel (a "woofer"). On the other hand, MPEG-2 AAC is also capable of delivering high-quality stereo

Table 14.2 MP3 compression performance

Sound quality	Bitrate (kbps)	Mode	Compression ratio
Telephony	8	Mono	96:1
Better than shortwave	16	Mono	48:1
Better than AM radio	32	Mono	24:1
Similar to FM radio	56–64	Stereo	26:1 to 24:1
Near-CD	96	Stereo	16:1
CD	112–128	Stereo	14:1 to 12:1

sound at bitrates below 128 kbps. It is the audio coding technology for the *DVD-Audio Recordable* (*DVD-AR*) format and is also adopted by XM Radio, one of the two main satellite radio services in North America. AAC was developed as a further compression and encoding scheme for digital audio to succeed MP3, and delivers better sound quality than MP3 for the same bitrate [13]. AAC is currently the default audio format for YouTube, iPhone and other Apple products plus iTunes, Nintendo, and PlayStation. It is also supported on Android mobile phones.

MPEG-2 audio can support up to 48 channels, sampling rates between 8 and 96 kHz, and bitrates up to 576 kbps per channel. Like MPEG-1, MPEG-2 supports three different "profiles", but with a different purpose. These are the *Main*, *Low Complexity* (*LC*), and the *Scalable Sampling Rate* (*SSR*). The LC profile requires less computation than the Main profile, but the SSR profile breaks up the signal so that different bitrates and sampling rates can be used by different decoders.

The three profiles follow mostly the same scheme, with a few modifications. First, an MDCT transform is carried out, either on a "long" window with 2,048 samples or a "short" window with 256 samples. The MDCT coefficients are then filtered by a *Temporal Noise Shaping* (*TNS*) tool, with the objective of reducing pre-masking effects and better encoding signals with stable pitch. The MDCT coefficients are then grouped into 49 scale factor bands, approximately equivalent to a good-resolution version of the human acoustic system's critical bands. In parallel with the frequency transform, a psychoacoustic model similar to the one in MPEG-1 is carried out, to find masking thresholds.

The Low complexity profile is the most widely used AAC profile and is more efficient than MP3 with a 30% increase in efficiency in terms of quality versus bitrate. It offers near-CD quality at very low bitrates such as 80 kbps for mono and 128 kbps for stereo audio input (44.1 kHz sampling frequency). It is mostly used for music development, vocal recordings and the like.

The Main profile uses a predictor. Based on the previous two frames, and only for frequency coefficients up to 16 kHz, MPEG-2 subtracts a prediction from the frequency coefficients, provided this step will indeed reduce distortion. Quantization for the Main profile is governed by two rules: keep distortion below the masking threshold, and keep the average number of bits used per frame controlled, using a bit reservoir. Quantization also uses scaling factors, used to amplify some of the scale factor bands and nonuniform quantization. MPEG-2 AAC also uses entropy coding for both scale factors and frequency coefficients.

For implementation, a nested loop is used for bit allocation. The inner loop adapts the nonlinear quantizer, then applies entropy coding to the quantized data. If the bit limit is reached for the current frame, the quantizer step size is increased to use fewer bits. The outer loop decides whether for each scale factor band the distortion is below the masking threshold. If a band is too distorted, it is amplified to increase the SNR of that band, at the price of using more bits.

In the SSR profile, a *Polyphase Quadrature Filter* (*PQF*) bank is used. The meaning of this phrase is that the signal is first split into four frequency bands of equal width, then an MDCT is applied. The point of the first step is that the decoder can decide to ignore one of the four frequency parts if the bitrate must be reduced.

14.2.5 MPEG-4 Audio

MPEG-4 AAC is another audio compression standard under ISO/IEC 14496. MPEG-4 audio integrates several different audio components into one standard: speech compression, perceptually based coders, text-to-speech, 3D localization of sound, and MIDI. MPEG-4 can be classified into MPEG-4 Scalable Lossless Coding (HD AAC) [14] and MPEG-4 (HE AAC) [14] . While MPEG-4 HD (High Definition) AAC is used for lossless high quality audio compression for High Definition videos, etc., MPEG-4 HE (High Efficiency) AAC is an extension of the Low complexity MPEG-2 AAC profile used for low bitrate applications such as streaming audio. MPEG-4 HE AAC has two versions: HE AAC v1, which uses only Spectral Band Replication (SBR, enhancing audio at low bit rates) and HE AAC v2, which uses SBR and Parametric Stereo (PS, enhancing efficiency of low bandwidth input). MPEG-4 HE AAC is also used for the digital radio standards DAB+, developed by the standards group WorldDMB (Digital Multimedia Broadcasting) in 2006, and in Digital Radio Mondiale, a consortium of national radio stations aimed at making better use of the bands currently used for AM broadcasting, including shortwave.

Perceptual Coders

One change in AAC in MPEG-4 is to incorporate a *Perceptual Noise Substitution* module, which looks at scale factor bands above 4 kHz and includes a decision as to whether they are noiselike or tonelike. A noiselike scale factor band itself is not transmitted; instead, just its energy is transmitted, and the frequency coefficient is set to zero. The decoder then inserts noise with that energy.

Another modification is to include a *Bit-Sliced Arithmetic Coding* (*BSAC*) module. This is an algorithm for increasing bitrate scalability, by allowing the decoder side to be able to decode a 64 kbps stream using only a 16 kbps baseline output (and steps of

1 kbps from that minimum). MPEG-4 audio also includes a second perceptual audio coder, a vector quantization method entitled *Transform-domain Weighted Interleave Vector Quantization* (*TwinVQ*). This is aimed at low bitrates and allows the decoder to discard portions of the bitstream to implement both adjustable bitrate and sampling rate. The basic strategy of MPEG-4 audio is to allow decoders to apply as many or as few audio tools as bandwidth allows.

Structured Coders

To have a low bitrate delivery option, MPEG-4 takes what is termed a *Synthetic/Natural Hybrid Coding* (*SNHC*) approach. The objective is to integrate both "natural" multimedia sequences, both video and audio, with those arising synthetically. In audio, the latter are termed *structured* audio. The idea is that for low bitrate operation, we can simply send a pointer to the audio model we are working with and then send audio model parameters.

In video, such a *model-based* approach might involve sending face animation data rather than natural video frames of faces. In audio, we could send the information that English is being modeled, then send codes for the base sounds (phonemes) of English, along with other assembler-like codes specifying duration and pitch.

MPEG-4 takes a *toolbox* approach and allows specification of many such models. For example, *Text-To-Speech* (*TTS*) is an ultra-low bitrate method and actually works, provided we need not care what the speaker actually sounds like. Assuming we went on to derive Face Animation Parameters from such low bitrate information, we arrive directly at a very low bitrate videoconferencing system. Another "tool" in structured audio is called *Structured Audio Orchestra Language* (*SAOL*, pronounced "sail"), which allows simple specification of sound synthesis, including special effects such as reverberation.

Overall, structured audio takes advantage of redundancies in music to greatly compress sound descriptions.

14.3   Other Audio Codecs

14.3.1 Ogg Vorbis

Ogg Vorbis [15] is an open-source audio compression format, part of the Vorbis project headed by Chris Montgomery of the Xiph.org foundation, which started in 1993. It was designed to replace existing patented audio compression formats by incorporating a variable bit rate (VBR) codec similar to MP3 with file sizes smaller compared to those of MP3 for the same bitrate and quality. It is targeted primarily at the MP3 standard, being more efficient even at low bit rates and with better quality audio at higher bitrates. Ogg Vorbis also uses a form of MDCT, specifically a forward adaptive codec. One of the major advantages of the Ogg Vorbis standard is its ability to be wrapped in other media containers, the most popular being Matroska and

Table 14.3 Comparison of MP3, MPEG-4 AAC, and Ogg vorbis

	MP3	MPEG-4 AAC	Ogg vorbis
File extension	.mp3	.aac, .mp4, .3gp	.ogg
Original name	MPEG-1 audio Layer 3	Advanced Audio Coding	Ogg
Developer	CCETT, IRT, Fraunhofer Society	Fraunhofer IIS, AT&T Bell Labs, Dolby, Sony Corp., and Nokia	Xiph.org Foundation
Released	1994	1997	v1.0 frozen May 2000
Algorithm	Lossy compression	Lossy compression	Lossy compression
Quality	Lower quality than AAC and Ogg	Better quality at same bit rate as MP3	Better quality and smaller file size than MP3 at same bit rates
Used in	Default standard for audio files	iTunes raised its popularity	Open-source platform

WebM. Ogg Vorbis is supported by many media players such as VLC, Mplayer, Audacity audio editing software, and most Linux distributions as well. It has limited native support in Windows and Mac OS but the Vorbis team have decoders available for various applications. Ogg Vorbis is gaining popularity with the gaming industry: Ubisoft uses the Ogg Vorbis format for its most recent game releases. Many popular browsers such as Firefox, Chrome, and Opera have native support for Ogg Vorbis. Table 14.3 compares the MP3, AAC, and Ogg vorbis standards.

Table 14.4 summarizes the target bitrate ranges and main features of other modern general audio codecs. They include many similarities to MPEG-2 audio codecs. Dolby Digital (AC-3) dates from 1992; it was devised to code multi-channel digital audio for 35mm movie film, placed alongside the optical analog audio channel. It is also used in HDTV audio and DVD-Video. AC-3 is a perceptual coder with 256 sample block length. The maximum bitrate for compressed 5.1 channel surround sound audio for 35mm film is 320 kbps (5.1 is one front left channel, one right front, one center channel, two surround channels, and a subwoofer). AC-3's predecessor, Dolby AC-2, was a transform-based codec.

Dolby Digital Plus (E-AC-3, or "Enhanced" AC-3) supports 13.1 channels. It is based on AC-3, with a low-loss and low-complexity conversion from E-AC-3 to AC-3. DTS (or Coherent Acoustics) is a digital surround system aimed at theaters; it forms part of the Blue Ray audio standard. WMA is a proprietary audio coder developed by Microsoft. MPEG SAOC [16], published in 2010, stands for "Spatial Audio Object Coding." It extends "MPEG Surround", which allows the addition of additional multi-channel side-information to core stereo data. MPEG SAOC processes "object signals" instead of channel signals, with not a great deal of extra bandwidth for the

Table 14.4 Comparison of audio coding systems

Codec	Bitrate kbps/channel	Complexity	Main application
Dolby AC-2	128–192	Low (encoder/decoder)	Point-to-point, cable
Dolby AC-3	32–640	Low (decoder)	HDTV, cable, DVD
Dolby Digital plus (Enhanced AC-3)	32–6,144	Low (decoder)	HDTV, cable, DVD
DTS: Digital surround	8–512	Low (for lossless audio extension)	DVD, entertainment, professional
WMA: Windows media audio	128–768	Low (low-bit-rate streaming)	Many applications
MPEG SAOC	As low as 48	Low (decoder/rendering)	Many applications

side-information. SAOC is aimed at such innovative usages as Interactive Remix, Karaoke, gaming, and mobile conferencing over headphones,

14.4 MPEG-7 Audio and Beyond

Recall that MPEG-4 is aimed at compression using objects. MPEG-4 audio has several interesting features, such as 3D localization of sound, integration of MIDI, text-to-speech, different codecs for different bitrates, and use of the sophisticated MPEG-4 AAC codec. However, newer MPEG standards are also aimed at "search": how can we find objects, assuming that multimedia is indeed coded in terms of objects?

MPEG-7 aims to describe a structured model of audio [17], so as to promote ease of search for audio objects. Officially called a method for *Multimedia Content Description Interface*, MPEG-7 provides a means of standardizing metadata for audiovisual multimedia sequences. MPEG-7 is meant to represent information about multimedia information.

The objective, in terms of audio, is to facilitate the representation and search for sound content, perhaps through the tune or other descriptors. Therefore, researchers are laboring to develop descriptors that efficiently describe, and can help find, specific audio in files. These might require human or automatic content analysis and might be aimed not just at low-level structures, such as melody, but at actually grasping information regarding structural and semantic content [18].

An example application supported by MPEG-7 is *automatic speech recognition* (*ASR*). Language understanding is also an objective for MPEG-7 "content". In theory, MPEG-7 would allow searching on spoken and visual events: "Find me the part where Hamlet says, 'To be or not to be.'" However, the objective of delineating a complete, structured audio model for MPEG-7 is by no means complete.

Nevertheless, low-level features are important. Useful summaries of such work [19,20] describe sets of such descriptors.

Further, standards in the MPEG sequence are mostly not aimed at further audio compression standardization. For example, MPEG-DASH (Dynamic Adaptive Streaming over HTTP) is aimed at streaming of multimedia using existing HTTP resources such as servers and content distribution networks, but is meant to be independent of specific video or audio codecs. We will examine it in more detail in Chap. 16.

14.5   Further Exploration

Good reviews of MPEG Audio are contained in the articles [9,10]. A comprehensive explication of natural audio coding in MPEG-4 appears in [21]. Structured audio is introduced in [22], and exhaustive articles on natural, synthetic, and SNHC audio in MPEG-4 appear in [23,24].

14.6   Exercises

1. (a) What is the threshold of quiet, according to Eq. (14.1), at 1000 Hz? (Recall that this equation uses 2 kHz as the reference for the 0 dB level.)
   (b) Take the derivative of Eq. (14.1) and set it equal to zero, to determine the frequency at which the curve is minimum. What frequency are we most sensitive to? Hint: One has to solve this numerically.
2. Loudness versus amplitude. Which is louder: a 1000 Hz sound at 60 dB or a 100 Hz sound at 60 dB?
3. For the (newer versions of the) Fletcher-Munson curves, in Fig. 14.1, the way this data is actually observed is by setting the y-axis value, the sound pressure level, and measuring a human's estimation of the effective perceived loudness. Given the set of observations, what must we do to turn these into the set of perceived loudness curves shown in the figure?
4. Two tones are played together. Suppose tone 1 is fixed, but tone 2 has a frequency that can vary. The *critical bandwidth* for tone 1 is the frequency range for tone 2 over which we hear *beats*, and a roughness in the sound. Beats are overtones at a lower frequency than the two close tones; they arise from the difference in frequencies of the two tones. The critical bandwidth is bounded by frequencies beyond which the two tones sound with two distinct pitches.

   (a) What would be a rough estimate of the critical bandwidth at 220 Hz?
   (b) Explain in words how you would set up an experiment to measure the critical bandwidth.

5. Search the web to discover what is meant by the following psychoacoustic phenomena:

   (a) Virtual pitch
   (b) Auditory scene analysis

    (c) Octave-related complex tones

    (d) Tri-tone paradox

    (e) Inharmonic complex tones

6. What is the compression ratio of MPEG audio if stereo audio sampled with 16 bits per sample at 48 kHz is reduced to a bitstream of 256 kbps?

7. In MPEG's polyphase filter bank, if 24 kHz is divided into 32 equal width frequency subbands

    (a) What is the size of each subband?

    (b) How many critical bands, at worst, does a subband overlap?

8. If the sampling rate f_s is 32 ksps, in MPEG Audio Layer 1, what is the width in terms of the frequency of each of the 32 subbands?

9. Given that the level of a *masking tone* at the 8th band is 60 dB, and 10 ms after it stops, the masking effect to the 9th band is 25 dB.

    (a) What would MP3 do if the original signal at the 9th band is at 40 dB?

    (b) What if the original signal is at 20 dB?

    (c) How many bits should be allocated to the 9th band in (a) and (b) above?

10. What does MPEG Layer 3 (MP3) audio do differently from Layer 1 to incorporate temporal masking?

11. Explain MP3 in a few paragraphs, for an audience of consumer-audio-equipment salespeople.

12. Implement MDCT, just for a single 36-sample signal, and compare the frequency results to those from DCT. For low-frequency sound, which does better at concentrating the energy in the first few coefficients?

13. Convert a CD-audio cut to MP3. Compare the audio quality of the original and the compressed version—can you hear the difference? (Many people cannot.)

14. For two stereo channels, we would like to be able to use the fact that the second channel behaves, usually, in a parallel fashion to the first, and apply information gleaned from the first channel to compression of the second. Discuss how you think this might proceed.

References

1. D.W. Robinson, R.S. Dadson, A re-determination of the equal-loudness relations for pure tones. B. J. Appl. Phys. **7**, 166–181 (1956)
2. H. Fletcher, W.A. Munson, Loudness, its definition, measurement and calculation. J. Acoust. Soc. Am. **5**, 82–107 (1933)
3. T. Painter, A. Spanias, Perceptual coding of digital audio. Proc. IEEE **88**(4), 451–513 (2000)
4. B. Truax, *Handbook for Acoustic Ecology*, 2nd edn. (Cambridge Street Publishing, 1999)
5. D. O'Shaughnessy, *Speech Communications: human and Machine*. (IEEE Press, 1999)

6. A.J.M. Houtsma, Psychophysics and modern digital audio technology. Philips J. Res. **47**, 3–14 (1992)

7. E. Zwicker, U. Tilmann, Psychoacoustics: matching signals to the final receiver. J. Audio Eng. Soc. **39**, 115–126 (1991)

8. D. Lubman, Objective metrics for characterizing automotive interior sound quality, in *Inter-Noise '92*, pp. 1067–1072 (1992)

9. D. Pan, A tutorial on MPEG/Audio compression. IEEE Multimed. **2**(2), 60–74 (1995)

10. S. Shlien, Guide to MPEG-1 audio standard. IEEE Trans. Broadcast. **40**, 206–218 (1994)

11. P. Noll, Mpeg digital audio coding. IEEE Signal Process. Mag. **14**(5), 59–81 (1997)

12. *Information technology—Generic coding of moving pictures and associted audio information, Part 7: advanced Audio Coding (AAC)*. Int. Standard: ISO/IEC 13818-7 (1997)

13. K. Brandenburg, MP3 and AAC explained, in *17th International Conference on High Quality Audio Coding*, pp. 1–12 (1999)

14. *Information technology—Coding of audio-visual objects, Part 3: audio*. Int. Standard: ISO/IEC 14496-3 (1998)

15. Vorbis audio compression (2013). http://xiph.org/vorbis/

16. J. Engdegård, B. Resch, C. Falch, O. Hellmuth, J. Hilpert, A. Hoelzer, L. Terentiev, J. Breebaart, J. Koppens, E. Schuijers, W. Oomen, Spatial audio object coding (SAOC)—The upcoming MPEG standard on parametric object based audio coding, in *Audio Engineering Society 124th Convention* (2008)

17. *Information technology—Multimedia content description interface, Part 4: audio*. Int. Standard: ISO/IEC 15938-4 (2001)

18. A.T. Lindsay, S. Srinivasan, J.P.A. Charlesworth, P.N. Garner, W. Kriechbaum, Representation and linking mechanisms for audio in MPEG-7. Signal Process.: Image Commun. **16**, 193–209 (2000)

19. P. Philippe, Low-level musical descriptors for MPEG-7. Signal Process.: Image Commun. **16**, 181–191 (2000)

20. M.I. Mandel, D.P.W. Ellis, Song-level features and support vector machines for music classification. in *6th International Conference on Music Information Retrieval* (2005)

21. K. Brandenburg, O. Kunz, A. Sugiyama, MPEG-4 natural audio coding. Signal Process.: Image Commun. **15**, 423–444 (2000)

22. E.D. Scheirer, Structured audio and effects processing in the MPEG-4 multimedia standard. Multimed. Syst. **7**, 11–22 (1999)

23. J.D. Johnston, S.R. Quackenbush, J. Herre, B. Grill, Review of MPEG-4 general audio coding, in *Multimedia Systems, Standards, and Networks*, ed. by A. Puri, T. Chen, pp. 131–155. (Marcel Dekker, Inc., 2000)

24. E.D. Scheirer, Y. Lee, J.-W. Yang, Synthetic audio and SNHC audio in MPEG-4, in *Multimedia Systems, Standards, and Networks*, ed. by A. Puri, T. Chen, pp. 157–177. (Marcel Dekker, Inc., 2000)

Part III

Multimedia Communications and Networking

Multimedia places great demands on networks and systems. Driven by an insatiable appetite for bandwidth on the Internet, advances in digital media compression technologies, and new user demands, multimedia communication and content sharing over the Internet have quickly risen to become a mainstream "killer" application over the past three decades. As well, we are witnessing a convergence of conventional telephone networks and television networks on the global Internet, and numerous new-generation multimedia-based applications have been developed over the Internet, e.g., Skype, YouTube, Netflix, Twitch, TikTok, etc.

The Internet, however, was not initially designed for multimedia content distribution and there are significant challenges to be addressed. Multimedia applications generally start playback before downloads have completed, i.e., in a *streaming* mode. In the early time, research attention was mostly focused on dedicated streaming protocols, such as the real-time transport protocol (RTP) and its control protocol (RTCP). There was also great effort toward multicast in the network layer as well as resource reservation protocols for large-scale multimedia content distribution.

Over the past two decades, content distribution networks (CDNs) and peer-to-peer (P2P) media streaming received substantial attention and were widely applied for both live and on-demand media streaming. Recently, web-based video streaming allows users to play videos directly from their web browsers with rich interactions, rather than having to download and install dedicated software.

Advances in wireless mobile networking and the emergence of sleek and smart portable devices are driving the revolution further. The dream of "anywhere and anytime" multimedia communication and content sharing has now become a reality. Meanwhile, advances in data centers and machine virtualization have catapulted the popularity of *cloud computing*. Attracted by the abundant resources in the cloud and the on-demand "pay-as-you-go" pricing model, an increasing number of multimedia services have been hosted on cloud computing platforms, e.g., Netflix, one of the leading video streaming service providers, reportedly making use of Amazon's cloud service. As well, Sony and Microsoft's game consoles are powered by cloud computing, which offload many computation-intensive multimedia processing tasks, e.g., 3D rendering, to remote servers, lifting the hardware and software constraints inherent in local consoles.

For delay-sensitive multimedia applications, aggregating all the resources in a centralized data center is not an ideal solution. Recent efforts toward fine-grained resource partition and distribution include *edge computing* (on geo-distribution) and *serverless computing* (on service abstraction).

This part examines the challenges and solutions for efficient multimedia communication and content sharing over computer networks, particularly over the wired Internet and wireless mobile networks. In Chap. 15, we look at the basic Internet service models and protocols for multimedia communications, and in Chap. 16 we go on to consider multimedia content distribution mechanisms. Chapter 17 further provides a quick introduction to the basics of wireless mobile networks and issues related to multimedia communication over such networks. In Chap. 18, we go on to examine cloud-assisted multimedia computing and content sharing, including recent advances in cloud gaming, edge computing, and serverless computing.

Network Services and Protocols for Multimedia Communications

Computer communication networks are essential to the modern computing environment that we know and have come to rely upon. Multimedia communications and networking share all major issues and technologies of computer communication networks. Indeed, the evolution of the Internet, particularly in the past three decades, has been largely driven by the ever-growing demands from numerous conventional and new generation multimedia applications. As such, multimedia communications and networking have become a very active area for research and industrial development.

This chapter will start with a review of the common terminologies and techniques in modern computer communication networks, specifically, the Internet, followed by an introduction to various network services and protocols for multimedia communications and content sharing, since they are becoming a central part of most contemporary multimedia systems. We also use Internet telephony as an example to illustrate the design and implementation of a typical interactive multimedia communication application.

15.1 Protocol Layers of Computer Communication Networks

It has long been recognized that network communication is a complex task that involves multiple levels of *protocols* [1–3]. Each protocol defines the syntax, semantics, and operations for a specific communication task. A widely used reference model for such a multilayer protocol architecture was proposed by the International Organization for Standardization (ISO) in 1984, called *Open Systems Interconnection* (OSI), documented by ISO Standard 7498. The OSI Reference Model has the following networking layers [4]:

© Springer Nature Switzerland AG 2021
Z.-N. Li et al., *Fundamentals of Multimedia*, Texts in Computer Science,
https://doi.org/10.1007/978-3-030-62124-7_15

1. **Physical Layer**. Defines the electrical and mechanical properties of the physical interface (e.g., signal level, specifications of the connectors, etc.); also specifies the functions and procedural sequences performed by circuits of the physical interface.
2. **Data Link Layer**. Specifies the ways to establish, maintain, and terminate a link, such as the transmission and synchronization of data frames, error detection and correction, and access protocol to the Physical layer.
3. **Network layer**. Defines the routing of data from one end to the other across the network, using circuit switching or packet switching. Provides such services as addressing, internetworking, error handling, congestion control, and sequencing of packets.
4. **Transport layer**. Provides end-to-end communication between *end systems* that support end-user applications or services. Supports either *connection-oriented* or *connectionless* protocols. Provides error recovery and flow control.
5. **Session layer**. Coordinates the interaction between user applications on different hosts, manages sessions (connections), such as completion of long file transfers.
6. **Presentation layer**. Deals with the syntax of transmitted data, such as conversion of different data formats and codes due to different conventions, compression, or encryption.
7. **Application layer**. Supports various application programs and protocols, such as File sharing (FTP), remote login (Telnet), e-mail (SMTP/MIME), Web (HTTP), network management (SNMP), and so on.

The OSI reference model was instrumental in the development of modern computer networks. Multimedia systems are generally implemented in the top three layers, but rely on the services from the underlying layers. The model, however, has never been fully implemented; instead, the competing and more practical TCP/IP protocol suite has become dominating. It is also the core protocols for the transport and network layers of today's Internet. For the data link layer, numerous Local Area Network (LAN) technologies have been developed and the IEEE 802 family of standards, particularly Ethernet and Wi-Fi, are dominating now.

Figure 15.1 compares the layers in the OSI model and the Internet (with TCP/IP being the core protocol suite). Figure 15.2 shows a typical home/office network setup nowadays, which, through an access network (ADSL or cable modem), is connected to an *Internet Service Provider* (ISP). The users inside the network are then able to access diverse multimedia services in the public Internet, and a firewall can protect them from malicious attacks. In the following sections, we present the details of different layers that are involved in such a networked system for multimedia communications.

15.2   Local Area Network (LAN) and Access Networks

For home or office users, the networks of direct use is generally a *Local Area Network* (LAN), which is restricted to a small geographical area, usually for a relatively small number of stations. The physical links that connect an end system inside a LAN

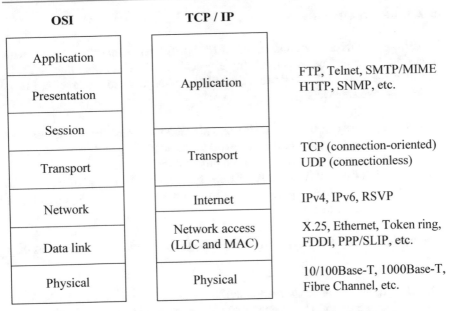

OSI	TCP / IP	
Application	Application	FTP, Telnet, SMTP/MIME HTTP, SNMP, etc.
Presentation		
Session	Transport	TCP (connection-oriented) UDP (connectionless)
Transport		
Network	Internet	IPv4, IPv6, RSVP
	Network access (LLC and MAC)	X.25, Ethernet, Token ring, FDDI, PPP/SLIP, etc.
Data link		
Physical	Physical	10/100Base-T, 1000Base-T, Fibre Channel, etc.

Fig. 15.1 Comparison of OSI and TCP/IP protocol architectures and sample protocols

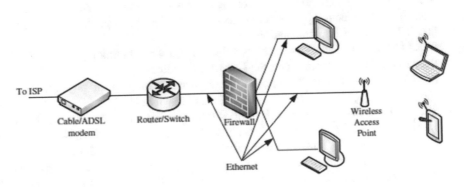

Fig. 15.2 A typical home/office network setup

toward the external Internet is referred to as the *Access Network*. It is also known as the "last mile" for delivering network services.

In this section, we describe the LAN services and representative wired LAN technologies, in particular, Ethernet. We then describe the typical network access technologies, including dialup, Digital Subscriber Line (DSL), Cable Networks, and Fiber-To-The-Home (FTTH), and their support to multimedia services.

15.2.1 LAN Standards

The IEEE 802 committee developed the IEEE 802 reference model for LANs, with a focus on the lower layers, namely, the Physical and the Data Link layers [5]. In particular, the Data Link layer's functionality is enhanced, and the layer has been divided into two sublayers:

- **Medium Access Control (MAC) layer**. This sublayer assembles or disassembles frames upon transmission or reception, performs addressing and error correction, and regulates access control to a shared physical medium.
- **Logical Link Control (LLC) layer**. This sublayer performs flow and error control and MAC-layer addressing. It also acts as an interface to higher layers. LLC is above MAC in the hierarchy.

Following are some of the important IEEE 802 subcommittees and the areas they define:

- **802.1 (Higher Layer LAN Protocols)**. It concerns the overall 802 LAN architecture, the relationship between the 802.X standards and wide area networks, as well as the interconnection, security, and management of LANs.
- **802.2 (LLC)**. The general standard for logical link control (LLC), which provides a uniform interface to upper layer protocols, masking the differences of various 802.X MAC layer implementations.
- **802.3 (Ethernet)**. It defines the physical layer and the data link layer's MAC of the wired Ethernet, in particular the CSMA/CD method.
- **802.11 (Wireless LAN)**. It defines the medium access method and physical layer specifications for wireless LAN (WLAN, also known as Wi-Fi).
- **802.16 (Broadband wireless)**. It defines the access method and physical layer specifications for broadband wireless networks. One product is WiMAX (Worldwide Interoperability for Microwave Access), which targets the delivery of last mile wireless broadband access as an alternative to cable and DSL. WiMax was also a candidate of 4G cellular networking, in competition with the LTE Advanced standard. It was, however, not was not successful in either market.

We next detail the Ethernet technology, which has become the *de facto* standard of wired LAN. We will describe wireless LAN technologies in Chap. 17.

15.2.2 Ethernet Technology

Ethernet is a LAN technology initially developed at Xerox PARC in 1970s [6]. It was inspired by ALOHAnet, an earlier random access network, and the idea was first documented in a memo by Robert Metcalfe. Ethernet was commercially introduced in 1980 and standardized in 1985, as IEEE 802.3. It soon defeated many other competing wired LAN technologies and has since become dominant in the market.

Preamble 7 bytes	Start of Frame Delimiter 1 bytes	MAC destination 6 bytes	MAC source 6 bytes	Type or Length 2 bytes	Payload Data 46-1500 bytes	CRC 4 bytes

Fig. 15.3 Ethernet frame structure

The basic Ethernet uses a shared bus. Each Ethernet station is given a 48-bit MAC address. The MAC addresses are used to specify both the destination and the source of each data packet, referred to as a *frame*. Figure 15.3 shows a typical Ethernet frame structure, which begins with a preamble and a start of frame delimiter, followed by an Ethernet header featuring source and destination MAC addresses. The middle section of the frame consists of the payload data including any headers for other protocols (e.g., IP) carried in the frame. The frame ends with a 32-bit *cyclic redundancy check* (CRC, see Sect. 17.3.1), which is used to detect data corruption in transit.

To send a frame, the recipient's Ethernet address is attached to the frame, which is then broadcast to everyone on the bus. On reception of a transmission, the receiver uses the destination address to determine whether the transmission is relevant to the station or should be ignored. Only the designated station will accept the frame, while others will ignore it. Note that, if two stations send frames simultaneously, a *collision* can happen. The problem is solved by *Carrier Sense Multiple Access with Collision Detection* (CSMA/CD) in medium access control. With CSMA/CD, a station that wishes to send a frame must first listen to the network (i.e., carrier sensing), wait until there is no traffic, and then send the frame. Obviously, multiple stations could be waiting and then send their messages at the same time, causing a collision. During frame transmission, the station compares the signal received with the one sent. If they are different, it detects a collision. Once a collision is detected, the station stops sending the frame, and the frame is retransmitted after a random delay.

For a LAN with multiple stations, often a *star* topology is used, in which each station is connected directly to a *hub* (and recently a *switch*). The hub is an active device and acts as a repeater. Every time it receives a signal from one station, it repeats, so that other stations will hear. Logically, this is still a bus, although it is physically a star network.

The maximum data rate for the early Ethernet is 10 Mbps, using unshielded twisted pairs. In its long life span, the Ethernet's physical layer has encompassed coaxial, twisted pair, and fiber optic physical media interfaces and speeds from 10 Mbps to 100 Gbps and beyond. The link layer has also evolved to meet the new bandwidth and market requirements. In 1989, *Ethernet switch* was introduced, which works differently from an Ethernet hub—in a switch, only the header of an incoming packet will be examined before it is either dropped or forwarded to another segment. This greatly reduces the forwarding latency and the processing load. The switched Ethernet has since been replacing the non-switched Ethernet given its bandwidth advantages, the improved isolation of devices from each other, and the ability to easily mix devices of different speeds. The latest switch can reach 400 Gbps.

These different generations of Ethernet technologies largely retain the same network protocol stack and interfaces and are, therefore, able to inter-connect and inter-operate. This is also a key reason for Ethernet's success, as opposed to other ad hoc or inflexible LAN technologies.

15.2.3 Access Network Technologies

An access network bridges the LAN in a home or office to the external Internet. To save cost for laying a new network line, an existing network that is already in the home is often used, in particular, the telephone or cable TV networks. Direct fiber optics connections have been popular nowadays for new buildings.

Dial-Up and Integrated Services Digital Network (ISDN)

Since the Public Switched Telephone Network (PSTN) is widely available in residential homes and offices, the very earlier Internet accesses are often using the telephone line to establish a dialed connection to an *Internet service provider* (ISP). Note that the traditional telephone lines carry analog voice signal only. To transmit digital data, a *modem* (modulator-demodulator) is needed between the computer and the telephone jack to modulate an analog carrier signal to encode digital information, and also demodulate a carrier signal to decode the transmitted information.

Modern dial-up modems typically have a maximum theoretical transfer speed of 56 kbps (using the V.90 or V.92 protocol), although in most cases 40–50 kbps is the norm. The connections usually have a latency as high as 300 ms or even more. Factors such as phone line noise, as well as the quality of the modem itself, play a large part in determining the connection speeds and delays. The low speed and relatively high delay make dial-up generally unsuitable for multimedia applications.

To overcome these limits, in the 1980s, the *International Telecommunication Union* (ITU) started to develop the Integrated Service Digital Network (ISDN) to meet the needs of various digital services in which digital data, voice, and sometimes video (e.g., in videoconferencing) can be transmitted [7].

Digital Subscriber Line (DSL)

DSL is the telephone industry's answer to the last mile challenge nowadays, which again makes use of existing telephone's twisted-pair wires to transmit modulated digital data signal [8]. Unlike traditional dial-up modems, which modulate bits into signals in the 300–3,400 Hz baseband (voice service), DSL modems modulate frequencies from 4,000 Hz to 1 MHz (and as high as 4 MHz), using *Quadrature Amplitude Modulation* (QAM). It then uses FDM (Frequency Division Multiplexing) to multiplex three channels:

- The high-speed (1.5–9 Mbps) downstream channel at the high end of the spectrum.
- A medium speed (16–640 kbps) duplex channel.
- A voice channel for telephone calls at the low end (0–4 kHz) of the spectrum.

DSL employs highly complex digital signal processing algorithms to overcome the inherent limitations of the existing twisted-pair wires. A notable technology is *Discrete Multi-Tone* (DMT), which, for better transmission in potentially noisy

Table 15.1 Different types of digital subscriber lines

Name	Meaning	Data rate	Mode
HDSL	High data rate digital subscriber line	1.544 Mbps or 2.048 Mbps	Duplex
SDSL	Single line digital subscriber line	1.544 Mbps or 2.048 Mbps	Duplex
ADSL	Asymmetric digital subscriber line	1.5–9 Mbps 16–640 kbps	Down Up
VDSL	Very high data rate digital subscriber line	13–55 Mbps 1.5–3 Mbps	Down Up
VDSL2	VDSL version 2	200–300 Mbps 100 Mbps	Down Up

channels (either downstream or upstream), sends test signals to all subchannels first. It then calculates the signal-to-noise ratio (SNR), to dynamically determine the amount of data to be sent in each subchannel. The higher the SNR, the more data sent. Theoretically, 256 downstream subchannels, each capable of carrying over 60 kbps, will generate a data rate of more than 15 Mbps.

Table 15.1 shows the evolution of various digital subscriber lines (*xDSL*). HDSL was an effort to deliver the T1 (or E1) data rate within a low bandwidth (196 kHz). It requires two twisted pairs for 1.544 Mbps or three twisted pairs for 2.048 Mbps. SDSL instead provides the same service as HDSL on a single twisted-pair line. ADSL (Asymmetrical DSL) adopts a higher data rate downstream (from network to subscriber) and lower data rate upstream (from subscriber to network). This asymmetric downstream and upstream bandwidth share well matches the traffic patterns of traditional client/server-based applications, e.g., the Web, but can have problems with such modern applications as peer-to-peer file sharing or two-way interactive voice or video conversation. More recently, the very-high-bit-rate DSL (VDSL) has been deployed and is still actively evolving; VDSL2-Vplus has achieved an uploading speed of 100 Mbps, thereby alleviating the problem of asymmetry.

Hybrid Fiber-Coaxial (HFC) Cable Networks

Besides telephone lines, another network access that is readily available in many homes is the Cable TV network. In such a network, optical fibers connect the core network with *Optical Network Units* (ONUs) in the neighborhood, each of which typically serves a few hundred homes through shared coaxial cables.

A *cable modem* can be used to provides bi-directional data communication via radio frequency channels on this *Hybrid Fiber-Coaxial* (HFC) network. Conforming to the Ethernet standard (with some modifications), it bridges Ethernet frames between the home LAN and the cable network. Technically, it modulates data to transmit it over the cable network and demodulates data from the cable network to receive it.

Traditionally, analog cable TV was allocated a frequency range of 50–500 MHz, divided into 6 MHz channels for NTSC TV in North America and 8 MHz channels in Europe. For HFC cable networks, the downstream is allocated a frequency range of 450–750 MHz, and the upstream is allocated a range of 5–42 MHz. For the downstream, a cable modem acts as a tuner to capture the QAM modulated digital stream. The upstream uses *Quadrature Phase-Shift Keying* (QPSK) [2] modulation, which is more robust in the noisy and congested frequency spectrum.

Today's cable modems have a download speed of 10–500 Mbps (typically 50 Mbps). The latest DOCSIS 4.0 (Data Over Cable Service Interface Specification) supports a data rate even up to 10 Gbps. It is, however, worth noting that the cable Internet access is shared among many neighboring homes, while the DSL based on a telephone line is dedicated. The cable service can slow down significantly if many people in the neighborhood access the Internet simultaneously. As such, in practice, cable's speed advantage over DSL is much less than the theoretical numbers suggest. In addition, both cable modem and DSL performance vary from one minute to the next depending on the pattern of use and traffic on the Internet, and DSL and cable service providers often implement so-called "speed caps" that limit the bandwidth or total monthly data of their services.

In most areas, both DSL and cable accesses are available, although some areas may have only one choice. These two technologies have dominated home Internet access around the world, and have very similar market shares. The competition is very tough, and they both try to provide better and richer services, particularly for multimedia applications. For example, with the advent of *Voice over Internet Protocol* (VoIP) telephony, cable modems have been extended to provide telephone service through Skype or even landline services, allowing customers who purchase the cable TV service to eliminate their plain old telephone service. On the other hand, many telephone companies are offering digital TV services through their networks, too. The convergence has made the *triple play* business model possible, that is, over a single broadband connection, provisioning two bandwidth-intensive services, high-speed Internet access and television, and the latency-sensitive telephone.

Fiber-To-The-Home (FTTH) or Neighborhood (FTTN)

Optical fibers can be laid to connect home networks to the core network directly. It replaces all or part of the conventional metal local loop used for last mile accesses, providing the highest bandwidth. For example, a 155 Mbps downstream can reach each of four homes through multiplexing over a 622 Mbps downstream that can be easily attained by a single fiber.

Since existing homes generally have only twisted pairs and/or coaxial cables, the implementation cost of pure Fiber-To-The-Home (FTTH) will be high, but many new high-rise buildings have already had built-in fiber accesses. Alternatively, the fiber can reach a node first (Fiber-To-The-Node or Neighborhood, FTTN) and then the nearby home users connect to this cabinet using traditional coaxial cable or twisted-pair wiring. As such, the combined Internet, TV, and telephone signal travels first over the fiber, and then over the existing telephone or cable wiring until it reaches the end-user's living space, where a VDSL or DOCSIS cable modem separates the signal and converts the data signal into the Ethernet format. The area served by the cabinet is usually less than one mile in radius and can contain several hundred customers.

Such fiber-based accesses are considered to be "future-proof" because the data rate of a connection is now only limited by the terminal equipment rather than the fiber, permitting long-term speed improvements by equipment upgrades before the fiber itself must be upgraded. It also offers good support for high-quality multimedia services.

In the United States, AT&T Fiber's FTTH is expected to reach 7 million homes by 2022. Its Internet 1000 plan offers connection speeds around 1 Gbps for both download and upload. This is sufficiently high for any type of home-based application nowadays or in the foreseeable future, thereby largely removing the last mile bottleneck for multimedia distribution to home users. As a matter of fact, with its services, a user can download 25 songs in 1 s, download a TV show in under 4 s, and download a 90 min HD movie in under 34 s.

Google Fiber offers similar services to end users. The high-speed connections also make rich multimedia services beyond the basic data plans possible; these include 1 TB (terabyte, or 10^{12} bytes) of Google Drive service and television service with a 2 TB DVR recorder that will record up to eight live television shows simultaneously.

15.3   Internet Technologies and Protocols

Through the access networks, the home and office users are connected to the external wide area Internet. The TCP/IP protocol suite plays the key roles in the Internet, interconnecting diverse underlying networks and serving diverse upper layer applications (see Fig. 15.1). For this reason, it is also known as the "narrow waist" of the Internet. TCP/IP were indeed developed before OSI, and have become the *de facto* standard for internetworking after their adoption by the Internet.

The *Internet Engineering Task Force* (IETF) and the *Internet Society* are the principal technical development and standard-setting bodies for the Internet. They publish *Request for Comments* (RFCs) that are authored by network engineers and scientists in the form of a memorandum describing methods, behaviors, research, or innovations applicable to the working of the Internet and networked systems.

15.3.1 Network Layer: IP

The network layer provides two basic services: *packet addressing* and *packet forwarding*. Point-to-point data transmission is readily supported within any LANs, and in fact, the LANs usually support broadcast. For a network-layer packet to be transmitted across different LANs or a *Wide Area Network* (WAN), *routers* are employed, which are network-layer devices that receive and forward packets according to their destination addresses. The forwarding is guided by *routing tables* that are collectively built and updated by the routers using *routing protocols*.

There are two common ways to move data through a network of links and routers, namely *circuit switching* and *packet switching*.

- **Circuit Switching**. The *public switched telephone network* (PSTN) is a good example of circuit switching, in which an end-to-end circuit must be established, which is dedicated for the duration of the connection at a guaranteed bandwidth. Although initially designed for voice communications, it was also used for data transmission in earlier ISDN networks.

  Circuit switching is preferable if the user demands a connection and/or more or less constant data rates, as in traditional voice communications and certain constant-bit-rate video communications. The establishment and maintenance of a circuit, however, can be costly, and for general data transfer of variable (sometimes bursty) rates, it can be inefficient given that the circuit and its resources are exclusively reserved.

- **Packet Switching**. Packet switching is used for many modern data networks, particularly today's Internet, in which data rates tend to be variable and sometimes bursty. Before transmission, data is broken into small *packets*, usually 1,000 bytes or less. The header of each packet carries necessary control information, such as the destination address, and the routers will examine the header of each individual packet and make individual forward decisions.

  Compared to circuit switching, the implementation of packet switching is simpler and, because the resources (e.g., bandwidth) are not exclusively reserved but shared among the packets, the network utilization can be much higher. This does come at a cost. Without a dedicated circuit, *store-and-forward* transmission is commonly used in a packet-switched network, which means that a packet must be received entirely and inspected before it is forwarded to the next hop. In addition to this store-and-forward delay, packets can suffer from queuing delay because if too many packets arrive, they need to be queued in a buffer of the router before they can be forwarded. If the buffer overflows when the link is severely congested, packet loss can happen.

For packet switching, two approaches are available to switch and route the packets: *datagram* and *virtual circuit*. In the former, each packet is treated independently, and no specific route is predetermined prior to the transmission; hence, the packets may be unknowingly lost or arrive out of order. It is up to the receiving station to detect and recover the errors and re-arrange the packets, say using TCP in the transport layer as we will see in the next section.

In virtual circuits, a route is predetermined through *request* and *accept* by all nodes along the route. It is a "circuit" because the route is fixed (once negotiated) and used for the duration of the connection; nonetheless, it is "virtual" because the "circuit" is only logical and not dedicated as in the true circuit switching. Sequencing (ordering the packets) is much easier in virtual circuits, and resources could be reserved along the virtual circuit too, providing guaranteed services.

The virtual circuit solution is seemingly more sophisticated and was considered as the technology for ensuring quality multimedia communications. A representative virtual-circuit network is ATM (*Asynchronous Transfer Mode*) [9], which was once believed to be a promising solution replacing the datagram-based Internet, moving toward better network traffic control and delivery, especially for multimedia content. It is, however, more complicated to implement, particularly for wide area networks, and the *Internet Protocol* (IP) (RFC 791, 2460) remains based on datagram, which means that it provides only a *Best Effort* service with no bandwidth, reliability, or delay guarantee.

As a datagram service, IP is *connectionless* and provides no end-to-end control. Every packet is treated separately and is not related to past or future packets. Hence, the packets can be received out of order and can also be dropped or duplicated. Packet fragmentation can also happen when a packet has to travel over a network that accepts only packets of a smaller size. In this case, the IP packets are split into the required smaller size, sent over the network to the next hop, and reassembled and resequenced afterward.

Each router maintains a *routing table*, which identifies for each packet the next hop that it should travel toward the destination. The routing tables are periodically updated through routing protocols with network topology information exchanged among the routers. The Internet is a loosely hierarchical network that is divided into a number of *Autonomous Systems* (ASes), each of which has one or more *gateways* for the nodes within the AS to communicate with those outside. Typical routing protocols within an AS include *Open Shortest Path First* (OSPF) and *Routing Information Protocol* (RIP), and among the gateways, the *Border Gateway Protocol* (BGP) has been widely used.

The IP protocol also provides global addressing of computers across all inter-connected networks, where every networked device is assigned a globally unique *IP address*. Application layer identifications, e.g., the URL (Uniform Resource Locator) of a server or client, can be mapped to the IP address of the server or client through the *Domain Name System* (DNS). In the current IPv4 (IP version 4) (see Fig. 15.4), the IP addresses are 32-bit numbers, usually specified using a *dotted decimal notation*. As an example, the web server of the authors' institution has a URL of www.sfu.ca and its IP address is 142.58.102.68 (=10001110 00111010 01100110 01000100 in binary format). The Internet addressing is undergoing a migration toward the 128-bit IPv6 (IP version 6), which offers a space of 2^{128} or approximately 3.4×10^{38} addresses. Much of the multimedia traffic, e.g., from YouTube, has now been carried over by IPv6 [10].

15.3.2 Transport Layer: TCP and UDP

The TCP (Transmission Control Protocol) and UDP (User Datagram Protocol) are two transport layer protocols used in the Internet to facilitate host-to-host (or end-to-end) communications.

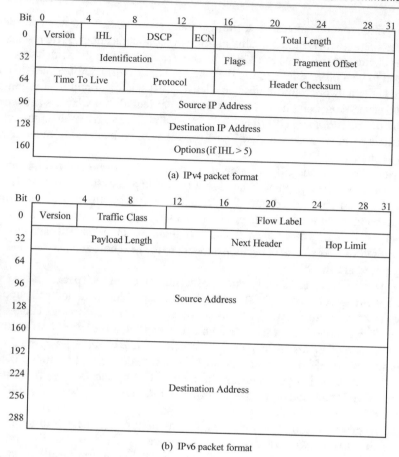

(a) IPv4 packet format

(b) IPv6 packet format

Fig. 15.4 Packet formats of IPv4 and IPv6

Transmission Control Protocol (TCP)

TCP (RFC 675, 793, 1122, 2581, 5681) offers a reliable byte pipe for sending and receiving application messages between two computers, regardless of the specific types of applications. It relies on the IP layer for delivering the data to the destination computer specified by its IP address.

TCP is *connection-oriented*: a connection must be established through a *3-way handshake* before the two ends can start communicating. For every TCP connection, both communicating ends allocate a buffer called a *window* to receive and send data. *Flow control* is established by only sending data in the window to the destination without overflowing its window. Since multiple application processes may use TCP/IP within one computer and the process may also establish multiple network connections, multiplexing/demultiplexing is needed by identifying connections using *port numbers*.

To ensure reliable transfer, TCP offers such services as message packetizing, error detection, retransmission, and packet resequencing. Each TCP datagram header contains the source and destination ports, sequence number, checksum, window field, acknowledgment number, and other fields, as illustrated in Fig. 15.5.

- The *source* and *destination ports*, together with the source and destination IP addresses in the network layer, are used for the source process to know where to deliver the message and for the destination process to know where to reply to the message. This 4-tuple ensures that a packet is delivered to a unique application process running in a particular computer. The port numbers range from 0 to 65535, and typical well-known port numbers include 80 for Web (HTTP), 25 for e-mail (SMTP), and 20/21 for FTP, to name but a few.
- As packets travel across the IP network, they can arrive out of order (by following different paths), be lost, or be duplicated. A *sequence number* reorders the arriving packets and detects whether any are missing. The sequence number is actually the byte count of the first data byte of the packet rather than a serial number for the packet.
- The *checksum* verifies with a high degree of certainty that the packet arrived undamaged, in the presence of channel errors. If the calculated checksum for the received packet does not match the transmitted one, the packet will be discarded and a retransmission will later be invoked. Details about Internet checksum calculation can be found at Sect. 17.3.1.
- The *window field* specifies how many bytes the destination's buffer can currently accommodate. This is typically sent with acknowledgment packets.
- *Acknowledgment* (ACK) packets have the *ACK number* specified—the number of bytes correctly received so far in sequence (corresponding to a sequence number of the first missing packet).

The source process sends packets to the destination process up to the window number and waits for ACKs before sending any more data. The ACK packet will arrive with the new window number information to indicate how much more data the destination buffer can receive. If an ACK packet is not received in a small time

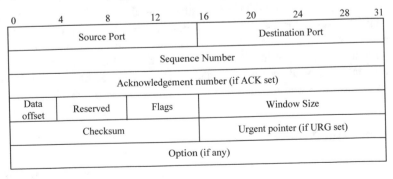

0	4	8	12	16	20	24	28	31
Source Port				Destination Port				
Sequence Number								
Acknowledgement number (if ACK set)								
Data offset	Reserved		Flags	Window Size				
Checksum				Urgent pointer (if URG set)				
Option (if any)								

Fig. 15.5 Header format of a TCP packet

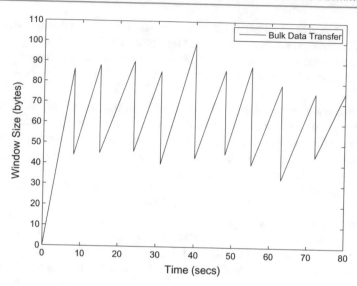

Fig. 15.6 Sawtooth behavior in TCP data transfer

interval, specified by *retransmission timeout* (RTO), the packet will be resent from the local window buffer.

TCP also implements a *congestion control* mechanism in response to network congestion, which can be observed by packet losses. TCP evolves over time with changes in different parts, particularly the congestion control algorithm. Reno, New Reno, and Sack are commonly used versions, all of which are mainly based on an *Additive Increase and Multiplicative Decrease* (AIMD) mechanism; that is, the sending rate, controlled by a sliding window, increases linearly when there is no congestion, but exponentially decreases when there is a packet loss, which indicates a potential congestion in the network.

The window-based AIMD has proven to be fair, robust, and efficient for multiple TCP flows competing for network resources; yet the variation of its transmission rate can be very high, leading to a well-known *sawtooth behavior*. As illustrated in Fig. 15.6, the TCP congestion window will grow linearly when there is no congestion, e.g., from 20 to 100 bytes over time, but when there is a packet loss, it can instantly reduce to 50 bytes (half of the window size that is before congestion), and the transmission rate is proportionally reduced too. While this is fine for general file transmission, it can be undesirable for many multimedia streaming applications that demand a relatively smooth transmission rate with a minimum threshold.

User Datagram Protocol (UDP)

UDP (RFC 768) is *connectionless* with no guarantee on delivery: if a message is to be reliably delivered, it has to be handled by its own application in the application layer. Essentially, the only thing UDP provides is multiplexing using port numbers

0	4	8	12	16	20	24	28	31
Source Port				Destination Port				
Length				Checksum				

Fig. 15.7 Header format of a UDP datagram

and error detection through a checksum. Even for multiplexing and demultiplexing, only the destination port number is used, which is less strict than TCP's 4-tuple does, but on other hand, more flexible for certain applications, e.g., multiparty audio/video conference where each participant expects to hear/see all others. In this scenario, using TCP's multiplexing/demultiplexing will require a connection to be established between every sender and receiver, which is simply too high a cost with many participants.

The UDP's packet format is illustrated in Fig. 15.7, whose header is of 8 bytes only, considerably shorter than that of TCP (20 bytes without option). Such a difference can be significant in many multimedia applications. For example, consider sending 64 kbps PCM-encoded voice, if the data chunks are collected every 20 ms, then each chunk is of 160 bytes. The header overhead of TCP is thus 12.5% and that of UDP is only 5%, not to mention there are header overhead in other layers.

Given the low header overhead and the removal of connection setup, UDP data transmission can be faster than TCP. It is, however, unreliable, especially in a congested network. Higher-level protocols can be used for retransmission, flow control, and congestion avoidance, and more realistically *error concealment* must be explored for acceptable Quality of Service (QoS).

TCP-Friendly Rate Control (TFRC)

Note that the sawtooth behavior of the window-based TCP congestion control is not well suited for media streaming, but an uncontrolled UDP flow can be too aggressive, which interferes with other flows, and easily starves an adaptive TCP flow competing for bandwidth. To avoid this, *TCP-Friendly Rate Control* (TFRC) (RFC 5348) has been introduced, which ensures a UDP flow to be reasonably fair when competing for bandwidth with TCP flows, where "reasonable" means its sending rate is within a factor of two of the sending rate of a TCP flow under the same conditions, i.e., as if the TCP flow is running over the same end-to-end path.

TFRC is generally implemented by estimating the equivalent TCP throughput over the same path using parameters that are observable by the sender or the receiver. RFC 5348 suggests the following equation for X_{Bps}, TCP's average sending rate in bytes per second:

$$X_{Bps} = \frac{s}{R \times \sqrt{2 \times b \times p/3} + (t_{RTO} \times (3 \times \sqrt{3 \times b \times p/8} \times p \times (1 + 32 \times p^2)))}$$

where s is the segment size in bytes (excluding IP and transport protocol headers), R is the round-trip time (RTT) in seconds, p is the loss event rate (between 0 and 1.0) of the number of loss events as a fraction of the number of packets transmitted, t_{RTO} is the TCP retransmission timeout value in seconds, and b is the maximum number of packets acknowledged by a single TCP acknowledgment.

Typically, t_{RTO} is set to $4R$ and $b = 1$. The TCP throughput equation can then be simplified as

$$X_{Bps} = \frac{s}{R \times (\sqrt{2 \times p/3} + 12 \times \sqrt{3 \times p/8} \times p \times (1 + 32 \times p^2))}$$

The parameters in the above equations are all known by the sender or can be estimated by the receiver and then feedback to the sender. The sender can then calculate the equivalent TCP throughput and accordingly control the sending rate of the UDP flow. TFRC co-exists well with TCP and other TFRC flows, but has a much lower variation of throughput over time compared with TCP, which makes it more suitable for media data with constant encoding rate, e.g., voice or constant-bit-rate (CBR) video, where a relatively smooth sending rate is the best match.

15.3.3 Network Address Translation (NAT) and Firewall

The 32-bit IPv4 addressing in principle allows $2^{32} \approx 4$ billion addresses, which seemed more than adequate. In reality, however, it has already largely been exhausted. In January 1995, IPv6 (IP version 6) was recommended as the *next generation IP* (IPng) by IETF. Figure 15.4 compares the packet formats of IPv4 and IPv6. Among the many improvements over IPv4, IPv6 adopts 128-bit addresses, allowing $2^{128} \approx 3.4 \times 10^{38}$ addresses. It is expected to settle the problem of IP address shortage for a long time.

Today we are still in the transition phase from IPv4 to IPv6. To solve the IPv4 address shortage, a practical solution is *Network Address Translation* (NAT) (RFC 4787). A NAT device, sitting behind a local private network and the external network, separates the local hosts from the external network. Each host on the LAN is assigned an internal IP address that cannot be accessed from the outside of the network. Instead, they all share a single public IP address that is kept by the NAT device, which typically maintains a dynamic *NAT table* that translates the addresses. To identify the multiple hosts behind the NAT, the port number in the transport layer is used.

When a local host sends out an IP packet with the internal address and a source port number, it goes through the NAT device, which changes the source IP address to the NAT device's public IP address and the source port number to a new port number that has not been associated with the public IP address. This record is kept by the NAT table, and an external destination host will see the public IP address and the new port number only. When a reply IP packet comes back from the external host, the destination address will be changed back to the internal IP address and the original source port number according to the NAT table, and the packet is then forwarded to the appropriate host.

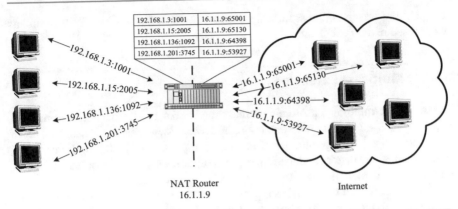

192.168.1.3:1001	16.1.1.9:65001
192.168.1.15:2005	16.1.1.9:65130
192.168.1.136:1092	16.1.1.9:64398
192.168.1.201:3745	16.1.1.9:53927

NAT Router
16.1.1.9

Internet

Fig. 15.8 An illustration of Network Address Translation (NAT). A single IP address of the NAT-enabled router (16.1.1.9) is effectively used by the four internal computers (on the left side) to communicate with the outside Internet by overwriting the port numbers

An example of NAT is shown in Fig. 15.8, where the PCs in a local area network behind the NAT device are of internal IP addresses 192.168.1.XXX, whereas the NAT device has a single public IP address of 16.1.1.9. To communicate with external hosts, a pair of (internal IP address : source port number) will be replaced by a pair of (public address : new port number). For example, (192.168.1.3:1001) is replaced by (16.1.1.9:65001), (192.168.1.15:2005) by (16.1.1.9:65130), and so on. Here the new port numbers 65001, 65130, ... are chosen from the unused port number space (associated with address 16.1.1.9), which in general is quite large, and therefore, many internal IP addresses can be supported.

While NAT alleviates the IP address shortage problem, it imposes fundamental restrictions on pair-wise connectivity of nodes, and may prohibit direct communication with one another. This is because it does not retain a host's original port number. For example, the default port number for a Web service is 80, which, however, can be arbitrarily changed by the NAT device, making a Web server behind the NAT hardly be accessible by external hosts. Whether communication is possible between two hosts depends on such factors as the transport protocol (UDP or TCP) and whether the hosts are located behind the same private network [11].

Similar penetration problem happens for a *firewall* [12], which is a software or hardware-based network security system that controls the incoming and outgoing network traffic based on a rule set. It has become an indispensable part for the safe operation of today's PCs and LANs given the vast threats from the insecure and untrusted public Internet. Yet it can block legitimate traffic too. For example, many firewalls blindly block any UDP-based traffic, making multimedia over UDP simply fail.

In today's Internet environment, over 50% of nodes are located behind NATs or firewalls. The connectivity constraints are a significant challenge to the viability for multimedia content distribution mechanisms over the Internet, particularly for peer-to-peer sharing. It is also one of the key motivations for HTTP-based streaming, which, using only the standard *Hyper Text Transfer Protocol* (HTTP) transactions, is capable of traversing most firewalls that let through the standard Web traffic, as we will see in the next chapter.

15.4   Multicast Extension

In network terminology, a *broadcast* message is sent to all nodes in a domain, a *unicast* message is sent to only one node, and a *multicast* message is sent to a set of specified nodes.[1] A large number of emerging applications, including Internet TV, online games, and distance education, require support for broadcast or multicast, i.e., simultaneous content delivery to a large number of receivers [13].

The initial design of TCP/IP supports one-to-one unicast communication only. Broadcast service is readily available in many LANs and also satellite-based networks; it is, however, simply not doable in the global Internet because it will cause a storm of data forward. Instead, multicast should be used. In the Internet environment, the primary issue for multicast is to determine at which layer it should be implemented. According to the *end-to-end argument*,[2] a functionality should be (1) pushed to higher layers if possible, unless (2) implementing it at the lower layer can achieve significant performance benefits that outweigh the cost of additional complexity. These two considerations can be conflicting for multicast, and in the past three decades, significant effort has been put to reconcile them, leading to multicast implementations in different layers.

15.4.1 Router-Based Architectures: IP Multicast

In his seminal work in 1989 [15], Deering argued that the second consideration should prevail and multicast should be implemented at the network layer. This view was widely accepted and, for much of the 1990s, the research and industrial community mainly focused on the router-based *IP Multicast* architecture, which was defined in RFC 1112 and was augmented in RFC 4604 and 5771.

IP multicast has open anonymous group membership. An IP multicast group address is used by a source (or sources for many-to-many communication) and its receivers to send and receive multicast messages. The source does not have to explicitly know its receivers, and a receiver can join or leave the multicast group at will. Recall that under IPv4, IP addresses are 32 bits. If the first 4 bits are 1110, the

---

[1] IPv6 also allows *anycast*, whereby the message is sent to any one of the specified nodes. This is useful for such services as selection from a cluster of server replicas.

[2] The end-to-end argument is a classic design principle of computer networking, first explicitly articulated by Saltzer et al. [14], which has since become a core principle of the Internet development. It states that application-specific functions should reside in the end hosts of a network rather than in intermediary network nodes provided they can be implemented "completely and correctly" in the end hosts. It ensures that the network core is simple, fast, and highly scalable.

message is an IP multicast message. It covers IP addresses ranging from 224.0.0.0 to 239.255.255.255, known as the Class D addresses. For example, if some content is associated with group 230.0.0.1, the source will send data packets destined to 230.0.0.1. Receivers for that content will inform the network that they are interested in receiving data packets sent to the group 230.0.0.1.

The *Internet Group Management Protocol (IGMP)* was designed to help the maintenance of multicast groups. Two special types of IGMP messages are used: Query and Report. The Query messages are multicast by routers to all local hosts, to inquire about group membership. The Report is used to respond to a query and to join groups. The routers periodically query group membership and declare themselves group members if they get a response to at least one query. If no responses occur after a while, they declare themselves nonmembers. IGMP version 2 also enforces a lower latency, so the membership is pruned more promptly after all members in the group leave. IGMP version 3 further supports source-specific multicast and introduces membership report aggregation.

Multicast routing is generally based on a shared tree: once the receivers join a particular IP multicast group, a multicast distribution tree is constructed for that group. For example, all data packets sent to the group 230.0.0.1 are distributed to routers that each has at least one receiver who joined 230.0.0.1 (i.e., each with at least one local group member), and each such router will further forward the packets to its local receivers.

One of the first trials of IP multicast was in March 1992, when the Internet Engineering Task Force (IETF) meeting in San Diego was streamed (audio only) on the Internet. Starting in the early 1990s, the *Multicast Backbone (MBone)* was built [16] and used for multicast services on the Internet [17,18]. Earlier applications, mostly multimedia-based, include vat for audio conferencing, vic and nv for video conferencing. Other application tools include wb for whiteboards in shared workspace and sdr for maintaining session directories on MBone.

Since many routers do not support multicast, MBone uses a subnetwork of routers (*mrouters*) that support multicast to forward multicast packets. As Fig. 15.9 shows, the mrouters, each being responsible for a local region (or so-called *island*), are connected with *tunnels*. Multicast packets are encapsulated inside regular IP packets for "tunneling", so that they can be sent to the destination through the islands.

IP multicast is a loosely coupled model that reflects the basic design principles of the Internet. It retains the IP interface and introduces the concept of open and dynamic groups, which greatly inspires later proposals. Given that the network topology is best-known in the network layer, multicast routing in this layer is also the most efficient. It remains a best-effort service, and attempts to conform to the traditional separation of routing and transport that has worked well in the unicast context. However, providing higher level features such as error, flow, and congestion control has been shown to be more difficult than in the unicast case. In general, UDP (not TCP) is used in conjunction with IP multicast, so as to avoid too many ACKs from TCP receivers. For reliable file sharing or replication, reliable multicast transport protocols [19,20] need to be implemented on top of UDP. For continuous

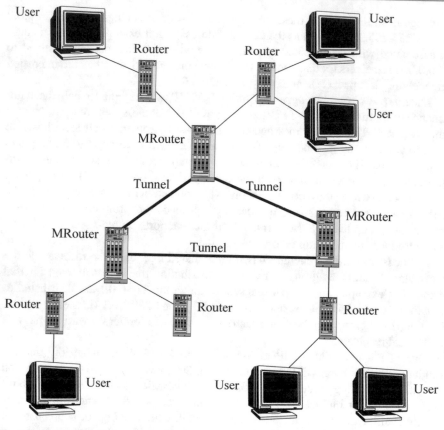

Fig. 15.9 Tunnels for IP multicast in MBone

streaming media, network and user heterogeneity should be accommodated, and for Video-on-Demand (VoD), asynchronous requests from subscribed users should be accommodated. These are not readily solved in IP multicast/UDP, either, and we will introduce solutions in the transport and application layers in the following sections and also the next chapter.

15.4.2 Non Router-Based Multicast Architectures

Today's IP multicast deployment remains limited in reach and scope. IP multicast calls for changes at the infrastructure level, i.e., in network routers. This introduces high complexity and serious scaling constraints. The flat topology of MBone, which has approximately 10,000 routes, is generally non-scalable [21]. The tunnel management is also very ineffective, that is, tunnels connecting islands can hardly be optimally allocated. Sometimes multiple tunnels are created over a single physical link, causing congestion. Beside technical obstacles, there are also economic and political concerns; in particular, there is a lack of incentive for network operators to install multicast-capable routers and to carry multicast traffic.

The placement of the multicast functionality was revisited in the late 1990s; researchers started to advocate moving multicast functionality away from routers toward the end systems [22]. In these approaches, multicast related features, such as group membership, multicast routing and packet duplication, are implemented at end systems, assuming only unicast IP service. The end systems participate in multicast communication via an *overlay network*, in the sense that each of its edges corresponds to a unicast path between two nodes in the underlying Internet.

Moving multicast functionality to end systems has the potential to address many of the problems associated with IP multicast. Since all packets are transmitted as unicast packets, deployment is easier and hence accelerated. Solutions for supporting higher layer features can be significantly simplified by leveraging well understood unicast solutions, and by exploiting application-specific intelligence.

Given that non-router based architectures push functionality to the network edges, there are several choices in instantiating such an architecture. On the one end of the spectrum is an *infrastructure-centric* architecture, where an organization that provides value-added services deploys proxies at strategic locations on the Internet. The end systems attach themselves to nearby proxies, and receive data using plain unicast. Such an approach is also commonly referred to as *Content Distribution Networks* (CDNs), and has been employed by companies such as Akamai. On the other end of the spectrum is a purely *application end-point* architecture, where functionality is pushed to the users (know as *peers*) actually participating in a multicast session. Administration, maintenance, responsibility for the operation of such a peer-to-peer system are distributed among the users, instead of being handled by a single entity.

While the application-layer solutions have the promise to enable ubiquitous deployment, they often involve a wide range of autonomous users that may not provide as good performance and easily fail or leave at will. It is impossible to completely prevent multiple overlay edges from traversing the same physical link, and thus some redundant traffic on physical links is unavoidable. Thus, the key challenge for application end-point architectures is to function, scale and self-organize with a highly transient population of users, without the need for a central server and the associated management overhead.

In the next chapter, we will detail the large-scale multimedia content distribution mechanisms over CDN, application-layer multicast, and general peer-to-peer networks.

15.5 Quality of Service (QoS) and Quality of Experience (QoE)

Fundamentally, multimedia network communication and traditional computer network communication are similar, since they both deal with data communications. However, challenges in multimedia network communications arise due to a series of distinct characteristics of audio/video data

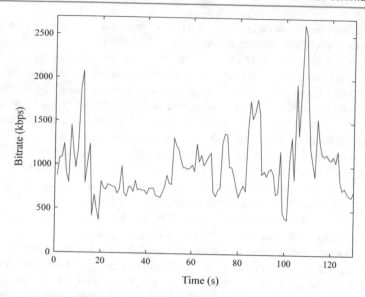

Fig. 15.10 The bitrate over time of an MPEG-4 video (Star Trek, 688 × 512 frame size)

- **Voluminous and Continuous**. They demand high data rates and often have a lower bound to ensure continuous playback. In general, a user expects to start playing back audio/video objects before they are fully downloaded. For this reason, they are commonly referred to as *continuous media* or *streaming media*.
- **Real-Time and Interactive**. They demand low startup delay and synchronization between audio and video for "lip sync". Interactive applications such as video conferencing and multiparty online gaming require two-way traffic, both of the same high demands.
- **Rate fluctuation**. The multimedia data rates fluctuate drastically and are sometimes bursty. In video-on-demand or Voice-over-IP (VoIP), no traffic most of the time but burst to high volume. In a variable-bit-rate (VBR) video, the average rate and the peak rate can differ significantly, depending on the scene complexity. For example, Fig. 15.10 shows the bitrate evolution of an MPEG-4 video stream, which has an average rate about 1 Mbps, but the minimum rate and maximum rate are 300 kbps and 2,600 kbps, respectively.

15.5.1 QoS and QoE for Multimedia Communications

Quality of Service (QoS) for multimedia data transmission depends on many parameters. We now list the most important ones below:

- **Bandwidth.** A measure of transmission speed over digital links or networks, often in kilobits per second (kbps) or megabits per second (Mbps).[3] As shown before, the data rate of a multimedia stream can vary dramatically, and both the average and the peak rates should be considered when planning for the bandwidth for transmission.

- **Latency (maximum frame/packet delay).** The maximum time needed from transmission to reception, often measured in milliseconds (msec, or ms). In voice communication, for example, when the round-trip delay exceeds 50 ms, *echo* becomes a noticeable problem; when the one-way delay is longer than 250 ms, *talker overlap* would occur, since each caller will talk without knowing the other is also talking.

- **Packet loss or error.** A measure (in percentage) of the loss or error rate of the packetized data transmission. The packets can get lost due to network congestion or garbled during transmission over the physical links. They may also be delivered late or in the wrong order. For real-time multimedia, retransmission is often undesirable, and therefore, alternative solutions like forward error correction (FEC), interleaving, or error-resilient coding are to be used.

  In general, for uncompressed audio/video, the desirable packet loss is $< 10^{-2}$ (lose every hundredth packet, on average). When it approaches 10%, the quality degradation becomes intolerable. For compressed multimedia data, the desirable packet loss is less than 10^{-7} to 10^{-8}. The error rate in modern wired communication links, in particular, fiber optics, can be quite low. For example, the Bit Error Rate (BER) objective for a fiber channel is 1 in 1012 (1 bit in 1,000,000,000,000 bits). At 2 Gbps, this equates to seven errors per hour. The BER, however, can be much worse in wireless links and is a key challenge for multimedia over wireless networks.

- **Jitter (*or delay jitter*).** A measure of smoothness (along time axis) of the audio/video playback. Technically, *jitter* is related to the variance of frame/packet delays. Figure 15.11 illustrates examples of high and low jitters in frame playbacks. A large buffer (jitter buffer) can be used to hold enough frames to allow the frame with the longest delay to arrive, so as to reduce playback jitter. However, this increases the latency and may not be desirable in real-time and interactive applications.

- **Sync skew.** A measure of multimedia data synchronization, often measured in milliseconds (msec). For a good *lip synchronization*, the limit of sync skew is ± 80 ms between audio and video. In general, ± 200 ms is still acceptable. For a video with voice, the limit of sync skew is 120 ms if video precedes voice and 20 ms if voice precedes video. The discrepancy is because we are used to have sound lagging image at a distance.

---

[3]For an analog signal, the bandwidth is generally measured in hertz, as in the fields of communications and signal processing. The network bandwidth and the frequency bandwidth can be linked by *Hartley's law* [23], which states "that the total amount of information that can be transmitted is proportional to frequency range transmitted and the time of the transmission."

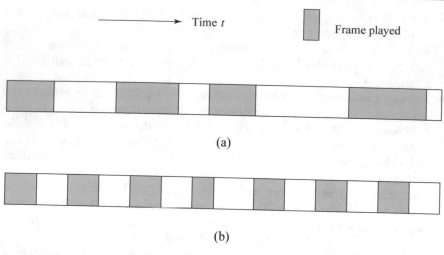

Fig. 15.11 Jitters in frame playback: **a** High jitter; **b** low jitter

Multimedia Service Classes

Unlike traditional file sharing and downloading applications that have largely uniform demands, there is a broad spectrum of multimedia data (from audio to image to video, and from low quality audio/video to medium quality and to high quality) and applications (one-way or two-way, interactive or non-interactive, realtime or non-realtime, and etc.). We now list a set of the typical multimedia applications of different QoS demands:

- Two-way traffic, low latency, and jitter, possibly with prioritized delivery, such as voice telephony and video telephony.
- Two-way traffic, low loss and low latency, with prioritized delivery, such as e-commerce applications.
- Moderate latency and jitter, strict ordering and sync. One-way traffic, such as streaming video; or two-way interactive traffic, such as web surfing and online gaming.
- No real-time requirement, such as downloading or transferring large files (movies). No guarantees for transmission.

Table 15.2 lists the general bandwidth/bitrate requirement for typical multimedia applications. Table 15.3 lists some specifications for tolerance to delay and jitter in digital audio and video of different qualities. As can be seen, the QoS demands of multimedia applications vary significantly, and therefore, the specific application demands must be taken into account in protocol and system design and deployment.

Table 15.2 Typical requirement on network bandwidth/bitrate

Application	Speed requirement
Telephone	16 kbps
Audio conferencing	32 kbps
CD-quality audio	128–192 kbps
Digital music	64–640 kbps
H. 261	64 kbps–2 Mbps
H. 263	< 64 kbps
H. 264	1–12 Mbps
MPEG-1 video	1.2–1.5 Mbps
MPEG-2 video	4–60 Mbps
MPEG-4 video	1–20 Mbps
HDTV (compressed)	>20 Mbps
HDTV (uncompressed)	>1 Gbps
4K (compressed)	>50 Mbps
4K (uncompressed)	>4 Gbps
8K (compressed)	>200 Mbps
8K (uncompressed)	>16 Gbps
MPEG-4 video-on-demand	250–750 kbps
Videoconferencing	384 kbps–2 Mbps

Table 15.3 Tolerance of latency and jitter in digital audio and video

Application	Average latency tolerance (ms)	Average jitter tolerance (ms)
Low-end videoconference (64 kbps)	300	130
Compressed voice (16 kbps)	30	130
MPEG-2 video (1.5 Mbps)	5	7
MPEG-2 audio (256 kbps)	7	9
HD/UHD video (>20 Mbps)	0.8	1

Quality of Experience (QoE): User Perceived QoS

Although QoS is commonly measured by the above objective parameters, it itself is a "collective effect of service performances that determine the degree of satisfaction of the user of that service", as defined by the *International Telecommunications Union* (ITU). In other words, it has everything to do with how a user *perceives* it, which is particularly true for services that involve multiple media and their interactions.

Quality of Experience (QoE) is a measure of the delight or annoyance of a user's experiences with a service, e.g., video streaming. More formally, ITU adopted the following definition in its Recommendation ITU-T P.10 (2016):

> The degree of delight or annoyance of the user of an application or service. It results from the fulfillment of his or her expectations with respect to the utility and / or enjoyment of the application or service in the light of the user's personality and current state.

Together with the perceptual nonuniformity we have studied in previous chapters, many issues of perception can be exploited in achieving the best QoE or user perceived QoS in networked multimedia. For example, in real-time multimedia, regularity is more important than latency (i.e., jitter and quality fluctuation are more annoying than slightly longer waiting), and temporal correctness is more important than the sound and picture quality (i.e., ordering and synchronization of audio and video are of primary importance). Humans also tend to focus on one subject at a time; a user's focus is usually at the center of a screen, and it takes time to refocus, especially after a scene change.

15.5.2 Internet QoS Architecture: IntServ and DiffServ

QoS policies and technologies enable such key metrics discussed in the previous section as latency, packet loss, and jitter to be controlled by offering different levels of service to different packet streams or applications. The conventional IP provides the "best-effort" service only, which does not differentiate among different applications. Therefore, it is hard to ensure QoS over the basic IP beyond expanding bandwidth. Unfortunately, in a complex and large-scale networks, abundant bandwidth is unlikely to be available everywhere (in practice, many IP networks routinely use oversubscription). Even if it is available everywhere, bandwidth alone cannot resolve problems due to sudden peaks in traffic, and there are always new networked applications demand higher and higher bandwidth, e.g., high-definition video and 3D/multiview video.

There have been significant efforts toward data networking with better or even guaranteed QoS. Pioneering works on the Internet QoS can be classified into two categories. *IntServ* or *integrated services* is an architecture that specifies the elements to guarantee QoS in fine-grains for each individual flow. The idea is that every router in the network implements IntServ, and every application that requires some kind of guarantees has to make an individual reservation in advance. In contrast to IntServ, *DiffServ* or *differentiated services* specifies a simple, scalable, and coarse-grained class-based mechanism for classifying and managing aggregated network traffic and providing specific QoS to different classes of traffic.

Integrated Service (IntServ) and Resource ReSerVation Protocol (RSVP)

In IntServ, *Flow Specs* describe what the resource reservation is for a flow, while the *Resource ReSerVation Protocol* (RSVP) [24] is used as the underlying mechanism to signal it across the network.

Flow specs include two components: First, what does the traffic look like? This is defined in *Traffic SPECification*, also known as TSPEC; Second, what guarantees does it need? This is defined in the service *Request SPECification*, also known as RSPEC.

RSVP is a setup protocol for Internet resource reservation, which targets a multicast setup (typical built on top of IP multicast) for general multimedia applications (unicast can be viewed as a special case of multicast). A general communication model supported by RSVP consists of m senders and n receivers, possibly in various multicast groups (e.g., in Fig. 15.12a, $m = 2$, $n = 3$, and the trees for the two multicast groups are depicted by the arrows—solid and dashed lines, respectively). In the special case of single-source broadcasting, $m = 1$; whereas in audio or video conferencing, each host acts as both sender and receiver in the session, that is, $m = n$.

The main challenges of RSVP are that many senders and receivers may compete for the limited network bandwidth, the receivers can be heterogeneous in demanding different contents with different QoS, and they can be dynamic by joining or quitting multicast groups at any time. To address these challenges, RSVP introduces Path and Resv messages. A Path message is initiated by the sender and travels toward the multicast (or unicast) destination addresses. It contains information about the sender and the path (e.g., the previous RSVP hop), so that the receiver can find the reverse path to the sender for resource reservation. A Resv message is sent by a receiver that wishes to make a reservation.

- **RSVP is receiver-initiated**. A receiver (at a leaf of the multicast tree) initiates the reservation request Resv, and the request travels back toward the sender but not necessarily all the way. A reservation will be merged with an existing reservation made by other receiver(s) for the same session as soon as they meet at a router. The merged reservation will accommodate the highest bandwidth requirement among all merged requests. The user-initiated scheme is highly scalable, and it meets the heterogeneous demands from the users.
- **RSVP creates only *soft state***. The receiver host must maintain the soft state by periodically sending the same Resv message; otherwise, the state will time out. There is no distinction between the initial message and any subsequent refresh message. If there is any change in reservation, the state will automatically be updated according to the new reservation parameters in the refreshing message. Hence, the RSVP scheme is highly dynamic.

Figure 15.12 depicts a simple network with two senders (S1, S2), three receivers (R1, R2, and R3), and four routers (A, B, C, D). Figure 15.12a shows that S1 and S2 send Path messages along their paths to R1, R2, and R3. In (b) and (c), R1 and R2 send out Resv messages to S1 and S2, respectively, to make reservations for S1 and S2 resources. From C to A, two separate channels must be reserved since R1 and R2 request different data streams. In (d), R2 and R3 send out their Resv messages to S1, to make additional requests. R3's request is merged with R1's previous request at A, and R2's is merged with R1's at C.

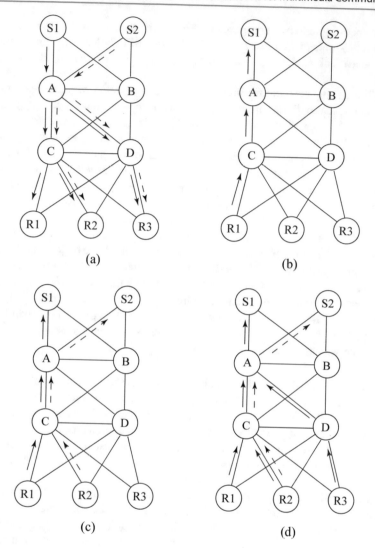

Fig. 15.12 A scenario of network resource reservation with RSVP: **a** Senders S1 and S2 send out their PATH messages to receivers R1, R2, and R3; **b** receiver R1 sends out RESV message to S1; **c** receiver R2 sends out RESV message to S2; **d** receivers R2 and R3 send out their RESV messages to S1

Any possible variation of QoS that demands higher bandwidth can be dealt with by modifying the reservation state parameters.

Differentiated Service (DiffServ)

As opposed to IntServ, DiffServ operates on the principle of traffic aggregation and classification, where data packets are placed into a limited number of *traffic classes*, rather than differentiating network traffic based on the requirements of individual flows. Each traffic class can be managed differently, ensuring preferential treatment for higher-priority traffic on the network.

In DiffServ, network routers implement *per-hop behaviors* (PHBs), which define the packet-forwarding properties associated with a class of traffic. In practice, the *Type of Service* octet in an IPv4 packet and the *Traffic Class* octet in an IPv6 packet can be used as the *DiffServ Code* (DS) to classify packets to enable their differentiated treatments (see Fig. 15.4).

The DS field contains a 6-bit *Differentiated services Code Point* (DSCP) value. In theory, a network could have up to 64 (i.e., 2^6) different traffic classes using different DSCPs. This gives a network operator great flexibility in defining traffic classes. Different PHBs may be defined to offer, for example, lower loss or lower latency for multimedia data than file transfer, or better service for audio than video, or even different services within a multimedia application data

- **Uncompressed audio.** PCM audio bitstreams can be broken into groups of every nth sample—prioritize and send k of the total of n groups ($k \le n$) and ask the receiver to interpolate the lost groups if so desired. For example, if two out of four groups are lost, the effective sampling rate is 22.05 kHz instead of 44.1 kHz. Loss is perceived as a change in sampling rate, not dropouts.
- **JPEG image.** The different *scans* in Progressive JPEG and different resolutions of the image in hierarchical JPEG can be given different services. For example, the best service for the scan with the DC and first few AC coefficients, and better service for the lower-resolution components of the hierarchical JPEG image.
- **Compressed video.** To minimize playback delay and jitter, the best service can be given to the reception of I-frames and the lowest priority to B-frames. In scalable video using layered coding, the base layer can be given a better service than the enhancement layers.

In practice, most networks use the following commonly-defined per-hop behaviors:

- **Default PHB**, which is typically the best-effort service.
- **Expedited Forwarding (EF)**, which is dedicated to low-loss and low-latency traffic. It is suitable for premium voice, video, and other real-time services, and is often given strict priority above all other traffic classes.
- **Assured Forwarding (AF)**, which achieves assurance of delivery under prescribed conditions. The traffic that exceeds the subscription rate faces a higher probability of being dropped if congestion occurs.
- **Class Selector PHBs**, which maintain backward compatibility with non-DiffServ traffic.

It is worth noting that the details of how individual DiffServ routers deal with the DS field, i.e., PHB, is configuration specific. For example, one implementation may divide the network traffic in Assured Forwarding into the following categories and allocate bandwidth accordingly:

- **Gold**: Traffic in this category is allocated 50% of the available bandwidth.
- **Silver**: Traffic in this category is allocated 30% of the available bandwidth.
- **Bronze**: Traffic in this category is allocated 20% of the available bandwidth.

Another implementation may have a different configuration or even completely ignore their differences.

Compared with IntSev, DiffServ has coarser control granularity (in aggregated classes, rather than individual flows), and is, therefore, simpler and scales well. They are, however, not necessarily exclusive to each other. In real-world deployment, IntServ and DiffServ may work together to accomplish the QoS targets with reasonable costs. In particular, RSVP can be applied to individual local flows within the network edge, and these flows are then aggregated with the DiffServ Code (DS) being added by QoS-aware Edge Devices. In the core network, there is no flow separation, where all packets of each specific class are treated equally by the PHBs. In other words, RSVP is tunneled in the core and only be visible and accommodated once the aggregated traffic arrived at the Edge Devices for the destination. Since IntServ is now confined within network edges, the costs for maintaining per flow states can be largely reduced, imposing minimum overhead to the high-speed core network.

15.5.3 Network Softwarization and Virtualization: SDN and NVF

IntServ and DiffServ represent the initial efforts toward improving the Internet QoS and have been implemented in many of today's Internet routers; however, their use in wide area networks have been limited. First, the complexity of maintaining these services in large-scale dynamic networks can be quite high, particularly for flow-based resource reservation protocols; Second, the scale and heterogeneity of Internet terminals and routers make a complete end-to-end QoS guarantee generally difficult, so for service differentiation. As such, it is difficult to predict the end-to-end behavior for a packet crossing multiple domains before reaching its destination, because it is up to all the service providers and their routers in the path to ensure that their policies will appropriately take care of the packet.

More importantly, in the traditional Internet design, the control plane and the data plane, as well as the software and hardware for them, are tightly coupled, making any change to the network core very difficult to be deployed. Recently, *Software-defined networking* (SDN) seeks to decouple the network control and forwarding functions, enabling network control to become directly programmable and the underlying infrastructure to be abstracted from applications and network services. Efficient and dynamic network configuration can then be realized to improve the network performance and monitoring [25].

The success of a global deployment of SDN greatly depends on the creation and agreement on standard application programming interfaces (APIs) between the control and data planes. The open source Ethane project at Stanford University is a pioneer toward this direction [26], which leads to the development of OpenFlow [27] and the NOX operating system for networks. OpenFlow has since been widely supported, with the following distinct features:

- **Highly programmable**. Since network control is decoupled from forwarding, its functions become highly programmable. Specifically, SDN enables network administrators to easily and quickly configure, manage, secure, and optimize network resources via standard and automated SDN programs, instead of proprietary software from different vendors.
- **Centrally manageable**. With the software-centric design, network intelligence is logically centralized in software-based SDN controllers. They work together to maintain a global view of the network, collectively acting as a single, logical switch for network traffic administration. Such abstraction allows administrators to dynamically adjust network-wide traffic flow, so as to meet rapidly changing needs.
- **Openly available**. SND, and OpenFlow in particular, advocates an open design. Given that instructions are provided by standardized software-empowered SDN controllers, instead of vendor-specific hardware and protocols, it greatly simplifies the network design, operation, management, as well as service migration for modern cloud or Internet of Things (IoT) applications.

Another recent effort toward making a more flexible and managcable Internet is *Network Functions Virtualization* (NFV) [28] . Extending the conventional machine virtualization concept to the network architecture, NVF seeks to virtualize the entire classes of network node functions into building blocks that may connect, or chain together, to create communication services. This will greatly reduce capital and operational expenditures, and accelerate service and product deployment. In particular, NFV adopts a central orchestration and management module that takes operator requests associated with each virtualized network function, translates them into the corresponding processing, storage, and network configuration, so as to execute the function. During the execution, the resource capacity and utilization of the function will be monitored and dynamically adjusted in realtime [29].

NFV is closely related to SDN, but does not necessarily depend on SDN. It can be implemented using existing networking and orchestration platforms, though unifying SDN and NFV in an ecosystem is clearly beneficial [30].

15.5.4 Rate Control and Buffer Management

SDN and NFV are still under active development and adoption, and it is worth noting they are solutions that improve the Internet, while not replace the Internet. Even with these advances in the network control plane, most of the time, a networked multimedia application still has to assume that the underlying network is of the best-effort service (or at least, without guaranteed QoS), and adaptive transmission and control are to be used [31].

A key concern here is rate fluctuation with multimedia data. As we have seen earlier, audio encoding is generally of *Constant Bit Rate* (CBR) during a talk, e.g., 64 kbps bitrate (8 kHz sampling frequency, 8 bits per sample). For video, CBR coding needs to maintain a constant bitrate at the source; yet variable distortions can be introduced given the scenes differ across frames. CBR coding is also less efficient than *Variable Bit Rate* (VBR) coding: to obtain comparable quality of coded media, the CBR bitrate is typically 15–30% higher than the mean VBR video bitrate.

To this end, VBR coding is often used. Usually, the more activities (motions in the video), the higher the required bitrate is. In this case, the typical bitrates for MPEG-1 (1.5 Mbps) and that for MPEG-2/4 (4 Mbps) are averages, and the real stream can have a low bitrate at one point and a much higher bitrate at another point (see Fig. 15.10). If the video is delivered through the network without any *work-ahead smoothing*, the required network throughput must be higher than the video's peak bitrate for uninterrupted playback.

To cope with the variable bitrate and network load fluctuation, buffers are usually employed at both sender and receiver ends [32]. A *prefetch buffer* can be introduced at the client side to smooth the transmission rate (reducing the peak rate). If the size of frame t is $d(t)$, the buffer size is B, and the number of data bytes received so far (at play time for frame t) is $A(t)$, then for all $t \in 1, 2, \ldots, N$, it is required that

$$\sum_{i=1}^{t} d(i) \leq A(t) \leq \sum_{i=1}^{t-1} d(i) + B \qquad (15.1)$$

If $A(t) < \sum_{i=1}^{t} d(i)$, we have inadequate network throughput, and hence buffer *underflow* (or *starvation*), whereas when $A(t) > \sum_{i=1}^{t} d(i) + B$, we have excessive network throughput and buffer *overflow*. Both are harmful to smooth, continuous playback. In buffer underflow, no data is available to play, and in buffer overflow, media packets must be dropped.

Figure 15.13 illustrates the limits imposed by the media playback (consumption) data rate and the buffered data rate. The transmission rates are the slopes of the curves. At any time, data must be in the buffer for smooth playback, and the data transmitted must be more than the data consumed. If the available bandwidth is as in Line II in the figure, at some point during playback, the data to be consumed will be greater than can be sent. The buffer will underflow, and playback will be interrupted. Also, at any point, the total amount of data transmitted must not exceed the total consumed plus the size of the buffer.

If the network available bandwidth is as in Line I and the media was sent as fast as possible without buffer considerations (as in normal file downloads), then toward the end of the video, the data received will be greater than the buffer can store at the time. The buffer will overflow and drop the extra packets. The server will have to retransmit the packets dropped, or these packets will be missing. This increases

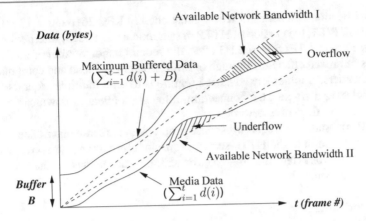

Fig. 15.13 The data that a client can store in the buffer assists the smooth playback of the media when the media rate exceeds the available network bandwidth

bandwidth requirements (and hence may cause underflow in the future). In many cases, such as multicast, no back channel is available.

To address this, we need to prefetch video data to fill the buffer and try to transmit at the mean video bitrate, or to keep the buffer full without exceeding the available bandwidth, which can be estimated as the TCP-friendly bandwidth as mentioned earlier. In either case, for video sections that require higher bandwidth than available, the data already in the buffer and the available network bandwidth should enable smooth playback without buffer underflow. If the data rate characteristics are known in advance, say for media stored in a server, it is possible to use the prefetch buffer more efficiently for the network. The media server can plan ahead for a transmission rate such that the media can be viewed without interruption and with minimized bandwidth [33].

15.6 Protocols for Multimedia Transmission and Interaction

We now review the protocols for multimedia communications. These protocols build on top of UDP or TCP and work with the best-effort Internet or with IntServ/DiffServ/SDN/NVF to provide quality multimedia data transmission, particularly in the streaming mode, and also enable various interactions between a media server and its clients.

15.6.1 HyperText Transfer Protocol (HTTP)

HTTP is a protocol that was originally designed for transmitting Web content, but it also supports transmission of any file type. The standards development of HTTP was monitored by both IETF and the World Wide Web Consortium (W3C), culminating in the publication of a series of RFCs, most notably RFC 2616 (June 1999), which defines HTTP/1.1, the version of HTTP in common use. In 2015, Internet Engineering Task Force (IETF) released HTTP/2, the second major version of the HTTP. It achieves improved efficiency through optimization of page load and compression of request headers. Another key advanced feature is to send multiple requests for data in parallel over a single TCP connection, allowing a client to download web files asynchronously from one server.

HTTP is a "stateless" request/response protocol, in the sense that a client typically opens a connection to the HTTP server, requests information, the server responds, and the connection is terminated—no information is carried over for the next request.

The basic request format is

```
Method URI Version
Additional-Headers

Message-body
```

The *Uniform Resource Identifier* (URI) identifies the resource accessed, such as the host name, always preceded by the token "http://" or "https://". A URI could be a URL, for example. It can also include query strings (some interactions require submitting data). Method is a way of exchanging information or performing tasks on the URI. Two popular methods are GET and POST. GET specifies that the information requested is in the request string itself, while the POST method specifies that the resource pointed to in the URI should consider the message body. POST is generally used for submitting HTML forms. Additional-Headers specifies additional parameters about the client. For example, to request access to this textbook's web site, the following HTTP message might be generated:

```
GET http://www.cs.sfu.ca/mmbook/ HTTP/1.1
```

The basic response format is

```
Version Status-Code Status-Phrase
Additional-Headers

Message-body
```

Status-Code is a number that identifies the response type (or error that occurs), and Status-Phrase is a textual description of it. Two commonly seen status codes and phrases are 200 OK when the request was processed successfully and 404 Not Found when the URI does not exist. For example, in response to the example request above for this textbook's URL, the web server may return something like

```
HTTP/1.1 200 OK Server:
[No-plugs-here-please] Date: Wed, 24 July 2019
20:04:30 GMT
Content-Length: 1045 Content-Type: text/html

<HTML>
...
</HTML>
```

Note that HTTP/2 has high-level compatibility with HTTP/1.1 in terms of methods, status codes, URIs, and header fields above.

HTTP builds on top of TCP to ensure reliable data transfer. It was not originally designed for multimedia content distribution, not to mention streaming media. Yet HTTP-based streaming has recently become popular, thanks to smart stream segmentation strategies and the abundant Web server resources available for HTTP data transfer, as we will examine in the next Chapter.

15.6.2 Real-Time Transport Protocol (RTP)

RTP, defined in RFC 3550, is designed for the transport of real-time data, such as audio and video streams. As we have seen, networked multimedia applications have diverse characteristics and demands; there are also tight interactions between the network and the media. Hence, RTP's design follows two key principles, namely *application-layer framing*, i.e., framing for media data should be performed properly by the application layer, and *integrated layer processing*, i.e., integrating multiple layers into one to allow efficient cooperation [34]. These distinguish RTP from other traditional application-layer protocols, such as the HTTP for Web transactions and FTP for file transfer, that each targets a single well-defined application. Instead, RTP resides in between the transport layer and the application layer, and bridges them for real-time multimedia transmission.

RTP usually runs on top of UDP, which provides an efficient (albeit less reliable) connectionless transport service. There are three main reasons for using UDP instead of TCP. First, TCP is a connection-oriented transport protocol; hence, it is more difficult to scale up in a multicast environment. From the very beginning, RTP had already targeted multicast streaming, with unicast being a special case only. Second, TCP achieves its reliability by retransmitting missing packets. As mentioned earlier, multimedia data transmission is loss-tolerant and perfect reliability is not necessary; the late arrival of retransmitted data may not be usable in real-time applications, either, and persistent retransmission would even block the data flow, which is undesirable for continuous streaming. Last, the dramatic rate fluctuation (sawtooth behavior) in TCP is often not desirable for continuous media.

TCP does not provide timing information, which is critical for continuous media. Since UDP has no timing information either, nor does it guarantee that the data packets arrive in the original order (not to mention synchronization of multiple sources),

RTP must create its own *timestamping* and *sequencing* mechanisms to ensure the ordering. RTP introduces the following additional parameters in the header of each packet:

- **Payload type** indicates the media data type as well as its encoding scheme (e.g., PCM audio, MPEG 1/2/4, H.263/264/265 audio/video, etc), so that the receiver knows how to decode it.
- **Timestamp** is the most important mechanism of RTP. The timestamp records the instant when the first octet of the packet is sampled, which is set by the sender. With the timestamps, the receiver can play the audio/video in proper timing order and synchronize multiple streams (e.g., audio and video) when necessary.
- **Sequence number** is to complement the function of timestamping. It is incremented by one for each RTP data packet sent, to ensure that the packets can be reconstructed in order by the receiver. This becomes necessary because, for example, all packets of one video frame can be set with the same timestamp, and timestamping alone becomes insufficient.
- **Synchronization source (SSRC) ID** identifies the sources of multimedia data (e.g., audio, video). If the data come from the same source (e.g., a translator or a mixer), they will be given the same SSRC ID, so as to be synchronized.
- **Contributing Source (CSRC) ID** identifies the source of contributors, such as all speakers in an audio conference.

Figure 15.14 shows the RTP header format. The first 12 octets are of a fixed format, followed by optional (0 or more) 32-bit Contributing Source (CSRC) IDs. Bits 0 and 1 are for the version of RTP, bit 2 (P) for signaling a padded payload, bit 3 (X) for signaling an extension to the header, and bits 4 through 7 for a 4-bit CSRC count that indicates the number of CSRC IDs following the fixed part of the header. Bit 8 (M) signals the first packet in an audio frame or last packet in a video frame, since an audio frame can be played out as soon as the first packet is received, whereas a video frame can be rendered only after the last packet is received. Bits 9 through 15 describe the payload type, Bits 16 through 31 are for sequence number, followed by a 32-bit timestamp and a 32-bit Synchronization Source (SSRC) ID.

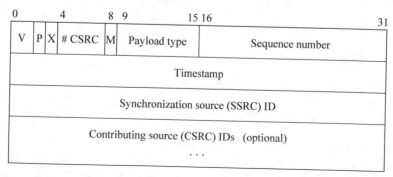

Fig. 15.14 RTP packet header

15.6.3 RTP Control Protocol (RTCP)

RTCP, also defined in RFC 3550, is a companion protocol of RTP. It monitors QoS in providing feedback to the source on the quality of data transmission and conveys information about the participants of a multicast session. RTCP also provides the necessary information for audio and video synchronization, even if they are sent through different packet streams.

RTCP provides a series of typical reports and is extensible, allowing for application-specific RTCP reports:

1. **Receiver report (RR)** provides quality feedback (number of last packet received, number of lost packets, jitter, and timestamps for calculating round-trip delays).
2. **Sender report (SR)** provides information about the reception of RR, number of packets/bytes sent, and so on.
3. **Source description (SDES)** provides information about the source (e-mail address, phone number, full name of the participant).
4. **Bye** indicates the end of participation.
5. **Application specific functions (APP)** provides for future extension of new features.

RTP and RTCP packets are sent to the same IP address (multicast or unicast) but on different ports. RTCP reports are expected to be sent by all participants, even in a multicast session which may involve thousands of senders and receivers. Such traffic will increase proportionally with the number of participants. Thus, to avoid network congestion, the protocol must include session bandwidth management, achieved by dynamically controlling the frequency of report transmissions. RTCP bandwidth usage should generally not exceed 5% of the total session bandwidth. Furthermore, 25% of the RTCP bandwidth should be reserved for media sources at all times, so that in large sessions new participants can identify the senders without excessive delay.

Note that, while RTCP offers QoS feedbacks, it does not specify how these feedbacks are to be used, but leaves the operations to the application layer. The rationale is that, as we have seen, the multimedia applications have highly diverse requirements (in bandwidth, delay, packet loss, and etc.), and therefore, no single set of operations can satisfy all of them. Instead, each application should customize its own operations with the feedbacks to improve QoS. This is quite different from TCP, which offers a uniform interface for a range of data applications with homogeneous QoS requirements, namely, delay or bandwidth insensitive, and perfect reliability. In the following sections and the next chapter, we will see more examples of the use of RTCP's QoS feedbacks.

15.6.4 Real-Time Streaming Protocol (RTSP)

The Real-Time Streaming Protocol (RTSP), defined in RFC 2326, is a signaling protocol to control streaming media servers. The protocol is used for establishing

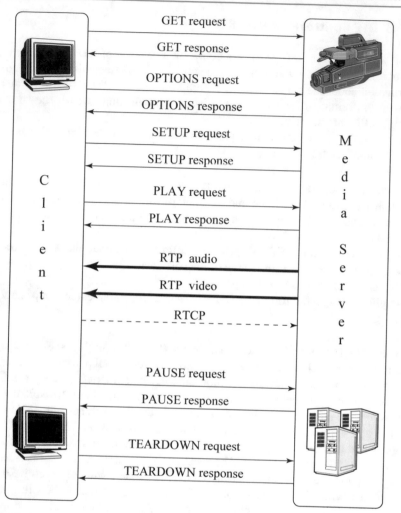

Fig. 15.15 A scenario of RTSP operations

and controlling media sessions between end points. Clients of media servers issue VCR-like commands, such as play, random-seek, and pause, to facilitate real-time control of playback of media files from the server. The transmission of streaming data itself is not a task of the RTSP protocol. Most RTSP servers use RTP in conjunction with RTCP for media stream delivery, although proprietary transport protocols are also possible.

Figure 15.15 illustrates a possible scenario of four typical RTSP operations

1. **Requesting presentation description**. The client issues a DESCRIBE request to the media server to obtain the presentation description, such as media types

(audio, video, graphics, etc.), frame rate, resolution, codec, and so on, from the server.

2. **Session setup**. The client issues a SETUP to inform the server of the destination IP address, port number, protocols, and TTL (for multicast). The session is set up when the server returns a session ID.

3. **Requesting and receiving media**. After receiving a PLAY, the server starts to transmit streaming audio/video data, using RTP. It is followed by a RECORD or PAUSE. Other VCR commands, such as FAST-FORWARD and REWIND are also supported. During the session, the client periodically sends an RTCP packet to the server, to provide feedback information about the QoS received (as described in Sect. 15.6.3).

4. **Session closure**. TEARDOWN closes the session.

15.7 Case Study: Internet Telephony

We now use a case of Internet Telephony to see the use of the protocols we have introduced, as well as introduce other important signaling protocols for multimedia communications.

As desktop/laptop computers and the Internet became readily available and more and more voice and data communications became digital, "voice over data networks," especially *Voice over IP* (VoIP), started to attract a great deal of interest in research and user communities. With ever-increasing network bandwidth and the ever-improving quality of multimedia data compression, *Internet telephony* [35] has become a reality.

The main advantages of Internet telephony over the *plain old telephone services* (POTS) are as follows:

- It provides great flexibility and extensibility in accommodating integrated services such as voicemail, video conversations, live text messages, and so on.
- It uses packet switching, not circuit switching; hence, network usage is much more efficient (voice communication is bursty and VBR-encoded).
- With the technologies of multicast or multipoint communication, multiparty calls are not much more difficult than two-party calls.
- With advanced multimedia data-compression techniques, various degrees of QoS can be supported and dynamically adjusted according to the network traffic, an improvement over the "all or none" service in POTS.
- Richer graphical user interfaces can be developed to show available features and services, monitor call status and progress, and so on.

As Fig. 15.16 shows, the transport of real-time audio (and video) in Internet telephony is supported by RTP (with its control protocol, RTCP), as described in Sect. 15.6.2. Streaming media is handled by RTSP and Internet resource reservation, if available, is taken care of by RSVP. Recently, new generations of Internet telephony,

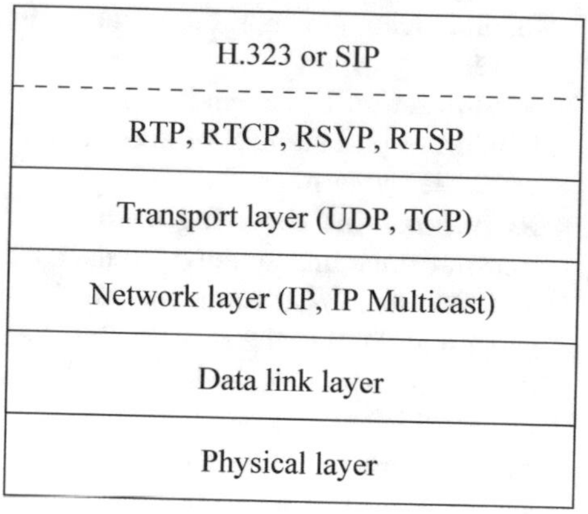

Fig. 15.16 Network
protocol structure for
Internet telephony

H.323 or SIP
RTP, RTCP, RSVP, RTSP
Transport layer (UDP, TCP)
Network layer (IP, IP Multicast)
Data link layer
Physical layer

most notably Skype, also uses the peer-to-peer and cloud computing technologies to achieve better scalability.

15.7.1 Signaling Protocols: H.323 and Session Initiation Protocol (SIP)

A streaming media server can be readily identified by a URL,, whereas acceptance of a call via Internet telephony depends on the callee's current location, capability, availability, and desire to communicate, which requires advanced signaling protocols.

The following are brief descriptions of the H.323 standard from ITU, and one of the most commonly used IETF standards, the Session Initiation Protocol (SIP).

H.323 Standard

H.323 [36,37] is an ITU standard for packet-based multimedia communication services. It specifies signaling protocols and describes terminals, multipoint control units (for conferencing), and gateways for integrating Internet telephony with *General Switched Telephone Network (GSTN)*[4] data terminals. The H.323 signaling process consists of two phases:

1. **Call setup**. The caller sends the *gatekeeper (GK)* a *Registration, Admission and Status (RAS) Admission Request (ARQ)* message, which contains the name and phone number of the callee. The GK may either grant permission or reject the request, with reasons such as "security violation" and "insufficient bandwidth".

---

[4]GSTN is a synonym for PSTN (public switched telephone network).

2. **Capability exchange**. An H.245 control channel will be established, for which the first step is to exchange capabilities of both the caller and callee, such as whether it is audio, video, or data; compression and encryption, and so on.

H.323 provides mandatory support for audio and optional support for data and video. It is associated with a family of related software standards that deal with call control and data compression for Internet telephony.

Signaling and Control

- **H.225**. Call control protocol, including signaling, registration, admissions, pack-etization and synchronization of media streams
- **H.245**. Control protocol for multimedia communications—for example, opening and closing channels for media streams, obtaining gateway between GSTN and Internet telephony
- **H.235**. Security and encryption for H.323 and other H.245-based multimedia terminals

Audio Codecs

- **G.711**. Codec for 3.1 kHz audio over 48, 56, or 64 kbps channels. G.711 describes PCM for normal telephony
- **G.722**. Codec for 7 kHz audio over 48, 56, or 64 kbps channels

Session Initiation Protocol (SIP)

SIP is IETF's recommendation (RFC 3261) for establishing and terminating sessions in Internet telephony. Different from H.323, SIP is a text-based protocol, and is not limited to VoIP communications—it supports sessions for multimedia conferences and general multimedia content distribution. As a client-server protocol, SIP allows a caller (the client) to initiate a request, which a server processes and responds to. There are three types of servers. A *proxy server* and a *redirect server* forward call requests. The difference between the two is that the proxy server forwards the requests to the next-hop server, whereas the redirect server returns the address of the next-hop server to the client, so as to redirect the call toward the destination.

The third type is a *location server*, which finds the current locations of users. Location servers usually communicate with the redirect or proxy servers. They may use finger, rwhois, *Lightweight Directory Access Protocol* (*LDAP*), or other protocols to determine a user's address.

SIP can advertise its session using e-mail, news groups, web pages or directories, or the *Session Announcement Protocol* (SAP). The *methods* (commands) for clients to invoke are

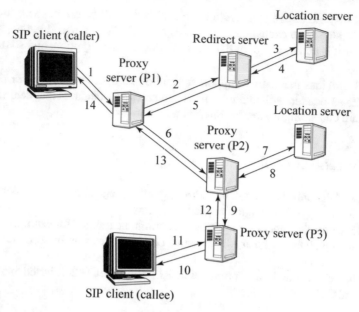

Fig. 15.17 A possible scenario of SIP session initiation

- INVITE—invites callee(s) to participate in a call.
- ACK—acknowledges the invitation.
- OPTIONS—inquires about media capabilities without setting up a call.
- CANCEL—terminates the invitation.
- BYE—terminates a call.
- REGISTER—sends user's location information to a registrar (a SIP server).

Figure 15.17 illustrates a possible scenario when a caller initiates a SIP session:

Step 1.	Caller sends an INVITE john@home.ca to the local Proxy server P1.
Step 2.	The proxy uses its Domain Name Service (DNS) to locate the server for
	john@home.ca and sends the request to it.
Steps 3, 4.	john@home.ca is not logged on the server. A request is sent to the nearby location server. John's current address, john@work.ca, is located.
Step 5.	Since the server is a redirect server, it returns the address john@work.ca to the proxy server P1.
Step 6.	Try the next proxy server P2 for john@work.ca.
Steps 7, 8.	P2 consults its location server and obtains John's local address, john_doe@my.work.ca.

Steps 9, 10. The next-hop proxy server P3 is contacted, which in turn forwards the invitation to where the client (callee) is.

Steps 11–14. John accepts the call at his current location (at work) and the acknowledgments are returned to the caller.

SIP can also use *Session Description Protocol* (SDP) (RFC 4566) to gather information about the users' media capabilities. As its name suggests, SDP describes multimedia sessions. The SDP descriptions are in a plain text format. They include the number and types of media streams (audio, video, whiteboard session, and etc.), destination address (unicast or multicast) for each stream, sending and receiving port numbers, and media formats (payload types). When initiating a call, the caller includes the SDP information in the INVITE message. The called party responds and sometimes revises the SDP information, according to its capability. Below we show an example session description, adapted from RFC 4566.

```
v=0
o=jdoe 2890844526 2890842807 IN IP4 10.47.16.5
s=SDP Seminar
i=A Seminar on the session description protocol
u=http://www.example.com/seminars/sdp.pdf
e=j.doe@example.com (Jane Doe)
c=IN IP4 224.2.17.12/127
t=2873397496 2873404696
a=recvonly
m=audio 49170 RTP/AVP 0
m=video 51372 RTP/AVP 99
a=rtpmap:99 H264/90000
```

This session description is being proposed to a receiving client (with username "jdoe") who was requesting a session from her host located at IPv4 address 10.47.16.5. The session is named "SDP Seminar", which has a more complete title "A Seminar on the session description protocol". It also contains a web-hosted PDF file and the description of one audio and one video that are part of this proposed session.

The media contents are both available on the same media server host, whose contact name is "Jane Doe", reachable by her indicated e-mail address. The two media streams are to be transported by the basic RTP Audio Video Profile (RTP/AVP) from an IPv4 multicast address 224.2.17.12 (Time To Live of up to 127 hops) with UDP ports 49170 and 51372 for audio and video, respectively. The audio has an RTP/AVP format 0 and the video has format 99, which the SDP server also defines and maps as being a "video/H264" media codec.

15.8   Further Exploration

General discussions on computer networks and data communications are given in the book by Kurose and Ross [3]. Stallings has an introductory book [38] on some of the recent advances in the networking field, including Software Defined Networking (SDN), Network Function Virtualization (NFV), Internet of Things (IoT), etc. The Request for Comments (RFCs) for many of the network protocols can be found on the website of IETF (Internet Engineering Task Force).

15.9   Exercises

1. What is the main difference between the OSI and TCP/IP reference models? Describe the functionalities of each layer in the OSI model and their relations to multimedia communications.
2. True or False.

   (a) ADSL uses cable modem for data transmission.
   (b) To avoid overwhelming the network, TCP adopts a flow control mechanism.
   (c) TCP flow control and congestion control are both window-based.
   (d) Out of order delivery won't happen with Virtual Circuit.
   (e) UDP has a lower header overhead than TCP.
   (f) Datagram network needs call setup before transmission.
   (g) The current Internet does not provide guaranteed services.
   (h) CBR video is easier for network traffic engineering than VBR video.

3. Consider multiplexing/demultiplexing, which is one of the basic functionalities of the transport layer.

   (a) List the 4-tuple that is used by TCP for demultiplexing. For each parameter in the 4-tuple, show a scenario that the parameter is necessary.
   (b) Note that UDP only uses the destination port number for demultiplexing. Describe a scenario where UDP's scheme fits better. *Hint: The scenario is very common in multimedia applications.*

4. Find out the IP address of your computer or smartphone/tablet. Is it a real physical IP address or an internal address behind a NAT?
5. Consider a NAT-enabled home network. (a) Can two different local clients access an external web server simultaneously? (b) Can we establish two web servers (both of port 80) in this network, which are to be accessed by external computers with the basic NAT setting? (c) If we want to establish only one web server (with port 80) in this network, propose a solution and discuss its potential problems.
6. What is the key difference between IPv6 and IPv4, and why are the changes in the IPv6 necessary ? Note that the deployment of IPv6 remains limited now.

Explain the challenges in the deployment and list two interim solutions that extend the lifetime of IPv4 before IPv6 is fully deployed.

7. Discuss the pros and cons of implementing multicast in the network layer or in the application layer. Can we implement multicast in any other layer, and how?

8. What is the relation between delay and jitter? Describe a mechanism to mitigate the impact of jitter.

9. Discuss at least two alternative methods for enabling QoS routing on packet-switched networks based on a QoS class specified for any multimedia packet.

10. Consider the conventional telephone network and today's Internet.

    (a) What are the key differences between the two types of networks? Why does the Internet become the dominating network now?

    (b) What are the key challenges for multimedia over the Internet?

11. Consider the *additive increase and multiplicative decrease* (AIMD) congestion control mechanism in TCP.

    (a) Justify that AIMD ensures fair and efficient sharing for TCP flows competing for bottleneck bandwidth.

    To facilitate your discussion, you may consider the simplest case with two TCP users competing for a single bottleneck. In Fig. 15.18, the throughputs of the two users are represented by the X-axis and the Y-axis, respectively. When the aggregated throughput exceeds the bottleneck bandwidth R, congestion will happen (in the upper right side of the figure), though it will be detected after a short delay given that TCP uses packet loss as the congestion indicator. For an initial throughput of the two users, say, x_0 and y_0, where $x_0 < y_0$, you can trace the throughput change with AIMD, and show that they will eventually converge to a fair and efficient share of the bottleneck bandwidth. *Hint: There is only one such point.*

    (b) Explain whether AIMD is suitable for multimedia streaming applications or not.

    (c) Explain the relation between AIMD and TCP-Friendly Rate Control (TFRC).

12. TCP achieves reliable data transfer through retransmission.

    (a) Discuss the possible overheads of retransmission.

    (b) List two applications that retransmissions are necessary.

    (c) List two applications that retransmissions are not necessary or not possible. Explain your answer.

13. Explain why RTP does not have a built-in congestion control mechanism, while TCP does. Also, note that RTSP is independent of RTP for streaming control, i.e., using a separate channel. This is known as *out-of-band*, because the data channel and control channel are separated. Is there any advantage or disadvantage in combining both of them into a single channel?

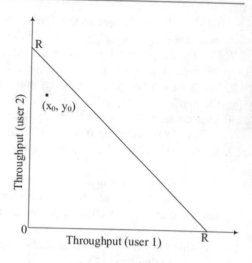

Fig. 15.18 Throughput of two TCP users sharing a bottleneck

14. Consider Fig. 15.12 that illustrates RSVP. In (d), receiver R3 decides to send an RSVP RESV message to S1. Assuming the figure specifies the complete state of the network, is the path reserved optimal for maximizing future network throughput? If not, what is the optimal path? Without modifying the RSVP protocol, suggest a scheme in which such a path will be discovered and chosen by the network nodes.

15. Consider a typical Internet telephony system of 64 kbps data rate with a sampling frequency of 8 KHz.

   (a) If the data chunks are generated every 20 ms, how many data samples are there in each data chunk, and what is the size of each chunk?
   (b) What is the header overhead when a data chunk is encapsulated into the RTP/UDP/IP protocol stack?
   (c) Assume there is only one caller and one callee, what is the bandwidth allocated to RTCP?

16. Specify on Fig. 15.13 the characteristics of *feasible* video transmission schedules. What is the *optimal transmission schedule*?

References

1. Andrew S. Tanenbaum, David J. Wetherall, *Computer Networks*, 5th edn. (Prentice Hall PTR, Upper Saddle River, New Jersey, 2012)
2. W. Stallings, *Data & Computer Communications*, 10th edn. (Prentice Hall, Upper Saddle River, New Jersey, 2013)
3. James F. Kurose, Keith W. Ross, *Computer Networking: A Top-Down Approach*, 7th edn. (Pearson, New York, 2016)

4. H. Zimmermann, OSI reference model—the ISO model of architecture for open systems interconnection. Commun. IEEE Trans. **28**(4), 425–432 (1980)
5. IEEE Standards for Local and Metropolitan Area Networks: Overview and Architecture. IEEE STD 802-1990 (1990)
6. J.F. Shoch, Y.K. Dalal, D.D. Redell, R.C. Crane, Evolution of the Ethernet local computer network. Computer **15**(8), 10–27 (1982)
7. M. Decina, E. Scace, CCITT recommendations on the ISDN: a review. Sel. Areas Commun. IEEE J. **4**(3), 320–325 (1986)
8. Jennie Bourne, Dave Burstein, *DSL: a Wiley Tech Brief (Technology Briefs Series)* (Wiley, Hoboken, 2002)
9. U.D. Black, *ATM, Volume III: internetworking with ATM*. (Prentice Hall PTR, Toronto, 1998)
10. Vaibhav Bajpai, Saba Ahsan, Jürgen Schönwälder, Jörg Ott, Measuring YouTube content delivery over IPv6. SIGCOMM Comput. Commun. Rev. **47**(5), 2–11 (2017). October
11. J. Rosenberg, J. Weinberger, C. Huitema, R. Mahy, STUN—Simple traversal of user datagram protocol (UDP) through network address translators (NATs) (2003)
12. Rolf Oppliger, Internet security: firewalls and beyond. Commun. ACM **40**(5), 92–102 (1997). May
13. J. Liu, S.G. Rao, B. Li, H. Zhang, Opportunities and challenges of peer-to-peer internet video broadcast. Proc. IEEE **96**(1), 11–24 (2008)
14. J.H. Saltzer, D.P. Reed, D.D. Clark, End-to-end arguments in system design. ACM Trans. Comput. Syst. **2**(4), 277–288 (1984). November
15. S. Deering, D. Cheriton, Multicast routing in datagram internetworks and extended LANs. ACM Trans. Comput. Syst. **8**(2), 85–110 (1990)
16. H. Eriksson, MBONE: the multicast backbone. Commun. ACM **37**(8), 54–60 (1994)
17. M.R. Macedonia, D.P. Brutzman, MBone provides audio and video across the internet. IEEE Comput. **27**(4), 30–36 (1994)
18. V. Kumar, *MBone: interactive Multimedia on the Internet* (New Riders, Indianapolis, 1995)
19. S. Paul et al., Reliable multicast transport protocol (RMTP). IEEE J. Sel. Areas Commun. **15**(3), 407–421 (1997)
20. B. Whetten, G. Taskale, An overview of reliable multicast transport protocol II. IEEE Netw. **14**, 37–47 (2000)
21. K.C. Almeroth, The evolution of multicast: from the MBone to interdomain multicast to internet2 deployment. IEEE Nctw. **14**, 10–20 (2000)
22. Y.-H. Chu, S.G. Rao, H. Zhang, A case for end system multicast. IEEE J. Sel. Areas Commun. **20**(8), 1456–1471 (2006). September
23. R.V.L. Hartley, Transmission of information. Bell Syst. Tech. J. (1928)
24. L. Zhang et al., RSVP: a new resource ReSerVation protocol. IEEE Netw. Mag. **7**(5), 8–19 (1993)
25. P.E. Veríssimo, C.E. Rothenberg, S. Azodolmolky, D. Kreutz, F.M.V. Ramos, S. Uhlig, Software-defined networking: a comprehensive survey. Proc. IEEE **103**(1), 14–76 (2015)
26. M. Casado, M.J. Freedman, J. Pettit, J. Luo, N. McKeown, S. Shenker, Ethane: taking control of the enterprise, in *Proceedings of ACM SIGCOMM*, p. 12 (2007)
27. N. McKeown, T. Anderson, H. Balakrishnan, G. Parulkar, L. Peterson, J. Rexford, S. Shenker, J. Turner, Openflow: enabling innovation in campus networks. SIGCOMM Comput. Commun. Rev. **38**(2), 69–74 (2008). March
28. R. Mijumbi, J. Serrat, J. Gorricho, N. Bouten, F. De Turck, R. Boutaba, Network function virtualization: state-of-the-art and research challenges. IEEE Commun. Surv. Tutor. **18**(1), 236–262 (2016)
29. B. Chatras, On the standardization of NFV management and orchestration APIs. IEEE Commun. Stand. Mag. **2**(4), 66–71 (2018)
30. M.S. Bonfim, K.L. Dias, S.F.L. Fernandes, Integrated NFV/SDN architectures: a systematic literature review. ACM Comput. Surv. **51**(6) (2019)

31. C. Liu, Multimedia over IP: RSVP, RTP, RTCP, RTSP, in *Handbook of Emerging Communications Technologies: the Next Decade*, ed. by R. Osso (CRC Press, Boca Raton, 2000), pp. 29–46

32. M. Krunz, Bandwidth allocation strategies for transporting variable-bit-rate video traffic. IEEE Commun. Mag. **35**(1), 40–46 (1999)

33. J.D. Salehi, Z.L. Zhang, J.F. Kurose, D. Towsley, Supporting stored video: reducing rate variability and end-to-end resource requirements through optimal smoothing. ACM SIGMETRICS **24**(1), 222–231 (1996). May

34. D.D. Clark, D.L. Tennenhouse, Architectural considerations for a new generation of protocols. SIGCOMM Comput. Commun. Rev. **20**(4), 200–208 (1990). August

35. H. Schulzrinne, J. Rosenberg, The IETF Internet telephony architecture and protocols. IEEE Netw. **13**, 18–23 (1999)

36. *Packet-based Multimedia Communications Systems*. ITU-T Recommendation H.323, November 2000 (earlier version September 1999)

37. J. Toga, J. Ott, ITU-T standardization activities for interactive multimedia communications on packet-based networks: H.323 and related recommunications. Comput. Netw. **31**(3), 205–223 (1999)

38. W. Stallings, *Foundations of Modern Networking: SDN, NFV* (IoT, and Cloud. Pearson, QoE, 2016)

Internet Multimedia Content Distribution

16

In the previous chapter, we have introduced the basic Internet infrastructure and protocols for real-time multimedia services. These protocol suites have been incorporated by client-side media players receiving streams from media servers over the Internet. The key functionality for multimedia data transfer is provided by the Real-Time Transport Protocol (RTP), including payload identification, sequence numbering for loss detection, and timestamping for playback control. Running on top of UDP, RTP itself does not guarantee Quality of Service (QoS), but relies on its companion, the RTP Control Protocol (RTCP), to monitor the network status and provide feedback for application-layer adaptation. The Real-Time Streaming Protocol (RTSP) coordinates the delivery of media objects and enables a rich set of controls for interactive playback.

Figure 16.1 shows a basic client/server-based multimedia media streaming system using the real-time protocol suite. It works fine for small-scale media content distribution over the Internet, in which media objects such as videos can be served by a single server to these users. Such an architecture has quickly become infeasible when more media contents are made available online and more users are network- and multimedia-ready.

There have been significant studies on efficient content distribution over the Internet, targeting a large number of users. Most of them were optimized for delivering conventional web objects (e.g., HTML pages or small images) or for file download. Streaming media, however, poses a new set of challenges [1–3]:

Huge size: A conventional static web object is typically in the order of 1–100K bytes. In contrast, a media object has a high data rate and a long playback duration, which combined yield a huge data volume. For example, a one-hour standard MPEG-1 video has a total volume of about 675 MB. Later standards have successfully improved the compression efficiency, but the video object sizes, even with the latest

© Springer Nature Switzerland AG 2021
Z.-N. Li et al., *Fundamentals of Multimedia*, Texts in Computer Science,
https://doi.org/10.1007/978-3-030-62124-7_16

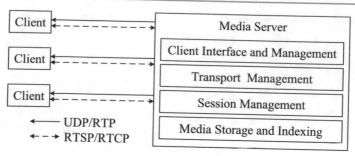

Fig. 16.1 A basic client/server-based media streaming system

H.266 compression, are still large, not to mention the new Ultra High Definition (UHD) and 3D videos.

Intensive bandwidth use: The streaming nature of delivery requires a significant amount of disk I/O and network bandwidth, sustaining over a long period.

Rich interactivity: The long playback duration of a streaming object also enables various client-server interactions. As an example, existing studies found that nearly 90% media playbacks are terminated prematurely by clients [4]. In addition, during a playback, a client often expects VCR-like operations, such as fast-forward and rewind. This implies the access rates might be different for different portions of a stream.

Many emerging applications, such as Internet TV and live event broadcast, further demand real-time multimedia streaming services with a massive audience, which can easily overwhelm the server. The scaling challenge for such multimedia content distribution is enormous. To reach 100 million viewers, delivery of TV quality video encoded in MPEG-4 (1.5 Mbps) may require an aggregate capacity of 150 Tbps. To put things into perspective, consider two large-scale Internet video broadcasts: the NBC's live telecast of Super Bowl XLIX in 2015, which has a record-high viewership within the United States with 114.4 million viewers, and the 2018 FIFA World Cup Final (France vs. Croatia), which has around 1 billion viewers worldwide. Even with low bandwidth Internet video of 500 Kbps, the NBC broadcast needs more than 57 Tbps server and network bandwidth, and the FIFA broadcast worldwide would be an order of magnitude higher. These can hardly be handled by any single server.

In this chapter, we discuss content distribution mechanisms that enable highly scalable multimedia content streaming, including proxy caching, multicast, content distribution networks, peer-to-peer, and web-based streaming.

16.1 Proxy Caching

To reduce client-perceived access latencies, as well as server/network loads, an effective means is to cache frequently used data at proxies close to clients. It also enhances the availability of objects and mitigates packet losses, as a local transmission is generally more reliable than a remote transmission. Proxy caching thus has become one of the vital components in virtually all web systems [5]. Streaming media, par-

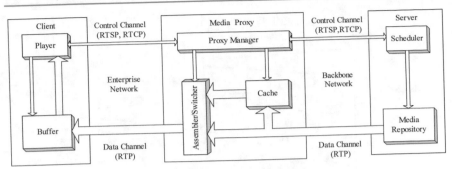

Fig. 16.2 A generic system diagram of proxy-cache-assisted media streaming using RTP/RTCP/RTSP

ticularly those pre-stored, could also benefit significant performance improvement from proxy caching, given their static nature in content and highly localized access interests.

Media caching, however, has many distinct focuses from conventional web caching [6]. On one hand, traditional web caching spends considerable effort to ensure that the copies at the origin servers and the proxy are consistent. Since the content of an audio/video object is rarely updated, such management issues are less critical in media caching. On the other hand, given the high resource requirements, caching each media object entirely at a proxy is hardly practical. It is necessary to decide which portions of which objects to be cached under cache space, disk I/O, and network I/O constraints, so that the benefit of caching outweighs the overhead for synchronizing different portions of a video stream in the proxy and in the server. A generic system diagram of proxy-cache-assisted media streaming is depicted in Fig. 16.2.

The proxy must reply to a client's PLAY request and initiate transmission of RTP and RTCP messages to the client for the cached portion, while request the uncached portion(s) from the server. Such fetching can be achieved through an RTSP Range request specifying the playback points, as illustrated in Fig. 16.3. The Range request also enables clients to retrieve different segments of a media object from multiple servers or proxies, if needed.

According to the selection of the portions to cache, we can classify existing algorithms into four categories: *sliding-interval caching*, *prefix caching*, *segment caching*, and *rate-split caching*.

16.1.1 Sliding-Interval Caching

This algorithm caches a sliding interval of a media object to facilitate consecutive accesses [7,8]. For illustration, given two consecutive requests for the same object, the first request may access the object from the server and incrementally store it into the proxy cache; the second request can then access the cached portion and release it after the access. If the two requests arrive close in time, only a small portion of

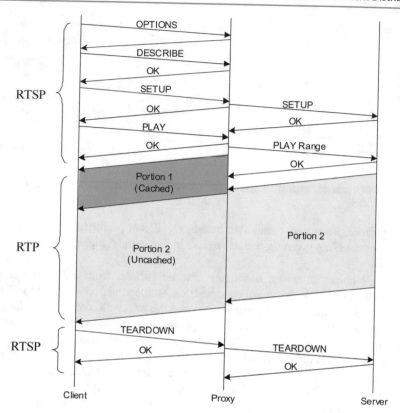

Fig. 16.3 Operations for streaming with partial caching

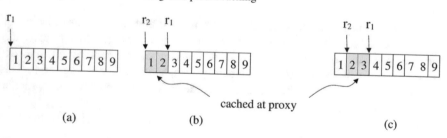

cached at proxy

(a)   (b)   (c)

Fig. 16.4 An illustration of sliding-interval caching. The object consists of 9 frames, each requiring one unit time to deliver from the proxy to a client. Requests 1 and 2 arrive at times 0 and 2, respectively. To serve request 2, only two frames need to be cached at any time instance. **a** Time 0: request 1 arrives; **b** time 1–2: frames 1 and 2 accessed by request 1 and cached; request 2 arrives; **c** time 2–3: frame 3 accessed by request 1 and cached; frame 1 read by request 2 and released

the media object needs to be cached at any time instance, and yet the second request can be completely satisfied from the proxy, as illustrated in Fig. 16.4. In general, if multiple requests for an object arrive in a short period, a set of adjacent intervals can be grouped to form a *run*, of which the cached portion will be released only after the last request has been satisfied.

Sliding-interval caching can significantly reduce the network bandwidth consumption and start-up delay for subsequent accesses. However, as the cached portion is dynamically updated with playback, the sliding-interval caching involves high disk bandwidth demands; in the worst case, it would double the disk I/O due to the concurrent read/write operations. To effectively utilize the available cache resources, the caching policy can be modeled as a two-constraint knapsack problem given the space and bandwidth requirements of each object [7], and heuristics can be developed to dynamically select the caching granularity, i.e., the run length, so as to balance the bandwidth and space usages. Given the memory spaces are large nowadays, it is also possible to allocate memory buffers to accommodate media data and thus avoiding the intensive disk read/write [8].

The effectiveness of sliding-interval caching diminishes with the increase of the access intervals. If the access interval of the same object is longer than the duration of the playback, the algorithm is degenerated to the unaffordable full-object caching. To address this issue, it is preferable to retain the cached content over a relatively long time period, and most of the caching algorithms to be discussed in the rest of this section fall into this category.

16.1.2 Prefix Caching and Segment Caching

This algorithm caches the initial portion of a media object, called *prefix*, at a proxy [9]. Upon receiving a client request, the proxy immediately delivers the prefix to the client and, meanwhile, fetches the remaining portion, the *suffix*, from the origin server and relays it to the client (see Fig. 16.5). As the proxy is generally closer to the clients than the origin server, the start-up delay for a playback can be remarkably reduced.

Segment caching generalizes the prefix caching paradigm by partitioning a media object into a series of segments, differentiating their respective utilities, and making caching decision accordingly (see Fig. 16.6). A salient feature of segment-based caching is its support to preview and such VCR-like operations as random access, fast-forward, and rewind. For example, some key segments of a media object (*hotspots*), as identified by content providers, can be cached [10]. When a client requests the object, the proxy first delivers the hotspots to provide an overview of the stream; the client can then decide whether to play the entire stream or quickly jump to

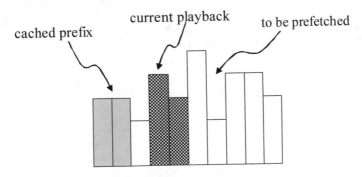

Fig. 16.5 A snapshot of prefix caching

Fig. 16.6 An illustration of
segment caching

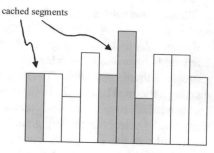

some specific portion introduced by a hotspot. Furthermore, in fast-forwarding and
rewinding operations, only the corresponding hotspots are delivered and displayed,
while other portions are skipped. As such, the load of the server and backbone network can be greatly reduced, but the client will not miss any important segments in
the media object.

The segments are not necessarily of the same length, nor predefined. One solution
is to group the frames of a media object into variable-sized segments, with the length
increasing exponentially with the distance from the start of the media stream, i.e., the
size of the segment i is 2^{i-1}, which consists of frames $2^{i-1}, 2^{i-1} + 1, ..., 2^i - 1$ [11].
The utility of a segment is calculated as the ratio of the segment reference frequency
over its distance from the beginning segment, which favors caching the initial segments as well as those with higher access frequencies. The proxy can also quickly
adapt to the changing access patterns of cached objects by discarding big chunks as
needed. If the access frequencies are not known in advance, segmentation should be
postponed as late as possible (called *lazy segmentation*), thus allowing the proxy to
collect a sufficient amount of access statistics to improve cache effectiveness [4].

16.1.3 Rate-Split Caching and Work-Ahead Smoothing

While all the aforementioned caching algorithms partition a media object along the
time axis, the rate-split caching (also known as *video staging*) [12] partitions a media
along the rate axis: the upper part will be cached at the proxy, whereas the lower part
will remain stored at the origin server (see Fig. 16.7). This type of partitioning is
particularly attractive for VBR streaming, as only the lower part of a nearly constant
rate has to be delivered through the backbone network. For a QoS network with
resource reservation, if the bandwidth is reserved based on the peak rate of a stream,
caching the upper part at the proxy significantly reduces the rate variability, which
in turn improves the backbone bandwidth utilization.

If the client has buffer capability (refer to Sect. 15.5.4 in the previous chapter),
work-ahead smoothing [13] can be incorporated into video staging to further reduce
the backbone bandwidth requirement.

Define $d(t)$ to be the size of frame t, where $t \in 1, 2, ..., N$, and N is the total
number of frames in the video. Similarly, define $a(t)$ to be the amount of data
transmitted by the video server during the playback time for frame t (for short, call
it at time t). Let $D(t)$ be the total data consumed and $A(t)$ be the total data sent at

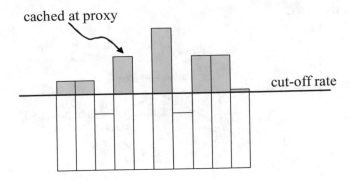

cached at proxy

cut-off rate

Fig. 16.7 An illustration of rate-split caching

time t. Formally, we have

$$D(t) = \sum_{i=1}^{t} d(i) \qquad (16.1)$$

$$A(t) = \sum_{i=1}^{t} a(i) \qquad (16.2)$$

Let the buffer size be B. At any time t, the maximum total amount of data that can be received without overflowing the buffer during time $1..t$ is $W(t) = D(t-1) + B$. Now it is easy to state the conditions for a server transmission rate that avoids buffer overflow or underflow

$$D(t) \le A(t) \le W(t) \qquad (16.3)$$

To avoid buffer overflow or underflow throughout the video's duration, Eq. (16.3) has to hold for all $t \in 1, 2, \ldots, N$. Define S to be the server transmission schedule (or plan), i.e., $S = a(1), a(2), \ldots, a(N)$. S is called a *feasible transmission schedule* if for all t, S obeys Eq. (16.3). Figure 16.8 illustrates the bounding curves $D(t)$ and $W(t)$ and shows that a constant (average)-bitrate transmission plan is not feasible for this video, because simply adopting the average bitrate would cause underflow.

When frame sizes $d(t)$ for all t are known ahead of the transmission time, the server can plan ahead to generate an optimal transmission schedule that is feasible and minimize the peak transmission rate [13]. Additionally, the plan minimizes schedule variance, optimally trying to smooth the transmission as much as possible.

We can think of this technique as stretching a rubber band from $D(1)$ to $D(N)$ bounded by the curves defined by $D(t)$ and $W(t)$. The slope of the total-data-transmitted curve is the transmission data rate. Intuitively, we can minimize the slope (or the peak rate) if, whenever the transmission data rate has to change, it does so as early as possible in the transmission plan.

As an illustration, consider Fig. 16.8 where the server starts transmitting data when the prefetch buffer is at state (a). It determines that to avoid buffer underflow at

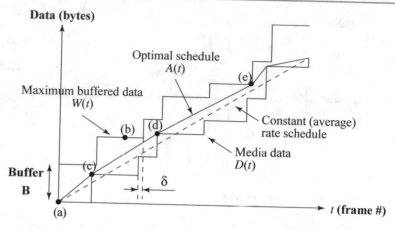

Fig. 16.8 The optimal smoothing plan for a specific video and buffer size. In this case, it is not feasible to transmit at the constant (average) data rate

point (c), the transmission rate has to be high enough to have enough data at point (c). However, at that rate, the buffer will overflow at point (b). Hence, it is necessary to reduce the transmission rate somewhere between points (c) and (b).

The earliest such point (that minimizes transmission rate variability) is point (c). The rate is reduced to a lower constant bitrate until point (d), where the buffer is empty. After that, the rate must be further reduced (to lower than the average bitrate!) to avoid overflow until point (e), when the rate must finally be increased.

Consider any interval $[p, q]$ and let $B(t)$ represent the amount of data in the buffer at time t. Then the maximum constant data rate that can be used without overflowing the buffer is given by R_{max}

$$R_{max} = \min_{p+1 \leq t \leq q} \frac{W(t) - (D(p) + B(p))}{t - p} \tag{16.4}$$

The minimum data rate that must be used over the same interval to avoid underflow is given by R_{min}

$$R_{min} = \max_{p+1 \leq t \leq q} \frac{D(t) - (D(p) + B(p))}{t - p} \tag{16.5}$$

Naturally it is required that $R_{max} \geq R_{min}$, otherwise no constant bitrate transmission is feasible over interval $[p, q]$. The algorithm to construct the optimal transmission plan starts with interval $[p, q = p + 1]$ and keeps incrementing q, each time recalculating R_{max} and R_{min}. If R_{max} is to be increased, a rate segment is created with rate R_{max} over interval $[p, q_{max}]$, where q_{max} is the latest point at which the buffer is full (the latest point in interval $[p, q]$ where R_{max} is achieved).

Equivalently, if R_{min} is to be decreased, a rate segment is created with rate R_{min} over interval $[p, q_{min}]$, where q_{min} is the latest point at which the buffer is empty.

Planning transmission rates can readily consider the maximum allowed network jitter. Suppose there is no delay in the receiving rate. At time t, $A(t)$ bytes of data

Table 16.1 Comparison of Proxy Caching Algorithms

		Sliding-interval caching	Prefix caching	Segment caching	Rate-split caching
Cached portion		Sliding intervals	Prefix	Segments	Portion of higher rate
VCR-like support		No	No	Yes	No
Resource demand	Disk I/O	High	Moderate	Moderate	Moderate
	Disk space	Low	Moderate	High	High
	Sync overhead	Low	Moderate	High	High
Performance improvement	Bandwidth reduction	High*	Moderate	Moderate	Moderate
	Start-up latency reduction	High*	High	High**	Moderate

* There is no reduction for the first request in a run
** Assume the initial segment is cached

have been received, which must not exceed $W(t)$. Now suppose the network delay is at its worst—$\delta\ sec$ maximum delay. Video decoding will be delayed by δ seconds, so the prefetch buffer will not be freed. Hence, the $D(t)$ curve needs to be modified to a $D(t - \delta)$ curve, as depicted in Fig. 16.8. This provides protection against overflow or underflow in the plan for a given maximum delay jitter.

We can either perform smoothing first and then select the cut-off rate for video staging, or select the cut-off rate and then perform smoothing. Empirical evaluation has shown that a significant bandwidth reduction can be achieved with a reasonably small cache space [12].

Table 16.1 summarizes the caching algorithms introduced above. While these features and metrics provide a general guideline for algorithm selection, the choice for a specific streaming system also largely depends on a number of practical issues, in particular, the complexity of the implementation. These algorithms are not necessarily exclusive to each other, and a combination of them may yield a better performance. For example, segment caching combined with prefix caching of each segment can reduce start-up latency for VCR-like random playback from any key-segment. If the cache space is abundant, the proxy can also devote certain space to assist workahead smoothing for variable-bit-rate (VBR) media [9].

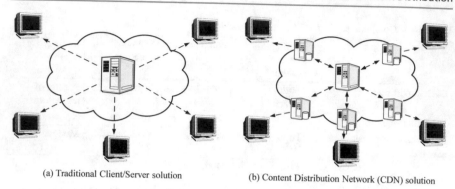

(a) Traditional Client/Server solution   (b) Content Distribution Network (CDN) solution

Fig. 16.9 Comparison between traditional single server and CDN

16.2   Content Distribution Networks (CDNs)

Caching is generally passive, in the sense that only if a user fetches an object would the object be cached at a proxy. In other words, a proxy needs time to fill up its cache space and there will be no immediate benefit for the first user accessing an object. A more proactive solution is a *Content Delivery Network* or *Content Distribution Network* (CDN) [14], which is a large geo-distributed system of servers deployed in data centers across the Internet; these servers replicate content from the origin server, pushing them to network edges closing to end users, so as to avoid middle-mile bottlenecks as much as possible (see Fig. 16.9). Originally for accelerating web accesses, this technology has rapidly evolved beyond facilitating static web content delivery. Today, CDNs serve a large fraction of the Internet data distribution, including both conventional web accesses and file download, and a new generation of applications like live streaming media, on-demand streaming media, and online social networks.

16.2.1 Request Routing and Redirection

A CDN provider hosts the content from content providers (i.e., CDN customers) and delivers the content to users of interest. This is done by mirroring the content in replicated servers and then building a mapping system accordingly. When a user types a URL into his/her browser, the domain name of the URL is translated by the mapping system into the IP address of a CDN server that stores a replica of content. The user is then redirected to the CDN server to fetch the content. This process is generally transparent to the user.

Figure 16.10 provides a high-level view of the request-routing in a CDN environment. The interaction flows are as follows:

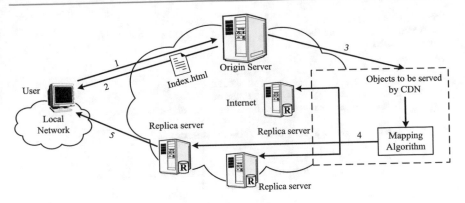

Fig. 16.10 A high-level view of request-routing in a CDN

Step 1. The user requests content from the content provider by specifying its URL in the web browser, and the request is directed to its origin server.

Step 2. When the origin server receives the request, it makes a decision to provide only the basic content (e.g., index page of the website), leaving others to CDN.

Step 3. To serve the high bandwidth demanding and frequently asked content (e.g., embedded objects fresh content, navigation bar, banner ads, etc.), the origin server redirects the user's request to the CDN provider.

Step 4. Using the mapping algorithm, the CDN provider selects the replica server.

Step 5. The selected server serves the user by providing the replicated copy of the requested object.

To assign the user to the best possible CDN server, the mapping system bases its answers on large amounts of historical and real-time data that have been collected and processed regarding the global network and server conditions. For performance optimization, the locations of the fewest hops or the highest server availability will be chosen. For cost-optimization, the least-expensive locations can be chosen instead. In real world, these two goals tend to align, as the replicated servers that are close to a user may have an advantage in both performance and cost. As an example, we use the Traceroute tools (`tracert` in MS Windows) to track the path between our institution and Hulu, a major video streaming service provider. The path tracking result is shown below.

```
tracert www.hulu.com
Tracing route to a1700.g.akamai.net [142.231.1.173]
over a maximum of 30 hops:

 1    1 ms    2 ms    1 ms    199.60.1.254
 2    3 ms    2 ms    1 ms    142.58.45.70
 3    1 ms   <1 ms   <1 ms    142.58.45.46
 4    1 ms   <1 ms   <1 ms    van-hcc1360-x-1-bby-sh1125-x-1.net.sfu.ca
                              [142.58.29.10]
 5    1 ms    1 ms    1 ms    ORAN-SFU-cr1.vantx1.BC.net
                              [142.231.1.45]
 6    2 ms    1 ms    1 ms    207.23.240.70
 7    1 ms    1 ms    1 ms    a142-231-1-173.deploy.akamaitechnologies.com
                              [142.231.1.173]
Trace complete.
```

It can be seen that the Hulu's web server we intended to reach is indeed an Akamai server (a1700.g.akamai.net), which is located in BCNet—a network that is close to our campus, while not in a network near Hulu's headquarter in California. This suggests that Hulu is using the CDN service provided by Akamai, one of the world's largest CDN providers, and the specific server offered to us is the nearest to us, which is also of low cost as our campus network is closely associated with BCNet.

The CDN provider gets paid by content providers, i.e., its customers. In turn, it pays ISPs, carriers, and network operators for hosting its servers in their data centers and for using their network resources. The amount of servers managed by a CDN provider can be very large. For example, Akamai maintains a network of 253,000 servers in 137 countries worldwide. This large overlay network of servers effectively reduces bandwidth costs and content access delays, and increases the global availability of content. It creates sizeable savings in capital and operational expenses, as the CDN customers no longer have to build their own large-size infrastructures that are not only expensive but also underutilized most of the time except during popular events. The massive amount of the monitoring data also enable CDN providers to effectively optimize the Quality of Experience of the end users through advanced data-driven and learning tools [15–17]. In addition, the CDN offers a content provider better protection from malicious attacks, because their large distributed server infrastructure can effectively absorb most of the attacking traffic.

16.2.2 Representative: Akamai Streaming CDN

For bandwidth-intensive streaming media, CDN provides better scalability by delivering the content over the last-mile from servers close to end users. The virtually unlimited resources from a large CDN also reduces the pressure on content providers to accurately predict capacity needs and enables them to gracefully absorb bursts of user demand. This is also one of the key reasons for the recent success of Cloud and Edge Computing as we will discuss in Chap. 18.

Fig. 16.11 Conceptual
relations among entry points,
reflectors, and edge servers
in Akamai's streaming CDN

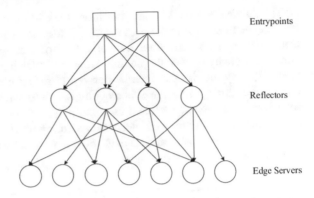

For large and comprehensive CDN operators, like Akamai, the service platform can comprise of multiple delivery networks, each being tailored to a specific type of content, e.g., static web content, dynamic news updates, or streaming media, to name but a few. At a high level, these delivery networks share a similar architecture, but the underlying technology and implementation of each system component may differ so as to best suit the specific type of content.

We now have a closer look at the Akamai's media streaming CDN, which has been widely used by such companies as Apple, Microsoft, and BBC for their video services [18]. So far Akamai has deployed approximately 253,000 servers in 137 countries around the world, delivering on average 50 Tbps of video data daily, with a record of more than 80 Tbps.

In this streaming CDN, once a live stream is captured and encoded, the stream is sent to an Akamai server, called *entry points*. To avoid having this single entry point becoming the single point of failure, multiple copies of the stream can be sent to additional entry points. If any entry point goes down, other copies can be used for recovery. The stream's packets are then transported from the entry point to a subset of *edge servers* that are close to end users.

Note that the transport system must simultaneously distribute thousands of live streams from their respective entry points to the subset of edge servers that are interested in the stream. To perform this task in a scalable fashion, an intermediate layer of servers called *reflectors* is used. Sitting between the entry points and the edge servers, each reflector can receive one or more streams from the entry points and then send those streams to one or more clusters of edge servers. This enables rapid replicating of a stream to a large number of edge clusters should the streaming event become extremely popular. The conceptual relations among entry points (sources), reflectors, and edge servers are shown in Fig. 16.11.

The use of reflectors also makes the content distribution more robust, because now there are multiple alternate paths between entry points and edge servers. If no single high-quality path is available between an entry point and an edge server, the system uses multiple link-disjoint paths that utilize different reflectors as intermediaries (see Fig. 16.11). Using the data forwarded along multiple paths, the edge servers can recover packet losses in individual paths, and forward the end users the best combined results.

The Akamai's servers residing in more than 1,600 of the world's networks also monitor the Internet in real time, gathering information about traffic, congestion, and trouble spots in the distribution network. A set of user agents will also continuously simulate users by repeatedly playing streams and testing their quality. A number of stream quality metrics can then be derived to reflect end-users' perception. These include the start-up time, the effective bandwidth to end users, the *stream availability*, which measures how often a user can play streams without failures, as well as the frequency and duration of interruptions during playback. Akamai uses this information to optimize routes and replicate data dynamically to deliver streams, offering end users high-quality experiences.

16.3   Broadcast/Multicast Video Distribution

Both proxy caching and CDN explore the temporal and geographical locality of users' interests in media objects. Such locality can also be explored through broadcast or multicast services to deliver the same content simultaneously to a massive amount of concurrent users. It works well for live media streaming. For media-on-demand services, the users' requests are asynchronous, and therefore, one single broadcast/multicast channel cannot serve the requests arriving at different times, even if they are for the same audio/video. In this section, we will introduce scalable broadcast/multicast solutions for media-on-demand with asynchronous requests and with heterogeneous users.

Note that there are subtle differences between broadcast and multicast as described in the previous chapter: the former is to all the destinations and the latter is to a group of destinations only. While broadcast is possible in air, cable networks, or local area networks, it simply cannot be carried over the global Internet. Nevertheless, we do not distinguish them here if the context is clear.

16.3.1 Smart TV and Set-Top Box (STB)

Among all possible Media-on-Demand services, the most popular is likely to be a subscription to the video: over high-speed networks, customers can specify the movies or TV programs they want and the time they want to view them. This will realize *Interactive TV* (iTV) or *Smart TV* that supports a growing number of activities, such as

- TV (basic, subscription, pay-per-view)
- Video-on-Demand (VoD)
- Information services (news, weather, magazines, sports events, etc.)
- Interactive entertainment (Online games, etc.)
- E-commerce (online shopping, stock trading, etc.)
- Digital libraries and distance education (e-learning, etc.)

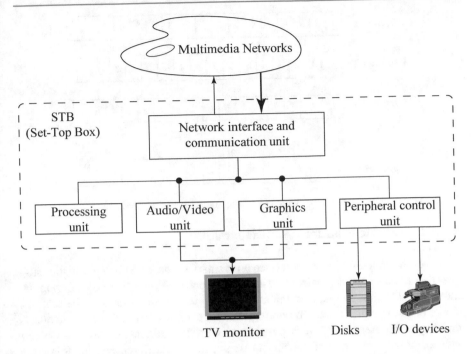

Fig. 16.12 General architecture of set-top box

The key differences between a smart TV and the conventional cable TV are (1) A smart TV invites user interactions; hence, the need for two-way traffic — downstream (content provider to user) and upstream (user to content provider), and (2) a smart TV is rich in information and multimedia services. With the penetration of *Digital Video Broadcasting* (DVB), the activities mentioned above all have emerged in today's *Multimedia Home Platform* (DVB-MHP).

To perform the above functions, a network-ready computer or a *Set-top Box* (STB) for a conventional TV set is required, which generally has the following components, as Fig. 16.12 shows:

- **Network interface and communication unit**, including a digital tuner, security devices, and a communication channel for basic navigation of Web and digital libraries as well as services and maintenance.
- **Processing unit**, including CPU, memory, and a special-purpose operating system for the STB.
- **Audio/video unit**, including audio and video decoders, Digital Signal Processor (DSP), buffers, and D/A converters.
- **Graphics unit**, supporting real-time graphics for animation and games.
- **Peripheral control unit**, including controllers for disks, audio and video I/O devices (e.g., digital video cameras), external memory card reader and writer, and so on.

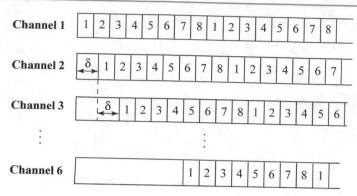

Fig. 16.13 Staggered broadcasting with $M = 8$ videos and $K = 6$ channels

16.3.2 Scalable Broadcast/Multicast VoD

Consider the Video-on-Demand service with smart TV users. Existing statistics suggest that most of the demands are usually concentrated on a few (10–20) popular movies or TV shows (e.g., new releases and top-ten movies/shows of the season). While one single multicast or broadcast channel cannot satisfy all the user requests arriving at different times, it is possible to smartly multicast or broadcast these videos, so that a number of clients can be put into the different groups following their requests [19].

One earlier solution is *Batching*, which, like sliding-interval caching, serves clients arriving close together in time using a single broadcast. An important quality measure of such a broadcast VoD service is the latency. We define the *access time* as the upper bound between the time of requesting a video and the time of actually consuming it. Apparently, the access time with batching increases with an increasing amount of client request aggregation.

Given the potentially high bandwidth of today's broadband networks and the low cost of local storage, it is conceivable that the video can be fed to the client in a relatively shorter time than its playback duration. This leads to the development of a series of *periodical broadcast* VoD solutions.

Staggered Broadcasting

For simplicity, we assume all videos are encoded using constant-bit-rate (CBR), are of the same length L (measured in time units), and will be played sequentially from beginning to end without interruption. The available high bandwidth W is divided by the playback rate b to yield the bandwidth ratio B. The bandwidth of the server is usually divided up into K logical channels ($K \geq 1$).

Assuming the server broadcasts up to M videos ($M \geq 1$), all can be periodically broadcast on all these channels with the start-time of each video staggered. This is referred to as *Staggered broadcasting*. Figure 16.13 shows an example of staggered broadcasting in which $M = 8$ and $K = 6$.

For staggered broadcasting, if the division of the bandwidth is equal among all K logical channels, then the access time for any video is $\delta = \frac{M \cdot L}{B}$. Note that the access time is actually independent of the value of K. In other words, the access time will be reduced linearly with an increased network bandwidth.

Pyramid Broadcasting

To improve the staggered broadcasting, *Pyramid broadcasting* [20] divides a video into segments of increasing sizes. That is, $L_{i+1} = \alpha \cdot L_i$, where L_i is the size (length) of Segment S_i and $\alpha > 1$. Segment S_i will be periodically broadcast on Channel i. In other words, instead of staggering the videos on K channels, the segments are now staggered. Each channel is given the same bandwidth, and the larger segments are broadcast less frequently.

Since the available bandwidth is assumed to be significantly larger than the video playback rate b (i.e., $B >> 1$), it is argued that the client can be playing a smaller Segment S_i and simultaneously be receiving a larger Segment S_{i+1}.

To guarantee continuous (noninterrupted) playback, the necessary condition is

$$playback\_time(S_i) \geq access\_time(S_{i+1}) \tag{16.6}$$

where the $playback\_time(S_i) = L_i$. Given the bandwidth allocated to each channel is $B/K \cdot b$, we have $access\_time(S_{i+1}) = \frac{L_{i+1} \cdot M}{B/K} = \frac{\alpha \cdot L_i \cdot M}{B/K}$, which yields

$$L_i \geq \frac{\alpha \cdot L_i \cdot M}{B/K} \tag{16.7}$$

Consequently,

$$\alpha \leq \frac{B}{M \cdot K} \tag{16.8}$$

The size of S_1 determines the access time for pyramid broadcasting. By default, we set $\alpha = \frac{B}{M \cdot K}$ to yield the shortest access time. The time drops exponentially with the increase in total bandwidth B, because α can be increased linearly.

A main drawback of the above scheme is the need for a large storage space on the client side, because the last two segments are typically 75–80% of the video size. Instead of using a geometric series, *Skyscraper broadcasting* [21] uses $\{1, 2, 2, 5, 5, 12, 12, 25, 25, 52, 52, ...\}$ as the series of segment sizes, to alleviate the demand on a large buffer.

Figure 16.14 shows an example of Skyscraper broadcasting with seven segments. As shown, two clients who made a request at time intervals (1, 2) and (16, 17), respectively, have their respective transmission schedules. At any given moment, no more than two segments need to be received.

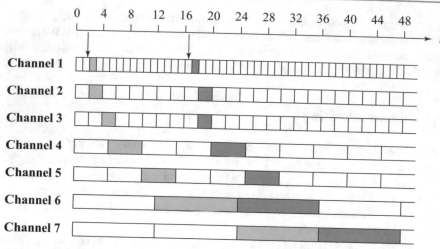

Fig. 16.14 Skyscraper broadcasting with seven segments

Harmonic Broadcasting

Harmonic broadcasting [22] adopts a different strategy. The size of all segments remains constant, whereas the bandwidth of channel i is $B_i = b/i$, where b is the video's playback rate. In other words, the channel bandwidths follow the decreasing pattern $b, b/2, b/3, \ldots b/K$. The total bandwidth allocated for delivering the video is thus

$$B = \sum_{i=1}^{K} \frac{b}{i} = H_K \cdot b \tag{16.9}$$

where K is the total number of segments, and $H_K = \sum_{i=1}^{K} \frac{1}{i}$ is the *harmonic number* of K.

Figure 16.15 shows an example of harmonic broadcasting. After requesting the video, the client is allowed to download and play the first occurrence of segment S_1 from channel 1. Meanwhile, the client will download all other segments from their respective channels.

Take S_2 as an example: it consists of two halves, S_{21} and S_{22}. Since bandwidth B_2 is only $b/2$, during the playback time of S_1, one-half of S_2 (say S_{21}) will be downloaded (prefetched). It takes the entire playback time of S_2 to download the other half (say S_{22}), just as S_2 is finishing playback. Similarly, by this time, two-thirds of S_3 has already been prefetched, and so the remaining third of S_3 can be downloaded just in time for playback from channel 3, which has a bandwidth of only $b/3$, and so on.

The advantage of harmonic broadcasting is that the Harmonic number grows slowly with K. For example, when $K = 30$, $H_K \approx 4$. If the video is 120 min long, this yields small segments—only 4 min (120/30) each. Hence, the access time for harmonic broadcasting is generally shorter than for Pyramid broadcasting, and the

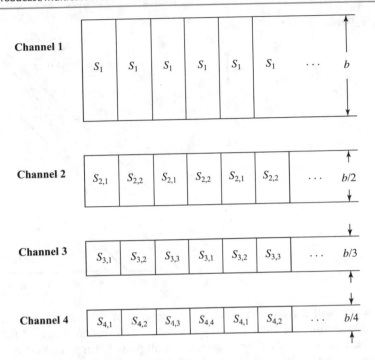

Fig. 16.15 Harmonic broadcasting

demand on total bandwidth (in this case $4b$) is modest. Its required buffer size at the client side is 37% of the entire video [22], which also compares favorably with the original pyramid broadcasting scheme.

However, the above Harmonic broadcasting scheme does not always work. For example, if the client starts to download at the second instance of S_1 in Fig. 16.15, then by the time it finishes S_1, only the second half of S_2—that is, S_{22}—is prefetched. The client will not be able to simultaneously download and play S_{21} from channel 2, since the available bandwidth is only half the playback rate.

An obvious fix to the above problem is to ask the client to delay the playback of S_1 by one slot, although it will double the access time.

Stream Merging

The above broadcast schemes are most effective when limited user interactions are expected—that is, once requested, clients will stay with the sequential access schedule and watch the video in its entirety.

Stream merging is more adaptive to dynamic user interactions, which is achieved by dynamically combining multicast sessions [23]. It still makes the assumption that the client's receiving bandwidth is higher than the video playback rate. In fact, it is common to assume that the receiving bandwidth is at least twice the playback rate, so that the client can receive two streams at the same time.

Fig. 16.16 Stream merging

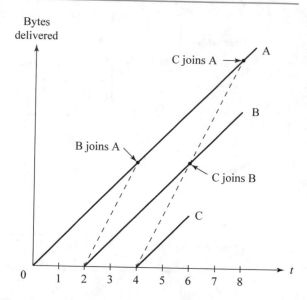

The server will deliver a video stream as soon as it receives the request from a client. Meanwhile, the client is also given access to a second stream of the same video, which was initiated earlier by another client. At a certain point, the first stream becomes unnecessary, because all its contents have been prefetched from the second stream. At this time, the first stream will merge with (or "join") the second.

As Fig. 16.16 shows, the "first stream" B starts at time $t = 2$. The solid line indicates the playback rate, and the dashed line indicates the receiving bandwidth, which is twice the playback rate. The client is allowed to prefetch from an earlier ("second") stream A, which was launched at $t = 0$. At $t = 4$, stream B joins A.

The technique of stream merging can be applied hierarchically [23]. As Fig. 16.16 shows, stream C, which started at $t = 4$, would join B at $t = 6$, which in turn joined A. The original stream B would have been obsolete after $t = 4$, since it joined A. In this case, it will have to be retained until $t = 6$, when C joins A.

A variation of stream merging is *piggybacking*, in which the playback rate of the streams is slightly and dynamically adjusted, to enable merging (piggybacking) of the streams.

16.3.3 Multi-rate Broadcast/Multicast for Heterogeneous Users

The Internet's intrinsic heterogeneity poses another challenge to multimedia broadcast/multicast. In traditional end-to-end adaptation schemes, the sender adjusts its transmission rate according to some feedback from its receiver. In a broadcast/multicast environment, this solution tends to be sub-optimal because there is no single target rate for a group of heterogeneous users.

It is thus necessary to use *multi-rate multicast*, in which the users in a multicast session can receive media data at different rates according to their respective bandwidths or processing capabilities [24]. From the viewpoint of a media source, multi-rate streams can be produced via two methods. The first is *information replication*; that is, the sender generates replicated streams for the same media content but at different rates. The second is *information decomposition*. A commonly used decomposition scheme is *cumulative layering*, in which a raw media sequence is compressed into some non-overlapping streams, or layers. The reconstructed quality is low if only one layer is decoded, but can be refined by decoding more layers. From the media compression's perspective, replication and decomposition can be implemented through transcoding and scalable audio/video coding (see Sect. 10.5.3), respectively. The remaining question is the efficient transmission of multi-rate video streams to a large group of heterogeneous users using the Internet multicast infrastructure.

Stream Replication

Stream replication can be viewed as a trade-off between single-rate multicast and multiple point-to-point connections. Its feasibility is well justified in a typical multicast environment where the bandwidths of the receivers usually follow some clustered distribution. This is because they use standard access interfaces, for example, a 1.5 Mbps ADSL, a 15 Mbps VDSL, and a 100 Mbps fiber access, or they might share some bottleneck links and hence experience the same bottleneck bandwidth. As a result, a limited number of streams can be used to match these clusters to achieve reasonably good fairness.

Due to its simplicity, stream replication has been advocated in many commercial video streaming products. For example, Netflix offers four data usage settings for its users: Low—0.3 GB/h; Medium—0.7 GB/h; High—up to 3 GB/h for HD, and 7 GB/h for Ultra HD. For YouTube, it originally offered videos at only one quality level, displayed at a resolution of 320 × 240 pixels using the Sorenson Spark codec (a variant of H.263), with mono MP3 audio. Later, 3GP format for mobile phones and high-quality mode of 480 × 360 pixels were added. Over time, a large set of options have been available on YouTube, as shown in Table 16.2, matching the demands from highly heterogeneous Internet and mobile users. The default video stream is encoded in H.264/MPEG-4 AVC format, with stereo AAC audio, and Google VP9 format has also been supported. Today, a video is typically available at multiple resolutions from 144p to 1080p HD to viewers; 1140p HD and 4K UHD are also available for certain videos.

Layered Multicast

For cumulative layered video (or scalable video), *Receiver-driven Layered Multicast* (RLM) [25] has been suggested, which takes advantage of the dynamic group concept in the IP multicast model. An RLM sender transmits each video layer over a separate multicast group. The number of layers, as well as their rates, is predetermined. Adaptation is performed only at the user's end by a probing-based scheme. Basically,

Table 16.2 A subset of YouTube media encoding options. Note that $2,560 \times 1,440$ (1440p) and $3,840 \times 2,160$ (2160p, or 4K UHD), as well as 360° videos have also been recently added

Itag value	Default container	Video encoding	Video profile	Video bitrate (Mbit/s)	Video resolution	Audio encoding	Audio bitrate (kbit/s)
13	3GP	N/A	MPEG-4 Visual	N/A	0.5	AAC	N/A
17	3GP	144p	MPEG-4 Visual	Simple	0.05	AAC	24
18	MP4	270p/360p	H.264	Baseline	0.5	AAC	96
22	MP4	720p	H.264	High	2-2.9	AAC	192
36	3GP	240p	MPEG-4 Visual	Simple	0.17	AAC	38
37	MP4	1080p	H.264	High	3-5.9	AAC	192
38	MP4	3072p	H.264	High	3.5-5	AAC	192
43	WebM	360p	VP9	N/A	0.5	Opus	128
44	WebM	480p	VP9	N/A	1	Opus	128
45	WebM	720p	VP9	N/A	2	Opus	192
46	WebM	1080p	VP9	N/A	N/A	Opus	192
82	MP4	360p	H.264	3D	0.5	AAC	96
83	MP4	240p	H.264	3D	0.5	AAC	96
84	MP4	720p	H.264	3D	2-2.9	AAC	152

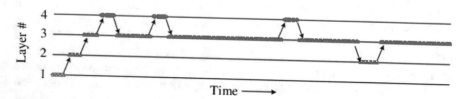

Fig. 16.17 An illustration of receiver-driven layered multicast [25]

a user periodically joins a higher layer's group to explore the available bandwidth. If packet loss exceeds a tolerable threshold after the join-experiment, i.e., congestion occurs, the user should leave the group; otherwise it will stay at the new subscription level.

Figure 16.17 shows an example of the layer joining/leaving behavior of a receiver. It started from layer 1 (the base layer), and then gradually joined enhancement layers 2, 3, and 4 as there was no congestion. After joining layer 4, however, congestion occurred and it had to leave this highest layer. It waited for a while, seeing no congestion, and re-joined layer 4. This triggered congestion again, and the receiver had to leave again, observing the network condition and planning for the next join-experiment. Note that the waiting time for the next join-experiment is longer than that of the previous one; such an *exponential backoff* ensures that the receiver will not be too aggressive in joining new layers and cause frequent congestion.

16.4 Application-Layer Multicast and Peer-to-Peer Streaming

Today the scope and reach of IP multicast remain limited, and many ISPs simply block or disable IP multicast due to various security and economic concerns [26]. The idea of using the application layer for multicast data forwarding came long time ago [27]. Though both application-layer multicast and IP multicast require intermediate nodes in the network topology to support the replication of data packets, the implementation in the application layer has much less demanding on end hosts, as compared to switches and routers in the Internet core [28]. This also motivated the design of general peer-to-peer overlay networks (i.e., over the Internet) for content sharing, including video streaming.

16.4.1 Application-Layer Multicast Tree

Figure 16.18 depicts an example of an application-layer multicast network that overlays the underlying Internet routers. When organizing the end hosts into an overlay for disseminating video streams, a series of important criteria must be considered for overlay construction and maintenance [29].

- *Overlay efficiency.* The overlay constructed must be efficient both from the network and the application perspectives. For multicast video, high bandwidth and low latencies are simultaneously required. However, for applications that are real-time but not interactive, a start-up delay of a few seconds can be tolerated.
- *Scalability and load balancing.* Since multicast sessions can scale to tens of thousands of receivers, the overlay must scale to support such large sizes, and the overhead associated must be reasonable even at large scales.
- *Self-organizing.* The overlay construction should take place in a distributed fashion and must be robust to dynamic changes in group membership. Further, the overlay must adapt to long-term variations in Internet path characteristics (such as bandwidth and latency), while being resilient to inaccuracies. The system must be self-improving in that the overlay should incrementally evolve into a better structure as more information becomes available.

Fig. 16.18 An illustration of application-level overlay network

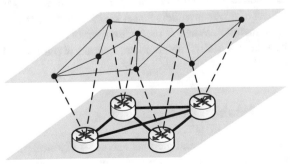

- *Node constraints.* Since the system relies on users contributing bandwidth, it is important to ensure that the total bandwidth a user is required to contribute does not exceed its inherent access bandwidth capacity. Also, a large fraction of users may stay behind NATs and firewalls—the connectivity restrictions posed by such users may severely limit the overlay capacity.

A number of proposals have emerged for application-layer multicast (also known as *overlay multicast, end-system multicast,* etc.) [29]. While they differ on a wide range of dimensions, earlier proposals were largely *push-based,* in which end-nodes are organized into structures (typically trees) for delivering data, with each data packet being disseminated using the same structure. Nodes on the structure have well-defined relationships, for example, the "parent-child" relationship in trees. Since all data packets follow this structure, it becomes critical to ensure the structure is optimized to offer good performance to all receivers. Furthermore, the structure must be maintained, as nodes join and leave the group at will—in particular, if a node crashes or otherwise stops performing adequately, all of its offspring in the tree will stop receiving packets, and the tree must be repaired. Finally, when constructing tree-based structures, loop avoidance is an important issue that must be addressed.

16.4.2 Representative: End-System Multicast (ESM)

The ESM system [30] employs a tree-based overlay protocol that is distributed, self-organizing, and performance-aware. The tree, rooted at the source, is optimized primarily for bandwidth, and secondarily for delay.

Group Management

Each ESM node maintains information about a small random subset of members, as well as information about the path from the source to itself. A new node joins the multicast session by contacting the source and retrieving a random list of members that are currently in the group. It then selects one of these members as its parent. Each node A also periodically picks one member (say B) at random, and sends B a subset of group members that A knows, along with the last timestamp it has heard for each member. When B receives a membership message, it updates its list of known members. Finally, members are deleted if their states have not been refreshed in a period.

Membership Dynamics

Dealing with graceful member leave is fairly straight-forward: the member continues forwarding data for a short period, while its children look for new parents using the parent selection method described below. This serves to minimize disruptions to the overlay. The members also send periodic control packets to their children to indicate existence.

Performance-Aware Adaptation

Each node maintains the application-level throughput it is receiving in a recent time window. If its throughput is significantly below the source rate, then it selects a new parent. One key parameter here is the *detection time*, which indicates how long a node must stay with a poor performing parent before it switches to another parent. The ESM system employs a default detection time of 5 s. The choice of this value has been influenced by the fact that a congestion control protocol is running on the data path (TCP or TFRC), and switching to a new parent thus requires going through a slow-start phase, which may take 1–2 s to get the full source rate. The protocol adaptively tunes the detection time because the nodes may not be capable of receiving the full source rate, there may be few good and available parent choices in the system, or the nodes may experience intermittent network congestion on links close to them.

Parent Selection

When a node (say A) joins the multicast overlay, or needs to make a parent change, it probes a random subset of nodes it knows. The probing is biased toward members that have not been probed or have a low delay. Each node B that responds provides the information about: (1) the throughput it is currently receiving, and delay from the source; (2) whether it is degree-saturated or not; and (3) whether it is a descendant of A. The probe also enables A to determine the round-trip time (RTT) to B. A waits for responses for a timeout period of 1 s, a large enough value of RTT, so as to maximize the number of responses received from members. From the responses A receives, it eliminates its descendants and the members that are saturated.

For each node B that has not been eliminated, A evaluates the performance (throughput and delay) it expects to receive if B were chosen as a parent. For example, the expected application throughput is the minimum of the throughput B is currently seeing and the available bandwidth of the path between B and A if the estimate is available. History of the past performance is maintained—if A has previously chosen B as parent, then it has an estimate of the bandwidth of the path between B and A. If the bandwidth to the nodes is not known, then A picks a parent based on delay. A identifies the node B that could best improve performance, and switches to the parent B either if the estimated application throughput is high enough for A to receive a higher quality stream, or if B maintains the same bandwidth level as A's current parent, but improves delay. The latter heuristic helps to increase tree efficiency by clustering nearby nodes.

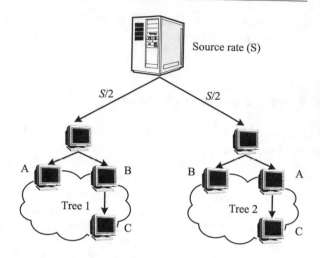

Fig. 16.19 A multi-tree application-layer multicast with two trees. Note that node A in Tree 1 and that in Tree 2 are physically the same nodes, so for node B or C

Multi-tree Structure

The tree-based designs are perhaps the most natural approach. One concern with them is that the failure of nodes, particularly those close to the root may disrupt the data delivery to a large number of users, and potentially result in poor transient performance. Furthermore, a majority of nodes are leaves in the structure, and their out-going bandwidth is not being utilized. More resilient structures, in particular, multi-tree [31,32], thus have been introduced.

In a multi-tree, the source encodes the stream into sub-streams and distributes each sub-stream along a particular overlay tree. The quality experienced by a receiver depends on the number of sub-streams that it receives. There are two key advantages of the multi-tree solution. First, the overall resiliency of the system is improved, as a node is not completely disrupted by the failure of an ancestor on a single tree. Second, the potential bandwidth of all nodes can be utilized, as long as each node is not a leaf in at least one tree.

Figure 16.19 illustrates how the multicast content is delivered with a multi-tree approach using two trees. The source distributes a stream rate $S/2$ over each tree, where S is the source rate. C receives $S/2$ from tree, with potentially different parents to reconstruct the original content. Nodes A and B each can contribute a bandwidth $S/2$, and allocate their bandwidth in Tree2 and Tree1, respectively. In a single tree approach, it is hard to utilize the contributions from these nodes. It can be seen that Akamai's streaming CDN that we examined earlier also uses a multi-tree solution (with reflectors), despite that the nodes there are dedicated replication servers.

16.4.3 Peer-to-Peer Mesh Overlay

Peer-to-peer (P2P) further extends the application-layer multicast paradigm by taking advantage of the ability of participating end hosts, or *peers*, in a multicast group to contribute their uplink bandwidth. It was first brought to the spotlight by the advent of Napster (1998) and Gnutella (2001). Later, the design philosophy in the highly popular BitTorrent software has converged with academic solutions in application-layer multicast, and a new generation of data-driven peer-to-peer streaming protocols on random mesh topologies emerged [29].

Data-driven or mesh overlay designs sharply contrast to tree-based application-layer multicast in that they do not construct and maintain an explicit structure for delivering data. The underlying argument is that, rather than constantly repair a structure in a highly dynamic peer-to-peer environment, we can use the availability of data to guide the data flow. In comparison, the tree-based application-layer multicast adopts a more rigid design [33,34], in that the structure of each tree needs to be actively managed as peers join and leave the session.

A naive approach to distribute data without explicitly maintaining a structure is to use *gossip algorithms* [35]. In a typical gossip algorithm, a node sends a newly generated message to a set of randomly selected nodes; these nodes do similarly in the next round, and so do other nodes until the message is spread to all. The random choice of gossip targets achieves resilience to random failures and enables decentralized operation. However, gossip cannot be used directly for video content distribution because its random push may cause significant redundancy with the high-bandwidth video. Without an explicit structure support, start-up, and transmission delays can be significant, too.

To handle this, mesh overlay adopts a *pull-based* technique for data dissemination. More explicitly, each node maintains a set of partners, and periodically exchange data availability information with the partners. The node may then retrieve the unavailable data from one or more partners, or supply the available data to partners. Redundancy is avoided, as the node pulls data only if it has not already possessed it. Since any data segment may be available at multiple partners, the overlay is robust to failures—departure of a node simply means its partners will use other partners to receive data segments. Finally, the randomized partnerships imply that the potential bandwidth available between the peers can be fully utilized. As a result, pull-based protocols are much simpler to design and more amenable to real-world implementations. It has the potential to scale with group size, as greater demand also generates more resources.

It is worth noting that CDN and peer-to-peer can work together as well. For example, in addition to using Akamai's own servers, Akamai can deliver content from other end-users' cache through the Akamai NetSession Interface, a download manager used to reduce download time and to increase quality.

There are common issues existing in both peer-to-peer file sharing and video streaming; for example, pricing for uploading/downloading and copyright protection. The key difference is the timing constraints that a streaming protocol must accommodate: if video segments do not arrive in time, they are not useful when it comes to the time of playing them back. Thus, an important component of the data-driven overlay is a scheduling algorithm, which strives to schedule the segments that must be downloaded from various partners to meet the playback deadlines.

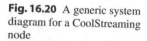

Fig. 16.20 A generic system diagram for a CoolStreaming node

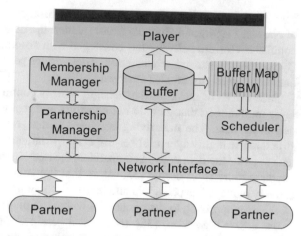

16.4.4 Representative: CoolStreaming

CoolStreaming [36] is the first large-scale data-driven peer-to-peer system being deployed in real world for video streaming. Other successful companies such as PPLive, PPStream, and UUSee also adopted mesh-based pull techniques to deliver live or on-demand media content to millions of users.

Figure 16.20 depicts the system diagram of a CoolStreaming node, which consists of three key modules: (1) a membership manager, which helps the node maintain a partial view of other overlay nodes; (2) a partnership manager, which establishes and maintains the partnership with other known nodes; (3) a scheduler, which schedules the transmission of video data. For each segment of a video stream, a CoolStreaming node can be either a receiver or a supplier, or both, depending dynamically on this segment's availability information, which is periodically exchanged between the node and its partners. An exception is the video source, which, as the *origin node*, is always a supplier. It could be a dedicated video server, or simply an overlay node that has a live video program to distribute.

Membership and Partner Management

Each CoolStreaming node has a unique identifier, such as its IP address, and maintains a membership cache (*mCache*) containing a partial list of the identifiers for active nodes in the overlay. In a basic node joining algorithm, a newly joined node first contacts the origin node, which randomly selects a *deputy node* from its mCache and redirects the new node to the deputy. The new node can then obtain a list of partner candidates from the deputy, and contacts these candidates to establish its partners in the overlay.

This process is generally viable because the origin node persists during the lifetime of streaming and its identifier/address is universally known. The redirection enables

Fig. 16.21 An illustration of partnerships in CoolStreaming with A being the origin node. The partnership is bidirectional, except for node A, which serves as a supplier only. For example, node F is a partner of nodes B, C, and E, and node E is a partner of nodes B, F, and H

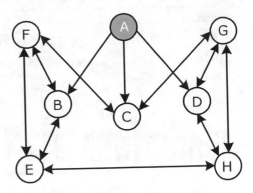

more uniform partner selections for newly joined nodes, and greatly minimized the origin node's load.

A key practical issue here is how to create and update the mCache. To accommodate overlay dynamics, each node periodically generates a membership message to announce its existence; each message is a 4-tuple $< seq\_num, id, num\_partner, time\_to\_live >$, where seq_num is the sequence number of the message, id is the node's identifier, $num\_partner$ is its current total number of partners, and $time\_to\_live$ records the remaining valid time of the message. Upon receiving the message of a new seq_num, the node updates its mCache entry for node id, or create the entry if not existing. The entry is a 5-tuple $< seq\_num, id, num\_partner, time\_to\_live, last\_update\_time >$, where the first four components are copied from the received membership message, and the fifth is the local time of the last update for the entry.

The following two events also trigger updates of an mCache entry: (1) the membership message is forwarded to other nodes through gossiping; and (2) the node serves as a deputy and the entry is included in the partner candidate list. In either case, $time\_to\_live$ is decreased by $current\_local\_time - last\_update\_time$. If the new value is less than or equal to zero, the entry will be removed; otherwise, $num\_partner$ will be increased by one in the deputy case.

Buffer Map Representation and Exchange

An example of partnership in an overlay is shown in Fig. 16.21. As said, neither the partnerships nor the data transmission directions are fixed. More explicitly, a video stream is divided into segments of a uniform length, and the availability of the segments in the buffer of a node can be represented by a *Buffer Map* (BM). Each node continuously exchanges its BM with the partners, and then schedules which segment is to be fetched from which partner accordingly.

Timely and continuous segment delivery is crucial to media streaming, but not to file download. In BitTorrent, the download phases of the peers are not synchronized, and new segments from anywhere in the file are acceptable. In CoolStreaming, the playback progress of the peers is roughly synchronized, and any segment downloaded

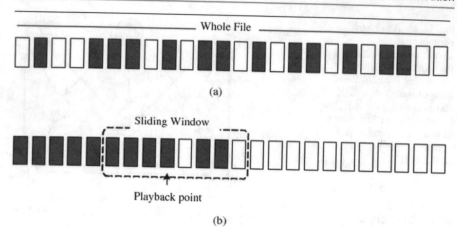

Fig. 16.22 Buffer snapshots of BitTorrent **a** and CoolStreaming **b**, where shaded segments are available in the buffer

after its playback time will be useless. A *sliding window* thus represents the active buffer portion, as shown in Fig. 16.22.

Suggested by experimental results, CoolStreaming adopts a sliding window of 120 segments, each of 1 s video. A BM thus consists of a bitstring of 120 bits, each indicating the availability of a corresponding segment. The sequence number of the first segment in the sliding window is recorded by another two bytes, which can be rolled back for extra long video programs (>24 h).

Scheduling Algorithm

Given the BMs of a node and its partners, a schedule is then generated for fetching the expected segments from the partners. For a homogeneous and static network, a simple round-robin scheduler may work well, but for a dynamic and heterogeneous network, a more intelligent scheduler is necessary. Specifically, the scheduling algorithm strikes to meet two constraints: the playback deadline for each segment, and the heterogeneous streaming bandwidth from the partners. If the first constraint cannot be satisfied, then the number of segments missing deadlines should be kept minimum, so as to maintain a continuous playback. This problem is a variation of the *Parallel machine scheduling*, which is known NP-hard. It is thus not easy to find an optimal solution, particularly considering that the algorithm must quickly adapt to the highly dynamic network conditions. CoolStreaming resorts to a simple heuristic of fast response time.

The heuristic first calculates the number of potential suppliers for each segment (i.e., the partners containing or to contain the segment in their buffers). Since a segment with less potential suppliers is more difficult to meet the deadline constraints, the algorithm determines the supplier of each segment starting from those with only one potential supplier, then those with two, and so forth. Among the multiple potential suppliers, the one with the highest bandwidth and enough available time is selected.

As an example, consider node F in Fig. 16.21, which has partners B, C, and E. Assume that a buffer map contains only four segments and they are 1000, 0010, 0011, and 0101 for nodes F, B, C, and E, respectively. That is, node F has only segment 1 available in its local buffer, but 2, 3, 4 missing. Among the three missing segments, segment 2 has only one supplier (node E) and segments 3 and 4 each has two suppliers (B,C for 3, and C,E for 4). As such, node F will schedule to fetch segment 2 first, from node E. It will then fetch segments 3 and 4. For segment 3, between the two potential suppliers B and C, the one with the higher bandwidth will be scheduled. The same strategy applies to segment 4.

Given a schedule, the segments to be fetched from the same supplier are marked in a BM-like bit sequence, which is sent to that supplier, and these segments are then delivered in order through the TCP-Friendly Rate Control (TFRC) protocol. There have been many enhancements to the basic scheduling algorithm in CoolStreaming, and existing studies have also suggested that the use of advanced *network coding* can possibly enable optimal scheduling [37,38].

Failure Recovery and Partnership Refinement

A CoolStreaming node can depart gracefully or accidentally due to an unexpected failure. In either case, the departure can be easily detected after an idle time of TFRC or BM exchange, and an affected node can quickly react through re-scheduling using the BM information of the remaining partners. Besides this built-in recovery mechanism, CoolStreaming also lets each node periodically establish new partnerships with nodes randomly selected from its local membership list. This operation serves two purposes: First, it helps each node maintain a stable number of partners in the presence of node departures; Second, it helps each node explore partners of better quality, e.g., those constantly having a higher upload bandwidth and more available segments.

16.5 Web-Based Media Streaming

Although peer-to-peer has proven to be highly scalable in video delivery, there are critical issues for peer-to-peer system deployment by content providers: (1) *Ease-of-use*. In peer-to-peer streaming, the users are usually required to install customized client software or plugins to be able to cache the video contents watched and exchange the contents with others—this is not user-friendly given that today's users are so familiar with using web browsers to consume Internet contents directly. (2) *Copyright*. In a peer-to-peer streaming system, the users exchange contents with each other autonomously—it is very difficult for the content providers to control the copyright in the video streaming.

Peer-to-peer also relies on peers' contribution to the system. In real world, there are many *free riders* who do not want to contribute their resources. Even if the peers are willing to contribute, the upload bandwidth of many peers are often constrained given the asymmetricity in such access networks as ADSL. Moreover, the data exchanged between peers need to traverse NAT in both directions through open ports, which is known to be difficult as we have discussed in the previous chapter. They are exposed to security threats too, and are often blocked by firewalls.

As the underlying protocol for web transactions, the *Hyper Text Transfer Protocol* (HTTP) is generally firewall-friendly because almost all firewalls are configured to support connections for web transactions. HTTP server resources are also widely available commodity, and therefore, supporting HTTP-based streaming for the massive audience can be cost-effective using the existing web infrastructure.

HTTP was not initially designed for streaming applications. It does not provide signaling mechanisms for interactive streaming control, and its underlying transport protocol, TCP, was not designed for continuous media, either. The key to support streaming with HTTP is to break the overall media stream into a sequence of small HTTP-based file downloads; each download includes one short chunk of an overall potentially unbounded stream. Using a series of the HTTP's GET commands, a user can progressively download the small files while playing those already being downloaded. Any damaged or delayed block will have limited impact, thus ensuring continuous playback. This process is illustrated in Fig. 16.23.

HTTP does not maintain session states on the server, either. Therefore, provisioning a large number of clients does not impose significant cost on server resources. This is quite different from RTP/RTCP/RTSP-based streaming that have to maintain per-session states. Yet each client can keep a record of its playback progress, and the progressive download also allows a client to seek to a specific position in the media stream by downloading the corresponding file, or more precisely, performing an HTTP's byte range request for the file, realizing similar functionalities offered by RTSP.

HTTP-based streaming has been implemented in commercial products. Today, representative online video providers, e.g., Netflix and YouTube is using HTTP to stream their videos to the users. Besides the superiorities we have mentioned earlier,

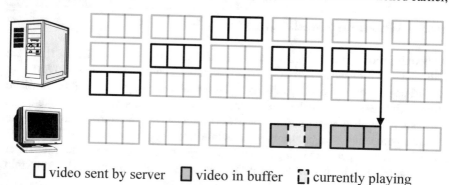

☐ video sent by server   ▨ video in buffer   ⌐⌐ currently playing

Fig. 16.23 An illustration of HTTP streaming

HTTP streaming is benefiting from the rapidly expanding capacity and dropping pricing of today's CDNs. Specifically, the emergence of such ad-based business models have boosted the importance of the users' Quality of Experience (QoE), as there is a crucial interplay between QoE and user engagement. As a consequence, content providers, with an objective of generating more revenue, are caring more about the visual quality and stability of their streaming service to maximize user engagement. Since today's CDNs are mainly designed and optimized to serve the web contents [39], HTTP streaming is capable of using the existing infrastructure to deliver high-quality media with low cost, better stability and security, and simpler interfaces [40]. More recently, to support real-time interactive multimedia application, WebRTC, which is built within the Web system but does not rely on HTTP, has been developed as well.

In the meantime, the HyperText Markup Language (HTML) also evolves to better support multimedia content distribution. The new version, HTML5, has added the new <video>, <audio> and <canvas> elements, as well as support for scalable vector graphics (SVG) content, so as to natively include and handle multimedia and graphical content.

16.5.1 Dynamic Adaptive Streaming over HTTP (DASH)

HTTP-based streaming, also under the name of *HTTP Adaptive Streaming (HAS)* or *HTTP Live Streaming (HLS)*, uses different manifest and segment formats under different implementations; hence, to receive the content from each server, a device must support its corresponding proprietary client protocol. There is a demand for standardization so that different devices can inter-operate. The heterogeneous networks and devices also require the media streaming to be dynamic and adaptive. To this end, the *Dynamic Adaptive Streaming over HTTP* (DASH) standard has been developed by the MPEG group. Work on DASH started in 2010; it became an international standard in November 2011, which was officially published as ISO/IEC 23009-1:2012 in April 2012 and was further revised in 2019 as ISO/IEC 23009-1:2019 [41].

DASH defines a set of implementation protocols across the servers, clients, and description files. In DASH, a video stream is encoded and divided into multiple segments, including initialization segments that contain the required information for initializing the media decoder, and media segments that contain the following data: (1) the media data, and (2) the stream access point, indicating where the client decoder can play. Subsegments can also be used such that a user can download them using the HTTP's partial GETS command, which includes a Range header field, requesting that only part of the entity be transferred.

A *Media Presentation Description* (MPD) describes the relation of the segments and how they form a video presentation, which facilitates segment fetching for continuous playback. A sample MPD file is shown below.

```
<MPD>
<BaseURL>http://www.baseurl_1.com</BaseURL>//Destination URL(s)
<Period>
   <AdaptationSet>//Video Set
      <Representation bandwidth="4190760" height="1080" width="1920">
         <SegmentInfo>... </SegmentInfo>//Quality_1
      </Representation>
      <Representation bandwidth="2073921" height="720" width="1280">
         <SegmentInfo>... </SegmentInfo>//Quality_2
      </Representation>
   </AdaptationSet>
   <AdaptationSet>//Audio Set
      <Representation bandwidth="127234" sampleRate="44100">
         <SegmentInfo>... </SegmentInfo>//Quality_1
      </Representation>
   </AdaptationSet>
</Period>
</MPD>
```

All BaseURLs are shown at the beginning of the MPD file. A client can analyze this part to acquire the destination URLs and then pull streaming data from the servers. In this simple MPD file, there is only one period, which consists of video and audio adaptation sets. There are two video sequences with different resolutions and bitrates, allowing the client to choose based on local and networking conditions. There is only one soundtrack (of 44.1 KHz sampling rate) for the video stream. In a more complex scenario, the audio set may have several soundtracks with different languages and bitrates, too.

Figure 16.24 illustrates the DASH-based streaming with the two video levels and one audio level. The client can use an adaption algorithm to choose the appropriate audio and video levels. During playback, the adaption algorithm will monitor the local and network status, so as to achieve the best possible QoS, e.g., request lower quality segments when the network bandwidth is low, and higher quality if enough bandwidth is available.

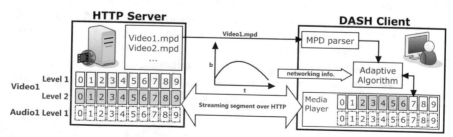

Fig. 16.24 A scenario of DASH-based streaming

DASH is available on Apple's Safari, Google's Chrome, and Microsoft's Edge, and has been used by such major streaming services as YouTube and Netflix. As compatible clients become available, it promises to be widely adopted in a wide range of devices and platforms. DASH is also codec agnostic. It allows seamless containerization of MPEG-4 as well as the improved HEVC video codec (*i.e.*, H.265) and Google VP9 .

As an open standard, the exact adaptation algorithm in DASH is left to specific implementations, where two key issues are to be addressed [42]:

- *Rate adaptation components.* DASH only defines the segmentation and the file description, and leaves rate adaptation for either the client or the server to implement. The client may utilize multi-path and multi-server approaches to receive video segments [43]. Such receiver-driven approaches, customized in the application layer, are highly flexible and scalable. On the other hand, the servers could also be able to adaptively change the bitrate for its clients, based on the perception of the client download speed and server load.
- *Rate adaptation strategies.* Rate adaptation strategies determine how different versions of segments are received by clients, to achieve such objectives as streaming stability, fairness, and high quality [44]. The rate adaptation strategies are conventionally based on predefined rules with parameters observed from the server or the client sides. Such rules are deterministic and tailored to specific network configurations. More recently, data-driven solutions that leverage advanced prediction or learning algorithms, together with the big data from the server, client, and network, have been developed [45–48]. Without using predefined rules, a self-learning client can dynamically adapt its behavior by interacting with the environment, so as to optimize the QoE under a vast range of highly dynamic network settings.

16.5.2 Common Media Application Format (CMAF)

The underlying transport layer protocol for HTTP streaming is TCP. Each request requires opening a separate connection using TCP to transfer requests and then closing it after completing the request. As such, the default HTTP streaming setup can incur a latency of 5 s or more, which can be a challenge for many live broadcast or interactive applications.

To reduce the latency, CMAF (Common Media Application Format) [49] optimizes the media data packaging and delivering process in HTTP streaming protocols, so as to achieve a lower latency, which is theoretically as low as 1 s, though practically around 2–3 s.

CMAF is not a protocol, but a standardized container that defines the segment format, as well as codecs and media profiles, which will be used to deliver the stream. Delivery and presentation are abstracted by a reference application model that accommodates a broad spectrum of practical implementations, including HTTP Live Streaming (HLS) and MPEG DASH (Dynamic Adaptive Streaming over HTTP).

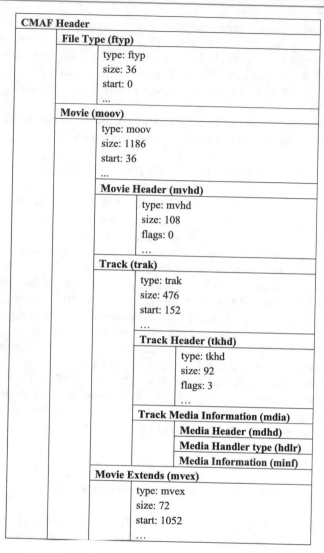

Fig. 16.25 A sample CMAF header

In CMAF, a track contains encoded media samples, including audio, video, and subtitles, which are stored in a container derived from the ISO Base Media File Format (ISO_BMFF). A sequence of one or more consecutive fragments from the same track forms a *segment*, which is further divided into *chunks*. To transmit each CMAF segment, a POST request containing a CMAF header is first sent to the media server. This header includes the information for initializing the track, as shown in Fig. 16.25.

Each chunk is then sent through HTTP 1.1 Chunked Transfer Encoding, as soon as its encoding and packaging are done. As such, a segment is progressively delivered

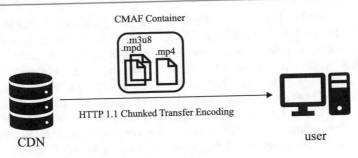

CMAF Container

HTTP 1.1 Chunked Transfer Encoding

CDN   user

Fig. 16.26 A CMAF-based streaming example

as each chunk is ready, which is much faster than loading and processing the entire segment.

To retrieve the segment, the player uses the manifest or playlist associated with a stream to establish a connection with the media and then makes a GET request.

Figure 16.26 shows a concrete example of CMAF-based streaming. The encoder wraps a single .mp4 file and a .mpd & .m3u8 file inside one CMAF container. Then, the data is pushed via HTTP 1.1 Chunked Transfer Encoding via the CDN to the end user's video player. The system no longer waits for a request because, with progressive delivery, the chunks are pushed to the player as soon as they are available.

Besides compatibility with HTTP streaming, the container design also eliminates the need for redundant storage of the same video file by using one file format that could be used by different devices. In this example, the different manifest (.mpd & .m3u8) use the same .mp4 files, making them platform independent. To synchronize video streams on different platforms, a server timestamp is incorporated, which, together with data buffering, enables the video players to play the same frame at the exact same time.

16.5.3 Web Real-Time Communication (WebRTC)

The more recent highly interactive applications, such as personal LiveChat/live cast and AR/VR streaming, have even tighter demands on the latency, preferably in the sub-second range. There is a need for a new streaming media protocol designed specifically to meet these real-time demands yet still being compatible with the web.

WebRTC [50] is an open source project and specification that was initiated by Google in 2011, targeting the high-quality real-time communication applications within browsers on both desktop and mobile platforms, as well as lightweight Internet-of-Things devices with limited capabilities. It has since been supported by many other companies, including Microsoft, Mozilla, Opera, Apple, etc., and has offered users the ability to communicate from within these Web and mobile computing platforms without the need for complicated plugins or additional hardware. Today, WebRTC is also being standardized through the World Wide Web Consortium (W3C) and the Internet Engineering Task Force (IETF).

Another related and once widely used protocol for this purpose is the Real-Time Messaging Protocol (RTMP) . It is, however, proprietary, relying on Adobe's Flash Media Server support, which is less popular in modern browsers now.

WebRTC works directly in a web browser without requiring additional plugins or downloading native apps. Yet unlike CMAF, it does not use HTTP to send media data. Instead, it establishes a connection using UDP and delivers encrypted media data over RTP. As such, WebRTC incurs a latency of 500 milliseconds or less. Also, since this process creates a direct connection between two web browsers, it follows a peer-to-peer communication structure. Multiple browsers can use WebRTC to communicate with one another, allowing for video conferencing between individuals and small groups.

Since UDP packs can be lossy and out-of-order, WebRTC uses NACK to retransmit the most critical packets or mitigate the impact through loss concealment. Readers can refer to Chap. 17 for a wide range of concealment techniques.

WebRTC exposes Javascript APIs to developers for writing real-time multimedia applications with diverse interactions. The key components includes

- `getUserMedia`: accesses the audio and video media data from storage or device;
- `RTCPeerConnection`: performs signal processing, codec handling, peer-to-peer communication, security, and bandwidth management;
- `RTCDataChannel`: enables bidirectional communication of data between peers with low latency.

Figure 16.27 shows an example of a simple LiveChat application that builds on WebRTC. Specifically, it starts from creating a connection through `RTCPeerConnection`, which exchanges necessary information using the Session Description Protocol (SDP) and DTLS (Datagram Transport Layer Security). This

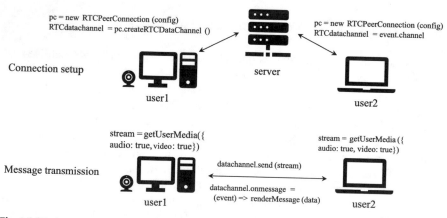

Fig. 16.27 A sample livechat application using WebRTC

signaling discovers where the two users are and how to connect. `RTCDataChannel` is then used for data communication with RTP. After setting up the connection, the sender can use `getUserMedia()` to generate a video stream from its camera, and send the stream via `datachannel.send()`. The receiver then uses the `datachannel.onmessage` handler to render the incoming messages. Voice and video are sent through separate channels and their bitrates are adaptable depending on the channel statistics. There is also a function, `getStats`, which gathers the statistics about the WebRTC session for media adaptation.

In summary, if latency is not very critical, e.g., more than 5 s or even 10 s is acceptable, DASH is preferred in terms of scalability, quality, and compatibility. It also works well with that existing caching infrastructure that largely focuses on the Web data. This latency can be further reduced by using CMAF, to around 2–3 s. If latency is critical and the application involves many real-time peer-to-peer interactions, WebRTC will be a good choice, with a latency around 0.5–1.0 s [51]. For instance, one of the most popular video conferencing tools, Zoom, started using WebRTC's DataChannels recently.

16.6 Exercises

1. Consider prefix caching with a total proxy cache size of S and N videos of user access probabilities $r_1, r_2, \ldots, r_N$, respectively. Assume that the *utility* of caching for each video is given by function $U(l_i)$ where l_i is the length of the cached prefix for video i. Develop an algorithm to optimize the total utility of the proxy. You may start from the simple case where $U(l_i) = l_i \cdot r_i$.

2. For the optimal work-ahead smoothing technique, how would you algorithmically determine at which point to change the planned transmission rate? What is the transmission rate?

3. Consider again the optimal work-ahead smoothing technique. It was suggested that instead of using every video frame, only frames at the beginning of statistically different compression video segments can be considered. How would you modify the algorithm (or video information) to support that?

4. Discuss the similarities and differences between proxy caching and CDN. Is it beneficial to utilize both of them in a system?

5. Discuss the similarities and differences between a CDN for web content distribution and that for multimedia streaming. What is the role of *reflectors* in Akamai's streaming CDN?

6. For Staggered broadcasting, if the division of the bandwidth is equal among all K logical channels ($K \geq 1$), show that the access time is independent of the value of K.

7. Given the available bandwidth of each user, $b_1, b_2, \ldots, b_N$, in a multicast session of N users, and the number of replicated video streams, M, develop a solution to allocate the bitrate to each stream, B_i, $i = 1, 2, \ldots, M$, so that the average *inter-receiver fairness* is maximized. Here, the inter-receiver fairness for user j

is defined as $max \frac{B_k}{b_j}$ where $B_k \le b_j, k = 1, 2, \ldots, M$, i.e., the video stream of the highest rate that user j can receive.

8. In a multicast scenario, too many receivers sending feedback to the sender can cause a *feedback implosion* that would block the sender. Suggest two methods to avoid the implosion and yet provide reasonably useful feedback information to the sender.

9. To achieve TCP-Friend Rate Control (TFRC), the Round-Trip Time (RTT) between the sender and the receiver must be estimated (see Sect. 15.3.2). In the unicast TFRC, the sender generally estimates the RTT and hence the TCP-friendly throughput, and accordingly controls the sending rate. In a multicast scenario, who should take care of this and how? Explain your answer.

10. In this question, we explore the scalability of peer-to-peer, as compared to the client/server. We assume that there is one server and N users. The upload bandwidth of the server is S bps, and the download bandwidth of user i is D_i bps, $i = 1, 2, \ldots, N$. There is a file of size M bits to be distributed from the server to all the users.

   (a) Consider the client/server architecture. Each user is now a client that is directly served by the server. Calculate the time to distribute the file to all the users.

   (b) Now consider the peer-to-peer architecture. Each user is now a peer, who can either download directly from the server or from other peers. Assume that the upload bandwidth of user i for serving other peers is U_i bps, $i = 1, 2, \ldots, N$. Calculate the time to distribute the file to all the users.

   (c) Using the results, explain in what conditions will peer-to-peer scale better (with more users) than client/server. Are these conditions naturally satisfied in the Internet?

11. Discuss the similarities and differences between peer-to-peer file sharing and peer-to-peer live streaming. How will such differences affect the implementation of a peer-to-peer living streaming? And how will they affect the calculation in the previous question.

12. Consider tree-based and mesh-based overlays for peer-to-peer streaming.

   (a) Discuss the pros and cons of each of them.

   (b) Why is the pull operation used in mesh-based overlays?

   (c) Propose a solution (other than those introduced in the book) to combine them toward a hybrid overlay. You may target different application scenarios, e.g., for minimizing delay or for multi-channel TV broadcast where some users may frequently change channels.

13. Consider Skype, a popular Voice-over-IP (VoIP) application that used to use peer-to-peer communication. The peers in Skype are organized into a hierarchical overlay network, with the peers being classified as *super peers* or *ordinary peers*,

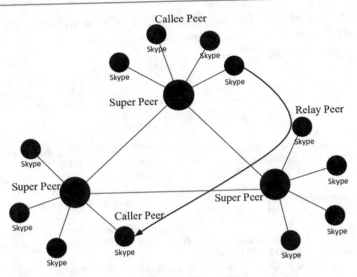

Fig. 16.28 An illustration of the Skype peer-to-peer network

as illustrated in Fig. 16.28. When two Skype users (caller and callee) need to set up a call, both the ordinary peers and the super peers can serve as relays.

(a) Skype generally uses UDP for audio streams but TCP for control messages. What kind of control messages are necessary for Skype peers, and why is TCP used?

(b) Explain the benefit of distinguishing super peers and ordinary peers.

(c) Besides one-to-one calls, Skype also supports multi-party conferences. How many copies of audio streams would be delivered in an N user conference, if each user needs to send its copy of stream to all others?

(d) Note that this number can be high. Skype reduces it by asking each user to send its stream to the conference initiator, who will combine the streams into one stream and then forward to each of the other users. How many streams are to be forwarded to the whole conference now? Discuss the pros and cons of this solution, and suggest improvements.

(e) Recently, Skype has migrated its service to the Microsoft Azure Cloud platform (See Chap. 18). Discuss the pros and cons of such migration.

14. One important reason that HTTP was not traditionally used for media streaming is that the underlying TCP has highly fluctuated transmission rate (the *saw-tooth behavior*), and during severe congestion or channel errors, it may persistently block the data pipe. Explain how DASH addresses these problems. Also, discuss other supports for streaming that are missing in the basic HTTP but addressed in DASH.

15. Identify two popular Internet multimedia applications that use DASH and WebRTC, respectively. Discuss how you find them out and why they are chosen for the particular application.

References

1. B. Li, Z. Wang, J. Liu, W. Zhu, Two decades of internet video streaming: a retrospective view. ACM Trans. Multimed. Comput. Commun. Appl. **9**(1s), 33, 1–33, 20 (2013)
2. D. Wu, Y.T. Hou, W. Zhu, Y.-Q. Zhang, J.M. Peha, Streaming video over the internet: approaches and directions. IEEE Trans. Circuits Syst. Video Technol. **11**(3), 282–300 (2001)
3. D. Wu, Y. Thomas Hou, Y.-Q. Zhang, Transporting real-time video over the internet: challenges and approaches. Proc. IEEE **88**(12), 1855–1877 (2000)
4. S. Chen, B. Shen, S. Wee, X. Zhang, Designs of high quality streaming proxy systems, in *INFO-COM 2004. Twenty-Third Annual Joint Conference of the IEEE Computer and Communications Societies*, vol. 3, pp. 1512–1521 (2004)
5. X. Jianliang, J. Liu, B. Li, Xiaohua Jia, Caching and prefetching for web content distribution. Comput. Sci. Eng. **6**(4), 54–59 (2004). July
6. J. Liu, X. Jianliang, Proxy caching for media streaming over the internet. IEEE Commun. Mag. **42**(8), 88–94 (2004)
7. R. Tewari, H.M. Vin, A. Dany, D. Sitaramy, Resource-based caching for web servers. in *Proceedings SPIE/ACM Conference on Multimedia Computing and Networking*, pp. 191–204 (1998)
8. S. Chen, B. Shen, Y. Yan, S. Basu, X. Zhang, SRB: shared running buffers in proxy to exploit memory locality of multiple streaming media sessions. in *Proceedings 24th International Conference on Distributed Computing Systems, 2004*, pp. 787–794 (2004)
9. S. Sen, J. Rexford, D. Towsley, Proxy prefix caching for multimedia streams. In *INFOCOM '99. Eighteenth Annual Joint Conference on IEEE Computer and Communications Societies*, vol. 3, pp. 1310–1319 (1999)
10. H. Fabmi, M. Latif, S. Sedigh-Ali, A. Ghafoor, P. Liu, L.H. Hsu, Proxy servers for scalable interactive video support. Computer **34**(9), 54–60 (2001)
11. K.-L. Wu, P.S Yu, J.L Wolf, Segment-based proxy caching of multimedia streams, in *Proceedings of the 10th International Conference on World Wide Web*, pp. 36–44. (ACM, 2001)
12. Z.-L. Zhang, Y. Wang, D.H.C. Du, D. Shu, Video staging: a proxy-server-based approach to end-to-end video delivery over wide-area networks. IEEE/ACM Trans. Netw. **8**(4), 429–442 (2000)
13. J.D. Salehi, Z.L. Zhang, J.F. Kurose, D. Towsley, Supporting stored video: reducing rate variability and end-to-end resource requirements through optimal smoothing. ACM SIGMETRICS **24**(1), 222–231 (1996). May
14. V.K. Adhikari, Y. Guo, F. Hao, V. Hilt, Z.-L. Zhang, M. Varvello, M. Steiner, Measurement study of Netflix, Hulu, and a tale of three CDNs. IEEE/ACM Trans. Netw. **23**(6), 1984–1997 (2015). December
15. A. Ganjam, J. Jiang, X. Liu, V. Sekar, F. Siddiqi, I. Stoica, J. Zhan, H. Zhang, C3: internet-scale control plane for video quality optimization, in *Proceedings of the 12th USENIX Conference on Networked Systems Design and Implementation, NSDI'15*, pp. 131—144, (USA, USENIX Association, 2015)
16. M.K. Mukerjee, D. Naylor, J. Jiang, D. Han, S. Seshan, H. Zhang, Practical, real-time centralized control for CDN-based live video delivery, in *Proceedings of the 2015 ACM Conference on Special Interest Group on Data Communication, SIGCOMM '15*, pp. 311—324. (New York, NY, USA, 2015)
17. J. Jiang, S. Sun, V. Sekar, H. Zhang, Pytheas: enabling data-driven quality of experience optimization using group-based exploration-exploitation. in *Proceedings of the 14th USENIX Con-

ference on Networked Systems Design and Implementation, NSDI'17, pp. 393—406. (USA, USENIX Association,2017)

18. E. Nygren, R.K. Sitaraman, J. Sun, The Akamai network: a platform for high-performance internet applications. SIGOPS Oper. Syst. Rev. **44**, 2–19 (2010). Aug
19. A. Hu, Video-on-demand broadcasting protocols: a compreshensive study, in *Proceedings IEEE INFOCOM* (2001)
20. S. Viswanathan, T. Imielinski, Pyramid broadcasting for video on demand service, in *IEEE Conference on Multimedia Computing and Networking*, pp. 66–77 (1995)
21. K.A. Hua, S. Sheu, Skyscraper broadcasting: a new broadcasting scheme for metropolitan video-on-demand systems, in *Proceedings of the ACM SIGCOMM*, pp. 89–100 (1997)
22. L. Juhn, L. Tseng, Harmonic broadcasting for video-on-demand service. IEEE Trans. Broadcast. **43**(3), 268–271 (1997)
23. D. Eager, M. Vernon, J. Zahorjan, Minimizing bandwidth requirements for on-demand data delivery. IEEE Trans. Knowl. Data Eng. **13**(5), 742–757 (2001)
24. B. Li, J. Liu, Multirate video multicast over the internet: an overview. Netw. IEEE **17**(1), 24–29 (2003)
25. S. McCanne, V. Jacobson, M. Vetterli, Receiver-driven layered multicast, in *Conference on Proceedings of the Applications, Technologies, Architectures, and Protocols for Computer Communications, SIGCOMM '96*, pp. 117–130 (1996)
26. C. Diot, B.N. Levine, B. Lyles, H. Kassem, D. Balensiefen, Deployment issues for the IP multicast service and architecture. IEEE Netw. **14**(1), 78–88 (2000)
27. S. Sheu, K.A Hua, W. Tavanapong, Chaining: a generalized batching technique for Video-On-Demand systems. in *Proceedings of the IEEE International Conference on Multimedia Computing and Systems* (1997)
28. M. Hosseini, D.T. Ahmed, S. Shirmohammadi, N.D. Georganas, A survey of application-layer multicast protocols. IEEE Commun. Surv. Tutor. **9**(3), 58–74 (2007)
29. J. Liu, S.G. Rao, B. Li, H. Zhang, Opportunities and challenges of peer-to-peer internet video broadcast. Proc. IEEE **96**(1), 11–24 (2008)
30. Y.-H. Chu, S.G. Rao, H. Zhang, A case for end system multicast. IEEE J. Sel. A. Commun. **20**(8), 1456–1471 (2006)
31. V.N. Padmanabhan, H.J. Wang, P.A. Chou, K. Sripanidkulchai, Distributing streaming media content using cooperative networking, in *Proceedings of the 12th International Workshop on Network and Operating Systems Support for Digital Audio and Video, NOSSDAV '02*, pp. 177–186 (ACM, New York, NY, USA, 2002)
32. M. Castro, P. Druschel, A.-M. Kermarrec, A. Nandi, A. Rowstron, A. Singh, Splitstream: high-bandwidth multicast in cooperative environments, in *Proceedings of the Nineteenth ACM Symposium on Operating Systems Principles, SOSP '03*, pp. 298–313. (ACM, New York, NY, USA, 2003)
33. V. Venkataraman, K. Yoshida, P. Francis, Chunkyspread: heterogeneous unstructured tree-based peer-to-peer multicast, in *Proceedings of the 5th International Workshop on Peer-to-Peer Systems (IPTPS)*, pp. 2–11 (2006)
34. N. Magharei, R. Rejaie, Y. Guo, Mesh or multiple-tree: a comparative study of live P2P streaming approaches, in *Proceedings of the IEEE INFOCOM* (2007)
35. P.T. Eugster, R. Guerraoui, A.M. Kermarrec, L. Massoulié, From epidemics to distributed computing. IEEE Comput. **37**, 60–67 (2004)
36. X. Zhang, J. Liu, B. Li, T.P. Yum, CoolStreaming/DONet: a data-driven overlay network for peer-to-peer live media streaming, in *INFOCOM 2005. 24th Annual Joint Conf. of the IEEE Computer and Communications Societies Proceedings of the IEEE*, vol. 3, pp. 2102–2111 (2005)
37. Z. Liu, C. Wu, B. Li, S. Zhao, UUSee: large-scale operational on-demand streaming with random network coding, in *Proceedings of the IEEE INFOCOM* (2010)
38. M. Wang, B. Li, R^2: random push with random network coding in live peer-to-peer streaming. IEEE J. Sel. Areas Commun. (2007)

39. G. Pallis, A. Vakali, Insight and perspectives for content delivery networks. Commun. ACM **49**(1), 101–106 (2006)

40. C. Ge, N. Wang, W.K. Chai, H. Hellwagner, QoE-assured 4K HTTP live streaming via transient segment holding at mobile edge, IEEE J. Sel. Areas Commun. (JSAC) (018)

41. ISO/IEC JTC 1/SC 29/WG 11 (MPEG). 23009-1: Dynamic adaptive streaming over HTTP (2019)

42. Y. Sani, A. Mauthe, C. Edwards, Adaptive bitrate selection: a survey. IEEE Commun. Surv. Tutor. **19**(4), 2985–3014 (2017)

43. S. Gouache, G. Bichot, A. Bsila, C. Howson, Distributed & adaptive HTTP streaming, in *Proceedings of the ICME* (2011)

44. S. Akhshabi, A.C Begen, Constantine Dovrolis. An experimental evaluation of rate-adaptation algorithms in adaptive streaming over HTTP, in *Proceedings of the ACM MMSys* (2011)

45. M. Claeys, S. Latré, J. Famaey, F. De Turck, Design and evaluation of a self-learning HTTP adaptive video streaming client. IEEE Commun. Lett. **18**(4), 716–719 (2014)

46. Y. Sun, X. Yin, J. Jiang, V. Sekar, F. Lin, N. Wang, T. Liu, B. Sinopoli, Cs2p: improving video bitrate selection and adaptation with data-driven throughput prediction, in *Proceedings of the 2016 ACM SIGCOMM Conference, SIGCOMM '16*, pp. 72–285. (ACM, 2016)

47. H. Mao, R. Netravali, M. Alizadeh, Neural adaptive video streaming with pensive, in *Proceedings of the Conference of the ACM Special Interest Group on Data Communication, SIGCOMM '17*, pp. 197–210. (Association for Computing Machinery, New York, NY, USA, 2017)

48. Z. Akhtar, Y.S. Nam, R. Govindan, S. Rao, J. Chen, E. Katz-Bassett, B. Ribeiro, J. Zhan, H. Zhang, Oboe: auto-tuning video ABR algorithms to network conditions, in *Proceedings of the 2018 Conference of the ACM Special Interest Group on Data Communication, SIGCOMM '18*, pp. 44–58. (ACM, New York, NY, USA, 2018)

49. ISO/IEC. 23000-19: multimedia application format (MPEG-A)—Part 19: common media application format (CMAF) for segmented media (2018)

50. C. Jennings, et al., WebRTC 1.0: real-time communication between browsers, in *W3C Candidate Recommendation* (2019)

51. B. Jansen, T. Goodwin, V. Gupta, F. Kuipers, G. Zussman, Performance evaluation of WebRTC-based video conferencing. SIGMETRICS Perform. Eval. Rev. **45**(3), 56–68 (2018)

Multimedia Over Wireless and Mobile Networks

The rapid developments in computer and communication technologies have made *ubiquitous computing* a reality. From cordless phones in the early days to later cellular phones, wireless mobile communication has been the core technology that enables anywhere and anytime information access and sharing. The new generation of smart mobile devices that emerged only in the recent years are driving the revolution further. Multimedia over wireless and mobile networks share many similarities as over the wired Internet; yet the unique characteristics of wireless channels and the frequent movement of users also pose new challenges that must be addressed.

17.1 Characteristics of Wireless Channels

Wireless radio transmission channels are far more error-prone than wire-line communications are. In this section, we briefly present the most common radio channel models to gain insight into the cause of errors and to classify the types of bit errors, the amount, and their patterns. More details can be found in [1–3].

Various effects cause radio signal degradation in the receiver side. They can be classified as short-range and long-range effects. Accordingly, path loss models are available for long-range atmospheric attenuation channels, and fading models are available for short-range degradation.

17.1.1 Path Loss

For long-range communication, the signal loss is dominated by atmospheric attenuation. Depending on the frequency, radio waves can penetrate the ionosphere (>3 GHz) and establish *line-of-sight* (LOS) communication, or for lower frequencies

© Springer Nature Switzerland AG 2021
Z.-N. Li et al., *Fundamentals of Multimedia*, Texts in Computer Science,
https://doi.org/10.1007/978-3-030-62124-7_17

reflect off the ionosphere and the ground, or travel along the ionosphere to the receiver. Frequencies over 3 GHz (which are necessary for satellite transmissions to penetrate the ionosphere, and have also been explored by 5G cellular standard) experience gaseous attenuations, influenced primarily by oxygen and water (vapor or rain).

The free-space attenuation model for LOS transmission is in inverse proportion to the square of distance (d^2) and is given by the Friis radiation equation

$$S_r = \frac{S_t G_t G_r \lambda^2}{(4\pi^2)d^2 L} \tag{17.1}$$

S_r and S_t are the received and transmitted signal power, G_r and G_t are the antenna gain factors, λ is the signal wavelength, and L is the receiver loss. It can be shown that if we assume ground reflection, attenuation increases to be proportional to d^4.

Another popular medium-scale (urban city size) model is the *Hata model*, which is empirically derived based on the Okumura path loss data in Tokyo. The basic form of the path loss equation in dB is given by

$$L = A + B \cdot \log_{10}(d) + C. \tag{17.2}$$

Here, A is a function of the frequency and antenna heights, B is an environment function, and C is a function depending on the carrier frequency. Again, d is the distance from the transmitter to the receiver.

For satellite communications that are mainly affected by rain, meteorological rainfall density maps can be used to communicate with the region. Attenuation is computed according to the amount of rainfall in the area on the given date.

17.1.2 Multipath Fading

Fading is a common phenomenon in wireless (and especially mobile) communications, in which the received signal power (suddenly) drops [1]. Signal fading occurs due to reflection, refraction, scattering, and diffraction (mainly from moving objects), as illustrated in Fig. 17.1. *Multipath fading* occurs when a signal reaches the receiver via multiple paths (some of them bouncing off buildings, hills, and other objects). Because they arrive at different times and phases, the multiple instances of the signal can cancel each other, causing the loss of signal or connection. The problem becomes more severe when higher data rates are explored.

For indoor channels, the radio signal power is generally lower, and there are more objects in a small place; some are moving. Hence, multipath fading is the main factor for signal degradation. In outdoor environments, refraction, diffraction, and scattering effects are also the important causes of signal degradation, mostly by the ground and buildings.

A multipath model probabilistically states the received signal amplitude, which varies according to whether the signals superimposed at the receiver are added

Fig. 17.1 An example of multipath

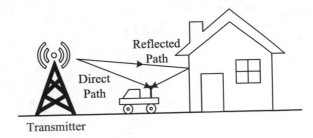

Transmitter

Fig. 17.2 The Gilbert-Elliot two-state Markov Chain model. State 0: Good; State 1: Bad

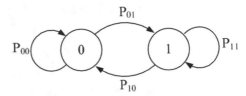

destructively or constructively. The *Doppler spread* of a signal is defined as the distribution of the signal power over the frequency spectrum (the signal is modulated at a specific frequency bandwidth). When the Doppler spread of the signal is small enough, the signal is coherent—that is, there is only one distinguishable signal at the receiver. This is typically the case for narrowband signals. When the signal is wideband, different frequencies of the signal have different fading paths, and a few distinguishable signal paths are observed at the receiver, separated in time.

For narrowband signals, the most popular models are *Rayleigh fading* and *Rician fading*. The Rayleigh fading model assumes an infinite number of signal paths with *non line-of-sight* (NLOS) to the receiver for modeling the probability density function P_r of received signal amplitude r:

$$P_r(r) = \frac{r}{\sigma^2} \cdot e^{\frac{-r^2}{2\sigma^2}} \tag{17.3}$$

where σ is the standard deviation of the probability density function. Although the number of signal paths is typically not too large, the Rayleigh model does provide a good approximation when the number of paths is over 5.

A Rayleigh fading channel can be approximated using a Markov process with a finite number of states, referred to as a *Finite State Markov Channel* [4]. The simplest form, known as the Gilbert-Elliot model [5], is with only two states, representing the *good* and the *bad* channel conditions. As illustrated in Fig. 17.2, state 0 has no error and state 1 is erroneous, and the wireless channel condition switches between them with transition probabilities P_{00} to P_{11}. It captures the short-term bursty nature of wireless errors, and has been widely used in simulations.

A more general model that assumes LOS is the Rician model. It defines a *K-factor* as a ratio of the signal power to the scattered power—that is, K is the factor by which the LOS signal is greater than the other paths. The Rician probability density function P_c is

Fig. 17.3 Rician PDF plot with K-factor = 0, 1, 3, 5, 10, and 20

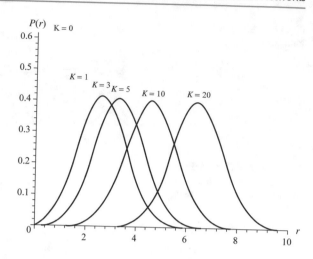

$$P_c(r) = \frac{r}{\sigma^2} \cdot e^{\frac{-r^2}{2\sigma^2} - K} \cdot I_0(\frac{r}{\sigma}\sqrt{2K}), \quad \text{where } K = \frac{s^2}{2\sigma^2} \tag{17.4}$$

As before, r and σ are the signal amplitude and standard deviation, respectively, and s is the LOS signal power. I_0 is a modified Bessel function of the first kind with 0 order. Note that when $s = 0$ ($K = 0$), there is no LOS, and the model thus reduces to a Rayleigh distribution. When $K = \infty$, the model reflects the popular *additive white Gaussian noise* (AWGN) conditions. Figure 17.3 shows the Rician probability density function for K-factors of 0, 1, 3, 5, 10, and 20, respectively, with a standard deviation of $\sigma = 1.0$.

For a wideband signal, the fading paths are more empirically driven. One way is to model the amplitude as a summation over all the paths, each having randomized fading. The number of paths can be 7 for a closed-room environment (six walls and LOS) or a larger number for other environments. An alternative technique of modeling the channel fading is by measuring the channel impulse response.

A similar technique is to use *rake receivers*, through which multiple radio receivers are tuned to signals with different phases and amplitudes, to recompose the transmission that is split to different distinguishable paths. The signal at each rake receiver is added up to achieve better SNR. To tune the rake receivers to the proper fading paths, a special *pilot channel* sends a well-known pilot signal, and the rake receivers are adjusted to recognize that symbol on each fading path.

17.2 Wireless Networking Technologies

Like wired networks, there is a large family of wireless networks, using different technologies to combat fading and pass loss and covering geographical areas of different sizes. In a wide-area cellular network, a field is covered by a number of *cells*. Each mobile terminal in a cell contacts its *access point (AP)* or *base station*

(BS), which serves as a gateway to the network. The access points themselves are connected through high-speed wired lines, or wireless networks or satellites that form a backbone *core network*. When a mobile user moves out of the range of the current access point, a *handoff* (or *handover*) is required to maintain the communication. The size of a cell is typically of 1,000 m in cities, but can be larger (macrocell) or smaller (microcell) depending on the location and the density of users. The whole network of cells collectively cover a city- or nation-wide area or even beyond, ensuring anywhere connection.

A *wireless local area network* (WLAN), on the other hand, covers a much shorter range, generally within 100 m. Given the short distances, the bandwidth can be very high while the access cost and power consumption can be low, making them ideal for use within a house or an office building. Many modern home entertainment systems are built around wireless local area networks. Public WLAN accesses have also been offered by many airports, shops, restaurants, or even city-wide.

In this section, we provide an overview of the different generations of wireless cellular networks and wireless local area networks.

17.2.1 Cellular Wireless Mobile Networks: 1G–5G

The first commercial cellular network was launched in Japan by Nippon Telegraph and Telephone (NTT) in 1979, initially in the metropolitan area of Tokyo. In the past 40 years, cellular wireless mobile networking has encompassed a rapid evolution from the first generation (1G) to today's advanced 5th generation (5G) that supports ultra-high bandwidth and ultra-low latency communication. They have become the foundation of today's mobile Internet, offering anytime and anywhere data access and sharing, particularly for multimedia content.

1G Cellular Analog Wireless Networks

The very early wireless communication networks were used mostly for voice communications, such as telephone and voice mail. The first-generation (1G) cellular phones used an analog technology with *frequency-division multiple access* (FDMA), in which each user is assigned a separate frequency channel during the communication. Its standards were *Advanced Mobile Phone System* (AMPS) in North America, *Total Access Communication System* (TACS), and *Nordic Mobile Telephony* (NMT) in Europe and Asia, respectively.

Figure 17.4 illustrates a sample geometric layout for an FDMA cellular system. A cluster of seven hexagon cells can be defined for the covered cellular area. As long as each cell in a cluster is assigned a unique set of frequency channels, interference from neighboring cells will be negligible. For clarity, the cells from the first cluster are marked with thicker borders.

Fig. 17.4 An example of geometric layout for an FDMA cellular system with a cluster size of seven hexagon cells

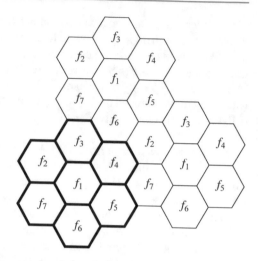

2G Cellular Networks: GSM and Narrowband CDMA

Starting from the second-generation (2G) wireless networks, digital technologies had replaced the analog technologies. The digital cellular networks adopted two competing technologies since 1993: *time division multiple access* (TDMA) and *code division multiple access* (CDMA). The *Global System for Mobile communications* (GSM) [6], which was based on TDMA, was the most widely used worldwide.

TDMA and GSM (Global System for Mobile Communications)

As the name suggests, TDMA creates multiple channels in multiple time slots while allowing them to share the same carrier frequency. In practice, TDMA is generally combined with FDMA —that is, the entire allocated spectrum is first divided into multiple carrier frequency channels, each of which is further divided in the time dimension by TDMA.

GSM was established by the *European Conference of Postal and Telecommunications Administrations* (CEPT) in 1982, with the objective of creating a TDMA-based standard for a mobile communication network capable of handling millions of subscribers and providing roaming services throughout Europe.

By default, the GSM network is circuit switched, and its data rate is limited to 9.6 Kbps, which is hardly useful for general data services (certainly not for multimedia data). *General Packet Radio Service* (GPRS), developed in 1999, supports packet-switched data over GSM wireless connections, so users are "always connected."

Code Division Multiple Access (CDMA)

Code division multiple access (CDMA) [7] is a major breakthrough in wireless communications. It is a *spread spectrum* technology, in which the bandwidth of a signal is spread before transmission. In its appearance, the spread signal might be indistinguishable from background noise, and so it has distinct advantages of being secure

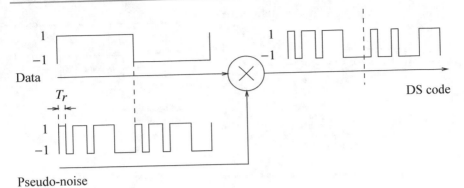

Pseudo-noise

Fig. 17.5 Spreading in direct sequence (DS) spread spectrum

and robust against intentional interference (known as *jamming*). Spread spectrum is applicable to digital as well as analog signals, because both can be modulated and "spread." The earlier generation of cordless phones and cellular phones, for example, used analog signals. However, it is the digital applications, in particular CDMA, that made the technology popular in modern wireless data networks.

The foundation of CDMA is *direct sequence (DS) spread spectrum*. Unlike FDMA, in which each user is supposed to occupy a unique frequency band at any moment, multiple CDMA users can make use of the same (and full) bandwidth of the shared wideband channel during the entire period of transmission! A common frequency band can also be allocated to multiple users in all cells—in other words, providing a reuse factor of $K = 1$. This has the potential to greatly increase the maximum number of users, as long as the interference from them is manageable.

As Fig. 17.5 shows, for each CDMA transmitter a unique *spreading code* is assigned to a direct sequence (DS) spreader. The spreading code (also called *chip code*) consists of a stream of narrow pulses called *chips*, with a bit width of T_r. Its bandwidth B_r is on the order of $1/T_r$.

The spreading code is multiplied with the input data by the DS spreader, as in Fig. 17.6. When the data bit is 1, the output DS code is identical to the spreading code, and when the data bit is 0 (represented by -1), the output DS code is the inverted spreading code. As a result, the spectrum of the original narrowband data is spread, and the bandwidth of the DS signal is:

$$B_{DS} = B_r \qquad (17.5)$$

The despreading process involves taking the product of the DS code and the spreading sequence. As long as the same sequence is used as in the spreader, the resulting signal is the same as the original data.

To separate the receivers for multiple access, i.e., CDMA, *orthogonal codes* can be used. As an example, consider spreading codes for two receivers: $(1, -1, -1, 1)$ and $(-1, 1, 1, -1)$, which are orthogonal to each other (in practice, the code length can be much longer); that is, their inner product is zero. Assume the data bit for receiver 1 is x and that for receiver 2 is y. The output DS code for receiver 1 is

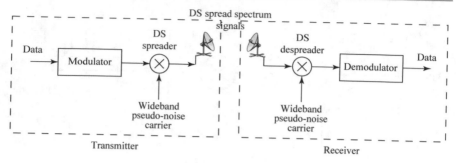

Fig. 17.6 Transmitter and receiver of direct sequence (DS) spread spectrum

$x \cdot (1, -1, -1, 1)$ and that for receiver 2 is $y \cdot (-1, 1, 1, -1)$. The sender combines them together, sending $x \cdot (1, -1, -1, 1) + y \cdot (-1, 1, 1, -1)$. The decoding results at receivers 1 and 2 using their respective codes will be

1)
$$(x \cdot (1, -1, -1, 1) + y \cdot (-1, 1, 1, -1)) \cdot (1, -1, -1, 1) = 4x \quad \text{(at receiver}$$

2)
$$(x \cdot (1, -1, -1, 1) + y \cdot (-1, 1, 1, -1)) \cdot (-1, 1, 1, -1) = 4y \quad \text{(at receiver}$$

which, after normalization by 4 (the spreading code length), become x and y for receivers 1 and 2, respectively. In other words, there is no interference between them.

Because T_r is small, B_r is much wider than the bandwidth B_b of the narrowband signal.

In practice, to support more users and achieve better spectrum utilization, nonorthogonal *Pseudo-random Noise* (PN) sequences can be used as codes. This is based on the observation that in general not all users are active in a cell. Since the effective noise is the sum of all other users' signals, as long as an adequate level of "average case" interference is maintained, the quality of the CDMA reception is guaranteed. Such a *soft capacity* makes CDMA much more flexible than TDMA or FDMA with hard capacity only, accommodating more users when necessary and alleviating the undesirable dropping of ongoing calls when reaching the capacity limit.

3G Cellular Networks: Wideband CDMA

The 2G cellular networks were mainly designed for voice communications with circuit switching and had very limited support for Internet data access, not to mention multimedia services. Starting from the third generation (3G), multimedia services have become the core issues for the cellular network development.

The 3G standardization process started in 1998, and was predominantly using *Wideband CDMA (WCDMA)*. The key differences in WCDMA air interface from a narrowband CDMA air interface are

- To support bitrates up to 2 Mbps, a wider channel bandwidth is allocated. The WCDMA channel bandwidth is 5 MHz, as opposed to 1.25 MHz for IS-95 and other earlier standards.
- To effectively use the 5 MHz bandwidth, longer spreading codes at higher chip rates are used. The chip rate specified is 3.84 Mcps, as opposed to 1.2288 Mcps.
- WCDMA supports variable bitrates, from 8 kbps up to 2 Mbps. This is achieved using variable-length spreading codes and time frames of 10 ms, at which the user data rate remains constant but can change from one frame to the other—hence bandwidth on demand.

The bandwidth made available by 3G networks gives rise to applications not previously available to mobile phone users. Examples include online maps, online gaming, mobile TV, and instant picture/video content sharing. The multimedia nature of these 3G wireless services also calls for a rapid development of new generations of handsets, where support for high-quality video, better software and user interface, and longer battery life are key factors. From the 3G era, these smartphones and tablets have greatly changed the way for people to interact with mobile devices and even their social behaviors.

4G Cellular Networks: LTE

IMT-Advanced, as defined by ITU, has been commonly viewed as the guideline for 4G standards [8]. An IMT-Advanced cellular system fulfills the following requirements:

- Based on an all-IP packet-switched network.
- Peak data rates of up to 100 Mbps for high mobility and up to 1 Gbps for nomadic/local wireless access.
- Dynamically share and use the network resources to support more simultaneous users per cell.
- Smooth handovers across heterogeneous networks.
- High quality of service for next generation multimedia support.

The pre-4G *3GPP Long-Term Evolution* (LTE) technology is often branded as 4G-LTE. The initial LTE releases had a theoretical capacity of up to 100 Mbps in the downlink and 50 Mbps in the uplink.

In September 2009, a number of proposals were submitted to ITU as 4G candidates, mostly based on two technologies:

- LTE Advanced standardized by the 3GPP.
- 802.16 m standardized by the IEEE (i.e., WiMAX).

Fig. 17.7 A 2 × 2 MIMO antenna system

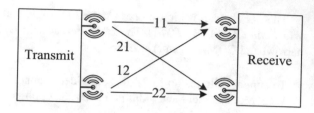

These two candidate systems have been commercially deployed: the first Mobile WiMAX network in South Korea in 2006, and the first LTE network in Oslo, Norway and Stockholm, Sweden in 2009. Today LTE Advanced has taken the place of WiMAX, and has been considered as the standard for 4G [9].

4G incorporates several new radio techniques to achieve higher rates and lower latencies than 3G [10]. They include *Space Division Multiplexing* via *multiple input/multiple output* (MIMO), *space time coding* (STC) using higher order of modulation and encoding schemes, sophisticated beam forming and beam directionality control, and intercell interference mitigation.

Using multiple sending and receiver antennas, MIMO creates multiple channels to carry user information, leading to higher capacity and less impact from interference. Figure 17.7 shows a typical 2 × 2 MIMO system. The beam-forming techniques temporarily improve gain and offer higher capacity. The properties of a beam are tuned or customized for a subscriber to achieve this capability for a limited duration. Space time coding improves the number of bits transmitted per Hz over the available bandwidth. These techniques collectively lead to higher capacity as required by advanced networks [11].

Additionally, techniques that reduce interference are also used to further boost the capacity, most notably *orthogonal frequency division multiplexing* (OFDM). While CDMA is well-suited for voice, OFDM can be a better transport mechanism for multimedia data. With a mix of technologies, backward compatibility is possible.

LTE Advanced also makes use of additional spectrums and multiplexing to achieve higher data speeds. It achieves peak download rates up to 299.6 Mbps and upload rates up to 75.4 Mbps depending on the user equipment category (e.g., with 4 × 4 MIMO antennas using 20 MHz of spectrum). Five different terminal classes have been defined from a voice centric class up to high-end terminals that support the peak data rates. It also enables lower data transfer latencies (theoretically < −5 ms latency for small packets, though practically much higher), and lower latencies for handover and connection setup time than with previous radio access technologies. Support for mobility is improved, too. Depending on the frequency band, it allows terminals to move at speeds up to 350 km/h (220 mph) or 500 km/h (310 mph). More importantly, through macro-diversity, also known as *group cooperative relay*, high bit rates are now available in a larger portion of a cell, especially to users in an exposed position in between several base stations. All these enables high-quality multimedia services with seamless mobility, even in such extreme scenario as today's high speed trains.

5G Cellular Networks and Beyond

4G is over 20 times faster than 3G, empowering today's high quality and real-time multimedia services. Despite the achievement, the past generations of cellular network technologies, up to 4G, were basically designed to meet the demands from voice and later video as well as Internet accesses. The current 4G networks have largely strained to their absolute bandwidth limits with little space to further improve.

Beyond bandwidth, a new era of innovative tools and technologies have accelerated the development of a wide range of new applications in recent years. These applications, coming from domains such as the ultra-high definition video, virtual and augmented reality, autonomous vehicles, and remote health and surgery, in general have multi-dimension quality of service (QoS) requirements, in terms of bandwidth, delay, and reliability. Also given that the number of connected devices is expected to reach tens or even hundreds of billions in the coming years, low-cost solutions for massive concurrent connections are expected.

To meet these demands, the newest generation cellular network, 5G NR (New Radio) [12–14], has begun its deployment in 2019. According ITU-R, 5G wireless systems will support three types of generic services:

- **Enhanced mobile broadband (eMBB).** The objective of this service is to maximize the data rate, while guaranteeing a moderate reliability, with packet error rate (PER) on the order of 10^{-3}. eMBB can be considered a direct extension of the 4G's broadband service. It is characterized by large payloads and stable connections with very high peak data rates, as well as moderate rates for users near the edge of a cell. Typical applications include high definition (HD) videos, virtual reality (VR), and augmented reality (AR), all with high data volume and hence bandwidth demand for transmission.
- **Ultra-reliable and low-latency communications (URLLC).** The objective of this service is to accommodate mission-critical low-latency transmissions of small payloads. The rate of a URLLC transmission is relatively low and the blocklength is short; the reliability requirement, however, can be very high, with a PER typically lower than 10^{-5}. It targets mainly latency-sensitive application scenarios, such as assisted and automated driving, and remote surgery.
- **Massive machine-type communications (mMTC).** This service accommodates a massive amount of Internet of Things (IoT) devices, which in general are inactive mostly of the time, and during the sporadically active times, send small data payloads only. It targets applications scenarios with high connection density, such as smart city and smart agriculture.

Table 17.1 shows the evolution from 3G to 5G in terms of bandwidth and latency. To achieve these goals, 5G has used a much wider and complex wireless spectrum, which includes all previous cellular spectrum and a large amount spectrum in the sub-6 GHz range and beyond. The sub-6 GHz 5G spans 450 MHz to 6 GHz, and the millimeter-wave 5G frequencies span 24.250–52.600 GHz. It also includes unlicensed spectrum. Additionally, there may be 5G spectrum in the 5,925–7,150 MHz

Table 17.1 Evolution from 3G to 5G

	3G	4G	5G
Deployment	2004	2006	2019
Theoretical Bandwidth	2 Mbps	200 Mbps	1 Gbps
Practical Bandwidth	144 Kbps	25 Mbps	200–400 Mbps
Latency	100–500 ms	20–30 ms	≤10 ms

range and 64–86 GHz range. Such a broad spectrum is utilized by advanced channel coding and multiple access technologies, e.g., nonorthogonal multiple access (NOMA). Two other important techniques in 5G are Massive MIMO and downsized cells.

- **Massive MIMO**. In 4G, MIMO is typically in a single user mode (SU-MIMO), in which both the base station and the user equipment have multiple antenna ports and antennas, and multiple data streams are transmitted simultaneously to the user equipment using the same time/frequency resources, doubling (2 × 2 MIMO), or quadrupling (4 × 4 MIMO) the peak throughput of the user.
  In a multiuser mode (MU-MIMO), the base station sends multiple data streams, one per user equipment, using the same time-frequency resources. It therefore increases the total cell throughput, i.e., cell capacity. The base station maintains multiple antenna ports, each of which is needed for a user equipment. Massive MIMO antennas further expands this design by a large number of individually-controlled antennas, each with embedded radio transceiver components. The number of antennas even exceeds the number of users. In practice, there can be 32 or even 64+ logical antenna ports in a 5G base station.
  Beamforming is then used on both mobile devices and networks' base stations to focus a wireless signal in a specific direction, rather than broadcasting to a wide area. In particular, 5G uses 3D Beamforming to create horizontal and vertical beams toward users, increasing the data rates for all users. As such, the throughput and capacity density in each sector can be greatly improved, so for the coverage and user experience.

- **Small cells**. Conventional base stations in the past generations of cellular networks can cover users 3–5 km away or even beyond 35 km with extended mode. Such coverage will remain valid in 5G when the identical frequency is used; however, for higher frequencies, e.g., millimeter wave (mmWave), the signals can hardly penetrate walls or buildings and therefore quickly decay beyond 500 m.
  Small cells use low power, short-range wireless transmission for base stations to cover a limited geographical area for indoor or outdoor applications. This concept, encompassing femtocells (10–50 m), pico cells (100–250 m), and micro cells (500 m–2.5 km), is developed in 4G but has become a necessity in 5G. Working together with regular base stations, they provide increased data capacity, and help service providers eliminate expensive rooftop systems and the associated installation or rental costs, which accelerates the deployment of 5G.

5G is the beginning of the promotion of digitalization from personal entertainment to industrial interconnection, and 6G, which is under active development, will continue this journey. The expansion of service scope for mobile networks enriches the telecom network ecosystem. The "one-size-fits-all" network paradigm employed in the past mobile networks (2G, 3G and 4G), however, becomes inflexible to support diversified multimedia services that have complicated and often contradicting demands. Even with the advanced technologies, performance metrics, e.g., bandwidth, latency, and the number of connections, can hardly be all optimized toward an ideal service that fits all. Such challenges are addressed by Network Slicing and mobile edge computing (MEC) in 5G and, more general, by an integration of modern communication, storage, and computation technologies, which will be the norm of future generation wireless systems (6G and beyond) [15]. We next introduce the network slicing; MEC will be discussed in Chap. 18 under the context of cloud/edge computing for multimedia services.

Network Slicing

In 5G, heterogeneous services are expected to coexist within a single physical network infrastructure. Their different QoS demands are accommodated through *network slicing* [16,17]. The basic idea is to partition the original network architecture into multiple logical and independent networks that are configured to effectively meet the various services requirements. For example, a low-latency slice, a maximum throughput/high bandwidth slice, and a slice enabling massive IoT deployment can be deployed in the same cell simultaneously. The operator provides separate resources for each of them with fine-grained QoS monitoring.

Network slicing was first introduced in 4G but never fully realized until the emergence of 5G. In essence, network slicing allows the creation of multiple virtual networks over a shared physical infrastructure, through the following two processes (see Fig. 17.8):

- **Virtualization**, which provides an abstract representation of the physical resources under a unified and homogeneous framework. With the framework, elementary network functions are to be used as building blocks to create network slices.
- **Orchestration**, which coordinates all the different network functions that are involved in the life-cycle of each network slice.

In this virtualized network scenario, physical components are secondary to logical self-contained partitions, i.e., slices, which allocate resources dynamically on a need basis. This could be achieved through software defined networking (SDN) and network function virtualization (NFV) [18]. Each slice will use some of the network's shared resources, while delivering a specific service with a specific service-level agreement to end users. As needs change, so can the allocated resources.

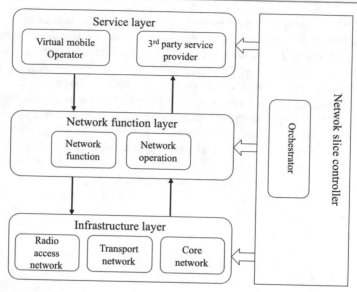

Fig. 17.8 A generic 5G network slicing framework

5G network slicing enables communication service providers to balance the disparate requirements between eMBB, URLLC, and mMTC applications. A derivative benefit is the ability to deploy the systems more quickly, because the functions do not need to be available as whole in the very beginning.

17.2.2 Wireless Local Area Networks (WLANs)

The increasing availability of such mobile computing devices as laptops and tablets brought about keen interest in *wireless local area networks* (WLANs), which potentially provide much higher throughput with much lower costs than the wide-area cellular wireless networks. The emergence lately of ubiquitous and pervasive computing [19] has further created a new surge of interest in WLANs and other short-range communication techniques.

Most of today's WLAN are based on the 802.11 family of standards (also known as Wi-Fi), developed by the IEEE 802.11 working group. They specify medium access control (MAC) and Physical (PHY) layers for wireless connectivity in a local area within a radius less than 100 m, addressing the following important issues:

- **Security**. Enhanced authentication and encryption, since the broadcast air over the air is more susceptible to break-ins.
- **Power management**. Saves power during no transmission and handles *doze* and *awake*.
- **Roaming**. Permits acceptance of the basic message format by different access points.

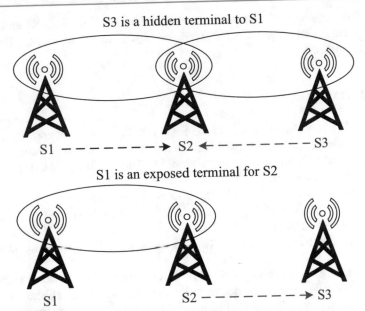

Fig. 17.9 An illustration of the hidden terminal problem. S2 is the access point; S1 and S3 are "hidden" to each other due to the long distance, but they can cause interference at S2

The initial 802.11 standard uses the 2.4 GHz radio band, which is the globally unlicensed *Industrial, Scientific and Medical* (ISM) short-range radio frequency band. As such, it faces interferences from both of its own users and many other wireless systems, e.g., cordless phones.

Similar to Ethernet, the basic channel access method of 802.11 is *carrier sense multiple access* (CSMA). However, *collision detection* (CD) in Ethernet is not employed. This is because of the unique *Hidden Terminal* problem in wireless communications. As shown in Fig. 17.9, wireless terminals S1 and S3 are at a far edge of the access point (S2)'s range. Recall that, unlike that of wired signals, the strength of a wireless signal decays very quickly with distance and a receiver can hear the signal only if its strength is above a certain threshold. As such, even if both S1 and S3 are "exposed" to access point S2, i.e., can hear S2 (and vice versa), they do not necessarily hear each other given the long distance. These two terminals are therefore "hidden" to each other—if they send packets simultaneously, the two packets will collide at S2, but neither S1 nor S3 can detect the collision.

To address the hidden terminal problem, 802.11 uses *collision avoidance* (CA); that is, during carrier sensing, if another node's transmission is heard, the current node should wait for a period of time for transmission to finish before listening again for a free communications channel. CSMA/CA can optionally be supplemented by the exchange of a Request to Send RTS packet sent by the sender, and a Clear to Send CTS packet sent by the intended receiver. This alters all the nodes within range of the sender, receiver or both, to not transmit for the duration of the intended transmission. For example, before sending a message to S2, S1 can first send an RTS request, and

S2 will then broadcast a CTS for S1, which will be heard by both S1 and S3. S1 can then send the message and S3 will temporarily refrain from sending, thus avoid potential collisions.

IEEE 802.11b/g

IEEE 802.11b is an enhancement of the basic 802.11. It uses DS spread spectrum and operates in the 2.4 GHz band. With the aid of new modulation technologies, it supports 5.5 and 11 Mbps in addition to the original 1 and 2 Mbps, and its functionality is comparable to Ethernet.

In North America, for example, the allocated spectrum for 802.11b is 2.400–2.4835 GHz. Regardless of the data rate (1, 2, 5.5, or 11 Mbps), the bandwidth of a DS spread spectrum channel is 20 MHz. Three nonoverlapped DS channels can be accommodated simultaneously, allowing a maximum of 3 access points in a local area.

IEEE 802.11g, an extension of 802.11b, is an attempt to achieve data rates up to 54 Mbps. It was designed to be downward compatible with 802.11b and hence still uses the 2.4 GHz band, but OFDM is used instead of DS spread spectrum.

IEEE 802.11a

IEEE 802.11a operates in the 5 GHz band and supports data rates in the range of 6–54 Mbps. It uses OFDMinstead of DS spread spectrum, too, and allows 12 nonoverlapping channels, hence a maximum of 12 access points in a local area.

Because 802.11a operates in the higher frequency (5 GHz) band, it faces much less radio interference, such as from cordless phones, than 802.11 and 802.11b. Coupled with the higher data rate, it has great potential for supporting various multimedia applications in a LAN environment.

802.11a products started shipping late, lagging 802.11b products due to the 5 GHz components being more difficult to manufacture. It was then not widely adopted in the consumer space given that the less-expensive 802.11b was already dominating the market. With the arrival of less expensive early 802.11g products on the market, which were backwards-compatible with 802.11b, the bandwidth advantage of the 5 GHz 802.11a in the consumer market was further reduced. It, however, does see penetration into enterprise network environments, which require increased capacity and reliability over 802.11b/g-only networks.

IEEE 802.11n, 802.11ac, and 802.11ax (Wi-Fi 6)

802.11n improves network performance over all the past 802.11 standards, achieving a significant increase in the maximum net data rate to 600 Mbps with the use of four spatial streams at a channel width of 40 MHz [20]. It builds on previous 802.11

standards by adding multiple-input multiple-output (MIMO), and frame aggregation to the MAC layer.

Channels operating with a width of 40 MHz are another feature incorporated into 802.11n; this doubles the channel width from 20 MHz in the previous 802.11 PHYs to transmit data, and provides twice the PHY data rate available over a single 20 MHz channel. It can be enabled in the 5 GHz mode, or within the 2.4 GHz mode if there is knowledge that it will not interfere with any other 802.11 or non-802.11 systems (such as cordless phones) using the same frequencies.

When 802.11g was released to share the band with existing 802.11b devices, it provided ways of ensuring coexistence between legacy and successor devices. 802.11n extends the coexistence management to protect its transmissions from legacy devices, including 802.11g/b/a, making its deployment much easier and smooth. It quickly replaced the existing 802.11a/b/g devices, offering much better support for multimedia over wireless.

802.11ac further increases the multi-station WLAN throughput to at least 1 Gbps and a single link throughput of at least 500 Mbps. This is accomplished by enhancing the air interfaces in 802.11n: wider radio bandwidth (up to 160 MHz) in the 5 GHz band, more MIMO streams (up to 8), MU-MIMO, and high-density modulation.

The latest standard, 802.11ax (also known as Wi-Fi 6), succeeds 802.11ac by introducing better power-control methods to avoid interference with neighboring networks. It also incorporates orthogonal frequency-division multiple access (OFDMA) and downlink MIMO and MU-MIMO, which further increases throughput. It offers real-world speeds that are roughly 30% faster than 802.11ac, with theoretical maximum transfer speeds up around 10 Gbps.

17.2.3 Bluetooth and Short-Range Technologies

It is known proximity-based services have constituted a considerable portion of the mobile data traffic. Such services enable geographically close users to directly exchange data. *Bluetooth* (named after the tenth-century king of Denmark, Harold Bluetooth) is a protocol intended for such short-range (called *piconet*) wireless communications [21].

Bluetooth uses *frequency hopping* (FH), a spread spectrum technology for data transmission, in the 2.4 GHz ISM short-range radio frequency band. Similar techniques have been used in the 802.11 WLAN. Bluetooth also employs a *master-slave* structure. One master may communicate with up to 7 slaves in a piconet; all devices share the master's clock. Packet exchange is based on a basic clock, defined by the master.

Bluetooth provides a secure and low-cost way to connect and exchange information between devices such as faxes, mobile phones, laptops, printers, Global Positioning System (GPS) receivers, digital cameras, and video game consoles. It was principally designed as a low-bandwidth technology. However, it permits moving or still pictures to be sent from a digital camera or mobile phone, at a speed of over 700 kbps, within a distance of 10 m.

Many other short or ultra short-range wireless communication protocols have been developed in recent years, including near field communication (NFC), Wi-Fi Direct, etc. Advanced technologies such as ultra-wideband (UWB) and cognitive radio are under active development as well. Batteryless backscatter communication has also been used to convey multimedia data, with prototype demonstration for 720p HD video at 10 fps up to 16 feet [22]. These technologies will facilitate direct data exchange between mobile or wearable devices or for Internet of Things (IoT) that consists of tiny devices or sensors.

17.3 Multimedia Over Wireless Channels

We have studied the evolution of 2G networks to high-capacity 3G/4G/5G networks as well as that of wireless local area networks. The main driving-force toward the new generation of higher speed wireless networks are from multimedia communications over wireless. Suggested multimedia applications range from streaming video, videoconferencing, online gaming, collaborative work, AR/VR to enhanced roadside assistance and online map guidance for drivers, to name but a few.

The characteristics of wireless handheld devices are worth keeping in mind when designing multimedia transmission over wireless, in particular video transmission [23]. First, both the handheld size and battery life limit the processing power and memory of the device. Thus, encoding and decoding must have relatively low complexity. On the other hand, the smaller screen sizes well accept relatively lower resolution videos, which helps reduce the processing time. This, however, is changing given the rapid adoption of high-resolution screens in mobile devices.

Second, due to memory constraints and reasons for the use of wireless devices, as well as billing procedures, real-time communication is likely to be required. Long delays before starting to see a video are either not possible or not acceptable.

Finally, wireless channels have much more interference than wired channels, with specific loss patterns depending on the environment conditions. The bitrate for wireless channels is also relatively lower, particularly during movement. This implies that even though a lot of bit protection must be applied, coding efficiency has to be maintained as well. And error-resilient coding is important.

Wireless communication standards generally expect the audio/video data be standard compliant, and most companies will concentrate on developing products using standards, in the interest of interoperability of mobiles and networks. The 3GPP/3GPP2 group has defined the following QoS parameters for wireless video-conferencing services [24,25].

- **Synchronization**. Video and audio should be synchronized to within 20 ms.
- **Throughput**. The minimum video bitrate to be supported is 32 kbps. Video rates of 128, 384 kbps, and above should be supported as well.
- **Delay**. The maximum end-to-end transmission delay is defined to be 400 ms.

- **Jitter**. The maximum delay jitter (maximum difference between the average delay and the 95th percentile of the delay distribution) is 200 ms.
- **Error rate**. A frame error rate of 10^{-2} or a bit error rate of 10^{-3} should be tolerated.

In this section, we are concerned mainly with sending multimedia data robustly over wireless channels, particularly for video communication, the natural extension to voice communication. We will introduce solutions for error detection, error correction, error-resilient entropy coding, and error concealment in the wireless network context, although most of these techniques are also applicable to other networks.

17.3.1 Error Detection

Error detection is to identify errors. caused by noise or other impairments during transmission from the sender to the receiver. Commonly used error detection tools include parity checking, checksum, and cyclic redundancy check (CRC) [26,27].

Parity Checking

With binary data, errors appear as bit flips. Parity checking adds a *parity bit* to a source bitstring to ensure that the number of *set bits* (i.e., bits with value 1) in the outcome is even (called *even parity*) or odd (called *odd parity*). For example, with even parity checking, a bit 1 should be appended to bitstring 10101000, and a bit 0 should be appended to 10101100.

This is a very simple scheme that can be used to detect any single or an odd number of errors on the receiver's side. An even number of flipped bits, however, will make the parity bit appear correct even though the data is erroneous.

Checksum

A checksum of an input message is a modular arithmetic sum of all the codewords in the message. The sender can append the checksum to the message, and the receiver can perform the same sum operation to check whether there is any error. It has been implemented in many network protocols, from data link layer, network layer to transport and application layers. The *Internet checksum* algorithm in these protocols works as follows (see more details in RFC 1071):

1. First pair the bytes of the input data to form 16-bit integers. If there is an odd number of bytes, then append a byte of zero in the end.
2. Calculate the 1's complement sum of these 16-bit integers. Any overflow encountered during the sum will be wrapped around to the lowest bit.

3. The result serves as the checksum field, which is then appended to the 16-bit integers.
4. On the receiver's end, the 1's complement sum is computed over the received 16-bit integers, including the checksum field. Only if all the bits are 1 will the received data be correct.

To illustrate this, let the input data be a byte sequence of $D_1, D_2, D_3, D_4, \ldots, D_N$. Using the notation $[a, b]$ for the 16-bit integer $a \cdot 256 + b$, where a and b are bytes, then the 16-bit 1's complement sum of these bytes is given by one of the following:

$$[D_1, D_2] +' [D_3, D_4] +' \cdots +' [D_{N-1}, D_N] \quad (N \text{ is even; no padding})$$

$$[D_1, D_2] +' [D_3, D_4] +' \cdots +' [D_N, 0] \quad (N \text{ is odd; append a zero})$$

As an example, suppose we have the following input data of four bytes: 10111011, 10110101, 10001111, 00001100. They will be grouped as 1011101110110101 and 1000111100001100.
The sum of these two 16 bit integers is

$$\begin{array}{r} 1011101110110101 \\ +1000111100001100 \\ \hline 0100101011000010 \end{array}$$

This addition has an overflow, which has been wrapped around to the lowest bit. The 1's complement is then obtained by converting all the 0s to 1s and all the 1s to 0s. Thus, the 1s complement of the above sum becomes 1011010100111101, which becomes the checksum.

The receiver will perform the same grouping and summation for the received bytes, and then add the received checksum too. It is easy to see that if there is no error, then the outcome should be 1111111111111111. Otherwise, if any bit becomes 0, then errors happen during transmission.

Cyclic Redundancy Check (CRC)

The basic idea behind *cyclic redundancy check* (CRC) is to divide a binary input by a keyword K that is known to both the sender and the receiver. The remainder R after the division constitutes the *check word* for the input. The sender sends both the input data and the check word, and the receiver can then check the data by repeating the calculation and verifying whether the remainder is still R. Obviously, to ensure that the check word R is fixed to r bits (zeros can be padded at the highest bits if needed), the keyword K should be of $r + 1$ bits.

CRC implementation uses a simplified form of arithmetic for the division, namely, computing the remainder of dividing with modulo-2 in GF(2) (Galois field with two elements), in which we have

$$0 - 0 = 0 + 0 = 0$$
$$1 - 0 = 1 + 0 = 1$$
$$0 - 1 = 0 + 1 = 1$$
$$1 - 1 = 1 + 1 = 0$$

In other words, addition and substraction are identical and both are equivalent to *exclusive OR* (XOR, $\oplus$). Multiplication and division are the same as in conventional base-2 arithmetic, too, except that, with the XOR operation, any required addition or substraction is now without carries or borrows. All these make the hardware implementation much simpler and faster.

Given the message word M, and the keyword K, we can manually calculate the remainder R using conventional long division, just with modulo-2 arithmetic. We also append r zeros to M before division, which makes the later verification easier, as we will see soon. For example, for $M = 10111$, and $K = 101$, we have

```
         10011
    101)1011100
         101
         ---
          110
          101
          ---
          110
          101
          ---
           11
```

Hence, $R = 11$, which is to be appended as the check word to the message.

It is not difficult to show that, in this case, $M \cdot 2^r \oplus R$ is perfectly divisible by K (which we leave as an exercise). Hence, instead of calculating the remainder on the receiver's side and comparing with the R from the sender, the receiver can simply divide $M \cdot 2^r \oplus R$ by K and check whether the remainder is zero or not. If it is zero, then there is not error; otherwise, error detected.

The keyword K indeed comes from a *generator polynomial* whose coefficients are the binary bits of the keyword K. For example, for $K = 100101$ in the binary format, its polynomial expression is $x^5 + x^2 + 1$. The keyword is therefore also called a *generator*. Choosing a good generator is a non-trivial job and there have been extensive studies [27]. For a well-chosen generator, two simple facts are known (more can be found in [27]):

1. If the generator polynomial contains two or more terms, all single-bit errors can be detected.
2. An r-bit CRC can detects all burst errors of length no more than r. Here, the burst means that the first and last bits are in error, and the bits in between may or may not be in error.

Standard generators of 8-, 12-,16-, and 32-bits have been defined in international standards. For example, the following 32-bit generator has been used in a number of

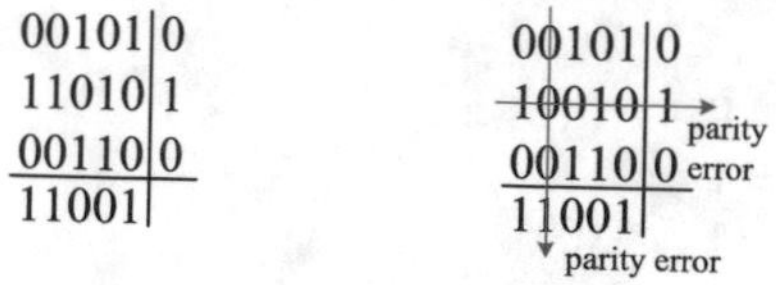

$$
\begin{array}{cc}
\begin{array}{c}
00101|0 \\
11010|1 \\
00110|0 \\
\hline
11001|
\end{array}
&
\begin{array}{c}
00101|0 \\
\overline{10010}|1 \;\rightarrow \\
00110|0 \;\text{error} \quad \text{parity}\\
\hline
1|001|
\end{array}
\end{array}
$$

parity error

Fig. 17.10 An example of two-dimensional even parity checking. Left: No error; Right: A single-bit error detected and corrected

link-layer IEEE protocols, in particular, the Internet since 1975:

$$
G_{CRC-32} = 100000100110000010001110110110111
$$

Note that it is of 33 bits (so that the remainder is of 32 bits). It is also the CRC generator used in MPEG-2, and the error message "CRC failed!" often appears for a scratched DVD that is hard to read by the disc player.

17.3.2 Error Correction

Once an error is detected, a retransmission could be used to recover the error, as such reliable transport protocols as TCP does. The back channel, however, is not always available, e.g., for satellite transmission, or can be quite expensive to create, e.g., in the broadcast or multicast scenarios. For real-time multimedia streaming, the delay for retransmission can be too long, making the retransmitted packet useless.

Instead, for real-time multimedia, *forward error correction* (FEC) is often used, which adds redundant data to a bitstream to recover some random bit errors in it [27]. Consider a simple extension to the parity checking, from one dimension to two dimensions [28]. We not only calculate the parity bit for each bitstring of M bits, but also group every M bitstrings to form a matrix and calculate the parity bit of each column of the matrix.

With this *two-dimensional parity checking*, we can both detect errors and correct errors ! This is because a bit error will cause a failure of a row parity checking and a failure of a column parity checking, which cross at a unique location—the flipped bit in the erroneous bitstring, as illustrated in the example in Fig. 17.10.

This is a very simple FEC scheme. It doubles the amount of parity bits, but the error correction capability is very limited; e.g., if two errors occur in a row, we will not able to detect them, not to mention correcting them.

There are two categories of practical error correction codes: *block codes* and *convolutional codes* [26,27]. The block codes apply to a group of bits, i.e., a block, at once to generate redundancy. The convolutional codes apply to a string of bits one at a time and have memory that can store previous bits as well.

Block Codes

The block codes take an input of k bits and append $r = n - k$ bits of FEC data, resulting in an n-bit-long string [29]. These codes are referred to as (n, k) codes. For instance, the basic ASCII words are of 7-bits; parity checking adds a single bit ($r = 1$) for any ASCII word, and so it is an (8,7) code, with eight bits in total ($n = 8$), of which 7 are data ($k = 7$). The *code rate* is k/n, or 7/8 in the parity checking case.

Hamming Codes

Richard Hamming observed that error correction codes operate by adding space between valid source strings. The space can be measured using a *Hamming distance*, defined as the minimum number of bits between *any* coded strings that need to be changed so as to be identical to another valid string.

To detect r errors, the Hamming distance has to be at least equal $r + 1$; otherwise, the corrupted string might seem valid again. This is not sufficient for correcting r errors however, since there is not enough distance among valid codes to choose a preferable correction. To correct r errors, the Hamming distance must be at least $2r + 1$.

This leads to the invention of first block code, the *Hamming(7, 4)-code*, in 1950. It encodes 4 data bits into 7 bits by adding three parity bits based on a generator matrix G, say

$$G = \begin{pmatrix} 1\ 1\ 0\ 1 \\ 1\ 0\ 1\ 1 \\ 1\ 0\ 0\ 0 \\ 0\ 1\ 1\ 1 \\ 0\ 1\ 0\ 0 \\ 0\ 0\ 1\ 0 \\ 0\ 0\ 0\ 1 \end{pmatrix}$$

Given an input data p (4 bits as a vector), the output code x is obtained by taking the product $G \cdot p$ and then performing modulo 2. As an example, for bits 1001, the input vector p is $(1, 0, 0, 1)^T$; the product will be vector $(2, 2, 1, 1, 0, 0, 1)^T$, and the encoded output x will be $(0, 0, 1, 1, 0, 0, 1)^T$ after modulo 2, or a 7-bit data block of 0011001.

The Hamming(7, 4)-code can detect and correct any single-bit error. To do this, a parity-check matrix H is used

$$H = \begin{pmatrix} 1\ 0\ 1\ 0\ 1\ 0\ 1 \\ 0\ 1\ 1\ 0\ 0\ 1\ 1 \\ 0\ 0\ 0\ 1\ 1\ 1\ 1 \end{pmatrix}$$

Similar to encoding, we take the product $H \cdot x$ with modulo 2, which yields a vector z of length 3. We can treat z as a 3-bit binary number. If it is zero, then there is no error; otherwise, it indicates the location of the error (by checking the corresponding column in H). For example, for $x = 0011001$, we have $z = 0$, i.e., no error; on the other hand, for $x' = 0111001$, we have $z = 010$, which corresponds to the second column of H, indicating bit 2 is erroneous.

Extended Hamming codes can detect up to two-bit errors or correct one-bit errors without detection of uncorrected errors. By contrast, the simple one-dimensional parity code can detect only an odd number of bits in error and cannot correct errors.

BCH and RS Codes

More powerful *cyclic codes* are stated in terms of generator polynomials of maximum degree equal to the number of source bits. The source bits are the coefficients of the polynomial, and redundancy is generated by multiplying with another polynomial. The code is cyclic, since the modulo operation in effect shifts the polynomial coefficients. The cyclic redundancy check (CRC) we have seen before belongs to this category, though it is mainly used for error detection. A widely used classes of cyclic error correction codes is the *Bose-Chaudhuri-Hocquenghem* (BCH) codes. The generator polynomial for BCH is also given over a Galois field (GF) and is the lowest-degree polynomial with roots of α^i, where α is a primitive element of the field and i goes over the range of 1 to twice the number of bits we wish to correct.

BCH codes can be encoded and decoded quickly using integer arithmetic. H.261 and H.263 use BCH to allow for 18 redundant bits every 493 source bits. Unfortunately, the 18 redundant bits will correct at most two errors in the source. Thus, the packets are still vulnerable to burst bit errors or single-packet errors.

An important subclass of BCH codes that applies to multiple packets is the *Reed-Solomon* (RS) codes. The RS codes have a generator polynomial over $GF(2^m)$, with m being the packet size in bits. RS codes take a group of k source packets and output n packets with $r = n - k$ redundancy packets. Up to r lost packets can be recovered from n coded packets if we know the erasure points.[1] Otherwise, as with all FEC codes, recovery can be applied only to half the number of packets, since error-point detection is now necessary as well.

In the RS codes, only $\lceil \frac{r}{2} \rceil$ packets can be recovered. Fortunately, in the packet FEC scenario, the packet itself often contains a sequence number and checksum or CRC in its header. In most cases, a packet with an error is dropped, and we can tell the location of the missing packet from the missing sequence number.

The RS codes are useful for both storage and transmission over networks. When there are burst packet losses, it is possible to detect which packets were received incorrectly and recover them using the available redundancy. If the video has scalability, a better use of allocated bandwidth is to apply adequate FEC protection on the base layer, containing motion vectors and all header information required to decode video to the minimum QoS. The enhancement layers can receive either less protection or none at all, relying just on resilient coding and error concealment. Either way, the minimum QoS is already achieved.

---

[1] Errors are also called *erasures*, since an erroneous packet can be useless, and has to be "erased."

A disadvantage of the block codes is that they cannot be selectively applied to certain bits. It is difficult to protect higher protocol-layer headers with more redundancy bits than for, say, DCT coefficients, unless they are sent explicitly through different packets. On the other hand, convolutional codes can do this, which makes them more efficient for data in which unequal protection is advantageous, such as videos.

Convolutional Codes

The convolutional FEC codes are defined over generator polynomials as well [26]. They are computed by shifting k message bits into a coder that convolves them with the generator polynomial to generate n bits. The rate of such code is defined to be $\frac{k}{n}$. The shifting is necessary, since coding is achieved using memory (shift) registers. There can be more than k registers, in which case past bits also affect the redundancy code generated.

After producing the n bits, some redundancy bits can be deleted (or "punctured") to decrease the size of n, and increase the rate of the code. Such FEC schemes are known as *rate compatible punctured convolutional* (RCPC) codes. The higher the rate, the lower the bit protection will be, but also the less overhead on the bitrate. A *Viterbi algorithm* with soft decisions decodes the encoded bit stream, although *turbo codes* are gaining popularity.

Given the limited network bandwidth, it is important to minimize redundancy, because it comes at the expense of bitrates available for source coding. At the same time, enough redundancy is needed so that the video can maintain required QoS under the current channel error conditions. Moreover, the data in a compressed media stream are of different importance. Some data are vitally important for correct decoding. For example, some lost and improperly estimated data, such as picture coding mode, quantization level, or most data in higher layers of a video standard protocol stack, will cause catastrophic video decoding failure. Others, such as missing DCT coefficients may be estimated or their effect visually concealed to some degree. As such, given certain channel conditions, different amount of FEC can be applied to these data to provide different level of protection, which is known as *unequal error protection (UEP)*.

RCPC puncturing is done after generation of parity information. Knowing the significance of the source bits for video quality, we can apply a different amount of puncturing and hence achieve UEP. Studies and simulations of wireless radio models have shown that applying unequal protection using RCPC according to bit significance information results in better video quality (up to 2 dB better) for the same allocated bitrate than videos protected using the RS codes.

Simplistically, the Picture layer in a video protocol should get the highest protection, the macroblock layer that is more localized will get lower protection, and the DCT coefficients in the block layer can get little protection, or none at all. This could be extended further to scalable videos in similar ways.

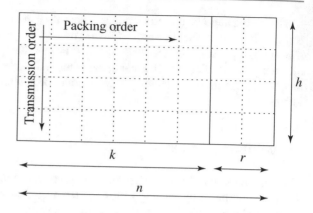

Fig. 17.11 Interleaving scheme for redundancy codes. Packets or bits are stored in rows, and redundancy is generated in the last r columns. The sending order is by columns, top to bottom, then left to right

The 3G networks have also incorporated data-type-specific provisions and recognize the video standard chosen for transmission; they can adaptively apply transport coding of the video stream with enough unequal redundancy suitable to the channel conditions at the time and QoS requested.

Packet Interleaving

It is also possible to use *packet interleaving* to increase resilience to burst packet loss. As Fig. 17.11 shows, the RS codes are generated for each of the h rows of k source video packets. Instead of transmitting with the original order, we can use the column-major order, so that the first packet of each of the h rows is transmitted first, then the second, and so on. Such an interleaving can effectively convert a burst loss to a series of smaller uniform losses across the original rows, which are much easer to handle given the enough redundancy in each row. In other words, we could tolerate more than r erasures with error correction and concealment.

It is worth noting that the interleaving does not increase bandwidth overhead but introduces additional delay.

17.3.3 Error-Resilient Coding

A video stream is either packetized and transmitted over a packet-switched channel or transmitted as a continuous bitstream over a circuit-switched channel, with the former being more popular nowadays. In either case, it is obvious that packet loss or bit error will reduce video quality. If a bit loss or packet loss is localized in the video in both space and time, the loss can still be acceptable, since a frame is displayed for a very short period, and a small error might go unnoticed.

However, digital video coding techniques involve variable-length codes, and frames are coded with different prediction and quantization levels. Unfortunately, when a packet containing variable bit-length data (such as DCT coefficients) is damaged, that error, if unconstrained, will propagate all the way throughout the stream.

This is called *loss of decoder synchronization*. Even if the decoder can detect the error due to an invalid coded symbol or coefficients out of range, it still cannot establish the next point from which to start decoding [30].

As we have learned in Chap. 10, this complete bitstream loss does not happen for videos coded with standardized protocol layers. The Picture layer and the group of blocks (GOB) layer or Slice headers have *synchronization markers* that enable decoder resynchronization. For example, the H.263 bitstream has four layers—the Picture layer, GOB layer, Macroblock layer, and Block layer. The Picture Layer starts with a unique 22-bit picture start code (PSC). The longest entropy-coded symbol possible is 13 bits, so the PSC serves as a synchronization marker as well. The GOB layer is provided for synchronization after a few blocks rather than the entire frame. The group of blocks start code (GBSC) is 17 bits long and also serves as a synchronization marker.[2] The macroblock and the Block layers do not contain unique start codes, as these are deemed high overhead.

Slice Mode

ITU standards after H.261 (i.e., H.263 to 266) support slice-structured mode instead of GOBs (see, for example, H.263 Annex K), where slices group blocks together according to the block's coded bit length rather than the number of blocks. The objective is to space slice headers within a known distance of each other. That way, when a bitstream error looks like a synchronization marker, if the marker is not where the slice headers should be it is discarded, and no false resynchronization occurs.

Since slices need to group an integral number of macroblocks together, and macroblocks are coded using VLCs, it is not possible to have all slices the same size. However, there is a minimum distance after which the next scanned macroblock will be added to a new slice. We know that DC coefficients in macroblocks and motion vectors of macroblocks are differentially coded. Therefore, if a macroblock is damaged and the decoder locates the next synchronization marker, it might still not be able to decode the stream.

To alleviate the problem, slices also reset spatial prediction parameters; differential coding across slice boundaries is not permitted. The ISO MPEG standards (and H.264 as well) specify slices that are not required to be of similar bit length and so do not protect against false markers well.

Other than synchronization loss, we should note that errors in prediction reference frames cause much more damage to signal quality than errors in frames not used for prediction. That is, a frame error for an I-frame will deteriorate the quality of a video stream more than a frame error for a P- or B-frame. Similarly, if the video is scalable, an error at the base layer will deteriorate the quality of a video stream more than in enhancement layers.

---

[2]Synchronization markers are always larger than the minimum required, in case bit errors change bits to look like synchronization markers.

Reversible Variable Length Code (RVLC)

Another useful tool to address the loss of decoder synchronization is *reversible variable length code* (RVLC) [31,32]. An RVLC makes instantaneous decoding possible both in the forward and backward directions. With the conventioanl VLC, a single-bit error can cause continuous errors in reconstructing the data even if no further bit error happens. In other words, the information carried by the remaining correct bits become useless. If we can decode from the reverse direction, then such information could be recovered. Another potential use of RVLC is in the random access of a coded stream. The ability to decode and search in two directions should halve the amount of indexing overhead with the same average search time as compared to the standard one-directional VLC.

A RVLC, however, must satisfy the prefix condition for instantaneous forward decoding (as we have seen in Chap. 7) and also a suffix condition for instantaneous backward decoding. That is, each code word must not coincide with any suffix of a longer code word. A conventional VLC, say, Huffman coding, satisfies only the prefix condition and can only be decoded from left to right.

As an example, consider the symbol distribution in Table 17.2. For input ACDBC, the Huffman coded bit stream (C_1 in Table 17.2) is 10010011101, which cannot be decoded instantaneously in the backward direction (right to left) because the last two bits 10 might be either symbol "C" or the suffix of "D."

To ensure both the prefix and the suffix conditions, we can use a VLC composed entirely of symmetrical code words, e.g., the second column (C_2) in Table 17.2. Each symmetric code is clearly reversible, and a bit stream formed by them is reversible too. For example, ACDBC will be coded as 0010101011101, which is uniquely decodable from both directions. Compared to Huffman coding (average code length of 2.21), this symmetric RVLC has a slightly longer average code length (2.44). More efficient asymmetric RVLC can also be systematically constructed (C_3 in the table), which has an average code length of 2.37. Though it is still higher than that of Huffman coding, the overhead is acceptable given the potential benefit of bi-directional decoding.

RVLC has been used MPEG-4 Part 3. To further help with synchronization, a data partitioning scheme in MPEG-4 groups and separates header information, motion vectors, and DCT coefficients into different packets and puts synchronization markers between them. Such a scheme is also beneficial to unequal protection.

Table 17.2 Huffman Code (C_1), Symmetric RVLC (C_2), and Asymmetric RVLC (C_3)

Symbol	Probability	C_1	C_2	C_3
A	0.32	10	00	11
B	0.32	11	11	10
C	0.15	01	101	01
D	0.13	001	010	000
E	0.08	000	0110	00100

Additionally, an adaptive intra-frame refresh mode is allowed, where each macroblock can be coded independently of the frame as an inter-or intra-block according to its motion, to assist with error concealment. A faster-moving block will require more frequent refreshing—that is, be coded in intra-mode more often. Synchronization markers are easy to recognize and are particularly well suited to devices with limited processing power, such as cell phones and mobile devices.

For interactive applications, if a back channel is available to the encoder, a few additional error control techniques are available with the feedback information. For example, according to the bandwidth available at any moment, the receiver can ask the sender to lower or increase the video bitrate (transmission rate control), which combats packet loss due to congestion. If the stream is scalable, it can ask for enhancement layers as well. Annex N of H.263+ also specifies that the receiver can notice damage in a reference frame and request that the encoder use a different reference frame for prediction—a reference frame the decoder has reconstructed correctly. Unfortunately, for many real-time streaming application with tight delay constraints or multicast/broadcast scenarios, such a backchannel for each receiver may not be available.

Error-Resilient Entropy Coding (EREC)

The main purpose of GOBs, slices, and synchronization markers is to re-establish synchronization in the decoder as soon as possible after an error. In Annex K of H.263+, the use of slices achieves better resilience, since they impose further constraints on where the stream can be synchronized. *Error-Resilient Entropy Coding* (EREC), further achieves synchronization after every *single* macroblock, without any of the overhead of the slice headers or GOB headers. It takes entropy-coded variable-length macroblocks and rearranges them in an error-resilient fashion. In addition, it can provide graceful degradation.

EREC takes a coded bitstream of a few blocks and rearranges them so that the beginning of all the blocks is a fixed distance apart. Although the blocks can be of any size and any media we wish to synchronize, the following description will refer to macroblocks in videos. The algorithm proceeds as in Fig. 17.12.

Initially, EREC slots (rows) of fixed bit-length are allocated with a total bit-length equal to (or exceeding) the total bit-length of all the macroblocks. The number of slots is equal to the number of macroblocks, except that the macroblocks have varying bit-length and the slots have a fixed bit-length (approximately equal to the average bit-length of all the macroblocks). As shown, the last EREC slot (row) is shorter when the total number of bits does not divide evenly by the number of slots.

Let k be the number of macroblocks, which is equal to the number of slots, l be the total bit-length of all the macroblocks, $mbs[\ ]$ be the macroblocks, $slots[\ ]$ be the EREC slots, the procedure for encoding the macroblocks is shown below.

EREC slots   Macroblocks

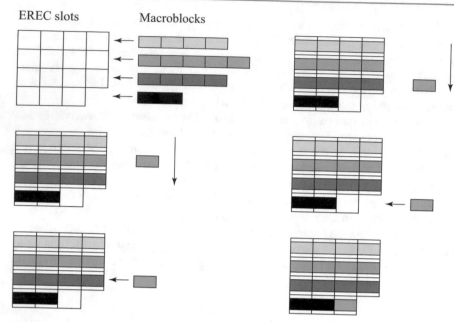

Fig. 17.12 Example of macroblock encoding using EREC

Procedure 17.1 (Macroblock encoding using EREC)

BEGIN
> $j = 0$;
> Repeat until $l = 0$
> {
> > for $i = 0$ to $k - 1$
> > {
> > > $m = (i + j) \bmod k$;
> > > // m is the macroblock number corresponding to slot i;
> > > Shift as many bits as possible (without overflow) from $mbs[i]$ into $slots[m]$;
> > > $sb = $ number of bits successfully shifted into $slots[m]$ (without overflow);
> > > $l = l - sb$;
> > }
> > $j = j + 1$;   // shift the macroblocks downwards
> }
END

The macroblocks are shifted into the corresponding slots until all the bits of the macroblock have been assigned or remaining bits of the macroblock don't fit into the slot. Then, the macroblocks are shifted down, and this procedure repeats.

The decoder side works in reverse, with the additional requirement that it has to detect when a macroblock has been read in full. It accomplishes this by detecting the end of macroblock when all DCT coefficients have been decoded (or a block end

Macroblocks   EREC slots

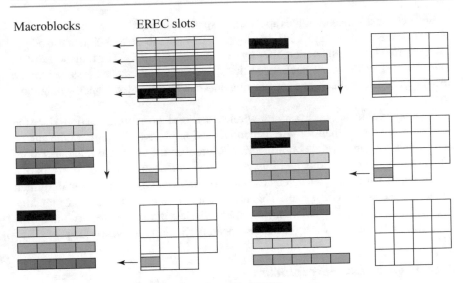

Fig. 17.13 Example of macroblock decoding using EREC

code). Figure 17.13 shows an example of the decoding process for the macroblocks coded using EREC in Fig. 17.12.

The transmission order of the data in the slots is row-major—that is, at first the data in slot 0 is sent, then slot 1, and so on, left to right. It is easy to see how this technique is resilient to errors. No matter where the damage is, even at the beginning of a macroblock, we still know where the next macroblock starts—it is a fixed distance from the previous one. In this case, no synchronization markers are used, so the GOB layer or slices are not necessary either (although we still might want to restrict spatial propagation of error).

When the macroblocks are coded using a data partitioning technique (such as the one for MPEG-4 described in the previous section) and also bitplane partitioning, an error in the bitstream will destroy less significant data while receiving the significant data. It is obvious that the chance for error propagation is greater for bits at the end of the slot than at the beginning. On average, this will also reduce visual deterioration over a nonpartitioned encoding. This achieves graceful degradation under worsening error conditions.

17.3.4 Error Concealment

Despite all the efforts to minimize occurrences of errors and their significance, errors can still happen unless with persistent retransmission, which, however, is not practical for continuous media with delay constraints. The residual error will be acoustically or visually annoying. *Error concealment* techniques are then introduced to approximate the lost data on the decoder side, so as to mitigate their negative audio or visual impact.

Error concealment techniques apply in the spatial, temporal, or frequency domain, or a combination of them [33,34]. For the case of video, these techniques use neighboring frames temporally or neighboring macroblocks spatially. The transport stream coder interleaves the video packets, so that in case of a burst packet loss, not all the errors will be at one place, and the missing data can be estimated from the neighborhood.

Error concealment is necessary for wireless audio/video communication, since the error rates are higher than for wired channels and might even be higher than can be transmitted with appropriate bit protection. Moreover, the error rate fluctuates more often, depending on various mobility or weather conditions. Decoding errors due to missing or wrong data received are also more noticeable on devices with limited resolution and small screen sizes. This is especially true if macroblock size remains large, to achieve encoding efficiency for lower wireless bitrates. Here, we summarize the common techniques for error concealment, particularly for video [35].

Dealing with Lost Macroblocks

A simple and popular technique for concealment can be used when DCT blocks are damaged but the motion vectors are received correctly. The missing block coefficients are estimated from the reference frame, assuming no prediction errors. Since the goal of motion-compensated video is to minimize prediction errors, this is an appropriate assumption. The missing block is hence temporally masked using the block in the reference frame.

We can achieve even better results if the video is scalable. In that case, we assume that the base layer is received correctly and that it contains the motion vectors and base layer coefficients that are most important. Then, for a lost macroblock at the enhancement layer, we use the motion vectors from the base layer, replace the DCT coefficients at the enhancement layer, and decode as usual from there. Since coefficients of less importance are estimated (such as higher frequency coefficients), even if the estimation is not too accurate due to prediction errors, the concealment is more effective than in a nonscalable case.

If the motion vector information is damaged as well, this technique can be used only if the motion vectors are estimated using another concealment technique (to be discussed next). The estimation of the motion vector has to be good, or the visual quality of the video could be inauspicious. To apply this technique for intra-frames, some standards, such as MPEG-2, also allow the acquisition of motion vectors for intra-coded frames (i.e., treating them as intra- as well as inter-frames). These motion vectors are discarded if the block has no error.

Combining Temporal, Spatial and Frequency Coherences

Instead of just relying on the temporal coherence of motion vectors, we can combine it with spatial and frequency coherences. By having rules for estimating missing block coefficients using the received coefficients and neighboring blocks in the same frame, we can conceal errors for intra-frames and for frames with damaged motion vector information. Additionally, combining with prediction using motion vectors will give us a better approximation of the prediction error block.

Missing block coefficients can be estimated spatially by minimizing the error of a smoothness function defined over the block and neighboring blocks. For simplicity, the smoothness function can be chosen as the sum of squared differences of pairwise neighboring pixels in the block. The function unknowns are the missing coefficients. In the case where motion information is available, prediction smoothness is added to the objective function for minimization, weighted as desired.

The simple smoothness measure defined above has the problem that it smooths edges as well. We can attempt to do better by increasing the order of the smoothing criterion from linear to quadratic or cubic. This will increase the chances of having both edge reconstruction and smoothing along the edge direction. At a larger computational cost, we can use an edge-adaptive smoothing method, whereby the edge directions inside the block are first determined, and smoothing is not permitted across edges.

Smoothing High-Frequency Coefficients

Although the human visual system is more sensitive to low frequencies, it would be disturbing to see a checkerboard pattern where it does not belong. This will happen when a high-frequency coefficient is erroneously assigned a high value. The simplest remedy is to set high-frequency coefficients to 0 if they are damaged.

If the frequencies of neighboring blocks are correlated, it is possible to estimate lost coefficients in the frequency domain directly. For each missing frequency coefficient in a block, we estimate its value using an interpolation of the same frequency coefficient values from the four neighboring blocks. This is applicable at higher frequencies only if the image has regular patterns. Unfortunately that is not usually the case for natural images, so most of the time the high coefficients are again set to 0. Temporal prediction error blocks are even less correlated at all frequencies, so this method applies only for intra-frames.

Estimating Lost Motion Vectors

The loss of motion vectors prevents decoding of an entire predicted block, so it is important to estimate motion vectors well. The easiest way to estimate lost motion vectors is to set them to 0. This works well only in the presence of very little motion. A better estimation is obtained by examining the motion vectors of reference macroblocks and of neighboring macroblocks. Assuming motion is also coherent, it is reasonable to take the motion vectors of the corresponding macroblock in the reference frame as the motion vectors for the damaged target block.

Similarly, assuming objects with consistent motion fields occupy more than one macroblock, the motion vector for the damaged block can be approximated as an interpolation of the motion vectors of the surrounding blocks that were received correctly. Typical simple interpolation schemes are weighted-average and median. Also, the spatial estimation of the motion vector can be combined with the estimation from the reference frame using weighted sums.

17.4 Mobility Management

Mobility is another distinct feature of wireless portable devices. The traditional TCP/UDP/IP networks were originally designed for communications between fixed ends. There are many issues that need to be resolved to support mobility, which has long been a research topic in the Internet community, particularly in recent years when the number of mobile terminals dramatically increases [36,37]. There is a broad spectrum of device and user mobility, in terms of both range and speed, as illustrated in Fig. 17.14.

The deep penetration of modern wireless accesses has made network connectivity anywhere and anytime a reality, which urges network operators/administrators to deploy mobility management protocols for ubiquitous accesses. From a network operator/administrator's view, a network usually covers a large geographical area (or administrative domain) consisting of several subnetworks. Mobility of a user in a network can be broadly classified into three categories:

- *Micro-mobility* (intra-subnet mobility), where movement is within a subnet.
- *Macro-mobility* (intra-domain mobility), where movement is across different subnets within a single domain.
- *Global mobility* (inter-domain mobility), where movement is across different domains in various geographical regions.

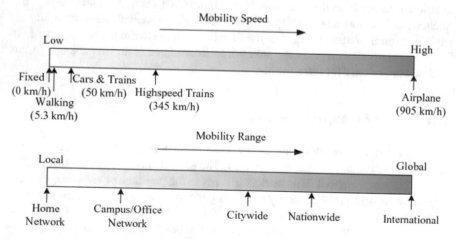

Fig. 17.14 Real-world mobility range and mobility speed

Global mobility involves longer timescales, where the goal is to ensure that mobile devices can re-establish communication after a move rather than provide continuous connectivity. Early studies on Mobile IP have addressed the simple scenario of global mobility that a computer is unplugged from a network, transported to another network, and then replugged. With the support of modern wireless mobile networks, such as 4G/5G and Wi-Fi, the mobility can be much more frequent with complex patterns. It is therefore important to ensure continuous and seamless connectivity during micro- and macro-mobility, together with secure authentication, authorization, and accounting. The short timescales here call for joint effort across multiple layers. This is further complicated with streaming media applications that expect un-interrupt data transfer during the movement.

To avoid interruption during communication, *handoff* (also known as *handover*) management is required, by which a mobile terminal keeps its connection active when it moves from one network access point to another. Another important function needed to support mobility is *location management*, which tracks the locations of mobile terminals, and provides such popular location-based services as searching nearby users or media content related to the locations of interest.

17.4.1 Network Layer Mobile IP

We start our discussion on mobility management from the network layer support for global mobility. The most widely used protocol for this purpose is Mobile IP, whose initial version was developed by IETF in 1996. IETF released Mobile IPv4 (RFC3220) and Mobile IPv6 (RFC6275) standards in 2002 and 2011, respectively. There are certain differences in their details, but the overall architectures and the high-level designs are similar for both versions.

The key support offered by Mobile IP is to assign a mobile host two IP addresses: a *home address* (HoA) that represents the fixed address of the *mobile node* (MN) and a *care-of-address* (CoA) that changes with the IP subnet to which the MN is currently attached. Each mobile node has a *home agent* (HA) in its home network, from which it acquires its HoA. In Mobile IPv4, the foreign network where the MN currently is attached should have a *foreign agent* (FA), which is replaced by an *access router* (AR) in Mobile IPv6. The mobile node obtains its CoA from its current FA or AR.

When the mobile node MN is in its home network, it acts like any other fixed node of that network with no special mobile IP features. When it moves out of its home network to a foreign network, the following steps are to be followed (see Fig. 17.15):

1. The MN obtains the CoA and informs its HA of the new address by sending a `Registration Request` message to the HA.
2. The HA, upon receiving the message, shall reply to the MN with a `Registration Reply` message. The HA keeps the binding record of the MN, which is transparent to a *correspondent node* (CN) that intends to communicate with the MN.

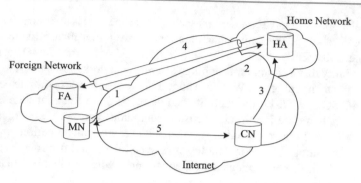

Fig. 17.15 The operations in mobile IP

Fig. 17.16 The data path in mobile IP

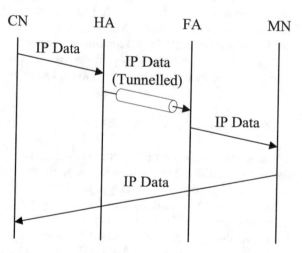

3. Once a packet from the CN to the MN arrives at the home network, the HA will intercept the packet.
4. The HA then forwards it to the FA by a tunnel that encapsulates the original packets (with the HoA in the headers) into packets with the CoA in the headers. Once the FA receives the tunneled packets, it removes the extra header and delivers it to the MN.
5. When the MN wishes to send data back to the CN, the packets are sent directly from the MN to the destination since the CN's IP address is known by the MN.

This data path for mobile IP is further illustrated in Fig. 17.16.

Such a simple implementation that involves MN, HA, and CN can cause *triangular routing*, that is, the communication between the MN and CN now has to go through the HA using tunneling. It can be quite efficient if the MN and CN are very close; in an extreme case, both MN and CN can be in the same network while the HA is far away. To alleviate triangular routing, the CN can also keep the mapping between the mobile's HoA and CoA, and accordingly send packets to the mobile directly, without going through the HA. In this case, the mobile node must update its CoA to CNs as well.

Even with route optimization, Mobile IP can still introduce significant network overhead in terms of increased delay, packet loss, and signaling when the MNs change their point of attachment to network frequently or the number of MNs grows dramatically. Hierarchical mobile IP (HMIP) (RFC 4140) is a simple extension that improves the performance by using a *mobility anchor point* (MAP) to handle the movement of a MN in a local region. The MN, if supporting HMIP, obtains a *regional CoA* (RCoA) and registers it with its HA as its current CoA; while RCoA is the locator for the mobile in Mobile IP, it is also its regional identifier used in HMIP. At the same time, the MN obtains a *local CoA* (LCoA) from the subnet it attaches to. When moving within the region, the MN only updates the MAP with the mapping between its RCoA and LCoA. It reduces the burden of the HA by reducing the frequency of updates. The shorter delay between the MN and the MAP also improves response time.

17.4.2 Link-Layer Handoff Management

Link layer handoff or handover occurs when a mobile devices changes its radio channels to minimize interference under the same access point or base station (called *intra-cell handoff*) or when it moves into an adjacent cell (called *intercell handoff*). For example, in GSM, if there is strong interference, the frequency or time slot can be changed for a mobile device, which remains attached to the same base station transceiver. For intercell handoff, there are two types of implementations, namely, *hard handoff* and *soft handoff*.

Hard Handoff

As illustrated in Fig. 17.17, a hard handoff is triggered when the signal strength from the existing BS perceived by the MN moving out of the cell is below a threshold before connecting to the new BS. The MN occupies only one channel at a time: the channel in the source cell is released and only then the channel in the target cell is engaged. Hence, the connection to the source is broken before or as the connection to the target is made. For this reason, hard handoff is also referred to as *break before make*. To minimize the impact of the event, the operations have to be short that causes almost no user-perceptible disruption to the session. In the early analog systems, it could be heard as a click or a very short beep; in modern digital systems, it is generally unnoticeable.

The implementation of hard handoff is relatively simple as the hardware does not need to be capable of receiving two or more channels in parallel. In GSM, the decision is done by the BS with mobile's assistance, which reports the signals strength back to the BS. The BS also knows the availability of channels in the nearby cells through information exchange. If the network decides that it is necessary for the mobile to hand off, it assigns a new channel and time slot to the mobile, and then informs the BS and the mobile of the change.

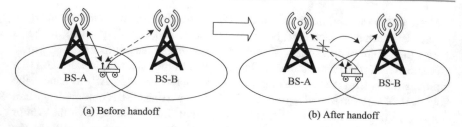

(a) Before handoff   (b) After handoff

Fig. 17.17 An illustration of hard handoff. Before handoff, the mobile node is connected to BS-A, but the signal strength becomes weaker when it is moving toward BS-B. At a certain time when the signal strength of BS-B is above a threshold (and that of BS-A becomes very weak), a hard handoff decision will be made, such that the connection to BS-A is first broken and the connection to BS-B is then established

The ongoing session however can be temporarily disrupted if the hard handoff fails. Reestablishment is then needed, which may be noticeable to the users and sometimes could fail, causing a session to be terminated abnormally. Also when the mobile stays between base stations, it can bounce back and forth, causing an undesirable *ping-pong* phenomenon.

Soft Handoff

In a soft handoff, the channel in the source cell is retained and used for a while in parallel with the channel in the target cell. In this case, the connection to the target is established before the connection to the source is broken. The interval, during which the two connections are used in parallel, may be brief or substantial. For this reason the soft handoff is also referred to as *make before break*, and is perceived by network engineers as a state of the call, rather than a instant event as in hard handoff.

One advantage of the soft handoff is that the connection to the source cell is broken only when a reliable connection to the target cell has been established, and therefore the chances that the call is terminated abnormally due to failed handoffs are lower. A soft handoff may involve using connections to more than two cells— connections to three, four or more cells can be maintained at the same time; the best of these channels can be used for the call at a given moment or all the signals can be combined to produce a clearer copy of the signal. Since fading and interference in different channels are not necessarily correlated, the probability of them taking place at the same moment in all channels is very low. Thus, the reliability of the connection becomes higher.

Soft handoff permits a smooth handoff that is critical to continuous media data flows. This advantage comes at the cost of more complex hardware and software in the device, which must be capable of receiving and processing several channels in parallel. This can be realized in CDMA or WCDMA through different transmission codes on different physical channels, so for 4G and 5G. In 5G, 0ms interruption time can be achieved by using intra-cell beam mobility and addition/release of small cell for dual connectivity operation, whereby simultaneous connections with the source cell and the target cell are maintained [38].

Vertical Handoff

A more interesting and complex handoff is between different types of networks, known as *vertical handoff* [39]. A typical example is between Wi-Fi and cellular networks, as the former is cheaper and fast and latter is of broader and ubiquitous coverage. Switching between them therefore combines their advantages [40].

A typical vertical handoff consists of three steps, namely, *system discovery, handoff decision*, and *handoff execution*. During the discovery phase, the mobile terminal determines which networks can be used. These networks may also advertise the supported data rates and the QoS parameters. In the decision phase, the mobile terminal determines whether the connections should continue using the current network or be switched to the target network. The decision may depend on various parameters or metrics including the type of the application (e.g., conversational or one-way streaming), the minimum bandwidth and delay required by the application, the transmit power, and the user's preferences. During the execution phase, the connections in the mobile terminal are rerouted from the existing network to the target network in a seamless manner.

The 3G networks support multimedia transmissions with a bitrate of 384 kbps for fast mobility to 2 Mbps for slow mobility; the 4G and 5G achieve even higher rates up to 100 Mbps and 1 Gbps, respectively. 5G also has a massive amount of antennas and limited network coverage due to higher radio channel attenuation. To enable smooth vertical handoff, besides the mobility solutions in the data link and network layers, additional support from the transport and application layers with cross-layer optimizations are also needed. For example, if the transport layer or application layer is aware of a potential handoff, then prefetching could be executed by the BS in the target cell, which can avoid the potential service interruption, thus enabling continuous streaming. In addition, software defined networking (SDN) and network functions virtualization (NFV) have been adopted in 5G, and their potentials are also under investigation.

17.5 Further Exploration

Rappaport [1], Goldsmith [2], and Tse and Viswanath [3] offer comprehensive and in-depth tutorials on the foundations of wireless communication. Wang et al. [33] give an in-depth discussion on error control in video communications. Recent advances on Wi-Fi standards can be found on Wi-Fi Alliance's website www.wi-fi.org, and that on cellular network standards can be found on 3GPP's website www.3gpp.org.

17.6　Exercises

1. In the implementations of TDMA systems such as GSM, an FDMA technology is still in use to divide the allocated carrier spectrum into smaller channels. Why is this necessary?

2. Consider the hard handoff and soft handoff for mobile terminals moving across cells.

   (a) Why is a soft handoff possible with CDMA? Is it possible with TDMA or FDMA?
   (b) Which type of handoff works better with multimedia streaming?

3. Identify the key technologies that drive the bandwidth expansion from 2G to 5G. Besides bandwidth expansion, what other advantages will be brought by 5G and how will they benefit multimedia applications.

4. Most of the schemes for channel allocation discussed in this chapter are fixed (or uniform) channel assignment schemes. It is possible to design a dynamic channel allocation scheme to improve the performance of a cellular network. Suggest such a dynamic channel allocation scheme.

5. The Gilbert-Elliot two-state Markov model has been widely used in simulations to characterize wireless errors, as illustrated in Fig. 17.2.

   (a) Given the state transition probabilities p_{00}, p_{11}, p_{10}, p_{01}, calculate steady-state probability P_0 and P_1 that the wireless channel is in state 0 and state 1, respectively.
   (b) Write a simple program to simulate the process. Run it for a long enough time and calculate the average length of error bursts. Discuss how it would affect multimedia data transmission.

6. Consider a wireless network whose signal does not decay dramatically, i.e., within the network range, the signal strength is always high enough. However, the signal can be blocked by physical barriers. Will this network have the hidden terminal problem? Briefly explain your answer.

7. Discuss the relationship between 5G and Wi-Fi 6 and their respective use cases for multimedia applications.

8. In today's networks, both the transport layer and link layer implement error detection mechanisms. Why do we still need error detection in the link layer given that the transport layer protocol, say TCP, assumes that the lower layers of a network is unreliable and seeks to guarantee reliable data transfer using error detection and retransmission? Hint: Consider the performance gain.

9. Discuss the error detection and correction capability of the two-dimensional parity check.

10. Calculate the Internet checksum of the following message: 10101101 01100001 10001000 11000001.

11. Consider cyclic redundancy check (CRC).

(a) Assume the keyword, K, is 1001, and the message M is 10101110. What is the width (in bits) of the CRC bits, R? What is the value of R? Please give detailed calculations.

(b) Prove that $M \cdot 2^r \oplus R$ is perfectly divisible by K, and verify it using the M, K, and R values above.

12. Discuss why interleaving increases the delay in decoding? Will interleaving be effective if the loss is uniformly distributed?

13. RVLC has been used in H.263+/MPEG-4 and beyond, which allows decoding of a stream in both forward and backward directions from a synchronization marker.

(a) Why is decoding from both directions preferred?

(b) Why is this beneficial for transmissions over wireless channels?

(c) What condition is necessary for the codes to be reversibly decodable? Are these two set of codes reversible: (00, 01, 11, 1010, 10010) and (00, 01, 10, 111, 110)?

(d) Why are RVLCs usually applied only to motion vectors?

14. Suggest two error concealment methods for audio streaming over wireless channels.

15. There is a broad spectrum of device and user mobility, in terms of both range and speed, as illustrated in Fig. 17.14. Discuss the challenges in the different mobility scenarios, and the potential solutions.

16. To alleviate triangular routing, a CN can also keep the mapping between the mobile's HoA and CoA, and accordingly encapsulate packets to the mobile directly, without going through the HA.

(a) In which scenario does this *direct routing* solution work best?

(b) Discuss any potential problem with the direct routing solution.

(c) Propose another solution that addresses the triangular routing problem. Discuss its pros and cons.

References

1. T.S. Rappaport, *Wireless Communications: principles and Practice*, 2nd edn. (Pearson Education, 2010)
2. A. Goldsmith, *Wireless Communications*. (Cambridge University Press, 2005)
3. D. Tse, P. Viswanath, *Fundamentals of Wireless Communication*. (Cambridge University Press, 2005)
4. H.-S. Wang, N. Moayeri, Finite-state Markov channel—A useful model for radio communication channels. Veh. Technol. IEEE Trans. **44**(1), 163–171 (1995)
5. E.N. Gilbert, Capacity of a burst-noise channel. Bell Syst. Tech. J. (1960)
6. M. Rahnema, Overview of GSM system and protocol architecture. IEEE Commun. Mag. 92–100 (1993)

7. A.J. Viterbi, *CDMA: principles of Spread Spectrum Communication*. (Addison Wesley Longman, 1995)
8. M.2134–Requirements Related to Technical Performance for IMT-Advanced Radio Interface(s). Technical report, ITU-R (2008)
9. A. Technologies, M. Rumney, *LTE and the Evolution to 4G Wireless: design and Measurement Challenges*. (Wiley, 2013)
10. M. Baker, From LTE-advanced to the future. Commun. Mag. IEEE **50**(2), 116–120 (2012)
11. C. Zhang, S.L. Ariyavisitakul, M. Tao, LTE-advanced and 4G wireless communications [Guest Editorial]. Commun. Mag. IEEE **50**, 102–103 (2012)
12. C. Wang, F. Haider, X. Gao, X. You, Y. Yang, D. Yuan, H.M. Aggoune, H. Haas, S. Fletcher, E. Hepsaydir, Cellular architecture and key technologies for 5G wireless communication networks. IEEE Commun. Mag. **52**(2), 122–130 (2014). February
13. 5G-PPP. 5G Vision: the next generation of communication networks and services. *European Commission White Paper* (2015)
14. M. Agiwal, A. Roy, N. Saxena, Next generation 5G wireless networks: a comprehensive survey. IEEE Commun. Surv. Tutor. **18**(3), 1617–1655 (2016)
15. Q. Bi, Ten trends in the cellular industry and an outlook on 6G. IEEE Commun. Mag. **57**(12), 31–36 (2019). December
16. P. Rost, C. Mannweiler, D.S. Michalopoulos, C. Sartori, V. Sciancalepore, N. Sastry, O. Holland, S. Tayade, B. Han, D. Bega, D. Aziz, H. Bakker, Network slicing to enable scalability and flexibility in 5G mobile networks. IEEE Commun. Mag. **55**(5), 72–79 (2017). May
17. H. Zhang, N. Liu, X. Chu, K. Long, A. Aghvami, V.C.M. Leung, Network slicing based 5G and future mobile networks: mobility, resource management, and challenges. IEEE Commun. Mag. **55**(8), 138–145 (2017). Aug
18. J. Ordonez-Lucena, P. Ameigeiras, D. Lopez, J.J. Ramos-Munoz, J. Lorca, J. Folgueira, Network slicing for 5G with SDN/NFV: concepts, architectures, and challenges. IEEE Commun. Mag. **55**(5), 80–87 (2017). May
19. J. Burkhardt, et al., *Pervasive Computing: technology and Architecture of Mobile Internet Applications*. (Addison Wesley Professional, 2002)
20. E. Perahia, R. Stacey, *Next Generation Wireless LANs: 802.11n and 802.11ac*. (Cambridge University Press, 2013)
21. L. Harte. *Introduction to Bluetooth*, 2nd edn. (Althos, 2009)
22. S. Naderiparizi, M. Hessar, V. Talla, S. Gollakota, J.R. Smith, Towards battery-free HD video streaming, in *Proceedings of the 15th USENIX Conference on Networked Systems Design and Implementation, NSDI'18*, pp. 233–247. (USA, USENIX Association, 2018)
23. G. Dimopoulos, I. Leontiadis, P. Barlet-Ros, K. Papagiannaki, P. Steenkiste, Identifying the root cause of video streaming issues on mobile devices, in *Proceedings of the 11th ACM Conference on Emerging Networking Experiments and Technologies, CoNEXT '15*. (ACM, New York, NY, USA, 2015)
24. Third Generation Partnership Project 2 (3GPP2). Video conferencing services—stage 1. *3GGP2 Specifications*, S.R0022 (2000)
25. Third Generation Partnership Project (3GPP). QoS for speech and multimedia codec. *3GPP Specifications*, TR-26.912 (2000)
26. A. Houghton, *Error Coding for Engineers*. (Kluwer Academic Publishers, 2001)
27. T.K. Moon, *Error Correction Coding: Mathematical Methods and Algorithms*. (Wiley-Interscience, 2005)
28. J.F. Kurose, K.W. Ross, *Computer Networking: a Top-Down Approach*, 7th edn. (Pearson, New York, 2016)
29. E.K. Wesel, *Wireless Multimedia Communications: networking Video, Voice, and Data*. (Addison-Wesley, 1998)
30. K.N. Ngan, C.W. Yap, K.T. Tan, *Video Coding For Wireless Communication Systems* (Marcel Dekker Inc, New York, 2001)
31. Y. Takishima, M. Wada, H. Murakami, Reversible variable length codes. IEEE Trans. Commun. **43**(2–4), 158–162 (1995)

32. C.W. Tsai, J.L. Wu, On constructing the Huffman-code-based reversible variable-length codes. IEEE Trans. Commun. **49**(9), 1506–1509 (2001)
33. Y. Wang, J. Ostermann, Y.Q. Zhang, *Video Processing and Communications*. (Prentice Hall, 2002)
34. M. Usman, X. He, M. Xu, K.M. Lam, Survey of error concealment techniques: research directions and open issues, in *2015 Picture Coding Symposium (PCS)*, pp. 233–238 (2015)
35. Y. Wang, Q.F. Zhu, Error control and concealment for video communication: a review. Proc. IEEE **86**(5), 974–997 (1998)
36. D. Le, F. Xiaoming, D. Hogrefe, A review of mobility support paradigms for the internet. Commun. Surv. Tutor. IEEE **8**(1), 38–51 (2006)
37. D. Saha, A. Mukherjee, I.S. Misra, M. Chakraborty, Mobility support in IP: a survey of related protocols. Netw. IEEE **18**(6), 34–40 (2004)
38. M. Tayyab, X. Gelabert, R. Jäntti, A survey on handover management: from LTE to NR. IEEE Access **7**, 118907–118930 (2019)
39. J. McNair, F. Zhu, Vertical handoffs in fourth-generation multinetwork environments. Wirel. Commun. IEEE **11**(3), 8–15 (2004)
40. J. Sommers, P. Barford, Cell versus WiFi: on the performance of metro area mobile connections, in *Proceedings of the 2012 ACM Conference on Internet Measurement Conference, IMC '12*, pp. 301–314. (ACM, New York, NY, USA, 2012)

Cloud Computing for Multimedia Services

18

The emergence of *cloud computing* [1] has dramatically changed the service models for modern computer applications. Utilizing elastic resources in powerful data centers, it enables end users to conveniently access computing infrastructure, platforms, and software provided by remote cloud providers (e.g., Amazon, Google, and Microsoft) in a pay-as-you-go manner or with long-term lease contracts. This new generation of computing paradigm, offering reliable, elastic and cost-effective resource provisioning, can significantly mitigate the overhead for enterprises to construct and maintain their own computing, storage, and network infrastructures. It has provided countless new opportunities for both new and existing applications.

Existing applications, from content sharing and file synchronization to media streaming, have experienced a leap forward in terms of system efficiency and usability through leveraging cloud computing platforms. These advances mainly come from exploiting the cloud's massive resources with elastic provisioning and pricing and with smart computational offloading.

On the other hand, start-up companies can easily implement their novel ideas into real products with minimum investment in the initial stage and expand the system scale without much effort later on. A representative is Dropbox, a typical cloud storage and file synchronization service provider, which, from the very beginning, has relied on Amazon's S3 cloud servers for file storage and used Amazon's EC2 cloud instances to provide such key functions as synchronization and collaboration among different users. Other examples include crowdsourced livecast service, Twitch.tv (later being integrated into Amazon's platform), and ultra short video sharing service, Tik Tok (using ByteDance's own cloud platform). We have also seen such new generation of multimedia services as cloud-based VoD and gaming that have emerged in the market and is actively changing the whole business model in these years. A prominent example is Netflix, a major Internet streaming video provider, which started migrating its infrastructure to Amazon's cloud platform in 2008, taking around

© Springer Nature Switzerland AG 2021
Z.-N. Li et al., *Fundamentals of Multimedia*, Texts in Computer Science,
https://doi.org/10.1007/978-3-030-62124-7_18

10 years till 2018 to completely eliminate its own infrastructure. It has since become one of the most important cloud users. In total, Netflix has over 1 petabyte of media data stored in Amazon's cloud. It pays by bytes for bandwidth and storage resources, so that the long-term costs become much lower than those with over-provisioning in self-owned servers. It has also embraced Google Cloud for certain workloads, mostly focused around intelligent user behavior analysis.

In the conventional cloud computing paradigm, an end user and the remote cloud data center are connected by the Internet, which typically incurs a delay in the hundred-millisecond level. This is not acceptable for many time-critical applications. There are also bandwidth cost and data privacy concerns on the long haul Internet transmission. As such, *Edge computing*, which pushes the function of cloud to the network edge that is closer to end users, has gained tremendous interest in recent years and has become an integral part of the modern cloud and 5G systems.

Serverless computing, as a new execution model for cloud applications, offloads the task of server provisioning and management at a fine granularity. In a serverless platform, developers only need to break up application codes into a collection of stateless functions and set events to trigger their executions; Platforms are responsible for handling every trigger and scaling precisely with the size of workloads with low overhead and fast response, which is quite suitable for bursty multimedia workloads.

In this chapter, we provide an overview of cloud computing, focusing on its impact on multimedia services. We discuss multimedia content sharing with cloud storage and multimedia computation offloading to the cloud. We then use cloud gaming as a case study to examine the role of the cloud in the new generation of interactive multimedia services. We also discuss the recent advances in edge computing and serverless computing.

18.1 Cloud Computing Overview

Cloud computing relies on sharing of resources to achieve coherence and economies of scale similar to a utility over a network, like the electricity grid does. As illustrated in Fig. 18.1, cloud users can run their applications on powerful server clusters offered by the cloud service provider, with system and development software readily deployed inside the cloud, mitigating the users' burden of full installation and continual upgrade on their local hardware/software. A cloud user can also store their data in the cloud instead of on their own devices, making ubiquitous data access possible.

At the foundation of cloud computing is the broader concept of resource virtualization and sharing. Focusing on maximizing the utilization of aggregated physical resources, cloud resources are shared by multiple users with dynamical on-demand allocation. For example, a cloud can serve European users during their business hours, while the same set of resources can later be used by North American users during their business hours. And each user sees its own dedicated virtual space.

The cloud services can be *public* or *private*. In a public cloud, the services and infrastructure are provided off-site over the public Internet. These clouds offer the

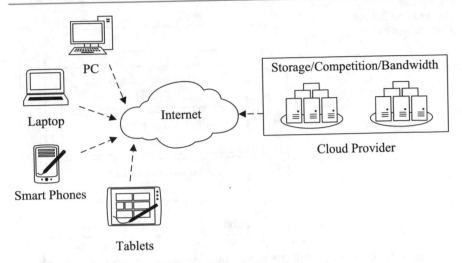

Fig. 18.1 A conceptual overview of cloud computing

greatest level of efficiency in resource sharing; however, they can be less secure and more vulnerable than private clouds. Unlike public clouds, in a private cloud, the services and infrastructure are maintained on a private network. These clouds offer the greatest level of security and control, though they require the company to still purchase and maintain the software and infrastructure. In either case, there are a common set of essential characteristics of cloud computing, as identified by the National Institute of Standards and Technology (NIST) [2]:

On-demand self-service. A user can unilaterally provision computing capabilities (e.g., server time and network storage) as needed without human interaction with each service provider;

Resource pooling and rapid elasticity. The provider's resources are pooled to serve multiple users, with different physical and virtual resources dynamically assigned and reassigned according to user demand. To a cloud user, the resources available for provisioning often appear unlimited and can be appropriated in any quantity at any time;

Measured service. Cloud systems automatically control and optimize resource use by leveraging a metering capability at an abstraction level appropriate to the type of service (e.g., storage, processing, bandwidth, and active user accounts). The resource usage can be monitored, controlled, and reported, providing transparency for both the provider and the users;

Broad network access. Persistent and quality network accesses are available to accommodate heterogeneous client platforms (e.g., mobile phones, tablets, laptops, and workstations).

In marketing, cloud services are mostly offered from data centers with powerful server clusters in three fundamental models (see Fig. 18.2): *infrastructure as a service* (IaaS), *platform as a service* (PaaS), and *software as a service* (SaaS), where IaaS is the most basic and each higher model abstracts from the details of the lower models. Network as a service (NaaS) and communication as a service (CaaS) have

Fig. 18.2 An illustration of
cloud service models

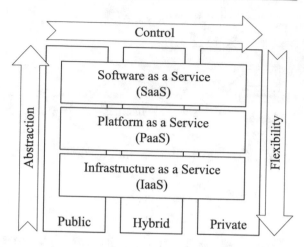

been recently added as part of the basic cloud computing models too, enabling a
telecommunication-centric cloud ecosystem.

Infrastructure as a Service (IaaS)

IaaS is the very basic and essential cloud service. Well-known examples of IaaS
include such infrastructure vendor environments as the Amazon's Elastic Compute
Cloud (EC2), which allow users to rent virtual machines on which to run applications,
and such cloud storage services as Amazon's Simple Storage Service (S3), which
allow users to store and retrieve data, at any time, from anywhere on the web.

In general, an IaaS provider offers a pool of computation resources, in the form of
physical machines, or more often, *virtual machines* (VMs), as well as other resources,
e.g., virtual-machine disk image library, data block or file-based storage, firewalls,
load balancers, IP addresses, virtual local area networks, and software bundles. For
wide-area connectivity, the users can use either the public Internet or dedicated virtual
private networks.

To deploy their applications, cloud users install operating-system images and their
application software on the cloud infrastructure. With machine virtualization, an IaaS
cloud provider can support a large numbers of users with its pool of hardware and
software, and scale services up and down according to users' varying requirements.
The cloud providers typically bill IaaS services on a utility computing basis, where
the cost reflects the amount of resources allocated and consumed.

Any virtualization system must ensure that each VM is given fair and secure access
to the underlying hardware. In state-of-the- art virtualization systems, this is often
achieved through the use of a software module known as a *hypervisor*, which works
as an arbiter between a VM's virtual devices and the underlying physical devices [3].
The use of a hypervisor brings many advantages, such as device sharing, performance
isolation, and security between running VMs. Having to consult the hypervisor each
time a VM makes a privileged call, however, introduces considerable overhead as

the hypervisor must be brought online to process each request. A self-contained VM also incurs substantial resource overhead and is slow in creation, as each VM runs an OS.

Recently, light-weight *container-based virtualization*   [4–6], e.g., Docker, has been advocated in such cloud-based video services as Netflix. Each container holds the components necessary to run the desired software, including files, environment variables, dependencies, and libraries. The host OS constrains the container's access to physical resources, such as CPU, storage, and memory, so a single container will not consume all of a host's physical resources. The use of container also facilitates the migration of service abstraction from *monolithic* to *microservices* [6,7]. Instead of being implemented as a whole, each microservice implements its own functionality and communicates with other microservices through language- and platform-agnostic APIs. This enables much finer-grained control over the cloud applications for both production and operation.

Platform as a Service (PaaS)

PaaS delivers development environments as a service, which typically includes the operating system, programming language execution environment, database, web server, etc. Applications can be built and ran on the PaaS provider's infrastructure and then delivered to end users via the Internet. As such, the cost and complexity of purchasing and managing the underlying hardware and software layers can be greatly reduced. Moreover, the underlying computation and storage resources can scale automatically to match the applications' demands. Google's App Engine is a typical example of PaaS. Besides IaaS, Microsoft Azure also supports PaaS that allows developers to use software developer kits (.NET, Node.js, PHP, Python, Java, and Ruby, etc.) and Azure DevOps to create and deploy applications.

Software as a Service (SaaS)

SaaS, probably the most widely used cloud service model to date, allows an application to run on the infrastructure and platforms offered by the cloud rather than on local hardware/software. As such, the user of an application does not have to heavily invest on its own servers, software, license, etc. SaaS is usually priced on a usage basis, or with a monthly or yearly flat fee per user. The price is scalable and adjustable if users are added or removed at any point.

SaaS greatly reduces IT operational costs by outsourcing hardware and software maintenance and support to the cloud provider. This enables the business to reallocate IT operations costs away from hardware/software spending and personnel expenses, toward meeting other goals. Besides cost saving and simplified maintenance and support on the user's side, cloud applications also enjoy superior scalability, which can be achieved by cloning tasks onto multiple virtual machines at runtime to meet changing work demands. A load balancer can distribute the workload over the virtual machines. Yet these are transparent to the cloud users, who see only the virtual machine allocated to itself. Google Apps and Microsoft Office 365 are typical examples of SaaS. As a matter of fact, in the fourth quarter of the fiscal year 2017, Office 365 revenue overtook that of conventional license sales for the first time.

Figure 18.2 illustrates the relations among different cloud service models, particularly in terms of their abstraction, control, and flexibility levels.

Amazon has played a leading role in the development of cloud computing by modernizing their data centers. In 2006, Amazon initiated a new product development effort to provide cloud computing to external customers, and launched the *Amazon Web Services* (AWS) on a utility computing basis, which has since become one of the most widely used cloud computing platforms. In 2019, AWS comprised more than 165 services spanning a wide range including computing, storage, networking, database, analytics, application services, as well as deployment, management, and development tools.

We next examine two representative services provided by Amazon's AWS, namely, S3 for storage and EC2 for computation, both of which have been widely used for supporting multimedia services.

18.1.1 Representative Storage Service: Amazon S3

Cloud storage has the advantage of being *always-on*, so that users can access their files from any device and can share their files with others who may access the content at an arbitrary time. With advanced storage management, cloud storage also provides a much higher level of reliability than local storage, yet with comparable or lower costs. All these are critical to media sharing, one of the most demanding storage services.

Amazon's S3 provides a web service interface that can be used to store and retrieve any amount of data, at any time, from anywhere on the web. It gives a developer access to the same highly scalable, reliable, secure, fast, inexpensive infrastructure that Amazon uses to run its own global network of web sites. The service aims to maximize the benefits of scale and to pass those benefits on to developers and end users. To this end, S3 is intentionally built with a minimal feature set with simple operations.

An S3 user can write, read, and delete objects each containing from 1 byte to 5 terabytes of data. Each object is stored in a *bucket* and retrieved via a unique, developer-assigned key. The bucket can be stored in one of several *regions*. The S3 user can choose a region to optimize for latency, minimize costs, or address regulatory requirements. As of 2019, it has distinct operations in 22 geographical regions: 7 in North America, 1 in South America, 5 in Europe and Africa, 1 in Middle-East, and 8 in Asia Pacific. The objects stored in a region never leave it unless transferred out, and, like a CDN routing, a network map is used to route the request. Figure 18.3 shows an example of a data object stored in the US West (Oregon) region.

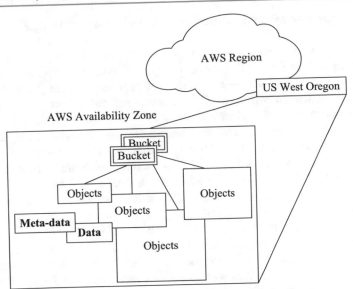

Fig. 18.3 An example of a data object stored in an Amazon AWS region (US West)

Table 18.1 Sample storage pricing of US east region (Year 2020)

	Standard storage	Intelligent tier (per GB)	Glacier storage (per GB)
First 50 TB/month	$0.023 per GB	$0.023	$0.004
Next 450 TB/month	$0.022 per GB	$0.022	$0.004
Over 500 TB/month	$0.021 per GB	$0.021	$0.004
Infrequent access/month	–	$0.0125	$0.004

S3 is built to be flexible so that protocols or functional layers can be easily added. The default download protocol is HTTP. A BitTorrent protocol interface is also provided to lower the costs for large-scale distribution.

The data stored in Amazon S3 is secure by default; only the bucket and object owners have access to the S3 resources created by them. It supports multiple access control mechanisms, as well as encryption for both secure transit and secure storage on disk. To increase durability, it synchronously stores the data across multiple facilities, and calculates checksums on all network traffic to detect corruption of data packets when storing or retrieving data. Unlike traditional storage systems that require laborious data verification and manual repair, S3 performs regular, systematic data integrity checks and is built to be automatically self-healing.

S3 also automatically archives objects to even lower cost storage options or perform recurring deletions, reducing the costs over an object's lifetime. The costs can be monitored and controlled by the users through the S3 APIs or Management Console. Table 18.1 shows a sample of the current pricing plan of the S3 services.

18.1.2 Representative Computation Service: Amazon EC2

Amazon's *Elastic Compute Cloud* (EC2) is a web service that provides resizable compute capacity in the cloud. It presents a virtual computing environment, allowing users to launch instances with a variety of operating systems, load them with customized application environment, and manage network access permissions. Different instance provisioning plans are available to meet a user' demands:

On-Demand Instances: On-Demand Instances let the user pay for compute capacity by hour with no long-term commitments. This frees the user from the costs and complexities of planning, purchasing, and maintaining hardware/software and transforms the commonly large fixed costs into much smaller variable costs. They also remove the need to buy safety net capacity to handle periodic traffic spikes;

Reserved Instances: Reserved Instances give the user the option to make a one-time payment for each instance it wants to reserve and in turn receive a significant discount on the hourly charge for that instance. There are three Reserved Instance types: *light, medium,* and *heavy utilization reserved*, which enable the user to balance the amount it pays upfront with effective hourly prices. A *Reserved Instance Marketplace* is also available, which provides users with the opportunity to sell the instances if their needs change (for example, want to move instances to a new AWS region, change to a new instance type, or sell capacity for projects that end before the reservation term expires);

Spot Instances: Spot Instances allow users to bid on unused EC2 capacity and run those instances for as long as their bid exceeds the current spot price. The spot price changes periodically based on supply and demand, and the user whose bids meet or exceed the price gains access to the available instances. If the user has flexibility in when the applications can run, using spot instances can significantly lower the costs.

An EC2 user has the choice of multiple instance types, operating systems, and software packages. It can select a configuration of memory, CPU, instance storage, and the optimal boot partition size that is optimal for specific choice of applications and operating systems, e.g., Linux distributions or Microsoft Windows Server. The user can also increase or decrease capacity within minutes, and has complete control of the instances with root access. Moreover, to achieve reliability, replacement instances can be rapidly and predictably commissioned. For each region, the current Service Level Agreement commitment is 99.95% availability.

The instance creation and configuration can be done through simple web service interfaces, as illustrated in Fig. 18.4. The user first selects a pre-configured *Amazon Machine Image* (AMI) template to boot up and run immediately (Step 1); or create an AMI containing the applications, libraries, data, and associated configuration settings. The user then chooses the expected instance type(s) (Steps 2 and 3), attaches storage, and set up network and security requirements, and starts, terminates, and monitors the instances using the web service APIs or a variety of management tools. Security and network access can also be configured on the instance.

Designed for use with other AWS modules, EC2 works seamlessly in conjunction with Amazon S3, and such other Amazon services as Relational Database Service (RDS), SimpleDB and Simple Queue Service (SQS) to provide a complete solution

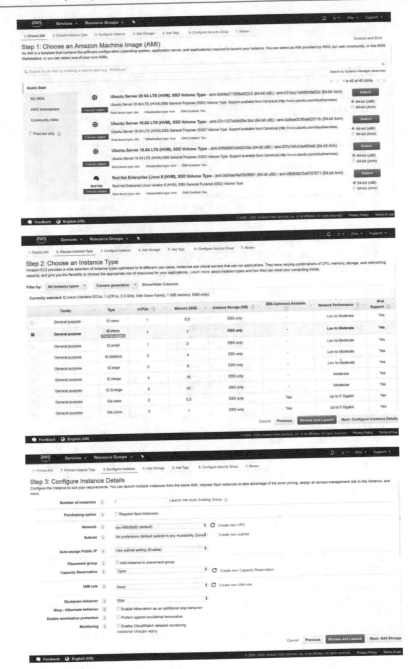

Fig. 18.4 Key steps in creating an Amazon EC2 instance

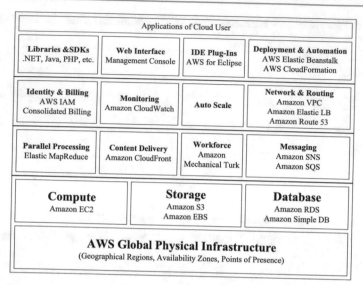

Fig. 18.5 Relations among the different components in Amazon Web Service (AWS)

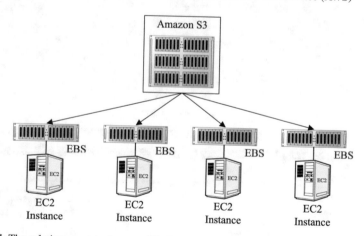

Fig. 18.6 The relations among Amazon S3, EC2, and the Elastic Block Store (EBS). EBS offers storage volumes from 1 GB to 1 TB that can be mounted as devices for EC2 instances, and the persistent storage is enabled by S3

for computing, query processing and storage across a wide range of applications, as Fig. 18.5 illustrates.

Specifically, an EC2 instance can be attached with a storage from the *Amazon Elastic Block Store* (EBS), which provides block level storage volumes from 1 GB to 1 TB. EBS provides the ability to create point-in-time snapshots of volumes, which can then be stored in S3 for long-term durability, as Fig. 18.6 illustrates.

18.2   Multimedia Cloud Computing

For multimedia applications and services over the Internet, there are strong demands for cloud computing, due to the massive storage and computation required for serving millions of wired Internet or mobile network users. In this new *multimedia cloud computing* paradigm [8], users can store and process their multimedia data in the cloud in a distributed manner, eliminating full installation of the media application software on local computers or devices.

Multimedia cloud computing shares many common characteristics with general-purpose cloud computing. Yet multimedia services are highly heterogeneous. There exist a wide array of types of media and associated services, such as voice over IP, video conferencing, photo sharing and editing, image search, and image-based rendering, to name but a few; the cloud should support different types of media and their services for massive user bases simultaneously. Besides service heterogeneity, different types of devices, such as smart TVs, PCs, and smartphones, have different capabilities for multimedia processing. The cloud should have adaptive capability to fit these types of devices, in terms of CPU and graphics processing unit (GPU), display, memory, and storage.

The multimedia cloud should also provide QoS provisioning to meet their distinct QoS requirements. There are two ways of providing QoS provisioning for multimedia: one is to add QoS to the current cloud-computing infrastructure and the other is to add a QoS middleware between the cloud infrastructure and the multimedia applications. The former focuses on the design and improvement within the cloud infrastructure. The latter focuses on improving cloud QoS in the middle layers, such as QoS in the transport layer and QoS mapping between the cloud infrastructure and media applications.

In summary, the heavy demands of multimedia data access, processing, and transmission would create bottlenecks in a general-purpose cloud. Today's cloud design has largely focused on allocating computing and storage resources through utility-like mechanisms, while QoS requirement in terms of bandwidth, delay, and jitter has yet to be addressed. To realize multimedia cloud computing, a synergy between the cloud and multimedia services becomes necessary; that is, *multimedia-aware cloud computing* with enhanced QoS support for multimedia applications and services, and *cloud-aware multimedia* that facilitates content storage, processing, adaptation, rendering, in the cloud with optimized resource utilization [8], as Fig. 18.7 illustrates.

18.3   Multimedia Content Sharing over Cloud

We first consider the use of the cloud for media sharing services. As we have seen in the previous chapters, representative media sharing services such as YouTube are developing extremely fast. In general, it is difficult if not impossible to predict the impact and the development of these new services in advance. The provision of resources is thus a great challenge, because any service with novel ideas, advanced techniques, and smart marketing strategies is possible to grow to the similar scale as

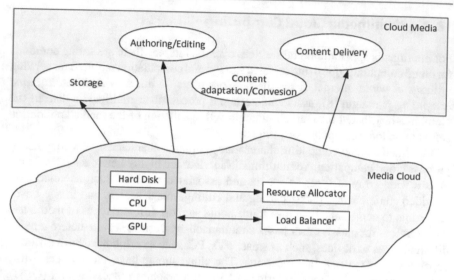

Fig. 18.7 Modules and their relations in multimedia cloud computing

YouTube. Yet there is a high possibility to fail, losing revenue and even being shut down.

Developers face a dilemma at the early stage of media sharing services. On one hand, to provision large enough resources at the beginning is costly and risky, and if the service is not as popular as expected to gain enough revenue, the resources would be wasted. On the other hand, starting the service small usually comes with scalability issues. New features and increasing user base will put high pressure on the insufficient infrastructure, which downgrades the quality of service.

The cloud, which offers reliable, elastic, and cost-effective resource provisioning with "pay-as-you-go" service, is clearly an elegant solution here, allowing designers to start a service small but easy to scale large. Besides starting a service from the cloud, a migration that moves the contents to the cloud is beneficial as well for existing media services, in the presence of scalability challenges.

Sharing is an essential part of cloud service. The request for easy and scalable sharing is the main reason that multimedia contents now occupy a large portion of cloud storage space. The always-on and centralized data centers can make one-to-many sharing highly efficient and synchronous—uploaded content can be readily shared to a large population instantly or lately. Sharing through a cloud (and edge) could also offer better QoS given that the connections with data centers are generally good, not to mention the firewall and network address translation (NAT) traversal problems commonly encountered in peer-to-peer sharing.

Figure 18.8 shows a generic framework that facilitates the migration of existing live media streaming services to a cloud-assisted solution. It is divided into two layers, namely, *Cloud Layer* and *User Layer*. The Cloud Layer consists of the live media source and dynamically leased cloud servers at the edge. Upon receiving a user's subscription request, the Cloud Layer will redirect this user to a properly selected

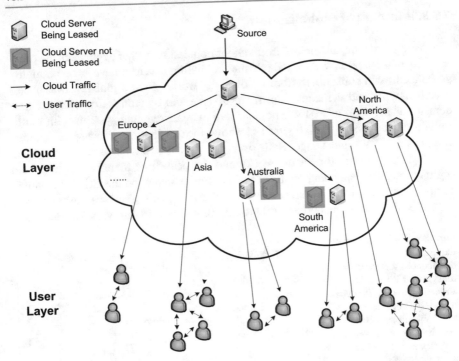

Fig. 18.8 A generic framework for migrating live media streaming service to the cloud

cloud server. Such a redirection, however, is transparent to the user, i.e., the whole Cloud Layer is deemed to be a single source server from a user's perspective. Since the user demands change over time, which are also location-dependent, the Cloud Layer will accordingly adjust the amount and location distribution of the leased servers at the edge. Intuitively, it will lease more server resources upon demand increase during peak times, and terminate leases upon decrease. The implementation of the User Layer can be flexible. They can be individual users purely relying on the Cloud Layer, or served by peer-to-peer or a CDN infrastructure, but seeking for extra assistance from the cloud during load surges. In other words, it can smoothly migrate diverse existing live streaming systems to the cloud.

There are, however, a number of critical theoretical and practical issues to be addressed in this generic framework. Though cloud services are improving, given the hardware, software, and network limits, latencies in resource provisioning remain exist, e.g., to start or terminate a virtual machine can take a few minutes in the current Amazon EC2 implementation. While such latencies have been gradually reduced with improved cloud/edge design, they can hardly eliminated. Therefore, the system must well predict when to lease new servers to meet the changing demands and when to terminate a server to minimize the lease costs. This can be done by a demand forecast algorithm for cloud users [9].

18.3.1 Impact of Globalization

The larger, dynamic, and non-uniform client population further aggravates the problem. To make it even worse, today's media sharing services have become highly globalized, with subscribers from all over the world. Such a globalization makes user behaviors and demands even more diverse and dynamic. Consider the user demand distribution of PPTV, a popular live media streaming systems with multi-million subscribers [10,11]. Figure 18.9 shows the distribution of two representative channels (CCTV3 and DragonBall) over one day. It is easy to see that they had attracted users from all over the world, and the peak time therefore shifted from region to region, depending on the timezone. For example, on the CCTV3 channel, the peak time of North America was around 20:00, while for Asian users, it was around 8:00. During the period 12:00 to 20:00, Asian users had very low demands,

Fig. 18.9 An illustration of the user demand distributions and variations of a popular live media streaming system (PPTV) on its two typical channels (CCTV3 and DragonBall). For ease of comparison, the user demands have been normalized by the corresponding maximum demand of each day. The time shown on x-axis is based on EST and spans slightly over one day, beginning from the night of the first day

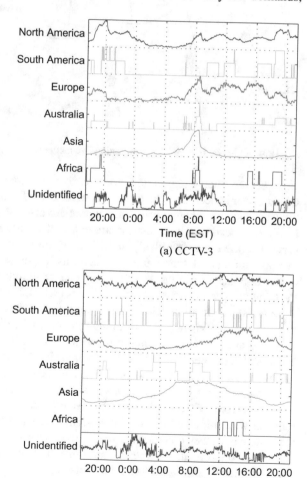

(a) CCTV-3

(b) DragonBall

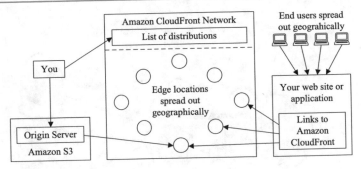

Fig. 18.10 An illustration of the CloudFront service, where a number of edge servers are geographically distributed, serving nearby cloud users

while European users generated most of their demands and the North American users also had moderate demands. Similar observations can also be found from the DragonBall channel, despite the fact that the streaming contents delivered on the two channels were completely different.

In this context, cloud, edge, and content distribution network (CDN) are to be integrated to serve the users with geo-distributed servers. This in fact a general trend in today's cloud development beyond highly centralized data centers. One example is Amazon's CloudFront, a cloud-based CDN that is integrated in AWS. Using a network of edge locations around the world, CloudFront caches copies of static content close to viewers, lowering latency when they download objects and offering high, sustained data transfer rates needed to deliver large popular objects to end users at scale. The requests for dynamic content can be carried back to the origin servers running in AWS, e.g., S3, over optimized network paths (see Fig. 18.10). These network paths are constantly monitored by Amazon, and the connections from CloudFront edge locations to the origin can be reused to serve dynamic content with the best possible performance.

18.3.2 Case Study: Netflix

One of the most successful migrations of media sharing applications to the cloud is Netflix, which now takes up a third of US download Internet traffic during peak traffic hours. Established in 1997, Netflix began to move away from its original core business model of mailing DVDs by introducing video-on-demand via the Internet in early 2007. The original Netflix digital video distribution was based on a few large Oracle servers with a Java front end, with DVD subscription being the main business. Later in 2008, it had suffered from storage data corruption bugs that took service down. The rapidly increased scale creates another challenge, which, however, can hardly be predicted when building private server clusters, not to mention the staff and skills needed to run a large and high-growth-rate data center infrastructure. Since then, Netflix started using Amazon's AWS for part of its services, and moved

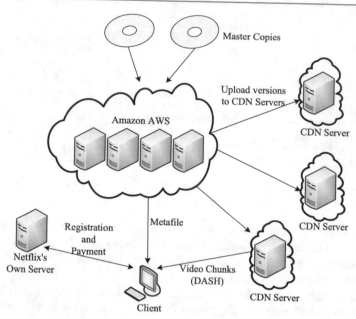

Fig. 18.11 The Netflix architecture during cloud migration. Its own servers were finally eliminated by 2018

its entire technology infrastructure to AWS in 2012. Till 2018, it has completely eliminating its own servers, including those for user management and billing.

To support the combination of huge traffic and unpredictable demand bursts, Netflix has developed a global video processing system using the AWS cloud. Figure 18.11 shows the architectural view of the cloud-based Netflix system, which includes the following key modules:

1. **Content Conversion**. Netflix purchases master copies of digital films from movie studios and, using the powerful EC2 cloud machines, converts them to over 50 different versions with different video resolutions and audio quality, targeting a diverse array of client video players running on desktop computers, smartphones, and even DVD/BluRay players or game consoles connected to television;
2. **Content Storage**. The master copies and the many converted copies are stored in S3. In total, Netflix has over 1 petabyte of data stored on Amazon;

These services are distributed across three AWS availability zones, even including the user registration and credit card payment. To serve worldwide users, the data are sent from S3 to content delivery networks (CDNs) that feed the local ISPs. Dynamic adaptive streaming over HTTP (DASH) is then available for streaming from the edge to end users with different demands on version, formats, and bit rates. The CDNs are used to include a combination of Akamai, Limelight, Level 3, etc., but have been replaced by OpenConnect, the company's proprietary CDN.

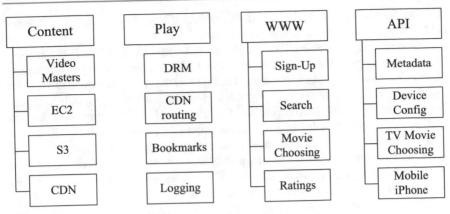

Fig. 18.12 The services integrated in Netflix and the cloud modules for them

Figure 18.12 shows different functional modules in Netflix and their relations with the cloud and CDNs.

Using Amazon's cloud and edge servers, Netflix can react better and faster to user demand with every increasing scale. It pays by bytes for computation, bandwidth and storage resources so that the long-term costs become much lower than those with over-provisioning in self-owned servers. Such costs, without the cloud, can be prohibitively high for dedicated content providers. Recently, Netflix further approaches container and microservice adoption in the already cloud-native infrastructure [12], achieving even lower overhead and cost with better responsiveness.

18.4   Multimedia Computation Offloading

Besides storage, computation is another rich resource offered by the cloud. Many computation-intensive tasks can now be migrated to the cloud, and the users do not have to maintain the ultra expensive high-performance servers or server clusters but just pay for the cost in an on-demand fashion.

Such *computation offloading* effectively expands the usability of local devices beyond their physical limits, which is particularly attractive for mobile devices [13–15]. Today's smartphones and tablets are increasingly penetrating into people's everyday life as efficient and convenient tools for communication and entertainment. The touch screen and all kinds of sensors provide even richer user experiences than desktop PCs do. Despite the fast development of such key components as CPU, GPU, memory, and wireless access technologies, and the effort toward unifying handheld and desktop computers, it remains widely agreed that mobile terminals will not completely replace laptop and desktop computers in the near future. Migrating popular PC software to mobile platforms or developing similar substitutes for them is still confined by their limited computation capability as well as the uniqueness of operating systems and hardware architectures. To make it even worse, battery, as the only power source of most mobile terminals, has seen relatively slow improvement in the past decade, which has become a major impediment in providing reliable and sophisticated mobile applications to meet user demands.

Combining the strength of the cloud and the convenience of mobile terminals thus becomes a promising route. An example is Apple's Siri service—after a piece of voice is recorded by an iPhone, a local recognizer will conduct speech recognition and decide whether to resort to the back end cloud to make an appropriate response. Other examples include MAUI [16] and CloneCloud [17]. The former enables fine-grained energy-aware offloading of mobile codes to a cloud based on the history of energy consumption. It achieves maximum energy savings of 90, 45, and 27%, and maximum performance speedups of roughly 9.5, 1.5, and 2.5 for face recognition, chess, and video game, respectively. CloneCloud uses function inputs and an offline model of runtime costs to dynamically partition applications between a weak device and the cloud. It reports speedups of 14.05, 21.2, and 12.43 for virus scanning, image search, and behavior profiling, respectively.

18.4.1 Requirements for Computation Offloading

Assembling local resources and remote clouds organically to make offloading transparent to end users requires non-trivial effort [14, 15].

First, *motivation for offloading*. In the very beginning, the major motivation to offload should be determined, to save energy locally, to improve computation performance, or both. This will serve as a guideline in the high-level design.

Second, *gain of offloading*. To understand the potential gain after offloading, a profiling or breakdown analysis of the application is needed to see whether the application can benefit from offloading. There is no incentive to resort to the cloud for a job that can be easily and efficiently executed locally.

Third, *decision of offloading*. The offloading decision can be made statically or dynamically. For static offloading, the related parameters need to be accurately estimated in advance, and the offloading strategy needs to be decided when developing the application. On the contrary, dynamic offloading monitors the runtime conditions and makes decisions accordingly, at the expense of higher overhead.

For multimedia applications, QoS requirements is also an important consideration. The critical QoS requirements (e.g., latency, image/video quality, computation accuracy) need to be respected so that offloading will not influence user experience. A simple solution for computation offloading is to move the whole computation engine to the remote cloud, then upload all the data required for computation to the cloud and download the computed results. While this is common in many implementations, complex enterprise applications are typically composed of multiple service components, which can hardly be migrated to the cloud in one piece and instantaneously.

This is further complicated with the wireless communications in mobile terminals. As compared to their wired counterparts, the mobile terminals are generally more resource-constrained; in particular, the wireless communication capacity and the battery capacity are their inherent bottlenecks [18, 19]. The energy trade-offs heavily depend on the workload characteristics, data communication patterns, and technologies used [20], all of which need to be carefully addressed. As such, offloading the whole computation module of an application to the remote cloud is not necessary, nor effective, if the data volume is large. This may not be a severe problem for users with high-speed wired network connection, but can dramatically reduce the benefit for mobile users.

18.4.2 Service Partitioning for Video Processing

Consider video encoding/compression/transcoding, an essential task in a broad spectrum of mobile multimedia applications. A local user uses his/her mobile terminal to capture video in real-time, expecting to encode or transcode the video and then stream it to others in real-time as well. Directly uploading the raw video without efficient encoding inevitably leads to high bandwidth cost and large transmission energy consumption. On the other hand, video encoding often incurs heavy computation, which results in high energy consumption as well. For example, to encode a video of 5 s (30 frames per second with resolution of 176×144 and pixel depth of 8 bits) using an H.264 encoder needs almost 1×10^{10} CPU cycles in total [21], or 2×10^9 CPU cycles per second on average, which means that a 2 GHz CPU is required for real-time encoding. Considering that the newest smartphones and tablets are equipped with high-definition cameras, the CPU workload can be 5–10 times higher than that in the above example.

Offloading the whole video compression task to the cloud, however, is not practical because it is identical to directly uploading the raw video data. The wireless transmission can be either too costly or simply impossible with limited bandwidth. It is known that motion-estimation is the most computation-intensive module, accounting for almost 90% of the computation. This module obviously should be the focus of offloading.

Yet it is not simple to decouple motion estimation from others given data dependency in coding, i.e., motion estimation of a frame depends on the data of the previous reference frame. A mobile terminal should upload the current video frame and the reference frame to the cloud for estimation and then download the estimated *motion vectors* (MVs) from the cloud and complete the remaining video encoding steps (e.g., DCT and entropy coding). While the MVs are of small volume, uploading the current frames essentially makes no difference as compared to upload the raw video.

One solution is to use *mesh-based motion estimation*, which partitions a frame into a coarse-grained mesh sketch and estimates one MV for each mesh node [22]. Regular triangular or rectangular meshes [23] have been commonly used given their simplicity, and both encoder and decoder can agree upon a mesh structure in advance. Unlike standard mesh-based motion estimation (see [22] for details), we need to use reversed mesh node selection and motion estimation, in which the mesh nodes are sampled on the P-frames and the MVs are calculated from the mesh nodes and the reference frame. As illustrated in Fig. 18.13, a set of regular mesh nodes are

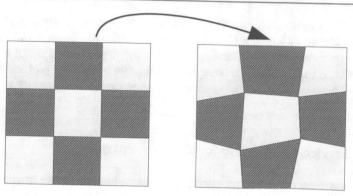

Fig. 18.13 Mesh-based motion estimation between two frames, using a regular mesh for the predictive frame

sampled on the P-frames of the macroblock(MB). Only the sampled mesh nodes and the reference frame are uploaded to the cloud. The MVs are then calculated on the cloud, and the result, to be sent back to the mobile terminal, is a set of MVs that describe the movements of the mesh nodes. Using this design, the most computation-intensive part of mesh-based motion estimation is offloaded to the cloud, while the local motion estimation for individual macroblocks within the mesh becomes much simpler. Compared to the standard mesh-based motion estimation, it loses the advantage of tracking the same set of mesh nodes over successive P-frames, yet saves more data transmission.

In summary, although it is very tempting and promising to leverage the much cheaper and more powerful resources on the cloud, the interaction between a mobile terminal and the cloud needs careful examination to avoid excessive transmission overhead. As such, partitioning tailored to specific applications is expected, as the example of motion estimation shows. The trade-off between the energy for computation and that for transmission can be found in many other applications that rely on computation offloading to extend the battery lifetime.

The closed-loop design with the remote cloud also introduces extra delays between mobile user and the cloud, which poses another challenge for interactive applications that will be discussed next.

18.5   Interactive Cloud Gaming

Recently, advances in cloud technology have expanded to allow offloading not only of traditional computation but also of such more complex tasks as high-definition 3D rendering, which turns the idea of *Cloud Gaming* into a reality [24,25]. Cloud gaming, in its simplest form, renders an interactive gaming application remotely in the cloud and streams the scenes as a video sequence back over the Internet to the player. A cloud gaming player interacts with the application through a *thin client*, which is responsible for displaying the video from the cloud rendering server as well as collecting the player's commands and sending the interactions back to the cloud.

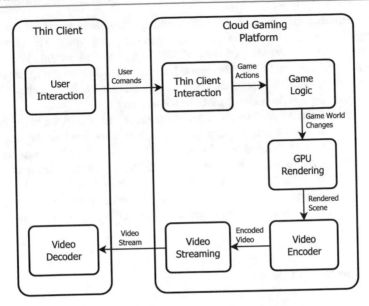

Fig. 18.14 A generic framework of cloud gaming

Figure 18.14 shows a high level architectural view of such a cloud gaming system with thin clients and cloud-based rendering.

Cloud gaming can bring great benefits by expanding the user base to the vast number of less-powerful devices that support thin clients only, particularly smartphones and tablets. As such, mobile users can enjoy high-quality video games without performing the computation-intensive image rendering locally. For example, the recommended system configuration for *Battlefield 3*, a popular first-person shooter game, is a quad-core CPU, 4 GB RAM, 20 GB storage space, and a high-quality graphics card with at least 1GB RAM. This is challenging for mobile devices of which the hardware is limited by their smaller size and thermal control. Furthermore, mobile devices have different hardware/software architecture from PCs, e.g., ARM rather than x86 for CPU (and often without GPU), lower memory frequency and bandwidth, and limited battery capacities. As such, the traditional console game model is not always feasible for such devices, which in turn become targets for cloud gaming. It also masks the discrepancy among different operating systems through such standard web development tools as HTML5, Flash, and JavaScript. It further reduces customer support costs since the computational hardware is now under the cloud gaming provider's full control, and offers better Digital Rights Management (DRM) since the codes are not directly executed on a customer's local device.

18.5.1 Workload and Delay in Cloud Gaming

As shown in Fig. 18.14, in cloud gaming, a player's commands must be sent over the Internet from its thin client to the cloud gaming platform. Once the commands reach the remote cloud, they are converted into appropriate in-game actions, which are interpreted by the game logic into changes in the game world. The game world changes are processed by the cloud system's GPU into a rendered scene. The rendered scene is then compressed by the video encoder, and sent to a video streaming module, which delivers the video stream back to the thin client. Finally, the thin client decodes the video and displays the video frames to the player.

To ensure interactivity, all of these serial operations must happen in the order of milliseconds. Intuitively, this amount of time, which is defined as the *interaction delay*, must be kept as short as possible in order to provide a rich experience to the cloud game players. There are, however, trade-offs: the shorter the player's tolerance for interaction delay, the less time the system has to perform such critical operations as scene rendering and video compression. Also, the lower this time threshold is, the more likely a higher network latency can negatively affect a player's experience of interaction.

Interaction Delay Tolerance

Studies on traditional gaming systems have found that different styles of games have different thresholds for maximum tolerable delay [26]. Table 18.2 summarizes the maximum delay that an average player can tolerate before the QoE begins to degrade [24]. As a general rule, the games that are played in the first person perspective, such as the shooter game Counter Strike, become noticeably less playable when actions are delayed by as little as 100 ms. This low delay tolerance is because such first person games tend to be action-based, and players with a higher delay tend to have a disadvantage [27]. In particular, the outcome of definitive game changing actions such as who "pulled the trigger" first can be extremely sensitive to the delay in an action-based *first person shooter* (FPS) game.

Third person games, such as *role playing games* (RPG), and many massively multi-player games, such as World of Warcraft, can often have a higher delay tolerance of up to 500 ms. This is because a player's commands in such games, e.g., use item, cast spell, or heal character, are generally executed by the player's avatar; there is often an invocation phase, such as chanting magic words before a spell is cast, and hence the player does not expect the action to be instantaneous. The actions must still be registered in a timely manner, since the player can become frustrated if the interaction delay causes them a negative outcome, e.g., they healed before an enemy attack but still died because their commands were not registered by the game in time.

Table 18.2 Delay tolerance in traditional gaming

Example game type	Perspective	Delay threshold (ms)
First person shooter (FPS)	First person	100
Role playing game (RPG)	Third person	500
Real-time strategy (RTS)	Omnipresent	1,000

The last category of games are those played in an "omnipresent" view, i.e., a top-down view looking at many controllable entities. Examples are *real-time strategy* (RTS) games like Star Craft and such simulation games as The Sims. Delays of up to 1,000 ms can be acceptable to these styles of games since the player often controls many entities and issues many individual commands, which often take seconds or even minutes to complete. In a typical RTS game, a delay of up to 1,000 ms for a build unit action that takes over a minute will hardly be noticed by the player.

Although there is much similarity between interaction delay tolerance for traditional gaming and cloud gaming, it is useful to stress the following critical distinctions. First, traditionally, the interaction delay was only an issue for multi-player online gaming systems, and was generally not considered for single player games. Cloud gaming drastically changes this: now all games are being rendered remotely and streamed back to the player's thin client. As such, we must be concerned with interaction delay even for a single player game. Also, traditional online gaming systems often hide the effects of interaction delay by rendering the action on a player's local system before it ever reaches the gaming server. For example, a player may instruct the avatar to move, and it immediately begins the movement locally; however, the gaming server may not receive the update on the position for several milliseconds. Since cloud gaming offloads its rendering to the cloud, the thin client no longer has the ability to hide the interaction delay from the player. Such visual cues as mouse cursor movement can be delayed by up to 1,000 ms, making it impractical to expect the player will be able to tolerate the same interaction delays in cloud gaming as they do in traditional gaming systems. The maximum interaction delay for all games hosted in a cloud gaming context should be no more 200 ms. Other games, specifically such action-based games as first person shooters likely require less than 100 ms interaction delay in order not to affect the player's QoE.

Video Streaming and Encoding

Cloud gaming's video streaming requirements are quite similar to live video streaming. Both cloud gaming and live video streaming must quickly encode/compress incoming video and distribute it to end users. In both cases, only a small set of the most recent video frames are of interest, and there is no need or possibility to access future frames before they are produced, implying encoding must be done with respect to very few frames.

Yet conventional live video streaming and cloud gaming have important differences. First, compared to live video streaming, cloud gaming has virtually no capacity to buffer video frames on the client side. This is because, when a player issues a command to the local thin client, the command must traverse the Internet to the cloud, be processed by the game logic, rendered by the processing unit, compressed by the video encoder and streamed back to the player. Given that this must all be done in under 100–200 ms, it is apparent that there is not much margin for a buffer. Live video streaming on the other hand can afford a buffer of hundreds of milliseconds or even a few seconds with very little loss to the QoE of the end user.

The sensitive real-time encoding needs of cloud gaming make the choice of video encoder of paramount importance for any cloud gaming provider. H.264 and 265 have been advocated, for both the very high compression ratio but the flexibility to well with stringent real-time demands.

18.5.2 Implementation and Deployment

Onlive and Gaikai are two industrial pioneers of cloud gaming. Boosting resource-limited users to play the games that used to be exclusive for high-end PCs and gaming consoles, both of them have seen great success with multi-million user bases and both have been acquired by Sony. They have since served as the technology foundation for Sony's cloud gaming on its Playstation gaming system.

Gaikai is implemented using two public clouds, namely Amazon EC2 and Lime-light. Figure 18.15 offers a practical view of Gaikai's workflow. When a user selects a game on Gaikai (*Step1* in Fig. 18.15), an EC2 virtual machine will first deliver the Gaikai game client to the user (*Step2*). After that, it forwards the IP addresses of game proxies that are ready to run the selected games to the user (*Step3*). The user will then select one game proxy to run the game (*Step4*). The game proxy starts to run the game, and the game screens will be streamed to the user via UDP (*Step5* and *Step6*). For multi-player online games, these game proxies will also forward user operations to game servers (mostly deployed by the game companies) and send the related information/reactions back to the users (*Step7*).

Onlive's workflow is quite similar, but is implemented with a private cloud environment. Using public clouds enables lower implementation costs and higher scalability; yet a private cloud may offer better performance and customization that could fully unleash the potentials of cloud for gaming.

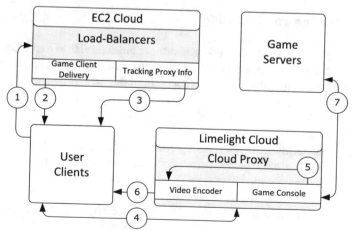

Fig. 18.15 The workflow of the Gaikai cloud gaming platform

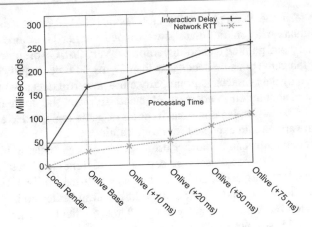

Fig. 18.16 Interaction delay in Onlive

Table 18.3 Processing time and cloud overhead

Measurement	Processing time (ms)	Cloud overhead (ms)
Local render	36.7	n/a
Onlive base	136.7	100.0
Onlive (+10 ms)	143.3	106.7
Onlive (+20 ms)	160.0	123.3
Onlive (+50 ms)	160.0	123.3
Onlive (+75 ms)	151.7	115.0

Figure 18.16 shows the extra latency brought by Onlive. For example, Onlive (+20 ms) indicates that an extra 20 ms is added on the network delay, bringing the total to 50 ms. The locally rendered copy has an average interaction delay of approximately 37 ms, whereas the Onlive baseline takes approximately four times longer at 167 ms to register the same game action. As is expected, for higher network latencies, the interaction delay increases. Impressively, the Onlive system manages to keep its interaction delay below 200 ms in many of our tests. This indicates that, for many styles of games, Onlive could provide acceptable interaction delays. However, when the network latency exceeds 50 ms, the interaction delays may begin to hinder the users' experience. Also, even with the baseline latency of only 30 ms, the system could not provide an interaction delay of less than 100 ms, the expected threshold for first person shooters.

Table 18.3 further gives the detailed breakdown for interaction processing and cloud overhead. The processing time is the amount of interaction delay caused by the game logic, GPU rendering, video encoding, etc; that is, it is the components of the interaction delay not explained by the network latency. The cloud overhead is the delay not caused by the core game logic or network latency, including the amount of delay caused by the video encoder and streaming system used in Onlive.

As can be seen, the cloud processing adds about 100–120 ms of delay in Onlive. This is better than earlier studies that show the cloud overhead is around 200 ms [28], suggesting that cloud gaming technology improves very fast. On the other hand, local rendering by the game console needs less than 37 ms. In other words, although the cloud, powered by data centers, is principally more powerful than any local console, the current implementation is not very efficient. To reach the optimal interaction delay threshold, better implementations in terms of game logic, video encoders and streaming software remain expected for cloud gaming.

The interaction paths can be further prolonged in a large-scale multi-user game, where the globally distributed users are served by different cloud data centers. The cloud providers, however, are often served by better network connections (e.g., near to the backbone networks and higher bandwidth) or even dedicated high-speed networks. If the cloud servers are smartly assigned to user clients, the latency may not change much and is still acceptable.

Cloud gaming is a rapidly evolving technology, with many exciting possibilities. Besides software and service providers, hardware manufacturers have also shown strong interests in cloud gaming, and have begun working on dedicated hardware solutions to address such prominent issues as fast and concurrent rendering and encoding of game scenes [29]. In 2015, Nvidia offered GeForce Now as its cloud-based game-streaming service. In 2018, Google unveiled Project Stream, which was formally announced at the 2019 Game Developers Conference as Stadia. It supports cloud-based gameplay at 4K at 60 fps and future plans for scaling to 8K at 120 fps. In the meantime, Microsoft unveiled Project xCloud, aiming to incorporate Microsoft Azure cloud services into cloud gaming. Sony also has the plan to use Azure as part of its underlying platform for cloud gaming.

18.6 Edge Computing and Serverless Computing for Multimedia

As we can see from the delay-sensitive multimedia applications above, aggregating all the resources in a centralized data center is not an ideal solution for cloud computing. Recent efforts toward fine-grained resource partition and distribution include *Edge computing* (on geo-distribution) and *Serverless computing* (on service abstraction).

18.6.1 Mobile Edge Computing

Edge computing is a distributed computing paradigm that expands the cloud-computing paradigm by bringing computation and data storage closer to the location where it is needed, i.e., network edge, so as to improve response times and save bandwidth [30–32]. From the conventional CDN's perspective, edge computing can also be viewed as an expansion that upgrades the distributed storage nodes with smart computing capability. Such expansion has been embraced by many of today's cloud and CDN service providers, under the names of AWS IoT Greengrass, CloudFront, Azure IoT Edge, Akamai Intelligent Edge, etc.

A typical edge node is made up of three components:

1. **Service module**, which includes containers that run native or third-party services locally at the edge node;
2. **Runtime**, which runs and manages the service modules deployed to the node;
3. **Cloud interface**, which offers the connection, interaction, and collaboration with the cloud data center.

Besides latency and traffic reduction, running tasks at the edge also enhances reliability as there is no single point of failure. For instance, the Akamai's Intelligent Edge includes 250,000 edge servers, deployed in thousands of locations around the world, which ingest 2.5 exabytes of data per year and interact with 1.3 billion devices. Moreover, it is easier for users to advocate cloud computing from privacy's perspective, as sensitive data can now be retained locally.

Edge computing is of particular importance to 5G multimedia applications [33, 34]. Even though 5G may reduce the latency to 10 ms or lower, the end-to-end latency to a cloud datacenter can be significantly longer as the data have to traverse the wired Internet. To seamlessly integrate it with cloud and fully unleash its potentials, processing data at the network edge becomes a necessity.

The concept of *Mobile Edge Computing*, also known as *multi-access edge computing* (MEC), initially presented by the European Telecommunications Standards Institute (ETSI) in 2015, is to bring storage and computing capability into cellular base stations, so as to run applications and performing related processing tasks closer to the cellular customers, without uploading or downloading a massive amount of data through the public Internet. MEC is a special case of the general edge computing paradigm and is mostly tied with the 5G network architecture (see Fig. 18.17 for a general view of the 5G MEC). It is considered as a key technology and architectural concept to enable the evolution to 5G.

Ultra-reliable and low-latency communications (URLLC) applications will especially benefit from MEC, which is also to be combined with network slicing for QoS assurance. For example, with the assistance of an edge server, augmented reality (AR) will be accelerated and timely feedback will help to modify a live representation of the world. Video streaming services will operate much more efficiently given the awareness of local network conditions [35]. For smart transportation, connected and automated cars can efficiently and reliably coordinate with other cars nearby in real time, in particular, within the coverage of the same base station.

The upgraded base stations will naturally offer the caching capability as well. The mMTC applications will also benefit from MEC, as the tiny connected devices in smart home and smart city can more easily access the computing resources at the edge than from the remote cloud data center, in terms of speed, cost, and reliability. Since the future 6G standard targets ultra high speed and low latency with even smaller cell sizes, MEC will play an important role as well to accommodate the vast amount of ubiquitous multimedia-empowered devices.

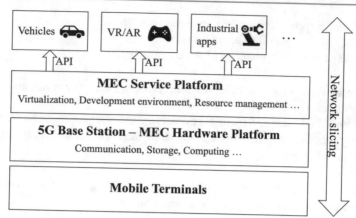

Fig. 18.17 An architectural view of mobile edge computing (MEC)

18.6.2 Serverless Computing for Video Processing

As a new execution model, *serverless computing* offloads the task of server provisioning and management from developers to platforms [36]. In a serverless platform, developers only need to break up application codes into a collection of stateless functions and set events to trigger their executions. Platforms are responsible for handling every trigger and scaling precisely with the size of workloads. The light-weight virtualization techniques used in serverless computing, represented by containerization, further enable function instances to spin up or down in milliseconds. Serverless computing has been implemented in commercial offerings and open source projects, such as AWS Lambda, Google Cloud Functions (GCF), and Apache OpenWhisk. They charge users at a fine-grained timescale (e.g., 100 ms), enabling truly "pay-as-you-go" pricing.

Given the advantages in scalability, start-up delay, and pricing, industry experts and researchers from academia have great interests in applying serverless computing in multimedia processing. Existing studies have shown its potentials toward executing massively parallel functions to speed up video processing [37], so do Amazon's practices.

Figure 18.18 plots the performance of a *transcoding* function deployed on two representative serverless platforms, namely, Amazon's AWS Lambda and Google's GCF, in a one day period. Figure 18.18a shows the I/O time variation of AWS Lambda. Increasing the memory size from 1,024 to 2,048 MB does not improve the download and upload latencies much, which indicates that for larger memory sizes, CPU power is not the bottleneck for I/O tasks. Figure 18.18b plots the variation of total duration. As the memory size decreases, the execution duration changes more and more dramatically. For example, the longest execution duration of 512 MB (56,640 ms) is 1.5 times as much as the shortest execution duration (37,735 ms), and the increment is even larger than the average execution duration of 2,048 MB (13,754 ms). Figure 18.18b plots the CPU types of VMs hosting the corresponding function instances. As can be observed, different CPU types correspond to differ-

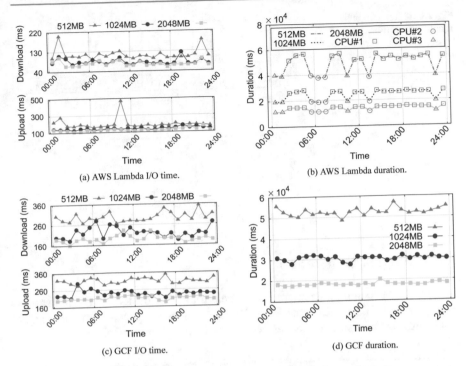

(a) AWS Lambda I/O time. (b) AWS Lambda duration.

(c) GCF I/O time. (d) GCF duration.

Fig. 18.18 Changes in execution time for the transcoding function in one day

ent performance levels. When the memory size is less than 1,152 MB, the CPU type of VMs hosting Lambda-MTCNN function instances is CPU#2; after that the CPU type of VMs is CPU#1. Clearly, the influence of the heterogeneous underlying infrastructures on performance cannot be ignored.

Figure 18.19 further plots the changes in execution duration of the *transcoding* task within one week. For AWS Lambda, the performance shows more drastic changes as the memory size decreases. The changes do not show a daily periodic pattern but are closely related to the heterogeneous underlying infrastructures. This implies that the scheduling and allocation of VMs are random and independent of time. In contrast, GCF exhibits a more stable execution duration regardless of the memory size.

In summary, the memory configuration for cost-efficient serverless functions is non-trivial [38]. The best memory configuration is influenced by the task type or even the video content. There is also opportunity for the combination with edge computing. While executing function codes closer to users improves the response speed, edge resource is usually limited compared to the data center. The trade-off between performance and cost in serverless edge computing will be an important consideration.

Fig. 18.19 Transcoding function execution duration changes within one week on different serverless platforms

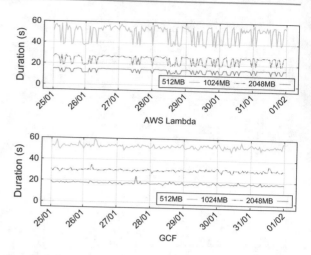

18.7 Further Exploration

Cloud and edge computing remains a new field for both industry and research community. Many of the related materials can be found as white papers from such major cloud and edge computing providers as Amazon, Google, and Microsoft.

18.8 Exercises

1. Discuss the relations and differences of the following systems: Cloud, Server Cluster, content distribution network (CDN), and Data Center.
2. Consider cloud-based video streaming and peer-to-peer video streaming

   (a) For each of them, what are the pros and cons?
   (b) Discuss a possible solution that combines these two. Discuss the benefit of this hybrid design and its potential issues.
   (c) Discuss the similarity and difference between a peer and an edge node if edge computing is used in conjunction with cloud.

3. Consider the cloud-based Netflix video-on-demand (VoD) service.

   (a) Describe the respective roles of Amazon EC2 and S3 in Netflix.
   (b) Netflix uploads the master videos to the cloud for transcoding. Why doesn't Netflix transcode the videos locally and then upload them to the cloud?
   (c) Why does Netflix still need a CDN service beyond S3?
   (d) What is the role of Edge computing in Netflix's design.

4. Is it always beneficial to offload computation to the cloud? List two application scenarios that simple computation offloading may not be cost-effective, and suggest possible solutions.

5. In this question, we try to quantify the cost savings of using cloud services. Without the cloud, a user has to purchase his or her own PC, say of price X. The value of the machine depreciates at a rate of $p\%$ per month, and when the value is below $V\%$, the machine is considered out-dated and the user has to purchase a new machine. On the other hand, using the cloud, the user doesn't have to buy his or her own machine, but leases from the cloud service provider with a monthly fee of C.

   (a) To make the cloud service cost-effective, how much should the provider set for the monthly lease fee C for a comparable cloud machine instance?
   (b) Are there any other real-world costs that can be included in the model, associated either with local machine or with the cloud?

6. Considering the energy consumption of a task with local computing at a mobile terminal and with offloading to the cloud. We assume the task needs C CPU cycles for computation. Let M and S be the computing speeds, in CPU cycles per second, of the mobile terminal and of the cloud machine, respectively. The local computing at the mobile terminal has energy consumption of P_M watts, and incurs no data exchange over the wireless interface. For offloading to the cloud, D bytes of data are to be exchanged over the wireless interface. We assume that the network bandwidth is B bps and the energy consumption of the air interface during transmitting or receiving is P_T watts.

   (a) Assume that the CPU of the mobile terminal consumes no energy when it is idle, nor docs the wireless interface of the terminal. What is the energy consumption of the mobile terminal if the task is executed locally, or offloaded to the cloud? Note that we don't consider the energy consumption in the cloud because the energy bottleneck of interest here is at the mobile terminal.
   (b) Under what condition does offloading to the cloud save energy?
   (c) What are other potential benefits with computation offloading, and under what conditions?

7. Besides the cost or energy savings, list two other advantages when using a cloud, as compared to building and maintaining a local infrastructure. Also list two disadvantages.

8. Consider cloud gaming, in which game scenes are rendered in the cloud and then streamed back to a thin client.

   (a) What are the benefits of using a cloud for gaming?
   (b) What types of games are most suitable for cloud gaming?

(c) Discuss the requirements for live video streaming and those for cloud gaming. How are they similar? What special requirements of cloud gaming make it more difficult?

(d) Suggest some solutions that can reduce the delay in cloud gaming.

References

1. M. Armbrust, A. Fox, R. Griffith, A.D. Joseph, R. Katz, A. Konwinski, G. Lee, D. Patterson, A. Rabkin, I. Stoica, M. Zaharia, A view of cloud computing. Commun. ACM **53**(4), 50–58 (2010). April
2. P. Mell, T. Grance, The NIST definition of cloud computing. Technical Report Special Publication 800-145, National Institute of Standards and Technology (NIST) (2011)
3. P. Barham, B. Dragovic, K. Fraser, S. Hand, T. Harris, A. Ho, R. Neugebauer, I. Pratt, A. Warfield, Xen and the art of virtualization. SIGOPS Oper. Syst. Rev. **37**(5), 164–177 (2003)
4. S. Soltesz, H. Pötzl, M.E. Fiuczynski, A. Bavier, L. Peterson, Container-based operating system virtualization: a scalable, high-performance alternative to hypervisors, in *Proceedings of the 2nd ACM SIGOPS/EuroSys European Conference on Computer Systems 2007, EuroSys '07*, pp. 275—287. (ACM, 2007)
5. S. Fu, J. Liu, X. Chu, Y. Hu, Toward a standard interface for cloud providers: the container as the narrow waist. IEEE Internet Comput. **20**(2), 66–71 (2016). March
6. B. Burns, B. Grant, D. Oppenheimer, E. Brewer, J. Wilkes, Borg, omega, and kubernetes. Commun. ACM **59**(5), 50–57 (2016). April
7. A. Jindal, V. Podolskiy, M. Gerndt, Performance modeling for cloud microservice applications, in *Proceedings of the ACM/SPEC International Conference on Performance Engineering, ICPE '19*, pp. 25—32. (ACM, New York, NY, USA, 2019)
8. W. Zhu, C. Luo, J. Wang, S. Li, Multimedia cloud computing. Signal Process. Mag. IEEE **28**(3), 59–69 (2011)
9. D. Niu, Z. Liu, B. Li, S. Zhao, Demand forecast and performance prediction in peer-assisted on-demand streaming systems, in *IEEE INFOCOM Mini-Conference* (2011)
10. Y. Huang, T. Fu, D. Chiu, J. Lui, C. Huang, Challenges, design and analysis of a large-scale P2P-VoD system, in *ACM SIGCOMM* (2008)
11. K. Xu, H. Li, J. Liu, W. Zhu, W. Wang, PPVA: a universal and transparent peer-to-peer accelerator for interactive online video sharing, in *IWQoS* (2010)
12. A. Leung, A. Spyker, T. Bozarth, Titus: introducing containers to the Netflix cloud. Queue **15**(5), 53–77 (2017). October
13. F. Liu, P. Shu, H. Jin, L. Ding, Y. Jie, D. Niu, B. Li, Gearing resource-poor mobile devices with powerful clouds: architectures, challenges, and applications. Wirel. Commun. IEEE **20**(3), 14–22 (2013)
14. X. Ma, Y. Zhao, L. Zhang, H. Wang, L. Peng, When mobile terminals meet the cloud: computation offloading as the bridge. Netw. IEEE **27**(5), 28–33 (2013)
15. K. Kumar, L. Yung-Hsiang, Cloud computing for mobile users: can offloading computation save energy? IEEE Comput. **43**(4), 51–56 (2010). April
16. E. Cuervo, A. Balasubramanian, D.-k. Cho, A. Wolman, S. Saroiu, R. Chandra, P. Bahl, Maui: making smartphones last longer with code offload, in *Proceedings of the 8th International Conference on Mobile Systems, Applications, and Services, MobiSys '10*, pp. 49–62. (ACM, New York, NY, USA, 2010)
17. B.-G. Chun, S. Ihm, P. Maniatis, M. Naik, A. Patti, Clonecloud: elastic execution between mobile device and cloud, in *Proceedings of the Sixth Conference on Computer systems, EuroSys '11*, pp. 301–314. (ACM, New York, NY, USA, 2011)

18. K. Kumar, J. Liu, L. Yung-Hsiang, B. Bhargava, A survey of computation offloading for mobile systems. Mob. Netw. Appl. **18**(1), 129–140 (2013)
19. H.T. Dinh, C. Lee, D. Niyato, P. Wang, A survey of mobile cloud computing: architecture, applications, and approaches. Wirel. Commun. Mob. Comput. pp. 1587–1611 (2011)
20. A.P. Miettinen, J.K. Nurminen, Energy efficiency of mobile clients in cloud computing, in *Proceedings of the 2nd USENIX Conference on Hot Topics in Cloud Computing, HotCloud'10.* (USENIX Association, Berkeley, CA, USA, 2010)
21. N. Imran, B.-C. Seet, A.M. Fong, A comparative analysis of video codecs for multihop wireless video sensor networks. Multimed. Syst. **18**(5), 373–389 (2012)
22. Y. Wang, J. Ostermann, Y.-Q. Zhang, *Video Processing and Communications*, vol. 5. (Prentice Hall, 2002)
23. M. Sayed, W. Badawy, A novel motion estimation method for mesh-based video motion tracking, in *IEEE International Conference on Acoustics, Speech, and Signal Processing, 2004. Proceedings of the (ICASSP'04)*, vol. 3, pp. 3–337. (IEEE, 2004)
24. M. Jarschel, D. Schlosser, S. Scheuring, T. Hossfeld, An evaluation of QoE in cloud gaming based on subjective tests, in *2011 Fifth International Conference on Innovative Mobile and Internet Services in Ubiquitous Computing (IMIS)*, pp. 330–335 (2011)
25. R. Shea, J. Liu, E.C.-H. Ngai, Y. Cui, Cloud gaming: architecture and performance. Netw. IEEE **27**(4), 16–21 (2013)
26. M. Claypool, Kajal Claypool, Latency and player actions in online games. Commun. ACM **49**(11), 40–45 (2006). November
27. M. Claypool, K. Claypool, Latency can kill: precision and deadline in online games, in *Proceedings of the First Annual ACM SIGMM Conference on Multimedia Systems, MMSys '10*, pp. 215–222. (ACM, New York, NY, USA, 2010)
28. K.-T. Chen, Y.-C. Chang, P.-H. Tseng, C.-Y. Huang, C.-L. Lei, Measuring the latency of cloud gaming systems, in *Proceedings of the 19th ACM International Conference on Multimedia, MM '11*, pp. 1269–1272 (2011)
29. Z. Zhao, K. Hwang, J. Villeta, Game cloud design with virtualized CPU/GPU servers and initial performance results, in *Proceedings of the 3rd Workshop on Scientific Cloud Computing Date, ScienceCloud '12*, pp. 23–30. (ACM, New York, NY, USA, 2012)
30. W. Shi, J. Cao, Q. Zhang, Y. Li, L. Xu, Edge computing: vision and challenges. IEEE Internet Things J. **3**(5), 637–646 (2016)
31. C. Li, Y. Xue, J. Wang, W. Zhang, T. Li, Edge-oriented computing paradigms: a survey on architecture design and system management. ACM Comput. Surv. **51**(2) (2018)
32. D. Wang, Y. Peng, X. Ma, W. Ding, H. Jiang, F. Chen, J. Liu, Adaptive wireless video streaming based on edge computing: opportunities and approaches. IEEE Trans. on Serv. Comput. (2018)
33. Y. Mao, C. You, J. Zhang, K. Huang, K.B. Letaief, A survey on mobile edge computing: the communication perspective. IEEE Commun. Surv. Tutor. **19**(4), 2322–2358 (2017)
34. N. Abbas, Y. Zhang, A. Taherkordi, T. Skeie, Mobile edge computing: a survey. IEEE Internet Things J. **5**(1), 450–465 (2018). Feb
35. C. Ge, N. Wang, W.K. Chai, H. Hellwagner, QoE-assured 4K HTTP live streaming via transient segment holding at mobile edge. IEEE J. Sel. Areas Commun. (JSAC) (2018)
36. E. Jonas, Q. Pu, S. Venkataraman, I. Stoica, B. Recht, Occupy the cloud: distributed computing for the 99%, in *Proceedings of the Symposium on Cloud Computing, SoCC '17*, pp. 445–451. (ACM, New York, NY, USA, 2017)
37. L. Ao, L. Izhikevich, G.M. Voelker, G. Porter, Sprocket: a serverless video processing framework, in *Proceedings of the ACM Symposium on Cloud Computing, SoCC '18*, pp. 263–274. (ACM, New York, NY, USA, 2018)
38. M. Zhang, Y. Zhu, C. Zhang, J. Liu, Video processing with serverless computing: a measurement study, in *Proceedings of the 29th ACM Workshop on Network and Operating Systems Support for Digital Audio and Video, NOSSDAV '19*, pp. 61–66. (ACM, 2019)

Part IV

Human-Centric Interactive Multimedia

The past decade has witnessed an explosion of new-generation social, mobile, and cloud computing technologies for multimedia processing and delivery. Rather than passively receiving media content as in the Web 1.0 era, Web 2.0 users actively participate in the online community for media production and sharing. Such popular Web 2.0-based social media sharing services as YouTube, Facebook, Twitter, Twitch, and TikTok have drastically changed the content distribution landscape, and indeed have become an integral part in people's daily life. With no doubt, online services in the future will focus on the user experience and participation with rich media and rich interactions.

The development on the coding algorithms and hardware for sensing, communication, and interaction also empower virtual reality (VR) and augmented reality (AR), providing better immersive experiences beyond 3D.

The sheer amount of user-generated content empowered by social media further demands automated multimedia data analysis and retrieval, so as to locate syntactically and semantically useful content. This calls for effective solutions to greatly extend the traditional text-based search, and to identify redundant or even pirated contents that indeed have been critical challenges for the management of media sharing services.

This part examines the challenges and solutions for the new generation of human-centric interactive multimedia in the Web 2.0 era. In Chap. 19, we examine the unique characteristics of social media sharing and their impact. In Chap. 20, we introduce the basic concepts of augmented reality and virtual reality as well as their applications. Chapter 21 further provides an introduction to multimedia content retrieval including the deep learning approach.

Online Social Media Sharing

19

Online social media, a group of Internet-based applications that build on the ideological and technological foundations of Web 2.0, allow the creation and exchange of user-generated contents [1]. These services combine rich graphical user interfaces (GUIs) and multimedia contents, and enable ubiquitous content generation and sharing and scalable communication. They have substantially changed the way that organizations, communities, and individuals communicate. Two distinct characteristics of Web 2.0 are considered as the key factors to the success of the new generation of social media [2]:

Collective intelligence. The base knowledge of general users of the Web is contributing to the content of the Web. The users have become the content providers and therefore the contents are richer and more dynamic.

Rich connections and activities. The users and their contents are linked to each other, creating an organic growth of connections and activities. A social network built out of the connections enables strong ties among the users as well as broad and rapid content propagation.

There are many types of online social media services, including user-generated content sharing (e.g., YouTube), online social networking (e.g., Facebook), question-and-answer (e.g., Quora), and collaboratively-edited encyclopedia (e.g., Wikipedia), to name but a few. In these social media services, the contents, as well as the users, have become interconnected, enabling convenient information *sharing* for feelings, activities, and location information, as well as resources, blogs, photos, and videos.

With the pervasive penetration of wireless mobile networks, the advanced development of smartphones and tablets, and the massive market of mobile applications, social media contents can now be easily generated and accessed at any time and anywhere. YouTube's own report for 2020 has suggested that 70% of the views come

© Springer Nature Switzerland AG 2021
Z.-N. Li et al., *Fundamentals of Multimedia*, Texts in Computer Science,
https://doi.org/10.1007/978-3-030-62124-7_19

from mobile. The crowdsourced livecast service, Twitch, has 2.2 million broadcasters monthly and 15 million daily active users, with around a million average concurrent users, and the mobile video clip sharing app, TikTok, has been opened by each of its US users eight times a day with an average daily usage of 52 min. The new trends on social media creation, deployment, and spreading, far beyond conventional media, have brought up numerous well-known Internet memes and celebrities that have no doubt been changing our daily life.

In this chapter, we present an overview of this rapidly evolving field, particularly on social media services for multimedia content sharing. We first look at important social media services, namely, user-generated media content sharing and online social networking. We closely examine the YouTube video sharing service. We then discuss media object propagation in social networks, and the user behaviors in sharing. We also discuss such novel services as crowdsourced livecast and mobile video clip sharing.

19.1   Representatives of Social Media Services

We now present the background for two keystones for social media sharing and their representative implementations.

19.1.1 User-Generated Content (UGC)

Having arisen in Web publishing and new media content production circles, *user-generated content* (UGC) plays a key role in today's social media services. It is used for a wide range of applications with different types of media, e.g., text, music, picture, and video, as well as a combination of open source, free software, and flexible licensing or related agreements to further reduce the barriers on collaboration, skill-building and discovery. For content generation and sharing, video data are arguably more difficult than such other types of media as text and pictures, given their large size, high bandwidth demand, and long playback duration.

In traditional video on-demand and live streaming services, videos are offered by enterprise content providers, stored in servers, and then streamed to users. A number of new generation video sharing websites, represented here by YouTube, offer users opportunities to make their videos directly accessible by others, by such simple operations in Web 2.0 as embedding and sharing.

Established in 2005, YouTube is so far the most significant and successful video sharing website. It allows registered users to upload videos, mostly short videos. The users can watch, embed, share, and engage with videos easily. As one of the fastest growing websites in the Internet, YouTube had served 100 million videos per day in 2006; by December 2013, more than 1 billion unique users visited YouTube each month, and over 6 billion hours of video were watched each month—that's almost an hour for every person on Earth, and 100 hours of new videos were uploaded every minute. In 2020, the monthly users of YouTube have further increased to 2 billion, with each visitor spending 11m 24 s per day; 500 hours of new videos have been uploaded every minute, and over 70% of the YouTube views are on mobile.

YouTube is also highly globalized [3]—it is localized in 91 countries and across 80 languages, with 85% of its traffic now comes from outside of the US (up from 80% in 2013). In fact, only 33% of the popular YouTube videos are in English. The success of similar sites (e.g., Vimeo and Tudou/YouKu) further confirm the mass market interest in UGC video sharing services.

YouTube is originally a video-on-demand service only, but started testing its own live streaming infrastructure, YouTube Live, in 2010. In the meantime, Twitch emerged as a highly popular crowdsourced livecast service, particularly on video game live streaming (e.g., individual gamecasts or eSports competitions), but also on music broadcasts, creative content, and more recently, "in real life" streaming. As of May 2018, it had 2.2 million broadcasters monthly and 15 million daily active users, with around a million average concurrent users.

19.1.2 Online Social Networking (OSN)

Online social networking (OSN) provides an Internet-based platform to connect people with social relations, e.g., friends, classmates, and colleagues in the real world, or people who simply share common interests.

Facebook, founded in 2004, is one of the dominating online social networking services on the Internet. As of 2020, Facebook had 2.5 billion monthly active users worldwide, and 1.66 billion of them log onto Facebook on a daily basis. It provides users with a platform to connect with friends, by updating status, uploading photos, commenting and "liking" others' posts, etc. Facebook has opened its API for developers to build thousands of applications and games, which makes it more enjoyable.

Another important social networking website, Twitter, is a representative of microblog, a simpler but much faster version of blog. It allows users to send text-based posts, called `tweets`, of up to 280 characters (doubled from 140 before 2017). Although short, the tweets can link to richer contents such as images and videos. By following friends or interested accounts, such as news providers, celebrities, brands, and organizations, Twitter users can obtain real-time notifications, and spread the posts by a `retweet` mechanism. Like Facebook, Twitter also opens its APIs, and there is a large collection of registered Twitter applications available, particularly for mobile users, making Twitter easier to access. As of 2019, there are 330 million monthly active users and 145 million daily active users on Twitter.

Both Facebook and Twitter support the sharing and propagation of such media objects as pictures, music, and videos among friends, although the media content may be hosted by external sites, e.g., YouTube. Nowadays 100 million hours of Facebook native videos are watched daily. Twitter also offered the Vine service earlier, which, available exclusively for mobile users, enables them to create and post video clips. A Vine user can create a short video clip up to six seconds long while recording through Vine's in-app camera. The clip can then be shared through

Twitter, Facebook, or other social networking services. Vine was later integrated directly into Twitter and, a few months before its discontinuation, TikTok (called Douyin in China) was released, with the added option of Duet, meaning that two different TikTok creators may collaborate at different times to create a final video. It has been a great success among the young generation, hitting one billion downloads globally.

19.2   User-Generated Media Content Sharing

Social media has greatly changed mechanisms of content generation and access, and also brings unprecedented challenges to server and network management. Understanding the distinct features of the user-generated media content is thus crucial to traffic engineering and to the sustainable development of these new generation of services. We now close examine two representatives, namely, YouTube for on-demand video and Twitch for livecast.

19.2.1 YouTube Video Format and Meta-Data

YouTube's video playback technology was first based on Adobe's Flash Player, and was later migrated to the HTML5 video player, whose benefits extend beyond web browsers to smart TVs and other streaming devices. YouTube used the H.263 video codec earlier, introduced "high quality" format with the H.264 codec for better viewing quality in late 2008, and later added the open VP9 codec , which gives higher quality video resolution with an average bandwidth reduction of 35%. In 2015, the support for 4K resolution was added, so for 360° video [4]. The support for 8K resolution was added in 2017, with a resolution up to $7,680 \times 4,320$ pixels.

A content generator can upload videos to YouTube in many formats, which are converted to several standard formats after uploading. It is well recognized that the use of uniform and easily-playable formats is critical to the success of YouTube (See Fig. 16.2 in Chap. 16 for a sample list of the audio/video formats offered by YouTube to date).

The video streaming to a YouTube viewer is then done by dynamic adaptive streaming over HTTP (MPEG-DASH), and WebRTC is also supported.

YouTube assigns each video a distinct 11-letter ID composed of the characters 0-9, a-z, A-Z, -, and _. Each video contains the following intuitive meta-data: *video ID, uploader, date added, category, length, number of views, number of ratings, number of comments*, and a list of *related videos*. The related videos are linked to other videos that have similar titles, descriptions, or tags, all of which are chosen by the uploader. A YouTube page only shows at most 20 related videos at once, but more can be displayed by scrolling down the list. A typical example of the meta-data is shown in Table 19.1.

Note that this video of year 2008 currently has 6 versions from 144 to 1080p, and the codec is VP9. That said, format conversion has been done lately.

Table 19.1 An example of the meta-data of a video https://www.youtube.com/watch?v=YiQu4gp oa6k

ID	YiQu4gpoa6k
Uploader	NewAgeEnlightenment
Date added	August 08, 2008
Category	Sports
Video length	270 s
Number of views	924, 691
Number of ratings	1, 039
Number of comments	212
Related videos	ri1h2_jrVjU, 0JdQlaQpOuU, ...

19.2.2 Characteristics of YouTube Video

There have been significant research efforts aimed at understanding the workloads of traditional media servers, for example, video popularity and access locality [5–7]. While sharing similar characteristics, many of the video statistics of these traditional media servers are quite different from YouTube-like sites, e.g., the video length distribution and user access pattern. More importantly, those videos are generally movies and TV programs that are not generated by ordinary users, nor are they connected by social relations.

Video Category

In YouTube, a category is selected by a user when uploading a video. Table 19.2 lists the number and percentage of the categories from a sample dataset of 5 million videos over a 1.5-year span [8]. We can see that the distribution is highly skewed: the most popular category is "Entertainment," at about 25.4%, and the second is "Music," at about 24.8%. These two categories of videos constitute half of the entire YouTube video collection, suggesting that YouTube is mainly an entertainment-like site. "Unavailable" are videos set to private, or videos that have been flagged as inappropriate content, for which the crawler can only get meta information from the YouTube API. "Removed" are videos that have been deleted by the uploader, or by a YouTube moderator (due to violation of the terms of use), but are still linked to by other videos.

Video Length

The length of YouTube videos is the most distinguishing difference from traditional video contents. Traditional servers contain a significant portion of long videos, typi-

Table 19.2 A sample set of YouTube video categories

Rank	Category	Count	Percentage (%)
1	Entertainment	1,304,724	25.4
2	Music	1,274,825	24.8
3	Comedy	449,652	8.7
4	People & Blogs	447,581	8.7
5	Film & Animation	442,109	8.6
6	Sports	390,619	7.6
7	News & Politics	186,753	3.6
8	Autos & Vehicles	169,883	3.3
9	Howto & Style	124,885	2.4
10	Pets & Animals	86,444	1.7
11	Travel & Events	82,068	1.6
12	Education	54,133	1.1
13	Science & Technology	50,925	1.0
14	Unavailable	42,928	0.8
15	Nonprofits & Activism	16,925	0.3
16	Gaming	10,182	0.2
17	Removed	9,131	0.2

Fig. 19.1 Histogram and cumulative distribution (CDF, the solid line) of YouTube video length

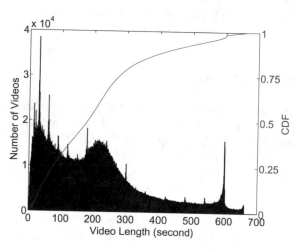

cally 1–2 hour movies (e.g., HPLabs Media Server [6] and OnlineTVRecorder [7]). YouTube, however, is mostly comprised of short video clips. Although YouTube has increased its initial 10-min length limit to 15 min and allows verified users to upload videos up to 12 hours and 128 GB, most of the user-generated videos remain quite short in nature.

Figure 19.1 shows the histogram and cumulative distribution function (CDF) of YouTube videos' lengths within 700 s, which exhibits three peaks. The first peak is within one minute, and contains 20.0% of the videos, which shows that YouTube is

Fig. 19.2 Length histograms
and cumulative distributions
for the four top categories

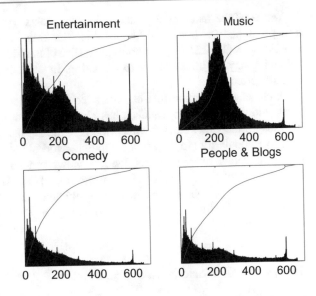

primarily a site for very short videos. The second peak is between 3 and 4 min, and contains about 17.4% of the videos. As shown in Fig. 19.2, this peak corresponds to the videos in the "Music" category, which is the second most popular category for YouTube. The third peak is near the maximum of 10 min, which is the earlier limit on the length of uploaded videos.

Figure 19.2 shows the video length distributions for the top four most popular categories. "Entertainment" videos have a similar distribution as the entire videos. "Music" videos have a high peak between three and four minutes (29.1%), which is the typical length range for music TVs. "Comedy" and "People & Blogs" videos have more videos within two minutes (53.1% and 41.7%, respectively), likely corresponding to "highlight" type of clips.

Access Patterns

Given that UGC video length is shorter by two orders of magnitude as compared to traditional movies or TV shows, YouTube's video content production is significantly faster with less effort [9]. The total number is in the billion order now, which is much higher than that of the traditional video services (e.g., only 412 in HPLabs' Media Server [6]), and this number is rapidly increasing with new user contributions. As such, the scalability challenges faced by YouTube-like social media services is indeed more significant.

It has been found that the 10% top popular videos account for nearly 80% of views, indicating that YouTube is highly skewed toward popular videos. This also implies that proxy caching can have high hit ratios since only a small portion of the videos will be requested frequently.

Yet YouTube users tend to abort the playback very soon, with 60% of videos being watched for less than 20% of their duration, which is particularly true for mobile users [10–12]. Furthermore, only 10% of the video are watched again on the following day [13]. As such, proxy caching, in particular, prefix caching [14], can be quite effective.

A closer examination that classifies YouTube videos into top videos, removed pirate videos, and random videos has also shown that copyrighted videos tend to get most of the views earlier, while videos in the top lists tend to experience sudden bursts of popularity [15,16]. In a campus network, however, the top popular videos do not contribute much to the total videos viewed on a daily basis, probably because the users are of closer relations in sharing video [13]. All these points suggest that YouTube users' viewing behaviors are highly diversified, affected by both the video quality (as in traditional video sharing) as well as their social relations (unique to social media).

19.2.3 Small-World in YouTube Videos

YouTube is a prominent social media service: there are communities and groups in YouTube, and thus videos are no longer independent from each other. Such social networking is unique to this new generation of video sharing services. It has been shown that, besides web searching, tracing related video is another top view source in YouTube-like UGC sites. There is a strong correlation between the number of views of a video and that of its top related videos [16], and this also provides more diversity on video views, helping users discover more videos of their own interest rather than the popular videos only.

The small-world network phenomenon is probably the most interesting characteristic for social networks. Milgram [17] initiated the study of small-world networks when investigating the phenomenon that people are linked by short chains of acquaintance (a.k.a., six degrees of separation). Such networks possess characteristics of both random graphs[1] and regular graphs[2] [19]. More formally, given the network as a graph $G = (V, E)$, the *clustering coefficient* C_i of a node $i \in V$ is the proportion of all the possible edges between neighbors of the node that actually exist in the graph, and the clustering coefficient of the graph, $C(G)$, is the average of the clustering coefficients of all nodes. The *characteristic path length* d_i of a node $i \in V$ is the average of the minimum number of hops it takes to reach all other nodes in V and the characteristic path length of the graph, $D(G)$, is then the average of the characteristic path lengths of all nodes. A small-world network has a large clustering coefficient like a regular graph, but also has a small characteristic path length like a random graph.

---

[1]A random graph is generated based on certain probability distributions. Purely random graphs, built according to the *Erdös-Rényi* (ER) model, exhibit a small characteristic path length (varying typically as the logarithm of the number of nodes) along with a small clustering coefficient [18].
[2]A regular graph is a graph where each vertex has the same degree.

Fig. 19.3 Small-world
characteristic of YouTube
videos

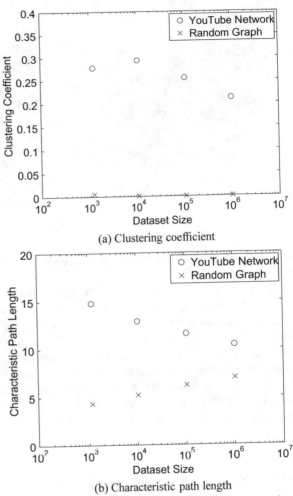

(a) Clustering coefficient

(b) Characteristic path length

The graph topology for the network of YouTube videos can be measured by using the related links in the YouTube dataset to form directed edges in a video graph. Figure 19.3a shows the clustering coefficient for the graph, as a function of the size of the dataset. The clustering coefficient is quite high (between 0.2 and 0.3), especially in comparison to random graphs (nearly 0). There is a slow decreasing trend in the clustering coefficient, showing that there is some inverse dependence on the graph size, which is common for small-world networks [20]. Figure 19.3b shows the characteristic path length for the graphs. It can be seen that the average diameter (between 10 and 15) is only slightly larger than the diameter of a random graph (between 4 and 8), which is quite good considering the still large clustering coefficient of these datasets. Moreover, as the size of graph increases, the characteristic path length decreases for the YouTube video graph, but increases for random graphs with the same number of nodes and average node degrees. This phenomena further verifies that the YouTube graph is a small-world network.

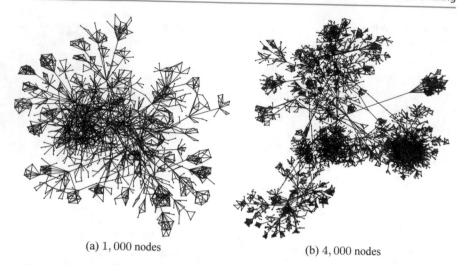

(a) 1, 000 nodes   (b) 4, 000 nodes

Fig. 19.4 Two sample graphs of YouTube videos and their links

The small-world characteristics of the video graph can also be observed from their visual illustrations (see Fig. 19.4 for two representatives of 1,000 and 4,000 nodes). The clustering behavior is very obvious in these two graphs, due to the user-generated nature of the tags, titles, and descriptions of the videos that are used by YouTube to find related ones. The results are similar to other real-world user-generated graphs that exist, yet their parameters can be quite different. For example, the graph formed by URL links in the World Wide Web exhibits a much longer characteristic path length of 18.59 [21]. This is likely due to the larger number of nodes (8×10^8 in the Web), but it also indicates that the YouTube network of videos is a much closer group.

19.2.4 YouTube from a Partner's View

YouTube displays advertisements on the webpages to monetize videos, and this has been the main source of YouTube's revenue. Besides user-generated videos, such companies and organizations as Electronic Arts, ESPN, and Warner Brothers are also providing their premium videos on YouTube now. To accommodate these content owners with copyrighted videos and popular channels, YouTube has introduced a *YouTube Partner Program*, which has largely improved the quality of YouTube videos, and has further increased YouTube's revenue. As of 2020, a qualified YouTube partner must reach the threshold of 1,000 subscribers and 4,000 valid public watch hours over the last 12 months.

The statistics of videos are of great potential value to the YouTube partners. For example, which videos are popular? And which external websites are referring more views? The partners can leverage these statistics to adapt their content deployment and user engagement strategies. To help the YouTube partners with this

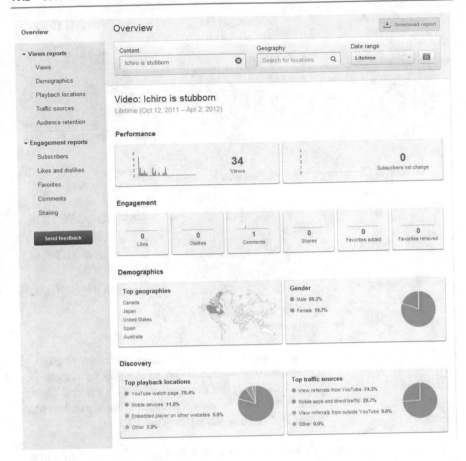

Fig. 19.5 YouTube insight dashboard

goal, YouTube introduced the *Insight Analytics* to provide various basic statistics on videos and channels. Figure 19.5 gives a snapshot of the Web-based Insight Analytics dashboard.

YouTube users have various means to reach YouTube videos. The last webpages where the viewers come from is called *referral sources*. Understanding referrals is essential for YouTube partners to adapt their user engagement strategy. We can classify the referral sources into four categories [22]:

Suggestion   The referral comes from YouTube's related video links;

Video Search   The referral comes from YouTube or search results, e.g., from Google;

YouTube Surfing   The referral comes from any YouTube pages (except for related video links and search results), including annotation links, YouTube channel pages, subscriber links, paid and unpaid YouTube promotion, etc.;

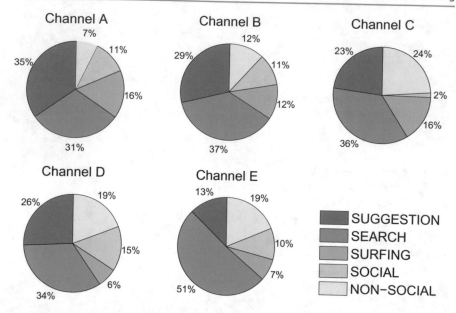

Fig. 19.6 Breakdown of the referral source

Social Referral	The referral source is a link on an external web page, or the video was embedded on an external web page;
Non-Social Direct	YouTube analytics does not identify a referral source, indicating that the viewer navigated directly to the video, e.g., by copying and pasting the URL.

Figure 19.6 shows the breakdown of the above five categories for four sample channels. It is clear that the breakdown percentages are channel-dependent. For example, one-third of the users reach Channel A videos from suggested videos, and one-third reach from search results; Channel B and Channel D are similar to Channel A; very few users reach Channel C videos from external sources; half of the users reach Channel E videos from search results.

This observation confirms that search results and related videos (Suggestion) are the top sources of views [15, 16]. Although Social Referral is not the top view source, the impact of external website referral cannot simply be ignored. There is a great chance that a user, attracted by an external referral, will watch more videos from the related video list. In other words, Social Referral can be considered as an introductive referral.

Table 19.3 lists the top-five external website sources for each channel. It does not disclose the specific names of the websites except such notable social networking service as Facebook, Twitter, and Reddit, and simply uses general descriptions. It again can be seen from the table that there are channel-dependency. No external website dominates the external referrals for Channel A, and the small percentage indicates that there is a great number of sources. In Channel E, over 60% of the

Table 19.3 Summary of top external websites referrers

	Channel A	Channel B	Channel C
1st	9.0% downloading site	16.2% Facebook	31.9% gaming wiki
2nd	4.4% Facebook	2.2% n/a	7.6% Facebook
3rd	2.6% forum	1.5% downloading site	5.3% gaming blog
4th	1.7% gaming site	1.2% n/a	5.1% gaming site
5th	1.5% gaming site	0.9% downloading site	3.7% Internet video site
	Channel D	**Channel E**	
1st	41.2% Reddit	62.4% Facebook	
2nd	9.9% Facebook	2.4% music streaming	
3rd	4.7% Twitter	2.0% music blog	
4th	2.0% blog	2.0% Twitter	
5th	1.7% entertainment site	1.6% music blog	

referrals are from Facebook, over 20 times greater than the second one. Facebook also dominates in Channel B, yet the percentage is not as high as in Channel E. Facebook is the second in both Channel C and Channel D, while the first ones have high percentage. In summary, YouTube videos have been sharing in many different portals and we will further study the propagation structures in online social networks later.

19.2.5 Crowdsourced Interactive Livecast

Crowdsourced livecast has become increasingly popular and seen great success in the past few years [23–26]. In this service, numerous amateur broadcasters can stream their own contents to viewers in their channels; examples include Twitch.tv,[3] Youtube Gaming,[4] Mixer,[5] Inke.tv,[6] etc. Besides watching the live content, a viewer can interact with the broadcaster and other viewers within the same channel. The crowdcast market is estimated to grow to more than 70 billion by 2021; in Facebook, people spend $3\times$ longer time watching live streaming compared to content that is no longer live.

Figure 19.7 shows a snapshot of the web interface of Twitch. The middle is the live streaming window, which is collaboratively generated by 4 distinct broadcaster players playing the game Mario. The viewer can follow the channel, or subscribe to the channel by pressing the button below the streaming window. The top right is the chatting window, where all registered viewers as well as the channel broadcasters can post messages.

---

[3] https://www.twitch.tv/.
[4] https://gaming.youtube.com/.
[5] https://www.mixer.com/.
[6] http://www.inke.cn/.

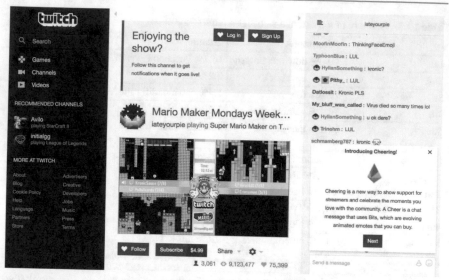

Fig. 19.7 An example of Twitch's web interface

Compared to professional video producers and distributors (e.g., TV channels), crowdsourced broadcasters are geo-distributed and highly heterogeneous in terms of their network/system configurations and therefore the generated video quality. The system is also extremely dynamic over time, for both the overall population and the distribution of viewers in each region. For instance, Twitch has live video content coming from contributing sources in more than 100 different countries with over 150 different resolutions, being shared by viewers all over the world. Although the total number of online viewers can exceed 2 million, it can be less than 200 thousand in the same day. The number of the broadcasters varies dramatically as well, not to mention that they can terminate their live streaming at any point of time.

As such, the source streaming will first need to be transcoded to different bitrates and then be delivered to viewers. Given the rich interactions between viewers and broadcasters as well as the diversified watching environments/preferences, viewers can have personalized quality of experience (QoE) demands (such as various preferences for streaming delays, channel switching latencies, and bitrates).

Consider the viewer interactions from a Twitch dataset of 300 channels for three popular gamecasts, namely, CS:GO, League of Legends (LOL), and FIFA. Figure 19.8 compares the daily average viewer numbers and the interaction message numbers of channels for the three games. Let an IA-ratio be defined as $\frac{\# \ of \ ave. \ interactions}{\# \ of \ ave. \ viewers}$. This ratio of FIFA is much larger than that of CS:GO, indicating the viewers of FIFA channels are more active with frequent interactions.

Figure 19.9 plots the viewing and interaction patterns over different time periods in a day for two typical channels from LOL and FIFA. It can be seen that the viewer and interaction numbers are relatively stable for the LOL channel, but not for the FIFA channel, which only attracts viewers with rich interactions in the prime time.

Fig. 19.8 The distribution of viewer numbers and interaction message numbers of channels in three games

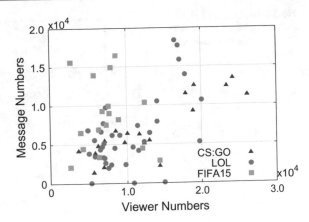

Fig. 19.9 The average viewing and interaction numbers of different time periods in a day for two gamecast channels

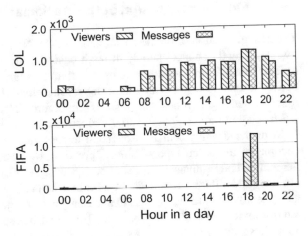

This suggests that the watching preference may vary with different time periods and/or different channels.

Finally, consider the interaction message number of each session from four popular channels, where a session is defined as the watching process between a viewer's join and departure. As illustrated in Fig. 19.10, the distribution is highly skewed, with 87% sessions never generating messages and more than 5% sessions having frequent messages (more than 10 messages per viewing). These viewers with frequent interactions generally prefer low streaming delay so as to communicate with the broadcaster and other active viewers in real time. In contrast, the silent viewers with less involvement are not so sensitive to streaming delay as long as the video and messages are delivered [24].

In summary, conventional streaming QoE definition considering the bitrate, startup delay, and rebuffer cannot fully characterize the interaction experience of viewers. Instead, the viewer QoE consists of not only the streaming QoE, but also the interaction QoE. The latter is affected by whether the streaming and the interactions are synchronous, and whether sending an interaction can be timely responded by the broadcaster and other active viewers [27].

Fig. 19.10 The CDF plot of interactive message numbers of each session in different channels

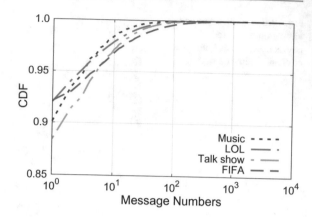

19.3   Media Propagation in Online Social Networks

The new generation of online social network (OSN) services, such as Facebook or Twitter, directly connect people through cascaded relations, and information thus spreads much faster and more extensively than through conventional web portals or newsgroup services, not to mention cumbersome emails [28]. As an example, Twitter first reported Tiger Woods' car crash 30 min before CNN, inverting the conventional 2.5-h delay of online blogging after mainstream news report [29].

With the development in broadband access and data compression, video has become an important type of object spreading over social networks, beyond earlier simple text or image object sharing [28,30]. Yet video objects, as richer media, possess quite different characteristics. From a data volume perspective, video objects are generally of much larger size than other types of objects; hence, most videos are fed from external hosting sites, e.g., YouTube, and then spread as URL links (together with titles and/or thumbnails). As a matter of fact, today's video sharing services and social networking services have become highly integrated [31,32]. YouTube enables automatic posting on Facebook and Twitter based on users' options, and the users can also share interesting videos on their social networking webpages. YouTube's own statistics revealed that 500 years of YouTube video are watched every day on Facebook, and over 700 YouTube videos are shared on Twitter each minute. Social-Baker's report further suggested that Facebook's native videos even have a higher Engagement Rate than YouTube links (0.22% on average compared to 0.10%).

From a social perspective, text diaries and photos often possess personal information, while videos are generally more "public." Together with shorter links, videos often spread more broadly than texts and images. Yet the sheer and ever-increasing data volume, the broader coverage, and the longer access durations of video objects also present significant challenges compared to other types of objects, not only to social networking service management, but also to network traffic engineering and to the resource provisioning of external video sites.

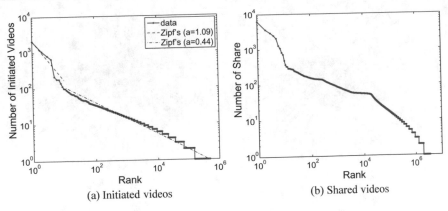

Fig. 19.11 Rank distributions of initiated videos and shared videos

19.3.1 Sharing Patterns of Individual Users

Since video object sharing involves both propagation over an OSN and accesses to the external video site, there are two critical questions to answer.

1. How often do users initiate video sharing?
2. How often do users further share a video upon receiving it?

Each initiator triggers the first share of a video. It has been found that, in a one-week dataset of 12.8 million video sharing and 115 million viewing events, 827 thousand initiating records can be extracted [33]. While this number is not small, it is only 6.5% out of the 12.8 million sharing records. This indeed reflects the pervasiveness and power of video spreading in social network.

The rank distribution of the initiators (in terms of the number of initiated videos) is plotted in Fig. 19.11a. It suggests that most users initiate few videos, but a few *active users* have initiated a remarkable number of videos. The most active user indeed has initiated over two thousand videos in one week.

Zipf's law [34] is usually used to a describe skewed distribution, which is a straight line in logarithmic scale. However, the data in Fig. 19.11a cannot be simply fitted by one Zipf line: the data after top-10 appear to be a straight line, but the top-10 data clearly differ from the rest. Yet they can be roughly fitted by another Zipf line. The distinction suggests the existence of two possible types of users with different initiating behaviors: (1) most of the initiators (over 99%) initiate only a few videos; and (2) a set of active initiators have much more friends and also initiate a much larger number of videos. The change at threshold 90 is actually quite sharp, showing clear distinction of the two types of initiators. These active initiators serve as hubs that draw much more attention than the general users and are worthy of particular attention in system optimization.

The distribution of the number of each user's shares is shown in Fig. 19.11b, which again indicates that there are some extremely active users sharing a great number of videos, although most of the users only share a small number of videos. There are

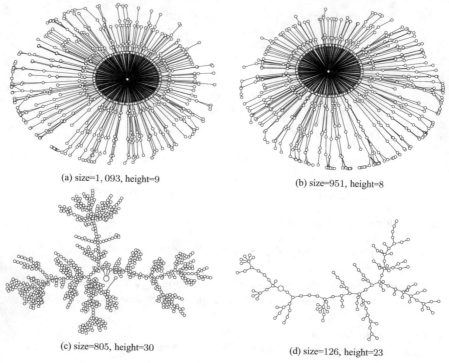

(a) size=1, 093, height=9   (b) size=951, height=8

(c) size=805, height=30   (d) size=126, height=23

Fig. 19.12 Illustration of spreading trees for popular videos

also users who have watched more than one thousand videos without sharing any, like *free-riders* in peer-to-peer systems.

The above observations suggest that the users have diverse activeness, and we can roughly distinguish three types of users.

- *Spreaders* (SU), a small number of users who initiate a lot of videos, and also have many friends, being hub-like. Some spreaders are non-personal accounts specifically interested in collecting and spreading interesting, funny, attractive contents, including videos; it is also possible that spreaders are bots, spreading videos in a spam manner;
- *Free-riders* (FU), who watch many videos without sharing any, which noticeably hinders video spreading;
- *Ordinary users* (OU), who sometimes initiate a few videos, watch some shared videos, and share some videos they watched.

19.3.2 Video Propagation Structure and Model

Figure 19.12 shows visual examples of two typical propagation structures: one type has a moderate depth, but limited branching—most of the branches are directly from the source, with no further branching, as shown by Fig. 19.12a, b; the other type

Table 19.4 Statistics and descriptions of the four videos (see Fig. 19.12)

Size	Height	Views	Length (s)	Category	Description
1093	9	34,531	123	News	A father picked up daughter from school by helicopter
951	8	14,281	60	Advt.	Earth hour promotion video
805	30	12,658	306	Music	Charity single "Children" by Chinese stars
126	23	1,431	235	Comedy	Funny lip sync video

branches frequently at different levels, and some branch can be very long, shown by Fig. 19.12c, d (the root node is enlarged for better visualization). Table 19.4 further lists the descriptions from each video's URL, revealing their content. We can see that the video propagation path and coverage are highly diverse, depending on both the video content itself and the user watching and sharing behavior.

There have been many studies on the propagation structure and model of message sharing through online social networks [28,35]. A widely used model that is aware of the users' status is the *epidemic model*, which describes the transmission of communicable disease through individuals [36]. Besides in epidemiology, it has also been recently used to model computer virus infections and information propagations such as news and rumors [37].

One classical epidemic model is the *SIR model* (Susceptible-Infectious-Recovered), first proposed by Kermack and McKendrick [38]. It considers a fixed population with three compartments: Susceptible (S), Infectious (I), and Recovered (R). The initial letters also represent the number of people in each compartment at a particular time t, that is, $S(t)$ represents the number of individuals not yet infected; $I(t)$ represents the number of individuals who have been infected and are capable of spreading the disease to those in the susceptible category; $R(t)$ represents the number of individuals who have been infected and then recovered. Given transition rate β from S to I and γ from I to R, the following equations can be derived:

$$\frac{dS}{dt} = -\beta SI, \quad \frac{dI}{dt} = \beta SI - \gamma I, \quad \frac{dR}{dt} = \gamma I$$

There is a natural mapping between conventional object sharing propagation in social networks and the compartments of the SIR model. For a particular object, all the users in the social network are Susceptible at the beginning; at a certain time, the users accessing the object are Infectious, indicating that they are able to infect others by sharing the object. They can be Recovered if they choose not to share. Yet for video spreading, the mapping is not complete:

1. A user can choose not to watch the received video, and likely not participate in the spreading as well. To differentiate these users and the users in R who have watched or directly shared the video, we categorize these users to a new compartment, Immune (Im).

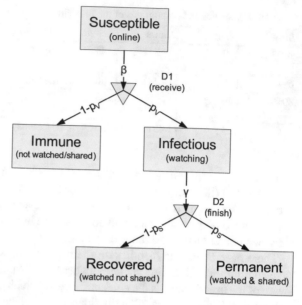

2. In the classical SIR model, the transition is time-dependent, i.e., at any time, there
   is a chance that the stage transits to the next one. While for video spreading in
   social networks, the transition of the stages depends on decisions at a certain time.
   For example, the user needs to choose watch or not, and share or not share. To
   address this problem, two temporary decision stage, D1 and D2 can be introduced.
3. It is necessary to differentiate the users who have shared the video and those who
   have not after watching the video. A new compartment, Permanent (P), can be
   introduced, indicating users who have shared the video, and otherwise Recovered.

The enhanced SI^2RP (Susceptible-Immune-Infectious-Recovered-Permanent)
model is illustrated in Fig. 19.13. For each video propagation process, the initiator is Infectious at the beginning.

The transition rate from S to D1 is β, and thus a Susceptible user will spend $1/\beta$
unit time to receive a shared video from a friend. The user then makes a decision
whether or not to watch the video. If the user is not interested in it and decides not to
watch or share, she/he is considered as Immune. The probability of the user watching
or directly sharing the video can be denoted as p_V.

If the user decides to watch the video, she/he becomes Infectious. The transition
rate from I to D2 is γ, indicating that the user will spend $1/\gamma$ time to finish watching
the video. The user then makes the second decision, whether or not to share the
video. If the user decides not to share, she/he becomes Recovered or Permanent
otherwise. The probability of a user deciding to share the video is denoted by p_S. The
transition rate β and γ can be inferred from measurement results, so are D_I, p_V, and
p_S in the model. These four probability distribution or probability characterize the
behavior of different types of users, namely, spreaders (SU), ordinary users (OU),
and free-riders (FU).

- An SU initiates video shares according to distribution D_I (An OU or FU does not initiate).
- A user watches videos shared by friends with probability p_V, which is based on the reception rate.
- After watching, an SU or OU shares the video with probability p_S, which is based on the share rate.

19.3.3 Video Watching and Sharing Behaviors

It has been found that, in social network sharing, compared to strangers, friends have relatively higher probability of reciprocal visits. When a content is uploaded by a friend, a user is more likely to browse. The users are also more active in viewing profiles than leaving comments, and consequentially, latent interactions cover a wider range of friends than visible interactions [39]. More importantly, most of the users are willing to share their resources to assist others with close relations, which naturally to leads to collaborative delivery [40].

Different from text or images that can be instantly viewed, a posted video will not be really watched until the recipient clicks the link. Upon receiving the video post, the recipient (friend or follower) has three options:

1. Watches the video, and thus the requirement of streaming quality, such as startup latency and playback continuity, should be satisfied.
2. Not to watch the live video, but download the video and expect to watch it later.
3. Shows no interest in the video. If the user does not want to watch the video now or later, she/he may not share the resources with other uploader's friends, either.

The coexistence clearly makes a system design more complicated. More specifically, there exist two types of friends interested in the posted video, namely, *streaming users* and *storage users*. The streaming users expect to watch the video immediately, and the storage users will download and watch the video at a different time, due to the presence of other concurrent events.

The streaming users might stop watching after a while if they find the video is out of their interest, even though the video is posted by friends. Such dynamics will affect the data delivery if they serve as relays for other users. On the other hand, the storage users who are downloading the video asynchronously do not have the concern of interest nor playback quality, until they start to watch the video. Hence such users are considered relatively stable.

19.4   Mobile Video Clip Sharing

Recently, the rapid development and penetration of mobile social networking have enabled the new generation video sharing services that use smart mobile terminals to instantly capture and share ultra-short video clips (usually of several seconds).

Many mobile apps, e.g., Twitter's Vine, Instagram, Snapchat, and TikTok, to name but a few, have incorporated such multimedia services and seen great acceptance, particularly by the youth community [11,12]. TikTok (known as Douyin in China) has become the 8th most-downloaded app in the past decade.

The instant video clips in these services are directly consumed at smart-terminals with specially designed mobile interfaces and operations. The expanded social relations and the distinct operations on the mobile terminals, particularly *screen scrolling*, have greatly increased the amount of videos available to watch, and in the meantime, shorten the time focusing on individual videos from tens of minutes to only a few seconds [41].

19.4.1 Mobile Interface Characteristics

In traditional video sharing and online social networking services, users need to click to view or link to one specific video, which only allows them to view one video each time/click. User experience is crucial to mobile instant video clip sharing. An instant video clip itself is of only several seconds long, thereby a mobile user can hardly tolerate a long delay, which would completely ruin the viewing experience. Mobile instant video clip sharing services instead return a playlist of video clips when a user touches the screen to view the updates for certain users, tags, or channels. As the user scrolls the smartphone/tablet's screen, video clips are seamlessly played from the generated list. Scrolling includes a series of user gestures, typically *click*, *drag* and *fling*, and the speed, acceleration, and continuity vary depending on the user's input.

Such a *batched view* implies that mobile users can watch a considerable amount of instant video clips within the playback time of one conventional video (e.g., from YouTube). A related new behavior is *passive view*. The media contents are arranged in order and a user has limited control over the order for playback. For two video clips of interest, if they are separated in the playlist, the user may have to download (and watch) all the video clips in between them. These videos of no particular interest are passively watched, and the resources for downloading and playing them will be consumed.

Although lack of common VCR controls (such as rewind and fast forwarding), batch/passive views with scrolling are effective in approaching successive instant video clips in the playlist, enabling users to find interesting contents more easily, and accelerating the propagation of popular videos. Yet, if not being handled properly, screen scrolling may ruin the viewing experience. In the worst case (e.g., downloading every instant video clip through a poor cellular connection), a vicious circle can be formed: the downloading of a just skipped video will take up the network resources and block the downloading of those of interest, which will in turn force the user to give up watching the target videos and scroll forward to search for other interesting videos.

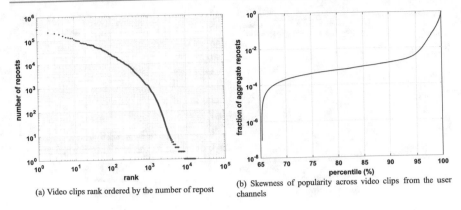

(a) Video clips rank ordered by the number of repost

(b) Skewness of popularity across video clips from the user channels

Fig. 19.14 Video popularity

19.4.2 Video Clip Popularity

For mobile video clip sharing, the number of reposts can be used to evaluate the video popularity. Figure 19.14 plots the number of reposts as a function of the ranks of the video clips by their popularity for a sample of 16 user channels, and Fig. 19.14 further plots the cumulative proportion of the total number of reposts versus the percentile of the video clips. The popularity of these sample video clips is extremely skewed: the top 5% video clips accounts for more than 99% reposts.

The popularity distributions for different generations of video sharing services show a trend of becoming more and more skewed over different generations of social media sharing (YouTube: 10–80%; OSN: 2–90% and 5–95%; Instant Video Clip: 2–95 and 5–99%). The YouTube result implies that, originally, users' interests across videos are not evenly distributed (biased toward popular videos); People tend to watch what others have watched, which is exaggerated when OSNs are introduced, as users in the same social group share common interests. On top of social networking, ubiquitous mobile accesses further leads to a more efficient and more extensive propagation of the video clips.

As the unpopular video clips move toward the bottom of the playlist, users hardly see them again. On the contrary, the popular video clips will be promoted to the popular section, and users can easily reach these posts. They become more and more popular, keeping on the top of the playlist and thereby being accessed more frequently. With the batched and passive views, the above process is accelerated and exacerbated. This extreme skewness suggests that identifying the popular videos and pre-fetching them could be beneficial.

19.4.3 Lifespan and Propagation

Figure 19.15 shows the average daily number of reposts after the video clips were created. As can be seen, the average number of reposts for the popular video clips monotonically decreases day by day. Even for many of the popular video clips, they

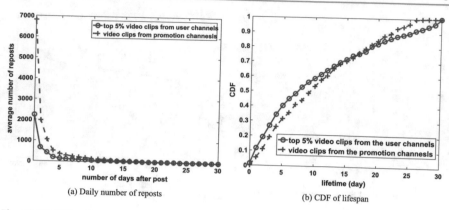

Fig. 19.15 Video lifespan and propagation

are most popular during the first day after the initial posts and are getting less and less popular afterwards. This fast decay feature of mobile instant video clips is quite unique: YouTube videos also reach the global peak immediately after introduction to the system, but decay much more slowly, while the requests for the new videos published in OSNs generally experience two or three days latency to reach the peak value, and then change dynamically with a series of unpredictable bursts [42].

Figure 19.15 plots the CDF of active lifespan of the popular video clips. It shows that more than half of the popular video reposts can only stay in active for less than 10 days. This result is quite different from the related observations on traditional video sharing services: some of YouTube videos can still get views even after one year since they were published [9]. The fast decay can be possibly explained by the mobile nature of ubiquitousness: As mobile users can upload and watch instant video clips at any time and anywhere, they can propagate very efficiently and extensively, and thus can reach the peak immediately; the frequent video watching and uploading from the mobile user further accelerate the fade of existing instant video clips, even for the popular ones.

The short lifespan and fast decay imply that popular contents are much more dynamic than those in other mobile VoD or video streaming applications. This introduces a dilemma for pre-fetching: on one hand, we would like to cache as many videos as possible to provide smooth watching experience; on the other hand, if the cached videos cannot be watched soon enough by the user, it becomes a huge waste for fetching them, as they will probably be flushed out by more recent feeds, having no chance to be viewed. As such, neither a simple download-and-watch scheme nor a naive pre-fetching/caching scheme would work efficiently, and a smart adaptive solution is expected. More importantly, it must work well with screen scrolling, a rich operation whose multiple factors, e.g., speed/acceleration, are to be considered.

19.5   Further Exploration

Research on the social relations and social graphs in the human society has a long history, so does that on disease propagation in epidemiology [36,38]. Online social media and social networking, however, appeared only in very recent years and are still undergoing rapid changes. Research in this field remains in an early stage and many exciting topics are to be explored [43].

19.6   Exercises

1. Find out a typical Web 1.0 application and a typical Web 2.0 application, and discuss their key differences.
2. Discuss the key differences between YouTube videos and the traditional movies/TV shows. How would they affect content distribution?
3. YouTube publishes statistics about its videos online. A recent one is as below:

   - More than 1.9 billion YouTube users are active monthly
   - More than 1 billion hours of YouTube videos are watched per day
   - More than 500 hours of video are uploaded to YouTube every minute
   - The majority of all users end up in the 25 to 44-year-old-age group
   - YouTube is available in more 90 countries, supporting 80 languages
   - Mobile makes up more than 50% of YouTube's global watch time.

   Check the latest statistics and estimate the monthly growth speed of YouTube. Suggest some reasons that make YouTube-like services expand so quickly and the potential challenges therein. Repeat these for two other social media services.
4. Is it beneficial to place all the content from a social media service in one server? If not, what are the challenges to place the content in multiple servers?
5. Discuss the propagation and consumption patterns of multimedia content in a social networking tool that you're familiar with.
6. Given a positive integer n and a probability value $0 \leq p \leq 1$, an *Erdös-Rényi* (ER) random graph $G(n, p)$ is with n vertices where each possible edge has probability p of existing. This is the most important class of random graphs.

   (a) Write a simple program to generate ER random graphs, and calculate their characteristic path lengths and clustering coefficients. Compare them with the YouTube video graph we have discussed earlier.
   (b) Discuss whether the graph formed by an online social network, say the graph of Facebook user accounts, is such a random graph or not. Hint: Think about the way that the edges are formed.

7. A simple model for information propagation is *gossip*. With gossip, a network node, upon receiving a message, will randomly forward it to the neighboring nodes with probability p.

Fig. 19.16 An example of
the tit-for-tat strategy

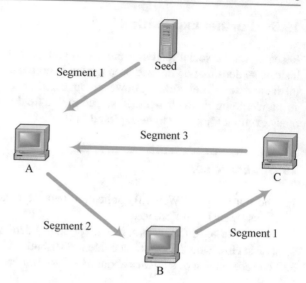

(a) Write a simple program to simulate the gossip algorithm in randomly generated
networks. A node may delay t time before forwarding. Discuss the impact of
p and t on the coverage and propagation speed of a message.

(b) Is it beneficial if the nodes can have different values of p ? If so, provide some
guidelines in the selection of p for each node.

(c) Is gossip suitable for modeling the propagation process of a picture shared in
a real-world social network, say Facebook? How about video?

8. In an online social network, a free rider only consumes videos but does not share
videos. Free riders also exist in peer-to-peer file sharing: they download data
segments from others, but never upload. BitTorrent adopts a *tit-for-tat* strategy to
solve the incentive problem, i.e., you get paid only if you contribute. As depicted
in Fig. 19.16, peers A, B, and C download different segments from each other.
This forms a feedback loop; for example, uploading segment 2 from A to B will
be feedback to A by the upload of segment 3 from C to A, which stimulates peer
A to cooperate.

(a) Discuss whether the tit-for-tat strategy works for video propagation with free
riders.

(b) For live video streaming with delay constraints, with tit-for-tat work?

9. For mobile video clip sharing, the input user gestures include click, drag, and
fling, where the latter two can cause screen scrolling. Once a gesture is given, the
following process of screen moving is predetermined. Given the fixed display size
of each clip (specifically, the fixed height), the motion of screen scrolling can be
modeled and calculated, and the details of the scrolling process can be obtained
accurately (e.g., how many videos are present, how long each video will stay

in the viewport). Although different operating systems have different technical details for implementation, the philosophy for animating the screen scrolling is generally the same, which is to gradually decelerate the scrolling speed until it reaches zero if there is no other finger touch detected during the deceleration. In the case of dragging, assume the screen scrolling speed will experience a uniform deceleration with a default deceleration $d = 2,000\,\text{pixels/s}^2$. Given the initial scrolling speed s_0 and the height of each instant video clip h, how many video clips will be covered by this drag gesture? When will the m-th video clip appears in viewport?

References

1. A.M. Kaplan, M. Haenlein, Users of the world, unite! The challenges and opportunities of social media. Bus. Horiz. **53**(1), 59–68 (2010)
2. Z. Wang, J. Liu, W. Zhu, Social-aware video delivery: challenges, approaches, and directions. IEEE Netw. **30**(5), 35–39 (2016)
3. V. Bajpai, S. Ahsan, J. Schönwälder, J. Ott, Measuring YouTube content delivery over IPv6. SIGCOMM Comput. Commun. Rev. **47**(5), 2–11 (2017)
4. J. Yi, S. Luo, Z. Yan, A measurement study of YouTube 360° live video streaming, in *Proceedings of the 29th ACM Workshop on Network and Operating Systems Support for Digital Audio and Video, NOSSDAV' 19*, pp. 49–54. (ACM, New York, NY, USA, 2019)
5. S. Acharya, B. Smith, P. Parnes, Characterizing user access to videos on the World Wide Web, in *Proceedings of the ACM/SPIE Multimedia Computing and Networking (MMCN)* (2000)
6. W. Tang, F. Yun, L. Cherkasova, A. Vahdat, *Long-term Streaming Media Server Workload Analysis and Modeling* (Technical report, HP Labs, 2003)
7. T. Hoßfeld, K. Leibnitz, A qualitative measurement survey of popular internet-based IPTV systems, in *Proceedings of the International Conference on Communications and Electronics (ICCE)*, pp. 156–161 (2008)
8. X. Cheng, J. Liu, C. Dale, Understanding the characteristics of internet short video sharing: a Youtube-based measurement study. IEEE Trans. Multimed. **15**(5), 1184–1194 (2013)
9. M. Cha, H. Kwak, P. Rodriguez, Y.-Y. Ahn, S. Moon, I Tube, You Tube, everybody Tubes: analyzing the world's largest user generated content video system, in *Proceedings of the 7th ACM SIGCOMM Conference on Internet Measurement (IMC '07)*, pp. 1–14 (2007)
10. A. Finamore, M. Mellia, M.M. Munafò, R. Torres, S.G. Rao, YouTube everywhere: impact of device and infrastructure synergies on user experience, in *Proceedings of the 2011 ACM SIGCOMM conference on Internet measurement (IMC '11)*, pp. 345–360 (2011)
11. S. Yarosh, E. Bonsignore, S. McRoberts, T. Peyton, YouthTube: youth video authorship on YouTube and Vine, in *Proceedings of ACM CSCW*, pp. 1423–1437 (2016)
12. F. Loh, F. Wamser, C. Moldovan, B. Zeidler, T. Houndefinedfeld, D. Tsilimantos, S. Valentin, From click to playback: a dataset to study the response time of Mobile YouTube, in *Proceedings of the 10th ACM Multimedia Systems Conference, MMSys '19*, pp. 267–272. (Association for Computing Machinery, New York, NY, USA, 2019)
13. P. Gill, M. Arlitt, Z. Li, A. Mahanti, YouTube traffic characterization: a view from the edge, in *Proceedings of the 7th ACM SIGCOMM Conference on Internet Measurement (IMC '07)*, pp. 15–28 (2007)
14. S. Sen, J. Rexford, D.F. Towsley, Proxy prefix caching for multimedia streams, in *Proceedings of the IEEE INFOCOM* (1999)

15. F. Figueiredo, F. Benevenuto, J.M. Almeida, The tube over time: characterizing popularity growth of YouTube videos, in *Proceedings of the fourth ACM International Conference on Web Search and Data Mining (WSDM '11)*, pp. 745–754 (2011)

16. R. Zhou, S. Khemmarat, L. Gao, The impact of YouTube recommendation system on video views, in *Proceedings of the 10th Annual Conference on Internet Measurement (IMC '10)*, pp. 404–410 (2010)

17. S. Milgram, The small world problem. Psychol. Today **2**(1), 60–67 (1967)

18. D.B. West, *Introduction to Graph Theory*, 2nd edn. (Oxford University Press, 2001)

19. D.J. Watts, S.H. Strogatz, Collective dynamics of "Small-World" networks. Nature **393**(6684), 440–442 (1998)

20. E. Ravasz, A.-L. Barabási, Hierarchical organization in complex networks. Phys. Rev. E **67**(2), 026112 (2003)

21. R. Albert, H. Jeong, A.-L. Barabási, The diameter of the World Wide Web. Nature **401**, 130–131 (1999)

22. X. Cheng, M. Fatourechi, X. Ma, J. Liu, Insight data of YouTube: from a partner's view. Technical Report. (Simon Fraser University, 2012)

23. H. Pang, Z. Wang, C. Yan, Q. Ding, L. Sun, First mile in crowdsourced live streaming: a content harvest network approach, in *Proceedings of the Thematic Workshops of ACM Multimedia*, pp. 101–109. (ACM, 2017)

24. X. Ma, C. Zhang, J. Liu, R. Shea, F. Di, Live broadcast with community interactions: Bottlenecks and optimizations. IEEE Trans. Multimed. **19**(6), 1184–1194 (2017)

25. B. Yan, S. Shi, Y. Liu, W. Yuan, H. He, R. Jana, Y. Xu, H.J. Chao, LiveJack: integrating CDNs and edge clouds for live content broadcasting, in *Proceedings of the ACM on Multimedia Conference*, pp. 73–81. (ACM, 2017)

26. C. Ge, N. Wang, W.K. Chai, H. Hellwagner, QoE-assured 4K HTTP live streaming via transient segment holding at mobile edge. IEEE J. Sel. Areas Commun. (JSAC)

27. F. Wang, C. Zhang, F. wang, J. Liu, Y. Zhu, H. Pang, L. Sun, Intelligent edge-assisted crowd-cast with deep reinforcement learning for personalized QoE, in *IEEE INFOCOM 2019—IEEE Conference on Computer Communications*, pp. 910–918 (2019)

28. D. Wang, Z. Wen, H. Tong, C.-Y. Lin, C. Song, A.-L. Barabasi, Information spreading in context, in *Proceedings of the 20th International Conference on World Wide Web (WWW '11)*, pp. 735–744 (2011)

29. J. Leskovec, L. Backstrom, J. Kleinberg, Meme-tracking and the dynamics of the news cycle, in *Proceedings of the 15th ACM SIGKDD International Conference on Knowledge Discovery and Data Mining*, pp. 497–506 (2009)

30. K. Dyagilev, S. Mannor, E. Yom-Tov, Generative models for rapid information propagation, in *Proceedings of the First Workshop on Social Media Analytics (SOMA '10)*, pp. 35–43 (2010)

31. A. Abisheva, V.R.K. Garimella, D. Garcia, I. Weber, Who watches (and shares) what on Youtube? And when? using Twitter to understand Youtube viewership, in *Proceedings of the 7th ACM International Conference on Web Search and Data Mining, WSDM '14*, pp. 593–602 (ACM, New York, NY, USA, 2014)

32. H. Yu, L. Xie, S. Sanner, Twitter-driven YouTube views: beyond individual influencers, in *Proceedings of the 22nd ACM International Conference on Multimedia, MM '14*, pp. 869–872. (ACM, New York, NY, USA, 2014)

33. X. Cheng, H. Li, J. Liu, Video sharing propagation in social networks: measurement, modeling, and analysis, in *Proceedings of the IEEE INFOCOM Mini-Conference* (2013)

34. G.K. Zipf, *The Psychobiology of Language*. (Psychology, 1999)

35. H. Kwak, C. Lee, H. Park, S. Moon, What is Twitter, a social network or a news media? In *Proceedings of the 19th International World Wide Web Conference on (WWW '10)*, pp. 591–600 (2010)

36. D.J. Daley, J. Gani, J.M. Gani, *Epidemic Modelling: an Introduction*, Cambridge Studies in Mathematical Biology. (Cambridge University Press, , 2001)

37. Z. Liu, Y.-C. Lai, N. Ye, Propagation and immunization of infection on general networks with both homogeneous and heterogeneous components. Phys. Rev. E **67**(1), 031911 (2003)

38. W.O. Kermack, A.G. McKendrick, A contribution to the mathematical theory of epidemics. Proc. R. Soc. Lond. Ser. A **115**(772), 700–721 (1927)
39. J. Jiang, C. Wilson, X. Wang, P. Huang, W. Sha, Y. Dai, B.Y. Zhao, Understanding latent interactions in online social networks, in *Proceedings of the 10th Annual Conference on Internet Measurement (IMC '10)*, pp. 369–382 (2010)
40. X. Cheng, J. Liu, Tweeting videos: coordinate live streaming and storage sharing, in *Proceedings of the 20th International Workshop on Network and Operating Systems Support for Digital Audio and Video, NOSSDAV '10*, pp. 15–20 (2010)
41. L. Zhang, F. Wang, J. Liu, Mobile instant video clip sharing with screen scrolling: measurement and enhancement. IEEE Trans. Multimed. **20**(8), 2022–2034 (2018)
42. H. Li, H. Wang, J. Liu, K. Xu, Video sharing in online social networks: measurement and analysis, in *Proceedings of ACM NOSSDAV*, pp. 83–88 (2012)
43. C. Kadushin, *Understanding Social Networks: Theories, Concepts, and Findings*. (Oxford University Press, 2012)

Augmented Reality and Virtual Reality 20

Imagine a world in which the physical objects surrounding us could be constructed or destructed on a whim, tinkered, and adjusted to fit the demands of the user. Such a world may never exist in the physical domain, but the worlds of *augmented reality* (AR) and *virtual reality* (VR) allow us to do just that. In this chapter, we examine this frontier of immersive multimedia and explore the enabling technologies and their impacts to the field, while simultaneously analyzing their challenges and restrictions.

Both augmented and virtual reality exist on a continuous scale often called the *virtuality continuum* [1] (Fig. 20.1). On one side is the completely real, reality: a real, physical world consisting solely of real, physical objects; on the opposing end is completely virtual, virtuality: a virtual world with virtual objects. The land between both of these extrema is called *mixed reality*, as aspects of the real and virtual begin to blend. In augmented reality, the physical reality that we live in is mixed with virtual objects, but not to the extent that the virtual aspects outnumber details of reality; in virtual reality, however, the physical reality is further removed, giving a fully immersive experience.

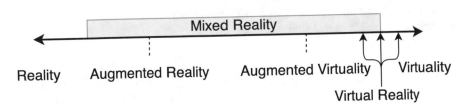

Fig. 20.1 Reality-virtuality continuum

Z.-N. Li et al., *Fundamentals of Multimedia*, Texts in Computer Science,
https://doi.org/10.1007/978-3-030-62124-7_20

Fig. 20.2 An example of what a user might see in an augmented reality app; from Keiichi Matsuda's short film hyperreality [4]. Used with the permission of Keiichi Matsuda, Keiichi Matsuda Ltd.

20.1   Defining Augmented Reality and Virtual Reality

The concepts of augmented reality and virtual reality have existed since the 1960s. In 1962, Morton Heilig created a very early virtual reality system, the Sensorama [2]; in 1968, Ivan Sutherland created the very first augmented reality system, the Sword of Damocles [3]. Sutherland also provided one of the earliest definitions of what constitutes augmented reality: a two-dimensional illusion of a three-dimensional object presented to the observer that visually changes in exactly the way that a real object would change when the user's head moves. Figure 20.2 shows a still image of what such an effect might look like (from [4], a short film on AR).

This broad definition was narrowed down by Ronald Azuma [5], who defined augmented reality to be a system that

- combines real and virtual objects in a real environment,
- aligns both real and virtual objects with each other,
- runs interactively in real time,
- is registered in three dimensions.

This revised version does successfully exclude films from being included, but also ends up excluding 2D overlays on top of live video. It also does not explicitly reference the visual illusory effect noted by Sutherland. The concept of a handheld computer with the works, e.g., cameras, sensors, touchscreen RGB displays, etc., was unthinkable at the time, either. Considering these factors, a modern AR system can be technically characterized by these factors:

- a predominantly visual system,
- that combines real and virtual objects in a real environment,

- while aligning the real and virtual together,
- running in real time,
- and allows for 6 degrees of freedom (position, orientation) of interaction with virtual objects.

This definition allows for the inclusion of 2D overlays while preserving the intent of having interactions in three dimensions as a user can physically move real objects with 6 degrees of freedom and virtually interact with objects with the same level of discretion. It also allows for auxiliary sensory augmentations (aural, tactile, gustatory, and/or olfactory) but mandates for a virtual object to visually exist first. This avoids including advanced sound systems capable of emitting spatialized sound. Furthermore, it makes no restrictions on the underlying hardware used to create the illusory effects to allow for future technologies yet to be discovered to be incorporated into the system.

In virtual reality, both the objects and the environment are virtualized. It therefore provides a simulated experience that can be similar to or completely different from the real world. This change, despite seemingly minor, affects how such a system can be constructed. As we no longer require a real environment, the nuances of object-environment alignment and supporting a viewfinder to allow the real environment to coexist with the objects are gone. This indeed simplifies the system as everything can now be self-contained in a closed system; the user's pose is now the ground truth as opposed to being relative to the real environment's orientation. For these reasons, an augmented reality system is technically capable of providing a virtual environment by ignoring and drawing over the user's physical surroundings.

One of the key differentiating factors of virtual reality is the existence of telepresence [6]: a sensation of being present in a distant elsewhere. This almost necessitates a *head-mounted display* (HMD) to provide a richness of presence and interactivity, leading to a categorization of display interfaces [1] that affect the perception of whether or not a system is augmented or virtual reality:

- Class 1—Monitor-based video displays upon which computer generated images are overlaid.
- Class 2—Video displays in Class 1, but with immersive HMD.
- Class 3—HMDs in Class 2, but with see-through capabilities rather than a video of the real environment.
- Class 4—Extension of Class 3 using a stereoscopic video of the real environment.
- Class 5—Completely graphical display environments where video reality is added.
- Class 6—Extension of Class 5 using partially immersive and real objects interact with the computer generated scene.

Despite appearing to be well delineated, the distinctions between each class rapidly break down when perception, orientation, real, virtual, and implementation are all taken into account. In practice, this also appears to be the case; in literature, the usage of augmented reality is restricted to primarily Class 3 displays (optical and head-mounted), but commercially successful implementations of augmented reality often utilize a Class 1 or Class 2 display (video, sometimes head-mounted), alongside most virtual reality systems.

In short, the medium of display is less relevant to whether a system is an augmented reality or virtual reality system than the actual perceived visual effects regardless of medium. An augmented reality system can be converted to a virtual reality system by a removal of the real environment, but the converse is not necessarily true as the removal of a virtual environment to incorporate the real would mean a fundamental overhaul (e.g., going from Class 5 displays to Class 3). In the remaining part of the chapter, we will focus on augmented reality systems, and many of the technical solutions apply to virtual reality systems as well.

20.2   Workflow of Augmented Reality

Figure 20.3 shows a typical augmented reality workflow, which can be broken down into three tasks: sensory data collection and processing, scene localization and alignment, and digital world generation and emission. These tasks can then be modeled after a source-sink data flow, transforming the low-level data into high-level semantics, which impact how the digital world is generated and emitted to the user's senses.

20.2.1 Sensory Data Collection

This step acts as an information source, generating a continuous stream of information that describes the real-world surroundings of the user. This information stream contains things such as a continuous sequence of images depicting what the user is currently observing, audio of the surroundings, and data describing the orientation

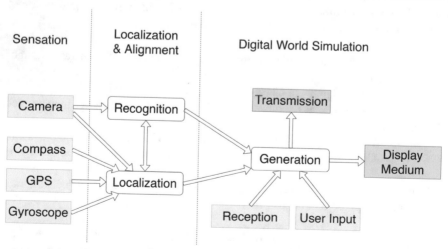

Fig. 20.3 General workflow model of augmented reality, where green nodes indicate a source, and red nodes indicate a sink

and position of the user's head. An augmented reality system attempts to fuse all of these together to extract an abstract representation of the user. In constrained environments and usage scenarios, there may be additional extrinsic sensory components, such as a mechanical linkage to determine the exact position relative to a room.

20.2.2 Localization and Alignment

Once the system extracts this digital representation of the user, it can begin to localize this user in the digital world. This process greatly varies depending on the application of the augmented reality, as there can exist multiple virtual worlds that are owned by one user and can only be acted upon by the owner. This happens in such applications as live translation, heads-up navigation, or digital personal assistants, etc. On the other end of the spectrum, we can have a single, globally-shared virtual world that all users exist in and can interact with. Examples include smart cities with digital signage and controls, or multiuser video games.

The general process of localization is to define a general mapping between the real-world coordinates to virtual objects. The mapping function must be surjective, as each digital object for each virtual world must match at least one physical coordinate. It does not have to be bijective, though bijectivity is highly desirable as it would guarantee perfect spatial alignment.

To aid in this localization task, the system can perform recognition on the physical surroundings, so as to adjust the outcome of the mapping function and better fit the estimate of where the digital representation of the user should be, or to trigger the placement of virtual objects based on a physical representation. Recognition can also be improved by utilizing knowledge of where the digital representation of the user currently is, with respect to other elements of the virtual world. Both steps can be used in tandem to compute a fixed-point estimate of the virtual coordinates of the user.

Accurate localization and alignment are computationally challenging, and are also one of the most active topics in AR/VR research.

20.2.3 World Generation and Emission

At this stage, the augmented reality system has the necessary information to begin rendering the two-dimensional perspective imagery that will be emitted to the display medium and presented to the user. The digital representation of the user has hopefully found a fixed-point positioning, and the conditionally displayed virtual objects are placed with respect to the recognition of some physical item. Secondary information, such as user input and in the case of multiuser augmented reality, received localization data of other users, are consolidated and reconciled with the existing virtual world. Again, the exact process varies depending on how the user input is observed by the system; for instance, with a dedicated directional controller, a visual or aural recognition-based controller, a spatial input controller, or their combinations.

Generally, the digital world is fully simulated at this point, and it advances one time-step forward. In multiuser environments, this simulated world is transmitted to other recipient users that exist in the same virtual world. Finally, the perspective of the digital representation of the user is rendered out and displayed to the user. Note that this stage is rife with many challenges: these displays need to have high refresh rates to minimize alignment error, high pixel densities to minimize pixelation, high per pixel intensity to be effective outdoors against the sun, while maintaining low power consumption to reduce or outright remove requiring a bulky battery.

20.3   Early Foundational Systems and Applications

Sutherland's system [3], a pioneer augmented reality system, was a relatively primitive optical HMD (head-mounted display) that used half-silvered mirrors to allow the user to view real-world objects along with the digital (Class 3). The technology behind it was truly decades ahead of its time. The system (Fig. 20.4) is driven by a mechanical linkage that measures the user's head position, which is then sent to an early proto-computer. This computer then outputs the matricies to translate and rotate the 3D objects in room-relative coordinates to a pipeline of three successive independent digital devices. The first device converts the object coordinates to be eye/device relative coordinates and feeds it to the next device. This second device then transforms the 3D coordinates into 2D coordinates and clips any vertices that are beyond the user's field-of-view and feeds the output to the last device, an analog display driver. It then generates the analog signals that are sent to half-silvered mirrors, which are finally viewed by the user.

Being a mechanical device that senses the position of the user, the analog positioning data, such as the amount of compression experienced by a spring, are to be translated into a digital format to be fed into a computer. This data are then be processed and turned into a projection matrix, encoding the position of the user and their viewing angle. When this projection matrix is multiplied against by a vector containing the position of a vertex, it converts the position to be relative to the pose of the user. In Sutherland's system, the computer both generated the coordinates of the digital objects and output them to the next device, i.e., the matrix multiplier.

As the name suggested, the matrix multiplier accepts two matrices as the input, and output as their resulting product. This is necessary because this system was built in 1968, one full year before the advent of Strassen's algorithm, the first published matrix multiplication algorithm with a sub-cubic computational complexity. To achieve an output of 30 frames per second (fps), a dedicated matrix processing chip was required, as "no available general-purpose computer would be fast enough to become intimately involved in the perspective computations required for dynamic perspective display" [[3], p. 759].

The clipping divider, for the same reason, was also required as a separate device. It takes the eye-relative coordinates and performs checks against the user's viewing cone. Anything that is outside of this cone would be "clipped" and removed. It does this by testing each vertex to determine if it lies outside of the viewing cone, as

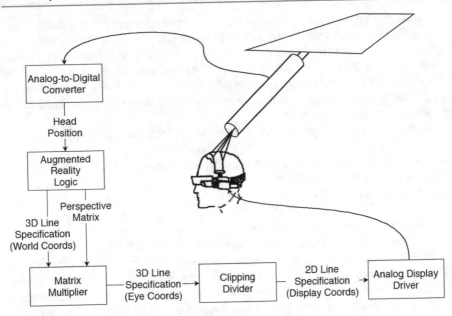

Fig. 20.4 A high-level breakdown of Sutherland's AR system [3]

defined by the perspective matrix, and any offending vertices are simply "moved" to the point of intersection with the edge of the cone. The clipping divider also performs depth testing, where vertices behind the viewing cone are removed, and converts the 3D eye-relative coordinates into 2D screen-relative coordinates.

The Sutherland's system has many limitations: it was heavy, bulky, and could only generate wireframe imagery; there was a confined area of movement allowed; and there was no mechanism for user interaction. Despite this, it is still a marvel of engineering that was considerably ahead of its time.

The first generation of augmented reality was severely limited in terms of power, cost, mobility, and standardization, which greatly narrowed the field of possible technological applications. Consequently, the vast majority of the literature during this period was focused on military and space-faring applications. The monumental leap from custom integrated circuits to a well-defined form factor for personal computers has then led to an explosion in AR applications. We present a non-exhaustive list of some prominent AR applications and usage scenarios in the early time.

Navigation Applications. One of the major motivating reasons for AR is in the domain of navigational assistance. In 1993, Loomis et al. utilized aural AR to create a navigational system for the visually impaired [7]. In 1995, Rekimoto and Nagao presented the NaviCam system to assist with indoor navigation by way of virtual signage [8]. In 1997, Feiner et al. created a mobile AR system using an optical HMD for highlighting nearby points of interest [9]. Narzt et al. further presented a system for navigating pedestrians and cars using mobile AR and a head-up display (HUD), and Tonnis et al. used AR projected onto a car's windshield to inform the user of potential hazards surrounding the car [10].

Military Applications. AR was widely used for combat and combat simulations. Following the development of the *Super Cockpit*, a battlefield AR system was developed by Julier et al. at the Naval Research Laboratory [11]. This mobile AR system would serve as the basis for a variety of combat research and training simulations as outlined in [12]. Exploratory applications of AR also included data visualization, e.g., Azuma et al. explore projecting reconnaissance data from unmanned aerial vehicles to ground troops [13].

Industry Applications. An AR system for maintenance and repair [14] was funded by the US Air Force. BMW [15], Volkswagan [16], Ford and a consortium of auto manufacturers [17], Boeing [18], and Airbus [19] all individually and independently experimented with AR systems for assisting their assembly lines and maintenance workers. Their applications ranged from assisting in manufacturing by overlaying instructions and schematics, cable management, to vehicle repairs by highlighting broken or blocking parts.

Outside of manufacturing, there are also applications in the field of medicine. In 1998, Fuchs et al. proposed an AR system for assisting in laparoscopic surgery [20]. This stereoscopic system presents a simulated cutaway view of where the laparoscopes will be inserted. In 1999, Navab et al. created an AR system to superimpose X-ray imagery onto the underlying physical appendage [21]. This idea was extended to other imaging devices like MRIs and also incorporated the visualization of tools hidden by skin and other tissues by Vogt in 2006 [22].

Education Applications. Evident by how manufacturers looked toward AR to assist front-line workers, AR can also be applied as an educational enhancement tool. Vlahakis et al. use AR as a virtual guide to offer personalized tours of archaeological sites through reconstruction of buildings and visualizations of everyday life using avatars [23]. Kaufmann et al. created an AR tool for mathematics and geometry education [24] and later use it as a case study toward collaborative AR education [25]. Schrier explored using AR games to assist in teaching history [26], while Klopfer et al. used them for environmental and museum education [27,28].

Another aspect that AR systems have been used for is as a collaboration tool. StudierStube is one of the first multiuser AR systems with a shared environment between users [29]. Rekimoto and Saitoh also used AR to create augmented surfaces that allowed virtual information to be presented [30]. A natural extension of this idea is to encompass an entire room, as proposed by Tamura in [31] to create AR meeting rooms.

Entertainment Applications. The 1st & Ten "yellow line" football broadcast system [32] and the FoxTrax "glowing puck" system [33] painted a virtual object within the broadcast media to highlight prominent and important features of the broadcast, such as where the hockey puck is, where the line of scrimmage is, or the driver of a race car. Aside such *pseudo-AR* from broadcasting, we have more traditional AR systems for entertainment like ARQuake [34] and Human Pacman [35]. There were also a number of indoor games such as AR hockey [36] and bowling [37]. The field of personal entertainment for AR has largely been enabled by the lowered barrier to AR as more devices come equipped with sufficient hardware to localize users.

20.4 Enabling Hardware and Infrastructure

From the mid 1990s to the late 2000s, three key advancements in hardware have greatly enabled the creation of augmented and virtual reality systems, including graphics processing unit (GPU), Global Positioning System (GPS), and the Internet and mobile networking.

20.4.1 Graphics Processing Unit (GPU)

To deliver an immersive experience, AR/VR applications typically require much higher resolution and refresh rate than broadcast videos, not to mention the rich interactions that demands real-time response. Full HD (1080p) and 60 Hz are the bottomline for today's HMDs and, ideally, 16K+ resolution and 16 KHz refresh rate for each eye are expected [38]. Such massive amount of data can hardly be handled by CPU alone, which has slowed down the development of early AR/VR systems. IBM's monochrome display adapter (MDA), introduced in the 1980s, was the first discrete video display card, rectifying the nuisance of building a custom chain of hardware devices to render computer generated imagery. Its modus operandi, not significantly different than how Sutherland's graphical pipeline worked, opens the era of modern graphic processing unit (GPU), which has since been the core of visualization applications, including AR/VR. In general, the graphical pipeline works by using a technique called *rasterization*. It is the transformation of a series of points and vectors into a discrete dot-line matrix that recreates the described shapes. This transformation can also be split into two parts, the geometry stage (top half of Fig. 20.5) and the pixelization stage (bottom half of Fig. 20.5) [39].

The geometry stage of the graphical pipeline accepts the input vertices, using screen-relative coordinates for 2D scenes and world-relative coordinates for 3D scenes. The corresponding projection matrix must also be supplied in the case of 3D rendering. The input list of vertices and optional matrices are transferred from the main system (CPU) memory to the device (GPU) memory, and then passed to the vertex shader.

The vertex shader is a programmable step that calculates the output positions for each supplied vertex coordinate, and can perform numerical operations on a per-

Fig. 20.5 The standard graphical pipeline

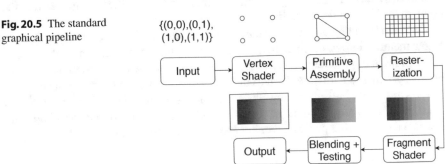

vertex basis. In 3D rendering, this is usually to multiply each vertex with the supplied view and projection matrices. The output, the actual vertices themselves, can then be passed to several other optional shaders before finally reaching the primitive assembly step. These optional shaders are generally reserved for specialized usages.

An intermediary step exists between the vertex shader and primitive assembly. To avoid assembling shapes outside of the screen, clipping is done immediately prior to primitive assembly in a manner that is identical to the clipping divider in Sutherland's system.

The primitive assembly step is where the vertices get transformed from a set of points into a connected geometric mesh. There are three well-defined primitives (point, line, triangle), and one removed primitive (quadrilateral). The primitive that is used to connect the vertices are supplied either in the input stage or in the vertex shader itself. The vertices are connected in the order that they are supplied. The example in Fig. 20.5 uses triangles, $(0, 0)$, $(0, 1)$, $(1, 0)$ to form one triangle, while $(0, 1)$, $(1, 0)$, $(1, 1)$ to form the other. If lines were used instead, each pair of the vertices forms a line segment, and in the case of points, each vertex is a point.

Rasterization is the final phase of the geometry stage. At this point, each primitive is broken down into multiple fragments, which essentially represent one pixel. There can be multiple fragments per pixel, but this is sufficient for explaining the next stage of the graphical pipeline. We begin to see the benefits of having many dedicated processing cores at this step, as each core can process one fragment.

The pixelization stage begins with the fragment shader. Much like the vertex shader, the fragment shader is a programmable step that acts on a per-fragment basis. The supplied programming instructs the GPU on how to color in each fragment. Lighting, texturing, and other material properties are all controlled by the fragment shader. The shader outputs the colored fragments to the penultimate step, blending and testing.

Blending and testing is not one specific step, but rather a collection of various operations that can be run to clean up the colored primitives. Recall that many fragments can exist at a single pixel, even within the same primitive. The majority of this step is to reconcile the different outputs that each of the fragments may have at that single pixel. For example, in a 2D space, one primitive may be behind another primitive such as two overlapping rectangles. Depth testing can be done to hide the deepest rectangle, and/or blending can be enabled to mix the colors together in the case of two transparent rectangles, although this depends on the configuration of the pipeline itself when it is initialized.

Once the blending and testing step is finished, the pixel-wise output is written to a final memory buffer, destined to a display device like a monitor, and marks the end of the process of producing one image frame. This pipeline is self-contained, and so highly specialized hardware, such as discrete video cards, can hardwire the pipeline process to achieve high efficiency. Since each step is independent of the previous, it is possible to begin rendering the next frame before the previous frame has completely finished the rendering process, thereby greatly speeding up frame rates.

20.4.2 Global Positioning System (GPS)

The Global Positioning System (GPS) was developed during the Cold War to provide a reliable navigation system, becoming fully operational in 1995. It includes a number of satellites in the medium Earth orbit of about 20,000 km up, acting as triangulation beacons for user localization. A receiver unit requires line-of-sight with at least 3 different satellites. Each of these satellites broadcast information beacons regarding their position (relative to the Earth) and on-board time. These beacons are sent at regular intervals, and the GPS receiver calculates the time difference from when the signal is sent to when it is received.

The accuracy of GPS was intentionally reduced to be accurate within 100 m in the beginning for non-military use. This limit was later turned off and improvements to GPS have now allowed for receivers to be accurate to within 30 cm. The accuracy can also be improved with Differential GPS (DGPS), which works by having *reference stations* that work identically to GPS receivers but with known fixed positions, or with Assisted GPS, which leverages cellular base stations to improve the startup performance.

One early influential system using GPS is Steven Feiner's touring machine [9]. It used a combination of DGPS and a gyroscope as its primary method for localization, something that carried over well into the 2010s and beyond. Figure 20.6 shows a picture of a user wearing and operating the system.

Feiner noted that there were three key issues in this system: the quality of the displays, the quality of user localization, and the loss of localization. In an outdoor environment, the Sun massively overpowers the brightness of most if not all displays, necessitating the use of neutral density filters, affect the perception of color. Similarly, the combined accumulated error of the magnetometer/inclinometer proved to be jarring, which was further compounded when the GPS signals lost line-of-sight. As a result, AR researchers began to investigate other alternative approaches toward

(a) Wearing and operating the Touring Machine.

(b) View through the optical display. The Buell and Fayweather buildings are behind St. Paul's Chapel.

Fig. 20.6 Feiner's touring machine [9]. Used with permission of Dr. Steven K. Feiner, Computer Graphics and User Interfaces Lab, Columbia University

user localization. One of these approaches is to solve the inverse problem: find the position and orientation of the virtual object and then display it relative to the user, rather than finding the user and displaying the object relative to the user.

20.4.3 Networking for Multiple Users

ARQuake was one of the first AR systems to support multiple simultaneous users [34]. It did so by cleverly utilizing an existing commercial application with multiuser support, Quake. Instead of using a traditional mouse and keyboard for device input, ARQuake translates the user's current pose into the corresponding mouse and keyboard inputs based on the user's previous pose. Localization was done using a hybrid approach as GPS was still relatively inaccurate at the time. Fiducial markers were placed onto buildings that were used by the system as a correction factor to align the virtual world positioning with the real world. This also meant that the virtual world in Quake must be designed to correlate with the real world, i.e., buildings in the Quake world need to match the buildings in the real world. The last step in converting Quake to run as an AR application was to remove the building textures in the Quake world so that they would render transparently to avoid occlusion of the real-world buildings (Fig. 20.7).

Fig. 20.7 User's perspective of ARQuake. The gridlines mark in-game walls. Used with permission of Dr. Thomas Wearable Computer Lab, University of South Australia [34]

A standalone AR system was also being designed to achieve full multiuser functionality and support [29]. The Studierstube system can be viewed as a full technological evolution of Sutherland's initial prototype. Rather than building AR around a single user, a client–server approach is used to connect users. In this system, there are two major and separate servers: a tracking server and an environment server. The tracking server acts as a relay server, maintaining the state of all tracked objects and users, and forwarding this data to each user. Users and objects are statically tracked using a magnetic sensor, in contrast to Sutherland's ultrasound sensors. Similarly, the environment server acts as a central database of all graphical models, which are distributed as needed to clients, who then independently render them locally. As a result, each client has its own environment database, and must be synchronized with the environment server. Exactly how this is achieved is not touched upon, although a simple policy of last action wins could be enacted.

The interaction delay of such a system is a combination of the synchronization of the environment server, the round-trip time of the slowest client, and the refresh frequency of the display. It is important that this interaction delay be minimized, although depending on the client-side implementation, it is possible to disguise it by assuming the client input always succeeds.

20.5 Modern Augmented Reality Systems and Applications

The new millennium has seen advances in miniature computers, including the smartphones and tablets, the integration of application-specific processors, and the commodification of computing. All these have embraced the development of new generation AR/VR systems and applications.

Prior to the ubiquity of the modern smartphone in 2007, there existed the personal digital assistant (PDA), also known as a handheld PC. These precursor devices were larger and bulkier than their smartphone cousins and also lacked their countless sensors, but otherwise had similar levels of functionality: telephony, touchscreen input, wireless capabilities, and occasionally a camera. In [40], researchers explored a prototype standalone AR system using a commercial, off-the-shelf personal digital assistant (PDA) device. For localization, they utilized an optical tracking approach with fiducial markers, which was far too CPU intensive for the hardware at the time (400 MHz).

One of the first commercially successful, mass-market consumer AR product with modern smartphones is a mobile game, Pokémon GO. Released in 2016, it became a massive worldwide phenomenon; at its peak, it was installed on about 10% of all US Android devices, with almost as many daily active users as Twitter. Pokémon GO's overnight success put AR in the public spotlight as it has demonstrated that consumers are ready to adopt AR, and that the state-of-the-art smartphone hardware is capable of supporting the fidelity required. It also marked the beginnings of a shift toward handheld, smartphone-based AR, rather than the HMDs originally incepted by Sutherland.

Pokémon GO's AR system functioned in a self-contained manner and did not rely on optical tracking; it used all the available sensors on a smartphone with a small minimal set of essential sensors: a compass, GPS, and gyroscope. Other sensors, like accelerometers, would be used in conjunction with the sensor basis to enhance the localization data. The phone's camera would be used to provide the background for the AR view, while the pose data would be used to render the correct pose for the virtual object. The end result is combined together and displayed to the user through the smartphone rather than an HMD.

A major consequence of processing all of this on the smartphone was that it required a massive amount of energy. A smartphone is an energy-constrained device, and thus quite often limits the usability to be around 30 min to an hour. This can be alleviated by cloud offloading [41]. Dedicated computing chips for image processing and motion processing have also been advocated. Apple and Google have heavily invested into creating powerful software development kits for their respective platforms with ARKit and ARCore, respectively. Other industry giants, such as Samsung and Snapchat, have also seen success with integrating AR into their camera-based applications that allow users to augment their faces with virtual decorations. Other AR development kits include ARToolkit and Vuforia.

These AR development kits are powered by three core components: a localization module, a scene graph module, and a physics engine. These components are provided through add-on platform services. The localization module makes use of the various sensors on the smartphone to perform *simultaneous localization and mapping* (SLAM). The resulting map consists of keypoints that represent physical features, as well as any planar surfaces that are detected. With this data, the scene graph and physics engine handle the generation of AR content that is displayed.

The other product borne from the commodification of computation is the lowered cost of producing fully customized microprocessors and application-specific integrated circuits. This has huge implications for AR and VR systems, as it enables purpose-built hardware that can work around issues in localization and rendering. Examples include NVIDIA's Pascal gaming GPU and the CloudXR software development kit that helps businesses create and deliver high-quality, wireless AR and VR experiences from any application based on OpenVR, a broadly used hardware and software interface.

The HoloLens system is one of the innovative systems of this period, as it features what Microsoft calls a *holographic processing unit* (HPU) as the powerhouse and centerpiece of the system (Fig. 20.8). It is a return to the roots of AR systems as it features dedicated integrated circuitry for different components, much like Sutherland's initial work. Unlike the prior works, the HoloLens also features a different method of combining light.

The principle mechanism of the HoloLens display is based around a concept known as *waveguided combining* (Fig. 20.9). An initial source of lightwaves (display) is emitted into a freeform prism with *total internal reflection* (TIR). This prism functions like a fiber optic cable that transports the light from the source to the exit (pupil). Surface gratings are made at the entrance and exit to guide the light to the desired TIR path. External sources of light (reality) pass through the "rear" of

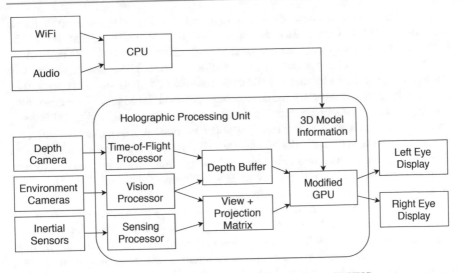

Fig. 20.8 A breakdown of HoloLens and its holographic processing unit (HPU)

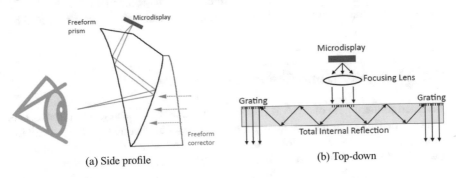

Fig. 20.9 How the HoloLens display functions

the prism. However, since the freeform prism will refract the light as it enters, an additional prism is necessary to counteract the distortion introduced by the curved backside of the TIR prism.

One of the main advantages of this display format is that it does not make any adjustments to the external lightwaves themselves. In contrast, previous approaches by Sutherland, Feiner, and Thomas all utilized half-silvered mirrors as the mechanism for combining light, resulting in a dimmed view of the user's surrounding environment. This natural dimming can be beneficial, however, as HoloLens' display does not target outdoor use due to the intensity of natural sunlight and skylight.

The HPU is also an interesting deviation from what previous works contended with. Recall from Sect. 20.4.1 that there were several issues with the contemporary graphical pipeline in regard to augmented reality. One such issue was the inability to directly manipulate buffers on the GPU with additional information regarding depth for occlusion. The HPU directly addresses this issue by integrating in a custom microprocessing chips that rectifies the problem.

HoloLens represents a new advancement in mixed reality systems, so do such new consumer HMDs as Oculus Rift and Valve Index for VR systems. Minecraft Earth, Half Life: Alyx, and Beat Saber are examples of different AR and VR applications that run using the new generation of consumer HMDs. There are, however, still numerous issues faced by these HMDs. As complete computer-in-a-headsets, they are entirely reliant on a battery for continuous operation, lasting for 2–4 h depending on usage. The display mechanism in HoloLens also requires an additional bracing points (the front visor) to mount an adjustable correctional prism. Consequently, the device as a whole is rather bulky and heavy, and rests atop a person's head. Contrast this with a smartphone: small, lightweight, monocular, and invariant to extrinsic lighting conditions, there is definitely an aspect of convenience and ease of use, both as an end-user and AR systems designer.

As a matter of fact, smartphone-based systems still feature a much wider selection of AR and VR applications due to their lower barrier of entry. Snapchat lenses are one of the most widely used AR application in this form factor for entertainment as users can digitally alter their faces and surroundings before sending them to their friends, while IKEA has launched an AR app that allows users to preview furniture in their homes with their smartphones. Mobile VR has also gained traction as Google Cardboard and Samsung Gear VR provide relatively cheaper headsets powered by smartphones.

20.6　Limitations and Challenges

The entertainment field of AR/VR has experienced astronomical growth during this time period. This growth was widely fuelled by the success of new generation smartphones, new HMD such as HoloLens, and the commoditization of computational power through cloud computing. The success of Pokémon GO inspired many other technology companies to pursue and incorporate AR into their applications. Apple, Samsung, and Snapchat have all incorporated facial processing with AR to allow users to augment their faces with virtual accessories, or to try a different face altogether.

Similar advancements were made in the consumer broadcasting sector as well. Crowdsourced livecasts are now commonly featuring more elements of pseudo-AR, e.g., both Riot Games and Valve Corporation have featured digital avatars from their games being brought to life in their broadcasts alongside human spectators and performers. Traditional broadcasting studios have also begun to incorporate these elements into their programming for weather report, soccer game broadcast, etc.

Given this shift toward consumer AR/VR, and the degree of hardware advancements, future AR/VR applications will be a reimagining of existing applications, but for a general audience. For example, the medical uses of AR might shift toward visualizing an individual's biometrics by displaying areas of interest, such as a broken bone or internal tissue damage. Likewise, future smart cities might utilize AR for dynamic signage to divert traffic in the event of congestion.

Despite these advancements, there are still many hurdles ahead. In this section, we present some of the pertinent obstacles, discuss how they affect the development of consumer grade systems, and the potential resolutions.

20.6.1 Color Perception

A distinct aspect of optical HMDs for AR is the issue of color perception in both indoor and outdoor environments. As these displays are see-through, what is ultimately perceived by the human eye is a combination of the background light (real-world environment) and the foreground display (virtual environment) [42]. This is further complicated by how the foreground light is generated, i.e., is the light source additive, where it is its own light source, or subtractive (e.g., LCD), where an ideal white light is filtered to produce the desired color? For the mass market, it is problematic as the same AR system can produce drastically different color perceptions when only the display is changed.

Consider the following scenario: a user is standing facing a red wall illuminated by a single perfectly white light, such that their entire point-of-view is filled by the wall. Naturally, the user's eyes receive only the light in the red wavelengths in this scenario. Suppose they equip an additive HMD, turn on the AR system, and request to view a model of Neptune, a notably blue planet. The rendered image is sent to the HMD, and it correctly displays the blue sphere on the optical display. However, what the user will perceive is not a blue sphere, but rather one that is the additive combination of the red and blue lights, that is, purple! Now suppose the user exchanges this display for another one that happens to use subtractive light. Without loss of generality, assume that this display is calibrated to reproduce colors under a perfectly white backlight. When the user turns on the AR system again, they will see a black sphere as the red light is filtered out by the display as it attempts to produce blue, the result of filtering red and green light from white. Figure 20.10 highlights this principle.

It is clear that there is a need for color correction on these displays, but it is not clear how this can be effectively carried out to maintain a truthful and accurate color reproduction [43–45]. In the above example, an additive color system will not be able to produce black, whereas a subtractive system will not be able to produce white, because both of them rely on an assumption that the "base" color will be either black or white, respectively. Realistically, a user will not be standing in front of a monochrome wall, the virtual objects will not be monochromatic, and the lighting conditions will not be idealized to be a single source and perfectly white (unless standing outside on a bright, perfectly clear day). These problems have been largely ignored on monitor-based display, due to simplifying the color model to a single input (camera) and a single output (monitor), compared to the multiple inputs that need to be separately considered and blended together.

Fig. 20.10 Color combining on an optical HMD. An example can be seen in Fig. 20.7; the color of the robot's face appears to be green, although it is actually yellow, and the bottom bar should be white, but is almost completely transparent

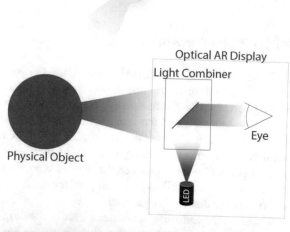

Fig. 20.11 A Pokémon being occluded by the table to appear to hide under it. This was done by hand-segmenting out the Pokémon and compositing with another image

20.6.2 Depth Perception

Another obstacle toward fully immersive AR is the occlusion of virtual objects by real objects. Currently, it is trivially easy to place virtual objects in front of real objects as we can draw atop the real-world due to our planar optical displays being the medium for AR. The converse, however, is extremely difficult to carry out for the same reason: to preserve the real object, it is now necessary to draw around the object, as close as possible to the boundary edge of the real object and the would-be virtual object (see Fig. 20.11).

There are multiple ways to achieve this outcome depending on the desired flexibility. Image segmentation, a well studied area of computer vision, can be used to isolate the different real-world objects, and can be used for simple occlusions of the nature of "the virtual object is behind the real object". For more complex queries such as "should the virtual object be behind or in front of the real object," there is a need for calculating the relative depth between real objects, and the relationship between the relative depth and virtual depth. To this end, there have been hardware and software solutions, each with their own advantages and disadvantages.

Hardware sensor solutions, such as a depth camera, are limited by the maximum distance and resolution that the depth can be discerned [46]. The further an object is from the sensor, the smaller it appears to the sensor, and the greater the noise. A simple implementation of a depth sensor is to use a calibrated infrared dot matrix to determine the relative depth and shape of the objects in view based on the distortion of the dots and how far each dot is from its neighbors. More complex implementations calculate the time it takes for one photon to hit an object and bounce back using the speed of light; these time-of-flight sensors offer greater fidelity and accuracy compared to the structured light sensors, but have a lower resolution due to the massive amount of data processing needed. Neither approach, however, is particularly well-suited for outdoor environments on a sunny day.

Software solutions have less constraints, but often require either multiple cameras or multiple images from different perspectives. A typical solution, *stereo triangulation* [47], uses two or more cameras, much like the human vision system. By finding the corresponding points of each image, rays can be traced from each camera, giving 3D intersection points that correspond to points from each image. In practice, finding the corresponding points is difficult for homogeneous regions of color or intensity, particularly when an object is in motion. Single camera techniques work under similar principles, but require a shift from image to image.

20.6.3 Localization

Localization is one of the most prominently discussed aspects of AR/VR throughout the ages in literature, but it is still a major obstacle. Holloway's analysis on registration errors [48] identifies various sources affecting AR objects to display themselves at incorrect locations. The two predominant sources of error were, and still are, attributed to localization error and localization delay. Hardware advancements have managed to significantly improve the accuracy of user pose localization as we have seen with GPS, going from about 100 m to 30 cm and below. What has not necessarily improved, however, is the polling rate of these devices in high-accuracy modes. Suppose, we have a magical sensor capable of perfect fidelity for any instantaneous measurement, but it is limited to at most 10–20 measurements per second or a corresponding update latency of 50–100 ms. At any given instantaneous measurement, our AR display will have perfect registration as there are no sources of error. However, any movement past this point in time will introduce a registration error as the true pose is now off from what was read, until our oracle sensor is able to provide a new reading.

Besides high accuracy for position and orientation data, a high update frequency is necessary, too, for two reasons: readings can always be accurately calibrated, but they may not be accurately forecasted, and the success of the kinetic depth illusion corresponds to the frequency of image updates. There will always be an inherent update latency due to the discrete nature of digital components; the best that can be done is an approximation of the analog by rapid polling, which remains a challenging for consumer AR/VR under budget constraints.

20.6.4 Information Presentation

As AR is the first of many steps toward fully immersive computing, it faces a rather unique challenge in finding a balance between preserving the view of the real-world, and the amount of virtual data that is displayed. Suppose, we have obtained a perfect AR system capable of recreating digital objects indistinguishable from the real. What is the maximal amount of the real-world that can be covered by virtual objects before it can no longer be considered AR rather than VR? Before users find the digital objects intrusive? Before users become overwhelmed by the amount of virtual objects? Hyperreality is a short concept film that succinctly highlights this problem [4]. Unfortunately, the boundary is highly subjective as some users may tolerate a higher degree of occlusion than others. Some early works have been conducted [49–51], but do not necessarily apply to the current and future form-factor shifts to smaller display devices.

Similarly, there is the question of how interactions are carried out between users and the virtual user interface. On an HMD, this will likely be using gesture-based recognition, but will these gesture-interactions be limited to just the user wearing the HMD, or should it accept input from any user? An argument could be made for either case: limiting it to solely the current user prevents malicious actors from interfering with the AR effect, but allowing nearby users to affect digital objects like a virtual keyboard reinforces the immersion and enhances the social aspect. Monitor-based displays will likely use the current pointer scheme, but the issue with this is allowing for depth control on the 2D plane. One potential resolution is to incorporate 2D gestures much like how modern smartphones have.

Regardless of the medium of display, what is clear is that the current WIMP (Windows, Icons, Menus, Pointers) paradigm of human–computer interaction is not entirely applicable. Recent works have been done in exploring new interaction paradigms better suited toward AR/VR [52,53].

20.6.5 Social Acceptance

In 2013, Google announced the development of a consumer grade optical HMD system called *Glass*. It was a pinnacle development of HMDs at the time: lightweight, unintrusive, and integrated with existing smartphones. While it was capable of being used as a medium for AR, Glass was initially marketed as a personal point-of-view recorder and hands-free digital assistant. Unfortunately, this was not well received at the time, and resulted in a widespread social rejection of the technology for societal and ethical reasons rather than the quality of the hardware.

Two predominant reasons for this rejection are the potential violations of privacy and safety concerns regarding the usage of the device. Despite the physical device itself being non-intrusive, the human operator can willfully decide to neglect their present immediate surroundings when focusing on the display. This has resulted in various recorded incidents where the operator was driving a vehicle, became inattentive to their physical surroundings, and collided with an avoidable object [54]. On the other hand, due to the device being so non-intrusive and innocuous, it was possible to record everything the wearer saw without requiring consent, either implicitly in an obvious recognition of a camera-like silhouette or explicitly via human interaction. The only outward indication of recording is a small light on the eyeware itself, and photographs can be captured silently.

For the general public to accept pervasive HMDs, a two-step approach may be needed. Had Google Glass not come with the inclusion of a camera and had its initial marketing campaign focus more toward the AR aspects, it would have a chance to be familiarized rather than demonized. Once this initial iteration has been accepted, only then would the inclusion of a camera be considered, as the public could properly weigh the benefits of AR versus the potential of misuse rather than immediately focusing on the negatives.

Regardless, one final aspect for consideration is the potential of motion sickness like symptoms. The effects of completely replacing a person's view with a digital one, like in VR, are not completely understood. Some users may experience general discomfort, visual fatigue, or nausea after prolonged exposure to a virtual environment. This may be another barrier toward general widespread adoption.

20.7 Further Exploration

AR/VR is under rapid development in recent years, and there is a long way to go toward truly immersive experience. The underlying technologies and the products available in the market are far from being mature but do evolve very fast. Related information can be found from conference/journal papers as well as whitepapers from Google, Microsoft, Apple, Samsung, NVIDIA, etc.

20.8 Exercises

1. In this chapter, we have explored different augmented reality and virtual reality systems. What are some advantages of this form of multimedia as compared to the forms we have seen?
2. Augmented reality systems integrate a wide array of engineering and science disciplines. List some of the different ones and how they contribute to AR and VR.
3. How many color channels will we need to display an AR image? VR?

4. You have been asked to design a new HPU chip that supports 4K resolution (3,840 × 2,160) for each eye at 60 fps.

   (a) What is the minimum size of the frame buffer necessary to support this resolution?
   (b) What is the minimum bandwidth that the serial bus needs to provide when connecting the chip and the displays if the bandwidth must be shared equally between the two displays?
   (c) What is it if the bandwidth is not shared between displays?
   (d) What if the resolution changes to be 1080p (1,920 × 1,080) at 240 fps?

5. What are some potential sources of error when trying to determine a user's location? Orientation?
6. Suppose you have been asked to write a localization algorithm for an AR system.

   (a) What are some requirements or constraints that you might ask about?
   (b) How would these change for a VR system?
   (c) In high-level pseudocode, write down your algorithm.
   (d) Suppose the augmented reality system supports multiple users simultaneously, in the same digital world. Does this affect your algorithm, and if so what changes would you make? If not, explain how your algorithm would handle one person located at the North Pole, and one person at the South Pole.

7. There will always be interaction delay in a mixed reality system. Where are some of the sources of this delay?
8. Describe some techniques that can be used to combat the interaction delay.
9. Discuss the tradeoffs between optical head-mounted displays (HMDs) and smartphone-based video displays for augmented reality.

References

1. P. Milgram, F. Kishino, A taxonomy of mixed reality visual displays. IEICE Trans. Inf. Syst. **77**(12), 1321–1329 (1994)
2. M.L. Heilig, Sensorama simulator. US Patent 3,050,870, 28 August 1962
3. I.E. Sutherland, A head-mounted three dimensional display, in *Proceedings of the December 9–11, 1968, Fall Joint Computer Conference, part I*, p. 757–764. (ACM, 1968)
4. K. Matsuda, Hyper-reality (2016)
5. R.T. Azuma, A survey of augmented reality. Presence: teleoperators Virtual Environ. **6**(4), 355–385 (1997)
6. J. Steuer, Defining virtual reality: dimensions determining telepresence. J. Commun. **42**(4), 73–93 (1992)
7. J.M. Loomis, R.G. Golledge, R.L. Klatzky, J.M. Speigle, J. Tietz, Personal guidance system for the visually impaired, in *Proceedings of the First Annual ACM Conference on Assistive Technologies*, pp. 85–91. (ACM, 1994)

8. J. Rekimoto, K. Nagao, The world through the computer: computer augmented interaction with real world environments, in *Proceedings of the 8th Annual ACM Symposium on User Interface and Software Technology*, pp. 29–36. (ACM, 1995)

9. S. Feiner, B. MacIntyre, T. Höllerer, A. Webster, A touring machine: prototyping 3D mobile augmented reality systems for exploring the urban environment. Pers. Technol. **1**(4), 208–217 (1997)

10. M. Tonnis, C. Sandor, G. Klinker, C. Lange, H. Bubb, Experimental evaluation of an augmented reality visualization for directing a car driver's attention, in *2005. Proceedings. Fourth IEEE and ACM International Symposium on Mixed and Augmented Reality*, pp. 56–59. (IEEE, 2005)

11. S. Julier, Y. Baillot, M. Lanzagorta, D. Brown, L. Rosenblum, Bars: battlefield augmented reality system, in *In NATO Symposium on Information Processing Techniques for Military Systems, Citeseer* (2000)

12. M.A. Livingston, et al., Military applications of augmented reality, in *Handbook of Augmented Reality*, pp. 671–706. (Springer, 2011)

13. R. Azuma, H. Neely, M. Daily, J. Leonard, Performance analysis of an outdoor augmented reality tracking system that relies upon a few mobile beacons, in *IEEE/ACM International Symposium on Mixed and Augmented Reality (ISMAR'06)*, pp. 101–104. (IEEE, 2006)

14. S.J. Henderson, S.K. Feiner, Augmented reality for maintenance and repair (ARMAR). Technical report. (Columbia University, New York, Department of Computer Science, 2007)

15. C. Sandor, G. Klinker, A rapid prototyping software infrastructure for user interfaces in ubiquitous augmented reality. Pers. Ubiquitous Comput. **9**(3), 169–185 (2005)

16. K. Pentenrieder, C. Bade, F. Doil, P. Meier, Augmented reality-based factory planning—An application tailored to industrial needs, in *Proceedings of the 6th IEEE and ACM International Symposium on Mixed and Augmented Reality*, pp. 1–9. (IEEE Computer Society, 2007)

17. W. Friedrich, D. Jahn, L. Schmidt, Arvika-augmented reality for development, production and service, in *ISMAR*, vol. 2002, pp. 3–4. (Citeseer, 2002)

18. D. Mizell, Boeing's wire bundle assembly project, *Fundamentals of Wearable Computers and Augmented Reality* (2001)

19. D. Willers, Augmented reality at Airbus, in *International Symposium on Mixed & Augmented Reality* (2006)

20. H. Fuchs, et al., Augmented reality visualization for laparoscopic surgery, in *International Conference on Medical Image Computing and Computer-Assisted Intervention*, pp. 934–943. (Springer, 1998)

21. N. Navab, A. Bani-Kashemi, M. Mitschke, Merging visible and invisible: two camera-augmented mobile C-arm (CAMC) applications, in *Proceedings of the 2nd IEEE and ACM International Workshop on Augmented Reality (IWAR'99)*, pp. 134–141. (IEEE, 1999)

22. S. Vogt, A. Khamene, F. Sauer, Reality augmentation for medical procedures: system architecture, single camera marker tracking, and system evaluation. Int. J. Comput. Vis. **70**(2), 179 (2006)

23. V. Vlahakis et al., Archeoguide: an augmented reality guide for archaeological sites. IEEE Comput. Graph. Appl. **22**(5), 52–60 (2002)

24. H. Kaufmann, D. Schmalstieg, Mathematics and geometry education with collaborative augmented reality, in *ACM SIGGRAPH Conference Abstracts and Applications*, pp. 37–41 (2002)

25. H. Kaufmann, *Collaborative Augmented Reality in Education*. (Vienna University of Technology, Institute of Software Technology and Interactive Systems, 2003)

26. K.L. Schrier, *Revolutionizing history education: using augmented reality games to teach histories*. Ph.D. thesis. (Massachusetts Institute of Technology, Department of Comparative Media Studies, 2005)

27. E Klopfer, K Squire, H. Jenkins, Environmental detectives: PDAs as a window into a virtual simulated world, in *Proceedings of the IEEE International Workshop on Wireless and Mobile Technologies in Education*, pp. 95–98. (IEEE, 2002)

28. E. Klopfer, J. Perry, K. Squire, M.-F. Jan, C. Steinkuehler, Mystery at the museum: a collaborative game for museum education, in *Proceedings of the Conference on Computer Support for*

Collaborative Learning: learning 2005: the next 10 years!, pp. 316–320. (International Society of the Learning Sciences, 2005)

29. Z. Szalavári, D. Schmalstieg, A. Fuhrmann, M. Gervautz, "Studierstube": an environment for collaboration in augmented reality. Virtual Real. **3**(1), 37–48 (1998)

30. J. Rekimoto, M. Saitoh, Augmented surfaces: a spatially continuous work space for hybrid computing environments, in *Proceedings of the SIGCHI conference on Human Factors in Computing Systems*, pp. 378–385. (ACM, 1999)

31. H. Tamura, Steady steps and giant leap toward practical mixed reality systems and applications, in *Proceedings of the International Status Conference on Virtual and Augmented Reality*, pp. 3–12 (2002)

32. M. Lake, When the game's on the line, the line's on the screen. New York Times (2000)

33. R. Cavallaro, The FoxTrax hockey puck tracking system. IEEE Comput. Graph. Appl. **2**, 6–12 (1997)

34. B. Thomas, B. Close, J. Donoghue, J. Squires, P. De Bondi, M. Morris, W. Piekarski, Arquake: an outdoor/indoor augmented reality first person application, in *Proceedings of the 4th International Symposium on Wearable Computers*, pp. 139–146. (IEEE, 2000)

35. A.D. Cheok, S.W. Fong, K.H. Goh, X. Yang, W. Liu, F. Farzbiz, Human Pacman: a sensing-based mobile entertainment system with ubiquitous computing and tangible interaction, in *Proceedings of the 2nd Workshop on Network and System Support for Games*, pp. 106–117. (ACM, 2003)

36. T. Ohshima, K. Satoh, H. Yamamoto, H. Tamura, Ar2 hockey: a case study of collaborative augmented reality, in *Proceedings of the Virtual Reality Annual International Symposium (VRAIS'98)*, p. 268. (IEEE Computer Society, 1998)

37. C. Matysczok, R. Radkowski, J. Berssenbruegge, AR-bowling: immersive and realistic game play in real environments using augmented reality, in *Proceedings of the ACM SIGCHI International Conference on Advances in Computer Entertainment Technology*, pp. 269–276. (ACM, 2004)

38. P. Lincoln, A. Blate, M. Singh, T. Whitted, A. State, A. Lastra, H. Fuchs, From motion to photons in 80 μs: towards minimal latency for virtual and augmented reality. IEEE Trans. Visualization Comput. Graph. **22**(4), 1367–1376 (2016)

39. T. Akenine-Möller, E. Haines, N. Hoffman, *Real-Time Rendering*, 4th edn. (CRC Press, 2018)

40. D. Wagner, D. Schmalstieg, First steps towards handheld augmented reality, in *Proceedings of the 7th IEEE International Symposium on Wearable Computers*, p. 127. (IEEE Computer Society, 2003)

41. R. Shea, A. Sun, S. Fu, J. Liu, Towards fully offloaded cloud-based AR: design, implementation and experience, in *Proceedings of the 8th ACM on Multimedia Systems Conference*, pp. 321–330. (ACM, 2017)

42. M.A. Livingston, J.H. Barrow, C.M. Sibley, *Quantification of contrast sensitivity and color perception using head-worn augmented reality displays*. Technical report. (Naval Research Lab, Washington DC, Virtual Reality Lab, 2009)

43. S.K. Sridharan, J.D. Hincapié-Ramos, D.R. Flatla, P. Irani, Color correction for optical see-through displays using display color profiles, in *Proceedings of the 19th ACM Symposium on Virtual Reality Software and Technology*, pp. 231–240. (ACM, 2013)

44. J.D. Hincapié-Ramos, L. Ivanchuk, S.K. Sridharan, P.P. Irani, Smartcolor: real-time color and contrast correction for optical see-through head-mounted displays. IEEE Trans. Visualization Comput. Graph. **21**(12), 1336–1348 (2015)

45. Y. Itoh, M. Dzitsiuk, T. Amano, G. Klinker, Semi-parametric color reproduction method for optical see-through head-mounted displays. IEEE Trans. Visualization Comput. Graph. **21**(11), 1269–1278 (2015)

46. S. Lee, Depth camera image processing and applications, in *19th IEEE International Conference on Image Processing*, pp. 545–548. (IEEE, 2012)

47. U.R. Dhond, J.K. Aggarwal, Structure from stereo—A review. IEEE Trans. Syst. Man Cybern. **19**(6), 1489–1510 (1989)

48. R. L. Holloway, *Registration Errors in Augmented Reality Systems*. Ph.D. thesis. (Citeseer, 1995)
49. S. Julier, M. Lanzagorta, Y. Baillot, L. Rosenblum, S. Feiner, T. Hollerer, S. Sestito, Information filtering for mobile augmented reality, in *Proceedings IEEE and ACM International Symposium on Augmented Reality (ISAR 2000)*, pp. 3–11. (IEEE, 2000)
50. B. Bell, S. Feiner, T. Höllerer, View management for virtual and augmented reality, in *Proceedings of the 14th Annual ACM Symposium on User Interface Software and Technology*, pp. 101–110. (ACM, 2001)
51. D. Kalkofen, E. Mendez, D. Schmalstieg, Interactive focus and context visualization for augmented reality, in *Proceedings of the 6th IEEE and ACM International Symposium on Mixed and Augmented Reality*, pp. 1–10. (IEEE Computer Society, 2007)
52. E. Normand, M. McGuffin, Enlarging a smartphone with AR to create a handheld vesad (virtually extended screen-aligned display), in *Proceedings of the IEEE International Symposium for Mixed and Augmented Reality* (2018)
53. M. Al-Kalbani, I. Williams, M. Frutos-Pascual, Analysis of medium wrap freehand virtual object grasping in exocentric mixed reality, in *IEEE International Symposium on Mixed and Augmented Reality (ISMAR)*, pp. 84–93. (IEEE, 2016)
54. B.D. Sawyer, V.S. Finomore, A.A. Calvo, P.A. Hancock, Google glass: a driver distraction cause or cure? Human Fact. **56**(7), 1307–1321 (2014)

Content-Based Retrieval in Digital Libraries

21

21.1 How Should We Retrieve Images?

Consider the image in Fig. 21.1 of a small portion of *The Garden of Delights* by Hieronymus Bosch (1453–1516), now in the Prado museum in Madrid. This is a famous painting, but we may be stumped in understanding the painter's intent. Therefore, if we are aiming at automatic retrieval of images, it should be unsurprising that encapsulating the semantics (meaning) in the image is an even more difficult challenge. A proper annotation of such an image certainly should include the descriptor "people." On the other hand, should this image be blocked by a "Net nanny" screening out "naked people" (as in [1])?

We know very well that web browsers have a web search button for multimedia content (usually images, or video for YouTube and its competitors), as opposed to text. For Bosch's painting, a text-based search will very likely do the best job, should we wish to find this particular image. Yet we may be interested in fairly general searches, say for scenes with deep blue skies and orange sunsets. By pre-calculating some fundamental statistics about images stored in a database, we can usually find simple scenes such as these.

In its inception, retrieval from digital libraries began with ideas borrowed from traditional information retrieval disciplines (see, e.g., [2]). This line of inquiry continues [3]. For example, in [4], images are classified into indoor or outdoor classes using basic information-retrieval techniques. For a training set of images and captions, the number of times each word appears in the document is divided by the number of times each word appears over all documents in a class. A similar measure is devised for statistical descriptors of the content of image segments, and the two information-retrieval-based measures are combined for an effective classification mechanism.

However, many multimedia retrieval schemes have moved toward an approach favoring multimedia content itself, either without regard to or reliance upon accom-

© Springer Nature Switzerland AG 2021
Z.-N. Li et al., *Fundamentals of Multimedia*, Texts in Computer Science,
https://doi.org/10.1007/978-3-030-62124-7_21

Fig. 21.1 How can we best characterize the information content of an image?

panying textual information, or at least textual-based search bolstered by multimedia evidence. This is commonly known as *content-based image retrieval (CBIR)*. Only recently has attention once more been placed on the deeper problem of addressing semantic content in images, of course also making use of accompanying text (possibly inserted when the media is archived). If data consists of statistical features built from objects in images and also of text associated with the images, each type of modality—text and image—provides semantic content omitted from the other. For example, an image of a red rose will not normally have the manually added keyword "red" since this is generally assumed. Hence, image features and associated words may disambiguate each other (see [5]).

In this chapter, however, we shall focus only on techniques and systems that make use of image features themselves, without text, to retrieve images from databases or from the web. The types of features typically used are such statistical measures as the color histogram for an image. Consider an image that is colorful—say, a Santa Claus plus sled. The combination of bright red and flesh tones and browns might be enough of an image signature to allow us to at least find similar images in our own image database (of office Christmas parties).

Recall that a color histogram is typically a three-dimensional array that counts pixels with specific red, green, and blue values. The nice feature of such a structure is that it does not care about the orientation of the image (since we are simply counting pixel values, not their orientation) and is also fairly impervious to object occlusions. A seminal paper on this subject [6] launched a tidal wave of interest in such so-called "low-level" features for images.

Other simple features used are such descriptors as *color layout*, meaning a simple sketch of where in a checkerboard grid covering the image to look for blue skies and orange sunsets. Another feature used is *texture*, meaning some type of descriptor typically based on an edge image, formed by taking partial derivatives of the image itself—classifying edges according to the closeness of spacing and orientation. An interesting version of this approach uses a histogram of such edge features. *Texture layout* can also be used. Search engines devised on these features are said to be *content-based*: the search is guided by image similarity measures based on the statistical content of each image.

Typically, we might be interested in looking for images similar to our current favorite Santa. A more industry-oriented application would typically be seeking a particular image of a postage stamp, say. Subject fields associated with image database search include art galleries and museums, fashion, interior design, remote sensing, geographic information systems, meteorology, trademark databases, criminology, and an increasing number of other areas.

The emerging deep learning technologies have brought state-of-the-art performances in feature matching. They were initially based on convolutional neural networks (CNNs) [7,8].

Many variations on the neural network configurations have since been developed, notably the *generative adversarial networks (GANs)*.

A more difficult type of search involves looking for a particular *object* within images, which we can term a *search-by-object* model. This involves a much more complete catalog of image contents and is a much more difficult goal. Generally, users will base their searches on *search by association* [9], meaning a first cut search followed by refinement based on similarity to some of the query results. For general images representative of a kind of the desired picture, a *category search* returns one element of the requested set, such as one or several trademarks in a database of such logos. Alternatively, the query may be based on a very specific image, such as a particular piece of art—a *target search*. There are also efforts to retrieve 3D shapes and objects [10,11].

Another axis to bear in mind in understanding the many existing search systems is whether the domain being searched is narrow, such as the database of trademarks, or wide, such as a set of commercial stock photos.

For any system, we are up against the fundamental nature of machine systems that aim to replace human endeavors. The main obstacles are neatly summarized in what the authors of the summary in [9] term the *sensory gap* and the *semantic gap*:

> The sensory gap is the gap between the object in the world and the information in a (computational) description derived from a recording of that scene.
>
> The semantic gap is the lack of coincidence between the information that one can extract from the visual data and the interpretation that the same data have for a user in a given situation.

Image features record specifics about images, but the images themselves may elude description in such terms. And while we may certainly be able to describe images linguistically, the message in the image, the semantics, is difficult to capture for machine applications.

21.2   Synopsis of Early CBIR Systems

The following provides examples of some early CBIR systems. It is by no means a complete synopsis. Most of these engines are experimental, but all those included here are interesting in some way. A good summary appears in [9].

- QBIC
  Query by Image Content (QBIC), developed by Niblack and colleagues [12,13] at IBM's Almaden Research Center in San Jose, was arguably the most famous search engine.
  One interesting feature in QBIC is the metric it uses for color histogram difference. The basic metric used for histogram difference is *histogram intersection*, basically an L_1-norm based measure. Instead of simple histogram intersection, the QBIC metric recognizes that colors that are *similar*, such as red and orange, should not have a zero intersection. Instead, a color-distance matrix A is introduced, with elements

$$a_{ij} = (1 - d_{ij}/d_{\max}) \tag{21.1}$$

  Here, d_{ij} is defined as a three-dimensional color difference (using Euclidean distance, or any other likely distance—sum of absolute values, say).
  Then a histogram-difference D^2 is defined as follows [14]:

$$D^2 = z^T A z \tag{21.2}$$

  Vector z is a histogram-difference vector (for vectorized histograms). For example, the histogram-difference vectors z would be of length 256 if we compared two-dimensional chromaticity histograms of size 16×16.
- Chabot
  Chabot was an early system from UC-Berkeley that aimed to include 500,000 digitized multiresolution images. Chabot uses the relational database management system POSTGRES to access these images and associated textual data. The system stores both text and color histogram data. Instead of color percentages, a "mostly red" type of simple query is acceptable.
- Blobworld
  Blobworld [15] was also developed at UC-Berkeley. It attempts to capture the idea of objects by segmenting images into regions. To achieve a good segmentation, an *expectation maximization (EM)* algorithm derives the maximum likelihood for a good clustering in the feature space. Blobworld allows for both textual and content-based searching. The system has some degree of feedback, in that it displays the internal representation of the submitted image and the query results, so the user can better guide the algorithm.
- WebSEEk
  A team at Columbia University developed several search engines, of which Web-SEEk was better known. It collects images (and text) from the web. The emphasis is on making a searchable catalogue with such topics as animals, architecture,

art, astronomy, cats, and so on. Relevance feedback is provided in the form of thumbnail images and motion icons. For video, a good form of feedback is also inclusion of small, short video sequences as animated GIF files.

- Photobook and FourEyes
  Photobook [16] was one of the earlier CBIR systems developed by the MIT Media Laboratory. It searches for three different types of image content (faces, 2-D shapes, and texture images) using three mechanisms. For the first two types, it creates an eigenfunction space—a set of "eigenimages." Then, new images are described in terms of their coordinates on this basis. For textures, an image is treated as a sum of three orthogonal components in a decomposition denoted as *Wold* features [17].
  With relevance feedback added, Photobook became FourEyes. Not only does this system assign positive and negative weight changes for images, but given a similar query to one it has seen before, it can search faster than previously.

- Informedia
  The Informedia (and later Informedia-II) Digital Video Library project at Carnegie Mellon University centers on "video mining." It was funded by a consortium of government and corporate sponsors. It uniquely combines speech recognition, image understanding, and natural language processing technologies. Features include video and audio indexing, navigation, video summarization and visualization, and search and retrieval of the video media.

- UC Santa Barbara Search Engines
  The Alexandria Digital Library (ADL) was a seasoned image search engine devised at the University of California, Santa Barbara. The ADL is concerned with geographical data: "spatial data on the web." The user can interact with a map and zoom into a map, then retrieve images as a query result type that pertains to the selected map area. This approach mitigates the fact that terabytes, perhaps, of data need to be stored for LANDSAT satellite images, say. Instead, ADL uses a multiresolution approach that allows fast browsing by making use of image thumbnails. Multiresolution images means that it is possible to select a certain region within an image and zoom in on it.

- MARS
  Multimedia Analysis and Retrieval System (MARS) was developed at the University of Illinois at Urbana-Champaign. The idea was to create a dynamic system of feature representations that could adapt to different applications and different users. Relevance feedback, with changes of weightings directed by the user, is the main tool used.

- Virage
  The visual information retrieval (VIR) image search engine operates on objects within images. Image indexing is performed after several preprocessing operations, such as smoothing and contrast enhancement. The details of the feature vector are proprietary; however, it is known that the computation of each feature is made by not one but several methods, with a composite feature vector composed of the concatenation of these individual computations.

21.3   C-BIRD—An Early Experiment

Let us consider the specifics of how image queries are carried out. To make the discussion concrete, we underpin our discussion by using the image database search engine devised by one of the authors of this text (see [18]). This system is called *content-based image retrieval from digital libraries* (*C-BIRD*), an acronym devised from *content-based image retrieval*, or *CBIR*.

The C-BIRD image database contains approximately 5,000 images, many of them keyframes from videos. The database can be searched using a selection of tools: text annotations, color histograms, illumination-invariant color histograms, color density, color layout, texture layout, and model-based search.

Although the system was developed in the early years, it still serves as a good example in illustrating the common techniques in CBIR based on image similarity. Moreover, it offers some unique features such as Search by Illumination Invariance, Feature Localization, and Search-by-Object Model.

Let's step through these options.

21.3.1 Color Histogram

In C-BIRD, features are precomputed for each image in the database. The most prevalent feature that is utilized in image database retrieval is the color histogram [6], a type of *global* image feature, that is, the image is not segmented; instead, every image region is treated equally.

A color histogram counts pixels with a given pixel value in red, green, and blue (RGB). For example, in pseudocode, for images with 8-bit values in each of R, G, B, we can fill a histogram that has 256^3 bins:

```
int hist[256][256][256];   // reset to 0
//image is an appropriate struct
//with byte fields red,green,blue

for i=0..(MAX_Y-1)
  for j=0..(MAX_X-1)
    {
      R = image[i][j].red;
      G = image[i][j].green;
      B = image[i][j].blue;
      hist[R][G][B]++;
    }
```

Usually, we do not use histograms with so many bins, in part because fewer bins tend to smooth out differences in similar but unequal images. We also wish to save storage space.

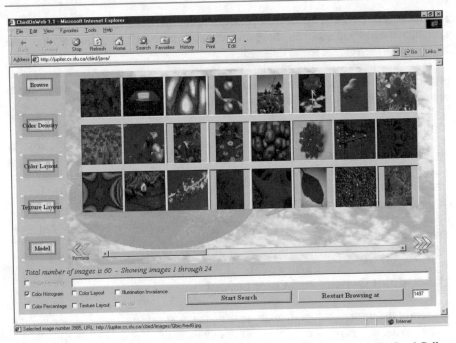

Fig. 21.2 Search by color histogram results. Some thumbnail images are from the Corel Gallery and are copyright Corel. All rights reserved

How image search proceeds is by matching the *feature vector* for the sample image, in this case, the color histogram, with the feature vector for every image in the database.

C-BIRD calculates a color histogram for each target image as a preprocessing step, then references it in the database for each user query image. The histogram is defined coarsely, with bins quantized to 8 bits, with 3 bits for each of red and green and 2 for blue.

For example, Fig. 21.2 shows that the user has selected a particular image—one with red flowers. The result obtained, from a database of some 5,000 images, is a set of 60 matching images. Most CBIR systems return as the result set either the top few matches or the match set with a similarity measure above a fixed threshold value. C-BIRD uses the latter approach and thus may return zero search results.

How matching proceeds in practice depends on what measure of similarity we adopt. The standard measure used for color histograms is called the *histogram intersection*. First, a color histogram $\mathbf{H}_i$ is generated for each image i in the database. We like to think of the histogram as a three-index array, but of course the machine thinks of it as a long vector—hence the term "feature vector" for any of these types of measures.

The histogram is *normalized*, so that its sum (now a `double`) equals unity. This normalization step is interesting: it effectively removes the *size* of the image. The reason is that if the image has, say, resolution 640×480, then the histogram entries sum to 307,200. But if the image is only one-quarter that size, or 320×240, the sum is only 76,800. Division by the total pixel count removes this difference. In fact, the normalized histograms can be viewed as *probability density functions* (*PDFs*). The histogram is then stored in the database.

Now suppose we select a "model" image—the new image to match against all possible targets in the database. Its histogram $\mathbf{H}_m$ is intersected with all database image histograms $\mathbf{H}_i$, according to the equation [6]

$$intersection = \sum_{j=1}^{n} \min(\mathbf{H}_i^j, \mathbf{H}_m^j) \qquad (21.3)$$

where superscript j denotes histogram bin j, with each histogram having n bins. The closer the intersection value is to 1, the better the images match. This intersection value is fast to compute, but we should note that the intersection value is sensitive to color quantization.

21.3.2 Color Density and Color Layout

To specify the desired colors by their density, the user selects the percentage of the image having any particular color or set of colors, using a color picker and sliders. We can choose from either conjunction (ANDing) or disjunction (ORing) a simple color percentage specification. This is a coarse search method.

The user can also set up a scheme of how colors should appear in the image, in terms of coarse blocks of color. The user has a choice of four grid sizes: $1 \times 1, 2 \times 2$, 4×4, and 8×8. Search is specified on one of the grid sizes, and the grid can be filled with any RGB color value—or no color value at all, to indicate that the cell should not be considered. Every database image is partitioned into windows four times, once for each window size. A clustered color histogram is used inside each window, and the five most frequent colors are stored in the database. Each query cell position and size corresponds to the position and size of a window in the image. Figure 21.3 shows how this layout scheme is used.

21.3.3 Texture Layout

Similar to the color layout search, this query allows the user to draw the desired texture distribution. Available textures are zero density texture, medium-density edges in four directions ($0°, 45°, 90°, 135°$) and combinations of them, and high-density texture in four directions and combinations of them. Texture matching is done by classifying textures according to directionality and density (or separation) and evaluating their correspondence to the texture distribution selected by the user in the texture block layout. Figure 21.4 shows how this layout scheme is used.

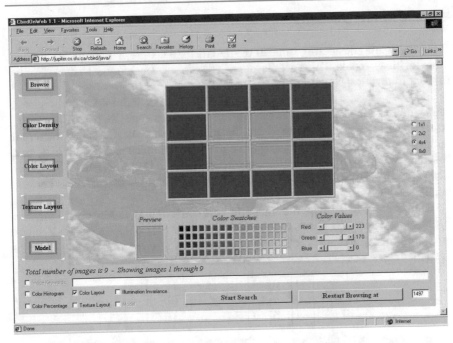

Fig. 21.3 Color layout grid

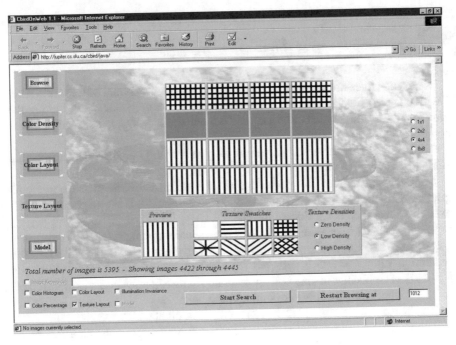

Fig. 21.4 Texture layout grid

Texture Analysis Details

It is worthwhile considering some of the details for a texture-based content analysis aimed at image search. These details give a taste of typical techniques systems employ to work in practical situations.

First, we create a texture histogram. A typical set of indices for comprehending texture is Tamura's [19]. Human perception studies show that "repetitiveness," "directionality," and "granularity" are the most relevant discriminatory factors in human textural perception [20]. Here, we use a two-dimensional texture histogram based on *directionality* ϕ and *edge separation* ξ, which is closely related to "repetitiveness." ϕ measures the edge orientations, and ξ measures the distances between parallel edges.

To extract an edge map, the image is first converted to luminance Y via $Y = 0.299R + 0.587G + 0.114B$. A *Sobel edge operator* [21] is applied to the Y-image by sliding the following 3×3 weighting matrices (*convolution masks*) over the image:

$$d_x : \begin{array}{|c|c|c|} \hline -1 & 0 & 1 \\ \hline -2 & 0 & 2 \\ \hline -1 & 0 & 1 \\ \hline \end{array} \qquad d_y : \begin{array}{|c|c|c|} \hline 1 & 2 & 1 \\ \hline 0 & 0 & 0 \\ \hline -1 & -2 & -1 \\ \hline \end{array} \qquad (21.4)$$

If we average around each pixel with these weights, we produce approximations to derivatives.

The edge magnitude D and the edge gradient ϕ are given by

$$D = \sqrt{d_x^2 + d_y^2}, \qquad \phi = \arctan \frac{d_y}{d_x} \qquad (21.5)$$

Next, the edges are thinned by suppressing all but maximum values. If a pixel i with edge gradient ϕ_i and edge magnitude D_i has a neighbor pixel j along the direction of ϕ_i with gradient $\phi_j \approx \phi_i$ and edge magnitude $D_j > D_i$, then pixel i is suppressed to 0.

To make a binary edge image, we set all pixels with D greater than a threshold value to 1 and all others to 0.

For edge separation ξ, for each edge pixel i we measure the distance along its gradient ϕ_i to the nearest pixel j having $\phi_j \approx \phi_i$ within $15°$. If such a pixel j doesn't exist, the separation is considered infinite.

Having created edge directionality and edge separation maps, C-BIRD constructs a 2D texture histogram of ξ versus ϕ. The initial histogram size is 193×180, where separation value $\xi = 193$ is reserved for a separation of infinity (as well as any $\xi > 192$). The histogram size is then reduced by three for each dimension to size 65×60, where joined entries are summed together.

The histogram is "smoothed" by replacing each pixel with a weighted sum of its neighbors and is then reduced again to size 7×8, with separation value 7 reserved for infinity. At this stage, the texture histogram is also normalized by dividing by the number of pixels in the image segment.

21.3.4 Search by Illumination Invariance

Illumination change can dramatically alter the color measured by camera RGB sensors, from *pink* under daylight to *purple* under fluorescent lighting, for example.

To deal with illumination change from the query image to different database images, each color-channel band of each image is first normalized, then compressed to a 36-vector [22]. Normalizing each of the R, G, and B bands of an image serves as a simple yet effective guard against color changes when the lighting color changes. A two-dimensional color histogram is then created using the *chromaticity*, which is the set of band ratios $\{R, G\}/(R + G + B)$. Chromaticity is similar to the chrominance in video, in that it captures color information only, not luminance (or brightness).

A 128×128–bin 2D color histogram can then be treated as an image and compressed using a wavelet-based compression scheme [23]. To further reduce the number of vector components in a feature vector, the DCT coefficients for the smaller histogram are calculated and placed in zigzag order, then all but 36 components are dropped.

Matching is performed in the compressed domain by taking the Euclidean distance between two DCT-compressed 36-component feature vectors. (This illumination-invariant scheme and the object-model-based search described next are unique to C-BIRD.) Figure 21.5 shows the results of such a search.

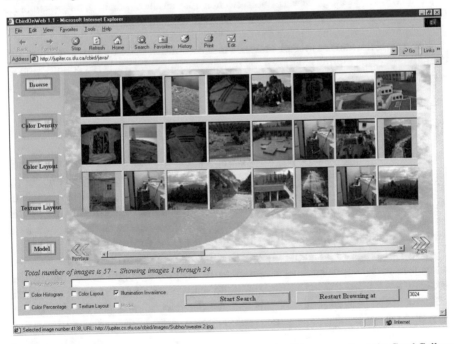

Fig. 21.5 Search with illumination invariance. Some thumbnail images are from the Corel Gallery and are copyright Corel.

Fig. 21.6 C-BIRD interface, showing object selection using an ellipse primitive. Image is from the Corel Gallery and is copyright Corel. All rights reserved

Several of the above types of searches can be done at once by checking multiple checkboxes. This returns a reduced list of images, since the list is the conjunction of all resulting separate return lists for each method.

21.3.5 Search-by-Object Model

The most important search type C-BIRD supports is the model-based object search. The user picks a sample image and interactively selects a region for object searching. Objects photographed under different scene conditions are still effectively matched. This search type proceeds by the user selecting a thumbnail and clicking the Model tab to enter Object Selection mode. An object is then interactively selected as a portion of the image; this constitutes an object query by example.

Figure 21.6 shows a sample object selection. An image region can be selected using primitive shapes such as a rectangle or ellipse, a magic wand tool that is basically a seed-based flooding algorithm, an active contour (a "snake"), or a brush tool, where the painted region is selected. All the selections can be combined with each other using Boolean operations such as union, intersection, or exclusion.

Once the object region is defined to a user's satisfaction, it can be dragged to the right pane, showing all current selections. Multiple regions can be dragged to the selection pane, but only the active object in the selection pane will be searched on.

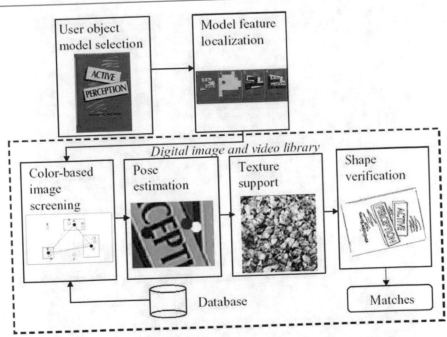

Fig. 21.7 Block diagram of object matching steps

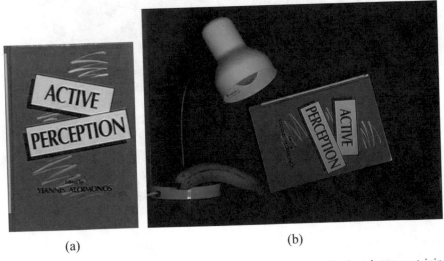

Fig. 21.8 Model and target images: **a** Sample model image; **b** sample database image containing the model book. Active Perception textbook cover courtesy Lawrence Erlbaum Associates, Inc.

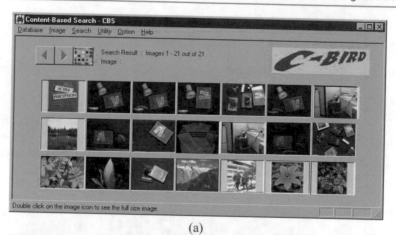

(a)

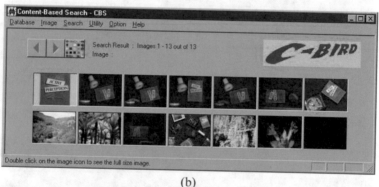

(b)

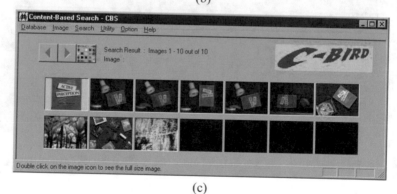

(c)

Fig. 21.9 Search result for the pink book model with illumination change support: **a** Search results using pose estimation only; **b** search results using pose estimation and texture support; **c** search results using GHT shape verification. Some thumbnail images are from the Corel Gallery and are copyright Corel. All rights reserved

The user can also control parameters such as flooding thresholds, brush size, and active contour curvature.

Details of the underlying mechanisms of this Search by Object Model are set out in [23] and introduced below as an example of a working system. Figure 21.7 shows a block diagram for how the algorithm proceeds. First, the user-selected model image is processed and its features are localized (details are discussed in [18]). Color histogram intersection, based on the reduced chromaticity histogram described in the previous section, is then applied as a first "screen." Further steps estimate the pose (scale, translation, rotation) of the object inside a target image from the database. This is followed by verification by the intersection of texture histograms and then a final check using an efficient version of a generalized Hough transform (GHT) for shape verification.

A possible model image and one of the target images in the database might be as in Fig. 21.8, where the scene in (b) was illuminated with a dim fluorescent light. Figure 21.9 shows some search results for the pink book in C-BIRD.

While C-BIRD is an experimental system, it does provide a proof in principle that the difficult task of search-by-object model is possible.

21.4 Quantifying Search Results

Generally speaking, some simple expression of the performance of image search engines are desirable. In information retrieval, *Precision* is the percentage of relevant documents retrieved compared to the number of all the documents retrieved, and *Recall* is the percentage of relevant documents retrieved out of all relevant documents. Recall and Precision are widely used for reporting retrieval performance for image retrieval systems as well. However, these measures are affected by the database size and the amount of similar information in the database. Also, they do not consider fuzzy matching or search result ordering.

In equation form, these quantities are defined as

$$
\begin{aligned}
Precision &= \frac{Relevant\ images\ returned}{All\ retrieved\ images} \\
Recall &= \frac{Relevant\ images\ returned}{All\ relevant\ images}
\end{aligned} \tag{21.6}
$$

Alternatively, they may also be written as

$$
\begin{aligned}
Precision &= \frac{TP}{TP + FP} \\
Recall &= \frac{TP}{TP + FN}
\end{aligned} \tag{21.7}
$$

where TP (True Positives) is the number of relevant images returned, FP (False Positives) is the number of irrelevant images returned, and FN (False Negatives) is the number of relevant images not returned.

In general, the more we relax thresholds and allow more images to be returned, the smaller the Precision, but the larger the Recall; and vice versa. Apparently, it is not quite meaningful to talk about either the Precision or Recall number by itself. Instead, they can be combined to provide a good measure, e.g., Precision when Recall is at 50%, Recall when Precision is at 90%, etc.

When multiple queries are involved, the numbers of the Precision and Recall values will again increase. To measure the overall performance of a CBIR system, the most common way is to summarize these values into a single value, i.e., the *mean average precision (MAP)*. The Average Precision of a single query q is defined as

$$AP(q) = \frac{1}{N_R} \sum_{n=1}^{N_R} Precision(n) \tag{21.8}$$

where $Precision(n)$ is the Precision value after the n-th relevant image was retrieved, and N_R is the total number of relevant images. The MAP is the mean of Average Precisions over all query images:

$$MAP = \frac{1}{N_Q} \sum_{q \in Q} AP(q) \tag{21.9}$$

where Q is the query image set, and N_Q is its size. The MAP has the advantage of reflecting both Precision and Recall oriented aspects and is sensitive to the entire ranking [24].

In the above definitions of Precision and Recall, one thing is missing: the value of TN (True Negatives). In the context of CBIR, if we are searching for dogs and the image database contains 100 relevant dog images, the fact that whether the database contains another 100 non-dog images or a million non-dog images, i.e., $TN = 100$ or $TN = 1,000,000$, often matters, because a larger TN tends to yield more distractions (noise).

Many other measures for the performance of CBIR systems have been devised, and among them a popular one is the *receiver operating characteristic (ROC)*. Given a database containing 100 relevant images (e.g., horses) and 1,000 other images (i.e., non-horses), if a CBIR system correctly retrieves 70 horse images from the 100 horse images, and incorrectly identified 300 "horse" images from the 1,000 other images, then the *True Positive Rate* $TPR = 70\%$ and the *False Positive Rate* $FPR = 30\%$. Obviously, $TPR = Recall$ as defined in Eq.21.7, and $FPR = 1 - TNR$, where TNR (True Negative Rate) is the percentage (70%) of the non-horses correctly identified. Figure 21.10 depicts the so-called ROC space, which plots TPR against FPR.

The 45° diagonal line in Fig. 21.10 indicates the performance of a random guess. In the above example, if we were to use a (fair) coin-flip to determine the outcome,

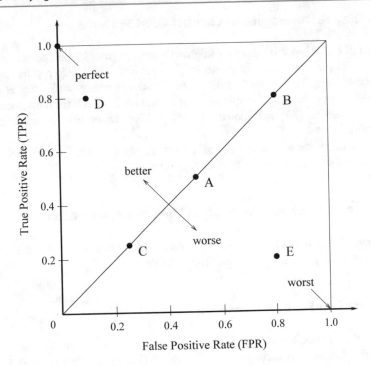

Fig. 21.10 The ROC space

then it will return a result of 50 true horse images and 500 false "horse" images from the two groups of images, i.e., $TPR = 50\%$ and $FPR = 50\%$, which is indicated by point A $(0.5, 0.5)$ on the diagonal line in the figure. A biased coin that is weighted to produce more positive outcomes (e.g., heads) may yield performance that is indicated by point B $(0.8, 0.8)$. Conversely, it may produce C $(0.25, 0.25)$.

Clearly, we would like to have CBIR systems that will produce results better than a coin-flip. Namely, their performance should be above the diagonal line. Point D $(0.1, 0.8)$ is an example of what would be a very good performance. In general, we aim to be as close as possible to the ultimately perfect point $(0, 1.0)$. Point E $(0.8, 0.2)$ indicates poor performance, and obviously $(1.0, 0)$ would be the worst.

If we measure the CBIR system at multiple TPR (or FPR) values, we will derive an *ROC curve*, which often provides a more comprehensive analysis of the system behavior. The perfect ROC curve would consist of two straight line segments, one vertical from $(0, 0)$ to $(0, 1.0)$ and one horizontal from $(0, 1.0)$ to $(1.0, 1.0)$, which indicates that there are no false positives or false negatives whatsoever. A normal ROC curve will be in the area between this ideal performance and the 45° diagonal line.

ROC is a general statistical measure for various classifiers. It has its applications in many disciplines, e.g., psychology, physics, medicine, and increasingly in machine learning and data mining.

21.5 Key Technologies in Current CBIR Systems

The field of CBIR has witnessed rapid growth and progress in the new millennium. Unlike in the early years, users are no longer satisfied with query results returning scenes containing "blue skies," or red blobs that may or may not be flowers. They are more and more looking for objects, people, and often search for human activities as well. Datta et al. [25] and Zhou et al. [26] provide comprehensive surveys of the CBIR field.

In this section, we will briefly describe some technologies and issues that are key to the success of current and future CBIR systems.

21.5.1 Robust Image Features and Their Representation

Many feature descriptors have been developed beyond the ones specified in MPEG-7. Here, we will only discuss *SIFT* and *Visual Words*.

Scale Invariant Feature Transform (SIFT)

In addition to global features such as color and texture histograms, local features characterized by scale invariant feature transform (SIFT) [27] have been developed because they are more suitable for searches emphasizing visual objects. SIFT has been shown to be robust against image noise; it also has a fair degree of invariance to translation, rotation, and scale change.

The process of extracting SIFT features from an image starts with building a multiresolution scale space, as shown in Fig. 21.11. At each scale, stacks of images are generated by applying Gaussian smoothing (filtering) operations. Each stack (the so-called octave) has $s + 3$ images—so $s = 2$ for the example shown in the figure. The standard deviation of the Gaussian filters in the stack increases by a factor of $2^{1/s}$. In this example, they are σ, $\sqrt{2}\sigma$, 2σ, $2\sqrt{2}\sigma$, and 4σ in the first octave. The third image from the top of the stack is used as the bottom image of the next stack by reducing its image resolution by half. The same Gaussian filtering process will continue, and hence at the next octave the Gaussian filters have standard deviations 2σ, $2\sqrt{2}\sigma$, 4σ, $4\sqrt{2}\sigma$, and 8σ. A simple operation of image subtraction, as indicated in the figure, generates difference of Gaussian (DOG) images.

Now, we are ready to talk about *Key points* for SIFT. In the DOG images, if a pixel's DOG value is the maximum or minimum when compared to its 26 neighbors in scale space (see Fig. 21.11), it is considered a possible Key point. Further, screening steps are introduced to make sure that the Key points are more distinct, e.g., are at the corners rather than distributed over a long edge. Then, the histogram of the edge (gradient) directions near the Key point are analyzed to yield a dominant direction θ (the so-called *canonical orientation*) for the Key point.

The local patch (16×16 pixels) at the Key point is now examined to produce a 128-dimensional SIFT descriptor. The patch is divided into 4×4 subwindows. In

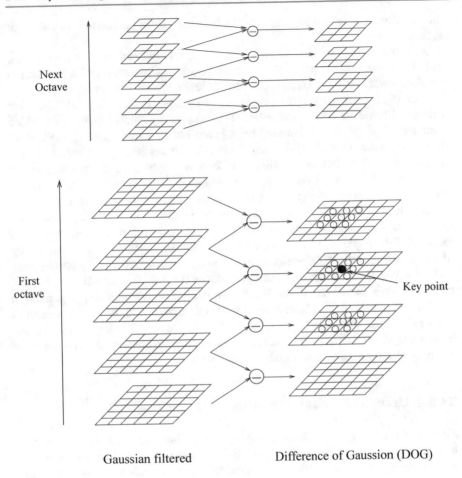

Fig. 21.11 The scale space and SIFT key point

each subwindow, a histogram of edge (gradient) directions is derived. The quantization is for every $45°$ so the histogram in each subwindow produces a vector of 8 dimensions. In total, we obtain a $4 \times 4 \times 8 = 128$ dimensional SIFT descriptor. Here, all the edge directions are calculated relative to the canonical orientation θ. Hence, this facilitates rotation invariance.

Visual Words

The *bag of words (BoW)* concept was originally a technique for document classification and text retrieval. As the name suggests, a mixed bag of words can be extracted from a query sentence and be used for a query. The stems of the words are used, e.g., "talk" for "talk," "talking," "talked," etc. In this way, the details of the word (e.g., tense, singular, or plural), the word order, and the grammar of the sentence are all ignored. The advantage is that this tends to be more robust against any variations of the text.

Analogously, bags of *Visual Words* can be extracted to represent image features. Fei-Fei and Perona [28] presented an earlier work in Computer Vision in which a bag of codewords is used to represent various texture features in images.

In CBIR, a common way of generating Visual Words is to use SIFT because of its good properties discussed above. This can be (a) object based, or (b) video frame based. If the search is based on a given object model, then clusters of its SIFT features can be used as the Visual Words describing the object. Commonly, the vector quantization method as outlined in Chap. 8 can be used to turn these Visual Words into codewords in a codebook. If the search is aimed at finding similar frames from a video or movie, then all SIFT features in the query video frame can be used to generate the Visual Words. Alternatively, Sivic and Zisserman [29] divided the video frame into dozens (or hundreds) of regions, and each region will produce a mean SIFT descriptor $\bar{x}_i$ and be used as a Visual Word. The clustering and matching of a large number of SIFT descriptors are shown to be computationally challenging.

Visual Words are rich in encapsulating essential visual features. However, compared to words from text, Visual Words are even more ambiguous. In general, a small-sized codebook will have limited discriminative power, not good enough to handle CBIR for large image and video databases. On the other hand, a large codebook also has its own problems, because the same feature contaminated by noise can easily be quantized to different codewords.

21.5.2 User Feedback and Collaboration

Relevance feedback, a well-known technique from Information Retrieval has been brought to bear in CBIR systems [30]. Briefly, the idea is to involve the user in a loop, whereby images retrieved are used in further rounds of convergence onto correct returns. The usual situation is that the user identifies images as good, bad, or don't care, and weighting systems are updated according to this user guidance.

The basic advantage of putting the user into the loop by using relevance feedback is that this way, the user need not provide a completely accurate initial query. Relevance feedback establishes a more accurate link between low-level features and high-level concepts, somewhat closing the semantic gap. As a result, the retrieval performance of the CBIR system is improved.

Besides mechanisms such as Relevance Feedback, it has been argued that users should play a more active role in CBIR systems. In the 2012 International Conference on Multimedia Retrieval (ICMR), the panelists (also the authors of [3]) again raised the question: Where is the User in Multimedia Retrieval? They pointed out the dominance of MAP, common in evaluations such as TRECVID, may have hindered the development of better and more useful multimedia retrieval systems. Although MAP has the merit of being objective and reproducible, it is unlikely that a single number will meet the needs of most users, who tend to have very different and dynamic tasks in mind.

One way to comprehend how people view images as similar is to study using user groups just what forms our basis of the perception of image similarity [31]. The function used in this approach can be a type of "perceptual similarity measure" and is *learned* by finding the best set of features (color, texture, etc.) to capture "similarity" as defined via the groups of similar images identified.

Another way to understand users is to talk to them and carefully analyze their search patterns. In other words, in addition to *content-based*, we need to be *context-based*, because the user's interpretation of the content is often influenced (or even determined) by the context.

Smith et al. [32] propose the first computer–human collaboration system for movie trailer creation, which is applied to the film "Morgan" released in 2016. The multimedia analysis algorithms retrieve and select 10 moments from the original movie as candidate scenes for a trailer, and then a professional film-maker is involved to complete the trailer creation. Specifically, the movie is first segmented into audio/visual snippets with visual shot-boundary detection and audio segmentation algorithms. The audio sentiment features are computed for each audio segment using OpenEAR model [33], which is trained on six emotion recognition datasets. Sentibank [34] is employed to extract visual sentiment representations from the keyframe of each shot. The visual sentiment features of all keyframes within the range of an audio segment are aggregated to form a composite visual sentiment feature. Besides, visual place attribute features are extracted with Places-CNN model [35]. After extracting the high-level audio and visual features, principal component analysis (PCA) is applied to project the features to a lower dimensional space, where the scenes with the highest response are selected as candidates for making the trailer. Finally, the film-maker is involved to create the trailer, by rearranging and cutting the shots, creating good transitions and overlaying the official soundtrack.

21.5.3 Other Post-processing Techniques

Beside User Feedback and Collaboration, other post-processing methods have been developed after the initial query results are in.

Spatial Verification

The quality of the query results can be verified and improved by *Spatial Verification*. Modern cameras often provide information about the location where pictures were taken, so it is trivial to check for consistency or relevance in certain query results if the user is only interested in images from certain locations, e.g., Paris or Rome.

In their paper [36], Philbin et al. go well beyond simple checks on geometric locations. They aim at verifying that image regions from the query image and the retrieved images are from the same object or scene region. It is argued that unlike words in Information Retrieval (e.g., "animal," "flower"), *Visual Words* inherently contain much more spatial information. For example, it is known from the theory of image geometry that two views of a rigid object are related by epipolar geometry, two views of a planar patch are related by a homograph, etc. They show that verifying mappings based geometric transformations can indeed improve the quality of results by spatial reranking.

Zhou et al. [37] describe a spatial coding method that records relative spatial information (e.g., left or right, above or below) of each pair of image features. A rotating spatial map is proposed that is more efficient than a simple x- and y-map.

Query Expansion

Another approach is to move the query toward positively marked content. *Query Expansion* proposed by Chum et al. [38] is such a method. Query Expansion is again a well-known method in Information Retrieval in which some combination of high-ranked relevant documents can be presented as a new query to boost the performance of the retrieval. The combination could simply be the averages (means) of the feature descriptors from the returned documents. The problem is that if one of the high-ranked document is a false-positive, then it will immediately harm the performance of the new query. Some robust statistical methods or even simply taking the median (instead of the mean) can help.

As mentioned above, the Visual Words used in CBIR often contain useful spatial information. Hence, high-ranked retrieved images can be verified before being used in forming a new query image. Chum et al. [38] use recursive average query expansion, which recursively generates new query images from all spatially verified returned images.

QA Paradigm

Question-answering (QA) tries to replace the large number of images or multimedia content in general, that is returned by a search query, by making use of media content, knowledge about the search domain, and also linguistic analysis to return hits based on users' natural-language questions. Since QA has traditionally been focused on text, bridging this technique over to multimedia content is referred to as MMQA [39]. Generally, MMQA attempts to combine traditional QA, based on textual metadata, with multimedia-oriented approaches if such are indeed more intuitive answers to users' queries.

21.5.4 Visual Concept Search

Search by Concept is another major step in closing the semantic gap. Typically, after local features are turned into words, the words are converted to semantic concepts with the aid of machine learning algorithms.

In interesting work presented by Wang et al. [40], image similarity is learned from 103 Flickr image groups on the Internet by adopting a support vector machine (SVM) classifier, with a histogram intersection kernel. The image groups range from *objects* such as Aquarium, Boat, Car, and Penguin, *scenes* such as Sunset and Urban to *concepts* such as Christmas and Smile. They show that such a system performs better than others that directly measure image similarity with simple low-level visual features such as color, texture, etc.

To assist in the development and testing of these systems, researchers and practitioners have developed many multimedia databases and benchmarks. The best known is perhaps the TREC Video Retrieval Evaluation (TRECVID) benchmark. TRECVID started in 2003, originally from the conference Text REtrieval Conference (TREC), co-sponsored by the National Institute of Standards (NIST) and the U.S. Department of Defense. Initially, TRECVID provided video data from professional sources such as broadcast news, TV programs, and surveillance systems, thus having limited styles and content. For example, the very typical head-shot of a news person, an overhead view of a variety of indoor scenes, etc. In the last decade, the benchmarks have expanded in terms of test data and objectives. For example, TRECVID 2013 evaluated the following tasks:

- Semantic indexing
- Interactive surveillance event detection
- Instance search
- Multimedia event detection
- Multimedia event recounting.

In the subsequent years, several new tasks were added such as the Video Hyperlink task (LNK) to address the issue of social media storytelling linking. The TRECVID 2019 [41] tasks included Ad-hoc Video Search, Activities in Extended Video, Instance Search, and Video to Text.

Myers et al. [42] reported a good performance for Multimedia Event Detection in their project, entitled SESAME. To start, multiple event classifiers were developed based on the bags of words from single data types, e.g., low-level visual features, motion features, and audio features; and high-level visual (semantic) concepts such as Birthday-party, Making-a-sandwich, etc. Various fusion methods (Arithmetic Mean, Geometric Mean, mean average precision (MAP) Weighted, Weighted Mean Root, Conditional Mixture Model, Sparse Mixture Model, etc.) were then tested. It was shown that some simple fusion methods, e.g., Arithmetic Mean, perform as well as more complex ones, or better.

21.5.5 Feature Learning with Convolutional Neural Networks

Convolutional neural networks (CNNs) [7] have brought a revolution in computer vision community with the powerful feature learning ability based on large-scale datasets. The AlexNet architecture [43] started to outperform the traditional mod-

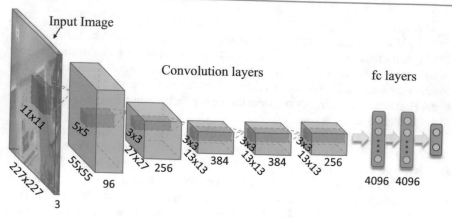

Fig. 21.12 An illustration of a typical convolutional neural network architecture

els based on hand-crafted features on ImageNet classification task in 2012. After that, a variety of CNN architectures have been proposed, such as VGGNet [44], GoogleNet [45], and ResNet [46]. They have been immensely successful and achieved state-of-the-art performances in various computer vision tasks, such as image classification and object detection. A convolutional neural network usually consists of multiple convolutional layers, and may also include fully-connected (fc) layers, pooling layers, normalization layers and activation functions such as rectified linear unit (ReLU). The weights of the convolution filters and fc layers are updated and learned using error back propagation during training. Figure 21.12 shows an illustration of the AlexNet architecture [43].

It has been shown that the generic feature descriptors extracted from the CNNs are very powerful and discriminative, which can be used to tackle a diverse range of visual recognition tasks [47]. The CNNs are usually trained on the ImageNet dataset for classification task and then used as a feature extractor, by feeding an image to the network and using the output of the upper layers (e.g., fc layers) as feature representation. In order to adjust the CNNs to learn more relevant features for other tasks using datasets that may have different image statistics, one strategy is to fine-tune the pre-trained CNN on the target dataset with more similar images at test time. The fine-tuning is achieved by first initializing the network weights with the pre-trained CNN and then retrain the model on the target dataset. These methods have been widely used in tasks including content-based image retrieval [8,48,49].

Babenko et al. [48] evaluate the performance of the generic CNN features, which are referred to as neural codes, in the image retrieval application. The authors employ the AlexNet architecture and use the outputs of the layer 5, 6, and 7, which are the last convolution layer and first two fully-connected layers, as the feature descriptors or neural codes for an input image. The performance of the neural codes are evaluated on four standard datasets (Oxford, Oxford 105K, Holidays, and UKB) and compared with traditional features. The results demonstrate that among all the layers, the features from the 6th layer generally perform the best. In comparison to other well-performing methods, the neural codes are in the same ball park. This is promising considering the fact that the CNN is trained on the ImageNet dataset for classification task, which is quite different from the retrieval task. The idea of fine-tuning is also tested in [48], where the authors collect the Landmarks dataset with 672 classes and 213,678 images. This dataset has similar image appearances with the landmark-type datasets (Holidays and Oxford). It is demonstrated that retraining the pre-trained CNN on the Landmarks dataset results in an improvement of retrieval performance on these two datasets.

Babenko et al. [48] further evaluate two different strategies for reducing the dimension of the features from the 6th layer, which is a vector with 4,096 entries. They first evaluate the performance of neural codes after being compressed using principal component analysis (PCA) to different dimensions. Experiments show that the neural codes can be compressed to 256 or even to 128 dimensions with only little loss of retrieval performance. The performance degrades gracefully after reducing feature dimension to be lower than 128. When comparing the neural codes and the traditional features compressed to the same dimension of 128, the neural codes achieve better results. This demonstrates that the neural codes are more resilient to compression and can provide better short-features than other hand-crafted descriptors. The authors further investigate a discriminative dimensionality reduction method by learning a low-rank projection matrix. The objective is to make the distance between the codes of similar images small and other images large. It is demonstrated that this dimension reduction method improves the performance of PCA-compressed neural codes even further for the extremely compressed 16-dimensional codes.

Besides using the generic features and fine-tuned features as in [48], Wan et al. [49] propose another two schemes for refining deep CNN features for content-based image retrieval. The first scheme is to use similarity learning based on the extracted features from the pre-trained model. Specifically, the similarity function of two samples is denoted by a bilinear form as: $S_W(\mathbf{x}_i, \mathbf{x}_j) = \mathbf{x}_i^T W \mathbf{x}_j$, where W is the matrix to be learned for computing similarity, $\mathbf{x}_i, \mathbf{x}_j$ are the CNN features. The similarity learning is based on a triplet constraint set containing the similar and dissimilar information of the image pairs, which is denoted as $\chi = \{(\mathbf{x}_i, \mathbf{x}_i^+, \mathbf{x}_i^-)\}$, $i = 1, \ldots, N$, where $(\mathbf{x}_i, \mathbf{x}_i^+)$ and $(\mathbf{x}_i, \mathbf{x}_i^-)$ represent a relevant image pair and an irrelevant image pair, respectively. The hinge loss for a triplet is defined as $l_W(\mathbf{x}_i, \mathbf{x}_i^+, \mathbf{x}_i^-) = \max\{0, 1 - S_W(\mathbf{x}_i, \mathbf{x}_i^+) + S_W(\mathbf{x}_i, \mathbf{x}_i^-)\}$. The loss is minimized by applying the Passive-Aggressive algorithm iteratively over all triplets to optimize W. The triplets are constructed by simply considering instances in the same class as relevant, and instances in different classes as irrelevant. Using the learned similarity measurement, this model is able to achieve better retrieval performance than using generic CNN features.

The second scheme proposed in [48] is to fine-tune the CNN with loss function based on cosine similarity. In this scheme, the CNN model is trained in an end-to-end manner by back-propagating the error to the previous layers and updating parameters of the entire model. Therefore, the feature representation is refined for the retrieval dataset. The cosine similarity of two features $\mathbf{x}_i, \mathbf{x}_j$ is defined as $S_{cos}(\mathbf{x}_i, \mathbf{x}_j) = \mathbf{x}_i^T \mathbf{x}_j / (\|\mathbf{x}_i\| \times \|\mathbf{x}_j\|)$. The loss function is defined similar to the hinge loss of the first scheme, and the triplet constraint set is also used during training. The experimental results demonstrate that refining the model by similarity guided retraining is able to achieve better performance than the first scheme.

21.5.6 Database Indexing

The scale of the image database has been growing explosively and the response time becomes a key issue in content-based image retrieval. Simple nearest neighbor search algorithm that matches the query with all images is considered too time-consuming. Even optimized search with structures such as K-D tree often does not meet the efficiency requirement for retrieval in large databases. Approximate nearest neighbor search methods are proposed to tackle this problem. It is usually required to perform special database organizing and indexing to enable fast approximate retrieval. We will briefly introduce two types of indexing techniques, *inverted indexing* and *hashing-based indexing*.

Inverted Indexing

The Inverted Indexing file structure can be considered as a matrix whose rows and columns denote images and visual words, respectively. To retrieve similar images from the database, only those images that contain the same visual words as the query are checked. This reduces the number of candidate images to be compared with the query and provides a substantial speed-up over the exhaustive search. The file structure is usually implemented by building a file list for each visual word. The file list of a visual word consists of the IDs of the images containing this visual word. Some other information such as Hamming Embedding [50], geometric statistics [51] can also be stored in the file list for verification or similarity measurement purposes.

The efficiency of inverted indexing has certain limitations and they begin to show up for very large datasets. The method has to increase the number of visual words or the list of files for each word will become too large for exhaustive search. Inverted Multi-index [52] is proposed to tackle this problem, by replacing the standard quantization inside the inverted index with the product quantization (PQ). The inverted multi-index is constructed as a multi-dimensional table following the PQ idea. The advantage of multi-index is that it produces finer subdivisions and returns shorter candidate lists with higher recall, which also improves the speed of the approximate nearest neighbor search. Zheng et al. [53] propose a coupled multi-index (c-MI) framework to perform feature fusion at indexing level. The features are coupled into a multi-dimensional inverted index with each dimension corresponding to one feature. The retrieval process votes for images that are similar in both SIFT and color attribute feature spaces.

Hashing-Based Indexing

Hashing techniques are widely adopted to achieve efficient retrieval for large datasets. These methods partition the database by the use of multiple hash functions to project the features into several buckets. During image retrieval, the query image is also projected by the hash functions and only the images within the same bucket are compared with the query. The locality sensitive hashing (LSH) is one of the representative hashing schemes [54]. The key intuition of LSH is to project the features with multiple hash functions, so that the probability of collision (falling within the same bucket) is much higher for features that are close to each other than for those that are far apart. Formally, a hash function $h(x)$ is called local sensitive, or (r_1, r_2, p_1, p_2)-sensitive, if the following two conditions hold for any features x, y with $p_1 > p_2$ and $r_1 < r_2$: "if $d(x, y) < r1$, then $P(h(x) = h(y)) > p_1$," "if $d(x, y) > r2$, then $P(h(x) = h(y)) < p_2$." The function $d(x, y)$ is some distance metric and P denotes probability. Kulis and Grauman [55] generate LSH to accommodate arbitrary kernel functions, which permits sub-linear time similarity search for a wide class of similarity functions. The hash functions can also be learned using deep convolutional neural networks [56].

21.6 Querying on Videos

Video indexing can make use of *motion* as the salient feature of temporally changing images for various types of queries. We shall not examine video indexing in any detail here but refer the reader to the excellent survey in [25].

In brief, since temporality is the main difference between a video and just a collection of images, dealing with the time component is first and foremost in comprehending the indexing, browsing, search, and retrieval of video content. A direction taken by the QBIC group [13] is a new focus on storyboard generation for automatic understanding of video—the so-called "inverse Hollywood" problem. In the production of a video, the writer and director start with a visual depiction of how the story proceeds. In a video understanding situation, we would ideally wish to regenerate this storyboard as the starting place for comprehending the video.

The first place to start, then, would be dividing the video into *shots*, where each shot consists roughly of the video frames between the on and off clicks of the Record button. However, transitions are often placed between shots—fade-in, fade-out, dissolve, wipe, and so on—so detection of shot boundaries may not be so simple as for abrupt changes.

Generally, since we are dealing with digital video, if at all possible we would like to avoid uncompressing MPEG files, say, to speed throughput. Therefore, researchers try to work on the compressed video. A simple approach to this idea is to uncompress just enough to recover the DC term, generating a thumbnail 64 times smaller than the original. Since we must consider P- and B-frames as well as I-frames, even generating a good approximation of the best DC image is itself a complicated problem.

Once DC frames are obtained from the whole video—or, even better, are obtained on the fly—many approaches have been used for finding shot boundaries. Features used have typically been color, texture, and motion vectors, although such concepts as trajectories traversed by objects have also been used [57].

Shots are grouped into *scenes*. A scene is a collection of shots that belong together and that are contiguous in time. Even higher level semantics exist in so-called "film grammar" [58]. Semantic information such as the basic elements of the story may be obtainable. These are (at the coarsest level) the story's exposition, crisis, climax, and denouement.

Audio information is important for scene grouping. In a typical scene, the audio has no break within a scene, even though many shots may take place over the course of the scene. General timing information from movie creation may also be brought to bear.

Text may indeed be the most useful means of delineating shots and scenes, making use of closed-captioning information already available. However, relying on text is unreliable, since it may not exist, especially for legacy video.

Different schemes have been proposed for organizing and displaying storyboards reasonably succinctly. The most straightforward method is to display a two-dimensional array of *keyframes*. Just what constitutes a good keyframe has of course been subject to much debate. One approach might be to simply output one frame every few seconds. However, action has a tendency to occur between longer periods of inactive story. Therefore, some kind of clustering method is usually used, to represent a longer period of time that is more or less the same within the temporal period belonging to a single keyframe.

Some researchers have suggested using a graph-based method. Suppose we have a video of two talking heads, the interviewer and the interviewee. A sensible representation might be a digraph with directed arcs taking us from one person to the other, then back again. In this way, we can encapsulate much information about the video's structure and also have available the arsenal of tools developed for graph pruning and management.

Other "proxies" have also been developed for representing shots and scenes. A grouping of *sets* of keyframes may be more representative than just a sequence of keyframes, as may keyframes of variable sizes. Annotation by text or voice, of each set of keyframes in a "skimmed" video, may be required for sensible understanding of the underlying video.

A *mosaic* of several frames may be useful, wherein frames are combined into larger ones by matching features over a set of frames. This results in set of larger keyframes that are perhaps more representational of the video.

An even more radical approach to video representation involves selecting (or creating) a *single* frame that best represents the entire movie! This could be based on making sure that people are in the frame, that there is action, and so on. In [59], Dufaux proposes an algorithm that selects shots and keyframes based on measures of motion-activity (via frame difference), spatial activity (via entropy of the pixel value distribution), skin-color pixels, and face detection.

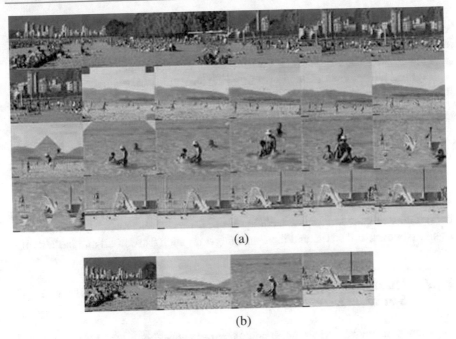

(a)

(b)

Fig. 21.13 Digital video and associated keyframes, beach video: **a** Frames from a digital video; **b** keyframes selected

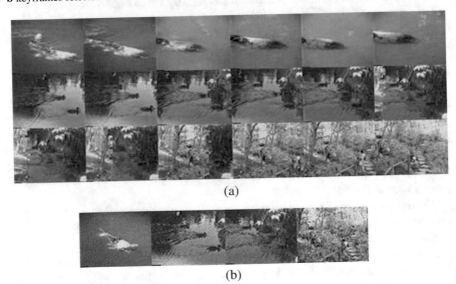

(a)

(b)

Fig. 21.14 Garden video: **a** Frames from a digital video; **b** keyframes selected

By taking into account skin color and faces, the algorithm increases the likelihood of the selected keyframe including people and portraits, such as close-ups of movie actors, thereby producing interesting keyframes. Skin color is learned using labeled image samples. Face detection is performed using a neural net.

Figure 21.13a shows a selection of frames from a video of beach activity (see [60]). Here, the keyframes in Fig. 21.13b are selected based mainly on color information (but being careful with respect to the changes incurred by changing illumination conditions when videos are shot).

A more difficult problem arises when changes between shots are gradual and when colors are rather similar overall, as in Fig. 21.14a. The keyframes in Fig. 21.14b are sufficient to show the development of the whole video sequence.

Other approaches attempt to deal with more profoundly human aspects of video, as opposed to lower level visual or audio features. Much effort has gone into applying data mining or knowledge-base techniques to *classifying* videos into such categories as sports, news, and so on, and then subcategories such as football and basketball.

21.7   Querying on Videos Based on Human Activity—A Case Study

Thousands of hours of video are being captured every day by CCTV camera, web camera, broadcast camera, etc. However, most of the activities of interest (e.g., a soccer player scores a goal) only occur in a relatively small region along the spatial and temporal extent of the video. In this scenario, effective retrieval of a small spatial/temporal segment of the video containing a particular activity from a large collection of videos is very important.

Lan et al.'s work [61] on activity retrieval is directly inspired by the application of searching for activities of interest from broadcast sports videos. For example, consider the scene of a field hockey game shown in Fig. 21.15. In terms of human activities, a variety of questions can be asked: Who is the attacker? What are the players in the bottom right corner doing? How many people are running? Which players are defending (marking) members of the opposing team? What is the overall game situation? Note that potential queries often involve social roles such as "defender," "attacker," or "man-marking." Lan et al. [61] present a model toward answering queries such as these.

The representation of human activity is a challenging, open problem. Much of the work focuses on recognition of low-level single-person actions (e.g., [62]). Lan et al. rely on low-level features and representations used by these methods to predict the actions of individuals, and build higher level models upon them. A model is presented that can be used to capture a variety of levels of detail in a unified framework. In addition to modeling of low-level actions (e.g., running or standing) and high-level events ("attack play," "penalty corner"), the social roles are also modeled. Social roles take into account inter-related people and are a complementary representation of the low-level actions typically used in the activity recognition literature. For example, a player engaged in the social role of "man-marking" is likely to have an opponent nearby. Further, the notions of low-level actions, social roles, and high-level events

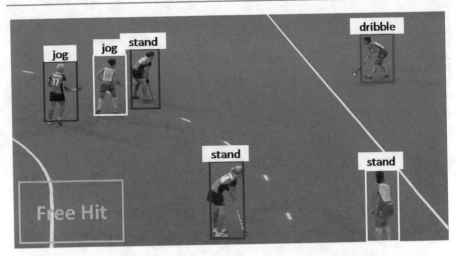

Fig. 21.15 An example of human activity in realistic scenes. Beyond the general scene-level activity description (e.g., Free Hit), we can explain the scene at multiple levels of detail such as low-level actions (e.g., standing and jogging) and mid-level social roles (e.g., attacker and first defenders). The social roles are denoted by different colors. In this example, we use magenta, blue and white to represent attacker, man-marking, and players in the same team as the attacker, respectively

naturally require a contextual representation—the actions and social roles of all the people in a scene are interdependent, and related to the high-level event taking place. The model captures these relationships, and allows flexible inference of the social roles and their dependencies in a given scene.

21.7.1 Modeling Human Activity Structures

Here we introduce the model in [61]. To illustrate, we describe an instantiation applicable to modeling field hockey videos.

We first describe the labeling. Assume an image has been pre-processed, so the location of the goal and persons in the image have been found. We separate the players into two teams according to their color histograms. Each person is associated with two labels: action and social role. Let $h_i \in \mathcal{H}$ and $r_i \in \mathcal{R}$ be the action and social roles of the person i, respectively, where $\mathcal{H}$ and $\mathcal{R}$ are the sets of all possible action and social role labels, respectively. Each video frame is associated with an event label $y \in \mathcal{Y}$, where $\mathcal{Y}$ is the set of all possible event labels.

The model is hierarchical, and includes various levels of detail: low-level actions, mid-level social roles, and high-level events. The relationships and interactions between these are included in the model. We define the score of interpreting a video frame I with the hierarchical event representation as

Fig. 21.16 Graphical illustration of the model. Different types of potentials are denoted by lines with different colors. Details of the potential functions are contained in Eq. 21.10

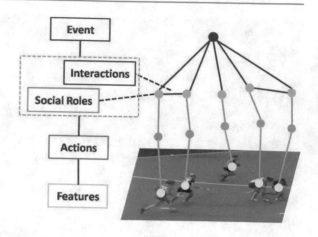

$$F_w(\mathbf{x}, y, \mathbf{r}, \mathbf{h}, I) = w^\top \Phi(\mathbf{x}, y, \mathbf{r}, \mathbf{h}, I) = \sum_j w_1^\top \phi_1(x_j, h_j)$$

$$+ \sum_j w_2^\top \phi_2(h_j, r_j) + \sum_{j,k} w_3^\top \phi_3(y, r_j, r_k) \tag{21.10}$$

Now we go over the model formulation using the graphical illustration shown in Fig. 21.16. The first term denotes standard linear models trained to predict the action labels of the persons in the scene (indicated by blue lines in Fig. 21.16). The second term captures the dependence between action labels and social roles (green lines in Fig. 21.16). The third term explores contextual information by modeling interactions between people in terms of their social roles under an event y (magenta and dark blue lines in Fig. 21.16). Social roles naturally capture important interactions, e.g., *first defenders* tend to appear in the neighborhood of an *attacker, man-marking* happens when there is a player from the opposing team.

The model parameters w are learned in a structured SVM framework [63]; for details the reader is referred to [61].

Inference

Given a test video, there are a variety of queries one might wish to answer. Using the proposed hierarchical model, one can formulate queries about any individual variable at any level of detail. For instance, one can query on the overall event label for the scene, or the social role label of a particular person. Figure 21.17 shows an overview of the testing phase.

For a given video and query variable q, the inference problem is to find the best hierarchical event representation that maximizes the scoring function $F_w(\mathbf{x}, y, \mathbf{h}, \mathbf{r}, I)$ while fixing the value of q to its possible values. For example, if q is the action of one person (one of the h_i), we would compute the maximum value of the scoring function

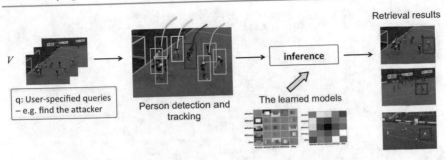

Fig. 21.17 An overview of the testing phase. Given a new video and a query, we first run the pre-trained person detector and tracker to extract the tracklet of each player. Then the tracklet features are fed into the inference framework. Finally, the retrieval results are obtained based on the inference scores computed using Eq. 21.10

F_w when fixing q to each possible action. We define the optimization problem as follows:

$$\max_{y,\mathbf{h},\mathbf{r}\backslash q} F_w(\mathbf{x}, y, \mathbf{h}, \mathbf{r}, I) = \max_{y,\mathbf{h},\mathbf{r}\backslash q} w^\top \Phi(\mathbf{x}, y, \mathbf{h}, \mathbf{r}, I) \qquad (21.11)$$

The score is used to represent the relevance of the instance (e.g., video frame) to the query. The goal of an activity retrieval system is to rank the data according to the relevance scores and return the top-ranked instances.

21.7.2 Experimental Results

The challenging *Broadcast Field Hockey Dataset* (developed in [61]) that consists of sporting activities captured from broadcast cameras is used to demonstrate the efficacy of the model. The model can carry out different inferences based on a user's queries. The method is directly applicable to multi-level human activity recognition. The goal is to predict events, social roles, or actions of each person. In this case, the query doesn't specify a particular event or social roles, but consists of more general questions, such as what is the overall game situation, or what are the social roles of each player. Figure 21.18 shows the visualizations of the predicted events and social roles.

21.8   Quality-Aware Mobile Visual Search

With the increasing popularity of mobile phones and tablets, *Mobile Visual Search* has attracted growing interest in the field of content-based image retrieval (CBIR). In this section, we present a novel framework for quality-aware mobile CBIR by Peng et al. [64]. On the mobile-client side, a query image is compressed to a certain quality level to accommodate the network condition and then uploaded onto a server with its quality level transferred as side information. On the server side, a set of features are extracted from the query image and then compared against the features

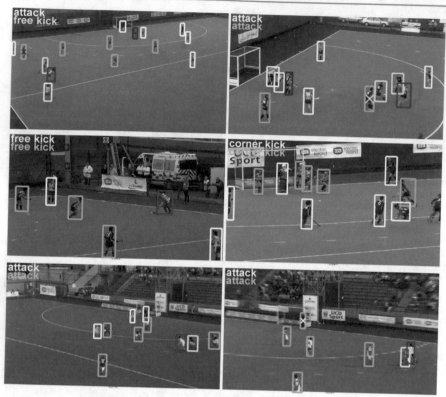

Fig. 21.18 Visualization of results on broadcast field hockey dataset. The ground truth event (white) and the predicted event are shown in the left corner of each image. Correct predictions are visualized in blue, otherwise yellow. Each bounding box is represented by a color, which denotes the predicted social roles. We use magenta, yellow, green, blue, and white to represent the social roles attacker, first defenders, defenders defend against space, defenders defend against person and other, respectively. The cross sign in the middle of a bounding box indicates incorrect predictions, and the ground truth social roles are indicated by the color of the cross sign

of the images in the database. As the efficacy of different features changes over query quality, we leverage the side information about the query quality to select a quality-specific similarity function that is learned offline using a *support vector machine (SVM)* method.

Mobile Visual Search enables people to look for visually similar products or find information about movies or CDs online by initiating a search request from a camera phone. Depending on which part of the visual search is performed on the mobile devices, there are several possible client–server architectures [65]:

1. A query image is transmitted to the server, and then feature extraction and retrieval are done on the server.
2. The mobile client extracts some features from the query image and uploads only the features to the server. The retrieval is performed on the server.

3. The mobile client maintains a cache of the image database. The retrieval is done locally on the client. Only if no match is found does the client sends a query to the server.

In each case, the system performance is constrained on the bandwidth, computation, memory, and power of mobile devices. Recently, there has been work focusing on designing compact descriptors for visual search. One representative work is the compressed histogram of gradients (CHoG) descriptor proposed by Chandrasekhar et al. which is shown to be highly discriminative at a low bitrate. Besides, there has been exploratory work by the MPEG committee toward defining a standard for visual search applications since 2011. This standardization initiative is referred to as "compact descriptors for visual search (CDVS)." It is evident that a low-bitrate descriptor can lead to shorter transmission latency, smaller memory overload, and potentially faster matching. Therefore, all the three aforementioned client–server architectures of mobile visual search can benefit from the advancement of compact-descriptor technology. Besides the great amount of research effort devoted to the design of visual descriptors, fusion methods for visual search have also attracted lots of attention in the CBIR community. Given that a descriptor is a set of characteristics of an image, such as color, shape, and texture, fusion techniques are shown to be effective in reducing the semantic gap of image retrieval based on feature similarity.

Here, we outline a framework for mobile visual search using a client–server architecture [64]. Specifically, a query image is compressed to a certain quality level on the mobile client and then uploaded to the server with its quality level transmitted as side information at the same time. A query quality-dependent retrieval algorithm based on the fusion of multiple features is then performed on the server. The motivations behind proposing such a framework are as follow:

- Although the computational capacity of mobile devices has become more and more powerful, there are several advantages to performing descriptor extraction on the server. Certainly, it eliminates the waiting time caused by computing descriptors on a mobile device of limited computing resources. More importantly, given the abundant computing resources on the server, it greatly relaxes the stringent constraint on the complexity and memory usage of the descriptor(s), which makes a fusion method computationally feasible in this framework.
- As bandwidth is also an important concern for visual search on wireless networks, the framework allows the client to compress a query image at a certain bitrate to accommodate the network condition.
- Since a specific descriptor is not equally important for images of different quality levels, the side information about the query quality could be leveraged to enhance the retrieval performance of the fusion method.

21.8.1 Quality-Aware Method

We outline a query quality-dependent fusion approach for mobile visual search. Specifically, it is based on the fusion of five common image features [66]:

- Tiny Images: This is the most trivial descriptor that compares images directly in color space after reducing the image dimensions drastically.
- Color Histograms: The histograms have 16 bins for each component of RGB color space, yielding a total of 48 dimensions.
- GIST: This descriptor computes the output of a bank of 24 Gabor-like filters tuned to 8 orientations at 4 scales. The squared output of each filter is then averaged on a 4×4 grid.
- Texton Histograms: For each image build a 512-dimensional histogram using a 512-entry universal texton dictionary.
- SSIM [67]: The self-similarity descriptors are quantized into 300 visual words by k-means. Unlike the descriptors mentioned before, SSIM provides a complementary measure of scene layout that is somewhat appearance invariant.

A query quality-dependent method for fusion image retrieval based on the five descriptors is then employed, as follows.

1. Categorize the query images into different quality levels based on the side information.[1]
2. For each "query image-retrieved image" pair, compute a 5-dimensional similarity-score vector $\mathbf{x}$ based on the five descriptors. The C-SVM formulation is adopted to learn the weights $\mathbf{w_k}$ to map $\mathbf{x}$ into a final score $s_f = \mathbf{w_k} \cdot \mathbf{x}$. Specifically, a set of positive (relevant) image pairs P is selected, and also a set of negative (irrelevant) image pairs N. For each image pair $p_i \in P \cup N$, compute a vector $\mathbf{x_i}$, and assign to $\mathbf{x_i}$ a label y_i ("1" for $p_i \in P$, and "-1" for $p_i \in N$). The weights $\mathbf{w}$ can be learned by solving the following optimization problem:

$$\min_{\mathbf{w},b,\xi} \quad \tfrac{1}{2}\mathbf{w}^T\mathbf{w} + C \sum_{i=1}^{l} \xi_i$$
$$\text{subject to} \quad y_i(\mathbf{w}^T \phi(\mathbf{x_i}) + b) \geq 1 - \xi_i, \qquad (21.12)$$
$$\xi_i \geq 0$$

This optimization is run for each quality level k to learn $\mathbf{w_k}$ using a query dataset of the corresponding quality.
3. At the test stage, the learned weight vector $\mathbf{w_k}$ is used to compute the final similarity score for a "query image-retrieved image" pair, with k being the quality level of the query image. The retrieved images are returned in decreasing order of the final similarity scores.

---

[1] For example, the JPEG standard uses a scalar to adjust a set of well-defined quantization tables.

Table 21.1 File size ranges of the images (386 × 256) from the Wang database compressed at different quality levels

Quality factor	100	75	50	30	20	15	10	8	5	3
Size range (KB)	7–56	6–38	4–24	4–18	3–14	3–12	3–9	3–8	3–6	3–5

21.8.2 Experimental Results

Datasets

Consider the Wang image database that contains 1,000 images of 10 classes (100 for each class). Images that belong to the same class are considered to be relevant. Ten copies of the Wang database are constructed at different quality levels. Specifically, images are compressed using JPEG compression with quality factor $k \in \{100, 75, 50, 30, 20, 15, 10, 8, 5, 3\}$. Let D_k be the database containing images at quality level k. D_{100} corresponds to the original Wang database, and D_3 contains images of the lowest quality. The file size ranges of the images after compression are listed in Table 21.1.

In the experiments, the query images can be from any quality level, whereas the images to be retrieved are always from D_{100}. To learn the weights $\mathbf{w_k}$ for quality level k, we selected 500 query images (50 for each class) from D_k, and pair them with the images in D_{100} for training. Let $(q_{k,i}, r_j)$ be such a pair, where $q_{k,i} \in D_k$ and $r_j \in D_{100}$. Let $c(\cdot)$ denote the class of an image. A pair $(q_{k,i}, r_j)$ is labeled positive or "1," if $c(q_{k,i}) = c(r_j)$. Otherwise, it is labeled negative or "−1." As there are 9 negative classes and only one positive class for each query image, we randomly select one-ninth of the negative pairs in order to balance the positive and negative samples during training. The remaining 500 images from each D_k that have not been selected for training are used for test. In the implementation, the dictionaries used to compute the descriptors are built using images from the SUN database [66]. We measure the similarity of two descriptors based on the histogram intersection distance and use a linear kernel for the SVM-based quality-aware fusion method.

Results

Let $s_f(q, r_n)$ be the final similarity score between a query image q and an image r_n in the database. The database images r_n are then sorted according to the similarity scores such that $s_f(q, r_n) \geq s_f(q, r_{n+1})$.

As the Precision and Recall values vary with the query images and the numbers of returned images, the mean average precision (MAP) is used instead, as defined in Eq. 21.9. The MAP results on the Wang database of 10 different quality levels are shown in Table 21.2. We can see that the performances of the Color Histograms, Tiny Images, and GIST do not change drastically as the image quality drops. On the contrary, the Texton Histograms and SSIM descriptors achieve the best performance

Table 21.2 MAP results of different quality levels

Query quality	100	75	50	30	20	15	10	8	5	3
Color histograms	0.3874	0.3874	0.3831	0.3835	0.3800	0.3775	0.3746	0.3752	0.3672	0.3626
Tiny images	0.3054	0.3052	0.3052	0.3047	0.3046	0.3043	0.3044	0.3038	0.3021	0.3021
GIST	0.3292	0.3276	0.3279	0.3264	0.3247	0.3228	0.3174	0.3125	0.2990	0.2824
Texton histograms	0.4260	0.4171	0.4256	0.4206	0.4135	0.4056	0.3843	0.3597	0.3119	0.2694
SSIM	0.4804	0.4805	0.4154	0.3656	0.3463	0.3202	0.2949	0.2777	0.2456	0.2178
Quality-aware fusion	0.5730	0.5726	0.5462	0.5330	0.5259	0.5168	0.5081	0.4975	0.4808	0.4701

near the higher end of the quality range (level 100, 75, and 50) and perform poorly near the lower end of the quality range (level 5 and 3).

It can also be observed that the quality-aware fusion method achieves significantly better MAPs than any individual descriptors on each quality level. Noticeably, the quality-aware fusion algorithm for the lowest-quality query images achieves comparable performance with the best-performing individual descriptor SSIM with the highest-quality query images. This strongly supports the advantages of fusing multiple descriptors for visual search.

The above discussion presented a quality-aware framework for Mobile Visual Search based on a query quality-dependent fusion method. The experimental results demonstrate the potential of taking into consideration the quality of query images to improve the performance of fusion image retrieval.

Current progressive coding techniques allow a mobile client to upload a bitstream that successively refines the reconstructed query image. In this case, the server can perform image retrieval using a query image of reduced quality, and then update the retrieved results as the query quality gets better and better.

21.9   Deep Incremental Hashing Network*

Hashing has been widely used for large-scale image retrieval because of its storage and search efficiency. Learning to hash aims to encode images with short binary hash codes that preserve the similarity of the original data in a supervised manner. With the development of deep learning techniques, recent works have found that hash functions learned by convolutional neural networks significantly outperform conventional hashing in many applications [56]. These deep supervised hashing methods usually require a large amount of data from different classes to train the deep model to achieve promising performance. With the explosive growth of the image databases, one practical challenge that arises for these methods is how to keep the model up to date. When a new set of images that belongs to a different category is added to the retrieval system, the update of the hash models often requires retraining on both the original and additional data, which is time-consuming. Besides, the hash codes of the original images would change if the model is updated and the new codes need to be generated by feeding all images into the new hash function.

In this section, we introduce the Deep Incremental Hashing Network (DIHN) proposed by Wu et al. [68] which addresses the aforementioned problem. DIHN learns the hash codes for the new images directly and keeps the codes of the original images unchanged. Simultaneously, a hash function is learned for query images using a subset of images from the entire database. The main contribution of the work in [68] are summarized as follows:

- A novel deep hashing framework, called deep incremental hashing network (DIHN), is proposed for learning hash codes in an incremental way. DIHN incrementally learns the hash codes for new images while holding those of the original ones invariant, which provides a flexible way for updating the modern image retrieval system.
- An incremental hashing loss is devised to preserve the similarities between images, which is used to train a deep convolutional neural network as the hash function only for query images. The hash loss incorporates the existing hash codes of the original images to train the hash function. Meanwhile, the hash codes for new images can be directly obtained during the optimization process.

21.9.1 Problem Definition

Assume we have a database containing m images, denoted as $D = \{d_i\}_{i=1}^m$. The set of class labels is denoted as $L = \{l_1, l_2, ..., l_c\}$, where each image d_i is associated with a label $l_j \in L$. Suppose the binary hash codes $B = \{b_i\}_{i=1}^m \in \{-1, +1\}^{k \times m}$ have already been learned for the existing images using some deep hashing methods, where k denotes the length of binary codes. Now, a set of n images $D' = \{d_i\}_{i=m+1}^{m \mid n}$ with a new class label set $L' = \{l_1', l_2', ..., l_{c'}'\}$ emerges. The goal is to learn the binary hash codes $B' = \{b_i\}_{i=m+1}^{m+n} \in \{-1, +1\}^{k \times n}$ for the new images while leaving B unchanged.

In addition, a hash function $h(\cdot)$ is needed for encoding the query images to their hash codes. The hash function is represented using a CNN model, which is trained with a set of images $Q = \{a_i\}_{i=1}^q$ containing sample images from both old and new images sets. The function maps Q to their corresponding hash codes $B_Q = \{b_{Q_i}\}_{i=1}^q \in \{-1, +1\}^{k \times q}$. The pairwise similarity information, denoted as $S \in \{-1, +1\}^{(m+n) \times q}$, is available during training. The first m rows of S represent the similarity between the images in D and Q, and the remaining n rows represent those between D' and Q. $S_{ij} = 1$ indicates that d_i and a_j are similar, while $S_{ij} = 0$ indicates they are dissimilar.

21.9.2 Descriptions of DIHN

The deep incremental hashing network (DIHN) has three inputs, i.e., the original database images D, the incremental database images D', and the images sampled from both the original and incremental databases Q, respectively. DIHN contains two important parts: *hash function learning* and *incremental hash code learning*. The hash function learning part exploits a deep CNN model to extract feature representations and learns the hash functions for query images based on the binary hash codes B_Q and images in Q. The incremental hash code learning part directly learns the hash codes B' for incremental database images D'. An incremental hashing loss function is proposed to preserve the similarities between database points. Note that the hash codes B for the original images D are already given and remain unchanged during the whole training process.

Deep Hash Function

The hash function in [68] is constructed by incorporating a CNN, forcing the output of its last fully-connected layer to have length k, which is the length of binary hash codes. The deep hash function $h(\cdot)$ is defined as

$$b_{Q_i} = h(a_i; \theta) = sign(f(a_i; \theta)) \tag{21.13}$$

where θ denotes the parameters of the CNN, and $f(\cdot)$ represents the function mapping from the input image a_i to the output of the last fully-connected layer.

Incremental Hashing Loss

The incremental hashing loss is proposed to preserve the semantic similarities between images, utilizing the pairwise similarity information S. The Hamming distances between the binary codes are used to measure similarity. The L2-norm loss is adopted to minimize the difference between the inner product of binary code pairs and the similarity S_{ij}. The incremental hashing loss in [68] is formulated as

$$\min_{B', B_Q} J(B', B_Q) = \sum_{i=1}^{m}\sum_{j=1}^{q}(b_i^T b_{Q_j} - kS_{ij})^2 + \sum_{i=m+1}^{m+n}\sum_{j=1}^{q}(b_i^T b_{Q_j} - kS_{ij})^2$$

$$s.t. \ B' = \{b_{m+1}, b_{m+2}, ..., b_{m+n}\} \in \{-1, +1\}^{k \times n}$$

$$B_Q = \{b_{Q_1}, b_{Q_2}, ..., b_{Q_q}\} \in \{-1, +1\}^{k \times q}$$

$$\tag{21.14}$$

The hash code b_i, $i = 1, ..., m$ in the first term is for the original images, which are given. The hash code b_i, $i = m + 1, ..., m + n$ in the second term is for the incremental images, which are directly learned during optimization. The hash code b_{Q_j} is learned from the Hash Function represented by the CNN, as shown in Eq. 21.13. In order to facilitate optimization, the $sign(\cdot)$ is replaced with its continuous relaxation function tanh($\cdot$). Substituting Eqs. 21.13–21.14 results in the following objective function:

$$\min_{B',\theta} J(B',\theta) = \sum_{i=1}^{m}\sum_{j\in\varphi}[b_i^T \tanh(f(d_j;\theta)) - kS_{ij}]^2 + \sum_{i=m+1}^{m+n}\sum_{j\in\varphi}[b_i^T \tanh(f(d_j;\theta)) - kS_{ij}]^2$$

$$+\lambda\sum_{j\in\varphi}[b_j - \tanh(f(d_j;\theta))]^2 + \mu\sum_{j\in\varphi}[\tanh(f(d_j;\theta))^T \mathbf{1}]^2$$

$$s.t. \; B' = \{b_{m+1}, b_{m+2}, ..., b_{m+n}\} \in \{-1, +1\}^{k\times n}$$

(21.15)

where φ is a subset of $\{1, 2, ..., m+n\}$ denoting the indices of the samples of Q in D and D'. There are terms added in this objective function. The first term is $\lambda\sum_{j\in\varphi}[b_j - \tanh(f(d_j;\theta))]^2$. It is added to minimize the gap between the codes learned from the CNN and the codes b_j for the database images, which are either fixed for the original images or directly learned from the objective for the incremental images. The second term $\mu\sum_{j\in\varphi}[\tanh(f(d_j;\theta))^T\mathbf{1}]^2$ encourages the numbers of -1 and $+1$ to be approximately equal among the codes. λ and μ are hyper-parameters, and $\mathbf{1}$ is the all-ones vector.

Optimization

The parameters θ and hash codes B' are learned by optimizing the objective function in an alternating strategy. Specifically, in θ-step, B' is fixed and the CNN is trained using back-propagation. While in B'-step, it is learned with θ fixed, using discrete cyclic coordinate descent (DCC) algorithm. The learning algorithm is summarized as follows.

Algorithm 21.1 (The learning algorithm for DIHN)

Input: original database set D; incremental database set D'; similarity matrix S; original hash codes B; code length k; maximum iteration number T.
Output: incremental hash codes B' and neural network parameter θ.
Initialize $B' \in \{-1, +1\}^{k\times n}$; neural network parameter θ.
Repeat
    for $i = 1 \rightarrow T$ **do**

    (a) Forward computation to compute $f(d;\theta)$ from the raw images;
    (b) Compute derivation of J with respect to $f(d;\theta)$.
    (c) Update the neural network θ with back-propagation.

    end
    Update B' according to (DCC) algorithm.
until convergence or reach maximum iterations

Table 21.3 Comparison of MAP w.r.t. different number-of-bits on two datasets. The best results for MAP are shown in bold, the second best results are underlined.

Methods	CIFAR-10				NUS-WIDE			
	12 bits	24 bits	36 bits	48 bits	12 bits	24 bits	36 bits	48 bits
DSH	0.6441	0.7421	0.7703	0.7992	0.7125	0.7313	0.7401	0.7485
DHN	0.6805	0.7213	0.7233	0.7332	0.7719	0.8013	0.8051	0.8146
DPSH	0.6818	0.7204	0.7341	0.7464	0.7941	0.8249	0.8351	0.8442
DQN	0.5540	0.5580	0.5640	0.5800	0.7680	0.7760	0.7830	0.7920
DSDH	0.7400	0.7860	0.8010	0.8200	0.7760	0.8080	0.8200	0.8290
ADSH	0.8898	<u>0.9280</u>	0.9310	0.9390	<u>0.8400</u>	0.8784	0.8951	0.9055
DIHN$_1$+ADSH	<u>0.8933</u>	0.9279	**0.9386**	0.9346	0.8357	<u>0.8862</u>	<u>0.8987</u>	0.9070
DIHN$_2$+ADSH	**0.8975**	**0.9294**	0.9293	0.9385	**0.8470**	**0.8926**	0.8964	<u>0.9103</u>
DIHN$_3$+ADSH	0.8916	0.9273	<u>0.9330</u>	**0.9456**	0.8353	0.8821	**0.9004**	**0.9115**
DIHN$_6$+ADSH	0.8784	0.9172	0.9293	<u>0.9401</u>	–	–	–	–
DIHN$_{11}$+ADSH	–	–	–	–	0.8120	0.8727	0.8880	0.8972

21.9.3 Experimental Results

Wu et al. [68] conduct extensive evaluations of the proposed method on two widely used datasets: CIFAR-10 and NUS-WIDE. The mean average precision (MAP) is used to evaluate the retrieval accuracy. We include some of the results in this section. For more details the reader is referred to [68].

The asymmetric deep hashing method (ADSH) [56] is adopted in the experiments to generate original binary hash codes B. The hash code of the incremental images are generated using DIHN. Table 21.3 shows the MAP performance comparisons on the two datasets. DIHN$_i$, ($i = 1, 2, 3, 6, 11$) means that the number of classes used as the incremental data is i. For normal incremental settings, i.e., $i = 1, 2, 3$, DIHN$_i$+ADSH can achieve competitive retrieval performance. For challenging settings with more incremental classes, i.e., $i = 6, 11$, the performance of DIHN$_i$+ADSH is a little inferior to that under the normal settings. In general, the DIHN achieves competitive retrieval performance.

In terms of time complexity, it is demonstrated that DIHN significantly decreases the training time, because it does not need to generate hash codes for the original images. For example, it takes about 28 min for DIHN$_3$+ADSH to achieve promising performance when $k = 12$. In comparison, ADSH requires three times (83 minutes) as much as DIHN$_3$+ADSH to achieve the similar performance.

21.10 Exercises

1. Devise a text-annotation taxonomy (categorization) for image descriptions, starting your classification using the set of Yahoo! categories, say.
2. Examine several website image captions. How useful would you say the textual data is as a cue for identifying image contents? (Typically, search systems use

word stemming, for eliminating tense, case, and number from words—the word *stemming* becomes the word *stem*.)

3. Suppose a color histogram is defined coarsely, with bins quantized to 8 bits, with 3 bits for each red and green and 2 for blue. Set up an appropriate structure for such a histogram, and fill it from some image you read. Template Visual C++ code for reading an image is on the text website, as `sampleCcode.zip` under "Sample Code."

4. Try creating a texture histogram as described in Sect. 21.3.3. You could try a small image and follow the steps given there, using MATLAB, say, for ease of visualization.

5. Describe how you may find an image containing some 2D "brick pattern" in an image database, assuming the color of the "brick" is yellow and the color of the "gaps" is blue. (Make sure you discuss the limitations of your method and possible improvements.)

   (a) Use color only.
   (b) Use edge-based texture measures only.
   (c) Use color, texture, and shape.

6. What are the typical components (layers) of a convolutional neural network (CNN)?

7. Describe the pre-training and fine-tuning procedure in a CNN for CBIR. What is the advantage of this procedure?

8. Describe in general the idea of the database indexing techniques: Inverted Indexing and Hashing-based Indexing.

9. The main difference between a static image and video is the availability of motion in the latter. One important part of CBR from video is motion estimation (e.g., the direction and speed of any movement). Describe how you could estimate the movement of an object in a video clip, say a car, if MPEG (instead of uncompressed) video is used.

10. Color is three-dimensional, as Newton pointed out. In general, we have made use of several different color spaces, all of which have some kind of brightness axis, plus two intrinsic-color axes.
    Let's use a *chromaticity* two-dimensional space, as defined in Eq. (4.7). We'll use just the first two dimensions, $\{r, g\}$. Devise a 2D color histogram for a few images, and find their histogram intersections. Compare image similarity measures with those derived using a 3D color histogram, comparing over several different color resolutions. Is it worth keeping all three dimensions, generally?

11. Suggest at least three ways in which audio analysis can assist in video retrieval-system-related tasks.

12. Implement an image search engine using low-level image features such as color histogram, color moments, and texture. Construct an image database that contains at least 500 images from at least 10 different categories. Perform retrieval tasks using a single low-level feature as well as a combination of features. Which

feature combination gives the best retrieval results, in terms of both Precision and Recall, for each category of images?

13. Another way of combining Precision and Recall is the F-score measure. The F-score is the harmonic mean of Precision P and Recall R, defined as

$$F = 2(P * R)/(P + R)$$

Experiment and determine how F behaves as P and R change.

References

1. M.M. Fleck, D.A. Forsyth, C. Bregler, Finding naked people. Eur. Congr. Comput. Vis. (2), 593–602 (1996)
2. C.C. Chang, S.Y. Lee, Retrieval of similar pictures on pictorial databases. Pattern Recogn. **24**, 675–680 (1991)
3. M. Worring, P. Sajda, S. Santini, D. Shamma, A.F. Smeaton, Q. Yang, Where is the user in multimedia retrieval? IEEE Multimed. **19**(4), 6–10 (2012)
4. S. Paek, C. L. Sable, V. Hatzivassiloglou, A. Jaimes, B. H. Schiffman, S.-F. Chang, K.R. McKeown, Integration of visual and text based approaches for the content labeling and classification of photographs, in *ACM SIGIR'99 Workshop on Multimedia Indexing and Retrieval*, pp. 423–444 (1999)
5. K. Barnard, D.A. Forsyth, Learning the semantics of words and pictures, in *Proceedings of the International Conference on Computer Vision*, pp. II, 408–415 (2001)
6. M.J. Swain, D.H. Ballard, Color indexing. Int. J. Comput. Vis. **7**, 11–32 (1991)
7. Y. LeCun, L. Bottou, Y. Bengio, P. Haffner, Gradient-based learning applied to document recognition. Proc. IEEE **86**(11), 2278–2324 (1998)
8. M. Tzelepi, A. Tefas, Deep convolutional learning for content based image retrieval. Neurocomputing **275**, 2467–2478 (2018)
9. A.W.M. Smeulders, M. Worring, S. Santini, A. Gupta, R. Jain, Content-based image retrieval at the end of the early years. IEEE Trans. Pattern Anal. Mach. Intell. **22**, 1349–1380 (2000)
10. J.W.H. Tangelder, R.C. Veltkamp, A survey of content based 3D shape retrieval methods. Multimed. Tools Appl. **39**, 441–471 (2008)
11. P. Huang, A. Hilton, J. Starck, Shape similarity for 3D video sequences of people. Int. J. Comput. Vis. **89**(2–3), 362–381 (2010)
12. M. Flickner et al., Query by image and video content: the QBIC system. IEEE Comput. **28**(9), 23–32 (1995)
13. W. Niblack, X. Zhu, J.L. Hafner, T. Breuel, D. Ponceleon, D. Petkovic, M.D. Flickner, E. Upfal, S.I. Nin, S. Sull, B. Dom, B.-L. Yeo, A. Srinivasan, D. Zivkovic, M. Penner, Updates to the QBIC system, in *Storage and Retrieval for Image and Video Databases*, pp. 150–161 (1998)
14. J. Hafner, H.S. Sawhney, W. Equitz, M. Flickner, W. Niblack, Efficient color histogram indexing for quadratic form distance functions. IEEE Trans. Pattern Anal. Mach. Intell. **17**, 729–736 (1995)
15. C. Carson, S. Belongie, H. Greenspan, J. Malik, Blobworld: image segmentation using expectation-maximization and its application to image querying. IEEE Trans. Pattern Anal. Mach. Intell. **24**(8), 1026–1038 (2002)
16. A. Pentland, R. Picard, S. Sclaroff, Photobook: content-based manipulation of image databases, in *Storage and Retrieval for Image and Video Databases (SPIE)*, pp. 34–47 (1994)
17. F. Liu, R.W. Picard, Periodicity, directionality, and randomness: Wold features for image modeling and retrieval. IEEE Trans. Pattern Anal. Mach. Intell. **18**, 722–733 (1996)

18. Z.N. Li, O.R. Zaïane, Z. Tauber, Illumination invariance and object model in content-based image and video retrieval. J. Vis. Commun. Image Rep. **10**, 219–244 (1999)

19. H. Tamura, S. Mori, T. Yamawaki, Texture features corresponding to visual perception. IEEE Trans. Syst. Man Cybern. **SMC-8**(6), 460–473 (1978)

20. A.R. Rao, G.L. Lohse, Towards a texture naming system: identifying relevant dimensions of texture, in *IEEE Conference on Visualization*, pp. 220–227 (1993)

21. D.A. Forsyth, J. Ponce, *Computer Vision: a Modern Approach*, 2nd edn. (Prentice Hall, 2012)

22. M.S. Drew, J. Wei, Z.N. Li, Illumination-invariant image retrieval and video segmentation. Pattern Recogn. **32**, 1369–1388 (1999)

23. M.S. Drew, Z.N. Li, Z. Tauber, Illumination color covariant locale-based visual object retrieval. Pattern Recogn. **35**(8), 1687–1704 (2002)

24. T. Deselaers, D. Keysers, H. Ney, Features for image retrieval: an experimental comparison. Inf. Retr. **11**(2), 77–107 (2008)

25. R. Datta, D. Joshi, J. Li, J.Z. Wang, Image retrieval: ideas, influences, and trends of the new age. ACM Comput. Surv. **40**(2), 5, 1–5, 60 (2008)

26. W. Zhou, H. Li, Q. Tian, Recent advance in content-based image retrieval: a literature survey (2017). arXiv preprint arXiv:1706.06064

27. D. Lowe, Distinctive image features form scale-invariant keypoints. Int. J. Comput. Vis. **20**(2), 91–110 (2004)

28. L. Fei-Fei, P. Perona, A Bayesian hierarchical model for learning natural scene categories, in *Proceedings of the IEEE Conference on Computer Vision and Pattern Recognition (CVPR)* (2005)

29. J. Sivic, A. Zisserman, Video Google: a text retrieval approach to object matching in videos, in *Proceedings of the Int.ernational Conference on Computer Vision (ICCV)* (2003)

30. Y. Rui, T. S. Huang, M. Ortega, S. Mehrotra, Relevance feedback: a power tool for interactive content-based image retrieval. IEEE Trans. Circ. Sys. Video Tech. **8**(5), 644–655 (1998)

31. B. Li, E. Chang, C.-T. Wu. DPF—A perceptual distance function for image retrieval, in *IEEE International Conference on Image Proceedings*, pp. II–597–II–600 (2002)

32. J.R. Smith, D. Joshi, B. Huet, W. Hsu, J. Cota. Harnessing AI for augmenting creativity: application to movie trailer creation, in *ACM International Conference on Multimedia*, pp. 1799–1808 (2017)

33. F. Eyben, M. Wöllmer, B. Schuller, OpenEAR—Introducing the Munich open-source emotion and affect recognition toolkit, in *2009 3rd International Conference on Affective Computing and Intelligent Interaction and Workshops*, pp. 1–6. (IEEE, 2009)

34. D. Borth, B. Ji, T. Chen, T. Breuel, S.F. Chang, Large-scale visual sentiment ontology and detectors using adjective noun pairs, in *ACM International Conference on Multimedia*, pp. 223–232. (ACM, 2013)

35. B. Zhou, et al., Learning deep features for scene recognition using places database, in *Advances in Neural Information Processing Systems*, pp. 487–495 (2014)

36. J. Philbin, O. Chum, M. Isard, J. Sivic, A. Zisserman, Object retrieval with large vocabularies and fast spatial matching, in *Proceedings of the EEE Conference on Computer Vision and Pattern Recognition (CVPR)* (2007)

37. W. Zhou, Y. Lu, H. Li, Y. Song, Q. Tian, Spatial coding for large scale partial-duplicate web image search, in *Proceedings of the ACM Conference on Multimedia (ACM Multimedia)* (2010)

38. O. Chum, J. Philbin, J. Sivic, M. Isard, A. Zisserman, Total recall: automatic query expansion with a generative feature model for object retrieval, in *Proceedings of the International Conference on Computer Vision (ICCV)* (2007)

39. T.-S. Chua, R. Hong, G. Li, J. Tang, From text question-answering to multimedia QA on web-scale media resources, in *Proceedings of the First ACM Workshop on Large-scale Multimedia Retrieval and Mining*, pp. 51–58 (2009)

40. G. Wang, D. Hoiem, D. Forsyth, Learning image similarity from Flickr group using fast kernel machines. IEEE Trans. Pattern Anal. Mach. Intell. **34**(11), 2177–2188 (2012)

41. G. Awad, TRECVID, et al., An evaluation campaign to benchmark video activity detection, video captioning and matching, and video search & retrieval. *Proceedings of TRECVID*, vol. 2019 (2019)

42. G.K. Meyers, et al., Evaluating multimedia features and fusion for examplar-based event detection. Mach. Vis. Appl. (2013)

43. A. Krizhevsky, I. Sutskever, G.E. Hinton, ImageNet classification with deep convolutional neural networks, in *Advances in Neural Information Processing Systems (NIPS)*, pp. 1097–1105 (2012)

44. K. Simonyan, A. Zisserman, Very deep convolutional networks for large-scale image recognition (2014). arXiv preprint arXiv:1409.1556

45. C. Szegedy, et al., Going deeper with convolutions, in *IEEE Conference on Computer Vision and Pattern Recognition (CVPR)*, pp. 1–9 (2015)

46. K. He, X. Zhang, S. Ren, J. Sun, Deep residual learning for image recognition, in *IEEE Conference on Computer Vision and Pattern Recognition (CVPR)*, pp. 770–778 (2016)

47. R. Sharif, et al., CNN features off-the-shelf: an astounding baseline for recognition, in *IEEE Conference on Computer Vision and Pattern Recognition Workshops*, pp. 806–813 (2014)

48. A. Babenko, A. Slesarev, A. Chigorin, V. Lempitsky, Neural codes for image retrieval, in *European Conference on Computer Vision*, pp. 584–599. (Springer, 2014)

49. J. Wan, et al., Deep learning for content-based image retrieval: a comprehensive study. In *ACM International Conference on Multimedia*, pp. 157–166. (ACM, 2014)

50. H. Jegou, M. Douze, C. Schmid, Hamming embedding and weak geometric consistency for large scale image search, in *European Conference on Computer Vision*, pp. 304–317. (Springer, 2008)

51. W. Zhou, H. Li, Y. Lu, Q. Tian, Large scale image search with geometric coding, in *ACM International Conference on Multimedia*, pp. 1349–1352. (ACM, 2011)

52. A. Babenko, V. Lempitsky, The inverted multi-index. IEEE Trans. Pattern Anal. Mach. Intell. **37**(6), 1247–1260 (2014)

53. L. Zheng, S. Wang, Z. Liu, Q. Tian, Packing and padding: coupled multi-index for accurate image retrieval, in *IEEE Conference on Computer Vision and Pattern Recognition (CVPR)*, pp. 1939–1946 (2014)

54. M. Datar, N. Immorlica, P. Indyk, V.S. Mirrokni, Locality-sensitive hashing scheme based on p-stable distributions, in *The Twentieth Annual Symposium on Computational Geometry*, pp. 253–262. (ACM, 2004)

55. B. Kulis, K. Grauman, Kernelized locality-sensitive hashing for scalable image search. IEEE Int. Conf. Comput. Vis. (ICCV) **9**, 2130–2137 (2009)

56. Q.Y. Jiang, W.J. Li, Asymmetric deep supervised hashing, in *AAAI Conference on Artificial Intelligence* (2018)

57. S.F. Chang et al., VideoQ: an automated content based video search system using visual cues. Proc. ACM Multimed. **97**, 313–324 (1997)

58. D. Bordwell, K. Thompson, *Film Art: an Introduction*, 9th edn. (McGraw-Hill, 2009)

59. F. Dufaux, Key frame selection to represent a video, in *International Conference on Image Processing*, pp. II, 275–278 (2000)

60. M.S. Drew, J. Au, Video keyframe production by efficient clustering of compressed chromaticity signatures. ACM Multimed. **2000**, 365–368 (2000)

61. T. Lan, L. Sigal, G. Mori, Social roles in hierarchical models for human activity recognition, in *Proceedings of the IEEE Conference on Computer Vision and Pattern Recognition (CVPR)* (2012)

62. C. Schuldt, I. Laptev, B. Caputo, Recognizing human actions: a local SVM approach, in *Proceedings of the International Conference on Pattern Recognition (ICPR)* (2004)

63. T. Joachims, Training linear SVMs in linear time, In *SIGKDD* (2006)

64. P. Peng, J. Li, Z.N. Li, Quality-aware mobile visual search, in *The 3rd International Conference on Integrated Information (IC-ININFO)* (2013)

65. B. Girod, V. Chandrasekhar, D.M. Chen, N.M. Cheung, R. Grzeszczuk, Y. Reznik, G. Takacs, S.S. Tsai, R. Vedantham, Mobile visual search. IEEE Signal Process. Mag. **28**(4), 61–76 (2011)

66. J. Xiao, J. Hays, K.A. Ehinger, A. Oliva, A. Torralba, Sun database: large-scale scene recognition from Abbey to Zoo, in *IEEE Conference on Computer Vision and Pattern Recognition (CVPR)*, pp. 3485–3492. (IEEE, 2010)
67. Z. Wang et al., Image quality assessment: from error visibility to structural similarity. IEEE Trans. Image Process. **13**(4), 600–612 (2004)
68. D. Wu, Q. Dai, J.Liu, B. Li, W. Wang, Deep incremental hashing network for efficient image retrieval, in *IEEE Conference on Computer Vision and Pattern Recognition (CVPR)*, pp. 9069–9077 (2019)

Index

© Springer Nature Switzerland AG 2021
Z.-N. Li et al., *Fundamentals of Multimedia*, Texts in Computer Science,
https://doi.org/10.1007/978-3-030-62124-7

Printed in the United States
by Baker & Taylor Publisher Services